Periodic Table of the Elements with the Gmelin System Numbers

1 H 2																1 H 2	2 He 1
3 Li 20	4 Be 26											5 B 13	6 C 14	7 N 4	8 O 3	9 F 5	10 Ne 1
11 Na 21	12 Mg 27											13 Al 35	14 Si 15	15 P 16	16 S 9	17 Cl 6	18 Ar 1
19 K 22 *	20 Ca 28	21 Sc 39	22 Ti 41	23 V 48	24 Cr 52	25 Mn 56	26 Fe 59	27 Co 58	28 Ni 57	29 Cu 60	30 Zn 32	31 Ga 36	32 Ge 45	33 As 17	34 Se 10	35 Br 7	36 Kr 1
37 Rb 24	38 Sr 29	39 Y 39	40 Zr 42	41 Nb 49	42 Mo 53	43 Tc 69	44 Ru 63	45 Rh 64	46 Pd 65	47 Ag 61	48 Cd 33	49 In 37	50 Sn 46	51 Sb 18	52 Te 11	53 I 8	54 Xe 1
55 Cs 25	56 Ba 30	57** La 39	72 Hf 43	73 Ta 50	74 W 54	75 Re 70	76 Os 66	77 Ir 67	78 Pt 68	79 Au 62	80 Hg 34	81 Tl 38	82 Pb 47	83 Bi 19	84 Po 12	85 At 8a	86 Rn 1
87 Fr 25a	88 Ra 31	89*** Ac 40	104 71	105 71													

* NH₄ 23 — NH_4 23

**Lanthanides 39

58 Ce	59 Pr	60 Nd	61 Pm	62 Sm	63 Eu	64 Gd	65 Tb	66 Dy	67 Ho	68 Er	69 Tm	70 Yb	71 Lu

***Actinides

90 Th 44	91 Pa 51	92 U 55	93 Np 71	94 Pu 71	95 Am 71	96 Cm 71	97 Bk 71	98 Cf 71	99 Es 71	100 Fm 71	101 Md 71	102 No 71	103 Lr 71

A Key to the Gmelin System is given on the Inside Back Cover

Gmelin Handbook of Inorganic Chemistry

8th Edition

Gmelin Handbook of Inorganic Chemistry

8th Edition

Gmelin Handbuch der Anorganischen Chemie

Achte, völlig neu bearbeitete Auflage

Prepared
and issued by

Gmelin-Institut für Anorganische Chemie
der Max-Planck-Gesellschaft
zur Förderung der Wissenschaften

Director: Ekkehard Fluck

Founded by

Leopold Gmelin

8th Edition

8th Edition begun under the auspices of the
Deutsche Chemische Gesellschaft by R. J. Meyer

Continued by

E. H. E. Pietsch and A. Kotowski, and by
Margot Becke-Goehring

Springer-Verlag Berlin Heidelberg GmbH 1990

Gmelin Handbook of Inorganic Chemistry

8th Edition

Th
Thorium

Supplement Volume A1a

Natural Occurrence. Minerals (Excluding Silicates)

With 14 illustrations

AUTHORS

Reiner Ditz, Bärbel Sarbas,
Peter Schubert, Wolfgang Töpper

EDITOR

Wolfgang Töpper

CHIEF EDITOR

Wolfgang Töpper

System Number 44

Springer-Verlag Berlin Heidelberg GmbH 1990

LITERATURE CLOSING DATE: 1987
IN SOME CASES MORE RECENT DATA HAVE BEEN CONSIDERED

Library of Congress Catalog Card Number: Agr 25-1383

ISBN 978-3-662-08912-5   ISBN 978-3-662-08910-1 (eBook)
DOI 10.1007/978-3-662-08910-1

© by Springer-Verlag Berlin Heidelberg 1990
Originally published by Springer-Verlag Berlin Heidelberg New York London Paris Tokyo in 1990
Softcover reprint of the hardcover 8th edition 1990

Preface

The present volume begins, in a summary fashion, with a description of the natural occurrence of the element Th. Especially stressed are those facts that are most important in understanding its geological distribution on Earth (as, e.g., mode of occurrence and distribution among minerals) and its behavior in minerals (as, e.g., diadochy and metamictization of minerals). Attached is a tabulation of the highest reported Th or ThO_2 content in minerals that normally (by their crystal-chemical formula) do not contain Th.

The main part of the present volume describes the minerals of Th. As can be seen from the crystal-chemical formulas, there are, in addition to a few minerals containing Th as the sole cation, a number of minerals that contain Th as an additional cation or as a diadochic element. In the case of diadochic substitution, the mineral may represent a Th-rich end member of a solid-solution series and, therefore, is described as a separate Th mineral, or may only sporadically contain higher amounts of Th (no mineral description is given).

The mineral descriptions, in this volume including oxides, carbonates, and phosphates/silicophosphates of Th, comprise the following topics:

— occurrence (comprising data on paragenesis, intergrowths, inclusions, and alteration; sometimes, for comparison, data on synthesis also);
— chemistry (comprising complete analysis, formula, diadochy, and the relationship between composition and type of geological occurrence or composition and physical properties);
— crystal form and structure;
— optical and other physical properties; and
— chemical and thermal behavior.

The silicates of Th and the deposits of Th will be described in the volume "Thorium" Suppl. Vol. A1b, that also contains a mineral index for both volumes.

Frankfurt am Main
May 1990

Wolfgang Töpper

Volumes published on "Radium and Actinides"

U   Uranium

List of Abbreviations

To save space in the text, and where it is appropriate, the following abbreviations are used for geographic names and the rare earth elements:

F.R.G.   Federal Republic of Germany
G.D.R.   German Democratic Republic
P.R.C.   People's Republic of China
R.S.A.   Republic of South Africa

RE   rare earth elements[*]
RE(Ce) cerian RE — comprises La to Eu[**]
RE(Y)   yttrian RE — comprises Gd to Lu and Y[**]

[*] Due to chemical similarities, summary data on the RE may also include those of Y. But note that this is not always clearly stated in the original papers.
[**] Provided that data for individual RE are given in the original paper, this subdivision is used throughout the present volume. Otherwise, the denomination of the original paper is used. But, note that there may be other subdivisions of the RE; refer to "Seltenerdelemente" A4, 1979, p. 38.

Table of Contents

1 Natural Occurrence

The parts of this chapter give only a brief outline (partly by only a single example) of the geochemical behavior of Th, especially of those facts that are most important in understanding its geological distribution on earth, see Chapter 1.2, pp. 3/7, and its occurrence in minerals, see Chapter 1.3, pp. 7/21. Largely used are the more recent papers, or compilations, or reference/text books.

For an outline of the geochemistry of Th, refer to the following books or papers, of which the last three take largely into account the types of deposit (for that, refer also to "Thorium" Suppl. Vol. A1b, Chapter Deposits):

Rankama, K.; Sahama, T. G., Geochemistry, Univ. Chicago Press, Chicago 1950, pp. 1/912, 570/3.
Goldschmidt, V. M., Geochemistry, Clarendon, Oxford 1954, pp. 1/730, 427/31.
Frondel, C. (Geol. Surv. Profess. Paper [U.S.] No. 300 [1956] 567/79; Proc. Intern. Conf. Peaceful Uses At. Energy, Geneva 1955 [1956], Vol. 6, pp. 568/77).
Adams, J. A. S.; Osmond, J. K.; Rogers, J. J. W. (Phys. Chem. Earth 3 [1959] 298/348).
Smyslov, A. A., Uran i Torii v Zemnoi Kore, Nedra, Leningrad 1974, pp. 1/232.
Kaplan, G. E.; Uspenskaya; T. A.; Zarembo, Yu. I.; Chirkov, I. V., Torii, Ego Syr'evy Resursy, Khimiya i Tekhnologiya, Atomizdat, Moscow 1960, pp. 1/224, 9/54.
Chirkov, I. V.; Kaplan, G. E.; Uspenskaya, T. A. (in: Nevskii, V. A., Trebovaniya Promyshlennosti k Kachestvu Mineral'nogo Syr'ya, Vol. 72, Spravochnik dlya Geologov, 2nd Ed., Gosgeoltekhizdat, Moscow 1961, pp. 3/83).
Olson, J. C.; Overstreet, W. C. (Geol. Surv. Bull. [U.S.] No. 1024 [1964] 1/61).

For an overview of properties (partly including diagnostic characteristics) of U- and Th-bearing minerals, refer to the following books or papers:

George, D'Arcy, Mineralogy of Uranium- and Thorium-Bearing Minerals, RMO-563 [1951] 1/198; N.S.A. 5 [1951] No. 833.
Getseva, R. V.; Savel'eva, K. T., Rukovodstvo po Opredeleniyu Uranovykh Mineralov, Gosgeoltekhizdat, Moscow 1956, pp. 1/260.
Soboleva, M. V.; Pudovkina, I. A., Minerals of Uranium: Reference Book, Washington 1961, pp. 1/455, AEC-tr-4487 [Russian original, Moscow 1957]; N.S.A. 15 [1961] No. 23739.
Heinrich, E. W., Mineralogy and Geology of Radioactive Raw Materials, McGraw-Hill, New York — Toronto — London 1958, pp. 1/654, 3/149.
Frondel, C., Systematic Mineralogy of Uranium and Thorium (Geol. Surv. Bull. [U.S.] No. 1064 [1958] 1/400).
Frondel, J. W.; Fleischer, M.; Jones, R. S., Glossary of Uranium- and Thorium-Bearing Minerals, 4th Ed. (Geol. Surv. Bull. [U.S.] No. 1250 [1967] 1/69).

For a classification scheme of radioactive minerals taking into account field and laboratory methods of identification, and one which is convenient for prospecting purposes, see Whittle, A. W. G. (J. Geol. Soc. Australia 2 [1955] 21/45, 25/30).

For Th recoils in geochemistry, refer to "Thorium" Suppl. Vol. A4, 1989, pp. 228/39. And for Th isotope geology (physics and chemistry of nuclides, natural distribution, geochronometry, etc.), refer to the relevant chapters in Rankama, K., Isotope Geology, Pergamon, London 1954, pp. 1/535, and Progress in Isotope Geology, Interscience Publ., New York — London 1963, pp. 1/705.

1.1 Chemical/Geochemical Characteristics of Thorium

Thorium is always tetravalent in minerals [1].

Atomic radius = 1.80 Å [2, p. 52]; ionic radius of Th^{4+} from different literature sources is 0.95 Å [3], 1.02 Å (Pauling, 1927; Ahrens 1952) or 1.10 Å (Goldschmidt, 1926; Wyckoff, 1948), see [2, p. 55]. Depending on coordination number, the ionic radius is given as 1.08, 1.12, and 1.17 Å for 6-, 8-, and 9-fold coordination, respectively [4].

First ionization potential I_1 = ~5.7 eV or ~131 kcal/mol, and electronegativity = 1.3 [2, pp. 49, 65] or 1.1 [5].

Thorium is radioactive, the predominating natural isotope is ^{232}Th [1]. An initial abundance in the earth's crust for ^{232}Th is given as 0.64 g × 10^{20} [6]; and its recent abundance represents 84.7% of the original content at the time of earth formation [7]. For the natural radioactivity of ^{232}Th and for other natural isotopes, see "Thorium", 1955, pp. 112/20 and [8]. According to their position in a crystal lattice, the Th isotopes behave differently during acid leaching, see [9].

Thorium is a lithophile [2, p. 184], [10], or pegmatophile [2, p. 184], or granitophile element [11]. Based on an investigation of the relationship between petrochemical dispersion (i.e., content in different types of rock) of Clarke content and the periodic system, Th belongs to the mainly salitic elements (that have their highest Clarke content in acidic rocks) [12].

References for 1.1:

[1] Rankama, K.; Sahama, T. G. (Geochemistry, Univ. Chicago Press, Chicago 1950, pp. 1/912, 570/3).
[2] Rösler, H. J.; Lange, H. (Geochemische Tabellen, Verlag Grundstoffind., Leipzig 1965, pp. 1/328).
[3] Mineev, D. A. (Geokhimiya **1960** 131/8; Geochem. [USSR] **1960** 156/66, 158).
[4] Whittaker, E. J. W.; Muntus, R. (Geochim. Cosmochim. Acta **34** [1970] 945/56, 952).
[5] Green, J. (Bull. Geol. Soc. Am. **64** [1953] (1001/12).
[6] Postnikov, D. V. (in: Garris, M. A.; Yusupov, B. M., Voprosy Izotopnoi Geokhronologii Urala i Vostochnoi Chasti Russkoi Platformy, Ufa 1972, pp. 17/20 from C.A. **82** [1975] No. 127756).
[7] Asimov, I. (J. Chem. Educ. **30** [1953] 398).
[8] Rogers, J. J. W.; Adams, J. A. S. (in: Wedepohl, K. H., Handbook of Geochemistry, Vol. 2, Springer, Berlin — Heidelberg — New York 1969, p. **90-B-1**).
[9] Starik, I. E.; Polevaya, N. I. (Tr. Radievogo Inst. Akad. Nauk SSSR **6** [1957] 104/18; C.A. **1958** 19407).
[10] Wilke, A. (in: Bischoff, G.; Gocht, W., Das Energiehandbuch, 2nd Ed., Vieweg, Braunschweig 1976, pp. 151/73, 151).
[11] De Wijs, H. J. (Geol. Mijnbouw **39** No. 3 [1960] 61/5).
[12] Smolin, P. P. (Dokl. Akad. Nauk SSSR **163** [1965] 212/5; C.A. **63** [1965] 12921).

1.2 Distribution and Abundance of Thorium in the Earth

Thorium and U have a wider distribution than is generally expected. For instance, their bulk content in granitic rocks is about thousand times that of Au, about one hundred times that of Ag, and about equal that of Sn [1]. Thorium in the earth's crust is about three times as abundant as U and a little less abundant than Pb [2], or it is three times as abundant as Sn [3], twice as abundant as Be and As, and nearly as abundant as Pb and Mo [4]; refer also to [5, pp. 189/91].

Thorium, like U, is generally concentrated in the later stages of magmatic crystallization [6, p. 427], see also [7, 33], and it behaves in fact much like the RE and U, with which elements it is frequently associated [3]. Concentration of Th is much slower than of U and occurs not only in alkaline, but also in acid and ultra-acid magmas, whereat the physico-chemical environment of Na-rich alkaline magmas is more suitable for an enrichment than that of K-rich acid magmas. During magmatic differentiation, Th may be present in the melt in completely moveable and invariable compounds as, for instance, complex carbonate or halogen compounds with alkali metals, or simple halogenides (especially with chlorine) [7]. For the complex formation of Th, refer also to [34]. Migration and fixation of U and Th are controlled by volatile activities during magmatic fractionation. For example, U is concentrated by F-complexing and Th is complexed with phosphates. At subsolidus temperature, U complexes are much less stable than Th counterparts [35]. And under hypogene conditions and at favorable pH and concentrations of CO_2 and alkalis, either a Th concentration with admixtures of U and RE, mixed U-Th ores, or pure U concentrations are formed from material of a single ore-generating (magma) chamber [8]. As follows from a study of Th and U distribution in volcanic rocks, both elements may be more prone to volatilization than the more abundant trace elements, and could have separated much earlier, perhaps in advance of the lava column (prior to extrusion), from the magma in a plutonic or hypabyssal environment [9].

A part of Th is brought into solution during weathering but is easily hydrolyzed (because of its very high ionization potential [6, pp. 427/31]) and accumulates in the hydrolyzates (hydroxides of Fe and Al [6, pp. 427/31], [10, 11]), which are notably rich in Th [12]. Under supergene conditions, Th may be leached by alkali carbonates and precipitated as a gel by changed acidity [13]; or an elevated migration is promoted by an alkali-carbonate or acid composition of solutions, by the presence of soluble soil acids, by an elevated temperature, by a moist warm climate, by the primary form of Th present in the metamict host mineral, and by other mechanisms [14].

As follows from a compilation of literature data, the elements Th, U, and K show a strong concentration towards the surface of the earth and are enriched in the continental crust (refer also to the next paragraphs). The mean abundance of Th, estimated by two independent methods, differs by a factor of two and is 5.3 to 11.4 ± 2 ppm (derived from geochemical data based on geological premises) or 2.5 to 4.6 ppm (derived from heat flow data) [15]; for details, see original papers cited in that reference. In contrast, the upper mantle (for a depth of ≦600 to 700 km) contains only about 0.175 ppm Th [16].

Based on different models of earth-crust (or lithosphere) composition (for the details, see original papers), the mean or Clarke concentration of Th varies between 8 and 12 ppm [5, p. 190], is 5.7 ppm Th [17], 4.8 ppm Th [18], or as low as 3.9 ppm Th [19]. See also, the lithosphere contains 11.5 ppm Th or 0.00050 atom per 100 Si (in sediments) [6, p. 75], or between 11 and 15 ppm Th [20]. For the content of [232]Th in the earth crust, see p. 2.

The content of Th in the upper (16 km thick) and lower layer (21 km thick) of the continental crust is calculated as follows for different areas and the surface heat flow given [15]:

part of the crust	content of ppm Th in		
	shield area with 1.0 µcal/cm$^2 \cdot$ s	continental area with	
		1.2 µcal/cm$^2 \cdot$ s	1.5 µcal/cm$^2 \cdot$ s
upper layer	4.00	5.28	7.48
lower layer	1.48	1.68	2.28

Based on the formula $A_x = A_o e^{-x/D}$ (with D = depth parameter, A_x and A_o = heat production rate at depth x and the surface, respectively), an average content of 4.43 ppm Th in a 33.3-km thick crust is calculated [21]. For the crust of the southwestern Australian Shield area is given: 20 ppm Th in the surface crust of 4.5 km thickness, 10 ppm Th in the surface young crust down to about 10 km, 2 ppm Th in medium- to high-pressure granulites at about 20 km depth, and 0.4 ppm Th at the base of the crust at about 35 km depth [22]. A mean content of 10.3 ± 4.2 ppm Th results from 32 composite samples prepared from 330 rocks collected in the Precambrian Canadian Shield [23].

The mean content of Th in the main types of lithospheric rock, derived from different authors, is given in [5, pp. 195, 197] as follows:

rock type	ppm Th	rock type	ppm Th
ultrabasic rocks	0.004 or 0.005	shale	12
basic rocks (basalts)	3 or 4	shale + clay	11
intermediate rocks		sandstone	1.7
syenite	13	carbonate	1.7
diorite	7	clay	7
acid rocks			
rich in Ca	8.5		
poor in Ca	17		
granite etc.	18		

Another compilation of literature data in [37, p. 312] gives as ranges for igneous rocks:

rock type	ultra-basic	basic		silicic		alkaline silicic pegmatite
		intrusive	extrusive	intrusive	extrusive	
ppm Th	low	0.5 to 5	0.5 to 10	1 to 25	9 to 25	1 to 2
ratio Th:U	—	3 to 4	3 to 7	2 to 6	4 to 7	variable (0.4 to 1.5)

And for residual rocks is given 8 to 132, mean 42 ppm Th for bauxites; 6 to 44, mean 24 ppm Th for bentonites [36, p. 325]. Computer-derived geochemical balances and element abundances from numerous literature data up to 1963 give as Th content: 11.4 ppm for igneous rocks, 13.1% for shale, 3.94 ppm for sandstone, 2.01 ppm for (terrestrial) carbonate, 0.2 ppm for evaporite, 13.1 ppm for oceanic clay, and 1.01 ppm for oceanic carbonate [24]. For further compilations of Th content in individual lithospheric rock types (partly also including data on individual upper-mantle material, as, e.g., eclogite), for instance, refer to [5, pp. 194/7], [21, 25, 26] and, especially for upper-mantle material, to [27, 28].

Besides vertical differences in Th content of the lithosphere (see pp. 3/4), there are also variations in lateral distribution and/or with time (stratigraphic succession) as is exemplified in the following:

From the ocean floor toward the continental Paleozoic fold belts, the amount of Th (together with a number of other elements) increases rapidly. This change is apparently affected by the distribution of host rocks (basic igneous rocks in the ocean floor and silicic igneous rocks in the continents) [29].

A compilation from literature data gives the following for changes with time of Th content in the earth:

	ppm Th	Ref.
present (young) upper crust	10.5	[18]
present (young) lower crust	1.95	[18]
post-Archean middle and lower crust	6.0	[30]
Archean upper crust	2.9	[18]
Archean middle crust[a]	8.4	[17]
Archean lower crust[b]	0.42	[17]
primitive earth mantle	0.0765 or	[32]
	0.070	[18]

[a] Represented by Lewisian amphibolite-facies gneisses. — [b] Represented by Lewisian granulite-facies gneisses.

Variations of the Th content within the stratigraphic succession of the continental crust are also observed. A large data set, which is globally representative and approaches mean element contents for crustal rocks of the five major stratigraphic units (their age is given in Gy = 10^9 a), gives the following Th content in $10^{-4}\%$; n = number of individual samples [31]:

absolute age stratigr. unit	3.8 to 3.4 Gy AR_1		3.4 to 2.5 Gy AR_{2-3}		2.5 to 1.6 Gy PR_1		1.6 to 0.55 Gy PR_{2-3}		0.55 to 0 Gy Ph	
	Th	n	Th	n	Th	n	Th	n	Th	n
komatiites	0.44	20	0.75	88	—	—	—	—	—	—
basic effusives	2.4	424	2.4	2760	2.2	788	3.5	3028	2.9	5723[b]
shales	7.3	57	8.6	1394	12.2	1507	13.8	4683[a]	11.8	6552[c]
granitoids	6.4	102	12.7	5615	21.9	3418	19.0	503	21.4	6969

[a] And 4 averaged samples of 1257 specimens. — [b] And 17 averaged samples of 284 specimens. — [c] And 1674 averaged samples of 28288 specimens.

The distribution of U, Th, and Pb between their terrestrial reservoirs will be influenced, primarily, by three factors: (1) element partitioning ratios; (2) time of crust formation; and (3) the degree to which pre-existing chemical differences are homogenized by recycling. An attempt to bring these factors into balance with the observational data is given in the plumbotectonics model in [37].

6

References for 1.2:

[1] Whittle, A. W. G. (J. Geol. Soc. Australia **2** [1955] 21/45, 22).
[2] Frondel, C. (Proc. Intern. Conf. Peaceful Uses At. Energy, Geneva 1955 [1956], Vol. 6, pp. 568/77; Geol. Surv. Profess. Paper [U.S.] No. 300 [1956] 567/79).
[3] Day, F. H. (The Chemical Elements in Nature, Harrap, London 1963, pp. 1/372, 354/61).
[4] Prakash, B.; Kantan, S. R.; Rao, N. K. (STI-PUB-15/22 [1962] 1/56, 12; N.S.A **16** [1962] No. 17974).
[5] Rösler, H. J.; Lange, H. (Geochemische Tabellen, Verlag Grundstoffind., Leipzig 1965, pp. 1/328).
[6] Goldschmidt, V. M. (Geochemistry, Clarendon, Oxford 1954, pp. 1/730).
[7] Turovskii, S. D. (Zap. Kirg. Otd. Vses. Mineral. Obshch. No. 1 [1959] 5/11, 5/6).
[8] Krol, O. F.; Gurkina, T. V. (Tr. Kaz. Nauchn. Issled. Inst. Mineral'n. Syr'ya No. 6 [1961] 117/22; C.A. **59** [1963] 1406).
[9] Gabelman, J. W. (AAPG Stud. Geol. No. 13 [1981] 23/36; C.A. **98** [1981] No. 129668).
[10] Krol, O. F. (Geol. Geokhim. Mineral. Mestorozhd. Redk. Elem. Kaz. **1966** 60/5; C.A. **67** [1967] No. 45906).

[11] Barretto, P. M. C.; Fujimori, K. (Chem. Geol. **55** [1986] 297/312, 311).
[12] Rankama, K.; Sahama, T. G. (Geochemistry, Univ. Chicago Press, Chicago 1950, pp. 1/912, 570/3).
[13] Burkser, E. S.; Kornienko, T. G.; Ushakova, A. M.; Faibishenko, I. Ya. (Ukr. Khim. Zh. **25** [1959] 334/6; C.A. **1959** 21453).
[14] Drozdovskaya, A. A.; Mel'nik, Yu. P. (Pitannya Geokhim. Mineral. Petrogr. **1963** 80/8 from C.A. **60** [1964] 15633).
[15] Heier, K. S. (Nature **208** [1965] 479/80).
[16] Lovering, J. F. (Nature **197** [1963] 138/40).
[17] Weaver, B. L.; Tarney, J. (Nature **310** [1984] 575/7).
[18] Taylor, S. R.; McLennan, S. M. (Phil. Trans. Roy. Soc. [London] A **301** [1981] 381/99, 384/5).
[19] Larimer, J. W. (Geochim. Cosmochim. Acta **35** [1971] 769/86, 772).
[20] Adams, J. A. S. (in: Nuclear Radiation in Geophysics, Academic Press Inc., New York 1962, pp. 1/17 from N.S.A. **17** [1963] No. 3187).

[21] Haack, U. (Earth Planet. Sci. Letters **62** [1983] 360/6).
[22] Lambert, I. B.; Heier, K. S. (Chem. Geol. **3** [1968] 233/8).
[23] Shaw, D. M. (Geochim. Cosmochim. Acta **31** [1967] 1111/3).
[24] Horn, M. K.; Adams, J. A. S. (Geochim. Cosmochim. Acta **30** [1966] 279/97, 286).
[25] Clark, S. P., Jr.; Peterman, Z. E.; Heier, K. S. (Geol. Soc. Am. Mem. No. 97 [1966] 521/41, 527/34, 536/8).
[26] Wedepohl, K. H. (Handbook of Geochemistry, Vol. 2, Springer, Berlin — Heidelberg — New York 1969, pp. **90-E-1/11, 90-K-1/5, 90-M-1/2**).
[27] Lovering, J. F.; Morgan, J. W. (Nature **197** [1963] 138/40).
[28] Wakita, H.; Nagasawa, H.; Ueda, S.; Kuno, H. (Geochem. J. [Tokyo] **1** [1967] 183/98).
[29] T'ung, Li; Chi-Lung, Jao (Dizhi Xuebao **45** No. 1 [1965] 82/92 from C.A. **63** [1965] 8062).
[30] Bailey, J. C. (Chem. Geol. **32** [1981] 139/54).

[31] Ronov, A. B.; Bredanova, N. V.; Migdisov, A. A. (Geokhimiya **1988** 180/98; Geochem. Intern. 25 No. 9 [1988] 24/42, 30/1).
[32] Anderson, D. L. (J. Geophys. Res. B **88** Suppl. [1983] 41/52, 45; C.A. **100** [1984] No. 195526).

[33] Taylor, S. R. (Phys. Chem. Earth **6** [1965] 133/213, 163).
[34] Mineev, D. A. (Geokhimiya **1960** 131/8; Geochem. [USSR] **1960** 156/66, 158/9).
[35] Wu, T.-W. (Chung-Kuo Ti Chih Hsueh Hui Chuan K'an No. 6 [1984] 219/28, 221).
[36] Adams, J. A. S.; Osmond, J. K.; Rogers, J. J. W. (Phys. Chem. Earth **3** [1959] 298/348).
[37] Zartman, R. E.; Doe, B. R. (Tectonophysics **75** [1981] 135/62).

1.3   Thorium in Minerals

1.3.1   Mode of Occurrence, Distribution Between Minerals, Diadochy, and Metamictization

Mode of Occurrence. Thorium occurs as an important constituent in relatively few minerals (for their description, see Chapter 2 in this volume and in "Thorium" Suppl. Vol. A1b), but there are minor amounts (down to traces) of Th in a large number of other minerals, refer to Table 1, pp. 16/8. Most of the earth's Th is contained chiefly in minerals in which it is a nonessential (occasional, or vicarious) constituent [1, p. 567].

Thorium is not known to occur in nature in elemental form [2] or as sulfide [3], arsenide, sulfosalt, and telluride [4]. With the exception of the few examples of Th minerals containing no other cation in their idealized formula (described in this volume and in "Thorium" Suppl. Vol. A1b), Th occurs chiefly as a minor constituent combined with oxygen and other elements to form (arranged by order of importance, refer also to p. 22) oxides, silicates, phosphates, carbonates, and fluorides [2]. The number of Th minerals is small compared to the number of U minerals, largely because Th does not form secondary minerals [5].

The most effective host minerals for Th are compounds of U, Zr, Nb, or Ce that are isostructural with Th compounds (such as zircon, uraninite, and monazite), or compounds in which mechanisms of coupled substitution are available for the housing of large, polyvalent ions (such as niobate-tantalates and monazite). Almost all these minerals are primary formations, and occur chiefly in pegmatites associated with alkalic or granitic igneous rocks. Only five of them (monazite, xenotime, allanite, zircon, and, to a lesser extent, pyrochlore) have a wide distribution as accessory minerals in igneous or metamorphic rocks [1, pp. 569/70]. Th-rich monazite rivals zircon as ubiquitous and important carrier of Th in common rocks [6, p. 310]. In placer deposits, potential source minerals for Th include monazite and several Th-bearing multiple (complex) oxide minerals. In veins or lodes, the principial Th-bearing minerals are fluorite, monazite, and uranothorite and, less commonly, allanite, brockite, and brannerite. Th-bearing pyrochlore is found in some carbonatites [7]. Occasionally, hypogene Th minerals show enrichment of Th as a result of supergene solutions and extraction of the more soluble constituents [8].

As follows from an evaluation of literature data on U and Th distribution in mineral matter of the earth's crust, there can be distinguished four groups: (1) slightly radioactive major rock-forming salic minerals (as quartz, K-feldspar, plagioclase, nepheline), which are composed mainly of the petrogenetic elements (Si, Al, O); (2) rock-forming melanocratic minerals with normal radioactivity (as biotite, amphibole, pyroxene), which, besides the petrogenetic elements, contain Mg, Fe, etc.; (3) somewhat radioactive accessory and ore minerals (as, e.g., apatite, eudialyte, fluorite, ilmenite, magnetite), whose crystal lattice contains elements with a relatively low Clarke content of $n \times 10^{-1}\%$ (P, Ti, Fe, etc.); and (4) highly radioactive minerals (as, e.g., titanite, allanite, monazite, zircon, loparite), whose crystal lattice contains elements with a Clarke content of $n \times 10^{-2}$ to $n \times 10^{-3}\%$ (Zr, Ce, La, Nb, Ta, etc.). By their Th:U ratio, these minerals may be divided into those displaying a low ratio of <1 to 2 (as, e.g., zircon, zirkelite, titanite, Y allanite), and those with a high ratio of >15 to 20 (as, e.g., allanite, monazite, thorianite) [9]. The major rock-forming minerals usually have Th:U ratios of about 3.5 (which correspond to the crustal average) [10]. Minerals of moderate to high radioactivity in igneous rocks, in decreasing order of frequency, are: zircon, allanite, monazite, xenotime, thorite, pyrochlore, and uraninite [11].

In rocks, Th and U may occur in various modes [12]: (1) as separate U and/or Th minerals (as, e.g., uraninite, thorianite, thorite, uranothorite, brannerite, or coffinite); (2) as iso-morphous admixture to other minerals (as, e.g., in the accessories zircon, apatite, monazite, xenotime, titanite, or allanite); (3) as microscopic and submicroscopic primary inclusions of U-Th minerals (detectable, partly only by their radiogram) in (mostly the dark-colored/mafic) rock-forming silicates; and (4) as microscopic to submicroscopic secondary inclusions forming, especially on grain boundaries and fissures, thin films or minute strings. The second-ary inclusions are formed by hydrothermal processes (mostly unrelated to the primary forma-tion of the host rock), and can easily be leached from the rock by acids.

See also, Th is found: in the major rock-forming minerals; in accessory minerals either as a major constituent (as, e.g., in thorite or thorianite) or by diadochic substitution (as, e.g., in epidote, allanite, titanite, apatite, zircon, monazite, or xenotime); or in an adsorbed form in intergranular spaces or in liquid inclusions [13]. As follows from leaching experiments, U and Th in monazite and "wiikite" samples (from the USSR?) occur only to the extent of several per cent as hydrolysis products or as other insoluble forms in lattice defects [14].

A compilation of literature data gives the following ranges in Th content for the major and accessory minerals of igneous rocks [6, p. 314]:

accessory mineral	ppm Th		rock-forming mineral	ppm Th
monazite	25000	to 2×10^5	biotite	0.5 to 50
allanite, in pegmatite	1000	to 20000	hornblende	5 to 50
allanite, in rock	500	to 5000	pyroxene	2 to 25
zircon, in pegmatite	100	to 2500	K-feldspar	3 to 7
zircon, in rock	50	to 4000	plagioclase	0.5 to 3.0
titanite	100	to 600	quartz	0.5 to 6.0
epidote	50	to 500	olivine	low
apatite, coarse aggregate	50	to 250 (?)		
apatite, in rock	20	to 150		
magnetite	0.3 to	20		
xenotime	low			

The following only shows examples of the mode of occurrence of Th in individual minerals (see below), or of the quantitative distribution of Th between individual minerals (see next paragraphs), for granitic rocks from various plutons:

The radioelements U and Th, detected by α-particle microradiography, in the Kyzyl-Tau granites, central Kazakhstan, occur (1) as separate radioactive minerals (uraninite, uranothorite), displaying long α tracks characteristic of ^{212}Bi (ThC); (2) as diadochic components in accessory minerals (allanite, apatite, monazite) without α tracks of ^{212}Bi; (3) as macro- and submicro-inclusions displaying a large density of α tracks; (4) as products of replacements (in secondary titanite, chloritized biotite, and magnetite, and in zircon, apatite, and monazite crystals); (5) as interstitial forms in veinlets located between (inter-grain) or transecting mineral grains (trans-grain); or (6) in dispersed forms [15]. For a number of granite plutons of Great Britain, the radioelement-bearing accessory minerals, besides apatite and zircon in all cases, are given below; for the compositional fields of U and Th (diagram wt% U vs. wt% Th) in radioelement-bearing accessories (of the named and/or other granite plutons), see original paper [23]:

location (pluton) and characteristics of rocks	radioelement-bearing accessory mineral
Loch Doon; diorite to granite	titanite, allanite, very rarely thorite, and (in a few extremely evolved low-Th rocks) monazite and xenotime
Skiddaw; greisenized granite	uraninite
Cairngorm	
less-evolved granite with 70 to 72% SiO$_2$	titanite, allanite
more-evolved granite with 75 to 77% SiO$_2$, slightly peraluminous	monazite, xenotime, thorite, uraninite, and complex Nb-Ta-U oxides/silicates
Carnmenellis; granite	monazite, uraninite

In accessory minerals of granite (from Kazakhstan?), a greater part of the mineral's total Th content occurs as mineral inclusion. Namely: 50 to 60% of the Th in magnetite (containing 53 ppm Th), 50% of the Th in ilmenite (containing 1.27 ppm Th), and 20 to 37% of the Th in zircon. The remaining Th, in each case, is incorporated diadochically in the crystal lattice [16].

Distribution Between Minerals. Up to 60% of the total Th and U in granites of the Megri pluton, Armenian SSR, is concentrated in accessory minerals (as zircon and its varieties, titanite, allanite, monazite, apatite, etc.). The remaining part is distributed in rock-forming minerals in dispersed form, in submicroscopic inclusions of U-Th minerals, and as a diadochic admixture [17]. In granitic rocks of the Susamyr batholite, central Tien Shan, 30 to 40% of the Th is distributed in rock-forming, the rest in accessory minerals (of which allanite is the main concentrator holding usually 50 to 70% of the rock's total Th). Thorium occurs (1) as molecular dispersion in light-colored rock-forming minerals (quartz, K-feldspar, and plagioclase); (2) as diadochic replacement of RE, Ca, and Zr in accessory minerals (allanite, zircon, titanite, and apatite) and, possibly, in dark-colored rock-forming minerals (biotite and hornblende); and (3) as microscopic and submicroscopic inclusions of Th and U-Th minerals of the uranothorite type [18]. In leucocratic granite of the Khantaus massif, Kazakhstan, the total Th content is distributed between the individual minerals as follows [19]:

	feldspar	quartz	biotite	magnetite	allanite	uranothorite
content of mineral in the rock, wt%	67.5	29.0	3.0	0.3	0.06	0.004
% of Th in mineral	12.3	5.5	16.4	1.8	3.6	54.5

In granitic rocks of the Kochkar region, Southern Urals, Th (diadochically together with RE for Ca) entered the lattice of rock-forming minerals (plagioclase) during the early and middle stages of magma differentiation. During a later enrichment of the magma in Th, the element becomes fixed chiefly in accessory monazite, holding, partly, 50 to 90% of the rock's total Th

[20]. Magnetite and biotite are the main Th carrier in granitoidic rocks of different age from 15 intrusions of the Chatkal-Kurama region, Uzbekistan. The content of Th, range and mean, in the individual minerals is given as follows (for details, see original paper); number of samples is given in parenthesis: 7.6 to 12.5, 8.5 ppm (21) in plagioclase; 4.7 to 13.0, 7.3 ppm (20) in K-feldspar; 5.5 to 15.3, 11.3 ppm (15) in quartz; 13.0 to 99, 33.4 ppm (25) in biotite; 10.1 to 14.5, 8.2 ppm[*] (3) in hornblende; and 13.8 to 135, 35.2 ppm (16) in magnetite [21].

As can be seen from the summarized data below (for the full details, see original paper), Th occurs in all analyzed minerals of rocks from different phases of formation of the Lovozero nepheline syenite massif, Kola Peninsula. Distribution of Th between the minerals is different, as is its behavior during crystallization of the magma. Thus, Th enters as a diadochic admixture in RE minerals (in urtites, lujavrites) during the magmatic stage, then accumulates in the remaining magma and enters in submicroscopic segregations of radioactive minerals (in foyaites and some eudialyte lujavrites) during the postmagmatic stage. For rocks of different phases of formation, the content in wt% of mineral in the rock (column a) and the percentage of total-rock Th contributed by the mineral (column b) is given as follows [22, pp. 47/51]:

minerals from	nepheline syenite of phase I		foyaites of phase II			
	a	b	a	b	a	b
feldspar + nepheline	89.7	82.9	77.8	47.0	92.3	39.5
acmite ± amphibole	4.7	4.5	10.5	1.7	4.1	1.8
titanite	2.7	10.5	—	—	—	—
eudialyte	—	—	10.0	25.9	0.1	0.3
others	—	—	0.2[a]	33.6	2.4[b]	1.2

minerals from	lujavrites of phase III			
	a	b	a	b
feldspar + nepheline	66.6	32.1	60.9	20.3
acmite ± amphibole	22.6	5.2	24.3	4.2
titanite	—	—	—	—
eudialyte	10.9	58.2	12.8	18.4
others	—	—	2.0[c]	3.6

[a] Loparite. — [b] Murmanite. — [c] Murmanite and lamprophyllite.

The distribution of Th (in percentage of total Th in rock) between rock-forming (essential) and accessory minerals and the portion not found in monomineralic fractions for different rock types of the Lovozero massif is given as follows [22, p. 55]:

---

[*] Given in the original paper, but 11.6 ppm is calculated from the individual data.

	ppm Th in rock	percentage of Th		
		in essential minerals	in accessory minerals	not found in mono-mineralic fraction
nepheline syenite, equigranular	7.6	87.4	10.5	2.1
foyaite	32.7	41.3	1.5	57.2
	23.2	50.9*)	59.5*)	—
acmite lujavrite	91	6.4	91.6	2.0
acmite-amphibole lujavrite	32.7	16.3	92.8	—
eudialyte lujavrite				
mesocratic	80	10.5	86.2	3.3
leucocratic	13.4	37.4	58.2	4.5
mesocratic	38	24.5	20.2	55.3

*) As given in the original paper.

Diadochy. The most important crystal-chemical factors affecting the incorporation of an element in minerals are: effective radius (inter-atomic distance), coordination number, form of the coordination polyhedron, structure of the outer electron shell, valency, and electro-negativity of the atom (refer also to [28]). The physico-chemical conditions of the mineral-forming process also have to be considered. Based on the effective radius and electro-negativity, there is a series of crystallochemical affinity for heterovalent substitution: Ca, Y, RE, Th, U [32].

The geochemistry of Th (and thus its distribution in minerals) is characterized by its crystal-chemical (diadochic) relationships to several elements with similar ionic radii, such as Ca^{2+}, RE^{3+}, and U^{4+}. A close and very important relationship exists between Th^{4+} and U^{4+} and, less importantly, with Ce^{3+} [24]. Therefore, a mutual substitution of U and Th in primary minerals is common, with the exception of low-temperature uranitite (pitchblende). Indeed, the geochemistry of both elements (in primary products, see below) is so closely linked that it is difficult to discuss one without the other [25]. By substituting Ca^{2+}, Y^{3+}, Zr^{4+}, and RE, U^{4+} and Th^{4+} preferentially enter the open lattices of accessory minerals [26]. Because of their larger ionic radii and higher valencies, neither U nor Th substitute to any degree in the structure of rock-forming minerals [25]; see also a compilation of Th contents in minerals given in "Thorium", 1955, pp. 9/10. But note that there are statements of a possibly diadochic occurrence of Th in rock-forming minerals, see p. 9. Thorium, like U, is diadochic chiefly with RE (especially Ce) and Zr, but it cannot be oxidized (it does not form secondary minerals, see p. 7, and so its geochemical behavior is departed from U); Th remains mainly in its few original minerals which are largely insoluble [27, p. 10]. Close geochemical relationships exist among Th, U, Ce, Zr, and Hf by similarities in their electronic structures. The similarities are illustrated by the isostructural relation of their dioxides and orthosilicates and by extensive mutual solid-solution series in these and other groups of compounds [1, pp. 567/8, 570/3]; refer also to Chapter 2.2 (Complex Oxides), pp. 75/236.

The explanation of the distribution of RE, U, and Th in minerals on the basis of (predomi-nantly the Goldschmidt) ionic radii alone (see above) is obviously insufficient, as very different values have since been given by other authors (see p. 2) and a mathematical determination of the degree of diadochy was developed (taking into account, besides ionic radius, a number of factors; for details, see [28]). Both facts lead to the following statements for Th: Y^{3+} and Th^{4+} ions are nearer in size than Y^{3+} and U^{4+} or Ce^{3+} and Th^{4+}; as follows from computed diadochy indices, substitution of La, Ce, and Pr by Th is easier than by U; conversely, substitution of Y and Er by U is easier than by Th; and Gd and Tb tend to be replaced by U and Th with approximately equal ease. The geochemical resemblance between the corresponding actinides and lanthanides is more than qualitative [29]. Note also, a study of geochemical series of elements shows that during the simultaneous entrance of Th and U into minerals, the natural genetic series of elements becomes complicated. The La and Ce, present in U minerals, are bound with Th. Therefore, in the common Th-U genetic series a part of the elements are bound with U, another only with Th, although the ionic radii of La and Ce, for instance, are sufficiently close to that of U [30]. Another investigation of diadochic substitution states that the radius ratio between substituting and substituted elements is of secondary importance. Instead, an essential condition for substitution is that the ratio of the atomic number and atomic weight of the participating elements is close to 1, 2, or 4 [31].

For details and examples of the diadochic behavior of Th, see "Thorium" 1955, pp. 6/9, comprising older, pre-1950 data; or "Seltenerdelemente" A4, 1979, pp. 87/96, comprising data up to 1977. Refer also to the relevant paragraphs in chapters covering chemistry of the individual minerals described in this volume and in "Thorium" Suppl. Vol. A1b. The most recent and exhaustive compilation of Th diadochy given in [33] comprises a list of minerals containing RE (and partly also Th, as can be seen from the crystal-chemical formulas) and gives types of substitution together with application to selected mineral groups. Thus, Th data are given, for instance, for fluorite, monazite, xenotime, zircon, apatite, niobates-tantalates-titanates, allanite, and chevkinite/perrierite.

Metamictization. Radioactive minerals sometimes are in the metamict state. This means although they exhibit crystalline form and habit, which should render them anisotropic, they are partly or totally isotropic/noncrystalline when examined by means of optical, X-ray, or differential thermal techniques [27, pp. 10/2], [34]; such minerals usually contain $>0.5\%$ U_3O_8 or ThO_2 in combination with RE and weak bases like Nb, Ta, and Ti [35]. As follows from compilations given in [36, 38, 56], besides a variety of other reasons (see original papers), complex diadochic substitution, radioactive element content (involving natural self-ir-radiation), and preceding hydratization of the minerals are generally emphasized as causes of metamictization or isotropization.

The characteristics of metamict minerals are given as follows:
— they are weakly to strongly radioactive [27, p. 12], [36];
— they are amorphous or nearly amorphous to X-ray diffraction [36], or yield a faint, fuzzy diffraction pattern [27, p. 12];
— they often are optically isotropic [27, p. 12], [36], and usually have a black or dark brown color and pitch-like luster [36];
— they have a conchoidal fracture and are without cleavage [27, p. 12], [36];
— they have an abnormally low specific gravity [27, p. 12], [36];
— they have originally weakly bonded structures [27, p. 12] with a low degree of chemical stability [36]; refer also to p. 13;
— they contain abnormal amounts of nonessential water [27, p. 12] which is lost during heat treatment [36];
— they may be focus of fractures radiating into surrounding mineral grains [27, pp. 22/3], [36], or may cause discoloration and/or pleochroic haloes in neighboring/enclosing minerals [27, pp. 20/1];
— they recrystallize during heating, usually at a definite temperature (below the fusion or sintering temperature [36]), with a considerable to minor evolution of heat (and light), to a crystalline polymorph (usually a polycrystalline aggregate) [27, p. 12], refer also to p. 13.

Additional changes of physical and structural properties with increasing degree of metamictization are as follows [36]:
— the isotropic index of refraction is usually lower than the indices for the nonmetamict equivalent;
— pleochroism diminishes;
— the mineral becomes less transparent;
— unit-cell dimensions of partially metamict specimens are somewhat larger than for non-metamict counterparts;
— the mineral often becomes thermoluminescent on heating (is pyrognomic [27, pp. 12, 22]);
— resistance to chemical attack, especially by acids, decreases; and
— oxidation of some cations occurs (especially of Fe^{2+} and U^{4+}).

Metamict minerals often display a complexity of chemical composition owing to large-scale heterovalent isomorphism within the various parts of the structure, and therefore the largest number of the isodesmic multiple oxides containing Nb, Ta, and/or Ti occurs in the metamict state; as also do numerous silicates (predominantly neso- and sorosilicates), but rarely phosphates [36]. This is corroborated by a complete systematic review of metamict minerals given in [37]; see also [38, p. 219], [51]. Note that the simple oxides thorianite and uraninite do not become metamict, apparently because of the inherent stability of their fluorite-type crystal structure [1, p. 569]. Factors influencing the metamict state of minerals are the stability of crystal lattice, content of radioactive elements, and geological age of their formation [39].

Metamictization results from the effects of irradiation by α particles originating from the decay of U and/or Th series radioisotopes, as is outlined in a short compilation of radiation-damage effects in natural crystals given in [38, pp. 214/6]. For additional data on radioactive disturbances in mineral structures, see [40]; more recent investigations on the metamict state (e.g., its nature, mechanism of formation, radiation damage) are given, for instance, in [41 to 50]. For quantitative data on heat production from stored energy in metamict minerals, see [55].

Metamict minerals can be recrystallized during heat treatment to a polycrystalline aggregate. But doubts remain that the newly acquired form is identical in structure with the primordial mineral [52], or that it is the natural low-temperature crystalline pre-metamict mineral, because under new equilibrium conditions imposed by prolonged high-temperature treatment, various internal physical and chemical changes may have occurred. Usually, the cooling is fast enough to freeze this artificial equilibrium state, hence one may measure a crystal structure merely belonging to the material at a high temperature [34]. Refer also to [53] for the behavior of metamict minerals on heating and the general problem of metamictization. Nevertheless, the DTA data obtained from heat treatment (done at best under defined conditions), together with the X-ray powder diffraction data of the newly formed products, are a powerful tool widely used to identify metamict minerals; refer, for instance, to [38, pp. 217/8], [54], and the relevant paragraphs on thermal behavior and/or crystal structure for the Th minerals described in this volume and in "Thorium" Suppl. Vol. A1b.

References for 1.3.1:

[1] Frondel, C. (Geol. Surv. Profess. Paper [U.S.] No. 300 [1956] 567/79; Proc. Intern. Conf. Peaceful Uses At. Energy, Geneva 1955 [1956], Vol. 6, pp. 568/77).
[2] Twenhofel, W. S.; Buck, K. L. (Geol. Surv. Profess. Paper [U.S.] No. 300 [1956] 559/66, 562; Proc. Intern. Conf. Peaceful Uses At. Energy, Geneva 1955 [1956], Vol. 6, pp. 562/7).

14

[3] Wilke, A. (in: Bischoff, G.; Gocht, W., Das Energiehandbuch, 2nd Ed., Vieweg, Braunschweig 1976, pp. 151/73, 151/2).

[4] Garside, L. A. (Bull. Nevada Bur. Mines Geol. No. 81 [1973] 1/121, 4).

[5] Nininger, R. D. (Minerals for Atomic Energy, Van Nostrand, New York 1954, pp. 1/367, 87).

[6] Adams, J. A. S. (Phys. Chem. Earth **3** [1959] 298/348).

[7] Staatz, M. H.; Olson, J. C. (Geol. Surv. Profess. Paper [U.S.] No. 820 [1973] 468/76, 471).

[8] Kelly, F. J. (Inform. Circ. U.S. Bur. Mines No. 8124 [1962] 1/38, 5).

[9] Smyslov, A. A. (Tr. Inst. Geol. Geofiz. Akad. Nauk SSSR Sibirsk. Otd. No. 286 [1975] 10/18, 272/83, 11; C.A. **84** [1975] No. 138633).

[10] Taylor, S. R. (Phys. Chem. Earth **6** [1965] 133/213, 163).

[11] Faul, H. (Nuclear Geology, Wiley, New York 1954, pp. 1/414, 84).

[12] Barthel, F.; Mehnert, K. R. (Neues Jahrb. Mineral. Abhandl. **114** [1970] 18/47, 24).

[13] Bajo, C.; Rybach, L.; Weibel, M. (Chem. Geol. **39** [1983] 299/318, 300).

[14] Starik, I. E.; Lazarev, K. F. (Radiokhimiya **1** No. 1 [1959] 60/5; Radiochem. [USSR] **1** [1959] 174; C.A. **1959** 18763).

[15] Baranov, V. I.; Tu, L.-T.; Korobkov, V. I. (Geokhimiya **1962** 411/9; Geochem. [USSR] **1962** 469/83).

[16] Trofimova, L. A.; Syromyatnikov, N. G.; Ivamova, E. I.; Turov, G. I. (Tr. Inst. Geol. Nauk Akad. Nauk Kaz. SSR No. 31 [1971] 168/77 from C.A. **80** [1974] No. 135861).

[17] Meliksetyan, B. M. (Izv. Akad. Nauk Arm. SSR Geol. Geogr. Nauki **14** No. 2 [1961] 21/41, 33, 38; C.A. **55** [1961] 17397).

[18] Leonova, L. L. (Geokhimiya **1962** 490/6; Geochem. [USSR] **1962** 571/8).

[19] Leonova, L. L.; Rehne, O. S. (Geokhimiya **1964** 778/93; Geochem. Intern. **1** [1964] 775/81).

[20] L'vov, B. K. (Tr. Vses. Nauchn. Issled. Geol. Inst. [2] **95** [1963] 13/44, 31, 37/40; C.A. **61** [1964] 15877).

[21] Magdiev, R. A.; Aizenshtat, V. I.; Gor'kovoi, O. P. (Zap. Uzbek. Otd. Vses. Mineral. Obshch. No. 29 [1976] 58/62; C.A. **87** [1977] No. 204538).

[22] Polyakov, A. I.; Kot, G. A. (Geokhimiya **1965** 73/85; Geochem. Intern. **2** [1965] 47/58).

[23] Webb, P. X.; Tindle, A. G.; Barritt, S. D. (Geophys. Res. Letters **14** [1987] 299/302).

[24] Goldschmidt, V. M. (Geochemistry, Clarendon, Oxford 1954, pp. 1/730, 427).

[25] Steacy, H. R.; Kaiman, S. (Short Course Handb. Mineral. Assoc. Can. **3** [1978] 107/40, 108).

[26] Wu, T.-W. (Chung-Kuo Ti Chih Hsueh Hui Chuan K'an No. 6 [1984] 219/28, 221).

[27] Heinrich, E. W. (Mineralogy and Geology of Radioactive Raw Materials, McGraw-Hill, New York — Toronto — London 1958, pp. 1/654).

[28] Vendel, M. (Magy. Tud. Akad. Musz. Tud. Oszt. Kozl. **23** [1958] 153/93; C.A. **1959** 2943. Acta Geol. [Budapest] **5** [1958] 301/433; C.A. **1959** 11916. Neues Jahrb. Mineral. Abhandl. **92** [1958] 184/202).

[29] Mineev, D. A. (Geokhimiya **1960** 131/8; Geochem. [USSR] **1960** 156/66, 157/9).

[30] Baskakov, M. P. (Vopr. Mineralog. Geokhim. Akad. Nauk Uzbek. SSR Inst. Geol. Geofiz. **1964** 56/61 from C.A. **62** [1965] 3840).

[31] Vlasov, K. A. (Dokl. Akad. Nauk SSSR **155** [1964] 1091/4; Dokl. Acad. Sci. USSR Earth Sci. Sect. **155** [1964] 162/4).

[32] Povarennykh, A. S. (in: Geokhimiya, Nauka, Moscow 1972, pp. 104/15, 104, 106, 108).

[33] Burt, D. M. (Rev. Mineral. **21** [1989] 259/307).

[34] Whittle, A. W. G. (J. Geol. Soc. Australia **2** [1954] 21/45, 36/8).

[35] Whittle, A. W. G. (Bull. Geol. Surv. South Australia No. 30 [1954] 126/51, 127).

[36] Mitchell, R. S. (Mineral. Record 4 [1973] 177/82).

[37] Bouška, V. (Acta Univ. Carolinae Geol. 1970 No. 3, pp. 143/69).

[38] Mitchell, R. S. (Mineral. Record 4 [1973] 214/23).

[39] Zhang, Peishan; Tao, Kejie (Zhongguo Xitu Xuebao 5 [1987] 1/7 from C.A. 107 [1987] No. 202264).

[40] Rankama, K. (Isotope Geology, Pergamon, London 1954, pp. 1/535, 125/9; Progress in Isotope Geology, Interscience, New York–London 1963, pp. 1/705, 63/80).

[41] Pachadzhanov, D. N.; Bandurkin, G. A. (Izv. Akad. Nauk Tadzh. SSR Otd. Geol. Khim. Tekhn. Nauk 1963 40/7; C.A. 61 [1964] 1615).

[42] Lipova, L. M. (Geokhimiya 1966 729/33; Geochem. Intern. 3 [1966] 549).

[43] Polezhaev, Yu. M. (Geokhimiya 1974 1648/53; Geochem. Intern. 11 [1974] 1157/61).

[44] Ewing, R. C. (Ceram. Glass Radioact. Waste Forms 1977 139/46; C.A. 90 [1978] No. 30882).

[45] Belyaeva, I. D. (in: Semenov, E. I.; Lebedeva, S. I., Mikrodiagnostika Redkometal'nykh Mineralov, Nauka, Moscow 1979, pp. 186/90).

[46] Ewing, R. C.; Haaker, R. F. (NUREG-CP-0005 [1979] 651/75 from C.A. 92 [1980] No. 44745).

[47] Yada, K.; Tanji, T.; Sunagawa, I. (Phys. Chem. Minerals 7 [1981] 47/52).

[48] Headley, T. J.; Ewing, R. C.; Haaker, R. F. (Nature 293 [1981] 449/50).

[49] Karioris, F. G.; Gowda, K. A.; Cartz, L.; Labbe, J. C. (J. Nucl. Mater. 108/109 [1982] 748/50; C.A. 97 [1982] No. 135495).

[50] Ewing, R. C.; Chakoumakos, B. C.; Lumpin, G. R.; Murakami, T. (MRS Bull. 12 No. 4 [1987] 58/66; C.A. 107 [1987] No. 185301).

[51] Noddack, W.; Jakobi, R. (Z. Anorg. Allgem. Chem. 284 [1956] 208/33, 210/3).

[52] Ruh, R.; Wadsley, A. D. (Acta Cryst. 21 [1966] 794/8).

[53] Pyatenko, Yu. A. (Geokhimiya 1970 1077/83; Geochem. Intern. 7 [1970] 758/63).

[54] Sidorenko, G. A. (Rentgenogr. Mineral. Syr'ya No. 1 [1959/62] 108/12 from C.A. 59 [1963] 8214).

[55] Kurath, S. F. (Am. Mineralogist 42 [1957] 91/9).

[56] Pabst, A. (Am. Mineralogist 37 [1952] 137/57, 137/42).

1.3.2 Content of Thorium as Accessory Element in Minerals

Owing (partly) to its diadochic behavior, Th occurs locally in higher amounts in other minerals that may be end-members of solid-solution series comprising a thorian and non-thorian component (as, e.g., zircon, coffinite, xenotime), or belong to the same mineralogical class/crystal-structure type as a definite Th mineral (as, e.g., loparite, pyrochlore, columbite, fergusonite). Without striving for completeness, and supplementing the glossary of U- and Th-bearing minerals given in [1], in the following compilation (Table 1 on pp. 16/8) only examples for both above cases are given from the more recent literature of minerals containing >1% Th or ThO_2. Preference is given to references containing a greater number of analyses, and the highest mentioned content is always given. Arrangement is according to the crystal-chemical classification of Strunz [2]; formulas are mainly after [2], partly after [3], or as given in the cited paper.

Table 1
Minerals Containing $> 1\%$ Th or ThO_2 as an Accessory Element.
Mineral species marked with an asterisk very often, or usually, contain Th.

mineral class; mineral species and formula	maximum Th content and reference
simple halogenides	
fluocerite, $(Ce,La)F_3$	1.56% ThO_2 [4]; 3.84% Th [5].
M_2O_3-type oxides	
lueshite, $NaNbO_3$	up to 2.9% Th[1]; 5% Th [6].
dysanalyte, $(Ca,Na,Ce)(Ti,Nb,Fe)O_3$	1.13% ThO_2 [7, p. 418].
loparite, $(Ce,Na,Ca)_2(Ti,Nb)_2O_6$	4.0% ThO_2 [7, p. 419]; 11.4% Th [1]; up to 13% ThO_2 [8, p. 33], [9], [7, p. 429] [a].
pyrochlores*, $(CaNa)_2(Nb,Ta)_2O_6(O,OH,F)$	up to 5.5% Th [1]; 5.8% ThO_2 [10]; 6.12% ThO_2 [11]; 6.30% [7, pp. 501/2], [12]; 7.23% ThO_2 [13]; 8.5% ThO_2 [14, 15]; 9.04% ThO_2 [16]; 11.1% ThO_2 [17]; 11.19% ThO_2 [8, pp. 12/23]; 11.28% ThO_2 [18]; 15.28% ThO_2 [19].
davidite, $(La,Ce)(Y,U,Fe^{2+})(Ti,Fe^{3+})_{20}(O,OH)_{38}$	1.18% ThO_2 [20].
MO_2-type oxides	
rutile, TiO_2	up to 3.64% ThO_2 [21, table Anhang A-3]; 5% Th [22].
columbite, $(Fe,Mn)(Nb,Ta)_2O_6$	2.45% ThO_2 [23]; 3.44% ThO_2 [24]; 4.88% ThO_2 [25].
samarskite, $(Y,Ce,U,Fe^{3+})_3(Nb,Ta,Ti)_5O_{16}$	up to 3.7% Th [1]; 5.72% ThO_2 [7, pp. 528/9]; 1 to 10% Th [26].
yttrotantalite, $(Y,U,Fe^{2+})(Ta,Nb)O_4$	up to 2.2% Th [1].
fersmite, $(Ca,Ce,Na)(Nb,Ti,Fe,Al)_2(O,OH,F)_6$	1.48 and 1.55% ThO_2 [27], [7, p. 483].
fergusonite*, $YNbO_4$	up to 6% Th [1]; 6.10% ThO_2 [28]; 6.80% ThO_2 [7, p. 431]; 6.8 to 8.4% Th [29]; 8.01% ThO_2 [30]; 10% Th [31]; $\sim$11% ThO_2 [32]; 12.88% ThO_2 [33].
formanite, $YTaO_4$	1.1% Th [1].
hydroxides	
"limonite"	up to 4.16% ThO_2 [34].
schoepite, $8[UO_2/(OH)_2] \cdot 8H_2O$	up to 25% ThO_2 [35].
clarkeite, $(Na,Ca,Pb)_2U_2(O,OH)_7$	2.2% Th [1]; 2.4% Th [36].
curite, $Pb_2U_5O_{17} \cdot 4H_2O$	$> 10\%$ ThO_2 [35]; 12.06% ThO_2 [37].
carbonates	
parisite, $Ca(Ce,La)_2(CO_3)_3F_2$	1.7% Th [38]; 5.36% ThO_2 [39].
synchisite, $Ca(Y,Ce)(CO_3)_2F$	1.62% ThO_2 [40].
ancylite, $SrCe(CO_3)_2OH \cdot H_2O$	1.81% ThO_2 [41].
phosphates	
xenotime, YPO_4	up to 2.2% Th [1]; 3.91% ThO_2 [42] [b]; 5.0% ThO_2 [43]; 5.0% Th [44]; 6.1% Th [45].

Table 1 (continued)

mineral class; mineral species and formula	maximum Th content and reference
rhabdophane, $(Ce,La)PO_4 \cdot H_2O$	6.04% ThO_2 [47]; 15.6% ThO_2 [48]; ~16% ThO_2 [49].
ningyoite, $(U,Ca,Ce)_2(PO_4)_2 \cdot 1\text{-}2\,H_2O$	1 to 2% Th [50].
crandallite, $CaAl_3(PO_4)_2(OH)_5 \cdot H_2O$	6.70% ThO_2 [51], [35].
apatite*, $Ca_5(PO_4)_3F$	15.46% ThO_2 [52][c].
arsenates	
gasparite-(Ce), $(Ce,RE)AsO_4$	1.95% ThO_2 [54][d].
nesosilicates	
zircon*, $ZrSiO_4$	~3% Th [55, 56]; up to 3.52 [57], 4% ThO_2 [58] or 5.3 [21, pp. 126/31], 6% Th [59], [60, table 13]; 7.35% ThO_2 [42]; up to 10% ThO_2 [62, 63]; 12.3% ThO_2 [64]; 13.1% Th [1]; up to 14.6% Th [65].
coffinite, $U(SiO_4)_{1-x}(OH)_{4x}$	3.06% ThO_2 [66]; 4.94% ThO_2 [42]; 5.99% ThO_2 [60, pp. 26/7]; 6.0% ThO_2 [64]; up to 7% ThO_2 [67].
tombarthite, $Y_4(Si,H)_4O_{12-x}(OH)_{4+2x}$	1.3 (dark) and 6.1% ThO_2 (light colored) [68].
neso-subsilicates	
titanite, $CaTiSiO_5$	9 to 10% ThO_2 [58].
lessingite, ~$(Ca,Ce,La,Nd)_5[(O,OH,F)/(SiO_4)_3]$	1.98% ThO_2 [69]; 3.42% ThO_2 [70].
rowlandite, $Y_4Fe^{2+}Si_4O_{14}F_2$ (?)	1.43% ThO_2 [72]; 2.69% ThO_2 [73].
stillwellite, $(Ce,La,Ca)BSiO_5$	1.80% ThO_2 [74].
gadolinite, $Y_2Fe^{2+}Be_2Si_2O_{10}$	up to 1.5% Th [1]; 1.65% ThO_2 [75]; 1.78 ThO_2 [76]; 2.30% ThO_2 [77]; 4.17% ThO_2 [78]; 4.26% ThO_2 [79].
uranophane, $(H_3O)_2Ca(UO_2)_2(SiO_4)_2 \cdot 3H_2O$	3.4% Th [80][e].
sorosilicates	
thalenite, $Y_3Si_3O_{10}(OH)(?)$	up to 1.6% ThO_2 [81].
hellandite, $(Ca,Y)_6(Al,Fe^{3+})Si_4B_4O_{20}(OH)_4$	up to 1.28% Th [1]; 1.46% ThO_2 [82].
mosandrite, $(Na,Ca,Ce)_3Ti(SiO_4)_2F$	1.24% ThO_2 [83]; 4.13% ThO_2 [84][f]; 21.58% ThO_2 [83][g].
allanite*, $(Ce,Ca,Y)_2(Al,Fe^{3+})_3(SiO_4)_3(OH)$	up to 4.35% Th [1]; 4.6% Th [85]; up to 5% Th [86]; 5.07% ThO_2 [87]; 5.60% ThO_2 [88, 89]; 7.36% ThO_2 [90]; up to 10% Th [45]; 11.93% Th [91]; 22.2 and 37.3% ThO_2 [92][h].
vesuvianite, $Ca_{10}Mg_2Al_4(SiO_4)_5(Si_2O_7)_2(OH)_4$	1.21% Th [93]; 1.40% ThO_2 [94]; up to 2.70% ThO_2 [95].
unclassified silicates	
laplandite, $Na_4CeTiPSi_7O_{22} \cdot 5H_2O$	1.32% ThO_2 [96].
sazhinite, $Na_2CeSi_6O_{14}(OH) \cdot n\,H_2O$	1.30% ThO_2 [97].
yftisite, $(Y,Dy,Er)_4(Ti,Sn)O(SiO_4)_2(F,OH)_6$	1.43% ThO_2 [98].

Table 1 (continued)

mineral class; mineral species and formula	maximum Th content and reference
undefined material	
"gummite"	up to 25% ThO_2 [61].
hydrocerite, $\sim (La,Ce,Th)_2(Si,P)_2O_7 \cdot 5H_2O$	13.3% Th [1]; 15.14% ThO_2 [47, 71].
unnamed metamict RE-Be-B silicate	2.31% ThO_2 [46].
unnamed silicotitanate of Th and RE	~ 4% Th [53].
"thucholite"	up to 48% Th in the ash [1].

Footnotes for Table 1:

[a] Variety irinite (or thorian loparite). — [b] Xenotime overgrowth on thorite. — [c] Th apatite. — [d] Mean of six analyses. — [e] Uranophane-like secondary mineral aggregate. [f] "Vudyavrite", $\sim Ce_4(Ti_2O_6)_3 \cdot n\ SiO_2 \cdot m\ H_2O$, as an amorphous product of alteration of metamict rinkite (lovchorrite) [84]; for vudyavrite, refer also to "Rare Earth Elements" A8, 1984, pp. 277/8. — [g] Vudyavrite. — [h] Two allanite-epidote samples intimately intergrown with zircon (containing <1% ThO_2); abundant apatite of the intergrowths contains only traces of radioelements; monazite is not a part of the intergrowth texture.

References for 1.3.2:

[1] Frondel, J. W.; Fleischer, M.; Jones, R. S. (Geol. Surv. Bull. [U.S.] No. 1250 [1967] 1/69).
[2] Strunz, H.; Tennyson, C. (Mineralogische Tabellen, 6th Ed., Geest & Portig, Leipzig 1977, pp. 1/621).
[3] Fleischer, M. (Glossary of Mineral Species, 5th Ed., Mineral. Record Inc., Tucson 1987, pp. 1/227).
[4] Chistyakova, M. B. (Tr. Mineral. Muzeya Akad. Nauk SSSR No. 23 [1974] 113/76, 138/9).
[5] Styles, M. T.; Young, B. R. (Mineral. Mag. **47** [1983] 41/6).
[6] Parker, R. L.; Sharp, W. N. (Geol. Surv. Profess. Paper [U.S.] No. 649 [1970] 1/24, 16).
[7] Vlasov, K. A. (Geochemistry and Mineralogy of Rare Elements and Genetic Types of Their Deposits, Vol. 2, Israel Progr. Sci. Transl., Jerusalem 1968, pp. 1/916 [Russian original, Moscow 1966]).
[8] Semenov, E. I. (Tipokhimizm Mineralov Shchelochnykh Massivov, Nedra, Moscow 1977, pp. 1/119).
[9] Haaker, R. F.; Ewing, R. C. (Sci. Basis Nucl. Waste Management **2** [1980] 281/8, 284/5).
[10] Kudrina, M. A. (Mineral'n, Syr'e Vses. Nauchn. Issled. Inst. Mineral'n. Syr'ya No. 7 [1963] 108/15 from C.A. **59** [1963] 15070).
[11] Kukharenko, A. A.; Orlova, M. P., Bulakh, A. G.; et al. (Kaledonskii Kompleks Ul'traosnovnykh, Shchelochnykh Porod i Karbonatitov Kol'skogo Poluostrova i Severnoi Karelii, Nedra, Moscow 1965, pp. 1/772, 357).
[12] Borodin, L. S.; Nazarenko, I. I. (Dokl. Akad. Nauk SSSR **115** [1957] 783/6; Dokl. Acad. Sci. USSR Earth Sci. Sect. **112/117** [1957] 745/8).
[13] Gold, D. P. (Intern. Mineral. Assoc. Papers Proc. 4th Gen. Meeting, New Delhi 1964 [1966], pp. 109/25, 122).
[14] Yusuf, S.; Saraswat, A. C. (Current Sci. [India] **46** [1977] 703/4 from C.A. **88** [1978] No. 39959).
[15] Petruk, W.; De'Alton, R. O. (Can. Mineralogist **13** [1975] 282/5).
[16] Bagdasarov, Yu. A. (Zap. Vses. Mineral. Obshch. **98** [1969] 395/406).

[17] Wimmenauer, W. (Eur. At. Energy Community — EURATOM Rept. No. 1827 [1964] 17/30, 26).
[18] Krivokoneva, G. K.; Sidorenko, G. A. (Geokhimiya 1971 187/200; Geochem. Intern. 8 [1971] 113/22, 114/5).
[19] Bonshtedt-Kupletskaya, E. M. (Zap. Vses. Mineral. Obshch. 95 [1966] 134/44).
[20] Gatehouse, B. M.; Grey, I. E.; Kelly, P. R. (Am. Mineralogist 64 [1979] 1010/7).

[21] Stupp, H. D. (Diss. Univ. Köln 1984, pp. 1/280, Table Anhang A-3).
[22] Eliseeva, O. P. (Tr. Inst. Geol. Rudn. Mestorozhd. Petrogr. Mineral. Geokhim. No. 27 [1960] 78/91, 89).
[23] Kudrin, V. A.; Kudrina, M. A.; Shuriga, T. N. (Geol. Mestorozhd. Redk. Elem. No. 25 [1965] 1/148, 85).
[24] Ellsworth, H. V. (Can. Geol. Surv. Econ. Geol. Ser. 11 [1932] 1/272, 260/1).
[25] Arkhangel'skaya, V. V. (Redkometal'nye Shchelochnye Kompleksy Yuzhnogo Kraya Sibirskoi Platformy, Nedra, Moscow 1974, pp. 1/128, 40/1).
[26] Cassedanne, J. P.; Baptista, A. (Anais Acad. Brasil. Cienc. 56 [1984] 71/83, 73).
[27] Makarochkin, B. A.; Es'kova, E. M.; Aleksandrov, V. B. (Dokl. Akad. Nauk SSSR 148 [1963] 179/82; Dokl. Acad. Sci. USSR Earth Sci. Sect. 148 [1963] 93/7).
[28] Butler, J. R.; Hall, R. A. (Mineral. Mag. 32 [1960] 392/407, 396).
[29] Smorchkov, I. W.; Tseitlin, S. G.; Batyreva, N. N. (Tr. Inst. Geol. Rudn. Mestorozhd. Petrog. Mineral. Geokhim. No. 99 [1963] 60/7).
[30] Kuo, C.-T.; Wang, I.-H.; Wang, H.-C.; Wang, C.-K.; Hou, H.-C. (Diqiu Huaxue 1973 No. 2, pp. 86/92 from Am. Mineralogist 60 [1975] 485).

[31] Lisitsina, G. A. (Tr. Inst. Geol. Rudn. Mestorozhd. Petrogr. Mineral. Geokhim. No. 27 [1960] 66/77, 71).
[32] Povilaitis, M. M.; Varshal, G. M. (Tipomorfizm Miner. 1969 196/209, 202).
[33] Makarochkin, B. A.; Mineev, D. A.; Aleksandrov, V. B. (Tr. Mineral. Muzeya Akad. Nauk SSSR No. 16 [1965] 252/8).
[34] Lovering, T. G. (Econ. Geol. 50 [1955] 186/95).
[35] Van Wambeke, L. (Bull. Soc. Fr. Mineral. Cristallogr. 95 [1972] 98/105).
[36] Frondel, C. (Geol. Surv. Profess. Paper [U.S.] No. 300 [1956] 567/79, 576/7).
[37] Sidorenko, G. A. (Rentgenograficheskii Opredelitel' Uranovykh i Uransoderzhashchikh Mineralov, Gosgeoltekhizdat, Moscow 1960, pp. 1/116, 35).
[38] Metz, M. C.; Brooking, D. G.; Rosenberg, P. E.; Zartman, R. E. (Econ. Geol. 80 [1985] 394/409).
[39] Kapustin, Yu. L. (Mineralogiya Karbonatitov, Nauka, Moscow 1971, p. 1/286, 184/5).
[40] Smith, W. L.; Smith, J.; Ross, D. R.; Levine, H. (Am. Mineralogist 45 [1960] 92/8).

[41] Efimov, A. F.; Kravchenko, S. M.; Vlasova, E. V. (Tr. Inst. Mineral. Geokhim. Kristallokhim. Redk. Elem. Akad. Nauk SSSR No. 16 [1963] 141/75, 173).
[42] Ixer, R. A.; Ashworth, J. R.; Pointer, C. M. (Geol. J. 22 [1987] 403/27, 414/5).
[43] Kinnaird, J. A. (J. African Earth Sci. 3 [1985] 229/51, 247).
[44] Meliksetyan, B. M. (Izv. Akad. Nauk Arm. SSR Geol. Geogr. Nauki 14 No. 2 [1961] 21/41, 31).
[45] Staatz, M. H. (Econ. Geol. 73 [1978] 512/23, 517).
[46] Kupriyanova, I. I.; Sidorenko, G. A.; Stolyarova, T. I. (Zap. Vses. Mineral. Obshch. 91 [1962] 573/81, 579).
[47] Semenov, E. I. (Mater. Mineral. Kol'sk. Poluostrova 1 [1959] 91/101, 97/9).
[48] Sveshnikova, E. V.; Semenov, E. I.; Khomyakov, A. P. (Zaangarskii Shchelochnoi Massiv Ego Porody i Mineraly, Nauka, Moscow 1976, pp. 1/80, 48/9).

20

[49] Bowie, S. H. U. (Chron. Mines d'Outre-Mer Rech. Min. **27** [1959] 304/8).
[50] Scharm, B.; Burda, J.; Hofreitr, V.; Sulovský, P.; Scharmova, M. (Casopis Mineral. Geol. **25** [1980] 113/24, 115/6).

[51] Van Wambeke, L. (Am. Mineralogist **56** [1971] 1366/84).
[52] Semenov, E. I. (Tr. Inst. Mineral. Geokhim. Kristallokhim. Redk. Elem. Akad. Nauk SSSR No. 9 [1962] 36/54, 48).
[53] Bonatti, S. (Atti 1st Conv. Geol. Nucl., Rome 1955, pp. 17/21; N.S.A. **14** [1960] No. 3701).
[54] Graeser, S.; Schwander, H. (Schweiz. Mineral. Petrogr. Mitt. **67** [1987] 103/13).
[55] Stussi, J. M. (Compt. Rend. D **271** [1970] 2255/8).
[56] Urunbaev, K. U. (Zap. Uzbek. Otd. Vses. Mineral. Obshch. No. 23 [1970] 146/51, 149).
[57] Yushkin, N. P.; Fishman, M. V.; Goldin, B. A. (Tr. Akad. Nauk SSSR Komi Filial Inst. Geol. No. 9 [1968] 143/62, 145).
[58] Willgallis, A. (Neues Jahrb. Mineral. Abhandl. **114** [1970] 48/60).
[59] Stupp, H. D.; Saager, R.; Thiel, K.; Vorwerk, R. (Appl. Mineral. Proc. 2nd Intern. Congr. Appl. Mineral. Minerals Ind., Los Angeles 1984 [1985], pp. 1109/27).
[60] Smits, G. (MINTEK-M 278 [1987] 1/41; C.A. **107** [1987] No. 10615).

[61] Lang, A. H. (Trans. Can. Inst. Mining Met. **53** [1950] 277/84).
[62] Dutra, C. V. (Bol. Soc. Brasil. Geol. **10** [1961] 25/37 from C.A. **56** [1962] 4396).
[63] Baretto, G. R.; Fujimori, K. (Chem. Geol. **55** [1986] 297/312).
[64] Parslow, G. R.; Brandstätter, F.; Kurat, G.; Thomas, D. J. (Can. Mineralogist **23** [1985] 543/51).
[65] Norton, D. A. (Am. Mineralogist **42** [1957] 492/505).
[66] Glatthaar, C. W.; Feather, C. E. (Appl. Mineral. Proc. 2nd Intern. Congr. Appl. Mineral. Minerals Ind., Los Angeles 1984 [1985], pp. 617/28).
[67] Speer, J. A.; Solberg, T. N.; Becker, S. W. (Econ. Geol. **76** [1981] 2162/75).
[68] Griffin, W. L.; Nilssen, B.; Jensen, B. B. (Norsk Geol. Tidsskr. **59** [1979] 265/71).
[69] Svyashin, N. V. (Tr. Inst. Geol. Ural'sk. Filial Akad. Nauk SSSR No. 70 [1965] 239/44).
[70] Voloshin, A. V. (Izv. Akad. Nauk Kaz. SSR Ser. Geol. **23** No. 6 [1966] 78/81; C.A. **66** [1967] No. 87465).

[71] Vlasov, K. A.; Kuz'menko, M. V.; Es'kova, E. M. (Lovozerskii Shchelochnoi Massiv, Akad. Nauk SSSR, Moscow 1959, pp. 1/623, 427/9).
[72] Kapustin, Yu. L. (Nov. Dannye Mineral. No. 33 [1986] 53/63, 60/1).
[73] Crook, W. W., III; Ewing, R. C.; Ehlmann, A. J. (Am. Mineralogist **63** [1978] 754/6).
[74] Dusmatov, V. D.; Efimov, A. F.; Semenov, E. I. (Dokl. Akad. Nauk SSSR **153** [1963] 913/5; Dokl. Acad. Sci. USSR Earth Sci. Sect. **153** [1963] 154/6).
[75] Aleksandrova, I. T.; Sidorenko, G. A. (Geol. Mestorozhd. Redk. Elem. No. 26 [1966] 66/90, 80/1).
[76] Harada, Z. (J. Fac. Sci. Hokkaido Imp. Univ. IV **3** [1936] 221/362, 303/4).
[77] Ganzeeva, L. V. (Redk. Elem. Syr'e Ekon. No. 7 [1972] 112/24, 123).
[78] Kudrina, M. A.; Kudrin, V. A. (Geol. Mestorozhd. Redk. Elem. No. 9 [1961] 120/6; C.A. **56** [1962] 6935).
[79] Kawai, T. (Nippon Kagaku Zasshi **81** [1960] 1220/1 from C.A. **1961** 7173).
[80] Rimsaite, J. (Geol. Surv. Can. Paper No. 81-23 [1982] 19/30, 22).

[81] Skorobogatova, N. V.; Kostin, N. E.; Sidorenko, G. A. (Geol. Mestorozhd. Red. Elem. No. 26 [1966] 106/17, 110/1).
[82] Oftedahl, J. (Tschermaks Mineral. Petrogr. Mitt. [3] **10** [1965] 125/9).
[83] Semenov, E. I.; Kazakova, M. E. (in: Semenov, E. I., Mineralogiya Shchelochnogo Massiva Ilimausak (Yuzhnaya Grenlandiya), Nauka, Moscow 1969, pp. 42/71, 47, 52/3).

[84] Slepnev, Yu. S. (Tr. Inst. Mineral. Geokhim. Kristallokhim. Redk. Elem. Akad. Nauk SSSR No. 13 [1962] 1/151, 84, 134).

[85] Leonova, L. L. (Geokhimiya 1962 490/6; Geochem. [USSR] 1962 571/8).

[86] Le Van Tiêt; Stussi, J. M. (Sci. Terre 18 [1973] 353/79).

[87] Kumskova, N. M.; Khvostova, V. A. (Geokhimiya 1964 660/71; Geochem. Intern. 1 [1964] 676/86).

[88] Lacroix, A. (Minéralogie de Madagascar, Vol. 1, Challamel, Paris 1922, pp. 1/624, 451/3).

[89] Azimov, P. T.; Smirnova, S. K. (in: Badalov, S. T.; Moisseva, M. I., Mineraly Uzbekistana, Vol. 3, Fan, Tashkent 1976, pp. 129/33; C.A. 91 [1979] No. 110251).

[90] Khvostova, V. A. (Tr. Inst. Mineral. Geokhim. Kristallokhim. Redk. Elem. Akad. Nauk SSSR No. 11 [1962] 1/120, 13/6, 24/36, 39/40, 64, 67).

[91] Gogel', G. N. (Aktsessornye Mineraly Granitoidov Tsentral'nogo Kazakhstana, Nauka, Alma-Ata 1966, pp. 1/181, 94/100).

[92] Oberli, F.; Sommerauer, J.; Steiger, R. H. (Schweiz. Mineral. Petrogr. Mitt. 61 [1981] 323/48, 330/2).

[93] Zykov, S. I.; Stupnikova, N. I.; Pavlenko, A. S. (Geokhimiya 1961 547/60; Geochem. [USSR] 1961 581/95, 588).

[94] Rodygina, V. G. (in: Tatarinov, P. M.; et al., Mineraly i Paragenezisy Mineralov Magmaticheskikh i Metasomaticheskikh Gornykh Porod, Nauka, Leningrad 1974, pp. 127/32; C.A. 84 [1976] No. 108487).

[95] Himmelberg, G. R.; Miller, T. P. (Am. Mineralogist 65 [1980] 1020/5).

[96] Es'kova, E. M.; Semenov, E. I.; Khomyakov, A. P.; Kazakova, M. E.; Sidorenko, O. V. (Zap. Vses. Mineral. Obshch. 103 [1974] 571/5).

[97] Es'kova, E. M.; Semenov, E. I.; Khomyakov, A. P.; Kazakova, M. E.; Shumskaya, N. G. (Zap. Vses. Mineral. Obshch. 103 [1974] 338/41).

[98] Pletneva, N. I.; Denisov, A. P.; Elina, N. A. (Mater. Mineral. Kol'sk. Poluostrova 8 [1971] 176/9).

2 Thorium Minerals

Out of the hitherto (end of 1986) known about 3200 mineral species, only a small number contains Th as an essential component or in greater amounts which are considered in the formula, refer to Fleischer, M. (Glossary of Mineral Species, 5th Ed., Mineralogical Record Inc., Tucson, Arizona, 1987, pp. 1/227). Th is only associated in considerable amounts with about 30 minerals, of which the following percentage belongs to individual mineral classes: oxide etc. 50%, silicate 27%, uranate 17%, and phosphate 6%, see Wilke, A. (in: Bischoff, G.; Gocht, W., Das Energiehandbuch, 2nd Ed., Vieweg, Braunschweig 1976, pp. 151/73, 152, 169).

Out of the Th-mineral species, partly representing endmembers of solid-solution series, described in this volume and in "Thorium" Suppl. Vol. A1b, the following numbers belong to the individual mineral class of

simple oxides	1	nesosilicates with an additional anion	2
complex oxides	10	boro-nesosubsilicates	2
carbonates	2	sorosilicates	4
phosphates	13	cyclosilicates	6
nesosilicates	3	unclassified and unnamed silicates	2+7

2.1 Simple Oxides

The dioxides of Th, U, and Ce have natural analogues in the mineral species thorianite (ThO_2), uraninite (UO_2), and cerianite (CeO_2). Inconsistently, they are comprised under the terms "uraninite group", see, for example, [1, p. 11], or "thorianite group", see, for example, [2, p. 110]. As first established by [3], the minerals are isostructural and crystallize in the cubic fluorite, CaF_2, lattice type.

According to literature data, the dioxides form a complete solid-solution series at elevated temperatures in synthetic material [4]. The related compound ZrO_2 may have the same structure [5], but only in form of its high-temperature polymorph, which is, however, not known in nature. The naturally stable monoclinic ZrO_2 phase, baddeleyite, shows a distorted CaF_2-type structure [4].

In the binary system ThO_2-UO_2, the complete solid-solution series is repeatedly proved for synthetic material; refer to the cited older literature in [1, pp. 47, 49], see also [6]. It is completely synthesized at 1350°C [7, p. 238], as well as between 1200 and 1000°C [8]. Exsolution into Th- and U-rich end members is observed at temperatures below 1000°C [7, p. 239]; for details, see [26, pp. 49, 58]. For corresponding natural samples, the continuity of the series is mostly assumed [9], [1, p. 49], [2, pp. 110, 116], as can also be deduced, for the part of the series with Th > U, from the composition of samples compiled in Table 2, pp. 48/54. But the complete isomorphic series was not found in nature [10]. Note that, formerly, miscibility gaps were reported for Th-rich thorianites with < 10% UO_2, for thorianites with 15.10 to 27.02% UO_2, and for uraninites with > 15% ThO_2 up to equal amounts of Th and U [11]. The latter gap was related to the above-mentioned exsolution phenomenon [26, p. 60]. For natural samples formed at temperatures below 700°C, a miscibility gap seems to exist, which is very narrow for samples from carbonatites. It is influenced not only by the formation temperature, but also by the RE content of the ThO_2-UO_2 phase [12]. A large central gap in the series is assumed by [4], or intermediate members are at least rare [1, p. 49]. But the finding of numerous thorian uraninites and uranian thorianites with 27.0 to 55.5% ThO_2 and

53.0 to 25.0% U_3O_8 in Quebec, Canada, extended the compositional range of known natural ThO_2-UO_2 compounds insofar as to suggest complete isomorphism for the solid-solution series [13].

Nevertheless, most natural material of the ThO_2-UO_2 system is close to one or the other of the end compositions, thorianite and uraninite [14]. The latter contains Th as a vicarious, but not as an essential constituent varying in content from 0 up to usually 14% ThO_2 [4]; compare also similar Th ranges deducible from important sample compilations of uraninite in [1, pp. 16/20], [2, pp. 121/2]. Thorium enters UO_2 in significant amounts only in uraninites of pegmatites and certain high-temperature deposits [14], whereas, especially the hydrothermally formed microcrystalline analogue pitchblende [1, p. 12] as well as coarsely crystalline uraninite of hydrothermal veins and of the well-known black-ore sandstone deposits of the Colorado Plateau are notably deficient in Th and Ce. This fact is ascribed to the geochemical association, rather than to a temperature dependency of the solid solubility [4].

Reviews of literature data suggest that complete solid solution exists among ThO_2 and stoichiometric CeO_2 and UO_2 [15] in binary and ternary systems, in both synthetic and natural material [14]. These statements must be restricted for natural samples, since CeO_2 or solid solutions with $Ce > (Th, U)$ are not known. Instead, Ce^{4+} reaches merely considerable amounts as solid solution in natural UO_2 and ThO_2 [4]. Exceptions are the few examples of the Ce-rich end member cerianite, see [16] and "Rare Earth Elements" A7, 1984, pp. 23/4. The isomorphic solubility up to 50% of CeO_2 or $LaO_{1.5}$ in ThO_2 and of $YO_{1.5}$ in UO_2 is experimentally determined, but in nature maximum RE contents of $\sim 8\%$ in thorianite (see also p. 46) and of $\sim 15.5\%$ in uraninite (see also "Rare Earth Elements" A7, 1984, p. 25) are observed [17]. For solid solution of $YO_{1.5}$ in ThO_2 in synthetic material, see also [18], and for a continuous synthetic series of $(Ce, U)O_2$, see [19].

The fact that Ce does not occur as an essential component, but is only dispersed in solid solution as a vicarious constituent in Th and U minerals, especially in their oxides, is attributed to its valence state. While Ce^{4+}, bearing a close crystallochemical resemblance to Th^{4+} or U^{4+}, is rare, Ce^{3+} has no mineralogically stable counterpart in Th, U, or Zr [4]. Trivalent RE such as La, Nd, Sm, and Gd, as well as Y, go into limited solid solution with ThO_2 in both natural and synthetic material [20]. The C-type ($=$ cubic) polymorphs of La_2O_3, Y_2O_3, etc. have a structure related to that of ThO_2, but with certain cation positions vacant. Therefore, series of ThO_2 with La_2O_3, Y_2O_3, etc. involve coupled entrance of vacancies to provide valence compensation [4].

The distribution of natural minerals in the oxidic Th-U-RE system is displayed in **Fig. 1**, p. 24.

Whereas solid solution among ThO_2 and stoichiometric CeO_2 and UO_2 is complete, the miscibility among these phases with ZrO_2 is limited [15]. In the ThO_2-UO_2-ZrO_2 system, ternary cubic synthetic phases were only found in the lower ZrO_2 range, between 17 and 1 mol% ZrO_2 at increasing ThO_2 content. Besides, only ZrO_2-rich phases with maximum UO_2 content of 8 mol% could be synthesized [22]. For a hypothetical phase diagram ThO_2-UO_2-ZrO_2 at temperatures below a possible solvus in the binary ThO_2-UO_2 system, see [26, p. 58]. In the binary system ThO_2-ZrO_2, a continuous solid-solution series is contradictory, because it shows a central miscibility gap between 25 and 75 mol% ZrO_2 [23]. Binary solid solutions of the oxides of Th, U, and Ce with ZrO_2 crystallize chiefly in fluorite-type structures and are restricted to very high formation temperatures [14]. Cubic ThO_2-ZrO_2 phases with 100 to 17.5 mol% ThO_2 were synthesized at 2000°C, but the ZrO_2 solubility in ThO_2 decreases to $\sim 1\%$ in solid

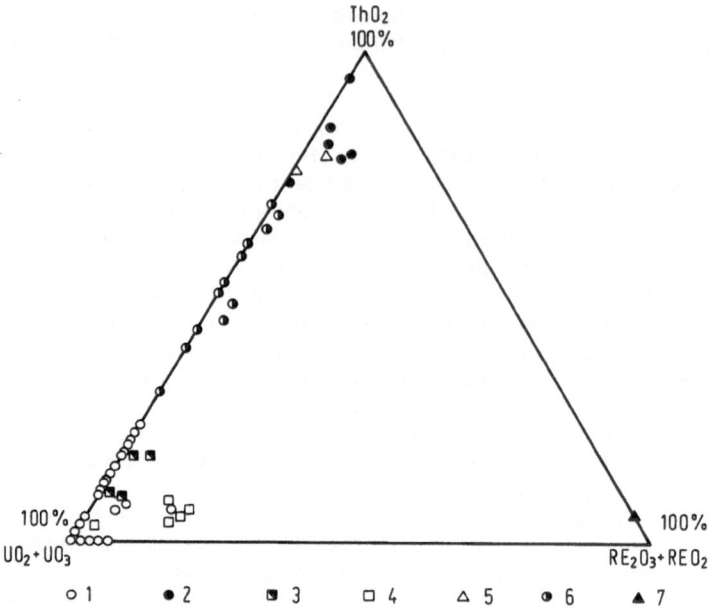

1 = uraninite; 2 = thorianite; 3 = bröggerite[a]; 4 = cleveite[b]; 5 = aldanite; 6 = uranothorianite; 7 = cerianite

Fig. 1. Compositional distribution in wt% of the oxide components of Th, U, and the rare earth
elements in naturally occurring minerals of the thorianite-uraninite group from [2, p. 111].

solution formed at lower temperatures [24]. Mutual solubility drops rapidly with decreasing
temperature, especially in the range between 1500 and 1100°C [14]. At still lower temperatures, between 1200 and 400°C, the maximum extent of ZrO_2 as solid solution in ThO_2, or
vice versa, reaches 2 mol%. Under hydrothermal conditions, below ~400°C a continuous,
although metastable, solid-solution series is observed [7, pp. 234, 237], [26, pp. 57/8]. In
natural ThO_2 and UO_2, Zr occurs only in very small amounts in solid solution. The restricted
solubility is related to the relatively small size of the Zr ion [4]. Note also that there is a close
paragenetic relationship of uranothorianite and uraninite with the orthosilicate of Zr (zircon),
but not with the Zr oxide, a fact which is often observed in early-metasomatic rocks of the
USSR [25].

From the above-mentioned oxides, only thorianite and its varieties are dealt with in the
following. Although Th is nearly always present in uraninite, the mineral should be described
exclusively within the group of U minerals, since uraninite with >15% ThO_2 is very rare. For
older data on uraninite, refer to "Uran" 1936, pp. 12/6, and especially for its Th content, see
"Thorium" 1955, pp. 8/9. An almost current and complete description of cerianite is given in
"Rare Earth Elements" A7, 1984, pp. 23/4; refer also to [1, pp. 53/5].

---

[a] The older varietal name bröggerite, originally introduced for RE-rich uraninite [1, p. 12],
was also applied to designate thorian uraninite [21], see also "Thorium" 1955, pp. 29/30, but
is abandoned in recent literature. — [b] Is a RE-rich uraninite, see "Rare Earth Elements" A7,
1984, p. 25.

References for 2.1:

[1] Frondel, C. (Geol. Surv. Bull. [U.S.] No. 1064 [1958] 1/400).
[2] Chukhrov, F. V.; Bonshtedt-Kupletskaya, E. M. (Mineraly Spravochnik, Vol. 2, Pt. 2, Nauka, Moscow 1965, pp. 1/341).
[3] Goldschmidt, V. M.; Thomassen, L. (Skr. Nor. Vidensk. Akad. Kl. I Mat. Naturvidensk. Kl. **1923** No. 2, pp. 1/48).
[4] Frondel, C. (Geol. Surv. Profess. Paper [U.S.] No. 300 [1956] 567/79; Proc. Intern. Conf. Peaceful Uses At. Energy, Geneva 1955 [1956], Vol. 6, pp. 568/77, 572/4).
[5] Bayer, G. (in: Wedepohl, K. H.; Handbook of Geochemistry II-5, Springer, Berlin 1974, p. **90-A-1**).
[6] Hund, F.; Niessen, G. (Z. Elektrochem. **56** [1952] 972/9).
[7] Mumpton, F. A.; Roy, R. (J. Am. Ceram. Soc. **43** [1960] 234/40).
[8] Slowinski, E.; Elliott, N. (Acta Cryst. **5** [1952] 768/70).
[9] George, D'Arcy (RMO-563 [1949] 1/198, 81/2; N.S.A. **5** [1951] No. 833).
[10] Rankama, K. K.; Sahama, T. G. (Geochemistry, Univ. Chicago Press, Chicago 1950, pp. 1/912, 572).

[11] Wells, R. C.; Fairchild, J. G.; Ross, C. S. (Am. J. Sci. [5] **26** [1933] 45/54, 47).
[12] Ramdohr, P. (Die Erzmineralien und ihre Verwachsungen, 4th Ed., Akademie Verlag, Berlin 1975, pp. 1/1277, 1131).
[13] Robinson, S. C.; Sabina, A. P. (Am. Mineralogist **40** [1955] 624/33, 631/2).
[14] Frondel, C. (Proc. 2nd Nucl. Eng. Sci. Conf., Philadelphia 1957, Vol. 2, pp. 305/12, 308).
[15] Haaker, R. F.; Ewing, R. C. (Sci. Basis Nucl. Waste Management **2** [1980] 281/8, 283).
[16] Graham, A. R. (Am. Mineralogist **40** [1955] 560/4).
[17] Semenov, E. I. (Mineralogiya Redkikh Zemel, Nauka, Moscow 1963, pp. 1/412, 48).
[18] Subbarao, E. C.; Sutter, P. H.; Hrizo, J. (J. Am. Ceram. Soc. **48** [1965] 443/6).
[19] Magneli, A.; Kihlborg, L. (Acta Chem. Scand. **5** [1951] 578/80).
[20] Brauer, G.; Gradinger, H. (Naturwissenschaften **38** [1951] 559/60).

[21] Fleischer, M. (Glossary of Mineral Species, 5th Ed., Mineral. Record Inc., Tucson 1987, pp. 1/227, 23).
[22] Brisi, C. (Atti. Accad. Sci. Torino Classe Sci. Fis. Mat. Nat. **94** [1959/60] 67/76, 67; C.A. **60** [1964] 6260).
[23] Ruff, O.; Ebert, F.; Woitinek, H. (Z. Anorg. Allgem. Chem. **180** [1929] 252/6, 256).
[24] Duwez, P.; Loh, E. (J. Am. Ceram. Soc. **40** [1957] 321/4).
[25] Brodin, B. V.; Sidorenko, G. A.; Shul'gin, A. S.; Dubinchuk, V. T. (Zap. Vses. Mineral. Obshch. **111** [1982] 198/209, 199/200, 202, 205/6).
[26] Mumpton, F. A. (Diss. Pennsylvania State Univ. 1958, pp. 1/90; Diss. Abstr. **19** [1958] 57).

2.1.1 Thorianite ThO_2

2.1.1.1 Name, Varieties, and Definitions

The first identification of the mineral in 1904 in placer deposits of the Balangoda district, Sri Lanka [1], was incorrectly interpreted as uraninite, but in the same year it was recognized as the new mineral species thorianite. It was named in allusion to its predominating Th composition, yielding 72.24 and 76.22% ThO_2 in the first two chemical analyses [2], see also

[3]. Later the thorianite definition was extended to the part of the ThO_2-UO_2 solid-solution series with Th > U (in atomic percent) [3]. Subsequently introduced compositional limits as, for example, $\leq 88\%$ ThO_2 and $\leq 46.5\%$ U [4] or 43.4 to 87.9% Th and $\leq 46.5\%$ U [5, p. 37], unfavorably restricted that definition. A more useful classification comprising the whole compositional spectrum of ThO_2-UO_2 mixtures uses the following names [6, pp. 680/1]:

mineral name	mol% ThO_2	composition between
thorianite	75 to 100	$UO_2 \cdot 3\,ThO_2$ and ThO_2
uranothorianite	25 to 75	$3\,UO_2 \cdot ThO_2$ and $UO_2 \cdot 3\,ThO_2$
uraninite and thorian uraninite	0 to 25	UO_2 and $3\,UO_2 \cdot ThO_2$

For intermediate phases of the ThO_2-UO_2 system between the end member thorianite and uraninite, several mineral names and definitions are developed. Thus, the variety name uranothorianite is used as a synonym for thorianite containing U in substitution for Th, see, for example, [5, p. 37]. The same definition is valid for uranoan (elsewhere written as uranian) thorianite, originally presented by [7]. For an exemplary discussion on the usage of these variety names with reference to the recommendation of the Committee on Nomenclature and Classification of Minerals of the Mineralogical Society of America (according to which, where possible, double words should be combined to single ones), see [8]. Originally introduced for the mineral occurrence at Easton, Pennsylvania, with a ThO_2:UO_2 ratio of approximately 1:1 [9], the name uranothorianite was subsequently suggested for varieties in which the content of U or U_3O_8 exceeds 15% [10 to 12]. However, the nomenclature for the thorianite minerals in the literature is all but uniform (see, e.g., Table 2, pp. 48/54). An example of confusional application of the aforementioned terms can be found in [13], where the names/spellings thorianite, uranothorianite, uranoan thorianite, and uranoanthorianite are used carelessly side by side without intentional discrimination.

Further, special names are given to varieties of thorianite from distinct origins, named in allusion to their type localities. Thus, aldanite was introduced for a mineral from the Aldan region, Eastern Siberia, by [14], that may be described as a Pb- and U-rich thorianite [6, pp. 676/7]. The original name was adopted later by [15] and [5, p. 37]. But its usage as a distinct mineral variety should be discontinued, since the typifying Pb is only of secondary, radiogenic origin, and therefore aldanite is just a very ancient thorianite [6, p. 683]. Mozambikite from the granite pegmatite of Muiane, Alto-Ligonha, Mozambique [18] is mentioned as another variety of thorianite [16, 17]. However, it is omitted in this volume, as it differs considerably from thorianite in chemical composition and crystallographic and physical properties, see the first description in [18] or "Rare Earth Elements" A7, 1984, p. 27. For its more probable agreement with thorite or thorogummite, see [19], [5, p. 38].

In spite of the inaccurate nomenclature of the "thorianite group" minerals (see the foregoing), in order to avoid confusion for the reader, the names given in the original papers are used throughout in the paragraphs of the following sections. Only minerals of the "thorianite group" with molar ratios Th:U > 1:1 are considered.

References for 2.1.1.1:

[1] Coomaraswamy, A. K. (Spolia Zeylan. **1** Pt. 4 [1904] 1 from [3]).
[2] Dunstan, W. R. (Nature **69** [1904] 510/1).
[3] Frondel, C. (Geol. Surv. Bull. [U.S.] No. 1064 [1958] 1/400, 47).

[4] Paone, J. (U.S. Bur. Mines Bull. No. 585 [1960] 863/72, 865).

[5] Frondel, J. W.; Fleischer, M.; Jones, R. S. (Geol. Surv. Bull. [U.S.] No. 1250 [1967] 1/69).

[6] Makarov, E. S.; Lipova, I. M. (Geokhimiya 1962 583/9; Geochem. [USSR] 1962 676/83).

[7] Palache, C.; Berman, H.; Frondel, C. (Dana's System of Mineralogy, 7th Ed., Vol. 1, Wiley, New York 1944, pp. 1/834, 621).

[8] Hanekom, H. J.; van Staden, C. M. v. H.; Smit, P. J.; Pike, D. R. (Mem. Geol. Surv. South Africa No. 54 [1965] 1/185, 66).

[9] Wells, R. C.; Fairchild, J. G.; Ross, C. S. (Am. J. Sci. [5] 26 [1933] 45/54, 47).

[10] George, D'Arcy (RMO-563 [1949] 1/198, 81/2; N.S.A. 5 [1951] No. 833).

[11] Nininger, R. D. (Minerals for Atomic Energy, Van Nostrand, New York 1954, pp. 1/367, 88).

[12] von Backström, J. W. (PEL-112 [1966] 1/19, 3; N.S.A. 21 [1967] No. 22051).

[13] Rimsaite, J. (Geol. Surv. Can. Paper No. 82-1B [1982] 253/66, 255, 259).

[14] Bespalov, M. M. (Sov. Geol. 1941 No. 6, pp. 105/7; Mineral. Abstr. 12 [1953/55] 460).

[15] Frondel, C. (Geol. Surv. Profess. Paper [U.S.] No. 300 [1956] 567/79; Proc. Intern. Conf. Peaceful Uses At. Energy, Geneva 1955 [1956], Vol. 6, pp. 568/77, 569).

[16] Strunz, H.; Tennyson, C. (Mineralogische Tabellen, 6th Ed., Geest and Portig, Leipzig 1977, pp. 1/621, 212).

[17] Ramdohr, P.; Strunz, H. (Klockmann's Lehrbuch der Mineralogie, 16th Ed., Enke, Stuttgart 1978, pp. 1/876, 548).

[18] Cotelo Neiva, J. M.; Correia Neves, J. M. (Intern. Geol. Congr. Rept. 21st Session Norden, Copenhagen 1960, Pt. 17, pp. 53/62).

[19] Fleischer, M. (Am. Mineralogist 45 [1960] 1316/7).

2.1.1.2 Occurrence, Paragenesis, Alteration, Intergrowths, and Inclusions

2.1.1.2.1 Occurrence and Paragenesis

General

The primary occurrence of the natural Th oxide mineral thorianite is related to the migration ability of ThO_2. Its solubility and deposition depends on temperature and pressure conditions and on the composition of the transporting solutions [1]. In pure water, ThO_2 has a very low solubility at 25°C and 1 atm, as well as at 400 and 500°C and 1 kbar (with 2×10^{-8} ion g/L for the latter, higher-temperature conditions) [2]. Such low solubility is reflected in a reduced abundance of the Th oxide (in contrast to the frequent occurrence of the U oxide uraninite) in relatively low-temperature, hydrothermal deposits [3]. Instead, thorianite predominantly forms in high-temperature deposits [4, p. 154], [5, pp. 10/1] as, for example, during the pegmatitic stage [6]. As a primary mineral thorianite is primarily formed in the hypogene zone (deeper part) of the earth's crust, where Th migrates preferentially in ascending acid solutions reaching concentrations of 10^{-4} ion g/L at pH 4.5, enhanced by transport in sulfate complexes; or in moderate alkaline carbonate waters where Th concentration reaches 10^{-5} ion g/L at pH 9 to 10. Precipitation starts by change of pH and/or composition of the solution, probably with the formation of metastable $Th(OH)_4$ which finally may convert to thorianite [7]; for details, see also [1]. In carbonated aqueous solutions (but not in solutions containing Cl, F, sulfate, or phosphate) the Th solubility increases with temperatures between 450 and 800°C at 1 kbar and with concentration of carbonates, which is attributed to the formation of Th complexes (as $[Th(CO_3)_n]^{4-2n}$) at high temperatures. Such conditions might cause Th to concentrate in solutions deriving from carbonate-rich magma (see the favorable occurrence of thorianite in carbonatites, refer to pp. 30/1). The low solubility of ThO_2 in aqueous solutions also explains the relatively high stability of thorianite during metamorphism, apart from carbonate environments [2]. Its very low solubility in the temperature and pH range of natural water characterizes thorianite as the most stable modification of Th oxides/hydroxides in the weathering profile of the earth's crust (for physicochemical details, see the original paper) [8], and causes its common accumulation in residual rocks [9, p. 165] and placer deposits [1, 10].

In the phase relation quadralateral SiO_2-ThO_2-UO_2-ZrO_2, the common primary occurrences of thorianite in quartz-free igneous and metamorphic rocks, particularly in carbonatites, alkaline rocks, and desilicified granitoids, are generally characterized by two assemblages: baddeleyite + uraninite-thorianite solid solution + thorite and baddeleyite + zircon + thorite + thorianite-uraninite solid solution [16]. If primary parageneses of thorianite (or baddeleyite) with quartz are, however, met in nature, a two-phase crystallization sequence must be assumed. The common existence of thorianite-zircon parageneses in nature can be considered in relation with a temperature dependence of stable associations ThO_2-$ZrSiO_4$ and $ThSiO_4$-ZrO_2 in the system ZrO_2-ThO_2-SiO_2; for details, see the original paper [149].

According to the favorable conditions of chemical transport and deposition, primary thorianite chiefly occurs in pegmatites [11 to 13], in pegmatite-like formations of intermediately basic rocks [14], and in carbonatites [14, 15]. Thorianite is also common in metasomatic rocks, such as occurrences in phlogopite-bearing rocks [15] and in so-called "pyrometasomatic deposits" (for explanation, see p. 33) [4, pp. 34, 244/5]. As an accessory mineral, thorianite is frequently found in K- and leucite-bearing (alkaline) massifs [15] and other alkaline rocks. Primary occurrences in granitic and basic intrusives, in volcanic and metamorphic rocks, and in a meteorite are of minor importance, see the following paragraphs on pp. 28/37. Best known and widespread is thorianite as detrital mineral [12], commonly found in placer deposits [11, 13, 17].

In general, thorianite rarely occurs in commercial quantities, except for its occurrence in the carbonatite of Palabora, R.S.A. [18], [19, p. 55]. Note also thorianite mineralizations of economic importance in phlogopite-pegmatite veins, granite pegmatites, and pyroxenites of southern Madagascar [20], especially in the region of Tranomaro [21]; in the commercially mined U-Th deposit of Ross Adams, Alaska [22]; and in the coastal-marine alluvial placers of Sri Lanka [23]; refer also to Chapter Deposits in "Thorium" Suppl. Vol. A 1b.

For a comprehensive compilation of worldwide occurrences of thorianite/uranothorianite, see [4]; particularly for occurrences in Canada, see also [5, 24, 25].

The following detailed description of extraterrestrial and terrestrial occurrences of thorianite is arranged for the latter with respect to the genetic aspect, i.e., beginning with magmatic plutonic (see pp. 29/30) and volcanic host rocks (see p. 30), through carbonatites (see pp. 30/1), pegmatites (see pp. 31/2), metasomatic host rocks and skarns (see pp. 33/4), and metamorphic host rocks (see pp. 34/5) to sedimentary and placer deposits (see pp. 35/7).

Extraterrestrial Occurrence

Thorianite occurs uniquely but as an important accessory mineral, together with baddeleyite, apatite, and niobian perovskite, in Ca-Al-rich inclusions of the B2 type from the Allende meteorite [26].

Terrestrial Occurrences, Magmatic Rocks

In connection with **basic plutonic magmatites**, thorianite is only reported from a gabbro in Tuva, USSR [27]; from a fenitized gabbro of the Gem Park complex, Colorado (see p. 33); and from the Zhaoanzhuang-type iron ore deposit in Archean ultrabasic rock at Wuyang, Henan Province, China [28].

Alkaline Plutonic Magmatites. Occurrences of accessory thorianite in alkaline intrusions are chiefly known from the Asiatic USSR: in Permian syenite of the Sandyk massif, northern Kirghizia [29]; in nepheline syenites of the post-Caledonian Balyktygkhem and the Variscan Dugda massifs, both eastern Tuva [30]; in syenites, nepheline- and leucite-bearing syenites (with an average thorianite concentration of 0.036%) of Permian intrusives, northern Tien Shan [31]; and in K- and leucite-bearing (alkaline) massifs of the Aldan region and the Synnyr Mountain Range, both Siberia [15]. Hypidiomorphic thorianite crystals, often forming monomineralic aggregates or intergrown with pyrochlore and loparite, occur in apatite-amphibole-acmite (aegirine) rocks, representing the final melanocratic products of an agpaitic crystallization series of complexly differentiated nepheline syenites of the Enisei Ridge, south-central Siberia [32]. Black thorianite crystals in association with giant feldspar phenocrysts, zircon, apatite, titanite, and other Th minerals (as thorite,uranothorite) concentrate in central early-intrusive parts of the Kyzyl-Ompul' syenite massif, northern Kirghizia, but are impoverished in later magmatic formations and in secondary mineralizations [29]. In a pegmatitic syenite of the same massif, thorianite reaches 0.00072 wt% of the total mineral content [33]. For thorianite in variably metasomatized alkaline magmatites, see pp. 32/3; and for thorianite as inclusion in steenstrupine from nepheline syenites of the Ilimaussaq batholith, southwestern Greenland, see p. 37.

Acidic Plutonic Magmatites. Most radioactive granitoids (with Th $\geq$ 40 to $50 \times 10^{-4}\%$ and U $\geq$ 5 to $10 \times 10^{-4}\%$) contain thorianite (and thorite) as highly radioactive minerals [34]. In the heterogeneous radioactive granitoids of the Grenville Structural Province in Canada, uranothorianite occurs especially in hybrid mineralized zones [35] and in contact zones between pyroxenite, granite pegmatite, and fluorite-carbonate veins, common in the Bancroft area. U-free thorianite is very rare and is found only in oxidation zones, where the original U was leached by oxidizing solutions [36]. Several thorianite and uranothorianite occurrences are known from marginal facies of epi- to mesozonal granitic plutons in Alaska. The occurrences are generally characterized as "felsic plutonic U deposits" in which numerous Th, U, and RE minerals are associated as disseminations in fissure veins and alkalic granite dikes in, or along, margins of alkalic and peralkalic granitic plutons, or in the granitic plutons themselves. Examples are: the Wheeler Creek alaskite and the Zane Hills monzonite to granodiorite plutons of the Purcell district, west-central Alaska; the porphyritic biotite syenite and alkali granite on Roy Creek, Mount Prindle, east-central Alaska (see also [37]); the granite and syenite pluton in the William Henry Bay (see also [38]) and the zoned, peralkaline to normal granite at Bokan Mountain (for details, see below), both southeastern Alaska [39, pp. 6, 29/30, 33, 65, 70]. In the Bokan Mountain area on Prince of Wales Island, Th-U mineralizations, with commonly metamict or partly metamict uranoan thorianite and uranothorite as chief ore minerals, are widespread and genetically related to a peralkaline granite boss. Originally, the ore minerals occur as syngenetic sparse disseminations or primary concentrations in predominantly riebeckite- and/or acmite (aegirine)-bearing granites; disseminations also occur in related pegmatites and aplites. The primary U-Th minerals formed during a late-magmatic stage and were possibly reworked by hydrothermal solutions leading to a subsequent epigenetic vein deposition of uranothorianite and other U-Th minerals (enriched to high-grade ore at Ross Adams) in the granite and adjacent pegmatites and aplites [40, pp. 1, 21, 34, 38, 46, 61/2, 67, 114]; see also [22, 41]. As an accessory mineral, uranothorianite is present in postmetamorphic granitic plutons, particularly in coarse-grained biotite- or biotite-amphibole-bearing granites (modally mainly monzogranites and quartz-monzonites), of the southern Appalachian Piedmont, southeastern USA [42]. Rare thoria-nite crystals (of 10 μm in size) accompany titanite in biotite of the accessory-rich gray biotite granite at Patka dip-head, Velence Mountains, Hungary [43]. Metamict thorianite, together with uraninite and gadolinite, are grouped with the late-magmatic autometamorphic accessory minerals concentrating in the late-magmatic granitoids, especially in the biotite granites (but also in aplites and pegmatites, see p. 31), of the Omchikandinsk massif, Verkhoyansk-Chukotka Fold Region, northeastern Siberia [44, pp. 32, 35/9, 42]. In granitoids of the Little Caucasus, USSR, thorianite occurs in the more alkaline granitoids (granosyenite and ademellite) and in alaskites [45]. Note also thorianite as partial fracture fillings in shattered magnetite of a potassium-alaskite xenolith in a diatreme of the Hopi Buttes volcanic field, Navajo County, Arizona. It presumably formed from high-temperature Th-U-rich volatile phases, which separated early from a volcanic magma in plutonic or hypabyssal environments and mineralized the country rocks after migration to higher crustal levels [46]. For exsolution phenomena in this thorianite, see p. 38. Hydrothermal uranothorianite and uraninite are equally distributed in a radioactive calc-alkaline granite from eastern Serbia, Yugoslavia; or are concentrated in associated cataclastic zones, where strong crushing and cracking of the granite enabled intense penetration by alkaline hydrothermal solutions of magmatic origin, which contain mainly U, less Th [47]. For thorianite as inclusion in hornblende or bastnaesite of granitic rocks from Zaire and from the USA, see p. 37.

Thorianite minerals are scarcely reported for **volcanic rocks**. Thus, black euhedral single crystals of uranothorianite (up to 1.5 mm across) with a distinct zonation and coatings of probably opal (see pp. 46 and 39, respectively) occur within the primary matrix of sanidinitic ejecta at Tre Croci near Viterbo, Latium, Italy. It is associated mainly with zircon, titanite, thorite, dark micas, betafite, and purple fluorite [48]. Thorianite, forming discrete tiny grains, is restricted to zones of extensive fluorphlogopite-pyrite replacements in the volcanic trachyte unit at the Rexspar U deposit, British Columbia, Canada. The U-Th mineralization (containing besides thorianite, also thorian uraninite and uranothorite) as well as the minerals of the hosting replacement zone, are all considered to be of primary, volcanogenic (not of secondary, hydrothermal) origin. They formed from deuteric volatile-rich fluids, syngenetically during late-stage magmatic differentiation and formation of the trachytic rocks [49]. In the Th-RE-rich ore body of Morro do Ferro, situated in the caldera of a volcanic tinguaite at Poços do Caldas, southeastern Brazil, thorianite, apart from its occurrence as secondary alteration product (see p. 39), also belongs to the primary minerals, which generally formed during volcanic emplacement or from later, hydrothermal activities [50].

Carbonatites. Thorianite in carbonatites of the USSR represents a very rare accessory mineral [9, p. 164], [51, p. 87]. It is generally found in early, unaltered, calcitic carbonatites (which form at relatively high temperatures in an alkaline environment under oxidizing conditions) [52]; or in early, dolomitized carbonatites [9, p. 78]. Thus, in calcitic carbonatites of the second stage in Siberia (Taimyr Peninsula [53]), it occurs as well-developed minute crystals (measuring 0.01 to 0.1 mm) usually dispersed in the rock, or sometimes it is connected with randomly orientated thin veinlets where it is associated with magnetite and apatite. Thorianite and pyrochlore do not occur together as if both minerals are mutually exclusive [51, p. 116]. In association with magnetite, apatite, phlogopite, and pyrochlore, thorianite occurs in actinolitized and dolomitized early carbonatite veins in melanocratic fenites at the southern margin of the Kovdor massif, Kola Peninsula. Its restriction to altered veins and its association with Th-free pyrochlore III suggest that thorianite is formed during recrystallization of Th-bearing pyrochlore II in the course of rock alteration (leaving thorianite unaffected by the secondary processes, see p. 38) [54, pp. 140, 162/3].

In the Gem Park complex, northern Wet Mountains, Colorado, accessory thorianite occurs in dolomite-barite-monazite-type carbonatite dikes [55, pp. 1, 12].

In the Loolekop carbonatite of the subvolcanic igneous complex of Palabora, northeastern Transvaal, R.S.A., thorianite/uranothorianite occurs usually in small amounts [56, p. 204], but may represent an economically important ore mineral, particularly concentrated in the younger transgressive carbonatite. In contrast, in the older banded carbonatite it forms a poor mineralization [19, pp. 55, 66, 93]. Irregularly distributed [57], it appears frequently as idiomorphic crystals ranging in size from 0.1 to 0.6 mm (or even up to 7mm [58]), but also forms irregular patches and veinlets [56, p. 204]; it may be distinctly zoned, owing to chemical inhomogeneity (see p. 46), and variably altered (see p. 38). Associated minerals, mainly carbonate gangue, magnetite, and baddeleyite, reveal color change or structural alteration, caused by radioactive blasting, in haloes around thorianite grains. Within the paragenetic mineral succession of the Loolekop carbonatite, uranoan thorianite formed, besides magnetite and baddeleyite, during an early, pegmatitic-pneumatolytic stage at relatively high temperatures well above 400°C [58]. Thorianite is also reported for a carbonatite at Kalkfeld, Namibia (South-West Africa) [59].

As a primary mineral, thorianite is chiefly found in **pegmatites** [12], usually associated with zircon, monazite, and beryl [11]. It is not nearly so common as uraninite in granitic pegmatites [4, p. 34], but is a typical mineral of rare-metal pegmatites [60], as, for example, in Nb-Ta-Sn pegmatites of Thailand (see p. 37), in rare-metal mineralized aplite-pegmatite at Jabal Sa'id, Hijaz region, Saudi Arabia [62], or in betafite- and columbite-bearing feldspathic aplites in syenite gneiss at Tambani, Malawi (Nyasaland) [61].

In America, thorianite occurs in deep-red hematitized microcline pegmatite at the contact of syenitic gneiss with metasediments at Cardiff Township, Bancroft area, Ontario [63], and in pegmatites and aplites, spacially and genetically related to an peralkaline granite boss in the Bokan Mountain area, Alaska, see p. 29. Together with thorite, uraninite, and further U minerals, it is reported from Precambrian pegmatite at Williams Quarry, Pennsylvania [64] and from pegmatites in Brazil (e.g., in the Amapá region, northern Brazil) [65].

In the USSR, thorianite occurs in alkaline pegmatites within the miaskitic intrusives of the Vishnevye Gory, Southern Urals [66, p. 90]. Among the multiphase granitoids of the Megri pluton, Little Caucasus, thorianite as a rare accessory mineral is restricted to granitoid veins of the early stage, especially to aplites and pegmatites, which are characterized by high U and Th contents [67]. Thorianite is accompanied by xenotime, monazite, zircon, and columbite in schlieric pegmatites derived as vein formations from late-intrusive, highly-differentiated, rare-element enriched leucocratic granites of the Yasmansk massif, Alai Mountain Range, Central Asia [68]. Metamict thorianite, together with uraninite and gadolinite, as a late-magmatic autometamorphic accessory mineral is mainly concentrated in late-magmatic aplite pegmatites and aplitic dikes along fractures of granites and near contacts to country rocks of the Omchikandinsk massif [44, pp. 32, 35/9, 42] and of other massifs of the Verkhoyansk-Chukotka Fold Region, northeastern Siberia [69, pp. 218/9, 225]. Thorianite with extraordinary amounts of Pb and U (aldanite) occurs, associated with monazite, ilmenite, zircon, rutile, and cassiterite, in Precambrian pegmatites of the Aldan region, Eastern Siberia [70].

Thorianite and calcite occur, unusually, as the only crystalline phases in carbocers (organic Ce substance) of natrolite-bearing pegmatites of the Khibiny tundra, Kola Peninsula, USSR [71].

The occurrences of thorianite in pegmatites within high-metamorphic series are well known: at various localities of Sri Lanka [72], see also [12, 73]; or in southern Madagascar, where thorianite, besides its predominance in pyroxenites (see pp. 34/5), also occurs in phlogopite-pegmatite veins and granite pegmatites at Soafia and Betroka [20], see also [74, 75], and together with spinel, corundum, and hibonite is found in pegmatitic plagioclasites in pyroxenites at North Vohimena and Andakota [76].

Terrestrial Occurrences, Metasomatic Rocks and Skarns

Occurrences of thorianite in metasomatic deposits are frequent, especially in the USSR and Canada (see the following paragraphs), and comprise multifold postmagmatic formations (from pneumatolytic to hydrothermal, including contact-metasomatic products) in various host rocks and alteration zones. It is generally stated that in postmagmatic, high-temperature, pneumatolytic-hydrothermal quartz-feldspar metasomatites and in medium-to-low tempera- ture albitites (albitized granites), U-ore formation is initially characterized by the paragenesis of thorianite, uraninite, and zircon in the early-metasomatic phase, sometimes followed by mixed Th-U oxides (as isolated segregations grown on zircon surfaces) or rarely by RE- bearing uranothorianite (as large inclusions in zircon). The initial association of zircon with the oxides and not with the orthosilicates of Th and U (as it is similarly observed for pneumatolytic- hydrothermal metasomatic deposits by [77]) is probably restricted to the very beginning of acid-alkaline alteration (which, in general, proceeds under relative alkaline reducing con- ditions and the availability of SiO_2) during postmagmatic metasomatism with beginning hydro- thermal U-ore formation in the original granitoids [78, pp. 198/200, 205/6]. For a second thorianite generation developing at increasing degree of metasomatism as inclusion in coffin- ite, see pp. 37/8. Thorianite/uranothorianite often also develop as postmagmatic pneumato- lytic-hydrothermal formations genetically related to alkaline and/or nepheline syenites: tho- rianite particularly in U-Th mineralizations within metasomatic skarn contact zones between metamorphic rocks and granites; uranothorianite in RE-Th-U mineralizations in skarns among quartzites, marbles, and granulites. As medium-to-low temperature hydrothermal formation, genetically related to subalkaline and alkaline granites and/or syenites, thorianite occurs in RE-U-Th mineralizations of the stockwork-impregnation type in alkaline granites, and uranothorianite in RE-Th-U mineralizations in pyroclastic alkaline rocks [79].

In various pyroxene-plagioclase-amphibole-rich host rocks of an undisclosed region of highly differentiated alkaline and subalkaline intrusions of the USSR, uranothorianite, together with zircon and its variety cyrtolite, occurs in high-temperature pneumatolytic-hydrothermal metasomatic U-Th ore mineralizations in fracture zones adjacent to syenitic dike-vein bodies. The ore is derived from residual postmagmatic solutions rich in Fe, Mg, U, Th, and Zr, which penetrated into rock fissures and reacted metasomatically with limestone. Apart from its formation during the above, late Mg-Fe-metasomatic phase, uranothorianite, together with cyrtolite, formed in Na-K-metasomatic quartz-feldspar rocks of the same region [80, pp. 69, 72/3, 76/8].

In phlogopite-diopside-olivine metasomatites of variable composition from the Kovdor massif, Kola Peninsula, USSR, accessory thorianite is restricted to strongly apatitized zones, in close association with baddeleyite (see also p. 37), zirkelite, and apatite. It shows a great variety of crystallographic forms (see pp. 59/60), even if it appears only as minute crystals up to 0.2 to 0.3 mm long. After segregation of the main mass of apatite and somewhat after formation of baddeleyite and zirkelite, thorianite was formed from SiO_2-undersaturated solutions and is closely related to the apatitization affected by postmagmatic solutions rich in P, CO_2, and minor amounts of rare elements [81, pp. 19/20, 26].

Thorianite (similar to that from Kovdor) in paragenesis with pyrochlore occurs in metasomatic rocks formed after melilite rocks of Cape Turii (a part of the Kola Peninsula). Thorianite, as an impregnation of fine rounded grains or as idiomorphic crystals, concentrates in calcite segregations of calcite-diopside-hornblende and calcite-garnet-vesuvianite rocks [82].

Besides pyrochlore, ilmenorutile, brannerite, bastnaesite, thorite, fluorite, and sulfides, uranothorianite occasionally occurs in mineralizations confined to albite-acmite (aegirine) zones in rare-metal albitites from strongly altered endo- and exocontact zones genetically related to foyaitic intrusions into calcareous country rocks [83]. Similarly, thorianite occurs in coarse-laminated albitite, in very strongly albitized alkaline pegmatite, and in albitized boston-ite, which all represent endo- and exocontact-hosted postmagmatic, pegmatitic, and hydro-thermal-metasomatic formations (with high content of volatile-rich minerals) of the nepheline syenite massif of Balyktygkhem, eastern Tuva, USSR [84]. Metasomatic thorianite occurs also: in albitized syenite of the Lakalii massif [85] and in acmite (aegirine) microclinite of an undisclosed alkali-metasomatic intrusion, both Eastern Siberia [86], see also [87]; and (together with zircon, fluorite, pyrochlore, and thorite) in sodalitized syenite of the Yasmansk massif, Alai Mountain Range, Central Asia, in which rare elements are typically concentrated in metasomatized rock varieties (or in late, highly differentiated granitoids, see p. 31) [68]. Thorianite, together with pyrochlore, occur as main radioactive minerals in tectonically in-duced, metasomatic-hydrothermal alteration zones (acting as structural traps for U-Th, Nb, and RE mineralizations) in nepheline and microcline syenites of the Elk massif, northeastern Poland [88, 89].

Within zones of fenitization, thorianite occurs in postmagmatic pneumatolytic-hydrother-mal carbonate and quartz-arfvedsonite veins and as replacement of monazite (see also p. 39) in fenitized granite pegmatites of the exocontact of miaskitic intrusions of the Vishnevye Gory, Southern Urals [66, pp. 86/7, 90]; or was deposited together with lueshite and pyrochlore in black serpentine dikelets and associated vermiculite border zones along small faults and fractures in gabbros and pyroxenites of the Gem Park complex, northern Wet Mountains, Colorado. There, thorianite formed during initial fenitic host rock alteration with subsequent introduction of fluids rich in Nb, RE, Sr, and Th [55, pp. 1, 10/1, 14, 17].

In a contact-metasomatic migmatitic zone (with pegmatites and schists in contact with granite) at the Row Property, Lake Charlebois area, northern Saskatchewan, Canada, thoria-nite is accompanied by mainly uraninite, minor molybdenite and pyrrhotite [24, p. 211].

Deposits of thorianite-uraninite in so-called pyrometasomatic formations (developed at high-hydrothermal activity) comprise several occurrences in marbles/skarns or meta-pyroxenites near contacts of granitic, syenitic, or pegmatitic intrusions of the Precambrian Grenville subprovince of the Canadian Shield in Ontario and Quebec, Canada [4, pp. 34, 244/5]. The intermediate members of the thorianite-uraninite series, uranian thorianite and high thorian uraninite, characteristically occur in the metasomatized impure marbles in a matrix of calcite, phlogopite, diopside, and accessory titanite, thorite/uranothorite, chon-drodite, and pyrite; whereas the more U-rich members are restricted to pyroxenites and other host rocks. The commonly twinned crystals of uranoan thorianite are either disseminated in typical salmon red calcite or concentrated in bands with phlogopite and diopside [90]. The same typical mineral association is described for thorianite occurrences in probably corresponding host rocks of the Bancroft area, eastern Ontario: in metasomatized zones in limestone [91]; in skarn zones in impure marbles [92]; or in micaceous marble [63]. Further, thorianite and uranoan thorianite occur as black cubes up to one quarter inch in size, commonly twinned, and often accompanied by uranothorite and uranophane in skarns of Quebec: mainly in pink calcite-diopside-mica rocks at Huddersfield and Grand Calumet townships, but also in scapolite-diopside rocks at Masham township and in the Camp zone of Huddersfield [93]. As is exemplified for the highly mineralized skarns of the Grenville marbles by the particular thorianite occurrence at the Yates U mine, Huddersfield, the genesis of the host rock and its minerals is explained by original Precambrian limestones which later were metamorphosed and mineralized by metasomatic fluids rich in silicates and elements such as F, U, and Th [94]; or the thorianite formation is related to an early phase of contact metamorphism of limestone by adjacent igneous rocks [95].

Probably genetically similar to the metasomatically reworked thorianite-bearing marbles of Canada, is the unique occurrence of thorianite in magnesian skarns formed at the contact zone between nepheline syenite and dolomite in Siberia (Synnyr Mountain Range, northeastern Lake Baikal region, USSR [15]) [96, p. 134]. Characterized as a calcic thorianite variety (see p. 47), it appears usually as irregular grains (up to 0.4 mm in size) associated with forsterite in calciphyres of the skarn rocks, see especially [97, p. 137], refer also to [96, p. 127], or, rarely, as fresh regular crystals (up to 0.3 mm) invariably in paragenesis with phlogopite in diopside-forsterite-phlogopite rocks of the Mg skarns, see especially [98], refer also to [97, p. 137]. The postmagmatic formation of thorianite (simultaneously with phlogopite, but after forsterite) is ascribed to both high temperature and high Ca concentration in the mineralizing medium [97, p. 142]; the Th content of the solution is attributed to a release during replacement of pyroxene and forsterite, which had higher rare-element contents with respect to the substituting newly-formed phlogopite [96, p. 134].

Terrestrial Occurrences, Metamorphic Rocks

Among the occurrences of thorianite minerals in metamorphic rocks, best known are the deposits of uranothorianite in Madagascar. They are confined to calco-magnesian rocks of high-metamorphic, granulite-facies pyroxenites, besides those in metamorphic cipolins (silicate marbles) and phlogopite pegmatites (see p. 32) [99], see also [21]. Of economic importance are deposits in an arched zone of 150 km length and up to 40 km width [100] in rocks of the Tranomaro group of the Precambrian Androyan System, within the interior bend of the Mandrare river [101]; for details on special localities, see [102, 103]. In the high-metamorphic Si-Al-rich country rocks of Madagascar, two major types of thorianite occurrence with specific paragenetic associations are differentiated. One type in Ca-Mg-rich banks within pyroxenites (of either scapolite-, plagioclase-, or calcite-dominated composition), pyroxenitic gneisses, and diopsidic cipolins has a mineral association with accessory hibonite, corundum, zircon, and occasionally thorogummite. The other type in elongated lenses and veins of phlogopite-bearing pyroxenites has an association with pargasite, phlogopite, and spinel [103]. Where uranothorianite occurs in both preceeding types, apatite and monazite are, in general, relatively rare [99]. For a dependence of the Th:U ratio in Madagascar thorianite on the type of hosting pyroxenite, see p. 45. Usually uranothorianite is irregularly scattered; it occurs more constantly and as a characteristic component only in phlogopite-bearing pyroxenites [102]. It generally appears as isolated euhedral crystals, their size varying with that of associated minerals [99], usually up to several mm [102]; it is coarser grained (mm to cm) in ore chimneys and lenses of pegmatitic habit in the center of calco-magnesian complexes, and forms fine-grained (<0.1 mm) xenomorphs in pyroxene, feldspar, and calcite [21]; it occurs as stringers in vein-like occurrences, or as nodules and aggregates [99], which reach cm to several dm in size in splendid crystal "cakes" of Morafeno or forms monomineralic masses several kg in weight at Ilapality [102].

Generally assumed for the uranothorianite of Madagascar is a metamorphic formation by crystallization syngenetically with the hosting Ca-Mg rocks [103], or during a regional granitization process [104]. Crystallization during granitization from Th and U, already present in the pyroxene-wernerite complex, may be favored by pegmatitic-hydrothermal conditions. This may have occurred simultaneously with the cessation of crystallization of phlogopite and apatite and of recrystallization of rock-forming silicates and calcite [99]. However, note also the opinion expressed in a common Mg-metasomatic origin of uranothorianite together with phlogopite in the metamorphic pyroxenites [105]. For other hypotheses, for example, of a primary magmatic thorianite formation, and for the related problem of a para- or orthogenic origin of the pyroxenites and of their Th content, see the discussions of older literature data in [99, 102]. A comprehensive review on the uranothorianite occurrence in southeastern Madagascar is given by [106].

A few other occurrences of thorianite in metamorphic rocks are: thorianite besides thorian monazite, zircon, cerianite, and kreittonite (a Zn spinel) in radioactive layers (probably a preserved fossil monazite placer) within high-grade metamorphic garnet-cordierite-silliman-ite gneiss near Bodenmais, Bavaria, F.R.G. [107]; thorianite together with thorite, thorian monazite, and florencite, all associated with chlorite, apatite, lazulite, and tourmaline, in various Th-RE-mineralized crystalline schists, Sopron Mountains, northwestern Hungary [108]; and thorianite/uranothorianite in serpentinized marble/limestone, contact-metamor-phically altered by intrusive pegmatites, of the Sherrer Quarry near Easton [109, pp. 48/9] and of the Williams Quarry [64], both Pennsylvania, USA.

Uranothorianite is hardly known from consolidated metasedimentary rocks. In the Bokan Mountain area, southeastern Alaska, the extensive hydrothermal U-Th mineralization, chiefly composed of uranothorite and uranoan thorianite (see also p. 29), is met only at a single locality (Cheri No. 1 prospect) in interstices of clastic metasedimentary country rocks adjacent to a peralkaline granite boss [40, pp. 1, 61/2]. From the USSR, in general, uranothorianite is reported as one of the main detrital ore minerals (besides zircon, thorite, uranothorite, allanite, ilmenite, hematite, and rutile) of radioactive Precambrian and Cambrian metasedimentary terrigeneous conglomerates (and gravellites) mineralized with rare elements (above all U, Th, Ta, Nb, RE, Zr). The rare elements (apart from U) are especially concentrated, probably due to erosion and weathering of the underlying granitic basement, in the basal strata of the conglomerates [110].

Terrestrial Occurrences, Sedimentary Rocks and Placers

Due to its high chemical stability (see pp. 27/8), thorianite/uranothorianite is widespread as a detrital mineral [12] occurring in alluvial deposits, black sands, or placers (generally together with zircon, ilmenite, thorite, and other minerals [11]), see below; refer also to "Thorium", 1955, p. 29. It is found world-wide in a large number of sand and gravel deposits of, for example, Sri Lanka, Madagascar, Siberia, the South Island of New Zealand, Alaska (especially in Au and Sn placers), California, Montana, and other regions [111]. As idiomorphic crystals it occurs in colluvial (slope wash) and diluvial placer deposits, probably of the USSR [112]; and in eluvial residual placers (in lateritic and kaolinitic weathering crusts above Th-mineral enriched migmatites, carbonatites, pegmatites, granites, syenites, and Sn-ore veins) of Madagascar, Malaysia, Nigeria, the Republic of South Africa, Indonesia, and Sri Lanka [113, pp. 30/1].

Most placer occurrences in the USA concentrate on Alaska, where uranothorianite occurs in northeastern Alaska in pre- and postglacial Au placers of the Chandalar region (with largest deposits on Little Squaw Creek) [39, p. 74], and in association with magnetite, hematite, and several metallic sulfides in high-stream gravels at Gold Bench on the South Fork of Koyukuk River [114, p. 8], [115]. In west-central Alaska, it occurs associated with Bi and Cu minerals in placers and placer concentrates of the Nixon Fork district, above all on Ruby and Eagle Creeks [116, pp. 10, 19], and in the magnetite-rich glacial Hogatza placer along Bear Creek at Zane Hills [117]. On the Seward Peninsula, central western Alaska, uranothorianite occurs, occasionally concentrated and accompanied by mainly zircon, titanite, and further Th minerals, in fluviatile (gravel) placer deposits at many localities in the northeastern Buckland-Kiwalik district, mainly on the slopes of the Granite Mountain syenite stock in the Peace River, Quartz Creek, and Sweepstakes Creek areas (for details, see [118], [119], and [120], respectively; note also large amounts of uranothorianite in black-sand concentrates of the same areas [121]), further on the south slope of Clem Mountain, and in the Hunter-Conolly Creek area [122], see also [123]. Uranothorianite and hydrothorite occur in mined creek and bench Au placers on Bonanza, Dime, and Sweepstakes Creeks, Koyuk district [39, p. 75]; and thorianite in placer deposits on the Kwiniuk River, Darby Mountains, all southeastern Seward Peninsula [124]. Refer also to comprehensive compilations of the above-mentioned and additional occurrences of thorianite/uranothorianite in (frequently heavy-mineral or gold-bearing) placer deposits of Alaska given in [125]; with preferential respect to occurrences on the Seward Peninsula and to older literature, see [126, pp. 5, 8/13]. Deduced from the mineral associations of thorianite/uranothorianite in the placer occurrences of Alaska, their potential source rocks are thought to be mainly granitic [117, 122, 124] and syenitic intrusives [126, p. 5], their contact zones to metasedimentary country rocks [116, pp. 10, 16], [124], or, if especially metallic sulfides, fluorite, bismuth, and silver are associated, postmagmatic hydrothermal formations in the intrusives [122], respectively metalliferous lodes as named in [114, p. 12], [115], [126, p. 5]; for local details, see the original papers.

Further occurrences of detrital thorianite in North America are known in black-sand concentrates from the Scott River near Helena, Montana [121]; in association with monazite, titanite, euxenite, allanite, and zircon in the heavy mineral fraction of black sands from the Sand Basin, Granite County, Montana [127]; and together with zircon, rutile, scheelite, sperrylite, gold, and platinum in black sands from the Fraser River, British Columbia, Canada [128].

In South America, detrital thorianite occurs as rolled grains in fluvial gravels of Brazil (e.g., in the Amapá region of northern Brazil) [65].

In Europe, it is only reported from gold-bearing alluvial deposits at Lemmenjoki, Inari, northern Lapland, Finland [129].

In Asia, thorianite occurs in black sands of a gold placer on the Boshogoch River, Transbaikalia [121], and aldanite in placers [27], respectively, in the non-magnetic fraction of gold washings of the Aldan region [70], all Eastern Siberia, USSR. Uranothorianite mostly as idiomorphic, scarcely altered (see p. 39) crystals, originally deriving from a local syenite massif, occurs in the heavy-mineral fraction of alluvial beach placers at Geyikli, south of Troy, Biga Peninsula, western Turkey [130]. Thorianite is found in alluvial deposits of Travancore, southern India [12]. It has been commercially mined from placers in Sri Lanka, where it occurs at many localities, mainly in the southern part of the island (see original paper) [131], especially in coastal-marine alluvial deposits, mainly in association with baddeleyite, thorian monazite, ilmenite, rutile, and zircon [23], as well as in alluvial deposits in the ravines of the highland interior [132], for example, associated commonly with zircon, ilmenite, thorite, and other heavy minerals (originally deriving from pegmatites) in stream gravels of many localities, notably in the Galle district, Southern Province, in the Balangoda district, and near Kondurugala, Sabaragamuwa Province [12, 121]. Together with thorite, thorianite occurs in washed gem gravels of the Southwest Group sediments of Sri Lanka [133].

From Africa, detrital thorianite is mainly reported to occur in alluvial deposits of Madagascar [12] (based on data from [74, 75]) and [76]; as very rare "radioactive black" in black sands of Rosetta, Nile delta, Egypt [134]. Uranothorianite is an occasional component of alluvial river sands of the Tambani region, Malawi (Nyasaland), south-central Africa [135].

2.1.1.2.2 Intergrowths, Inclusions, and Alteration

Intergrowth of thorianite and baddeleyite, epitaxially grown together with their (100) planes, are reported from metasomatic phlogopite-rich rocks of the Kovdor massif, Kola Peninsula [81, p. 20]; and intergrown with loparite and pyrochlore is thorianite of alkaline rocks of the Enisei Ridge, Eastern Siberia, both USSR [32].

Inclusions of Thorianite in Other Minerals. As an inclusion in rock-forming minerals, thorianite occurs: in Ca-rich hornblende of granite from Zaire (Belgian Congo) [144]; in feldspar of syenite from Kyzyl-Ompul', Kirghizia [29]; and in amphibole, acmite (aegirine), apatite, and biotite of alkaline rocks from the Enisei Ridge, Siberia [32]. As commonly xenomorphous fine-grained inclusions thorianite occurs in feldspar, pyroxene, and calcite of high-metamorphic Ca-Mg-facies rocks of southern Madagascar [21], especially in plagioclase of plagioclase-dominated pyroxenites, or in diopside and calcite of calcitic pyroxenites and diopsidic cipolins; it crystallizes also in phlogopite, pargasite, and diopside of phlogopitic pyroxenites [103]. Inclusions of uranothorianite in diopside of metamorphic limestone from Huddersfield, Quebec, show prominent reaction rims [95]. Thorianite occurs as an inclusion in phlogopite of metasomatic phlogopite-rich rocks from an area of Mg skarns in Siberia [96, p. 128], [97, p. 137], [98], and in the Kovdor massif, Kola Peninsula, where it also occurs as inclusions bound to microcracks in apatite [81, p. 20]. Very rare are inclusions of uranothorianite in cryptocrystalline, sometimes pegmatitic minerals of quartz-feldspar metasomatites from an undisclosed locality of the USSR [80, p. 77].

In (more) accessory minerals, uranoan thorianite is present as an inclusion: in bastnaesite of the Redstone granite, Westerly, Rhode Island [145]; in baddeleyite of the Loolekop carbonatite, R.S.A., causing radioactive haloes and attendant isotropism in the host mineral [58]; and in steenstrupines of lujavrite (for their alteration, see p. 38), of a recrystallized border zone between lujavrite and naujaite [146], of recrystallized naujaites (associated in the inclusions with britholite), and of late, partly hydrothermal acmite (aegirine)- or analcime-rich veins in naujaite (sometimes associated in the inclusions with britholite and astrophyllite), all of the Ilimaussaq batholith, southwestern Greenland [136, pp. 4, 9, 13, 21/3, 30]. Thorianite occurs as an inclusion in monazite from rocks of the tin belt, Thailand. Its size (rarely > 10 μm) seems to vary proportionally with that of the host mineral. The close relation of the thorianite inclusions to Dy and Y concentrations speaks for their generation by simple exsolution. Idiomorphic thorianite inclusions of up to 100 μm in size are oriented parallel to the planes (100) and (010) of large magmatic monazite of Nb-Ta-Sn pegmatites. They show Si rims, suggesting a diadochic replacement of SiO_4^{4-} for PO_4^{3-} effective as charge compensation during Th^{4+} incorporation [147].

Small inclusions of thorianite may occasionally account for the analyzed Th content of some accessory minerals, as it is observed for Th-rich britholite (see "Thorium" Suppl. Vol. A 1b) [148], as well as, in case of metasomatic granitoids, for zircon or coffinite. In those rocks, early-metasomatic thorianite or, rarely, RE-bearing uranothorianite occur as inclusions in zircon and are followed with increasing degree of metasomatism by a second generation of thorianite. This second generation, after replacements of zircon by uranothorite and of uranothorite by coffinite, forms as paragenetic inclusion in coffinite (containing 2 to 7.5% Th) and causes the chemical heterogeneity of the latter. The inclusions of the second generation fix the released Th, since they originate as an accessory phase during substitution of $ThSiO_4$ by $USiO_4 + ThO_2$ proceeding as diffusion process (without SiO_2 introduction), rather than as a metamict decomposition of thorite relicts in coffinite. Finally, the thorianite inclusions of the second generation are replaced by a U-free thorite phase [78, pp. 199/206].

As **inclusions of other minerals in thorianite** are observed: zircon and apatite in peripheral zones of thorianite crystals of syenite from the Kyzyl-Ompul' massif, Kirghizia [29]; small amounts of rutile for gabbro of Tuva [27]; microscopically tiny crystals of baddeleyite and zirkelite (and probably impurities of apatite and thorite, as is deduced hypothetically from the chemical thorianite composition, see p. 47) for phlogopite-bearing metasomatites of Kovdor, Kola Peninsula [81, p. 20]; or occasionally forsterite for calciphyres (but not for the associated phlogopite-rich metasomatites) of Mg skarns, Siberia [97, p. 137], see also [96, p. 127]. Included in thorianite of the Loolekop carbonatite, R.S.A., may be apatite, calcite [19, p. 67], baddeleyite [56, p. 204], magnetite, carbonate, and, as a characteristic feature, finely scattered galena, which apparently resulted from a reaction of thorianite-derived radiogenic Pb with S from the sulfide mineralization stage of the ore formation. Large crystals of uranoan thorianite of the carbonatite show a distinct zonation which may partly be caused by an increase of the galena inclusions inwards from the margin [58].

Alteration of Thorianite. Since pure ThO_2 is relatively stable under natural conditions of weathering and metamorphism, see pp. 27/8, the high-thorian thorianites are generally little affected by alteration, whereas varieties rich in U are commonly altered [4, p. 34], [11, 111, 121], a fact which is generally attributed to the higher content of oxidizable U [113, p. 19].

Only few alterations of thorianite referable to the postmagmatic stage are known: From the syenite of Kyzyl-Ompul', Kirghizia, which is subjected to postmagmatic metamorphism causing in primary thorianite chemical contamination, intense corrosion phenomena (cavernation), and marginal alteration zones with, apparently, secondary thorogummite [29]. From the Ilimaussaq batholith, southwestern Greenland, where inclusions of thorianite and britholite in steenstrupine react with each other during assimilation of their host mineral in late lujavrites. This reaction path hypothetically proceeded with formation of thorite and monazite according to the reaction: britholite $((Ca,Ce,Na)_5[(Si,P)O_4]_3(OH,F))$ + thorianite $((Th,Ce)O_2) \rightarrow$ monazite $(Ce[(P,Si)O_4])$ + thorite $((Th,Ce)SiO_4)$ + Ca^{2+} + Na^+ + H_2O [136, p. 30]. And from a granitic xenolith in a diatreme of the Hopi Buttes volcanic field, Arizona, where thorianite contains spheroidal blebs of pitchblende which are thought to be due to selective U exsolution [46]. Thorianite from dolomitized and actinolitized carbonatite of the Kola Peninsula is mainly not affected by the alteration processes, apart from a single case of growth of a red ochrous product similar to ferrithorite [54, pp. 162/3]. In the Loolekop carbonatite, R.S.A., are observed one case of clusters of baddeleyite crystals replacing and growing from uranoan thorianite, and a different case of baddeleyite forming the matrix of brecciated uranoan thorianite [58]. However, samples of the same locality frequently show shrinkage and cataclastic fractures cemented by sulfides, magnetite, and gangue minerals [14, 58].

More commonly for thorianite minerals are alteration processes as weathering and hydratization, during which various supergene U minerals develop [11] (especially gummite-like material (thorian gummite) [121] or thorogummite [4, p. 34], see also [111]), or ill-defined poorly-crystallized metamictic earthy substances are formed, which may occasionally remain unidentified, see, for example, [56, p. 206], [109, pp. 50, 52/3]. Gummite proper, generally known only as an alteration product of uraninite, occurs solely as the probable product of final stages of oxidation and hydration of uranothorianite from fluviatile placer deposits of the Seward Peninsula, Alaska [122]. The secondary products sometimes form coatings around the primary thorianite minerals [111], as is observed for thorianite from Tuva and aldanite from Aldan, Siberia [27]; for thorianite, often covered by thorogummite, from Kirghizia [29], or for zoned uranothorianite, sometimes coated by amorphous silica (probably opal), from Latium, Italy [48]. See also brownish red and golden yellow secondary rims (probably metamict, slightly hydrated thorianite) around primary thorianite cores from the Omchikandinsk massif, northeastern Siberia [44, pp. 42, 44], [69, p. 225], and reaction rims with probably gummite-like material around thorianite, but not if in contact with apatite, at the Loolekop carbonatite [137]. Due to secondary oxidation in pyroxenites of Madagascar, thorianite and diopside may be covered by ferrugineous coatings [103].

Special alterations referred to secondary processes during placer formation occur for detritic, but idiomorphic, uranothorianite from colluvial and diluvial placer deposits, probably of the USSR. Within the crystals, U is partially oxidized (for the reason, see p. 55) and redistributed from lattice positions into internal narrow U-enriched fissures, with major deposition as uraninite [112]. Only scarce alteration is mentioned for usually homogeneous idiomorphic uranothorianite from beach placers of the Biga Peninsula, western Turkey, for which the rare, irregularly distributed worm-like alteration zones and zonally altered crystals with seams of pyrite (probably pyritized microorganisms) are explained by the activity of microorganisms during sedimentation causing reducing conditions under which uranothorianite behaves inertly and remains stable [130]. For uranothorianite from the same locality, other alteration phenomena gradually increase during mineral transport in surface water from the source rock region towards final deposition in nearby alluvial beach placers (5 km apart). Close to the source rock, alteration, corrosion, or weathering are weak, new formations are rare; but already common are lamellation and small irregular fissures (entailing susceptibility to mechanical disintegration and chemical weathering) owing to volume change due to radioactive decay. More remote from the source rock, grains of uranothorianite are edge-rounded and corrosion is frequent, microscopically visible as crater-like cavities caused by removal of exposed grain parts, or as redeposition and as new formation of secondary nuclei; gradual dissolution of the crystal lattice and precipitation of gel-like orangite are evident. In the area of deposition (beach, shelf, and adjacent sea bottom), the mineral grains are always penetrated by cracks and holes; small fragments reach the grain size of schluff. For a gradual Th enrichment accompanying the alteration in uranothorianite, see p. 55 [138].

Alteration and Replacement by Thorianite. Lately formed thorianite replaces monazite in postmagmatic pneumatolytic-hydrothermal mineralizations formed during the stage of albitization in fenitized granite pegmatites at the exocontact of the miaskitic intrusions of Vishnevye Gory, Southern Urals [139, 140], see also [66, p. 86]. In the Ilimaussaq complex, southwestern Greenland, accessory thorianite of green albite-acmite (aegirine)-bearing veins (cutting various country rocks adjacent to the alkaline intrusions) is suggested to be an alteration product of thorite or thorogummite [141]. Especially in the Kvanefjeld area with mainly lujavritic nepheline syenites, thorianite also occurs as an alteration product of steenstrupine [142]; or small yellow-green thorianite(-like) grains are important constituents of pseudomorphs after eudialyte, which is associated with steenstrupine [143].

Other occurrences of secondary thorianite (without given source mineral) are: Zr-rich thorianite (besides mainly thorbastnaesite) in zones of alteration or weathering in the Morro do Ferro ore body, southeastern Brazil [50]; and scarce Zn- and W-enriched thorianite in postmagmatically strongly altered syenite of Kyzyl-Ompul', Kirghizia [29].

References for 2.1.1.2.1 and 2.1.1.2.2:

[1] Zhidikova, A. P.; Kolesov, G. M.; Khodakovskii, I. L. (Geokhimiya **1980** 821/6).
[2] Legret, M.; Poty, B. (5th Réun. Ann. Sci. Terre, Rennes 1977, p. 313).

[3] Frondel, C. (Proc. Intern. Conf. Peaceful Uses At. Energy, Geneva 1955 [1956], Vol. 6, pp. 568/77; Geol. Surv. Profess. Paper [U.S.] No. 300 [1956] 567/79, 572).

[4] Heinrich, E. W. (Mineralogy and Geology of Radioactive Raw Materials, McGraw-Hill, New York 1958, pp. 1/654).

[5] Lang, A. H. (Geol. Surv. Can. Econ. Geol. Rept. No. 16 (Interim Account) [1952] 1/173).

[6] Getseva, R. V.; Savel'eva, K. T. (Rukovodstvo po Opredeleniyu Uranovykh Mineralov, Gosgeoltekhizdat, Moscow 1956, pp. 1/260, 125).

[7] Drozdovskaya, A. A.; Mel'nik, Yu. P. (Geokhimiya 1968 402/10; Geochem. Intern. 5 [1968] 362).

[8] Kopeikin, V. A. (At. Energiya SSSR 56 [1984] 221/3; Soviet At. Energy 56 [1984] 242/5).

[9] Kapustin, Yu. L. (Mineralogiya Karbonatitov, Nauka, Moscow 1971, pp. 1/286).

[10] Page, L. R. (Intern. Geol. Congr. Rept. 21st Sess. Norden, Copenhagen 1960, Pt. 15, pp. 149/64, 155).

[11] Chukhrov, F. V.; Bonshtedt-Kupletskaya, E. M. (Mineraly Spravochnik, Vol. 2, Pt. 2, Nauka, Moscow 1965, pp. 1/341, 116).

[12] Frondel, C. (Geol. Surv. Bull. [U.S.] No. 1064 [1958] 1/400, 52).

[13] Ramdohr, P.; Strunz, H. (Klockmanns Lehrbuch der Mineralogie, 16th Ed., Enke, Stuttgart 1978, pp. 1/876, 548).

[14] Ramdohr, P. (Die Erzmineralien und ihre Verwachsungen, 4th Ed., Akademie Verlag, Berlin 1975, pp. 1/1277, 1131/2).

[15] Semenov, E. I. (Typochimizm Mineralov Shchelochnykh Massivov, Nedra, Moscow 1977, pp. 1/119, 57/9).

[16] Speer, J. A. (Rev. Mineral. 5 [1980] 113/35, 129/30).

[17] Palache, C.; Berman, H.; Frondel, C. (Dana's System of Mineralogy, 7th Ed., Vol. 1, Wiley, New York 1944, pp. 1/834, 620/2).

[18] von Backström, J. W. (PEL-112 [1966] 1/19, 3, 17; N.S.A. 21 [1967] No. 22051).

[19] Hanekom, H. J.; van Staden, C. M. v. H.; Smit, P. J.; Pike, D. R. (Mem. Geol. Surv. South Africa No. 54 [1965] 1/185).

[20] Holmes, A.; Besairie, H. (Mem. Inst. Sci. Madagascar D 6 [1954] 191/9, 193/4).

[21] Moreau, M. (Rappt. CEA-R-Fr. Commis. Energ. At. No. 1685 [1960] 1/19, 2/4).

[22] MacKevett, E. M., Jr. (Proc. 2nd U.N. Intern. Conf. Peaceful Uses At. Energy, Geneva 1958, Vol. 2, pp. 502/8; Mining Eng. 11 [1959] 915/9).

[23] Kogan, B. I.; Rozhanets, A. V. (Redk. Elem. Syr'e Ekon. No. 3 [1969] 127/31).

[24] Lang, A. H.; Griffith, J. W.; Steacy, H. R. (Geol. Surv. Can. Econ. Geol. Rept. No. 16, 2nd Ed. [1962] 1/324).

[25] Traill, R. J. (Geol. Surv. Can. Paper No. 69-45 [1970] 1/649, 562/3).

[26] Lovering, J. F.; Wark, D. A.; Sewell, D. K. B. (Lunar Planet. Sci. 10 [1979] 745/7).

[27] Makarov, E. S.; Lipova, I. M. (Geokhimiya 1962 583/9; Geochem. [USSR] 1962 676/83, 677).

[28] Yu, S.; Li, S.; Liu, K.; Zhuang, L.; Li, Z. (Zhongguo Dizhi Kexueyuan Yichang Dizhi Kuangchan Yanjiuso Sokan No. 5 [1982] 1/22 from C.A. 100 [1984] No. 54749).

[29] Turovskii, S. D. (Zap. Vses. Mineral. Obshch. 87 [1958] 369/70).

[30] Zykov, S. I.; Stupnikova, N. I.; Pavlenko, A. S.; Tugarinov, A. I.; Orlova, L. P. (Geokhimiya 1961 547/60; Geochem. [USSR] 1961 581/95, 585, 594).

[31] Turovskii, S. D.; Kokarev, G. N. (in: Smorchkov, I. E.; Aktsessornye Mineraly Izverzhennykh Porod, Moscow 1968, pp. 130/4).

[32] Grebennikova, O. T. (Nauchn. Tr. Irkutsk. Gos. Nauchn. Issled. Inst. Redkikh Tsvetn. Metal. No. 14 [1966] 97/100; C.A. 70 [1969] No. 70075).

[33] Leonova, L. L.; Pogiblova, L. S. (Geokhimiya **1961** 901/6; Geochem. [USSR] **1961** 999/1004).
[34] Smyslov, A. A. (Tr. Inst. Geol. Geofiz. Akad. Nauk SSSR Sibirsk. Otd. No. 286 [1975] 10/8, 272/83, 11/2).
[35] Rimsaite, J. (Geol. Surv. Can. Paper No. 81-23 [1982] 19/30, 24, 28).
[36] Rimsaite, J. (Geol. Surv. Can. Paper No. 82-1B [1982] 253/66, 255, 259).
[37] Burton, P. J. (Diss. Univ. Alaska, Fairbanks 1981, pp. 1/72, 3, 59/61).
[38] Eakins, G. R. (Geol. Rept. Alaska Div. Geol. Geophys. Surv. No. 44 [1975] 1/62, 8, 15; C.A. **85** [1976] No. 35590).
[39] Nokleberg, W. J.; Bundtzen, T. K.; Berg, H. C.; Brew, D. A.; Grybeck, D.; Robinson, M. S.; Smith, T. E.; Yeend, W. (Geol. Surv. Bull. [U.S.] No. 1786 [1987] 1/104).
[40] MacKevett, E. M., Jr. (Geol. Surv. Bull. [U.S.] No. 1154 [1963] 1/125).

[41] MacKevett, E. M., Jr. (Bull. Geol. Soc. Am. **70** [1959] 1796).
[42] Speer, J. A.; Becker, S. W.; Farrar, S. S. (Mem. Virginia Polytech. Inst. State Univ. Dept. Geol. Sci. No. 2 [1980] 137/48, 137, 142, 145).
[43] Pantó, G. (Acta Geol. Acad. Sci. Hung. **19** [1975] 59/93, 61, 69).
[44] Nekrasov, I. Ya.; Ipat'eva, I. S. (Mater. Geol. Polezn. Iskop. Yakutsk. ASSR No. 5 [1961] 32/50; C.A. **58** [1963] 363).
[45] Meliksetyan, B. M. (Aktsessornye Miner. Magmat. Metamorf. Porod **1982** 52/66, 54/5).
[46] Gabelman, J. W. (AAPG Stud. Geol. No. 13 [1981] 23/36, 32, 35).
[47] Ristic, M. (Proc. Intern. Conf. Peaceful Uses At. Energy, Geneva 1955 [1956], Vol. 6, pp. 634/40, 638/9).
[48] De Casa, C. G.; Della Ventura, G. C.; Parodi, G. C.; Stoppani, F. S. (Riv. Mineral. Ital. **1987** 97/104, 101, 104).
[49] Preto, V. A. (CIM Bull. [Montreal] **71** [1978] 82/8, 82, 85/8; C.A. **90** [1979] No. 90212).
[50] Barretto, P. M. C.; Fujimori, K. (Chem. Geol. **55** [1986] 297/312, 297, 299, 307, 309/11).

[51] Gaidukova, V. S.; Zdornik, T. B. (Geol. Mestorozhd. Redk. Elem. No. 17 [1962] 86/117).
[52] Kapustin, Yu. L. (Geokhimiya **1966** 1311/21; Geochem. Intern. **3** [1966] 1054/64, 1054/5, 1063).
[53] Saltykova, V. S. (Tr. Inst. Mineral. Geokhim. Kristallokhim. Redk. Elem. No. 2 [1959] 189/207, 190).
[54] Kapustin, Yu. L. (in: Vlasov, K. A.; Mineralogiya i Geticheskie Osobennosti Shchelochnykh Massivov, Nauka, Moscow 1964, pp. 135/94).
[55] Parker, R. L.; Sharp, W. N. (Geol. Surv. Profess. Paper [U.S.] No. 649 [1970] 1/24).
[56] Russell, H. D.; Hiemstra, S. A.; Groeneveld, D. (Trans. Geol. Soc. South Africa **57** [1954] 197/208).
[57] Pike, D. R. (Proc. 2nd U.N. Intern. Conf. Peaceful Uses At. Energy, Geneva 1958, Vol. 2, pp. 91/6, 94).
[58] Forster, I. F. (Trans. Geol. Soc. South Africa **61** [1958] 359/65).
[59] Verwoerd, W. J. (Rep. Suid-Afrika Dept. Mynwese Geol. Opname Bull. Handbook No. 6 [1967] 1/452, 295, 305).
[60] Ginzburg, A. I.; Gorzhevskii, D. I. (Izv. Akad. Nauk SSSR Ser. Geol. **1957** No. 6, pp. 14/29, 21).

[61] Bowie, S. H. U. (Chron. Mines Outre-Mer Rech. Miniere No. 271 [1959] 304/8).
[62] Hackett, D. (J. African Earth Sci. **4** [1986] 257/67, 265/6).
[63] Kennedy, I. (Mineral. Rec. **10** [1979] 153/8, 156/7).
[64] Adams, J. W.; Arengi, J. T.; Parrish, I. S. (GJBX-166 [1980] 1/463, Pt. II, Pennsylvania, Locality 3; C.A. **94** [1981] No. 50447).

42

[65] Pereira, W.; Atalla, L. T.; Abrão, A. (IPEN-Pub-69 [1985] 1/128, 100; C.A. **104** [1986] No. 133122).

[66] Es'kova, E. M.; Zhabin, A. G.; Mukhitdinov, G. N. (Mineralogiya i Geokhimiya Redkikh Elementov Vishnevykh Gor, Nauka, Moscow 1964, pp. 1/318).

[67] Meliksetyan, B. M. (Izv. Akad. Nauk Arm. SSR Geol. Geogr. Nauki **14** No. 2 [1961] 21/41, 21, 29; C.A. **1961** 17397).

[68] Ifantopulo, T. N. (in: Mineev, D. A., Osobennosti Raspredeleniya Redkikh Elementov v Izverzhennykh Gornykh Porodakh, Akad. Nauk SSSR, Moscow 1970, pp. 86/96, 86/9, 95/6).

[69] Nekrasov, I. Ya. (Tr. Yakutsk. Filiala Akad. Nauk SSSR Ser. Geol. **12** [1962] 1/334).

[70] Bespalov, M. M. (Soviet Geol. **1941** No. 6, pp. 105/7; Mineral. Abstr. **12** [1955] 460).

[71] Loskutov, A. V.; Polezhaeva, L. I. (Mater. Mineral. Kol'sk. Poluostrova **1968** No. 6, pp. 276/81, 276, 279; C.A. **73** [1970] No. 68378).

[72] Holmes, A. (Proc. Geol. Assoc. Can. **7** Pt. 2 [1955] 81/106, 89, 99/100).

[73] Coomaraswamy, A. K. (Spolia Zeylan. **1** Pt. 4 [1904] 1 from Neues Jahrb. Mineral. Geol. Paläontol. **1906** 165).

[74] Lacroix, A. (Bull. Soc. Franc. Mineral. **37** [1914] 176/80).

[75] Lacroix, A. (Minéralogie de Madagascar, Vol. 1, Challamel, Paris 1922, pp. 1/624, 245/7).

[76] Delbos, L. (Compt. Rend. **244** [1957] 214/5).

[77] Nevskii, V. A.; Ginzburg, A. I.; Kozlova, P. S.; Ontoev, D. O.; Apel'tsin, F. R.; Kupriyanova, I. I.; Kudrin, V. S.; Epshtein, E. M. (Geologiya Postmagmaticheskikh Torievo-redkometal'nykh Mestorozhdenii, Atomizdat, Moscow 1972, pp. 1/407 from [78, p. 199]).

[78] Brodin, B. V.; Sidorenko, G. A.; Shul'gin, A. S.; Dubinchuk, V. T. (Zap. Vses. Mineral. Obshch. **111** [1982] 198/209).

[79] Belova, L. N.; Vol'fson, F. N.; Kazanskii, V. I.; Paverov, N. P.; Lukin, L. I.; Mel'nikov, I. V.; Nevskii, V. A.; Omel'yanenko, B. I.; Rybalov, B. L.; Smorchkov, I. E.; Sonyushkin, F. P.; Tishkin, A. I.; Khoroshilov, L. V.; Chentsov, I. G. (Probl. Geol. Miner. Mestorozhd. Petrol. Mineral. **1** [1969] 150/74, 165/8; C.A. **73** [1970] No. 37242).

[80] Trofimov, A. A.; Apollonova, G. N. (Izv. Vysshikh Uchebn. Zavedenii Geol. Razvedka **19** No. 1 [1976] 69/78).

[81] Krasnova, N. I.; Kartenko, N. F.; Rimskaya-Korsakova, O. M.; Firyulina, V. V. (Mineral. Geokhim. **1967** No. 2, pp. 19/27; C.A. **68** [1968] No. 116291).

[82] Bulakh, A. G.; Mazalov, A. A. (Am. Mineralogist **59** [1974] 378/80).

[83] Apel'tsin, F. R. (Geol. Mestorozhd. Redk. Elem. No. 14 [1961] 105/15, 108/10).

[84] Pavlenko, A. S.; Vainshtein, E. E.; Shevaleevskii, I. D. (Geokhimiya **1957** 351/67, 351, 356/8).

[85] Portnov, A. M. (Geol. Rudn. Mestorozhd. **9** No. 6 [1967] 74/5 and C.A. **68** [1968] No. 80378).

[86] Pavlenko, A. S. (Geokhim. Shchelochnogo Metasomatoza **1963** 7/73 from [87, p. 105]; C.A. **60** [1964] 7820).

[87] Pavlenko, A. S.; Orlova, L. P.; Akhmanova, M. V.; Tobelko, K. I. (Zap. Vses. Mineral. Obshch. **94** [1965] 105/13, 105, 112/3).

[88] Bareja, E.; Kubicki, S. (Kwart. Geol. **27** [1983] 215/23, 218, 223).

[89] Kubicki, S. (Biul. Inst. Geol. No. 347 [1984] 49/54).

[90] Robinson, S. C.; Sabina, A. P. (Am. Mineralogist **40** [1956] 624/33, 627/8, 632/3).

[91] Robinson, S. C.; Abbey, S. (Can. Mineralogist **6** [1957/61] 1/14, 2).

[92] Robinson, S. C. (Can. Mineralogist **6** [1957/61] 513/21, 515).

[93] Shaw, D. M. (Geol. Rept. Quebec Dept. Nat. Resour. No. 80 [1958] 1/52, 1, 15, 30/1, 35/6, 39/40, 43/4).

[94] Leavitt, D. L. (Mineral. Rec. **12** [1981] 359/63).
[95] Stevenson, J. S.; Stevenson, L. S. (Trans. Roy. Soc. Can. IV [3] **49** [1955] 105/10).
[96] Andreev, V. A. (Tr. Buryatsk. Kompleks. Nauchn. Issled. Inst. Akad. Nauk SSSR Sibirsk. Otd. No. 22 [1966] 121/34).
[97] Andreev, G. V.; Maslov, V. M. (Tr. Buryatsk. Kompleks. Nauchn. Issled. Inst. Akad. Nauk SSSR Sibirsk. Otd. No. 22 [1966] 136/42).
[98] Andreev, G. V. (Tr. Buryatsk. Kompleks. Nauchn. Issled. Inst. Akad. Nauk SSSR Sibirsk. Otd. No. 12 [1963] 97/9).
[99] Moreau, M. (Chron. Mines Outre-Mer Rech. Min. No. 279 [1959] 315/8).
[100] Roubault, M. (Proc. Intern. Conf. Peaceful Uses At. Energy, Geneva 1955 [1956], Vol. 6, pp. 152/61, 159/60).

[101] Murdock, T. G. (U.S. Bur. Mines Inform. Circ. No. 8196 [1963] 1/147, 103/4).
[102] Behier, J. (Trav. Bur. Geol. Repub. Malgache **89** [1958] 1/146, 129/34).
[103] Moreau, M. (Ann. Geol. Madagascar No. 33 [1963] 197/201).
[104] de la Roche, H.; Marchal, J. (Trav. Bur. Geol. Haut Commissariat Madagascar Dependences **76** [1956] 1/60, 17/8; C.A. **1956** 13358).
[105] de la Roche, H. (Territ. Madagascar Doc. Bur. Geol. No. 140 [1958] 1/11 from C.A. **1959** 5989).
[106] Roubault, M. (Géologie de l'Uranium, Masson, Paris 1958, pp. 1/462, 358/67).
[107] Teuscher, E. O. (Spec. Publ. Soc. Geol. Appl. Miner. Deposits No. 2 [1982] 460/8, 464, 467; C.A. **99** [1983] No. 91334).
[108] Fazekas, V.; Kósa, L.; Selmeczi, B. (Foldt. Kozl. **105** [1975] 297/308, 305).
[109] Wells, R. C.; Fairchild, J. G.; Ross, C. S. (Am. J. Sci. [5] **26** [1933] 45/54).
[110] Shcherbin, S. S. (in: Vol'fson, F. I., Geologicheskye Voprosy Genezisa Endogennykh Uranovykh Mestorozhdenii, Nauka, Moscow 1968, pp. 50/64, 50, 60, 62/3; C.A. **71** [1969] No. 103950).

[111] Nininger, R. D. (Minerals for Atomic Energy, Van Nostrand, New York 1954, pp. 1/367, 88/9).
[112] Egorov, B. L. (Zap. Vses. Mineral. Obshch. **99** [1970] 493/7).
[113] Kaplan, G. E.; Uspenskaya, T. A.; Zarembo, Yu. I.; Chirkov, I. V. (Torii i Ego Syr'evye Resursy, Khimia i Tekhnologiya, Atomizdat, Moscow 1960, pp. 1/224).
[114] White, M. G. (U.S. Geol. Surv. Circ. No. 195 [1952] 1/12).
[115] Wedow, H., Jr.; et al. (U.S. Geol. Surv. Circ. No. 248 [1953] 1/15, 3, 5).
[116] White, M. G.; Stevens, J. M. (U.S. Geol. Surv. Circ. No. 279 [1953] 1/19).
[117] Staatz, M. H. (Geol. Surv. Profess. Paper [U.S.] No. 1161 A-BB [1981] Y1/Y2).
[118] West, W. S. (U.S. Geol. Surv. Circ. No. 250 [1953] 28/31).
[119] Killeen, P. L.; White, M. G. (U.S. Geol. Surv. Circ. No. 250 [1953] 15/20, 15/8).
[120] Gault, H. R.; Black, R. F.; Lyons, J. B. (U.S. Geol. Surv. Circ. No. 250 [1953] 1/10, 5/6).

[121] George, D'Arcy (RMO-563 [1949] 1/198, 81/2; N.S.A. **5** [1971] No. 833).
[122] West, W. S.; Matzko, J. J. (U.S. Geol. Surv. Circ. No. 250 [1953] 21/7).
[123] Eakins, G. R.; Forbes, R. B. (TID-27593 [1976] 1/396, 103/9).
[124] West, W. S. (U.S. Geol. Surv. Circ. No. 300 [1953] 1/7, 5).
[125] Cobb, E. H. (Geol. Surv. Bull. [U.S.] No. 1374 [1973] 1/213).
[126] Bates, R. G.; Wedow, H., Jr. (U.S. Geol. Surv. Circ. No. 202 [1953] 1/13).
[127] Heinrich, E. W.; Conrad, M. A. (Am. Mineralogist **45** [1960] 459/64, 460, 462/3).
[128] Thompson, R. M. (Am. Mineralogist **39** [1954] 525/8).
[129] Hytonen, K.; Kallio, P. (Geologi **26** [1974] 103; C.A. **83** [1975] No. 30829).
[130] Andaç, M.; Mücke, A. (Bull. Mineral Res. Explor. Inst. Turkey No. 85 [1975] 150/2).

44

[131] Herath, J. W. (Econ. Bull. Sri Lanka Geol. Surv. Dept. No. 2 [1975] 1/72, 21, 24).

[132] Anonymous (Bibl. Ser. Intern. At. Energy Ag. No. 4 [1962] 1/134, 22; C.A. **58** [1963] 7754).

[133] Rupasinghe, M. S.; Senaratne, A.; Dissanayake, C. B. (J. Gemmology **20** [1986] 177/84, 177, 182).

[134] Gindy, A. R. (Econ. Geol. **56** [1961] 436/41).

[135] Davidson, C. F. (Proc. Intern. Conf. Peaceful Uses At. Energy, Geneva 1955 [1956], Vol. 6, pp. 207/9).

[136] Buchwald, V.; Sørensen, H. (Medd. Grønland **162** No. 11 [1961] 1/35).

[137] Bouwer, R. F. (Diss. Univ. Cape Town 1957 from [19, p. 67]).

[138] Newesely, H.; Andaç, M. (Naturwissenschaften **65** [1978] 153).

[139] Zhabin, A. G.; Mukhitdinov, G. N.; Kazakova, M. E. (Tr. Inst. Mineral. Geokhim. Kristallokhim. Redk. Elem. No. 4 [1960] 51/72, 55/7, 69/70).

[140] Zhabin, A. G.; Kazakova, M. E. (Dokl. Akad. Nauk SSSR **134** [1960] 164/7).

[141] Hansen, J. (Medd. Grønland **181** No. 8 [1968] 1/47, 2, 19, 27).

[142] Makovicky, M.; Makovicky, E.; Nielsen, B. L.; Karup-Møller, S.; Sørensen, E. (RISO-R-416 [1980] 1/186, 39, 48).

[143] Bondam, J.; Sörensen, H. (Proc. 2nd U.N. Intern. Conf. Peaceful Uses At. Energy, Geneva 1958, Vol. 2, pp. 555/9, 556/7).

[144] Ledoux, A. (Ann. Soc. Geol. Belg. **40** [1912/13] 177/210, 186, 190).

[145] Smith, W. L.; Cisney, E. A. (Am. Mineralogist **41** [1956] 76/81, 78, 80).

[146] Sørensen, H. (Medd. Grønland **167** No. 1 [1962] 1/251, 92/3, 96).

[147] Pluhar, E. (Berlin. Geowiss. Abhandl. A **12** [1979] 1/53, 24/6).

[148] Dubinchuk, V. T.; Sidorenko, G. A. (Konst. Svoistva Mineral. No. 12 [1978] 18/26, 23).

[149] Mumpton, F. A. (Diss. Pennsylvania State Univ. 1958, pp. 1/90, 37/40, 63; Diss. Abstr. **19** [1958] 57).

2.1.1.3 Chemistry

Analyses. Thorianite is composed essentially of ThO_2 and forms with the isomorphous mineral uraninite a more or less complete solid-solution series from almost pure ThO_2 to UO_2 [1]. Natural miscibility gaps within the series, as reported in 1933 for thorianite with < 10% UO_2 and between 15.10 and 27.02% UO_2 [2, p. 47], do not exist. Rather, the continuity of the series, proved for synthetic compounds (see, e.g. [3, 4]), is now also corroborated by natural minerals. This is particularly demanded by [5, p. 682], [6], and is also evident (for the part of the series with Th > U) from the chemical compositions of thorianite minerals compiled in the Tables 2 (complete/almost complete analyses) and 3 (partial analyses, comprising contents of Th, U, and sometimes Pb), see pp. 48/54 and 54/5, respectively. For important compilations of older analyses, see [1, 7, 8], [9, p. 48], [10, p. 168], [11, pp. 84/5], [12, pp. 114/5], [13, pp. 314, 378/9]; see also "Thorium", 1955, p. 29.

Although being the Th-richest mineral, with a theoretical composition of 100% ThO_2 = 87.88% Th [12, p. 114], pure thorianite is hitherto not found in nature [5, p. 681]. The Th content varies in natural samples due to multifold admixtures and isomorphic substitutions for Th, particularly by U (see the variety of uranothorianite) [12, pp. 110, 114], refer also to pp. 46/7. The highest Th content ever determined is 94.85% ThO_2 for thorianite from a carbonatite, Kola Peninsula, USSR [14, pp. 78, 164/5]. The Th-poorest sample, which still satisfies the restriction to samples with molar Th:U ratios > 1:1 (see p. 26), is uranothorianite with 38.47% ThO_2 from metamorphic limestone, Easton, Pennsylvania (but note strong effects of alteration and impurities for the analyzed specimen; see analysis No. 42 in Table 2, p. 53) [2, pp. 47, 50].

Samples with equal amounts of Th and U are reported especially from placers of the Seward Peninsula, Alaska [15], and from sanidinitic ejecta, Latium, Italy [16], but see also Tables 2 and 3, pp. 48/54 and 54/55. For a statistical evaluation of the ThO_2 content of more than 50 thorianite minerals, see "Thorium" Suppl. Vol. A 1b.

A compilation of thorianite analyses for 24 localities of Madagascar shows considerable variation in contents of Th (50 to 75%) and U (5 to 25%), but the summary content of Th + U (varying between 75 and 80%) is remarkably constant [17]; for similar values of samples from Madagascar, see [18 to 21]. This holds also for other analyses given in Table 2, pp. 48/54. The great range in Th and U contents entails wide variation of the Th:U ratios, which is assumed to be a general characteristic for thorianite [22]. The variation of the ratio (from 12.5 to 2.3) in the samples from Madagascar is tentatively related to genetic conditions, since thorianites with higher Th:U ratios occur in more alkaline, phlogopite-bearing pyroxenites than in common or feldspathic pyroxenites [17]. Note also the evidence for a preferential occurrence of Th-rich thorianites in phlogopite-bearing, calcite-garnet-vesuvianite-bearing, and other early-metasomatic rocks (as, e.g., on the Kola Peninsula), whereas uraniferous varieties, occasionally enriched in U to compositions intermediate between ThO_2 and UO_2, are bound to later carbonatites (as, e.g., at Loolekop, R.S.A.). Such compositional changes for thorianite agree with the general geochemical characteristic of Th and U, namely a gradual decrease of the Th:U ratio with time in natural evolutionary series of rocks and minerals [23].

Thorianite from Canada is generally given to contain 75 to 93% ThO_2 and up to 10% U_3O_8, uranothorianite 40 to 75% ThO_2 and 10 to 40% U_3O_8 [24]. But 15% U_3O_8 is usually regarded as a useful point of separation between both varieties [22], see also [25, p. 3]. As can be seen from Table 2, this discrimination is not consequently used in the literature.

Thorianite is generally grouped as simple oxide of the **formula** type AX_2 [49]. According to the main chemical components, textbooks and reviews give for thorianite the general formulas ThO_2 [50, 51] or $(Th, U)O_2$ [52 to 54], meaning a strong idealization, since all natural members of the thorianite-uraninite series are chemically rather complicated [55]. Considering the significant contents of RE and/or Pb (see the following pages and Tables 2 and 3, pp. 48/54 and 54/5), a general formula $(Th, U, RE, Pb)O_x$ seems to be more appropriate (but note that Pb is only a secondary component, see p. 46). Its application leads to the following formulas for individual samples [5, pp. 680, 682]:

thorianite from Madagascar	$(Th_{0.899}U_{0.066}RE_{0.008}Pb_{0.027})O_{1.99}$ [*)]
thorianite from Sri Lanka	$(Th_{0.711}U_{0.243}RE_{0.008}Pb_{0.037})O_{2.10}$
thorianite from Tuva, USSR	$(Th_{0.705}U_{0.267}RE_{0.008}Pb_{0.020})O_{2.13}$
aldanite from Aldan region, USSR	$(Th_{0.700}U_{0.168}RE_{0.003}Pb_{0.129})O_{1.96}$
aldanite from Aldan region, USSR	$(Th_{0.690}U_{0.155}RE_{0.006}Pb_{0.149})O_{1.93}$

[*)] Indices for U, Pb, and O are incorrectly given in the original paper.

The chemical formula for the Th-richest natural thorianite (see analysis No. 1 on p. 48) is given as $(Th_{0.96}RE_{0.03}U_{0.01})_{\Sigma 1.00}O_{1.985}$ [14, p. 164]. Detailed formulas, which summarize almost completely the chemical analyses, are for two calcic thorianites from skarn of Siberia $(Th_{0.79}Ca_{0.29}Si_{0.03}Pb_{0.01}U_{0.01})O_{2.00}$ and $(Th_{0.79}Ca_{0.25}Si_{0.07}Pb_{0.01}U_{0.01})O_{2.00}$ [32, p. 139], and for thorianite of the Allende meteorite $(Th_{0.689}U_{0.167}Ca_{0.099}Al_{0.071}Nb_{0.025}Dy_{0.013})O_2$ [35].

For an older thorianite formula, Th_4O_7, introduced in analogy to the known compounds Th_4S_7 or U_4O_7, see [56].

Diadochic and Trace Elements. Among the various elements occurring in substitution for Th in thorianite, U is apparently always present in considerable amounts [9, p. 47], since both elements show complete solid solution, see p. 44. Uranium is originally present as U^{4+} in the thorianite structure. The content of U^{6+} is probably secondary, owing to oxidation of U^{4+} with a concomitant entry of O into the vacant 8-fold position of the mineral structure, as occurs in partly oxidized uraninite [50]. For oxidation (and partial leaching) of U due to mineral alteration, see also p. 55. U-rich crystals are often zoned [54] with U concentrated in the core, as is mentioned for large crystals of uranothorianite from the Loolekop carbonatite ($UO_2:ThO_2$ ratio in the cores 1:2, in marginal zones 1:3) [57], and for uranothorianite from sanidinitic ejecta, Latium, in which the ThO_2 content progressively increases from UO_2-enriched cores towards outer zones [16]. For an inhomogeneous U and Th distribution among individual particles taken from two thorianite samples from Sri Lanka, see Table 3, p. 54/5.

Lead is also nearly constantly present in thorianite, usually in amounts of several percent [9, p. 49]. It is assumed that Pb enters homogeneously into the structure of thorianite minerals, and is probably present in the bivalent state [5, pp. 682/3]. Note for synthetic ThO_2 crystals, that Pb^{2+} occupying the Th site is more stable than tetravalent Pb at the elevated temperature required during crystal growth [58]. As Pb in natural samples is presumably of radiogenic origin [1, 59, 60], [9, p. 49], [11, p. 83], [26, p. 25], it may occur as thorian or uranian Pb; its proportions depend on the original amounts of Th and U in the mineral, and on the mineral's age [12, p. 114]. Origination of Pb by radioactive decay is particularly concluded from the observation that Pb-rich thorianite varieties, especially aldanites (refer also to p. 26), are the most ancient (here Precambrian) of an investigated series of samples [5, p. 683]. For more details and a discussion of older literature data on Pb in thorianite (with main respect to its origin, atomic weight, isotopic composition, content, ratios with other radioactive elements, and to its usage for age determination of thorianite), see [13, pp. 203/4, 212, 214, 236/7, 376/7, 407]. Note also a release of Pb during alteration of thorianite, see p. 55. For Pb present as PbS after reaction with sulfur in porous, schlieric and cloudy zones [61], respectively as galena inclusions [57], in uranoan thorianite from Palabora, see p. 38.

Also a product of radioactive decay processes is He, which occurs uniformly dispersed in thorianite [62]. It is usually present [59, 60], [12, p. 114] in amounts of 3.5 to 10.5×10^{-6} cm^3/g of the mineral [11, pp. 81, 83], see also [9, p. 49]. Large amounts of He gas occur especially in more ancient thorianite minerals [62]. For older data on He in thorianite concerning its origin, content, production rate, and its usage for age determination of thorianite minerals, see [63]; refer also to [13, pp. 197/9]. Note that in addition to Pb and He, the other products of radioactive decomposition of Th, U, and Ac may also occur as admixtures in thorianite [11, p. 81].

The rare earth elements, chiefly Ce or the Ce-group elements, generally substitute for Th [1, 59, 64], [9, pp. 48/9], [11, p. 83], in amounts up to 8.04% RE_2O_3, see "Seltenerdelemente" A4, 1979, p. 111; refer also to Table 2, pp. 48/54. Besides Th, the RE are major components of thorianite (as is indicated by spectral analyses) from rocks of the alkaline complex of the Vishnevye Gory, Southern Urals [65, 66]. However, note that almost complete isomorphic series between ThO_2 and RE oxides are only observed for synthetic products [9, p. 53], [67]. The strong predominance of light RE in thorianite is clearly evident from the relative RE distributions given in the original papers for thorianite: from carbonatite of Siberia (with 0.19% ΣRE_2O_3) and of the Kola Peninsula (with 0.12% ΣRE_2O_3) [68, p. 1060], similar data in [14, p. 143]; from metasomatite, Kola Peninsula (sample No. 2 in Table 2) [26, p. 25]; and from placers of Madagascar (sample No. 8 in Table 2) [5, p. 678] and of Sri Lanka [69, p. 406], similar data for both placer occurrences in [67]. Although the light RE prevail in the thorianite from Sri Lanka, even if the whole RE spectrum from La to Lu is present [67], thorianite, in comparison with other RE-incorporating phases, is grouped under minerals with no selective, but a complex RE composition, with relatively uniform proportions of the lanthanoids in [69, pp. 406/7, 410]. For the thorianites from carbonatites, the Ce-group predominance (observable also for associated RE-bearing carbonatite minerals) is attributed to the highly alkaline character of the Ce-group elements, which are concentrated by alkaline solutions commonly taking part in carbonatite formation [68, pp. 1060/4]. A particular abundance of Dy in thorianite, as is generally assumed by [67], is uniquely known from a meteoritic thorianite which contains 1% Dy_2O_3 as the sole RE [35]. Those RE capable of tetravalence, such as Ce, Pr, and Tb, usually are by far more frequent in thorianite minerals than the trivalent RE elements [67], [26, p. 26]. For synthetic ThO_2 successfully doped with tetravalent, less so with trivalent, lanthanides, magnetic resonance studies proved that, for example, Tb^{4+} substitutes for Th at the cubic Th^{4+} site [70].

Some thorianites show Zr contents [9, p. 49], [12, p. 110], [60]; refer also to analysis Nos. 2 and 4 in Table 2, p. 48, and to a secondary thorianite with high Zr concentrations (deduced from qualitative X-ray analysis) from Morro do Ferro, Brazil [71]. The highest Zr content hitherto determined is 3.68% ZrO_2 for an older analysis of thorianite (with 72.24% ThO_2) from gem washings, Balangoda, Sri Lanka [72]. Generally, the role of Zr in the analyses is uncertain [60], but some authors are convinced of an isomorphous substitution for Th [59], since ZrO_2 and ThO_2 are isostructural (fluorite-type lattice) and form partial solid-solution series in synthetic material [9, p. 49], see also "Thorium" 1955, p. 354. The Zr content in the above-mentioned thorianite sample No. 2 from Kovdor is partially explained by isomorphic substitution, but also attributed to inclusions of baddeleyite [26, pp. 23/5]; note also baddeleyite as an important contaminant in uranoan thorianite from Palabora [41]. The elevated Zr content of the above-mentioned sample No. 4 from Sri Lanka is probably due to inclusions of zircon [72].

Major amounts of CaO can be incorporated into thorianite [64], mainly $\leq$ 1%, but also up to a few percent, see Table 2, pp. 48/54. Higher amounts are solely reported for calcic thorianite from magnesian skarns, Siberia (analysis Nos. 7 and 9 in Table 2), for which the chemical formulas (see p. 45) suggest a replacement of 20% of the Th ions by Ca [32, p. 139]. But for the same samples, Ca is also attributed to mechanical admixtures [31]. The elevated content of Ca in the Kovdor thorianite (analysis No. 2 in Table 2) is explained by isomorphous substitution for Th, but also by inclusions of apatite, indicated by the simultaneous presence of P [26, p. 23].

Among the other elements which may occur in thorianite, Fe and Si are noted in several analyses, see Table 2. The role of Fe, usually appearing as Fe^{3+}, is, in general, uncertain [9, p. 49], [60], and may be due to isomorphous substitution [26, p. 23], [59]. The SiO_2 content of the Kovdor thorianite is completely attributed to thorite inclusions (see recalculation of the analysis in the original paper) [26, pp. 23/5]. Further elements observed as traces in thorianite are, for example, Mg, Mn, Zn, Nb, Ar, N [11, p. 81], Cu, Al, Sc, Y, Ti [12, p. 114/5], and, beyond them, also Ta, K, Na, see footnotes on p. 54 for Table 2.

The chemical composition of thorianite, including the various diadochic replacements, especially by U, Ca, and the RE (see the preceding pages), affect the mineral's crystallographic and physical properties, particularly the lattice constant, cell volume, formula units, density, refractive index, and sometimes the color; for details, see the corresponding Sections 2.1.1.4 and 2.1.1.5 on pp. 58/67 and 67/73. Such influences on the mineral properties are occasionally used to deduce the chemical composition of thorianite, see, for example, [23], [73], or [74].

Table 2

Chemical Composition and Density of Thorianite Samples (with Molar Ratios $Th:U > 1:1$) from Various Types of Occurrence. For the footnotes, see p. 54. Analysis numbers marked with an asterisk indicate samples whose contents are recalculated from atomic percentages given in the original paper.

	content in % in sample number						
	1	2	3	4	5	6	7
ThO_2	94.85	93.65	93.02	92.44	92.00	91.10	90.28
U_3O_8	2.12	0.48	—	3.46	6.10	6.50	—
UO_3	—	—	—	—	—	—	1.00
UO_2	—	—	4.73	—	—	—	—
PbO	—	0.08	1.80	0.26	0.30	0.15	0.98
ΣRE_2O_3	1.66	0.46[a)]	—	1.57[b)]	—[c)]	—[d)]	0.32
SiO_2	—	1.15	—	0.44	0.40	0.32	1.70
CaO	—	0.65	—	0.08	—	—	6.0
Fe_2O_3	—	0.86	0.29	0.14	—	1.32	—[e)]
l.o.i.	—	—	—	0.04	1.80	—	—
others	1.20[f)]	2.51[g)]	—	1.46[h)]	—	—	—
sum	99.83	99.84	99.84	99.89	100.60	99.39	100.28
D in g/cm³	8.89[i)]	9.29 ± 0.01	9.33	—	—	—	9.107

No. 1 = Thorianite from early carbonatite, Kola Peninsula, USSR [14, pp. 78, 164/5]. – No. 2 = Thorianite from apatitized phlogopite-diopside-olivine metasomatite, Kovdor massif, Kola Peninsula, USSR; for contents recalculated after subtraction of assumed apatite, respectively thorite and baddeleyite admixtures, see the original paper [26, pp. 19, 23, 25]. – No. 3 = Thorianite from alluvium, province of Betroka, southern Madagascar [27]. – No. 4 = Thorianite from calcitic carbonatite (Taimyr Peninsula [28]), Siberia, USSR [29]. – Nos. 5 and 6 = Thorianite from pyroxenite or cipolino, Madagascar [30]. – No. 7 = Thorianite from Mg skarn (Synnyr Mountain Range, northeastern Lake Baikal region [31]). Siberia [32, pp. 136/8, 140].

Table 2 (continued)

	8	9	10	11	12*	13	14
ThO$_2$	89.91	89.64	89.00	88.20	88.12	84.48	79.70
U$_3$O$_8$	6.86	–	6.70	6.70	6.56	13.00	17.90
UO$_3$	–	1.63	–	–	–	–	–
UO$_2$	–	–	–	–	–	–	–
PbO	2.40	1.45	0.07	2.08	2.27	0.42	0.90
ΣRE$_2$O$_3$	0.42	0.26	–$^{j)}$	–$^{j)}$	–	–$^{d)}$	–$^{j)}$
SiO$_2$	–	0.90	0.12	0.53	–	0.40	0.32
CaO	–	7.0	–	–	–	–	–
Fe$_2$O$_3$	–	–$^{e)}$	–	–	–	–	–
l.o.i.	–	–	4.40	2.27	–	1.22	1.10
others	–	–	–	–	–	–	–
sum	99.59	100.88$^{k)}$	100.29	99.78	96.95	99.52	99.92
D in g/cm^3	9.09	9.107	–	–	–	–	–

content in % in sample number

No. 8 = Thorianite from Madagascar [5, pp. 678, 680]. – No. 9 = Thorianite from Mg skarn (Synnyr Mountain Range, northeastern Lake Baikal region [31]), Siberia, USSR [32, pp. 136/8, 140]. – Nos. 10, 11, and 13, 14 = Thorianite from pyroxenite or cipolino, Madagascar [30]. – No. 12 = Thorianite from pegmatite, Betroka, southern Madagascar [33, pp. 191, 194].

Table 2 (continued)

	15*	16*	content in % in sample number 17	18	19	20	21
ThO_2	79.02	76.28	75.2	74.20	73.10	70.58	70.54
U_3O_8	16.31	13.76	–	–	23.70	18.03	25.67
UO_3	–	–	–	–	–	–	–
UO_2	–	–	18.6	14.10	–	–	–
PbO	2.73	2.48	–	–	0.22	11.09	3.14
ΣRE_2O_3	–	–	1.0[l)	6.30[m)	–[j)	0.21	0.51
SiO_2	–	–	–	0.80	0.49	–	–
CaO	–	–	2.3	–	–	–	–
Fe_2O_3	–	–	–	3.10	–	–	–
l.o.i.	–	–	–	–	2.17	–	–
others	–	–	2.9[n)	–	–	–	–
sum	98.06	92.52	100.0	98.50	99.68	99.91	99.86
D in g/cm^3	–	–	–	–	–	9.10	9.50

No. 15 = Thorianite from phlogopite pegmatite vein, Soatia, southern Madagascar [33, pp. 191, 193]. – No. 16 = Thorianite probably from pegmatite; average of eight samples from the Balangoda district, Sri Lanka [34]. For the individual analyses, see [13, pp. 378/9]. – No. 17 = Thorianite (assigned only on the basis of chemical composition) from Ca-Al-rich inclusions of type B, Allende meteorite [35]. – No. 18 = Thorianite probably from the USSR [11, pp. 84/5]. – No. 19 = Thorianite from pyroxenite or cipolino, Madagascar [30]. – No. 20 = Uranothorianite (aldanite) from placer, Aldan region, Siberia [5, pp. 677/8, 680]. – No. 21 = Thorianite from placer, Balangoda, Sri Lanka [5, pp. 677/8, 680].

Table 2 (continued)

	content in % in sample number						
	22	23	24	25	26	27	28
ThO_2	70.07	69.30	69.30	68.33	67.50	64.68	64.30
U_3O_8	16.84	–	27.88	26.72[o]	29.50	–	–
UO_3	–	14.90	–	–	–	–	20.40
UO_2	–	–	–	–	–	29.80	–
PbO	12.73	11.20	1.56	0.97	0.87	–	12.50
ΣRE_2O_3	0.31	3.70[p]	0.53	–	–[j]	–	0.70[p]
SiO_2	–	0.24	–	1.08	0.50	4.77	0.50
CaO	–	–	–	0.41	–	–	–
Fe_2O_3	–	–	–	0.23	–	1.29	–
l.o.i.	–	–	–	–	1.70	–	–
others	–	0.66[f]	–	0.59[q]	–	0.12[r]	1.60[f]
sum	99.95	100.00	99.27	98.33	100.07	100.66	100.00
D in g/cm³	8.80	8.92[i]	9.00	–	–	–	8.92[i]

No. 22 = Uranothorianite (aldanite) from Aldan region, Siberia [5, pp. 678, 680]. – No. 23 = Aldanite from pegmatite, Aldan region, Siberia [36]. – No. 24 = Thorianite from gabbro, Tuva, USSR [5, pp. 677/8, 680]. – No. 25 = Thorianite probably from the People's Republic of China; given are calculated mean values from two analyses performed by different methods [37]. – No. 26 = Thorianite from pyroxenite or cipolino, Madagascar [30]. – No. 27 = Uranothorianite from placer, probably of the USSR [38]. – No. 28 = Aldanite from pegmatite, Aldan region, Siberia [36].

Table 2 (continued)

			content in % in sample number				
	29	30*	31*	32	33	34	35
ThO₂	62.77	61.85	61.0	59.9	58.80	55.82	51.80
U₃O₈	29.30	28.63	30.9	16.6	–	–	25.70
UO₃	–	–	–	–	–	–	–
UO₂	–[s)]	–	–	–	37.93	26.65	–
PbO	0.63	3.12	1.1	9.61	–	13.40	12.43
ΣRE₂O₃	0.50	–	–	4.6[t)]	–	4.12[t)]	3.82[t)]
SiO₂	0.66	–	–	–	1.55	–	–
CaO	0.66	–	–	–	–	–	–
Fe₂O₃	3.94	–	–	–	0.90	–	–
l.o.i.	0.70	–	–	–	–	–	–
others	0.21[u)]	–	–	–	–	–	–
sum	98.71	93.60	93.0	90.71[v)]	99.18	99.99	93.75[v)]
D in g/cm³	–	–	–	9.1 ± 0.3[i)]	–	–	9.1 ± 0.3[i)]

No. 29 = Thorianite from syenite, Kyzyl-Ompul' massif, northern Kirghizia, USSR [39]. – No. 30 = Thorianite probably from pegmatite, Sri Lanka; average of ten samples [34]. For the individual analyses, see [13, pp. 378/9]. – No. 31 = Thorianite from nepheline syenite, Dugda massif, eastern Tuva, USSR [40]. – No. 32 = Uranoan thorianite (surface material) from carbonatite, Loolekop, R.S.A.; for contents recalculated to Σ = 100%, see original paper [41]. – No. 33 = Thorianite probably from the USSR [11, pp. 84/5]. – No. 34 = Uranoan thorianite from carbonatite, Palabora, R.S.A. [25, p. 10]. – No. 35 = Uranoan thorianite (drilling sample taken at depth) from carbonatite, Loolekop, R.S.A.; for contents recalculated to Σ = 100%, see original paper [41].

Table 2 (continued)

	content in % in sample number						
	36	37	38	39	40*	41	42
ThO$_2$	49.7	48.20	46.4	46.13	45.69	44.72	38.47
U$_3$O$_8$	36.3	33.42[o]	40.0	–	46.10	43.87[o]	–
UO$_3$	–	–	–	42.74	–	–	33.15
UO$_2$	–	–	–	–	–	–	4.44
PbO	8.6	9.63	8.0	1.49	9.32	2.57	5.21
ΣRE$_2$O$_3$	–[ae]	–	–[ae]	–	–	–	2.49[w]
SiO$_2$	–	0.48	–	–[y]	–	1.02	–[x]
CaO	–	0.12	–	–[z]	–	1.29	0.97
Fe$_2$O$_3$	–	0.34	–	–	–	0.08	1.57[aa]
l.o.i.	–	–	–	–	–	–	–
others	–	0.42[ab]	–	–	–	0.279[ac]	13.09[ad]
sum	94.6	92.61	94.4	90.36	101.11	93.83	99.39
D in g/cm^3	–	–	–	–	–	–	6.68

No. 36 = Uranian thorianite from metasomatized impure marble, Huddersfield, Quebec, Canada [6]; for further, partial analyses of samples of the same area, see Table 3; for analyses of samples in composition intermediate between thorianite and uraninite, see original paper. – No. 37 = Thorianite probably from the People's Republic of China; given are calculated mean values from two analyses performed by different methods [37]. – No. 38 = Uranian thorianite from metasomatized impure marble, Dungannon, Ontario, Canada [6]. – No. 39 = Uranothorianite from pyroxene-plagioclase-rich metasomatite, USSR [42]. – No. 40 = U-thorianite from granitic rock, Grenville Structural Province, Bancroft area, Ontario, Canada [43]. – No. 41 = Thorianite probably from the People's Republic of China; given are calculated mean values from two analyses performed by different methods [37]. – No. 42 = Uranothorianite from metamorphosed, serpentine-rich limestone, Easton, Pennsylvania [2, pp. 47/8, 50].

Footnotes for Table 2:

a) Comprises 0.03% La_2O_3, 0.29% Ce_2O_3, 0.02% Pr_2O_3, 0.10% Nd_2O_3, 0.01% Sm_2O_3, and 0.01% $(Eu + Gd)_2O_3$. — b) Comprises 1.05% Ce_2O_3 and 0.52% La_2O_3. — c) Separate spectrographic analyses showed <200 ppm La, 500 to 800 ppm Ce, 200 to 500 ppm Yb, 300 ppm Lu, and 500 to 1000 ppm Y. — d) Separate spectrographic analyses showed <200 ppm La, 500 to 800 ppm Ce, 200 to 500 ppm Yb, 100 to 200 ppm Lu, and 500 to 1000 ppm Y. — e) Separate semiquantitative X-ray spectrometry showed 0.03 to 0.10% Fe.

f) No individual elements given. — g) Comprises 2.35% ZrO_2, 0.14% P_2O_5, and 0.02% Y_2O_3. — h) Comprises 0.70% ZrO_2, 0.34% Al_2O_3, 0.03% TiO_2, 0.02% MgO, 0.03% Nb_2O_5, 0.01% Ta_2O_5, 0.27% Y_2O_3, and 0.06% H_2O^+. — i) Given density is not clearly assignable to chemically analyzed sample. — j) Separate spectrographic analyses showed <200 ppm La, 500 to 800 ppm Ce, 200 to 500 ppm Yb, and 500 to 1000 ppm Y.

k) Erroneously given as 100.17% in the original paper. — l) Comprises Dy_2O_3. — m) Comprises Ce_2O_3. — n) Comprises 1.5% Al_2O_3 and 1.4% Nb_2O_5. — o) Recalculated from atomic percent.

p) Originally given as ΣRE; CeO_2 absent. — q) Comprises 0.20% Al_2O_3, 0.16% MgO, 0.07% MnO, 0.04% TiO_2, 0.04% K_2O, and 0.08% Na_2O. — r) Comprises TiO_2. — s) Separate semiquantitative X-ray spectrometry showed 0.5 to 5.0% Pb. — t) Originally given as rare earth oxides.

u) Comprises 0.05% TiO_2, 0.06% MgO, and 0.10% H_2O^-. — v) Low summary content is attributed to restricted sample pureness. — w) Originally given as rare earths. — x) Included in "others". — y) Separate semiquantitative X-ray spectrometry showed 0.5 to 1.0% Ca.

z) Separate semiquantitative X-ray spectrometry showed <0.001% Fe. — aa) Erroneously given as R_2O_3 in the original paper; corrected to Fe_2O_3 by [9, p. 48]. — ab) Comprises 0.13% Al_2O_3, 0.21% MgO, 0.02% MnO, and 0.06% TiO_2. — ac) Comprises 0.06% Al_2O_3, 0.09% MgO, 0.003% MnO, 0.006% TiO_2, 0.03% K_2O, and 0.09% Na_2O. — ad) Comprises 0.53% MgO, 0.31% MnO, 4.39% H_2O^+, and 7.86% insoluble including SiO_2.

ae) Small amounts of rare earth elements are indicated.

Table 3
Partial Chemical Analyses of Thorianite.

sample and host rock, locality				Ref.
uranian thorianite from	55.5 % ThO_2	25.0 % U_3O_8	6.8% PbO	[6]
metasomatized impure marble,	41.0 % ThO_2	39.8 % U_3O_8	8.4% PbO	[6]
Quebec, Canada	40.1 % ThO_2	36.8 % U_3O_8	7.7% PbO	[6]
thorianite from skarn, Huddersfield,				
Quebec, Canada	62 % ThO_2	28 % U_3O_8		[44]
uranothorianite from monazite				
sand, undisclosed locality;				
one analysis	58.5 % ThO_2	26.4 % U_3O_8		[45]
mean of 2 analyses	75.75% ThO_2	13.65% U_3O_8		[45]
mean of 2 analyses	54.25% ThO_2	38.3 % U_3O_8		[45]
mean of 2 analyses	48.4 % ThO_2	35.35% U_3O_8		[45]
thorianite from Hogatza placer,				
Alaska	47.3 % ThO_2	36.8 % UO_2		[46]

Table 3 (continued)

sample and host rock, locality				Ref.
thorianite (No. T_1) from Sri Lanka*)	47.6 % ThO_2	31.6 % UO_2		[47]
same sample, particle 1	47.7 % Th	29.8 % U	3.5% Pb	[48, p. 1159]
same sample, particle 2	67.3 % Th	11.1 % U	2.6% Pb	[48, p. 1159]
thorianite (No. T_2) from Sri Lanka*)	51.9 % ThO_2	26.9 % UO_2		[47]
same sample, particle 1	50.4 % Th	29.6 % U	3.6% Pb	[48, p. 1159]
same sample, particle 2	50.0 % Th	26.7 % U	3.5% Pb	[48, p. 1159]
same sample, particle 3	56.7 % Th	24.0 % U	3.4% Pb	[48, p. 1159]

*) Reliability of the analysis is $\pm 2\%$.

Secondary **alteration of** thorianite minerals may influence their **chemical composition**. During weathering, thorianite is occasionally hydrated [12, p. 116], [59], entailing small amounts of secondary water in some analyses [9, pp. 48/9], [60]; see also analysis Nos. 4, 29, and 32 in Table 2, and especially U-rich varieties [1]. Alteration of thorianite often causes partial or total oxidation of primary U^{4+} to U^{6+} [2, p. 46], [12, p. 116], [59, 60]. Self-oxidation of U^{4+} to U^{6+} is assumed for some placer thorianites (see p. 39) owing to release of O due to radioactive decay of Th to Pb in the mineral (schematically as $ThO_2 \rightarrow PbO + \frac{1}{2} O_2 (+ 7 \text{ He})$). This leads to a partial redistribution of U into crystal fissures [38]. Increasing alteration during transport and deposition of uranothorianite from placers of western Turkey (see p. 39) is accompanied by a strong relative Th enrichment (the Th:U ratio in the mineral increases from originally $\sim 1:1$ towards finally $3:1$) due to redox-potential controlled elevated solubility of U^{6+} [75]. Leaching of U^{6+} is also observed for thorianite from postmagmatically altered alkaline rocks [76] and for ultimately U-free thorianite from oxidation zones in granitoids of the Grenville Structural Province, Canada [77]. Also the radiogenic Pb tends, in general, to be released from thorianite, either during thermal metamorphism or by circulating ground water [78], or during weathering or other secondary processes [79]. Note, for instance, reduced Pb contents (and simultaneously increased amounts of Be, Ti, Mg, Fe, Al, Ca, and Si) in strongly altered thorianite from Kirghizia [39]. Note also that Pb loss affects age determinations of thorianite if carried out on Pb isotopes [78].

Natural α decay of Th and/or U in thorianite evidently does not enhance its dissolution with subsequent leaching of the radioactive elements, as it is contrarily proved for many other Th-U-bearing accessory minerals in crystalline rocks [80]. However, isotopic fractionation occurs owing to preferential leaching of short-lived ^{228}Th, but not of long-lived ^{230}Th and ^{234}U (tested in bicarbonate-carbonate solvents after mineral irradiation imitating natural α decay). The process of preferential leaching causes localized disorder in the lattice, but is strikingly reduced by natural self-annealing in thorianite subsequent to radiation damage [47, 81], [48, pp. 1158, 1160/3].

For secondarily formed thorianites with enriched contents of Zr or of Zn and W, see p. 39.

References for 2.1.1.3:

[1] George, D'Arcy (RMO-563 [1949] 1/198, 81/2; N.S.A. **5** [1951] No. 833).

[2] Wells, R. C.; Fairchild, J. G.; Ross, C. S. (Am. J. Sci. [5] **26** [1933] 45/54).

[3] Hund, F.; Niessen, G. (Z. Elektrochem. **56** [1952] 972/9, 972).

56

[4] Slowinski, E.; Elliott, N. (Acta Cryst. **5** [1952] 768/70).

[5] Makarov, E. S.; Lipova, I. M. (Geokhimiya **1962** 583/9; Geochem. [USSR] **1962** 676/83).

[6] Robinson, S. C.; Sabina, A. P. (Am. Mineralogist **40** [1956] 624/33).

[7] Hintze, C. (Handbuch der Mineralogie, Vol. 1, Pt. 2, Veit & Co., Leipzig 1904/1915, pp. 1209/2674, 1670).

[8] Doelter, C. (Handbuch der Mineralchemie, Vol. 3, Pt. 1, Steinkopff, Dresden and Leipzig 1918, pp. 1/965, 224/6).

[9] Frondel, C. (Geol. Surv. Bull. [U.S.] No. 1064 [1958] 1/400).

[10] Nag-Chowdhury, B. D.; Das, S.; Dasgupta, A. (Proc. Natl. Inst. Sci. India **10** [1944] 167/74).

[11] Soboleva, M. V.; Pudovkina, I. A. (AEC-tr-4487 [1961] 1/455; N.S.A. **15** [1961] No. 23739).

[12] Chukhrov, F. V.; Bonshtedt-Kupletskaya, E. M. (Mineraly Spravochnik, Vol. 2, Pt. 2, Nauka, Moscow 1965, pp. 1/341).

[13] Holmes, A. (Bull. Natl. Res. Council [U.S.] No. 80 [1931] 124/459).

[14] Kapustin, Yu. L. (Mineralogiya Karbonatitov, Nauka, Moscow 1971, pp. 1/286).

[15] West, W. S.; Matzko, J. J. (U.S. Geol. Surv. Circ. No. 250 [1953] 21/7, 23).

[16] De Casa, C. G.; Della Ventura, G. C.; Parodi, G. C.; Stoppani, F. S. (Riv. Mineral. Ital. **1987** 97/104, 101, 104).

[17] Behier, J. (Trav. Bur. Geol. Repub. Malgache **89** [1958] 1/146, 129/30).

[18] Behier, J. (Ann. Geol. Madagascar No. 29 [1960] 1/78, 70/1).

[19] Moreau, M. (Chron. Mines Outre-Mer Rech. Min. No. 279 [1959] 315/8).

[20] Moreau, M. (Rappt. CEA-R-Fr. Commis Energ. At. No. 1685 [1960] 1/19, 5).

[21] Murdock, T. G. (U. S. Bur. Mines Inform. Circ. No. 8196 [1963] 1/147, 105).

[22] Nininger, R. D. (Minerals for Atomic Energy, Van Nostrand, New York 1954, pp. 1/367, 88/9).

[23] Bulakh, A. G.; Mazalov, A. A. (Am. Mineralogist **59** [1974] 378/80).

[24] Lang, A. H.; Griffith, J. W.; Steacy, H. R. (Geol. Surv. Can. Econ. Geol. Rept. No. 16, 2nd Ed. [1962] 1/324, 39).

[25] von Backström, J. W. (PEL-112 [1966] 1/19; N.S.A. **21** [1967] No. 22051).

[26] Krasnova, N. I.; Kartenko, N. F.; Rimskaya-Korsakova, O. M.; Firyulina, V. V. (Mineral. Geokhim. **1976** No. 2, pp. 19/27; C.A. **68** [1968] No. 116291).

[27] Lacroix, A. (Bull. Soc. Franc. Mineral. **37** [1914] 176/80).

[28] Saltykova, V. S. (Tr. Inst. Mineral. Geokhim. Genezisa Mestorozhd. Redk. Elem. No. 2 [1959] 189/207, 190).

[29] Gaidukova, V. S.; Zdornik, T. B. (Geol. Mestorozhd. Redk. Elem. No. 17 [1962] 86/117, 116).

[30] Moreau, M. (Ann. Geol. Madagascar No. 33 [1963] 197/201).

[31] Semenov, E. I. (Typochimizm Mineralov Shchelochnykh Massivov, Nedra, Moscow 1977, pp. 1/119, 59).

[32] Andreev, G. V.; Maslov, V. M. (Tr. Buryatsk. Kompleks. Nauchn. Issled. Inst. Akad. Nauk SSSR Sibirsk. Otd. No. 22 [1966] 136/42).

[33] Holmes, A.; Besairie, H. (Mem. Inst. Sci. Madagascar Ser. D **6** [1954] 191/9).

[34] Holmes, A. (Proc. Geol. Assoc. Can. **7** Pt. 2 [1955] 81/106, 101).

[35] Lovering, J. F.; Wark, D. A.; Sewell, D. K. B. (Lunar Planet. Sci. **10** [1979] 745/7).

[36] Bespalov, M. M. (Sov. Geol. **1941** No. 6, pp. 105/7; Mineral. Abstr. **12** [1953/55] 460).

[37] Luo, A. (Yanshi Kuangwu Ji Ceshi **5** [1986] 67/9).

[38] Egorov, B. L. (Zap. Vses. Mineral. Obshch. **99** [1970] 493/7).

[39] Turovskii, S. D. (Zap. Vses. Mineral. Obshch. **87** [1958] 369/70).

[40] Zykov, S. I.; Stupnikova, N. I.; Pavlenko, A. S.; Tugarinov, A. I.; Orlova, L. P. (Geokhimiya **1961** 547/60; Geochem. [USSR] **1961** 581/95, 585).

[41] Russell, H. D.; Hiemstra, S. A.; Groeneveld, D. (Trans. Proc. Geol. Soc. South Africa **57** [1954/55] 197/208, 197, 204).

[42] Trofimov, A. A.; Apollonova, G. N. (Izv. Vysshikh Uchebn. Zavedenii Geol. Razvedka **19** No. 1 [1976] 69/78, 69, 74/5).

[43] Rimsaite, J. (Geol. Surv. Can. Paper No. 81-23 [1982] 19/30, 24).

[44] Shaw, D. M. (Geol. Rept. Quebec Dept. Nat. Resour. No. 80 [1958] 1/52, 40).

[45] Kember, N. F. (Analyst [London] **77** [1952] 78/85, 85).

[46] Staatz, M. H. (Geol. Surv. Profess. Paper [U.S.] No. 1161 A-BB [1981] Y1/Y2).

[47] Eyal, Y.; Fleischer, R. L. (Nature **314** [1985] 518/20).

[48] Eyal, Y.; Fleischer, R. L. (Geochim. Cosmochim. Acta **49** [1985] 1155/64).

[49] Winchell, A. N. (Elements of Optical Mineralogy, Pt. II, 4th Ed., Wiley, New York 1951, pp. 1/551, 65).

[50] Frondel, C. (Proc. Intern. Conf. Peaceful Uses At. Energy, Geneva 1955 [1956], Vol. 6, pp. 568/77; Geol. Surv. Profess. Paper [U.S.] No. 300 [1956] 567/79, 568/9).

[51] Strunz, H.; Tennyson, C. (Mineralogische Tabellen, 6th Ed., Geest and Portig, Leipzig 1977, pp. 1/621, 211/2).

[52] Frondel, J. W.; Fleischer, M.; Jones, R. S. (Geol. Surv. Bull. [U.S.] No. 1250 [1967] 1/69, 37).

[53] Paone, J. (U.S. Bur. Mines Bull. No. 585 [1960] 863/72, 865).

[54] Ramdohr, P.; Strunz, H. (Klockmanns Lehrbuch der Mineralogie, 16th Ed., Enke, Stuttgart 1978, pp. 1/876, 548).

[55] Rankama, K. K.; Sahama, T. G. (Geochemistry, Univ. Press, Chicago 1950, pp. 1/912, 571/2).

[56] Wasserstein, B. (Nature **174** [1954] 1004/5).

[57] Forster, I. F. (Trans. Geol. Soc. South Africa **61** [1958] 359/65, 360).

[58] Kolopus, J. L.; Finch, C. B.; Abraham, M. M. (Phys. Rev. [3] B **2** [1970] 2040/5, 2044).

[59] Heinrich, E. W. (Mineralogy and Geology of Radioactive Raw Materials, McGraw-Hill, New York 1958, pp. 1/654, 34).

[60] Palache, C.; Berman, H.; Frondel, C. (Dana's System of Mineralogy, Vol. 1, 7th Ed., Wiley, New York 1944, pp. 1/834, 621).

[61] Ramdohr, P. (Die Erzmineralien und ihre Verwachsungen, 4th Ed., Akademie-Verlag, Berlin 1975, pp. 1/1277, 1131/2).

[62] Barnes, R. S.; Mazey, D. J. (J. Nucl. Energy **5** [1957] 1/3).

[63] Strutt, R. J. (Proc. Roy. Soc. [London] A **84** [1911] 379/88).

[64] Haaker, R. F.; Ewing, R. C. (Sci. Basis Nucl. Waste Management **2** [1980] 281/8, 283).

[65] Zhabin, A. G.; Mukhitdinov, G. N.; Kazakova, M. E. (Tr. Inst. Mineral. Geokhim. Kristallokhim. Redk. Elem. No. 4 [1960] 51/72, 69).

[66] Es'kova, E. M.; Zhabin, A. G.; Mukhitdinov, G. N. (Mineralogiya i Geokhimiya Redkikh Elementov Vishnevykh Gor, Nauka, Moscow 1964, pp. 1/318, 159).

[67] Semenov, E. I. (Mineralogiya Redkikh Zemel, Izv. Akad. Nauka, Moscow 1963, pp. 1/412, 48/50).

[68] Kapustin, Yu. L. (Geokhimiya **1966** 1311/21; Geochem. Intern. **3** [1966] 1054/64).

[69] Semenov, E. I.; Barinskii, R. L. (Geokhimiya **1958** 314/33; Geochem. [USSR] **1958** 398/419).

[70] Baker, J. M.; Chadwick, J. R.; Garton, G.; Hurrell, J. P. (Proc. Roy. Soc. [London] A **286** [1965] 352/65, 352/3).

58

[71] Barretto, P. M. C.; Fujimori, K. (Chem. Geol. **55** [1986] 297/312, 310).
[72] Dunstan, W. R. (Nature **69** [1904] 510/1).
[73] Grebennikova, O. T. (Nauchn. Tr. Irkutsk. Gos. Nauchn. Issled. Inst. Redk. Tsvetn. Metal. No. 14 [1966] 97/100; C.A. **70** [1969] No. 70075).
[74] Loskutov, A. V.; Polezhaeva, L. I. (Mater. Mineral. Kol'sk. Poluostrova **6** [1968] 276/81, 281; C.A. **73** [1970] No. 68378).
[75] Newesely, H.; Andaç, M. (Naturwissenschaften **65** [1978] 153).
[76] Mineeva, I. G. (Geokhimiya **1965** 443/55, 451/5; Geochem. Intern. **2** [1965] 284).
[77] Rimsaite, J. (Geol. Surv. Can. Paper No. 82-1B [1982] 253/66, 255).
[78] Robinson, S. C.; Loveridge, W. D.; Rimsaite, J.; Van Peteghem, J. (Can. Mineralogist **7** [1963] 533/46, 542/5).
[79] Polanski, A. (Geochimia Izotopów, Wydawn. Geol., Warzawa 1961, pp. 1/392; Geochemistry of Isotopes, Office of Technical Services, U.S. Dept. of Commerce, Washington 1965, pp. 1/473, 291).
[80] Petit, J. C.; Langevin, Y.; Dran, J. C. (Geochim. Cosmochim. Acta **49** [1985] 871/6).

[81] Eyal, Y.; Fleischer, R. L. (Mater. Res. Soc. Symp. Proc. **44** [1985] 687/91).

2.1.1.4 Crystal Form and Structure

Crystal Form

Thorianite crystallizes in the cubic/isometric crystal system, hexoctahedral class [1 to 3], [4, p. 211]; see also "Thorium" 1955, p. 29.

Thorianite is originally found as rough crystals, usually cubes, or as embedded grains [3], isolated or in aggregates (see, e.g., [5, 6]), interstitial [7], or as impregnation of fine rounded grains [8]. Primary crystals are euhedral/idiomorphic (see, e.g., [9, p. 204], [10], [11, p. 66], [8, 12]) or hypidiomorphic [6], but may also occur in irregular forms (see, e.g., [9, p. 204], [7, 10], [11, p. 66]), with curved [13] and corroded faces, or as deformed crystals [5], or as fragments [10, 14]. The cubes become worn on the edges when subjected to wave action [15] so that they commonly appear as more or less rounded grains in alluvial deposits [3], see also [1, 16, 17]. But note also idiomorphic crystals of uranothorianite from colluvial and diluvial placers of the USSR and from beach placers of western Turkey, see pp. 35 and 36, respectively. The crystal size of thorianite varies widely from a fraction of 1 mm up to 5 cm [18, p. 82 (Engl. transl.)], but it characteristically occurs as small crystals [19, 20], usually in the mm range; for individual data, see Section 2.1.1.2.1, pp. 27/37.

Habit. The appearance of thorianite is almost identical to that of uraninite [15]. Crystals are usually cubic [1, 2, 16, 17], less commonly octahedral, sometimes pseudotetragonal [19]. Thorianite from carbonatite dikes of Colorado invariably appears as well-formed equidimensional dodecahedral crystals (measuring ⅛ to ¼ mm across) [21]. Aldanites from pegmatites and gold washings of the Aldan region, Eastern Siberia, show both cubic and octahedral habits [22]. Characteristically, simple cubes occur, which are modified by, in most cases weakly developed, faces of an octahedron [17] (see examples in [8, 10, 14, 23]) or of a dodecahedron (see, e.g., [14]), respectively a rhombododecahedron (see, e.g., [8, 24]). For cubo-octahedral crystals of uranothorianite from Madagascar, the octahedral faces are never sufficiently developed to give the crystals a real octahedral habit [5]. Among the investigated intermediate mineral members of the thorianite-uraninite series from metasomatized impure marbles of Canada, the cubic forms of the uranian thorianites and high-thorian uraninites are, in contrast to the low-thorian uraninites, not modified by the octahedron [25]. A structural-geometric analysis of crystal growth forms of thorianite from the Gulinskii intrusion, Eastern Siberia, provided evidence of an evolution during the mineral growth process from crystals with cubo-octahedral habit, typical for early formations, to cubic crystals [26].

The most common **crystal form** for thorianite minerals is the simple cube $\{100\}$, sometimes modified by $\{111\}$ or $\{113\}$ [3] (see also [1]), or by $\{110\}$ [16]. The forms $\{210\}$ and $\{211\}$ are also given, especially for crystals from carbonatites [19]. A distinction is given according to which cubes of thorianite often occur with fine facets of an octahedron (111) and an tetragontrioctahedron ($=$ trapezohedron) (113); cubes of uranothorianite occur in some crystals with traces of facets of an octahedron, trapezohedron, and a pentagonal dodecahedron (note that the latter would lower the symmetry); and aldanite occurs rarely with faces of an octahedron [18, pp. 81/2, 85 (Engl. transl.) and pp. 77/8, 81/2 (Russ. orig.)]. Among samples from Madagascar, the cube and the cubo-octahedron (but never $\{311\}$) are observed as the most common crystal forms for uranothorianite [5], whereas thorianite cubes are provided either with small octahedral faces [10], or also with trapezohedral faces (311) on their edges [27], see also [28]. Thorianite from magnesian skarns in Siberia shows cubic and trigonal forms [7], uranothorianite from placers in Alaska mainly cubic, sometimes octahedral forms [29]. A variety of forms are given for cubic and octahedral aldanites from the Aldan region, Eastern Siberia: $\{111\}$, $\{200\}$, $\{220\}$, $\{311\}$, $\{331\}$, $\{420\}$, $\{422\}$, $\{511\}$, $\{531\}$, and $\{442\}$ [22]. The most detailed crystallographic description is presented for 14 individual thorianite crystals from metasomatic rocks of the Kovdor massif, Kola Peninsula. Due to variable growth conditions (see original paper), various types of crystal morphology developed, see **Fig. 2**, p. 60. Besides the most frequent cubo-octahedral combinations (b and c in the figure), there are forms compressed (d to f) or stretched (g to j) along one crystallographic axis. One extreme case of a pseudorhombohedral crystal (k) displays only octahedral faces. The dominating crystal faces of the Kovdor thorianites are (100) and (111); less frequently (110), (210), and (211) occur, all restricted to smallest grains or very fine fragments; one crystal (a) shows simply (111) [30, pp. 21/2].

Twinning after the spinel law is rather common for thorianite [18, p. 81 (Engl. transl.)], [19] and is useful as a characteristic feature for mineral identification [31, pp. 34/5], especially for discrimination from usually untwinned uraninite, see, for example, [25]. The twinned cubes generally occur as interpenetration twins after (111) [3, 16], with [111] as twin axis [25]; (011) as twin plane is only given by [17]. A small thorianite crystal from Betroka, Madagascar, is considered to be a penetration twin formed by rotation (of 60° around a 3-fold axis) of two crystals with the face combination p (100), a^1 (111), and a^3 (311); for a detailed figure, see original paper [28]; see also [27]. Note also polysynthetic twins after the cube for uranothorianite from Madagascar [5]. Twinning characterizes thorianite/uranothorianite from Canada [20], where the twins occur in high-metamorphic/metasomatic rocks of Ontario and Quebec, for details see [14, 25, 32 to 35]. Twins are also described for aldanites from pegmatites and gold washings of the Aldan region, Eastern Siberia [22]; for uranothorianite from placers in Alaska [29], and, rarely, for thorianite from metasomatites of the Kovdor massif, Kola Peninsula, USSR [30, p. 21]. Thorianite twins of an unclear character are found in alkaline rocks of the Enisei Ridge, Siberia [6].

Standard reference books generally give for thorianite a poor **cleavage** parallel to (001) [1, 3, 16], [31, p. 34], or imperfect parallel to (100) [18, p. 81 (Engl. transl.)], [19]. Only in thin sections, a cleavage parallel to (111) is occasionally observable [36]. Data for individual natural thorianite occurrences uniquely exist for the cleavage parallel to (100) and are only reported for samples from Madagascar [28] and from the Loolekop carbonatite, R.S.A., where crystals show a good cleavage [11, p. 67], [37], see also [36]. Note for synthetic ThO_2 a good, but not perfect, cubic cleavage parallel to the planes (100), (010), and (001) [38].

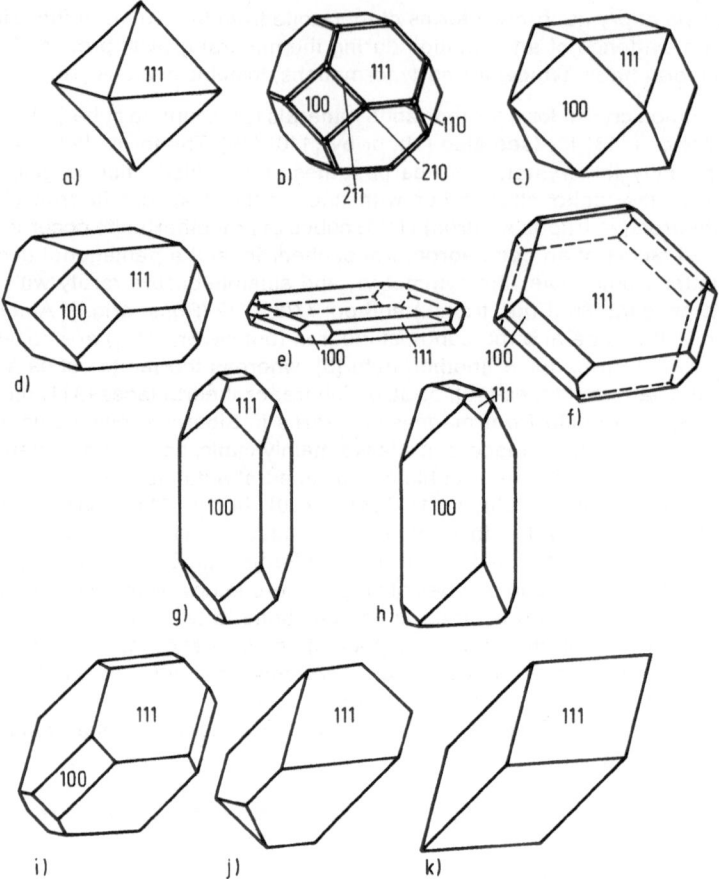

Fig. 2. Morphologies of thorianite crystals from metasomatic rocks of the Kovdor massif, Kola Peninsula, USSR, from [30, p. 22].

Crystal Structure

The **space group** of thorianite is $O_h^5 - Fm3m$ (No. 225) [4, p. 211], see also [3, 39], [40, p. 400]; for synthetic ThO_2, see also [41, pp. 42/3] and "Thorium" 1955, p. 206.

The theoretical **number of formula units** (Z) is 4 for ThO_2 [3, 39], [4, p. 211]. But Z is <4 when calculated for a series of natural thorianite minerals of Sri Lanka (3.74), Madagascar (3.64), and the USSR (3.64, 3.54, and 3.51). This is explained by vacant lattice sites caused by the presence of U^{6+} in the mineral structure [10].

The **unit-cell dimension** a_0 of natural thorianite is variable, as it depends on changing $ThO_2:UO_2$ ratios in the minerals [4, p. 212], or on the degree of substitution of Th by U and other elements [42, p. 135], [3]. Therefore, it is used to distinguish roughly between the mineral members of the isomorphous ThO_2-UO_2 group [43, p. 23], [44]. The variations of a_0 for thorianite given in the standard reference literature spread over different ranges, with maxima up to 5.62 Å [40, p. 399] and minima down to 5.50 Å [42, p. 135], but mostly cover the range from

5.60 to 5.54 Å as given in [4, p. 212]. For lattice constants of individual thorianite minerals, arranged by decreasing ThO_2 contents, see Table 4. Additionally to these data, note also two high values of $a_0 = 5.617 \pm 0.001$ Å[*] for a Th-rich thorianite from alkaline rocks, Enisei Ridge, Siberia [6], and of $a_0 = 5.603 \pm 0.005$ Å for a thorianite from altered melilite rocks, Turii Peninsula, northwestern USSR, which is, as indicated by its physical properties, a thorian end member of the ThO_2-UO_2-CeO_2 mineral group [8]. Other lattice constants of individual thorianite minerals given in the literature are omitted if the samples are not chemically analyzed.

For comparison, note lattice constants (all measured at room temperature) of synthetic ThO_2 with $a_0 = 5.5997$ Å [46] or several values between 5.5953 and 5.5952 Å [47, pp. 45/6], see also [41, pp. 42/3], [48, 49]; for compiled values from older literature, including samples of the ASTM file, see [46], see also [3].

Table 4
Unit-Cell Constants a_0 and ThO_2 Contents for Natural Thorianite Minerals.
For a complete sample description, see Tables 2 and 3, pp. 48/54 and 54/5, respectively. The number of chemical analysis given after the sample locality refers to the corresponding number in Table 2.

host rock and/or locality; analysis number	a_0 in Å	% ThO_2	Ref.
metasomatite, Kovdor, USSR; No. 2	5.619 ± 0.005	93.65	[30, p. 23]
skarn, Siberia; No. 7	5.57[a]	90.28	[50]
Madagascar; No. 8	5.60	89.91	[10]
skarn, Siberia; No. 9	5.57[a]	89.64	[50]
pegmatite, Sri Lanka; No. 16	5.573 ± 0.003	76.28	[51]
placer, Aldan region, Siberia; No. 20	5.58	70.58	[10]
placer, Sri Lanka; No. 21	5.58	70.54	[10]
Aldan region, Siberia; No. 22	5.58	70.07	[10]
gabbro, Tuva, USSR; No. 24	5.57	69.30	[10]
pegmatite, Aldan region, Siberia; No. 23	5.578 ± 0.009	69.30	[22]
Sri Lanka	5.575 ± 0.003	68.55[b]	[51]
placer, USSR; No. 27	5.54 to 5.56	64.68	[24]
pegmatite, Aldan region, Siberia; No. 28	5.539 ± 0.005	64.30	[22]
carbonatite, Republic of South Africa; No. 32	5.582 ± 0.003	59.9	[9, pp. 204, 206]
metasomatized marble, Quebec[c]	5.564 ± 0.002	55.5	[25]
carbonatite, Republic of South Africa; No. 35	5.588 to 5.591	51.80	[9, pp. 204, 206]
metasomatized marble, Quebec; No. 36	5.547 ± 0.002	49.7	[25]
metasomatized marble, Ontario; No. 38	5.553 ± 0.002	46.4	[25]
metasomatite, USSR; No. 39	5.50	46.13	[52]
metasomatized marble, Quebec[c]	5.539 ± 0.002	41.0	[25]
metasomatized marble, Quebec[c]	5.540 ± 0.002	40.1	[25]
metamorphic limestone, Pennsylvania; No. 42	5.505	38.47	[44]

[a] Note also $a_0 = 5.58 \pm 0.02$ Å for a thorianite sample (without detailed chemical analysis) from the same locality, given in [7]. — [b] Chemical analysis from [53]. — [c] For sample description and chemical analysis, see Table 3, pp. 54/5.

[*] Value, originally given in kX units, is converted to Å using the conversion factor 1.00202 ± 0.00004, see [45].

Within the system ThO_2-UO_2, the unit-cell dimension of synthetic material decreases linearly from pure ThO_2 ($a_0 = 5.595$ Å) to pure UO_2 ($a_0 = 5.468$ Å) [3], as is particularly shown by continuous solid-solution series which give, for samples of the part of the series with mole ratios ThO_2:$UO_2 > 1$ (as defined for thorianite on p. 26), ranges of a_0 from 5.5953 to 5.5017 Å [*] [48], or from 5.587 to 5.526 Å [54], or from 5.586 to 5.528 Å [55]. The lattice contraction is explained by incorporation of U^{4+} (into ThO_2), which has a smaller ionic radius than Th^{4+} [48] (0.97 and 1.02 Å, respectively [4, p. 30]), occurring without alteration of the structure type [43, p. 23].

For natural thorianite-uraninite minerals, the general decrease of the lattice constant with decreasing Th content is also evident. This is shown, for example, from investigations of samples from Ontario and Quebec, Canada (comprising uraninites, thorian uraninites, and uranian thorianites) [25], and from Sri Lanka, Madagascar, and the USSR (comprising uraninite, uranothorianite, aldanites, and thorianites) [10]; refer also to Table 4, p. 61. But in natural material, this simple relation is complicated by the presence of other elements in solid solution, particularly of RE and radiogenic Pb; by partial oxidation of U^{4+} to U^{6+} (see below); by radiation damage in the lattice; and by other factors [25]. A decrease of a_0 upon the entry of U into thorianite is generally assumed by [56]. The elevated lattice constant of $a_0 = 5.60 \pm 0.01$ Å for thorianite (heated at 1000°C) from the alkaline complex of the Vishnevye Gory, Southern Urals, is referred to its high RE content [57, 58]. The variation of a_0 (5.52 to 5.59 Å) for a series of eight thorianites from Madagascar (with a ThO_2 range from 67.50 to 92.00%, see analysis Nos. 5, 6, 10, 11, 13, 14, 19, and 26 in Table 2, pp. 48/51) is attributed to substitution of U and RE for Th. But, even if the U-rich thorianites have the lowest a_0 values, there is yet no simple arithmetic relation between the lattice constant and the Th:U ratio [59]. The influence of U and Pb on a_0 is questioned for the above-mentioned thorianite minerals from Sri Lanka, Madagascar, and the USSR [10]. A general tendency of a decrease of the lattice parameter caused by increasing admixtures of U, Zr, Pu, Ce, Pr, Eu, Yb, and Y compounds into ThO_2 is corroborated by synthetic material; refer to the corresponding diagrams, compiled from literature data, in [60]; for the influence of Y, see especially [61]. No correlation exists between the unit-cell edge and the specific gravity [44].

Several thorianite minerals from Sri Lanka, Madagascar, and the USSR show a scatter from the experimentally proved (see above) correlation between a_0 and the ThO_2 content towards higher a_0 values in the high-Th range (see also [6]). Similar results on synthetic samples (examined by [54]) are explained by variable degrees of equilibrium attained during sample fusion [10]. Note, for natural uraninite and thorian uraninite samples from Ontario and Quebec, a similar spread of a_0 in the high-U range, which is tentatively attributed to variable U^{6+}:U^{4+} ratios [25].

Natural ThO_2-UO_2 minerals have, in general, higher a_0 values than corresponding synthetic oxides, as is particularly indicated by the already repeatedly mentioned sample series from [10, 25]. It could be shown that heating (at 1000°C in air for 10 min) of the samples from Canada reduces their lattice constants by about half a percent resulting in a closer approach to linear correlation with the mineral composition (U:Th ratio); for a_0 values of heated and unheated material, see original paper; data for the unheated uranian thorianites are also given in Table 4, p. 61 [25]. A comparable decrease of the unit-cell edge occurred upon heating (at 810°C for ~7.5 h) of two thorianite minerals from Sri Lanka (T_1 and T_2, for chemical analyses, see Table 3, p. 55): a_0 for T_1 (5.544 Å) and T_2 (5.553 Å) were reduced by 0.019 and 0.012 Å, respectively [62]. The dependency of the unit-cell edge (and the molar volume) on

---

[*] See footnote, p. 61.

different heating temperatures is particularly investigated for synthetic ThO_2 in the temperature range from 8.5 to 787.2°C, resulting in a variation of a_0 from 5.59475 to 5.63050 Å; for the deduced thermal linear expansion coefficient, see below; for further details, see the original paper [47, pp. 45/6]. The shrinkage of the unit cell, as observed upon sample heating to 800°C in air or in vacuum for natural uraniferous thorianite and uraninite, obviously implies an oxidation of U^{4+} (with ionic radii of 1.05 or 0.97 Å) to smaller U^{6+} cations (0.80 Å). Shrinkage under vacuum conditions additionally suggests that in the O_2-rich lattice the excess oxygen occupies the available interstitial positions ($\frac{1}{2},0,0$) in 6-fold coordination with U. When pure thorianite is heated no lattice shrinkage occurs, since Th^{4+} ions do not react because of their reluctance to donate electrons [51]. During oxidation of synthetic mixed ThO_2-UO_2 crystals at temperatures between 100 and 1000°C, the unit cell contracts as long as the average U valency increases from 4.0 to 5.0, but begins to expand on further valency increase up to 5.5 [63].

The **cell volume** for synthetic thorianite (thoria) is given as $175.16_9 \times 10^{-24}$ cm^3; the molar volume as 26.38 ± 0.01 cm^3/mol [41, p. 66], see also [64], or as 26.378 ± 0.005 cm^3/mol, all values measured at 25°C. As the lattice constant, the cell or molar volume also expands with increasing temperature. The linear and volume thermal expansion coefficients of ThO_2 are, respectively, $\alpha_a = 0.6216 \times 10^{-5} + 3.541 \times 10^{-9}$ T $- 0.1125$ T^{-2} and $\alpha_v = 1.865 \times 10^{-5} + 10.96 \times 10^{-9}$ T $- 0.3375$ T^{-2} (with T = temperature from 298 to 1073 K) [47, pp. 39, 45/6]. The entry of U in the thorianite lattice causes a reduction of the cell volume [56], as is also deducable from the preceding data on a_0.

The **X-ray powder diffraction pattern** of thorianite closely resembles that of uraninite and cerianite. The d values of the interplanar spacings vary, according to the lattice constant, with variable degree of substitution of U and other elements for Th. Therefore, d spacings are distinctive for thorianite samples with relatively low U content [42, pp. 135, 141], [3]. For experimental diagrams giving the variation of the (200) and (111) spacings for the series ThO_2-UO_2 and ThO_2-ZrO_2, see [79]. From a sample series of thorianite-uraninite minerals from Madagascar and the USSR it is deduced that the d values become larger with increasing Th content, but the relative line intensities reveal no anomalies due to changing chemical composition. The latter fact is explained by the similarity of the atomic scattering power of the heavy elements U, Th, and Pb [10]. However, thorianite-uraninite minerals from Canada indicate, with increasing Th content, an intensification of the (222) reflection relative to those for the (311) and (400) planes [25]. In addition to the use for mineral identification within the thorianite-uraninite group (see, e.g., [14]), X-ray data also prove the face-centered crystal structure for thorianite (see, e.g., [10, 22] and also below). The usually observed completeness of the patterns and the sharpness of the lines indicate that thorianite, in general, is well crystallized; see also p. 71. X-ray data for a Th-rich thorianite from alkaline rocks of the Turii Peninsula, northwestern USSR, characterize the mineral to be slightly metamict [8]. A weak diffraction pattern of (partly) metamict/hydrated thorianite from biotite granite of the Omchikandisk massif, northeastern Siberia, was only obtained after sample heating at 700 to 900°C for 3 h. Compared to reference patterns, by far fewer diffraction lines appeared (see original paper), owing to the mineral alteration processes [65, 66]. See the original papers for further d values of heated samples, given for two thorianites from Sri Lanka (heated at 800°C; 30 min in vacuum or 15 min in a muffle furnace) [51], and for a RE-rich thorianite (heated at 1000°C) from the alkaline complex of the Vishnevye Gory, Southern Urals [57]. The X-ray diffraction pattern of the ThO_2 mineral remains virtually unaffected during experimental heavy-ion bombardment (of up to 5000 Ar$^+$ ions/nm^2 at 3 or 4 MeV, imitating natural irradiation, see also pp. 65 and 71) owing to the mineral's stabile fluorite-type structure [67], see also [68]. In the following table, the interplanar spacings of natural thorianite minerals from various localities are compiled and compared with reference data on synthetic ThO_2; selectively chosen are the three d values with always highest relative intensities (given in parentheses); samples are arranged with decreasing d value of the, in most cases, strongest (111) reflection:

source	d value (in Å) for plane			% ThO_2	Ref.
	(111)	(220)	(311)		
Kola Peninsula, USSR[a]	3.26(10)	1.96(6)	1.680(8)	—[b]	[13]
Madagascar	3.25(9)	1.99(10)	1.70(7)[c]	89.91	[10]
Kovdor massif, Kola Peninsula	3.245(10)	1.983(9)	1.694(9)	93.65	[30, pp. 24/5]
Turii Peninsula, Kola Peninsula	3.234(80)	1.9807(41)	1.690(30)[c]	very high[d]	[8]
Enisei Ridge, Siberia	3.22(9)	1.978(10)	1.688(10)	very high[e]	[6]
—[f]	3.22(10)	1.960(9)	1.679(9)	—[g]	[43, p. 25]
Balangoda, Sri Lanka[h]	3.216(10)	1.964(10)	1.675(10)	—[b]	[40, p. 400]
Siberia	3.210(10)	1.972(10)	1.682(10)	90.28/89.64	[50]
Aldan region, Siberia	3.20(8)	1.97(9)	1.68(9)[c]	70.07	[10]
Balangoda, Sri Lanka[i]	3.18(10)	1.953(7)[j]	1.667(8)[c]	—[b]	[39]
synthetic ThO_2[k]	3.234(100)	1.980(58)	1.689(64)	pure	[46]
synthetic ThO_2	3.22(10)	1.982(10)	1.685(10)	pure	[43, p. 26]
synthetic ThO_2[l]	3.21(10)	1.971(8)	1.682(9)	pure	[42, p. 135]

[a] No plane indices given in the original paper. — [b] No chemical composition given. — [c] Plane indexed as (113). — [d] As indicated by physical properties (e.g., n, D, a_0). — [e] As indicated by spectrographic analysis and a_0.
[f] Uranothorianite monocrystal of unspecified origin. — [g] Given as uranothorianite with $U \gg 1\%$. — [h] d values of this thorianite sample are often used as reference data by Russian authors. — [i] Calculated d values (see original paper) are, mostly, slightly higher than the given measured values. — [j] Plane indexed as (022).
[k] ASTM card No. 4-0556; for two older ThO_2-patterns from the ASTM file (card Nos. 1-0731 and 2-1278), see the original paper cited. — [l] Reference sample of the U.S. Geological Survey standard file.

The **crystal structure** of thorianite is a face-centered cubic lattice of the fluorite (CaF_2)-type, as is given in the general reference literature [19, 56, 69, 70] and individually determined by [6, 10, 22, 59]; see also "Thorium" 1955, p. 9. An antifluorite-type lattice for thorianite is specified by [36]; a probable perovskite structure (deduced from X-ray analysis) is only demanded for a calcic thorianite from Mg skarns of Siberia [7, 71]. Thorianite is isostructural with uraninite and cerianite [19, 56, 69], and also with ZrO_2 [72]. It is attributed to the general structure type $A^{[8]}O_2$ [56], which implies an 8-fold cubic coordination for the Th ion and a 4-fold tetrahedral coordination for O [73]; see also [69]. Atomic positions are given for 4 Th in (a) as 0,0,0 and for 8 O in (c) as $\frac{1}{4},\frac{1}{4},\frac{1}{4}$ [39].

The chemical bonding in thorianite is of a mixed ionic-covalent type (with 25% δ character), characterized by directed polar δ bonds with f^4d^3s hybridization of Th and sp^3 hybridization of O [73, 74]. As follows from IR absorption spectra (see p. 69), a strengthened Th-O bond in thorianite is developed due to electron hybridization in the $f^2d^2sp^3$ orbit [75, p. 57]. Because of considerable differences in the electronegativity between the main cation Th (given as

165 kcal/g-atom) and the oxygen anion (given as 530 kcal/g-atom), bonding in thorianite is considered to be strong [76]. Effective charges are $Q_{Th} = +2.52$ and $Q_O = -1.26$ [73], or $+2.6$ and -1.3, respectively [74].

The ideal structure of the cation-oxygen polyhedra may be considerably altered by incorporation of isomorphous elements and by variable mineral formation temperatures (note the influence on IR spectra, see p. 69) [75, pp. 58/9]. Natural irradiation by α decay in thorianite (see metamictization, p. 71) probably entails localized submicroscopic loss of the atomic order in the thorianite lattice (as deduced from leaching data, see p. 55). But the overall structure of thorianite remains generally crystalline, probably owing to a steady-state process of radiation damage and recovery, during which subsequent natural self-annealing heals the damaged structure [62, 77, 78].

References for 2.1.1.4:

[1] Palache, C.; Berman, H.; Frondel, C. (Dana's System of Mineralogy, Vol. 1, 7th Ed., Wiley, New York 1944, pp. 1/834, 620/2).
[2] Winchell, A. N. (Elements of Optical Mineralogy, Pt. 2, 4th Ed., Wiley, New York 1951, pp. 1/551, 65).
[3] Frondel, C. (Geol. Surv. Bull. [U.S.] No. 1064 [1958] 1/400, 47/53).
[4] Strunz, H.; Tennyson, C. (Mineralogische Tabellen, 6th Ed., Geest and Portig, Leipzig 1977, pp. 1/621).
[5] Behier, J. (Ann. Geol. Madagascar No. 29 [1960] 1/78, 70).
[6] Grebennikova, O. T. (Nauchn. Tr. Irkutsk. Gos. Nauchn. Issled. Inst. Redk. Tsvetn. Metal. No. 14 [1966] 97/100; C.A. **70** [1969] No. 70075).
[7] Andreev, G. V. (Tr. Buryatsk. Kompleks. Nauchn. Issled. Inst. Akad. Nauk SSSR Sibirsk. Otd. No. 12 [1963] 97/9).
[8] Bulakh, A. G.; Mazalov, A. A. (Am. Mineralogist **59** [1974] 378/80).
[9] Russell, H. D.; Hiemstra, S. A.; Groeneveld, D. (Trans. Geol. Soc. South Africa **57** [1954] 197/208).
[10] Makarov, E. S.; Lipova, I. M. (Geokhimiya **1962** 583/9; Geochem. [USSR] **1962** 676/83).

[11] Hanekom, H. J.; van Staden, C. M. v. H.; Smit, P. J.; Pike, D. R. (Mem. Geol. Surv. South Africa No. 54 [1965] 1/185).
[12] De Casa, C. G.; Della Ventura, G. C.; Parodi, G. C.; Stoppani, F. S. (Riv. Mineral. Ital. **1987** 97/104, 101, 104).
[13] Kapustin, Yu. L. (in: Vlasov, K. A.; Mineralogiya i Geticheskie Osobennosti Shchelochnykh Massivov, Nauka, Moscow 1964, pp. 135/94, 162/3).
[14] Stevenson, J. S.; Stevenson, L. S. (Trans. Roy. Soc. Can. IV [3] **49** [1955] 105/10).
[15] Nininger, R. D. (Minerals for Atomic Energy, Van Nostrand Reinhold, New York 1954, pp. 1/367, 88/9).
[16] George, D'Arcy (RMO-563 [1949] 1/198, 81/2; N.S.A. **5** [1951] No. 833).
[17] Getseva, R. V.; Savel'eva, K. T. (Rukovodstvo po Opredeleniyu Uranovykh Mineralov, Gosgeoltekhizdat, Moscow 1956, pp. 1/260, 124/5).
[18] Soboleva, M. V.; Pudovkina, I. A. (AEC-tr-4487 [1961] 1/455; N.S.A. **15** [1961] No. 23739).
[19] Chukhrov, F. V.; Bonshtedt-Kupletskaya, E. M. (Mineraly Spravochnik, Vol. 2, Pt. 2, Nauka, Moscow 1965, pp. 1/341, 113/4).
[20] Lang, A. H.; Griffith, J. W.; Steacy, H. R. (Geol. Surv. Can. Econ. Geol. Rept. No. 16, 2nd Ed. [1962] 1/324, 45).

66

[21] Parker, R. L.; Sharp, W. N. (Geol. Surv. Profess. Paper [U.S.] No. 649 [1970] 1/24, 17).
[22] Bespalov, M. M. (Sov. Geol. **1941** No. 6, pp. 105/7; Mineral. Abstr. **12** [1953/55] 460).
[23] Gaidukova, V. S.; Zdornik, T. B. (Geol. Mestorozhd. Redk. Elem. No. 17 [1962] 86/117, 116).
[24] Egorov, B. L. (Zap. Vses. Mineral. Obshch. **99** [1970] 493/7).
[25] Robinson, S. C.; Sabina, A. P. (Am. Mineralogist **40** [1956] 624/33).
[26] Evzikova, N. Z. (Diss. Izd. LGU, 1963, Author's Abstract from [192, pp. 20/1]).
[27] Lacroix, A. (Bull. Soc. Franc. Mineral. **37** [1914] 176/80).
[28] Lacroix, A. (Minéralogie de Madagascar, Vol. 1, A. Challamel, Paris 1922, pp. 1/624, 245/7).
[29] West, W. S.; Matzko, J. J. (U.S. Geol. Surv. Circ. No. 250 [1953] 21/7, 23).
[30] Krasnova, N. I.; Kartenko, N. F.; Rimskaya-Korsakova, O. M.; Firyulina, V. V. (Mineral. Geokhim. **1967** No. 2, pp. 19/27; C.A. **68** [1968] No. 116291).

[31] Heinrich, E. W. (Mineralogy and Geology of Radioactive Raw Materials, McGraw-Hill, New York 1958, pp. 1/654).
[32] Kennedy, I. (Mineral. Rec. **10** [1979] 153/8, 156).
[33] Satterly, J. (Ann. Rept. Ontario Dept. Mines **65** VI [1956] 1/181, 15).
[34] Shaw, D. M. (Geol. Rept. Quebec Dept. Nat. Resour. No. 80 [1958] 1/52, 15, 36).
[35] Leavitt, D. L. (Mineral. Rec. **12** [1981] 359/63).
[36] Ramdohr, P. (Die Erzmineralien und ihre Verwachsungen, 4th Ed., Akademie Verlag, Berlin 1975, pp. 1/1277, 1131/2).
[37] Forster, I. F. (Trans. Geol. Soc. South Africa **61** [1958] 359/65, 360).
[38] Campbell, W. B.; Hurst, V. J.; Moody, W. E. (J. Am. Ceram. Soc. **42** [1959] 262).
[39] Berry, L. G.; Thompson, R. M. (Geol. Soc. Am. Mem. No. 85 [1962] 1/281, 184).
[40] Mikheev, V. I. (Rentgenometricheskii Opredelitel' Mineralov, Gosudarst. Nauch.-Izdatel. Lit. po Geol. i Okhrane Nedr., Moscow 1957, pp. 1/868).

[41] Robie, R. A.; Bethke, P. M.; Toulmin, M. S.; Edwards, J. L. (Geol. Soc. Am. Mem. No. 97 [1966] 27/74).
[42] Frondel, C.; Riska, D.; Frondel, J. W. (Geol. Surv. Bull. [U.S.] No. 1036-G [1956] 91/153).
[43] Sidorenko, G. A. (Rentgenograficheskii Opredelitel' Uranovykh i Uransoderzhashchikh Mineralov, Gosgeoltekhizdat, Moscow 1960, pp. 1/116).
[44] Arnott, R. J. (Am. Mineralogist **35** [1950] 386/400, 388, 390).
[45] Jerrard, H. G.; McNeill, D. B. (A Dictionary of Scientific Units, Chapman and Hall, London 1963, pp. 1/197, 128).
[46] Swanson, H. E.; Tatge, E. (NBS-C-No. 539 [1953] 57/8).
[47] Skinner, B. J. (Am. Mineralogist **42** [1957] 39/55).
[48] Hund, F.; Niessen, G. (Z. Elektrochem. **56** [1952] 972/9, 973).
[49] Frondel, C. (Am. Mineralogist **40** [1955] 876/84, 882/3).
[50] Andreev, G. V.; Maslov, V. M. (Tr. Buryatsk. Kompleks. Nauchn. Issled. Inst. Akad. Nauk SSSR Sibirsk. Otd. No. 22 [1966] 136/42).

[51] Mukherjee, B. (Nature **186** [1960] 464/5).
[52] Trofimov, A. A.; Apollonova, G. N. (Izv. Vysshikh Uchebn. Zavedenii Geol. Razvedka **19** No. 1 [1976] 69/78, 75).
[53] Nag-Chowdhury, B. D.; Mousuf, A. K. (Proc. Natl. Inst. Sci. India **12** [1946] 341/9, 347).
[54] Trzebiatowski, W.; Selwood, P. W. (J. Am. Chem. Soc. **72** [1950] 4504/6).
[55] Slowinski, E.; Elliot, N. (Acta Cryst. **5** [1952] 768/70).
[56] Semenov, E. I. (Typokhimizm Mineralov Shchelochnykh Massivov, Nedra, Moscow 1977, pp. 1/119, 57).

[57] Zhabin, A. G.; Mukhitdinov, G. N.; Kazakova, M. E. (Tr. Inst. Mineral. Geokhim. Kristallokhim. Redk. Elem. No. 4 [1960] 51/72, 64/5, 69/70).

[58] Es'kova, E. M.; Zhabin, A. G.; Mukhitdinov, G. N. (Mineralogiya i Geokhimiya Redkikh Elementov Vishnevykh Gor, Nauka, Moscow 1964, pp. 1/318, 159/60).

[59] Moreau, M. (Ann. Geol. Madagascar No. 33 [1963] 197/201).

[60] Chukhrov, F. V.; Lapin, V. V.; Ovsyannikova, N. I. (Mineraly Spravochnik. Diagrammy Fazovykh Ravnovesii, Vol. 2, Nauka, Moscow 1974, pp. 1/490, 45, 94, 106, 174, 187/8, 190/1).

[61] Subbarao, E. C.; Sutter, P. H.; Hrizo, J. (J. Am. Ceram. Soc. 48 [1965] 443/6).

[62] Eyal, Y.; Fleischer, R. L. (Geochim. Cosmochim. Acta 49 [1985] 1155/64).

[63] Anderson, J. S.; Edington, D. N.; Roberts, L. E. J.; Wait, E. (J. Chem. Soc. 1954 3324/31, 3324).

[64] Bulakh, A. G.; Bulakh, K. G. (Fiziko-Khimicheskie Svoistva Mineralov i Komponentov Gidrotermal'nykh Rastvorov, Nedra, Leningrad 1978, pp. 1/167, 78).

[65] Nekrasov, I. Ya.; Ipat'eva, I. S. (Mater. Geol. Polez. Iskop. Yakutsk. ASSR No. 5 [1961] 32/50, 42/4).

[66] Nekrasov, I. Ya. (Tr. Yakutsk. Filiala Akad. Nauk SSSR Ser. Geol. No. 12 [1962] 1/334, 224/5).

[67] Cartz, L.; Karioris, F. G.; Fournelle, R. A.; Gowda, K. A.; Ramasami, K.; Sarkar, G.; Billy, M. (Sci. Basis Nucl. Waste Management 3 [1981] 421/7).

[68] Karioris, F. G.; Gowda, K. A.; Cartz, L. (Radiat. Effects Letters Sect. 58 [1981] 1/3).

[69] Frondel, C. (Proc. Intern. Conf. Peaceful Uses At. Energy, Geneva 1955 [1956], Vol. 6, pp. 568/77; Geol. Surv. Profess. Paper [U.S.] No. 300 [1956] 567/79).

[70] Ramdohr, P.; Strunz, H. (Klockmanns Lehrbuch der Mineralogie, 16th Ed., Enke, Stuttgart 1978, pp. 1/876, 135/6).

[71] Andreev, G. V. (Tr. Buryatsk. Kompleks. Nauchn. Issled. Inst. Akad. Nauk SSSR Sibirsk. Otd. No. 22 [1966] 121/34, 128).

[72] Bayer, G. (in: Wedepohl, K. H., Handbook of Geochemistry II-5, Springer, Berlin 1974, p. 90-A-1).

[73] Zuev, V. V. (Mineral. Sb. [Lvov] 21 [1967] 206/10; C.A. 70 [1969] No. 89521).

[74] Zuev, V. V. (Obogashch. Rud 15 No. 1/2 [1970] 99/102; C.A. 73 [1970] No. 80715).

[75] Povarennykh, A. S. (Konst. Svoistva Mineral. No. 13 [1979] 53/78).

[76] Povarennykh, A. S. (Zap. Vses. Mineral. Obshch. 85 [1956] 593/7).

[77] Eyal, Y.; Fleischer, R. L. (Nature 314 [1985] 518/20).

[78] Eyal, Y.; Fleischer, R. L. (Mater. Res. Soc. Symp. Proc. 44 [1985] 687/91).

[79] Mumpton, F. A. (Diss. Pennsylvania State Univ. 1958, pp. 1/90, 12; Diss. Abstr. 19 [1958] 57).

2.1.1.5 Physical Properties

Optical Properties

Diaphaneity. Thorianite is commonly opaque [1, 2] or nearly opaque [3]. Translucent are only the smallest grains [4, 5] and also crystals [6, 7], or fine fragments [8], [9, pp. 81, 85]. Transparent thorianite is known solely from the Vishnevye Gory, Southern Urals [10, 11], or for very fine splinters [12].

The characteristic **color** is black [1] (see also [2, 12 to 14]), which may frequently fade to grayish and brownish tones [14, 15], such as dark gray [1, 2, 12, 13], [9, p. 82], gray [16, p. 34], brownish black [2, 12], brown-black [13], or dark reddish brown [2]. A distinction between brown-black thorianite, black uranothorianite, and tarry or blue-black aldanite is made by [9, p. 81]. Additional colors, mentioned for individual thorianite occurrences, are pitch black to greenish [17], greenish black [18, 19], deep emerald green [5], dark brown to nearly black [20], reddish brown [6, 21], resinous brown, brown, and yellow [22, 23], [24, p. 137], dark yellow-brown and amber yellow [25, pp. 75/6], golden yellow [26], transparent yellow [10, 11], and light orange [4, 7]. Note that the lighter colors are usually observed for fine translucent mineral fragments and splinters, such as yellow [9, p. 85], reddish or red-brown [1, 13], brownish or brown-red [8], [9, pp. 82/3], but also dark green [16, p. 34]; or they appear because of mineral alteration, in the colors brown, cinnamon gray, and sometimes yellow-brown owing to corrosion [9, p. 82], or brownish black and yellowish brown owing to weathering of originally blackish crystals [1]. Thorianite crystals with a bronze tarnish are sometimes reported [2, 14] or, especially for aldanites, with reddish coatings [17, 27], or with brownish red and golden yellow rims around black isotropic crystal cores of partly metamict/hydrated late-magmatic thorianite [18, 19]. Note also yellow to yellow-brown alteration zones along fracture planes in black uranothorianite crystals from Easton, Pennsylvania [28].

In general, an intensification of the color towards black owing to the entry of U into thorianite is assumed [29]. Thus, the light, reddish brown color of the Kovdor thorianite is attributed to its low content of U, but also of other chromatophores (see analysis No. 2 in Table 2, p. 48), and to its tiny grain size [6].

The color variability of natural thorianite is comparable to the colors observed for synthetic material of the ThO_2-UO_2 system, which change from white for pure ThO_2 through yellowish, greenish, and brownish tones (of various shades, see original paper) finally towards black, which is reached at concentrations of $< \sim 52$ mol% ThO_2 [30, p. 973].

In transmitted light (under the microscope), natural thorianite is usually not totally opaque, but shows similar, even fainter, colors than macroscopically, in general with brownish [1 to 3], [9, p. 83] and greenish tones [1, 2, 14]. Still lighter tints are observed particularly in thin sections of very Th-rich samples, such as yellowish brownish [6], orange-yellow [21], or light brown and light gray [4].

The **luster** of thorianite, as given in standard reference literature, is adamantine; semimetallic [1]; submetallic (especially when fresh) [2]; semimetallic to adamantine [9, p. 81], [13] or pitchy [8]; submetallic to horny [12], greasy [14, 15], [16, p. 34], or dull [14]; also pitchy [31]; sometimes resinous [1]. Individual occurrences additionally show: metallic luster for uranothorianite from Latium, Italy [32]; brilliant luster for thorianite from carbonatite, Kola Peninsula, USSR [33]; and semimetallic, on fracture planes slightly adamantine, luster for brilliant black uranothorianite crystals from Madagascar [34]. See also adamantine luster on fracture planes, but only vitreous to dull luster on crystal faces, of thorianite from northern Kirghizia [35]. Vitreous luster is especially known for calcic thorianite [22, 33], [24, p. 137], pitchy luster for aldanite [9, p. 81], [17, 27]. Occasionally, also weak lusters occur, such as dull for thorianite from Colorado [7], and greasy, or resinous on fresh surfaces, for thorianite from the Enisei Mountain Range, southern Siberia [21]. But note that weak lusters are generally considered to result from alteration, as, for example, dull to greasy [1] and resinous or horn-like [2].

The **streak** is usually black, gray, or greenish, see, e.g., [13, 14]. Only Russian samples display also brownish colors, for instance, brownish yellow [24, p. 137], light yellow-brown [21], and light brown [6]. Especially for aldanites greenish brown [17] and bluish brown streak colors are observed [9, p. 88].

Thorianite is **isotropic**, see, for instance, [1, 12]. According to [2], sometimes it should reveal weak anomalous birefringence. (But no literature on individual occurrences proves that.)

The mean **refractive index** is n = 2.20, see "Thorium" 1955, p. 29; refer also to [1, 8, 13, 31]. A value n = 2.20 is also determined for many individual occurrences, see, for instance, [20, 21, 33]. General ranges for n are given as between 2.35 and 2.20 [14], 2.35 and 2.12 [16, p. 34], or 2.2 and 1.8 [9, pp. 83, 413]. The lower values usually are attributed to low U or high RE contents [29]. This is confirmed by refractive indices of n = 2.14 ± 0.01 for Th-rich thorianite from Kovdor (analysis No. 2 in Table 2, p. 48) [6], of n = 1.8 for Th-rich thorianite from Madagascar [36] (chemical analysis in [37]), or of n = 1.777 to 1.781 for RE-rich thorianite from the Vishnevye Gory, Southern Urals, USSR [10, 11]. The low refractive index n = 2.10 for calcic thorianite from Siberia (analysis Nos. 7 and 9 in Table 2, pp. 48/9) [22, 23] is attributed to Ca substitution [24, p. 141]. Note also that the conclusion for thorianite from the Turii Peninsula, northwestern USSR, to be a thorian end-member of the ThO_2-UO_2-CeO_2 system, is drawn from its n = 2.14 and other properties (see pp. 61 and 64) [4]. For a partly metamict/hydrated thorianite from biotite granite, Omchikandinsk massif, northeastern Siberia, n varies between 1.86 and 1.91. Differently colored rim zones of those samples show lower values of 1.716 to 1.680, which increased to 1.78 after heating for 3 to 4 h. The possibility that they might be caused by a secondarily formed Th mineral is argued [18, 19].

The **reflectivity** (R) of pure thorianite material (with n_{Na} = 2.129) is calculated to be ~13% [8]. Without comment, R = 13.5% is given (as general?) in [1]. For an unoriented sample of an undisclosed origin, R (measured in air) is 13.7% at 610 nm (white light). The values for other wavelengths are: 19% at 470 nm (blue), 15.7% at 520 nm (green), 14.2% at 575 nm (yellow), 15.7% at 600 nm (orange), and 13.5% at 700 nm (red) [38]. Uranothorianites from the USSR show R = 16% [39] and between 13.5 and 16% (depending on the U content) [25, p. 75]. The reflectivity of thorianite generally increases with the content of U (see also [1]), so that it is always lower than for uraninite [8]. Large zoned crystals of uranoan thorianite from the Loolekop carbonatite, R.S.A., compared to its U-enriched cores, have a lower reflectivity in Th-rich marginal zones (see also p. 46) [40]. Samples of the same locality are characterized by a gray reflection color, which distinguishes them from associated magnetite which they resemble otherwise [41].

Thorianite from Loolekop shows **internal reflections** of brown-yellow and brown-red colors, especially visible when embedded in oil [8]. In large zoned crystals (see above), they are slightly yellow-brown in the U-rich cores and strongly reddish brown in the Th-rich crystal margins [40].

Absorption Spectra. Thorianite absorps at low frequencies in the IR spectral range in a manner similar to the oxides of other high-valent elements. The shift to the long-wave range, however, is not as big as expected theoretically. Important absorption bands occur at wavenumbers between 100 and 1100 cm^{-1} with strongly pronounced peaks at 430 and 368 cm^{-1}, and fainter ones at 680, 535, 180, and 130 cm^{-1}. Isomorphous element substitution, as well as variable mineral formation temperatures may affect the IR spectral curve. The positions of the absorption bands may shift and their intensities change [42]. In the visible wavelength range between 4500 and 7000 Å, pure synthetic ThO_2 has no discrete absorption bands, but is characterized by an intense continuous absorption band, whose long-wavelength edge gradually shifts towards longer wavelengths with increasing U content. This band is attributed to resonance of U ions between different oxidation states. With increasing U concentration, four discrete bands with maxima at 5100, 5400, 6000, and 6500 Å appear.

Maximum light absorption occurs in the 10% UO_2 sample. Further U increase causes a fading out of the peaks, with exception of the band at 6500 Å which shifts to shorter wavelengths. It is tentatively concluded that these absorptions are due to electronic transitions between various states of the U^{4+} ion [43].

Differential absorption of γ rays in the low-energy range between 20 and 610 keV shows, for thorianite, a characteristic single intensity line at 240 keV, which is caused by $Th_B = {}^{212}Pb$ and is clearly distinctive from peaks due to U [44].

Neither **luminescence** [1], [9, p. 82] nor **fluorescence** [2] is reported for thorianite.

Other Physical Properties

Thorianite is brittle [1, 2, 12, 31], [9, p. 83]. The Mohs **hardness** H_M varies/decreases with the degree of alteration [1, 2, 12, 14]. The most frequently mentioned general range for H_M is 5 to 7 [14, 15], [16, p. 34], 6.5 to 7 especially for unaltered material [2]; the most common mean value is 6.5 [3, 12, 45]. Other ranges given between 5 and 6 [8] seem to be too low, those between 6.5 and 7.5 [1], [9, pp. 81, 83, 413] are probably too high, since H_M is theoretically calculated to be 7 [46]. The microhardness of thorianite ranges from 920 to 1235 kg/mm^2 (at a load of 100 g) [47], or from 872.8 to 1138 kg/mm^2. The mean of 1020 kg/mm^2 (calculated for 10 measurements of the latter range), corresponds to $H_M = 7.049$ (converted by means of Khrushchov's formula) [48]. Samples from unspecified localities in the USSR give lower values, as 713 to 845 kg/mm^2 for aldanite [9, p. 82], and 600 to 720 kg/mm^2 [25, p. 75] or ≤ 600 kg/mm^2 for uranothorianite [39].

The **fracture** of thorianite is uneven to subconchoidal [1, 12 to 14] or conchoidal [2].

Literature data on the general range of the **specific gravity** of thorianite minerals diverge widely. It is usually > 9 g/cm^3 [14, 15], but considerably less when highly altered [14] (note also [1, 2], [9, p. 83]), as, for example, 6.68 g/cm^3 for strongly altered uranothorianite from Easton, Pennsylvania [28], or 4.2 to 4.3 g/cm^3 for central zones of partly metamict/hydrated thorianite from the Omchikandinsk massif, northeastern Siberia [18]. The most comprehensive range of D covers values between 8.07 and 9.7 g/cm^3 [9, p. 413], but also values up to 10 g/cm^3 may be reached [8, 31]. $D = 10$ g/cm^3 is calculated for samples with $a_0 = 5.5952$ Å [1] and $a_0 = 5.596$ Å [2]. The following specific gravities are given for individual samples: 10.5 to 8 for thorianites from various host rock types of the Bancroft area, Ontario [49]; 9.8 for probably Th-rich thorianite of the Enisei Mountain Range, southern Siberia [21]; ≤ 9.43 [50], 9.32, and 8.98 for individual thorianites from Sri Lanka [51]; 9.2 as a mean for uranothorianites from several placers, Seward Peninsula, Alaska [52]; always > 9 (one sample with 9.18) for uranothorianites from Madagascar [34]; 8.89 and 8.4 for two thorianites, see [4] and [33], a range from 8.1 to 6.9 for several placer uranothorianites [39], and 5.35 for a RE-rich thorianite [11], all from the USSR. For further values in the range between 9.50 and 8.80 (6.68), see Table 2, pp. 48/54. For pure synthetic ThO_2, D is 10.011 [53] or 10.013 ± 0.001 g/cm^3, both measured at 25°C [54]; $D = 9.991$ g/cm^3 (at 26°C) is calculated for a synthetic ThO_2 sample (later adopted into the ASTM file as card No. 4-0556) [55].

There is an essentially linear variation in specific gravity with substitution of U for Th (up to 10.9 g/cm^3 for pure UO_2). However, in natural material the variation is obscured by the presence of RE and of other elements contained in solid solution, and by alteration [2]. The specific gravity generally increases with the content of U, but decreases owing to the entry of RE [29]. The low value of D for calcic thorianite (sample Nos. 7 and 9 in Table 2, pp. 48/9) from magnesian skarn, Siberia, is attributed to Ca substitution, because the measured value of 9.107 g/cm^3 conforms well with 9.15 g/cm^3 calculated for a theoretical thorianite composition of 93% ThO$_2$ + 7% CaO [24, pp. 137, 141]. For a discussion of probable reasons (type of bonding/lattice, ratio U^{4+}:U^{6+}, conditions of sample fusion) causing the difference between measured and calculated densities of synthetic members of the ThO$_2$-UO$_2$ system, see [30, pp. 974/5]. No correlation of D with the mineral variety nor with the lattice constant is found by [50].

Magnetic Properties. Whereas thorianite is generally characterized as diamagnetic [2] and uranothorianite as nonmagnetic [1], the few individually analyzed samples from the Enisei Mountain Range, southern Siberia, from the Bancroft area, Ontario, and from the Yates U mine, Quebec, are always nonmagnetic, see [21], [49], and [56], respectively. Only aldanite from the Aldan region, Eastern Siberia, is weakly magnetic [17].

Strong **radioactivity** is a characteristic property of thorianite [1]; see also [3], [9, pp. 83, 85]. Geiger probe readings for highly radioactive thorianite from the Yates U mine, Quebec, give ⩾ 0.5 mr/h [57]. The radioactivity of thorianite from phlogopite-bearing metasomatites, Kovdor massif, Kola Peninsula, is somewhat lower than usual, which is explained by its low U content (see sample No. 2 in Table 2, p. 48) [6]. Despite its very strong radioactivity, thorianite is always well crystallized [16, p. 12] and not metamict [2], [16, p. 34].

Metamictization. Almost alone among radioactive minerals, thorianite and uraninite do not become metamict, apparently because of the inherent stability of their fluorite-type crystal structure [58].

Detailed X-ray investigations on thorianite, in general, proved the high crystallinity of the mineral despite its natural irradiation by α decay. Apart from the above explanation by the mineral's stabile structure, this is also attributed to the process of natural self-annealing which heals probable radiation damage [59 to 61]. Leaching experiments on thorianite proved the character of the damaged structure to be only temporary [62]. The lack of a tendency for metamictization is also proved by thermal investigations [63] and by experimental heavy-ion (Ar$^+$) bombardments imitating the natural irradiation, see p. 63.

Only few examples of metamict thorianite are reported: Amorphous or cryptocrystalline thorianite occurs in at least one deposit, probably in Madagascar [64]. X-ray data characterize high-thorian thorianite from altered melilite rocks of the Turii Peninsula, northwestern USSR, to be slightly metamict [4] and late-magmatic thorianite from granitoids of the Verkhoyansk-Chukotka Fold Region, northeastern Siberia, to be metamict because a diffraction pattern was obtained only after heating of the sample [18, 19]. The U-Th minerals including uranoan thorianite of the radioactive mineralization in the Bokan Mountain area, Canada, are generally said to occur in the metamict or partly metamict state [65]. A metamict structure is also evident from a radioactive mineral, tentatively identified as uraninite-thorianite, which occurs as small black cubes in gravel deposits of the Sweepstakes Creek area, northeastern Seward Peninsula, Alaska [66].

References for 2.1.1.5:

[1] Chukhrov, F. V.; Bonshtedt-Kupletskaya, E. M. (Mineraly Spravochnik, Vol. 2, Pt. 2, Nauka, Moscow 1965, pp. 1/341, 113/7).
[2] Frondel, C. (Geol. Surv. Bull. [U.S.] No. 1064 [1958] 1/400, 51).
[3] Winchell, A. N. (Elements of Optical Mineralogy, Pt. 2, 4th Ed., Wiley, New York 1951, pp. 1/551, 65).
[4] Bulakh, A. G.; Mazalov, A. A. (Am. Mineralogist **59** [1974] 378/80).
[5] Heinrich, E. W.; Conrad, M. A. (Am. Mineralogist **45** [1960] 459/64, 460).

72

[6] Krasnova, N. I.; Kartenko, N. F.; Rimskaya-Korsakova, O. M.; Firyulina, V. V. (Mineral. Geokhim. **1967** No. 2, pp. 19/27, 22/3; C.A. **68** [1968] No. 116291).
[7] Parker, R. L.; Sharp, W. N. (Geol. Surv. Profess. Paper [U.S.] No. 649 [1970] 1/24, 17).
[8] Ramdohr, P. (Die Erzmineralien und ihre Verwachsungen, 4th Ed., Akademie Verlag, Berlin 1975, pp. 1/1277, 1131/2).
[9] Soboleva, M. V.; Pudovkina, I. A. (AEC-tr-4487 [1961] 1/455; N.S.A. **15** [1961] No. 23739).
[10] Zhabin, A. G.; Mukhitdinov, G. N.; Kazakova, M. E. (Tr. Inst. Mineral. Geokhim. Kristallokhim. Redk. Elem. No. 4 [1960] 51/72, 69).

[11] Es'kova, E. M.; Zhabin, A. G.; Mukhitdinov, G. N. (Mineralogiya i Geokhimiya Redkikh Elementov Vishnevykh Gor, Nauka, Moscow 1964, pp. 1/318, 159).
[12] Palache, C.; Berman, H.; Frondel, C. (Dana's System of Mineralogy, Vol. 1, 7th Ed., Wiley, New York 1944, pp. 1/834, 621).
[13] Getseva, R. V.; Savel'eva, K. T. (Rukovodstvo po Opredeleniyu Uranovykh Mineralov, Gosgeoltekhizdat, Moscow 1956, pp. 1/260, 124/5).
[14] George, D'Arcy (RMO-563 [1949] 1/198, 81/2; N.S.A. **5** [1951] No. 833).
[15] Nininger, R. D. (Minerals for Atomic Energy, Van Nostrand Reinhold, New York 1954, pp. 1/367, 88/9).
[16] Heinrich, E. W. (Mineralogy and Geology of Radioactive Raw Materials, McGraw-Hill, New York 1958, pp. 1/654).
[17] Bespalov, M. M. (Sov. Geol. **1941** No. 6, pp. 105/7; Mineral. Abstr. **12** [1953/55] 460).
[18] Nekrasov, I. Ya.; Ipat'eva, I. S. (Mater. Geol. Polez. Iskop. Yakutsk. ASSR No. 5 [1961] 32/50, 42, 44).
[19] Nekrasov, I. Ya. (Tr. Yakutsk. Filiala Akad. Nauk SSSR Ser. Geol. No. 12 [1962] 1/334, 225).
[20] Gaidukova, V. S.; Zdornik, T. B. (Geol. Mestorozhd. Redk. Elem. No. 17 [1962] 86/117, 116).

[21] Grebennikova, O. T. (Nauch. Tr. Irkutsk. Gos. Nauchn. Issled. Inst. Redk. Tsvetn. Metal. No. 14 [1966] 97/100; C.A. **70** [1969] No. 70075).
[22] Andreev, G. V. (Tr. Buryatsk. Kompleks. Nauchn. Issled. Inst. Akad. Nauk SSSR Sibirsk. Otd. No. 12 [1963] 97/9).
[23] Andreev, V. A. (Tr. Buryatsk. Kompleks. Nauchn. Issled. Inst. Akad. Nauk SSSR Sibirsk. Otd. No. 22 [1966] 121/34, 128).
[24] Andreev, G. V.; Maslov, U. M. (Tr. Buryatsk. Kompleks. Nauchn. Issled. Inst. Akad. Nauk SSSR Sibirsk. Otd. No. 22 [1966] 136/42).
[25] Trofimov, A. A.; Apollonova, G. N. (Izv. Vysshikh Uchebn. Zavedenii Geol. Razvedka **19** No. 1 [1976] 69/78).
[26] Ledoux, A. (Ann. Soc. Geol. Belg. **40** [1912/13] 177/210, 190).
[27] Makarov, E. S.; Lipova, I. M. (Geokhimiya **1962** 583/9; Geochem. [USSR] **1962** 676/83, 677).
[28] Wells, R. C.; Fairchild, J. G.; Ross, C. S. (Am. J. Sci. [5] **26** [1933] 45/54, 50).
[29] Semenov, E. I. (Typokhimizm Mineralov Shchelochnykh Massivov, Nedra, Moscow 1977, pp. 1/119, 57).
[30] Hund, F.; Niessen, G. (Z. Elektrochem. **56** [1952] 972/9).

[31] Ramdohr, P.; Strunz, H. (Klockmanns Lehrbuch der Mineralogie, 16th Ed., Enke, Stuttgart 1978, pp. 1/876, 548).
[32] De Casa, C. G.; Della Ventura, G. C.; Parodi, G. C.; Stoppani, F. S. (Riv. Mineral. Ital. **1987** No. 2, pp. 97/104, 101, 104).
[33] Kapustin, Yu. L. (in: Vlasov, K. A., Mineralogiya i Geneticheskie Osobennosti Shchelochnykh Massivov, Nauka, Moscow 1964, pp. 135/94, 162/3).

[34] Behier, J. (Ann. Geol. Madagascar No. 29 [1960] 1/78, 70).

[35] Turovskii, S. D. (Zap. Vses. Mineral. Obshch. **87** [1958] 369/70).

[36] Lacroix, A. (Minéralogie de Madagascar, Vol. 1, Challamel, Paris 1922, pp. 1/624, 245/7).

[37] Lacroix, A. (Bull. Soc. Franc. Mineral. **37** [1914] 176/80).

[38] Gray, I. M.; Millman, A. P. (Econ. Geol. **57** [1962] 325/49, 337).

[39] Egorov, B. L. (Zap. Vses. Mineral. Obshch. **99** [1970] 493/7).

[40] Forster, I. F. (Trans. Geol. Soc. South Africa **61** [1958] 359/65, 360).

[41] Hanekom, H. J.; van Staden, C. M. v. H.; Smit, P. J.; Pike, D. R. (Mem. Geol. Surv. South Africa No. 54 [1965] 1/185, 66).

[42] Povarennykh, A. S. (Konst. Svoistva Mineral. No. 13 [1979] 53/78, 57/9).

[43] Gruen, D. M. (J. Am. Chem. Soc. **76** [1954] 2117/20).

[44] Charlet, J. M. (Bull. Soc. Franc. Mineral. **91** [1968] 151/65, 157).

[45] Cooper, P. F.; MacNevin, A. A.; Winward, K. (Mineral. Ind. New South Wales **1979** Nos. 38, 40, 44, pp. 1/80, 17).

[46] Zuev, V. V. (Mineral. Sb. [Lvov] **21** [1967] 206/10; C.A. **70** [1969] No. 89521).

[47] Young, B. B.; Millman, A. P. (Bull. Inst. Mining Met. No. 689 [1964] 437/66, 451).

[48] Povarennykh, A. S.; Lebedeva, A. D. (Konst. Svoistva Mineral. No. 4 [1970] 121/8, 122/3).

[49] Satterly, J. (Ann. Rept. Ontario Dept. Mines **65** VI [1956] 1/181, 15).

[50] Arnott, R. J. (Am. Mineralogist **35** [1950] 386/400, 387).

[51] Dunstan, W. R. (Nature **69** [1904] 510/1).

[52] West, W. S.; Matzko, J. J. (U.S. Geol. Surv. Circ. No. 250 [1953] 21/7, 23).

[53] Robie, R. A.; Bethke, P. M.; Toulmin, M. S.; Edwards, J. L. (Geol. Soc. Am. Mem. No. 97 [1966] 27/74, 66).

[54] Skinner, B. J. (Am. Mineralogist **42** [1957] 39/55, 46).

[55] Swanson, H. E.; Tatge, E. (NBS-C-539 [1953] 57/8).

[56] Stevenson, J. S.; Stevenson, L. S. (Trans. Roy. Soc. Can. IV [3] **49** [1955] 105/10).

[57] Leavitt, D. L. (Mineral. Rec. **12** [1981] 359/63).

[58] Frondel, C. (Geol. Surv. Profess. Paper [U.S.] No. 300 [1956] 567/79; Proc. Intern. Conf. Peaceful Uses At. Energy, Geneva 1955 [1956], Vol. 6, pp. 568/77, 570).

[59] Eyal, Y.; Fleischer, R. L. (Nature **314** [1985] 518/20).

[60] Eyal, Y.; Fleischer, R. L. (Mater. Res. Soc. Symp. Proc. **44** [1985] 687/91).

[61] Eyal, Y.; Fleischer, R. L. (Geochim. Cosmochim. Acta **49** [1985] 1155/64).

[62] Petit, J. C.; Langevin, Y.; Dran, J. C. (Geochim. Cosmochim. Acta **49** [1985] 871/6).

[63] Noddack, W.; Jakobi, R. (Z. Anorg. Allgem. Chem. **284** [1956] 208/33, 224/5).

[64] Davidson, C. F. (Mining Mag. [London] **94** [1956] 197/208, 198).

[65] MacKevett, E. M., Jr. (Geol. Surv. Bull. [U.S.] No. 1154 [1963] 1/125, 67/9, 114).

[66] Gault, H. R.; Black, R. F.; Lyons, J. B. (U.S. Geol. Surv. Circ. No. 250 [1953] 1/10, 5/6).

2.1.1.6 Chemical and Thermal Behavior

The solubility of thorianite in acids is characteristically less than that of uraninite. It generally dissolves in HNO_3 or H_2SO_4, but not in HCl. The reaction is accompanied by emission of gas, principally He [1, 2], [3, pp. 49, 52], [4, pp. 82/3]. Dissolution in concentrated HNO_3 and in dilute H_2SO_4 may occur with gel precipitation [4, p. 82]. When treated with HNO_3, thorianite from Quebec leaves insoluble residues; but it dissolves completely, even if very slowly, in a mixture of HF and $HClO_4$ [5, p. 538]. The dissolution of ThO_2 in acids depends on its previous heat treatment: low-burned oxide (<600°C) generally dissolves readily, but if heated to higher temperatures, the dissolution becomes increasingly difficult. Hot HNO_3 dissolves all ThO_2 types, high-burned oxide requiring the longest time of reaction [6]. Decomposition of thorianite also occurs by heating in sulfuric acid with ammonium sulfate or by prolonged sintering with sodium peroxide at 460 ± 20°C for 60 min [7]. Uranothorianite from Loolekop, R. S. A., behaves negatively to standard etch reactions [8].

In aqueous solutions, thorianite dissolves only slightly, following the equation $ThO_2 + 4\,H^+ \rightarrow Th^{4+} + 2\,H_2O$. For constructed solubility curves (in the pH range from 3 to 11) and for related values of the electroneutrality point, of free energies, of solution constants, and of activities of the various Th species involved in the dissolution process of thorianite in aqueous solution at various temperatures, see the original paper [9]. A thorough investigation of thoria solubility in aqueous solutions of $NaClO_4$ (with a constant ionic force of $\mu = 0.1$) between 17 and 25°C and at 50°C during 20 to 800 days gives the following results: Within the experimental errors, temperature does not influence the solubility between pH 1.1 and 7.4, but the mode of solution varies with pH. Between pH 1.1 and 3.0, ThO_2 dissolves with formation of a positively charged hydroxide according to $ThO_2 + H^+ + H_2O \rightleftharpoons Th(OH)_3^+$, and between pH 3.0 and 7.4 an electroneutral complex forms according to $ThO_2 + 2\,H_2O \rightleftharpoons Th(OH)_4^0$. The corresponding preliminary solution constants are $\log K_1^0 = 3.30 \pm 0.07$ and $\log K_2^0 = 6.66 \pm 0.06$. For a diagram of $-\log \Sigma m_{Th}$ vs. pH (Σm_{Th} comprising solubilities and activities for all Th species involved), see the original paper [10]. For isotopic fractionation of Th and U due to dissolution of thorianite in bicarbonate/carbonate solvents, see p. 55.

Thorianite is infusible before the blowpipe but disintegrates or cracks at higher temperatures [2], [4, p. 83], emitting a strong white glow resulting from Ce [3, p. 49].

The melting point of ThO_2 is 3300 ± 100°C. During heating up to 1000°C, thorianite should, in general, not show any thermal anomalies [3, pp. 49, 51]. Uranothorianite from placers of the USSR shows two weak endothermic effects at 740 and 900°C, and a weak exothermic effect at 810°C (taken from figure 2 in the original paper). The weak broad exothermic band between 350 and 700°C is attributed to oxidation of U^{4+} [11]. Slow heating (at a rate of 280°C/h) of thorianite from Sri Lanka results in sample fragmentation that begins with breaking into small pieces at 700°C and culminates in explosive disintegration at 950°C. The thermal instability is not referred to as a thermal shock, but attributed to pressure build-up by originally dispersed gas which accumulates by diffusion into existing flaws in the mineral, along which spontaneous fracturing occurs. The evolved gas is mainly composed of He, also of CO_2 and H_2O, and traces of H_2, O_2, and Ar [12]. Note also a weight loss of 1.1 and 0.7 rel% for two thorianite samples heated at 810°C (for ~7.5 h), which is comparable to the probable amount of He accumulated during the minerals' age [13]. Heating of thorianite may also cause volatilization of Pb [5, pp. 542/3]. For reduction of the unit-cell dimensions of heated thorianite, refer to pp. 62/3.

References for 2.1.1.6:

[1] Palache, C.; Berman, H.; Frondel, C. (Dana's System of Mineralogy, Vol. 1, 7th Ed., Wiley, New York 1944, pp. 1/834, 621).

[2] Chukhrov, F. V.; Bonshtedt-Kupletskaya, E. M. (Mineraly Spravochnik, Vol. 2, Pt. 2, Nauka, Moscow 1965, pp. 1/341, 116).

[3] Frondel, C. (Geol. Surv. Bull. [U.S.] No. 1064 [1958] 1/400).

[4] Soboleva, M. V.; Pudovkina, I. A. (AEC-tr-4487 [1961] 1/455; N.S.A. **15** [1961] No. 23739).

[5] Robinson, S. C.; Loveridge, W. D.; Rimsaite, J.; Van Peteghem, J. (Can. Mineralogist **7** [1962/63] 533/46).

[6] Rodden, C. J.; Warf, J. C. (Natl. Nucl. Energy Ser. Div. VIII **1** [1950] 160/207, 164).

[7] Doležal, Z.; Povondra, P.; Šulcek, Z. (Decomposition Techniques in Inorganic Analysis, Iliffe Books, London 1968, pp. 1/224, 65, 163 [Czechish original, Prague 1966]).

[8] Bouwer, R. F. (Diss. Univ. Cape Town 1957 from Hanekom, H. J.; van Staden, C. M. v. H.; Smit, P. J.; Pike, D. R., Mem. Geol. Surv. South Africa No. 54 [1965] 1/185, 66).

[9] Kopeikin, V. A. (At. Energiya SSSR **56** [1984] 221/3; Soviet At. Energy **56** [1984] 242/5).

[10] Zhidikova, A. P.; Kolesov, G. M.; Khodakovskii, I. L. (Geokhimiya **1980** 821/6, 822/3; C.A. **93** [1980] No. 135162).

[11] Egorov, B. L. (Zap. Vses. Mineral. Obshch. **99** [1970] 493/7).

[12] Barnes, R. S.; Mazey, D. J. (J. Nucl. Energy **5** [1957] 1/3).

[13] Eyal, Y.; Fleischer, R. L. (Geochim. Cosmochim. Acta **49** [1985] 1155/64, 1157).

2.2 Complex Oxides

As can be seen from the following sections, complex or multiple oxides containing Nb, Ta, and Ti form a great number of individual minerals with a variable chemical composition, which:

(1) constitute, partly, solid-solution series;

(2) mainly are in the metamict state; and

(3) have very similar, or nearly identical physical properties.

Many of these complex oxides contain Th as an essential (see Sections 2.2.1 and 2.2.2), or as an additional component not expressed by the formula and sometimes amounting to >1% ThO_2 (see pp. 16/8).

Whereas most of the complex silicates, in which U and Th enter the crystal lattice by substitution, have corresponding nonradioactive and therefore non-metamict analogs (refer to "Thorium" Suppl. Vol. A1b, Chapter 2.5), the complex oxides (i.e., niobotantalates and titanates) do not possess such analogs [1].

The general formula of the complex oxides, or niobate-tantalate-titanates, can be expressed by $A_mB_nO_{2(m+n)}$, with m:n between 1:1 and 1:2, where A = RE, U, Ca, Th, Fe^{2+}, Na, Mn, Zr and B = Nb, Ta, Ti, Fe^{3+}, Sn, W(?), Zr(?) [2]. Or their composition may be given by $A_xB_yO_z$, where A = RE, U, Ca, Th, Fe^{2+}, Na; B = Nb, Ta, Ti, Fe^{3+}; and O = O, OH, and F [3]. An empirical formula $A_{n-x}B_pX_q$ is proposed for isometric (cubic) titanium-tantalum niobates, where x is the value determining the deficiency in the atomic amounts of the A-group elements; x ranges from 2.0 to 0.5 and shows a definite dependence on the contents of Ti, Zr, U, Th, and H_2O. The usual minerals with an increased A-cation deficiency are metamict [4]. Whereas in titano-niobo-tantalates of the general formula $A_xB_yZ_z$ the B positions are occupied by highly stable cations (such as Nb, Ta, Ti, Fe^{3+}, etc.), the A positions may or may not show deficiencies (due to various processes, see below), and thus can be used to divide the minerals into different groups. There are primary and secondary deficiencies (a distinction firstly given by [5]). The former may be caused by: (a) coupled substitutions in which the entry of higher valency cations is not fully balanced by the corresponding entry of lower valency cations; (b) simple substitutions involving the entry of cations of the same valency but of a size too large for the lattice positions to be filled (such cations are mainly Ba, Pb, K); (c) depletion of cations during crystallization; and (d) alteration (= primary alteration) by later residual solutions. Secondary deficiencies are essentially due to the weathering processes (= secondary alteration), which are in general much more intense in hot and humid climates. Apart from hydration, alteration processes selectively leach, by decreasing importance, Na, RE, Ca, and U. In extreme cases, titano-niobo-tantalates are practically reduced during alteration to a mixture of oxides and hydroxides. The change in chemical composition during alteration affects also the physical properties of the minerals (e. g., density and refractive index decrease, hardness is approximately proportional to the grade of alteration), and causes marked changes in the X-ray diffraction patterns. For the detailed discussion, see the original paper [6]; see also examples given for individual mineral species in the paragraphs of this chapter.

The pH of a suspension in water gives an indication of the chemical composition of titano-tantalo-niobates chiefly by the predominant cation in group A. Thus, minerals with Ca having been replaced by U, RE, and Th (as obruchevite, hatchettolite, ellsworthite, betafite, and Ti-betafite) give a pH of 7.0 to 7.8 [11].

In conformity with the complicated chemistry, nomenclature and classification of the complex oxides containing Nb, Ta, and Ti is nonuniform and based on a diversity of criteria, see, for example, [2, 4, 7 to 10].

The chemical formulas for the minerals described in the paragraphs of the following sections are based on chemical grounds, and are given in accordance with the proposals of the International Mineralogical Association (see [12]), or of the literature cited. No attempt is made to assign the minerals described to a specific nomenclature; the names given in the original paper are used throughout. Likewise, except a few cases with a more general background, only papers which give information for the behavior of thorium are cited.

References for 2.2:

[1] Orcel, J.; Fauquier, D. (Chemistry Earth Crust Trans. Geochem. Conf. Celebrating 100th Anniv. Birth Acad. V. I. Vernadskii, Moscow 1964, Vol. 2, pp. 340/9, 341 [Russian original, Moscow 1964]).

[2] Palache, C.; Berman, H.; Frondel, C. (Dana's System of Mineralogy, 7th. Ed., Vol. 1, Wiley, New York 1946, pp. 1/834, 745/7).

[3] Steacy, H. R.; Kaiman, S. (Short Course Handb. Mineral. Assoc. Can. **3** [1978] 107/40, 113/4).

[4] Ginzburg, A. I.; Gorzhevskaya, S. A.; Erofeeva, E. A.; Sidorenko, G. A. (Geokhimiya **1958** 486/500; Geochem. [USSR] **1958** 615/36).

[5] Borodin, L. S.; Nazarenko, I. I. (Dokl. Akad. Nauk SSSR **115** [1957] 783/6; Dokl. Acad. Sci. USSR Earth Sci. Sect. **112/117** [1957] 745/8).

[6] Van Wambeke, L. (Neues Jahrb. Mineral. Abhandl. **112** [1969/70] 117/49).

[7] Barsanov, G. P. (Tr. Mineral. Muzeya Akad. Nauk SSSR No. 10 [1959] 3/16, 11/6).

[8] Ginzburg, A. I.; Gorzhevskaya, S. A.; Erofeeva, E. A.; Sidorenko, G. A. (Geologiya Mestorozhdenii Redkikh Elementov, Vol. 10: Titano-Tantalo-Niobaty I, Nedra, Moscow 1960, pp. 1/168).

[9] Gorzhevskaya, S. A.; Sidorenko, G. A. (Geologiya Mestorozhdenii Redkikh Elementov, Vol. 23: Titano-Tantalo-Niobaty II, Nedra, Moscow 1964, pp. 1/116).

[10] Bonshtedt-Kupletskaya, E. M. (Zap. Vses. Mineral. Obshch. **95** [1966] 134/44).

[11] Erofeeva, E. A. (in [8, pp. 47/60, 52]).

[12] Fleischer, M. (Glossary of Mineral Species, 5th Ed., Mineralogical Record Inc., Tucson, Arizona, 1987, pp. 1/227).

2.2.1 Complex Oxides with Ti (and Sometimes Nb and/or Fe)

2.2.1.1 Yttrocrasite $(Y, Th, Ca, U)(Ti, Fe^{3+})_2(O, OH)_6$

The mineral description is given in "Rare Earth Elements" A7, 1984, pp. 44/5; for additional information, see [1], refer also to [2]. A complete wet chemical analysis is given in "Seltenerdelemente" A1, 1938, p. 89. An electron microprobe analysis of yttrocrasite (from the type locality or from a nearby location, see [3]) gives: CaO 1.71, MnO 0.14, Fe_2O_3 1.63, TiO_2 49.56, WO_3 1.79, PbO 0.38, Ta_2O_5 0.06, Nb_2O_5 0.02, ThO_2 9.12, U_3O_8 2.36, RE_2O_3 29.05, H_2O 5.07%; sum = 100.89%. The indexed X-ray powder diffraction data are given as follows [4]:

hkl	$d_{obs.}$	$d_{calc.}$	I[*)]
200	6.455	6.431	90
110	4.331	4.310	90
310	3.121	3.126	30
211	2.954	2.944	20
002	2.401	2.403	20
102	2.369	2.363	20
510	2.241	2.241	30
600	2.147	2.144	100
420	1.865	1.863	30
022	1.658	1.656	10
912	1.189	1.186	5
11.10	1.132	1.133	10
414	1.094	1.093	10

[*)] Intensities are visually estimated.

Because there are inconsistencies, especially concerning analytical and X-ray diffraction data, it is questionable whether yttrocrasite is a valid member of the euxenite group [3], see also [5].

References for 2.2.1.1:

[1] Palache, C.; Berman, H.; Frondel, C. (Dana's System of Mineralogy, 7th. Ed., Vol. 1, Wiley, New York 1946, pp. 1/834, 793).
[2] Soboleva, M. V.; Pudovkina, I. L. (AEC-tr-4487 [1957] 1/455, 410/1; N.S.A. **15** [1961] No. 23739).
[3] Peacor, D. R.; Simmons, W. B., Jr.; Essene, E. J.; Heinrich, E. W. (Am. Mineralogist **67** [1982] 156/69, 163/4).
[4] Crook III, W. W. (Am. Mineralogist **62** [1977] 1009/11).
[5] Crook III, W. W. (Mineral. Rec. **8** [1977] 88/90).

2.2.1.2 Thorutite-Brannerite Series

Thorutite and brannerite, ideally $ThTi_2O_6$ and UTi_2O_6, respectively, are isostructural [1, 2] and form a series of solid solution. Up to a content of about 13% ThO_2, the isomorphous replacement between U and Th in brannerite is complete [3, p. 123], as is corroborated by chemical analyses of natural brannerites indicating, that for a considerable number of samples, Th^{4+} is diadochically incorporated for U [3, pp. 115/7], [4], [5].

As thorutite and brannerite usually occur in the metamict state, their identification is based predominantly on their physical properties prior to, or after heating at high temperatures (recrystallization), see the relevant paragraphs on pp. 79 and 101/2, respectively. But it is without certainty to assume that the recrystallized samples are identical with the original structure of the mineral. Therefore, one has build up a stock of data derived from synthetic products which may serve as diagnostic criteria [6].

Minerals of the betafite type (refer to Section 2.2.1.4, pp. 131/48) are somewhat similar in appearance to those of the brannerite type, but may be readily distinguished chemically by a considerable content of Ta and Nb, and optically by relative low refractive index [7].

References for 2.2.1.2:

[1] Ruh, R.; Wadsley, A. D. (Acta Cryst. **21** [1966] 974/8).
[2] Szymanski, J. T.; Scott, J. D. (Can. Mineralogist **20** [1982] 271/9).
[3] Povilaitis, M. M. (Zap. Vses. Mineral. Obshch. **92** [1963] 113/23).
[4] Povilaitis, M. M. (Tr. Mineral. Muzeya Akad. Nauk SSSR No. 12 [1961] 64/79).
[5] Chernysheva, L. V.; Kupriyanova, I. I.; Sidorenko, G. A. (Sb. Kratk. Soobshch. Mineral. Geokhim. No. 1 [1968/70] 83/8).
[6] Shabalin, B. G. (Petrol. Mineral. i Rudoobraz. v Predelakh Ukr. Shchita, Kiev 1984, pp. 13/7, 16; Ref. Zh. Geol. **1985** 1 V 235).
[7] Whittle, A. W. G. (J. Geol. Soc. Australia **2** [1955] 21/45, 34/5).

2.2.1.2.1 Thorutite $(Th, U, Ca)Ti_2(O, OH)_6$

Thorutite from an undisclosed syenite massif of the USSR was described in 1958. It is named in allusion to its composition [1]. Synonymous is the name smirnovite [2]. There are two allotropic forms of synthetic $ThTi_2O_6$ stable below and above 1300°C, α-$ThTi_2O_6$ and β-$ThTi_2O_6$, respectively [3].

Thorutite occurs as idiomorphous grains scattered in the nepheline syenite, or in druses in strongly sericitized microcline-nepheline veins (cutting the syenite) in association with thorite, zircon, calcite, and subordinate barite and galena [1].

Synthetic $ThTi_2O_6$ has been formed from the oxides at 1610°C (in vacuum) during 1 h by solid-state reaction [4] and below 1300°C by a flux method [5]. Jointly precipitated hydroxides of Th and Ti yield homogeneous transparent grains of $ThTi_2O_6$ after heating at 1073 to 1273 K for 3 to 4 h; the grains, up to 6 µm in size and without a definable habit, aggregated into blocks of 140 to 300 µm in diameter [6].

The chemical analysis of thorutite is given in Table 5, p. 92. Whereas the U and Ca contents are considered as an isomorphous replacement for Th, the oxides of Fe, Al, and Si belong to mechanical admixtures. The small content of $(Nb, Ta)_2O_5$ has to be taken as a chemical admixture without significance for the mineral formula. According to its loss during heating in a closed tube, water belongs to the formula [1].

The crystals of thorutite are metamict; have a short-prismatic habit; are up to 2 cm in length and 0.5 to 1 cm wide; and are black with a fatty luster. Fracture is conchoidal. In thinnest fragments they are translucent with dark brown color. Streak is brown. Thorutite is isotropic, with a refractive index >2.1. Mohs' hardness is 5 to 6. Specific gravity is 5.82 g/cm^3 [1].

Synthetic $ThTi_2O_6$ is monoclinic, space group C2/m (No. 12) [4] or C2/c (No. 15) [5]. The lattice constants, number of formula units, and specific gravity for $ThTi_2O_6$ synthesized by various methods are as follows:

| | synthetic $ThTi_2O_6$ formed from | | |
	oxides by solid-state reaction [4]	hydroxides by calcination [6]	oxides by flux method [5] *)
a	9.822 ± 0.005 Å	0.981 ± 0.02 nm	10.808 ± 0.002 Å
b	3.824 ± 0.002 Å	0.380 ± 0.02 nm	8.580 ± 0.002 Å
c	7.036 ± 0.005 Å	0.703 ± 0.02 nm	5.196 ± 0.002 Å
β	118.84° ± 0.05°	118.79° ± 0.08°	115°15′ ± 0.5′
Z	2	2	4
$D_{calc.}$	6.08 g/cm^3	6.13 g/cm^3	6.462 ± 0.006 g/cm^3
$D_{exp.}$	6.0 ± 0.5 g/cm^3	5.98 ± 0.5 g/cm^3	—

*) α-$ThTi_2O_6$.

The crystal structure of $ThTi_2O_6$, see **Fig. 3**A, p. 80, is related to the structure of the anatase form of TiO_2 and contains layers of somewhat distorted Ti-O octahedra (with Ti-O distances ranging from 1.83 to 2.20 Å) connected by the Th atoms in interlayer octahedral positions. All Th are located in mirror planes normal to the b axis at two levels y = 0 and 1/2. The Ti-O octahedra with Ti atoms at y = 1/2 are joined by a common edge, and those with Ti atoms at y = 0 are united at opposite ends by additional edge-sharing. These build up into infinite zig-zag sheets centered about the (100) planes. There are two Th-O distances of 2.36 Å and four of 2.35 Å, but the octahedron is flattened; two more oxygen atoms further away at 2.96 Å are probably unbonded [4]. The same structural arrangement is given for low-temperature α-$ThTi_2O_6$, see Fig. 3B. The Ti-O distances are between 1.87 and 2.04 Å; for the Th-O distances are given 2.28, 2.36, and 2.45 Å, two other oxygen atoms are at 2.69 Å [5].

For the X-ray data (d values) of thorutite, see [1].

After heating at 950°C, the IR spectrum of thorutite shows strong absorptions at 565, 410, and 280 cm^{-1}, weaker absorptions are at 670, 220, 140, and 96 cm^{-1} [7]. For synthetic $ThTi_2O_6$ are given: strong absorptions at 565, 412, and 268 cm^{-1}, together with a weaker one at 364 cm^{-1}, which are in accord with the structural features [8]; or absorptions at 715, 545, 405, 272, and 210 cm^{-1} [6].

After calcination at about 1000°C for 1 h, the black mineral thorutite becomes gold in color and a X-ray powder diagram is developed. The same gold color appears during heating in a closed tube, accompanied by loss of water [1]. Synthetic $ThTi_2O_6$ prepared in vacuum at 1610°C, when heated at 1000°C in air for a further period of 12 h, turned from black to light buff without any obvious change of its X-ray powder pattern [4].

A)

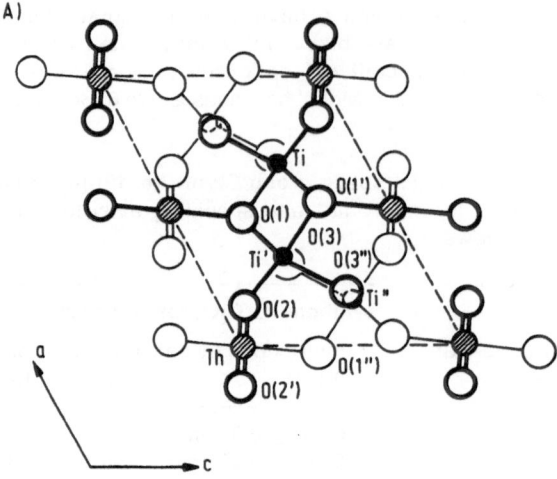

B)

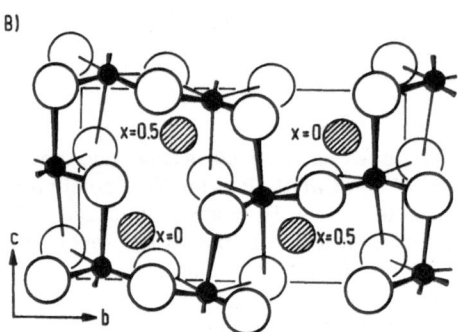

Fig. 3. Structure of synthetic $ThTi_2O_6$, A from [4] and B from [5].
Black circles are Ti and Th atoms are hatched.

References for 2.2.1.2.1:

[1] Gotman, Ya. D.; Khapaev, I. A. (Zap. Vses. Mineral. Obshch. **87** [1958] 201/2).
[2] Chukhrov, F. V.; Bonshtedt-Kupletskaya, E. M. (Mineraly Spravochnik, Vol. 2, Pt. 3, Nauka, Moscow 1967, pp. 1/676, 441).
[3] Perez y Jorba, M.; Mondange, H.; Collongues, R. (Bull. Soc. Chim. France **1961** 79/81).
[4] Ruh, R.; Wadsley, A. D. (Acta Cryst. **21** [1966] 974/8).
[5] Loye, O.; Laruelle, P.; Harari, A. (Compt. Rend. C **266** [1968] 454/6).
[6] Shabalin, B. G. (Petrol. Mineral. i Rudoobraz. v Predelakh Ukr. Shchita, Kiev 1984, pp. 13/7; Ref. Zh. Geol. **1985** 1 V 235).
[7] Povarennykh, A. S. (Konst. Svoistva Mineral. **13** [1979] 53/78, 75/6).
[8] Shabalin, B. G.; Povarennikh, A. O. (Dopov. Akad. Nauk Ukr. RSR B **1982** No. 5, pp. 30/4).

2.2.1.2.2 Brannerite $(U, Ca, Y, Ce)(Ti, Fe)_2O_6$

Brannerite from a gold placer near the head of Kelly Gulch, western Custer County, Idaho, was described in 1920. It is named after John Casper Branner (1850 – 1922), formerly President of Leland Stanford University, California [1]; see also [2,3]. Synonymous are cordobaite [4] and lodochnikite [5, 6]. Thorian brannerite has been named (needlessly) absite [7].

An orthorhombic polymorph, orthobrannerite, was described in 1978 [8].

Due to its nonuniform microscopic habit (refer to pp. 81/5), its metamict state (refer to p. 99), and its heterogeneous chemical composition (refer to pp. 89/97), brannerite is hard to define as a mineral species, see p. 89 and [9].

Detailed compilations for occurrence and the properties of brannerite (with and without Th content) are given in [10 to 15]. Note that in the following paragraphs only papers are cited which deal with brannerites containing $> 1\%$ ThO_2, or which reveal that it contains Th.

References for 2.2.1.2.2:

[1] Hess, F.; Wells, R. (J. Franklin Inst. **189** [1920] 225/37).
[2] Spencer, L. S. (Mineral. Mag. **19** [1920/22] 336).
[3] Palache, C.; Berman, H.; Frondel, C. (Dana's System of Mineralogy, 7th Ed., Vol. 1, Wiley, New York 1946, pp. 1/834, 774/5).
[4] George, D'Arcy (RMO-563 [1949] 1/198, 78/9; N.S.A. **5** [1971] No. 833).
[5] Gotman, Ya. D. (Zap. Vses. Mineral. Obshch. **87** [1958] 197/200).
[6] Fleischer, M. (Am. Mineralogist **43** [1958] 1007).
[7] Whittle, A. W. G. (Bull. Geol. Surv. South Australia No. 30 [1954] 126/51, 142).
[8] X-Ray Laboratory, Peking Inst. Uranium Geol. and X-Ray Labor., Wuhan Geol. College (Acta Geol. Sinica **1978** 241/51 from Am. Mineralogist **64** [1979] 656).
[9] Krishna Rao, N.; Rao, G. V. U. (J. Geol. Soc. India **24** [1983] 489/501, 489/90).
[10] Povilaitis, M. M. (Tr. Mineral. Muzeya Akad. Nauk SSSR No. 12 [1962] 64/79).

[11] Povilaitis, M. M. (Zap. Vses. Mineral. Obshch. **92** [1963] 113/23).
[12] Povilaitis, M. M.; Yakovlevskaya, T. A.; Knyazeva, D. N.; Belyaeva, I. D. (Zap. Vses. Mineral. Obshch. **97** [1968] 150/61).
[13] Chukhrov, F. V.; Bonshtedt-Kupletskaya, E. M. (Mineraly Spravochnik, Vol. 2, Pt. 3, Nauka, Moscow 1967, 1/676, 433/40).
[14] Ferris, C. S.; Ruud, C. D. (Colo. School Mines Quart. **66** No. 4 [1971] 1/35).
[15] Korolev, K. G.; Miguta, A. K.; Polyakova, U. M.; Rumyancheva, G. V. (Mineralogiya, Geologicheskie i Fizikokhimicheskie Osobennosti Obrazovaniya Uranotitanatov, Nedra, Moscow 1979, pp. 1/144, 29/42, 67/70, 72/90).

2.2.1.2.2.1 Occurrence, Paragenesis, Intergrowths, Inclusions, and Alteration

As can be seen from the following paragraphs, **occurrence and paragenesis** of brannerite as a primary mineral is connected with alkaline granitic rocks, pegmatites, and attendant pneumatolytic-hydrothermal or metasomatic formations. As a secondary product, it occurs mainly with ancient quartz-pebble formations.

Accessory brannerite and aeschynite, the latter partially or completely replacing the former, are confined to areas rich in hornblende, biotite, and ilmenite of grayish rose, porphyric granite from the Borovskii pluton, northern Kazakhstan. In the electromagnetic fraction of two heavy-mineral concentrates ("shlich") therefrom, brannerite forms idiomorphic, long prismatic grains, 0.1 to 0.5 mm in size, and sometimes with a distinct striation parallel to the prism [1].

Large subhedral to euhedral, or broken, grains of chemically zoned orthobrannerite (or uranium titanate-brannerite? as named in [2]) occur in biotite granite of the Lawler Peak, west-central Arizona [3, pp. 141, 165]; for the zonation, see pp. 86/7. Textural studies of thin sections of the rocks from Lawler Peak as well as from Dells Granite plutons show that, despite euhedral forms, brannerite and its accompanying accessory minerals generally occur in the interstitial places between the rock-forming minerals [4, pp. 356, 359/61].

Accessory thorian brannerite of the Crockers Well prospect, South Australia, occurs in sodic granitic rocks (mainly trondhjemite and alaskite) and in steeply dipping veins and fractures of the following types, cutting the sodic granitic rocks and their surrounding sodic felsic gneisses: (1) An early generation of massive, coarse grained pegmatoid veins (several centimeters thick); (2) phlogopite-rich veins, commonly schistose; and (3) "dry" fractures containing thin selvedges of phlogopite, muscovite, and chlorite. In addition, local phlogopite-rich breccias (forming diatreme-like bodies and dikes, ranging from < 1 cm to > 40 m across) also contain disseminated thorian brannerite. At Crockers Well, the brannerite, commonly forming aggregates with niobian rutile, has an approximate composition of $U^{4+}_{0.64}Th_{0.20}$-$Pb_{0.20}Y_{0.04}Fe^{3+}_{0.13}Ca_{0.16}Si_{0.10}Ti_{2.05}O_6$ and occurs in two forms: a greenish translucent type generally found within 20 to 40 m below the earth's surface (originally named "absite", see [5, 6]), and a dark brown opaque semimetallic type found at greater depths. The differences in physical and chemical properties are ascribed to weathering or supergene processes forming the more hydrated, near-surface type. Only the small portion of the granitic brannerite probably crystallized directly from the sodic granitic melt (which itself has been generated by anatexis of the surrounding gneisses during high-grade regional metamorphism). In the significant zones of fracture-controlled U-Th mineralization, thorian brannerite formed contemporaneously with an F-rich assemblage (of, e. g., fluor-bearing phlogopite, fluorapatite, fluorite, tourmaline, etc.) from saline fluids evolved during the crystallization of the sodic granitic rocks (in a manner more akin to porphyry copper and stockwork Mo deposits) [7, pp. 8/10, 14/6]. Note also that an extremely radioactive mineral, almost certainly brannerite, has been found only as eluvial pieces up to 2 inches in diameter at the Comfort lease, Cloncurry-Mount Isa area, Queensland, Australia. The mineral is believed to occur in lenses of sodic migmatite within biotite schists of the Argylla Series (for a systematic search was without result in a dike of unstressed granite pegmatite) concordantly intrusive into the schists [8].

Prismatic crystals of fully metamict orthobrannerite, up to 8 × 12 mm in size, were found in the weathering residue of a biotite-pyroxene syenite, Yunan, and in an alkalic lamprophyre, Szechuan, both P.R.C. [9]. Th-containing brannerite together with Th-rich zircon, radioactive rutile, titanite, and barite is observed in siliceous parts of a quartz syenite near Prilep, Topolchane district, Yugoslavia [10].

Brannerite, forming large (up to 15 cm) well-developed monoclinic and orthogonal prismatic crystals, occurs in U-bearing potassic pegmatite dikes and masses in highly metamorphosed Cambrian sediments (now gneiss and amphibolite) of Sierra Albarrana, Cordoba, Spain [11]. There it is a primary mineral together with uraninite, monazite, beryl, tourmaline, and muscovite [12]. Brannerite occurs in radioactive pegmatites, associated with the higher part of metasediments of the Wollaston Group, middle Foster Lake area, northern Saskatchewan, Canada [13]; in pegmatite of Rajputana, India, together with magnetite, apatite, tourmaline, etc. [14]; and in the heavy-mineral fraction of borehole samples from a pegmatitic complex of the northern Bals batholith, Romania [15]. Probably genetically related to granitic pegmatite dikes, brannerite (with Th more abundant than usual) was found in more or less open fissures in greenschist near Åmot, Modum district, southeast Norway [16]. Brannerite (randomly distributed and occasionally associated with slightly radioactive smoky quartz) occurs in mesothermal quartz veins cutting quartz monzonite, West Walker River region, Mono county, California. The veins are considered to be related genetically to nearby pegmatite and aplite veins [17].

Brannerite occurs in quartz-albite veins with an U-Th-RE mineralization (the I and L vein system) transecting peralkaline granite and its late-stage phases, Bokan Mountain, southeastern Alaska [18]. Together with muscovite, fluorite, beryl, and pyrite, brannerite has formed at early stages of mineralization, prior to Mo-W ores, in veinlike quartz-microcline bodies of pneumatolytic-hydrothermal origin in leucocratic granite of an undisclosed locality, USSR. Exceptionally, idiomorphic brannerite is found in the altered country rock within 1 to 2 cm from the veinlike bodies, at places where the latter thin out [19, pp. 64/5]. Brannerite, together with uraninite, coffinite, Mo sulfides, sphalerite, etc. occurs in a metasomatic U-Mo ore deposit in felsite porphyries of an undisclosed locality, USSR [20].

Coarse brannerite occurs in thick quartz veins, Wernecke Mountains, and together with uraninite, on the margins of pinch-and-swell quartz veins paralleling the bedding of chloritic metasedimentary rocks west of Kiwi Lake, Wernecke Mountains, Yukon Territory, Canada [21].

Irregular grains (occasionally up to 4 mm in diameter) of brannerite (lodochnikite) occur together with barite in the central portion of muscovite-containing microcline-nepheline veins cutting through syenitic rocks of an undisclosed locality, USSR [22]. Lodochnikite (containing 5.36% ThO_2 and 1.82% CeO_2) occurs in the central zone of small (0.5 to 1.5 cm) barite-siderite veinlets cutting across an altered nepheline syenite vein of an undisclosed locality, USSR [23].

Irregular grains of brannerite (0.02 to 0.1 mm in size) occur in metasomatic veinlets, or veinlike segregations composed mainly of quartz, pyrite, and sericite in a series of Devonian sediments with intercalated tuffaceous material and albitophyre, cut through by intrusive acid rocks, of an undisclosed U deposit, USSR. Adjacent to the veins, the hydrothermally altered country rock also contains brannerite as irregular platy, or rarely rhomboidal segregations of 0.2 to 0.3 mm in size and composed of fine-striped individual brannerite crystals of 0.01 mm in length. The latter type of brannerite, predominating quantitatively in the deposit, is formed by replacement of rutile (which developed from the accessory ilmenite of the country rock) in a medium with large concentrations of S^{2-} ions. Both types of brannerite are intergrown with pyrite (see pp. 85/6). Although the presence of sericite (large amounts), chlorite, tetrahedrite, sphalerite, and galena in the ore is indicative of medium temperature conditions, the increased content of Th and RE in brannerite is characteristic of relatively high-temperature formation (refer also to p. 89) [24, pp. 54/5, 58], [25]. Brannerite in pinacoidal form or in irregular, elongated crystals (of 8 × 3 mm in size) occurs, together with smaller quantities of olive-green to colorless tourmaline and fluorapatite, in the lower parts of a zonal quartz vein enveloped by leucogranite, which is formed by metasomatic alteration of granite gneiss, Woława Góra near Kowary (Schmiedeberg), Sudetes, southwest Poland [26].

Brannerite, together with thorite, uraninite, monazite, and other minerals, occurs in greisen, Kazakhstan, which is formed metasomatically from alumo-silicate rocks of the normal series [27]. Brannerite is found as discrete black grains, a few of which appear to be euhedral, within calcite, muscovite, and tourmaline, as well as irregular masses up to 1.2 cm in size in a skarn zone bordering the Deloro granitic intrusion near Madoc township, Ontario, Canada.

Native Au is almost invariably associated with the brannerite; and uraninite is associated with and, in part, is veining the brannerite (see p. 87) [28]. Brannerite together with coffinite, uraninite, and native Au is found in siderite ore of the Hüttenberger Erzberg, Carinthia, Austria [29].

Idiomorphic crystals of brannerite(?), up to 0.25 mm in length, occur together with calcite, dolomite, and muscovite in small fluorite veins (up to 0.3 m thick) in dolomitized limestone near a massif of alkalic syenite (and monzonite) of an undisclosed locality, USSR. The brannerite(?) and rare grains of thorite (orangite), rutile, brookite, corundum, etc. are found within the fluorite [23].

Brannerite occurs in silicified breccias in metasedimentary rocks, Yukon Territory, Canada [21]; and, together with copper minerals, as a high-intensity hydrothermal (epigenetic) deposition in the Archean phyllites and younger schists intruded by granite at Khandela, Sikar district, Rajasthan, India [30]. A few grains of thorian brannerite occur in an extremely altered (originally metamorphic?) rock composed of clay material, minute green flakes of graphite, well-crystallized quartz, and amorphous silica at the Dead Horse Bay uranium prospect, York Peninsula, South Australia [31].

Metamict brannerite occurs as lenticular or roughly spherical nodules of two distinct modes of occurrence in a probable Precambrian light-gray granite gneiss of the San Bernardino Mountains, California: Smaller nodules (weighing 1 to 20 g) lie along the foliation of the gneiss, locally concentrated along a definite lamina, and seem to consist entirely of brannerite surrounded by a thin layer of biotite; larger nodules (weighing up to 200 g) are found in tabular shear zones and contain brannerite together with grains of rutile (submicroscopic and megascopic). The larger nodules not only are surrounded by biotite, but contain small plates and sheaves of that mineral, indicating that brannerite may have replaced biotite. No bodies of rock that commonly is called pegmatite occur in the brannerite area. It seems that the San Bernardino brannerite has formed at great depth under conditions of high pressure and temperature [32].

Disseminated brannerite grains (60 to 250 μm in size), enveloped by sulfides (see p. 87), occur together with uraninite in a stratabound U deposit hosted in metasediments (now sillimanite-bearing quartz-feldspar-biotite gneiss) of the Karpinka Lake uranium prospect, northern Saskatchewan, Canada. Brannerite is considered to have formed by adsorption of U on TiO_2 sponges created by alteration of primary Fe-Ti oxides [33].

Brannerite, partly displaying a complicated internal structure (see, e.g., [34, 35] and below) and often containing Th, is one of the major U minerals in Precambrian slightly metamorphosed quartz-pebble conglomerates of Canada, South Africa, and Australia (see "Rare Earth Elements" A7, 1984, pp. 47/8), where its formation is considered to be secondary (see, e.g., [36]) or as an authigenic constituent formed, partially, by adsorption of U on titania collectors including altered Fe-Ti oxides (see, e.g., [37, p. N10]; refer also to [38, 39], [40, pp. 31/2], and [41]). Thus, in the Blind River-Elliot Lake region, Ontario, brannerite occurs typically as ovoid aggregates, about 0.3 mm in diameter, together with anatase or rutile [42]; or forms ovoid red-brown to black grains showing bladed rutile surrounded by a uranium oxide and RE oxides [43]. There are three, texturally and chemically distinct types of U-Ti grains ("brannerite") in the Elliot Lake area: Type-I grains usually display the rounded shape of detrital particles and contain variable amounts (within the same polished section, but even more with the location) of quartz, anatase, rutile, brannerite, often galena, and sometimes pyrrhotite. Type-II grains are limited in number and have a somewhat cheesy texture, with numerous pits and holes in the polished section. Although this type of "brannerite" is similar to type-I in content of UO_2 and TiO_2, the contents of CaO are high, those of SiO_2 low (see Table 5, pp. 92/6). One grain of type-III brannerite displays a patchy, irregular, concentric zoning (which may reflect a particular type of antecedent mineral) with U- and Ti-bearing phases present at the periphery, like the type-I grains [40, pp. 15/6]. At four localities of the Blind River region, near the contact of the Mississagi Formation and underlying pre-Huronian rocks, brannerite, partly altered (see p. 87), is observed together with Th-RE-rich uraninite and monazite in the pyritized matrix of quartz-pebble conglomerates. All three primary minerals occur as discrete grains averaging about 100 μm in diameter; and their relative proportions vary to such an extent that any one of the three may predominate, and one or both of the others may be almost absent [44]. The observed U-Ti phases in the Precambrian quartz-pebble conglomerates from the Elliot Lake area and from the Pongola basin, South Africa, range from U-free leucoxene/rutile to U-enriched brannerite. However, the presence of brannerite must be regarded with reservation, for metamict brannerite can only be verified by X-ray diffraction methods after prolonged heating (see pp. 101/2). The redistribution and subsequent adsorption of U and Ti phases during diagenesis and/or metamorphism of the conglomerates do not lead to formation of authigenic brannerite, but rather result in microcrystalline leucoxene/rutile admixtures containing U in varying amounts [45]; see also [46]. Hydrothermal alterations (albitization, chloritization, and carbonatization) of the conglomeratic U ores, Pronto Mine, Ontario, markedly redistributed the radioactive minerals, generally concentrating them near footwall contacts, and changed the habits of both brannerite and uraninite. The former becomes increasingly euhedral and less metamict, thus clear and birefringent. Relatively large boxwork euhedra of brannerite appear. Veinlets of albite-brannerite, of calcite-uraninite, of uraninite, and of brannerite-uraninite cut the newly-formed albitites and chloritites. In the new carbonate rocks, euhedral brannerite and uraninite become conspicuous, but the radioactive minerals generally avoid the dolomite-barite-rich rocks, and are absent in carbonate pseudomorphs after quartz pebbles [47]. Brannerite of the Witwatersrand quartz-pebble conglomerates, R.S.A., occurs as felts of needle- or lath-shaped crystals grading sometimes into irregular patches of massive brannerite. Both types show inclusions of sulfides (see p. 86) and are in close association with patches of rutile [37, pp. N7/N9]. For Witwatersrand brannerite pseudomorphous after uraninite (so-called "uraninite ghosts"), see p. 86. Brannerite-like material, as irregularly shaped masses, occurs interstitial to other groundmass constituents of the conglomerates from the Dominion Reef near Klerksdorp, western Transvaal, R.S.A. [48]. The main difference for brannerite formation in quartz-pebble conglomerates of the Elliot Lake area and of eastern Transvaal (Farm Gunsteling) is that the former has been generated from detrital titanomagnetite containing segregation lamellas of ilmenite, and the latter, possibly, from nonsegregated Ti-containing magnetite, ilmenite, or rutile [46].

Intergrowths of thorian brannerite (absite) and davidite have been identified in a few places of the Crockers Well prospect, South Australia [6].

Intergrowths of coffinite, brannerite, and pitchblende occur in veins of metasomatic Fe-U deposits, USSR [50].

Intergrowths of brannerite and pyrite, partly formed simultaneously, occur in metasomatic veinlets, or veinlike segregations in a series of Devonian sediments of an undisclosed U deposit, USSR. Whereas the brannerite of the veinlets is in nearly all cases surrounded by thin rims of pyrite, the brannerite forming xenoliths, composed of idiomorphic tiny crystals in the altered rock adjacent to the veinlets, is connected and intergrown with quartz-sericite masses, and the pyrite occurs on the periphery of those masses or, sometimes, occurs as parallel lamellas, intersecting themselves at distinct angles within the brannerite. These lamellas resemble grid-like, sagenitic rutile crystals (see below) [24, p. 57].

Abundant **inclusions** of galena occur in brannerite from the Witwatersrand, R.S.A. [51]; and both needles and massive parts of the Witwatersrand brannerite are speckled with minute inclusions of galena. The larger ones sometimes display cubic outlines. The massive brannerite is often characterized by lath-shaped inclusions of pyrrhotite [37, p. N7]. Small inclusions of pyrrhotite and radiogenic galena generally occur in brannerite of the Algoma district [43]; quartz, galena, pyrrhotite, or pyrite are present in many brannerite grains of the Blind River uranium ores, both Ontario, Canada [52].

Tiny inclusions of rutile, anatase(?), and radiogenic galena are abundant in the dark brown brannerites of Crockers Well, South Australia, but are absent in the green varieties found within 20 to 40 m below earth's surface [7, p. 10]. Minute dust-like inclusions of anatase also occur in Witwatersrand brannerite as a product of reconstitution after radioactive breakdown of U in the mineral lattice [37, p. N7]. And most brannerite of the Blind River uranium ores contains intimately admixed rutile or anatase [52].

Relicts of lamellar crystals of sagenitic rutile, and sometimes irregularly shaped rutile, are included within brannerite xenoliths, composed of idiomorphic tiny crystals, from altered rock adjacent to metasomatic veinlets, or veinlike segregations in a series of Devonian sediments of an undisclosed U deposit, USSR. With increasing distance from the veinlets, the portion of rutile increases and at some distance completely supersedes brannerite, indicating that the latter formed pseudomorphously after rutile which itself is, partly together with pyrite, pseudomorphous after ilmenite (see p. 83) [24, p. 57].

Except for small, generally rounded grains of anatase or rutile scattered throughout brannerite from skarn near Madoc township, Ontario, Canada, this mineral appears to be relatively uniform. But at higher magnifications, some brannerites show wispy masses of slightly higher reflectivity, particularly adjacent to some of the uraninite veinlets cutting through [28].

Plagioclase (often saussuritized) and dark minerals are included in brannerite from fissures in greenschist near Åmot, southeast Norway [16].

For inclusions of brannerite(?) in fluorite, refer to p. 84.

Alterations. Brannerite pseudomorphous after detrital uraninite grains ("uraninite ghosts") occurs in quartz-pebble conglomerates of the Witwatersrand, R.S.A. The central portions of the "ghosts" are usually leached, with the resulting cavities filled, in part, by reconstituted chalcopyrite, pyrrhotite, and galena. Often the grains are also surrounded by aureoles of authigenic sulfides [37, pp. N7/N9]. Brannerite-like material replaces uraninite of the Dominion Reef conglomerates, western Transvaal, either as a distinct border surrounding, or fingering into the grains [48].

Pseudomorphous after rutile seems to be some of the brannerite occurring, together with uraninite, in the margins of pinch-and-swell quartz veins in chloritic metasedimentary rocks west of Kiwi Lake, Yukon Territory, Canada [21]. Brannerite rimming and apparently replacing rutile occurs in the quartz-pebble conglomerates of the Quirke Ore Zone, Elliot Lake area, Ontario, Canada [41].

Together with coffinite and quartz, brannerite replaces, due to a post-depositional mobility of U and Th, a prismatic detrital heavy mineral of the Quirke Ore Zone conglomerate [41].

Uranium titanate (brannerite?) crystals from the Lawler Peak granite, west-central Arizona, contain as many as four chemically distinct internal zones (for its individual composition, see

p. 91) corresponding to areas of different color, some grains are almost exclusively zone A material and may represent the original material: Zone A is black, lustrous, and isotropic, forming the center of most grains; it consists of highly uraniferous titanate with 40 to 50% UO_2, 3 to 6% ThO_2, and 0.7 to 3% PbO. Zone B is brown, lustrous, and isotropic, occurring outside of, or intergrown with zone A; it contains 20 to 25% UO_2, 5 to 8% ThO_2, and 4 to 6% PbO. The composition of both zones A and B, which appear to be totally metamict, can be expressed by a brannerite formula $(U, Ca, Th, RE) (Ti, Si, Fe)_2O_6$. Together they comprise over 80% of most grains. Zone C is pale, yellowish brown and is probably a multiphase aggregate, with anatase predominating, and forms a thin layer on zones A and B, or may be intergrown with them. Zone C is highly enriched in Ti (54 to 64% TiO_2) and contains 2 to 7% UO_2, 4 to 6% ThO_2, and 1 to 3% PbO. It is not clear whether U and Th are substituted in the anatase, or are in an intergrown metamict U-Th-Ti-bearing phase. Zone D is rust-brown and also probably a multiphase aggregate, and sometimes it is present as an irregular outer zone on zone C. Relative to zone C, it is enriched in Fe, Si, and Pb, but depleted in Ti; whereas U and Th contents in both zones are similar [2]. Brannerite from biotite granite has suffered complete alteration to a TiO_2 mineral (anatase) with a relatively wide range of unusual contaminants, and a brannerite from hydrothermally altered muscovite granite has, for the most part, broken down to anatase and other species, both Lawler Peak [3, pp. 165, 198].

Brannerite from veinlike quartz-microcline bodies in leucocratic granite of an undisclosed locality, USSR, was altered in various degree to secondary pseudocollomorphic products (mainly anatase) which may form rims (envelopes) on the crystals, traverse it as veinlets, or build up complete pseudomorphs. Compared with unaltered brannerite, the secondary product is enriched in Th (4.98 against 1.34% ThO_2), Ti, Fe, RE, and water, whereas U is diminished (and probably is fixed with metatorbernite) [19, pp. 68/9, 77/8]. Discrete brannerite grains of the pyritized matrix of quartz-pebble conglomerates, Blind River area, Ontario, Canada, were partially altered to a mixture of anatase and thorogummite. All stages of alteration occur, from narrow rims around homogeneous brannerite to separate grains composed only of the beige alteration product. As follows from the chemical compositions, the alteration process, likely the result of weathering, results in an enrichment of Th relative to U [44]. Note also that for marginal parts of brannerite from the Blind River-Elliot Lake district an enrichment in Th is mentioned, and that with marginally large changes in Th content, the source rock has been primarily different in Th content [49].

Brannerite from grayish rose porphyric granite, Borovskii pluton, northern Kazakhstan, is partially or completely replaced by aeschynite, which itself is converted into gelatinous decomposition products. Alteration by aeschynite occurs either synchronously to, or after the formation of brannerite [1].

Brannerite from the stratabound Karpinka Lake uranium prospect, northern Saskatchewan, Canada, is enveloped by ovoids of sulfides, in general pyrrhotite. However, composed rims consisting of euhedral to subhedral pyrite surrounded by pyrrhotite and chalcopyrite also occur [33].

A polished section shows much of the brannerite from skarn near Madoc township, Ontario, Canada, to be brecciated and cemented by calcite and veined and replaced by native Au [28].

References for 2.2.1.2.2.1:

[1] Platonov, A. N. (Sb. Nauchn. Rab. Nauchn. Issled. Sekt. Kiev. Gos. Univ. No. 2 [1964] 110/5, 110/3).
[2] Silver, L. T.; Williams, I. S.; Woodhead, J. A. (GJBX-45-81 [1980] 1/380, 219/57).

[3] Silver, L. T.; Woodhead, J. A.; Williams, I. S.; Chappell, B. W. (GJBX-7-84 [1984] 1/412).
[4] Silver, L. T.; Woodhead, J. A.; Williams, I. S. (Uranium Explor. Methods Rev. NEA/IAEA R & D Programme Proc. Symp., Paris 1982, pp. 355/67).
[5] Whittle, A. W. G. (Bull. Geol. Surv. South Australia No. 30 [1954] 126/51, 142).
[6] Campana, B.; King, D. (Bull. Geol. Surv. South Australia No. 34 [1958] 1/133, 54/5).
[7] Ashley, P. M. (Mineral. Deposita 19 [1984] 7/18).
[8] Lawrence, L. J. (Proc. Roy. Soc. Queensland 66 No. 5 [1954] 69/76, 69/71).
[9] X-Ray Laboratory, Peking Inst. Uranium Geol. and X-Ray Labor., Wuhan Geol. College (Acta Geol. Sinica 1978 241/51 from Am. Mineralogist 64 [1979] 656).
[10] Vukasović, M.; Mihailović-Vlajić, N. (Radovi Sek. Istraž. Nukl. Drugih Mineral. Sirovina No. 2 1963 67/8; Ref. Zh. Geol. 1964 No. 10 V 135).

[11] Arribas, A. (Bol. Real. Soc. Espan. Hist. Nat. Secc. Geol. 65 [1967] 157/70, 163).
[12] Arribas, A. (Intern. Geol. Congr. Rept. 21st Session Norden, Copenhagen 1960, Pt. 15, pp. 98/108, 100/1).
[13] Ray, G. E. (Saskatchewan Dept. Mineral Resour. Rept. No. 208 [1981] 1/33, 25/6).
[14] Umamaheswararao, G. V.; Krishnaswamy, R. (J. Sci. Ind. Res. [India] A 15 [1956] 401/2 from C.A. 1960 5349/50 and Ref. Zh. Geol. 1957 No. 9389).
[15] Parcalabescu, I. D.; Nistor, I. (Mine Petrol Gaze 35 [1984] 497/503 from C.A. 103 [1985] No. 40103).
[16] Van Autenboer, T.; Skjerlie, F. J. (Norges Geol. Unders. No. 200 [1957] 5/7).
[17] Pabst, A. (Am. Mineralogist 39 [1954] 109/17, 109).
[18] Staatz, M. H. (Econ. Geol. 73 [1978] 512/23, 512, 517).
[19] Povilaitis, M. M. (Tr. Mineral. Muzeya Akad. Nauk SSSR No. 12 [1961] 64/79).
[20] Kazantsev, V. V. (in: Petrov, R. P., Tekstury Strukt. Uranovykh Rud Endogen. Mestorozhd., Atomizdat, Moscow 1977, pp. 99/110, 195/202 from C.A. 89 [1978] No. 166363).

[21] Bell, R. T.; Jones, L. D. (Geol. Surv. Can. Paper No. 79-1A [1979] 397/9).
[22] Gotman, Ya. D. (Zap. Vses. Mineral. Obshch. 87 [1958] 197/200).
[23] Il'inskii, G. A. (Vestn. Leningr. Univ. Geol. Geogr. No. 6 1961 134/8, 134/5).
[24] Zavarzin, A. V. (Geol. Rudn. Mestorozhd. 6 [1961] 54/8, 54/5, 58).
[25] Davidson, C. F. (Mining Mag. 106 [1962] 92/3).
[26] Lis, J.; Stępniewski, M.; Sylwestrzak, H. (Biul. Inst. Geol. No. 193 [1965] 207/28, 214, 227).
[27] Plyushchev, E. V. (Tr. Inst. Geol. Geofiz. Akad. Nauk SSSR Sibirsk. Otd. No. 286 [1975] 161/6, 272/83, 161, 164).
[28] Steacy, H. R.; Plant, A. G.; Boyle, R. W. (Can. Mineral. 12 [1973/74] 360/3).
[29] Meixner, H. (Carinthia II 78 [1968] 96/115, 98).
[30] Dar, K. K. (J. Geol. Soc. India 5 [1964] 112/20, 115).

[31] Hiern, M. N. (South Australia Dept. Mines Mining Rev. No. 107 [1957] 80/1).
[32] Hewett, D. F.; Stone, J.; Levine, H. (Am. Mineralogist 42 [1957] 30/8, 33/4, 37/8).
[33] Williams-Jones, A. E.; Sawink, M. J. (Econ. Geol. 80 [1985] 1927/41, 1931, 1938).
[34] Rudnitskaya, L. S. (in: Bezgubov, A. I.; Byvshikh, Yu. I.; Dement'ev, P. K.; et al., Uran v Drevnikh Konglomeratakh, Gosatomizdat, Moscow 1963, pp. 129/40, 131/3).
[35] Kulik, N. A. (in: Mitropol'skii, A. S., Geokhimiya i Mineralogiya Radioaktivnykh Elementov Sibiri, Nauka, Novosibirsk 1970, pp. 17/53, 30/40).
[36] Robertson, J. A. (Geol. Surv. Profess. Paper [U.S.] No. 1161-A-BB [1981] U1/U23).
[37] Schidlowski, M. (Geol. Surv. Profess. Paper [U.S.] No. 1161-A-BB [1981] N1/N29).
[38] Ferris, C. S.; Ruud, C. O. (Colo. School Mines Quart. 66 No. 4 [1971] 1/35).
[39] Krishna Rao, N.; Rao, G. V. U. (J. Geol. Soc. India 24 [1983] 489/501).
[40] Theis, N. J. (Geol. Surv. Can. Bull. No. 304 [1979] 1/50, 31/2).

[41] Robinson, A.; Spooner, E. T. C. (J. Geol. Soc. London **141** [1984] 221/8, 222/4).
[42] Roscoe, S. M.; Steacy, H. R. (Proc. 2nd U.N. Intern. Conf. Peaceful Uses At. Energy, Geneva 1958, Vol. 2, pp. 475/83, 482).
[43] Robertson, J. A. (Ontario Dept. Mines Geol. Rept. No. 57 [1968] 1/162, 91/2).
[44] Traill, R. J. (Can. Mining J. **75** [1954] 63/8).
[45] Saager, R.; Stupp, H. D. (TMPM Tschermaks Mineral. Petrogr. Mitt. [3] **32** [1983] 83/102).
[46] Stupp, H. D. (Diss. Univ. Köln 1984, pp. 1/280, 94/9).
[47] Heinrich, E. W. (in: Amstutz, G. C.; El Goresy, A.; Frenzel, G.; et al., Ore Genesis: The State of the Art, Springer, Berlin – Heidelberg – New York 1982, pp. 16/24, 22).
[48] Glatthaar, C. W.; Feather, C. E. (Appl. Mineral. Proc. 2nd Intern. Congr. Appl. Minerals Miner. Ind., Los Angeles 1984 [1985], pp. 617/28, 621).
[49] Ramdohr, P. (Z. Deut. Geol. Ges. **130** [1979] 439/58, 442).
[50] Kozyr'kov, V. D. (in: Petrov, R. P.; Tekstury Strukt. Uranovykh Rud Endogen. Mestorozhden., Atomizdat, Moscow 1977, pp. 60/70, 195/202 from C.A. **89** [1978] No. 166270).
[51] Feather, C. E. (Geol. Surv. Prof. Paper [U.S.] No. 1161-A-BB [1981] Q1/Q23, Q13).
[52] Roscoe, S. M. (Can. Mining J. **80** No. 7 [1959] 65/6).

2.2.1.2.2.2 Chemistry

At present, there are two views of the chemical nature of brannerite and related minerals: They may be complex oxides of a number of elements, or may be titanates of U with a complex mode of diadochic substitution [1, pp. 60/2]. Chemical analyses of thorutite-brannerite series samples, see Table 5, pp. 92/7, most satisfactorily correspond to an AB_2O_6-type formula [2, pp. 67/70]. As follows from literature data, members of this series rich in Th are genetically related to derivatives of alkaline magmas [3, p. 157]; see also pp. 78 and 81/5.

Calculations for brannerite from various types of occurrence show that cations with an ionic radius between 1.32 and 0.79 Å may be contained in group A, and those between 0.79 and 0.57 Å in group B of the above formula type. Ions with a radius close to 0.79 Å (as Mg and Fe^{2+}) may enter group A in some cases, or group B in others [4]. Group-A cations comprise U^{6+}, U^{4+}, Ca, RE, Th, Pb, sometimes Sc; and group-B cations are Ti, Fe^{2+}, Fe^{3+}, partly Nb, Ta, and U^{4+}, but also Mg, Mn, Sn. An insignificant surplus of group-B cations for some analyzed material may be caused by admixed secondary products enriched in Ti [2, pp. 67/70]; see also [5]. For a short discussion of the influence of ionic radius, ionic potential, and electronegativity of Th, U^{4+}, Zr, Fe^{3+}, and RE on formation and stability of natural titanate minerals (as, e. g., brannerite and thorutite), see [1, pp. 62/4]. Differences in the $U^{4+}:U^{6+}$ ratio for brannerite may be related to the degree of oxidation of U^{4+} near the earth's surface [3, p. 155]. Brannerites from low-temperature formations seen to be characterized by low content of Th [6], [7, p. 141] or by insignificant contents of Th and RE [8].

For contents (variation and mean) of Y_2O_3 and/or $RE_2O_3 + Y_2O_3$ in brannerites, see "Seltenerdelemente" A4, 1979, pp. 9, 14/5.

For a recalculation into components (as, e. g., $ThTi_2O_6$ and $(Th,U)O_2$, besides others) of seven analyses of brannerite given in the literature, see [35].

Due to the negative correlation between U_3O_8 and TiO_2 contents for primary brannerites from pegmatites and for secondary brannerites to rutile-leucoxene aggregates from conglomerates (Blind River, Witwatersrand, and eastern Transvaal), which together form a continuous series, it is proposed to set arbitrarily the lower limits for brannerite at 26% U_3O_8 and 32% TiO_2, corresponding to primary brannerite. Except one, all above brannerites with U_3O_8 > 26% show a U_3O_8 : TiO_2 ratio < 0.6. The diagram U_3O_8 : ThO_2 vs. ThO_2 reveals a great variation in Th values for the investigated U-Ti phases (brannerite and rutile-leucoxene aggregates) [7, p. 96, 136/43].

Electron microanalyses for more homogeneous brannerite from a skarn zone bordering a granitic intrusion near Madoc township, Ontario, Canada, gives 45 to 50% U, 1.2 to 2.2% Th, 3.0 to 3.4% Ca, 19.5 to 22.0% Ti, and 1.5 to 2.0% Fe, with U and Ti varying antipathetically [9]. Brannerite from Dzhida, Central Asia, USSR, has 40.5% U, 6.9% Th, 1.94% Pb [10]. Chemical analysis of thorian brannerite from a brecciated shear zone in adamellite, Crockers Well, South Australia, gives 21.33% U, 8.196% Th, and 2.228% Pb [11].

The low totals for UO_2, ThO_2, PbO_2, FeO, TiO_2, CaO, and SiO_2 contents (summing up to an average of 94.9%, variation is between 90.1 and 103.3%) in brannerite grains from different Witwatersrand reefs and recovery plants, R.S.A., probably result from metamictization and voids in the crystal lattice structure; and possibly also result from the fact that Fe may be present as Fe_2O_3 instead of FeO and that Pb may be a sulfide rather than an oxide. The SiO_2 content of the grains is probably derived from the abundant authigenic silica of the reefs, having entered the brannerite structure either at the time of formation or later when the grains become metamict and somewhat porous. The UO_2 : ThO_2 ratio varies extensively and may have been inherited from the source uraninite. The contents, variation and mean, for the above mentioned elements and the UO_2 : ThO_2 ratio in 30 grains of brannerite are as follows; for details, see original paper [12]:

	content in %		content in %
UO_2	26.2 to 49.5; 36.0	TiO_2	11.5 to 52.5; 31.9
ThO_2	0.2 to 5.4; 2.7	FeO	0.7 to 14.8; 0.5
PbO_2	0.3 to 37.8; 11.7	SiO_2*)	1.0 to 17.3; 4.1
CaO	<0.1 to 1.1; 0.5	UO_2 : ThO_2	6.4 to 165.0; 13.3

*) Determined only for 21 samples.

For brannerite (as defined in the paper) from conglomerates of Gunsteling, eastern Transvaal, R.S.A., and of Quirke Mine No. 2, Elliot Lake area, Canada, the following contents, variation and mean, are given; for details and for analyses of rutile-leucoxene aggregates, see original paper. Anomalously high contents for some samples of PbO are caused by numerous inclusions of galenite, high FeO contents by pyrite or pyrrhotite [7, pp. 139/40 and Table A-3]:

	content in % for brannerite from	
	Gunsteling; 32 analyses	Quirke; 20 analyses
U_3O_8	23.46 to 39.04; 35.34	27.90 to 40.82; 33.05
ThO_2	0.60 to 1.47; 1.04	0.61 to 2.68; 1.47
PbO	1.97 to 9.96; 4.62	0.74 to 6.00; 2.34
CaO	1.32 to 2.95; 2.29	0.73 to 3.38; 1.40
TiO_2	31.63 to 44.02; 38.33	34.33 to 52.34; 44.45
FeO	0.53 to 8.28; 1.20	0.77 to 2.29; 1.37
SiO_2	4.75 to 11.63; 7.75	4.77 to 18.44; 6.97

The content of major elements in some brannerite samples from quartz-pebble conglomerates of the Blind River region, Ontario, Canada, varies from 33.8 to 51.8% U_3O_8, 0.3 to 5.0% ThO_2, 1.1 to 6.5% (REE + Y) oxides, 2.7 to 2.9% CaO, 32.4 to 39.0% TiO_2, and 2.4 to 2.5% FeO [13]. Compared with 40 to 52% U_3O_8, 0.3 to 9.15% ThO_2, and 32 to 40% TiO_2 for brannerite samples other than Blind River area, variously colored brannerite from different mines of the township of Elliot Lake, Ontario, shows the following contents [14]:

mine	color	% U_3O_8	% ThO_2	% TiO_2
Nordic	reddish brown[a]	32.6	2.6	30.3
Nordic	darker[b]	32.2	1.8	30.5
Denison[c]	black	41	6.1	20
Denison[c]	dark brown	36	1.8	30
Quirke[c]	dark brown	24	1.7	37
Quirke[c]	brown	20	2.2	30
Quirke[c]	cream	6	1.2	27

[a] Altered. — [b] More glassy, less altered. — [c] Samples are from an individual layer of homogeneous conglomerate [15].

Analyses of eight brannerite samples from a stratabound U deposit in sillimanite-bearing quartz-feldspar-biotite gneiss of the Karpinka Lake uranium prospect, Saskatchewan, Canada, show 51.5 to 60.7% UO_2, 1.0 to 2.5% ThO_2, 0.4 to 2.4% PbO, 0.1 to 1.1% CaO (7 samples only), 30.2 to 38.1% TiO_2, 1.0 to 2.3% FeO, 3.4 to 10.1% SiO_2, and 0.6 to 6.1% Al_2O_3 (7 samples only). The very low Pb content is puzzling, except one suggests that brannerite is probably far more susceptible to Pb loss than uraninite [16].

Chemically zoned orthobrannerite (or uranium titanate-brannerite?, see p. 82) from the Lawler Peak granite, west-central Arizona, suggests recrystallization with expulsion of Pb and Th from the central zone A of the crystal. The outer zone B appears to be a continuous band of lower reflectivity around the central zone, but the Th content of the outer zone is rather spotty. The few high-Th spots in the inner zone coincide with less reflective spots. Besides the zones A and B, plots of U vs. Th content and of U vs. Th: U ratio reveal an additional zone C, made up predominantly of anatase, and an outermost zone D, high in Fe content [17, pp. 141/9]; for the zonation, refer also to pp. 86/7. Microprobe analyses for individual zones, are given as follows [18]:

	content in % in zone				
	A	A	B	C	D
ThO_2	2.8	5.9	6.4	6.5	3.6
UO_2	50.4	41.6	22.2	3.9	5.2
PbO_2	1.2	2.6	5.3	2.3	5.0
CaO	3.8	3.0	3.4	1.6	2.8
Y_2O_3	1.0	1.0	0.7	0.4	0.3
SiO_2	3.0	3.3	9.5	4.1	14.7
TiO_2	32.1	34.7	38.3	63.4	23.4
FeO	1.3	1.2	1.9	5.4	27.2
MnO	0.5	0.4	0.4	0.2	0.4
ZrO_2	0.1	0.2	0.8	0.5	1.1
Nb_2O_5	1.0	0.9	1.4	2.6	1.5

Table 5

Chemical Composition and Density of Thorutite and Brannerite Samples from Various Types of Occurrence.

n.d. = not determined; tr. = traces. For the footnotes, see p. 97.

	content in % in sample number										
	1	2	3	4	5	6	7	8	9	10	11
ThO_2	54.10	12.92	12.81	9.25	9.18	9.15	8.49	8.41	8.04	7.81	7.59
UO_2	0.14	–	–	37.10	36.51	tr.	37.31	–	45.22	9.63	10.16
UO_3	1.43	–	31.83	–	–	26.47	–	–	–	31.83	32.67
U_3O_8	–	37.40	–	–	–	–	–	40.95	–	–	–
PbO	–	1.30	2.79	2.75	2.31	1.75	3.04	0.58	0.58	1.23	0.72
CaO	1.07	3.40	–	2.31	2.29	2.70	2.07	–	2.27	3.07	3.04
ΣRE_2O_3	–	2.24	5.14	1.20	1.89	9.75	1.88	tr.	1.29	tr.	tr.
TiO_2	36.10	38.20	35.13	33.25	31.32	39.60	31.32	37.96	34.23	37.38	36.68
FeO	1.10	–	–	1.11	1.42	1.28	0.79	–	1.64	4.47	5.17
Fe_2O_3	–	0.16	1.37	–	–	2.73	–	–	–	–	–
H_2O^-	–	0.06	2.54	–	–	0.05	–	–	–	–	–
H_2O^+	0.94[a]	–	7.41	–	–	2.35	–	1.14[a]	–	0.35[a]	0.52[a]
others	3.86[b]	3.20[c]	0.70[d]	13.03[e]	15.09[f]	–	15.11[g]	–	6.75[h]	–	–
sum	97.74	98.98	99.72	100.00	100.01	96.11	100.01	88.46	100.02	95.77	96.55
D in g/cm^3	5.82	5.57	~4.1	–	–	4.76	–	4.02	–	5.40	5.04

No. 1 = Thorutite from a nepheline syenite massif, USSR [19]. – No. 2 = "Lodochnikite?" from a fluorite vein in a dolomitized limestone near an alkaline syenite massif, USSR [20]. – No. 3 = "Absite" from fissures in granitoid rocks, Crockers Well, South Australia [21]; specific gravity from [22]. – Nos. 4 and 5 = "Absite" from Crockers Well, South Australia [23]. – No. 6 = Brannerite from granite gneiss, San Bernardino county, California [24]. – No. 7 = "Absite" from Crockers Well, South Australia [23]. – No. 8 = Brannerite [25]; refer also to [2, pp. 68/9]. – No. 9 = Brannerite (center), Mono county, California [23]. – Nos. 10 and 11 = Brannerite [25]; refer also to [2, pp. 68/9].

Table 5 (continued)

						content in % in sample number					
	12	13	14	15	16	17	18	19	20	21	22
ThO_2	6.97	6.78	5.58	5.20	5.12	5.10	5.0	~5	4.7	4.5	4.18
UO_2	46.20	8.87	47.21	35.66	14.10	20.65	8.2	35.51	46.2[o]	29.9	13.62
UO_3	—	31.43	—	15.21	39.48	24.16	32.0	15.01	—	11.4	26.86
U_3O_8	—	—	—	—	—	—	—	—	—	—	—
PbO	0.17	2.60	0.49	0.65	0.27	0.07	—	0.65	2.1	0.1	1.22
CaO	1.74	2.41	1.10	n.d.	0.24	0.46	2.8	—	1.2	4.1	3.45
ΣRE_2O_3	1.92	5.08	1.99	2.16	0.13	4.76	2.1	~2.06	1.6	3.7	3.24
TiO_2	34.65	35.22	34.46	35.09	37.77	39.82	32.9	35.48	32.0	43.8	41.72
FeO	1.45	—	1.99	1.36	—	—	2.4	1.09	2.6	—	—
Fe_2O_3	—	2.48	—	0.43	0.48	2.07[n]	—	0.99	—	1.8[n]	0.58
H_2O^-	—	—	—	n.d.	—	—	—	n.d.	—	—	—
H_2O^+	—	2.00[a]	—	2.61	—	0.12[a]	2.6[a]	2.66	—	1.0[a]	2.08[a]
others	6.84[i]	3.19[j]	7.18[k]	1.35[l]	1.91[m]	2.90[o]	—	1.18[p]	4.6[r]	—	2.33[b]
sum	99.92	100.06	100.00	99.72	99.50	100.11	88.0	99.63	95.7	100.3	99.28
D in g/cm³	—	5.41	—	5.70	5.88	~5.62	5.43	—	—	—	5.12

No. 12 = Brannerite (edge), Mono county, Californa [23]. – No. 13 = Brannerite (end), Mono county, California [23]. – No. 15 = Brannerite (sample No. 212 with chonchoidal fracture and strong pitchy glance) from a quartz vein of the lower horizon of a Mo-W deposit, USSR [3, pp. 150, 155]. – No. 14 = Brannerite from a quartz vein of the lower horizon of a Mo-W deposit, USSR [3, pp. 150, 155]. – No. 16 = Brannerite (oral communication I. D. Starynkevich-Borneman) [2, pp. 68/9]. – No. 17 = Brannerite (oral communication E. F. Kuznetsova) [2, pp. 68/9]. – No. 18 = Brannerite from a mesothermal quartz vein cutting quartz monzonite, Walker River, Mono county, California [26]. – No. 19 = Brannerite (sample No. 212a with uneven fracture and dull luster) from a quartz vein of the lower horizon of a Mo-W deposit, USSR [3, pp. 150, 155]. – No. 20 = Brannerite (A zone) from altered muscovite granite, Lawler Peak, Arizona [17, p. 177]. – No. 21 = Brannerite (oral communication E. F. Kuznetsova) [2, pp. 150, 155]. – No. 22 = Brannerite [25], refer also to [2, pp. 68/9].

Table 5 (continued)

	content in % in sample number										
	23	24	25	26	27	28	29	30	31	32	33
ThO_2	4.1	3.96	3.58	3.41	3.12	3.0	3.0	2.93	2.64	2.58	2.5
UO_2	10.3	16.45	24.68	23.52	41.07	–	48.6[q]	38.81	45.01	31.23	52.0
UO_3	33.5	35.41	–	21.63	9.21	–	–	–	–	–	6.0
U_3O_8	–	–	–	–	–	39.0	–	–	–	–	–
PbO	0.2	0.13	0.48	0.35	–	–	1.4	0.41	0.95	0.46	tr.
CaO	2.9	–	0.38	2.99	0.80	–	2.5	2.29	1.06	1.57	0.1
ΣRE_2O_3	3.9	–	10.22	5.57	4.94	3.6	–	4.68	0.82	8.43	–
TiO_2	39.0	42.05	27.75	31.92	36.12	32.5	34.7	28.16	27.91	26.77	36.0
FeO	2.9	–	1.34	2.55	–	–	0.1[q]	1.87	0.98	2.36	0.7
Fe_2O_3	–	0.25	–	–	1.27[n]	3.0	–	–	–	–	–
H_2O^-	–	–	–	–	–	1.4	–	–	–	–	–
H_2O^+	2.0[a]	0.04[a]	–	4.51[a]	–	6.5	–	–	–	–	1.0[a]
others	1.4[f]	1.82[u]	31.49[v]	3.64[w]	3.59[x]	1.2[y]	7.2[z]	20.85[aa]	20.64[ab]	26.62[ac]	–
sum	100.2	100.11	100.02	100.09	100.12	90.2	98.9	100.00	100.01	100.02	98.3
D in g/cm^3	5.42	5.88	–	5.43	~5.83	4.2	–	–	–	–	5.45

No. 23 = Brannerite from a gold placer, Kelly Gulch, Custer county, Idaho [26]. – No. 24 "Lodochnikite" from a microcline-nepheline vein cutting syenitic rocks, USSR [27]. – No. 25 = "Brannerite" form placer, Idaho [23]. – No. 26 = Brannerite (from aplite [28]) [25], refer also to [2, pp. 68/9]. – No. 27 = Brannerite (oral communication E. F. Kuznetsova) [2, pp. 68/9]. – No. 28 = Brannerite from Torrens Gorge, South Australia [28]. – No. 29 = Brannerite (central part) from muscovite-biotite granite, Lawler Peak, Arizona [17, p. 112]. – No. 30 = "Brannerite" from placer, Idaho [23]. – No. 31 = Brannerite phase of a composite grain from conglomerate, Rio Algom-Denison ore reef, Ontario [23]. – No. 32 = "Brannerite" from placer, Idaho [23]. – No. 33 = Brannerite from pegmatite, Lodrino, Canton Tessin, Switzerland [29].

Table 5 (continued)

95

	content in % in sample number										
	34	35	36	37	38	39	40	41	42	43	44
ThO_2	2.35	1.79	1.76	1.75	1.70	1.59	1.45	1.34	1.34	1.34	1.29
UO_2	none	49.40	30.79	42.06	46.08	54.94	0.91	10.00	40.20	37.28	39.12
UO_3	42.36	–	16.49	–	–	–	42.86	38.19	–	–	–
U_3O_8	–	–	–	–	–	–	–	–	–	–	–
PbO	2.36	0.64	–	0.59	1.33	1.28	3.52	0.74	1.37	0.50	0.01
CaO	2.80	0.79	0.60	0.85	0.51	0.78	3.37	1.75	0.52	0.34	2.17
ΣRE_2O_3	1.25	0.72	5.62	1.84	0.58	–	2.80	0.50	1.63	1.52	2.05
TiO_2	31.80	24.54	31.21	27.70	27.98	35.45	37.20	40.09	33.08	38.74	35.48
FeO	–	0.59	–	0.89	0.59	1.45	–	2.90	1.01	1.58	1.43
Fe_2O_3	–	–	4.11	–	–	–	0.91	0.03	–	–	–
H_2O^-	–	–	–	–	–	–	–	n.d.	–	–	–
H_2O^+	–	–	1.69[a]	–	–	–	–	2.77	–	–	–
others	13.25[ad]	21.54[ae]	6.72[af]	24.32[ag]	21.24[ah]	7.51[ai]	6.12[aj]	1.72[ak]	20.86[al]	18.67[am]	18.47[an]
sum	96.17	100.01	99.17	100.00	100.01	103.00	99.14	100.03	100.01	99.97	100.02
D in g/cm³	–	–	5.56	–	–	–	5.1	5.49	–	–	–

No. 34 = Brannerite [25], refer also to [2, pp. 68/9]. – No. 35 = Brannerite phase of a composite grain from conglomerate, Rio Algom-Denison ore reef, Ontario [23]. – No. 36 = Brannerite from greisen, USSR [30]. – Nos. 37 and 38 = Brannerite phase of a composite grain from conglomerate, Rio Algom-Denison ore reef, Ontario [23]. – No. 39 = Brannerite from paragneiss (metamorphosed arkoses), mean for 8 analyses, Karpinka Lake uranium prospect, Saskatchewan [16]. – No. 40 = Brannerite from pegmatite, Rajputana, India [31], see also [32]. – No. 41 = Brannerite from a quartz vein of the upper horizon [3, pp. 150, 155] of a Mo-W deposit, USSR [2, p. 67]. – No. 42 = Brannerite phase of a composite grain from conglomerate, Rio Algom-Denison ore reef, Ontario [23]. – Nos. 43 and 44 = Brannerite phase of a composite grain from conglomerate, Rio Algom-Denison ore reef, Ontario [23].

Table 5 (continued)

	content in % in sample number									
	45	46	47	48	49	50	51	52	53	54
ThO_2	1.25	1.24	1.20	1.2	1.19	1.18	1.15	1.10	1.03	1.0
UO_2	32.55	32.30	—	2.4[q]	31.65	33.24	53.15	38.81	28.75	2.3
UO_3	—	—	—	—	—	—	—	—	31.08	38.5
U_3O_8	—	—	51.76	—	—	—	—	—	—	—
PbO	1.82	4.55	2.34	0.2	0.76	0.64	0.90	1.03	0.17	4.0
CaO	2.86	4.80	2.74	0.6	0.53	1.03	0.33	5.52	0.44	1.2
ΣRE_2O_3	3.47	3.61	1.10	—	0.88	3.42	0.80	1.58	0.28	4.2
TiO_2	34.57	35.47	32.45	64.2	44.35	42.27	20.18	36.12	35.35	33.4
FeO	1.56	2.03	3.11	8.3	1.13	1.50	0.89	1.97	—	—
Fe_2O_3	—	—	—	—	—	—	—	—	0.40	4.9
H_2O^-	—	—	—	—	—	—	—	—	0.76	2.0
H_2O^+	—	—	2.35[a]	—	—	—	—	—	1.11	7.6
others	21.91[ap]	16.02[aq]	0.16[ar]	18.0[as]	19.51[at]	16.71[au]	22.60[av]	13.87[aw]	0.85[ax]	1.3[ay]
sum	99.99	100.02	97.21	94.9	100.00	100.00	100.00	100.00	100.22	100.4
D in g/cm^3	—	—	5.17	—	—	—	—	—	5.46	4.16

Nos. 45 and 46 = Brannerite of type II from conglomerate, Rio Algom-Denison ore reef, Ontario [23]. – No. 47 = Brannerite from pegmatite, Fuenteovejuna, Cordoba, Spain [33]. – No. 48 = Brannerite (rim) from altered muscovite granite, Lawler Peak, Arizona [17, p. 177]. – No. 49 = Brannerite phase of a composite grain from conglomerate, Rio Algom-Denison ore reef, Ontario [23]. – Nos. 50 and 51 = Brannerite phase of a composite grain from conglomerate, Rio Algom-Denison ore reef, Ontario [23]. – No. 52 = Brannerite of type II from conglomerate, Rio Algom-Denison ore reef, Ontario [23]. – No. 53 = Orthobrannerite from alkalic rocks (syenite and lamprophyre), People's Republic of China [34]. – No. 54 = Brannerite from a clay-filled crevice within gneiss and migmatite, Thackaringa, New South Wales, Australia [28].

Footnotes for Table 5:

a) Comprises H_2O^- and H_2O^+. — b) Comprises 1.50% Al_2O_3, 0.44% SiO_2, 1.12% Nb_2O_5, 0.08% Ta_2O_5, and 1.72% l.o.i. — c) Comprises l.o.i. — d) Comprises 0.38% ZrO_2, 0.19% Sc_2O_3, and 0.13% P_2O_5. — e) Comprises 0.55% Al_2O_3, 0.66% SiO_2, and water.

f) Comprises 0.41% Al_2O_3, 0.52% SiO_2, and water. — g) Comprises 0.25% Al_2O_3, 1.11% SiO_2, and water. — h) Comprises 0.47% Al_2O_3, 0.52% SiO_2, 0.32% Nb_2O_5, and water. — i) Comprises 0.56% Al_2O_3, 0.50% SiO_2, and water. — j) Comprises 1.07% Al_2O_3, 0.50% SiO_2, 0.22% MgO, and 1.4% $(Ta,Nb)_2O_5$.

k) Comprises 0.47% Al_2O_3, 0.20% SiO_2, and water. — l) Comprises 0.12% Al_2O_3, 0.18% SiO_2, and 1.05% Nb_2O_5. — m) Comprises 1.45% SiO_2 and 0.46% $(Nb,Ta)_2O_5$. — n) Comprises total iron. — o) Comprises 0.46% Al_2O_3, 0.10% SiO_2, 0.82% $(Nb,Ta)_2O_5$, 0.10% MgO, 0.32% K_2O, 0.12% Na_2O, and 0.98% l.o.i.

p) Comprises 0.12% Al_2O_3 and 1.06% Nb_2O_5. — q) Cation oxidation state not determined. — r) Comprises 1.1% SiO_2, 1.2% Nb_2O_5, 0.5% Ta_2O_5, and 1.8% Y_2O_3. — s) Comprises 2.17% SiO_2 and 0.16% insoluble. — t) Comprises 0.6% SiO_2, 0.2% ZrO_2, 0.3% BaO, 0.1% SrO, and 0.2% CO_2.

u) Comprises 0.12% Al_2O_3, 0.21% SiO_2, 0.14% $(Nb,Ta)_2O_5$, 0.68% MgO, 0.12% S, and 0.55% l.o.i. — v) Comprises 0.62% Al_2O_3, 1.14% SiO_2, 17.6% Nb_2O_5, and water. — w) Comprises 2.12% Al_2O_3, 1.07% SiO_2, and 0.45% MgO. — x) Comprises 0.79% SiO_2, 0.15% $(Nb,Ta)_2O_5$, 0.12% SnO, and 2.53% l.o.i. — y) Comprises 1.1% SnO_2 and 0.1% ZrO_2.

z) Comprises 2.9% SiO_2, 0.3% MnO, 1.0% Nb_2O_5, 1.4% Y_2O_3, 0.5% each Ce_2O_3 and Nd_2O_3, and 0.3% each Sm_2O_3 and Gd_2O_3. — aa) Comprises 0.57% Al_2O_3, 0.64% SiO_2, 9.92% Nb_2O_5, and water. — ab) Comprises 0.68% Al_2O_3, 9.94% SiO_2, 1.02% Nb_2O_5, and water. — ac) Comprises 0.27% Al_2O_3, 0.94% SiO_2, 15.15% Nb_2O_5, and water. — ad) Comprises 2.85% SiO_2 and 10.4% l.o.i.

ae) Comprises 0.89% Al_2O_3, 10.06% SiO_2, 0.22% Nb_2O_5, and water. — af) Comprises 0.23% Al_2O_3, 1.87% SiO_2, 1.52% MgO, and 3.10% $(Nb,Ta)_2O_5$. — ag) Comprises 0.93% Al_2O_3, 10.49% SiO_2, 0.52% Nb_2O_5, and water. — ah) Comprises 1.4% Al_2O_3, 10.58% SiO_2, and water. — ai) Comprises 1.83% Al_2O_3, 5.41% SiO_2 for seven samples, and 0.27% MnO for three samples.

aj) Comprises 1.24% Al_2O_3, 0.52% SiO_2, 0.03% V_2O_5, and 4.33% l.o.i. — ak) Comprises 0.6% Al_2O_3, 0.1% SiO_2, 0.79% Nb_2O_5, and 0.23% MnO. — al) Comprises 0.91% Al_2O_3, 9.19% SiO_2, 0.66% Nb_2O_5, and water. — am) Comprises 0.77% Al_2O_3, 7.97% SiO_2, 0.81% Nb_2O_5, and water. — an) Comprises 0.71% Al_2O_3, 5.02% SiO_2, and water.

ap) Comprises 2.16% Al_2O_3, 6.34% SiO_2, 0.04% S, and water. — aq) Comprises 0.98% Al_2O_3, 4.55% SiO_2, 0.56% S, and water. — ar) Comprises SiO_2. — as) Comprises 0.6% Al_2O_3, 1.6% SiO_2, 0.4% ZrO_2, 14.8% Nb_2O_5, 0.4% Ta_2O_5, 0.1% MgO, and 0.1% P_2O_5. — at) Comprises 0.78% Al_2O_3, 6.9% SiO_2, 2.23% Nb_2O_5, and water.

au) Comprises 1.05% Al_2O_3, 4.36% SiO_2, 0.88% Nb_2O_5, and water. — av) Comprises 1.58% Al_2O_3, 11.18% SiO_2, 0.07% Nb_2O_5, and water. — aw) Comprises 0.75% Al_2O_3, 3.56% SiO_2, 0.16% Nb_2O_5, and water. — ax) Comprises 0.65% SiO_2, 0.19% Nb_2O_5, and 0.01% Ta_2O_5. — ay) Comprises 0.2% ZrO_2, 0.2% CO_2, 0.8% SO_2, and 0.1% P_2O_5.

References for 2.2.1.2.2.2:

[1] Strelkin, M. F.; Zheleznyak, N. N.; Zavarzin, A. V.; Kashcheev, L. P.; Nikol'skii, A. L (Vopr. Prikl. Radiogeol. No. 2 [1967] 60/89).
[2] Povilaitis, M. M. (Tr. Mineral. Muzeya Akad. Nauk SSSR No. 12 [1961] 64/79).
[3] Povilaitis, M. M.; Yakovlevskaya, T. A.; Knyazeva, D. N.; Belyaeva, I. D. (Zap. Vses. Mineral. Obshch. **97** [1968] 150/61).
[4] Povilaitis, M. M. (Zap. Vses. Mineral. Obshch. **92** [1963] 113/23, 118).

98

[5] Chukhrev, F. V.; Bonshtedt-Kupletskaya, E. M. (Mineraly Spravochnik, Vol. 2, Pt. 3, Nauka, Moscow 1967, pp. 1/676, 436).

[6] Frenzel, G.; Ottemann, J.; Kurtze, W. (Neues Jahrb. Mineral. Abhandl. **124** [1975] 75/102, 95).

[7] Stupp, H. D. (Diss. Univ. Köln 1984, pp. 1/280).

[8] Rakovich, F. I. (Dopov. Akad. Nauk Ukr. RSR B **1976** No. 11, pp. 976/80).

[9] Steacy, H. R.; Plant, A. G.; Boyle, R. W. (Can. Mineral. **12** [1973/74] 360/3).

[10] Narbutt, K. I.; Labutina, I. P.; Shuba, I. D.; Kardakov, K. A.; Samoilov, G. P. (Tr. Inst. Geol. Rudn. Mestorozhd. Petrogr. Mineral. Geokhim. Akad. Nauk SSSR No. 28 [1959] 122/37, 130).

[11] Greenhalgh, D.; Jeffery, P. M. (Geochim. Cosmochim. Acta **16** [1959] 39/57, 45/6).

[12] Feather, C. E. (Geol. Surv. Profess. Paper [U.S.] No. 1161-A-BB [1981] Q1/Q23, Q13/Q15, Q19).

[13] Traill, R. J. (Can. Mining J. **75** [1954] 63/8).

[14] Robertson, J. A. (Ontario Dept. Mines Geol. Rept. No. 57 [1968] 1/162, 92).

[15] Roscoe, S. M. (Can. Mining J. **80** No. 7 [1959] 65/6).

[16] Williams-Jones, A. E.; Sawiuk, M. J. (Econ. Geol. **80** [1985] 1927/41, 1933/4).

[17] Silver, L. T.; Woodhead, J. A.; Williams, I. S.; Chappell, B. W. (GJBX-7-84 [1984] 1/412).

[18] Silver, L. T.; Williams, I. S.; Woodhead, J. A. (GJBX-45-81 [1980] 1/380, 137).

[19] Gotman, Ya. D.; Khapaev, I. A. (Zap. Vses. Mineral. Obshch. **87** [1958] 201/2).

[20] Il'inskii, G. A. (Vestn. Leningr. Univ. **1961** No. 6, pp. 134/8, 135).

[21] Whittle, A. W. G. (Bull. Geol. Surv. South Australia No. 30 [1954] 126/51, 143).

[22] Whittle, A. W. G. (J. Geol. Soc. Australia **2** [1955] 21/45, 33).

[23] Theiss, N. J. (Geol. Surv. Can. Bull. No. 304 [1979] 1/50, 47/8, 50).

[24] Hewett, D. F.; Stone, J.; Levine, H. (Am. Mineralogist **42** [1957] 30/8, 36).

[25] Soboleva, M. V.; Pudovkina, I. A. (AEC-tr-4487 [1957] 1/455, 399; N.S.A. **15** [1961] No. 23739).

[26] Pabst, A. (Am. Mineralogist **39** [1954] 109/17, 112).

[27] Gotman, Ya. D. (Zap. Vses. Mineral. Obshch. **87** [1958] 197/200).

[28] Rayner, E. O. (New South Wales Dept. Mines Tech. Rept. **5** [1957/60] 63/102, 77).

[29] Bianconi, F.; Simonetti, A. (Schweiz. Mineral. Petrogr. Mitt. **47** [1967] 887/934, 903).

[30] Korolev, K. G.; Miguta, A. K.; Polyakova, U. M.; Rumyancheva, G. V. (Mineralogiya, Geologicheskie i Fiziko-Khimicheskie Osobennosti Obrazovaniya Uranotitanatov, Nedra, Moscow 1979, pp. 1/144, 35/6).

[31] Umamaheswararao, G. V. (J. Sci. Ind. Res. [India] A**15** [1956] 401/2 from C.A. **1960** 5349/50).

[32] Dar, K. K. (J. Geol. Soc. India **5** [1964] 112/20, 115).

[33] George, D'Arcy (RMO-563 [1949] 1/198, 78; N.S.A. **5** [1971] No. 833).

[34] X-Ray Laboratory, Peking Inst. Uranium Geol. and X-Ray Labor., Wuhan Geol. College (Acta Geol. Sinica **1978** 241/5 from Am. Mineralogist **64** [1979] 656).

[35] Schröcke, H. (Neues Jahrb. Mineral. Abhandl. **106** [1966] 1/54, 12).

2.2.1.2.2.3 Crystal Form and Structure

Crystal Form. Brannerite is monoclinic [1], [2, p. 894], [3, p. 137] with $a:b:c = 2.580:1:1.889$, $\beta = 119°30'$ [4, p. 150], which correlates well with $2.625:1:1.848$ for synthetic UTi_2O_6 [5] or with $1.186:1:1.182$, $\beta = 114°22'$ for "lodochnikite" [3, p. 137]. Orthobrannerite has $a:b:c = 0.636:1:0.555$ or, after heat treatment, $0.632:1:0.543$ [6].

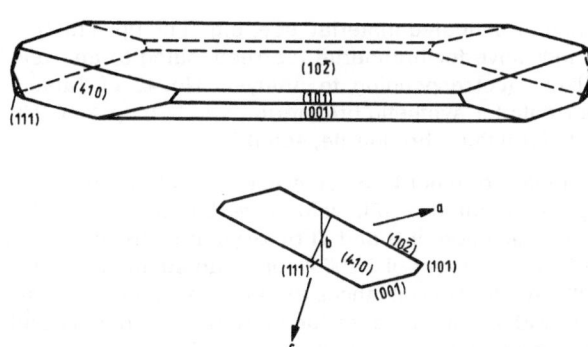

Fig. 4. Crystal form of brannerite from pegmatite, Lodrino, Tessin, Switzerland, from [2, p. 896].

The crystal morphology of brannerite samples from pegmatite comprises the forms $(10\bar{2})$, (101), (001), (410), and (111) for Lodrino, Tessin, Switzerland [2, pp. 895/6, 932/3], see **Fig. 4**, and (110), (130), (010), and (201) for Fuenteovejuna, Cordoba, Spain [1]. Brannerite from an undisclosed Mo-W deposit, USSR, shows a (100), b (010), l (120), n (210), y (310), v (510), d (111), k ($\bar{1}$12), and ($\bar{2}$12) [4, pp. 150, 152/3]; for a figure, see "Rare Earth Elements" A7, 1984, p. 52. "Lodochnikites" from barite-siderite veins in altered nepheline syenite, and from fluorite veinlets in dolomitized limestone near an alkaline syenite massif, both USSR, have as essential forms, in order of decreasing frequency, n (210), a (100), b (010), W (012), and u (111). Additionally, there have been observed 16 other forms; for the details (including figures of crystal forms) and for the measured and calculated spherical coordinates (φ, ϱ), see original paper [3, pp. 136/8]. For a comparison of equivalent crystal forms given in the latter reference and in [5], see [4, p. 152]. Crystals of orthobrannerite are bounded by (001), (120), (021), (110), and (140); some crystals show additionally (210), (180), and a form with uncertain indices [6].

The habit of brannerite is often prismatic or needle-like, see pp. 81/5. The latter habit seems to be related to intermediate or lower-temperature formations [7].

Twinning with c axis contained in the plane of intergrowth was observed for brannerite from Sierra Albarrana, Cordoba, Spain [8].

Distinct cleavage parallel to one direction is mentioned for brannerite from microcline-nepheline veins cutting syenitic rocks, USSR [9].

Natural brannerite is metamict and, therefore, **crystal structure** data can be obtained only after heating the sample [10]; refer also to pp. 101/2. X-ray and electron diffraction data indicate that natural brannerite contains an insignificant part of particles up to 5×10^{-7} cm in size, the majority is dispersed in nearly molecular state. Heating, especially between 710 and 840°C, increases the particle size and generates a diffraction pattern [11]. But the intensity of its diffraction lines may vary considerably due to variations in chemical composition [12].

Synthetic brannerite, UTi_2O_6, has a symmetry and unit-cell size compatible with the morphological form development of metamict brannerite (as is recorded by [13, p. 110]) [5], and also with X-ray powder diffraction data of thermally recrystallized material [14]. For compilations of powder diffraction data, partly including a discussion or taken from the literature, of heated and/or unheated material, see, e.g., [13, pp. 113/7], [11, 15 to 17]. For d values viewed to be indicative for brannerite (corresponding to analysis No. 13 of Table 5, p. 93) and "Iodochnikite" (corresponding to analysis No. 42 of Table 5, p. 95), see [18]. Indexed X-ray powder data for synthetic brannerite, see [19, pp. 273, 276]; and for indexed neutron diffraction data for natural brannerite, see [20].

Synthetic, stoichiometric brannerite is isostructural with thorutite, see pp.79/80, and is monoclinic, space group C2/m (No. 12), with a = 9.8123, b = 3.7697, c = 6.9253 Å, β = 118.957(6)°. The coordination of U and Ti by oxygen is distorted octahedral. There are two bond distances of 2.252 and four of 2.296 Å for U; an additional pair of short nonbonded U-O distances has 2.824 Å. The Ti-O distances are between 1.854 and 2.104 Å, and are within about one standard deviation the same as for thorutite. For details and for figures of the coordination polyhedra, see original paper [19, pp. 271/5].

As can be seen from a diagram composition vs. lattice parameters given for synthetic $(U^{4+},Th)Ti_2O_6$ in [21], with increasing content of Th the parameters b and c increase (from 3.750 to 3.818 Å and from 6.890 to 6.988 Å, respectively), whereas the parameter a and the angle β are only slightly diminished (from 9.78 to 9.76 Å and from 118°40′ to 118°30′, respectively).

References for 2.2.1.2.2.3:

[1] George, D'Arcy (RMO-563 [1949] 1/198, 78/9; N.S.A. **5** [1971] No. 833).

[2] Bianconi, F.; Simonetti, A. (Schweiz. Mineral. Petrogr. Mitt. **47** [1967] 887/934).

[3] Il'inskii, G. A. (Vestn. Leningr. Univ. **1961** No. 6, pp. 134/8).

[4] Povilaitis, M. M.; Yakovlevskaya, T. A.; Knyazeva, I. D.; Belyaeva, I. D. (Zap. Vses. Mineral. Obshch. **97** [1968] 150/61).

[5] Patchett, J. E.; Nuffield, E. W. (Can. Mineralogist **6** [1957/61] 483/9, 485, 487).

[6] X-Ray Laboratory, Peking Inst. Uranium Geol. and X-Ray Labor., Wuhan Geol. College (Acta Geol. Sinica **1978** 241/51 from Am. Mineralogist **64** [1979] 656).

[7] Chernikov, A. A. (Sov. Geol. No. 4 [1981] 85/98, 88).

[8] Arribas, A. (Bol. Real. Soc. Espan. Hist. Nat. Secc. Geol. **65** [1967] 157/70, 163/4).

[9] Gotman, Ya. D. (Zap. Vses. Mineral. Obshch. **87** [1958] 197/200).

[10] Chukhrov, F. V.; Bonshtedt-Kupletskaya, E. M. (Mineraly Spravochnik, Vol. 2, Pt. 3, Nauka, Moscow 1967, pp. 1/676, 441).

[11] Povilaitis, M. M. (Tr. Mineral. Muzeya Akad. Nauk SSSR No. 12 [1961] 64/79, 72/7).

[12] Povilaitis, M. M. (Zap. Vses. Mineral. Obshch. **92** [1963] 113/23, 113).

[13] Pabst, A. (Am. Mineralogist **39** [1954] 109/17).

[14] Kaiman, S. (Can. Mineralogist **6** [1957/61] 389/90).

[15] Zavarzin, A. V. (Geol. Rudn. Mestorozhd. **1961** No. 6, pp. 54/8).

[16] Lis, J.; Stępniewski, M.; Sylwestrzak, H. (Biol. Inst. Geol. No. 193 [1965] 207/28, 215/8).

[17] Feather, C. E. (Geol. Surv. Profess. Paper [U.S.] No. 1161-A-BB [1981] Q1/Q23, Q17/Q18).

[18] Sidorenko, G. A. (Rentgenograficheskii Opredelitel' Uranovykh i Uransoderzhashchikh Mineralov, Gosgeoltekhizdat, Moscow 1960, pp. 1/116, 84/5).

[19] Szymanski, J. T.; Scott, J. D. (Can. Mineralogist **20** [1982] 271/9).

[20] Ivanova, O. A.; Mel'nikov, I. V.; Gorshkov, A. I.; Vampilov, M. V. (Izv. Akad. Nauk SSSR Ser. Geol. **1982** No. 2, pp. 63/71, 70).

[21] Sidorenko, G. A. (Kristallokhimiya Mineralov Urana, Atomizdat, Moscow 1978, pp. 1/217, 38/9).

2.2.1.2.2.4 Optical and Other Physical Properties. Chemical and Thermal Behavior

In addition to the data for **optical properties** (such as luster, color, streak, refractive index, reflectivity) of brannerites, partly containing Th, given in "Rare Earth Elements" A7, 1984, pp. 54/6, there are the following statements:

"Lodochnikite" from microcline-nepheline veins cutting syenitic rocks, USSR, is black with fatty luster, translucent in very thin fragments with brown color. Its streak is brown-black; reflection color gray with a reflectivity of 19%. Refractive index is 2.16, after calcination 2.19 [1].

Orthobrannerite from alkaline rocks, People's Republic of China, is black with adamantine luster; streak is brown to black. It is grayish white under reflected light with reddish brown internal reflections, and has a reflectance of 15 to 17%. Refractive index is 2.328. It is non-fluorescent [2].

The IR spectrum of brannerite shows strong absorption at 548, 384, and 258 cm^{-1}; weaker absorption occurs at 210, 122, and 88 cm^{-1} [3]. Orthobrannerite has a strong broad band at 546 cm^{-1} and several weak H_2O and OH bands [2].

In addition to the data for **other physical properties** (such as hardness, microhardness, fracture, density, and magnetism) of brannerites, partly containing Th, given in "Rare Earth Elements" A7, 1984, pp. 56/8, there are the following statements:

Microhardness for fully metamict brannerite is between 515 and 532 kg/mm^2 (load 100 to 250 g); after heating to 1000°C it rises to 542 to 546 kg/mm^2 (load 50 to 100 g) [2].

Density varies between 4.0 and 5.9 g/cm^3 as can be seen in Table 5, pp. 92/6.

Due to its radioactivity, around brannerite inclusions in vein-like quartz-microcline bodies in leucogranite of an undisclosed Mo-W occurrence, USSR, the neighboring minerals become discolored: quartz turns to dark smoky; pale rose fluorite to black violet; and microcline to brownish red [4].

Natural brannerite is metamict, see, for example, [5, 6] and thus gives no, or only weak, X-ray diffraction data, which become distinct after heating, see, for example, [7, 8].

In addition to the data for **chemical and thermal behavior** of brannerites, partly containing Th, given in "Rare Earth Elements" A7, 1984, p. 58, there are the following statements:

Brannerite from Åmot, Modum district, Norway, is soluble in both HCl and HNO_3 [9]. Orthobrannerite from the People's Republic of China is not attacked by HCl and H_2SO_4, but dissolves in warm H_2SO_4 and H_3PO_4 [2]. Oxidative leaching of brannerite-type minerals at 40 and 60°C in a solution containing 10 g H_2SO_4 and 3 g $Fe_2(SO_4)_3$ per liter starts at 40°C with a low rate, leaving TiO_2 and $PbSO_4$ (anglesite) as a residue; the latter originates from galena inclusions [10]. The rate of dissolution at 90°C of an ideal free crystal of brannerite suspended in an alkaline carbonate solution (containing 20 g $NaHCO_3$, 35 g Na_2CO_3, and 40 g Na_2SO_4 per liter) is calculated to be 1.33 µm^3/s for a leaching time between 2 and 48 h. A free crystal of maximum dimension 89 µm will be completely dissolved in 48 h [11]. For the use of hydrofluoric acid as a diagnostic (a 50% acid solution dissolves brannerite within 2 to 3 s) and structural etching medium (to reveal, e.g., zonation, twinning, cleavage), see [12].

Caused by its chemical composition (nearly all U occurs as UO_2; water is not detectable), the heating curve of brannerite from Lodrino, Tessin, Switzerland, displays only one exothermic peak at 676°C; the absence of endothermic peaks is due to missing water. The estimated low content of 1% H_2O (see analysis No. 33 of Table 5, p. 94) is caused by entrance of water (adsorption) during the grinding of the sample [13]. Orthobrannerite shows two weak endothermic peaks at 100° and 340°C and a strong peak at 620°C (due to its recrystallization) [2]. For a compilation of heating curves for other brannerite samples, see [14] or [15].

Whether the heating process merely returns the brannerite from its metamict state, or whether it actually synthesizes it, is debatable. Synthesis of brannerite can be done only in inert atmospheres at temperatures $>2000°C$ [16]. Whereas heating in vacuum of metamict brannerite from Mo-W ores of an undisclosed locality, USSR, leads to recrystallization of the (primary?) brannerite structure between 600 and 900°C. Heating in air produces, besides brannerite, the phase U_3O_8 at 800°C and, in addition, phase X (corresponding to $U^{6+}Ti_2O_7$, orthorhombic) at 900°C. Other specimens, devoid of Th, furthermore give rutile. It seems that Th can stabilize the brannerite structure [17, 18]. Orthobrannerite seems to have regained its original crystal structure after heat treatment [2].

References for 2.2.1.2.2.4:

[1] Gotman, Ya. D. (Zap. Vses. Mineral. Obshch. **87** [1958] 197/200).
[2] X-Ray Laboratory, Peking Inst. Uranium Geol. and X-Ray Labor., Wuhan Geol. College (Acta Geol. Sinica **1978** 241/51 from Am. Mineralogist **64** [1979] 656).
[3] Povarennykh, A. S. (Konst. Svoistva Mineral. No. 13 [1979] 53/78, 75/6).
[4] Povilaitis, M. M. (Tr. Mineral. Muzeya Akad. Nauk SSSR No. 12 [1961] 64/79, 65).
[5] Chukhrov, F. V.; Bonshtedt-Kupletskaya, E. M. (Mineraly Spravochnik, Vol. 2, Pt. 3, Nauka, Moscow 1967, pp. 1/676, 433).
[6] Palache, C.; Berman, H.; Frondel, C. (Dana's System of Mineralogy, 7th Ed., Vol. 1, Wiley, New York 1946, pp. 1/834, 774).
[7] Pabst, A. (Am. Mineralogist **39** [1954] 109/17).
[8] Patchett, J. E.; Nuffield, E. W. (Can. Mineralogist **6** [1957/61] 483/9).
[9] Van Autenboer, T.; Skjerlie, F. J. (Norges Geol. Unders. No. 200 [1957] 5/7).
[10] Smits, G. (Appl. Mineral. Proc. 2nd Intern. Congr. Appl. Mineral. Minerals Ind., Los Angeles 1984 [1985], pp. 599/616 from C.A. **103** [1985] No. 108167).

[11] Szymanski, J. T.; Scott, J. D. (Can. Mineralogist **20** [1982] 271/9, 275/6, 278).
[12] Sergeev, V. N. (Izv. Sibirsk. Otd. Akad. Nauk SSSR Geol. Geoliz. No. 1 [1958] 103/5).
[13] Bianconi, F.; Simonetti, A. (Schweiz. Mineral. Petrogr. Mitt. **47** [1967] 887/934, 902/4).
[14] Adler, H. A.; Puig, J. A. (Am. Mineralogist **46** [1961] 1086/96, 1091, 1093).
[15] Povilaitis, M. M.; Yakovlevskaya, T. A.; Knyazeva, D. N.; Belyaeva, I. D. (Zap. Vses. Mineral. Obshch. **97** [1968] 150/61, 158).
[16] Robertson, D. S.; Steenland, N. C. (Econ. Geol. **55** [1960] 659/94, 680).
[17] Krivokoneva, G. K. (Dokl. Akad. Nauk SSSR **194** [1970] 1168/71; Dokl. Acad. Sci. USSR Earth Sci. Sect. **194** [1970] 137/40).
[18] Krivokoneva, G. K. (Zap. Vses. Mineral. Obshch. **101** [1972] 554/67).

2.2.1.3 Zirkelite $(Ca, Th, Ce)Zr(Ti, Nb)_2O_7$

There occurs much confusion in the literature regarding composition, formula, and crystal structure of zirkelite and zirconolite, see, for instance, discussion on nomenclature in [1, 2]. Following Pudovkina et al. [3] and Hogarth [4], both mineral names are synonymous. But crystal structure and chemical investigations by Mazzi and Munno [2] showed that zirkelite and zirconolite are polymorphs identical in chemical formula (refer to pp. 109/10) and have similar crystal structures; nevertheless they differ in the stacking of identical pairs of layers of polyhedra: zirkelite is trigonal and zirconolite is monoclinic (refer also to [5] and pp. 124/6). But a listing of zirconolite with the pyrochlore-betafite-microlite series by [6] is excluded by its monoclinic symmetry [4]. A third polymorph, polymignite, is orthorhombic and (according to chemical principles suitable for a subdivision of Chapter 2.2, refer to pp. 75/6) is described on pp. 232/4.

Zirkelite (of composition $(Ca,Fe,Th,U)_2(Ti,Zr)_2O_5$ [7]) from a heavy-mineral concentrate obtained from weathered jacupirangite, São Paulo, Brazil, was described in 1895 and is named after Ferdinand Zirkel (1838−1912), formerly Professor of Mineralogy, University of Leipzig, Germany [8]. A more recent reinvestigation of the Brazilian zirkelite gives the formula $(Ca,Th,RE)_{1.1}Zr_{1.0}(Ti,Nb,Fe)_{1.9}O_{7.0}$ [3]. Another zirkelite (of composition $(Ca,Ce,Y,Fe)(Ti,Zr,Th)_3O_7$ [7]) from placers (gem gravels) of Ceylon was described in 1913 [9]; refer also to [10].

Zirconolite (containing only 0.58% ThO_2) from a nepheline-pyroxene rock formed metasomatically in the exterior zone of a pyroxenite massif (= Afrikanda, Kola Peninsula [11]), USSR, was described in 1956 and, because of differences in composition as compared with zirkelite, is named in analogy to other Zr-containing minerals (as, e.g., zircon) [12]. But note that a more recent investigation showed zirconolite to be the Nb-free variety of zirkelite [3].

Structural information for the above minerals is frequently difficult or impossible to obtain because of small grain size and metamict nature of investigated material [13], and most samples have not been investigated by X-ray analysis [14]. Therefore, and as problems over nomenclature have not been fully resolved [13], in the following sections and paragraphs the mineral name is given as used in the relevant paper.

References for 2.2.1.3:

[1] Busche, F. D.; Prinz, M.; Keil, K.; Kurat, G. (Earth Planet. Sci. Letters **14** [1972] 313/21, 313/5).

[2] Mazzi, F.; Munno, R. (Am. Mineralogist **68** [1983] 262/76).

[3] Pudovkina, Z. V.; Dubakina, L. S.; Lebedeva, S. I.; Pyatenko, Yu. A. (Zap. Vses. Mineral. Obshch. **103** [1974] 368/72).

[4] Hogarth, D. D. (Am. Mineralogist **62** [1977] 403/10).

[5] White, T. J. (Am. Mineralogist **69** [1984] 1156/72).

[6] Parker, R. L.; Fleischer, M. (Geol. Surv. Profess. Paper [U.S.] No. 612 [1968] 1/43, 31).

[7] Strunz, H.; Tennyson, C. (Mineralogische Tabellen, 6th Ed., Geest & Portig, Leipzig 1977, pp. 1/621, 189).

[8] Hussak, E.; Prior, G. T. (Mineral. Mag. **11** [1895] 80/8).

[9] Blake, G. S.; Smith, G. F. (Mineral. Mag. **16** [1913] 309/16).

[10] Palache, C.; Berman, H.; Frondel, C. (Dana's System of Mineralogy, 7th Ed., Vol. 1, Wiley, New York 1946, pp. 1/834, 740).

[11] Bulakh, A. G.; Il'inskii, G. A.; Kukharenko, A. A. (Zap. Vses. Mineral. Obshch. **89** [1960] 261/73, 262).

[12] Borodin, L. S.; Nazarenko, I. I.; Rikhter, T. L. (Dokl. Akad. Nauk SSSR **110** [1956] 845/8).

[13] Fowler, M. B.; Williams, C. T. (Mineral. Mag. **50** [1986] 326/8, 326).

[14] Gieré, R. (Contrib. Mineral. Petrol. [Berlin] **93** [1986] 459/70, 459/60).

2.2.1.3.1 Occurrence, Paragenesis, Intergrowths, Inclusions, and Alteration

Zirkelite is a peculiar mineral characterized by an extensive isomorphous replacement related to conditions of formation (see pp. 110/4) and its **occurrence and paragenesis** was reported from a wide variety of terrestrial rocks as well as from lunar samples, where it is a common accessory phase [1, pp. 459/60]. In view of its relatively common occurrence in late-stage mesostasis areas of lunar basalts, it is probable, however, that terrestrial zirconolite is not as uncommon as it seems. It is likely to exist also in similar environments in terrestrial basic intrusive and extrusive rocks, but because of its small grain size, was invariably overlooked [2]. Up to now, terrestrial zirkelite is found exclusively in alkalic-ultrabasic rocks and carbonatites, being typical for the latter [3, p. 109]. Or it occurs especially (1) in rocks of the alkalic-ultrabasic formation and associated carbonatites, where it is formed in the postmagmatic phase during autometasomatic alteration of the rocks; (2) in the exocontact of nepheline syenite massifs, within metasomatically altered basic effusive rocks and granite gneiss, where it is connected with pegmatoid veinlets and veins; and (3) in placer deposits. Lunar zirkelite (mainly poor in Th, see p. 111) is scattered as tiny inclusions (up to 10 µm) in the interstice of mare basalt minerals (e. g., samples from Apollo 11, 12, 15, and 17 missions), or associatated with lately-formed minerals in the groundmass and in fragments of breccias (e. g., Luna 20 and Apollo 14). Despite a wide variation in geological conditions, formation of zirkelite is characterized as a crystallization during last stages of magmatic or metasomatic processes [4]; refer also to the examples given in the following paragraphs.

Dark reddish brown anhedral zirconolite grains (between 2 and 60, average 15 µm in size) occur together with baddeleyite and apatite, mutually forming intergrowths, and with hydrothermal minerals (as chlorite, uralite, biotite, kaersutite) in mesostasis (groundmass) areas of ultrabasic cumulates of the Rhum layered pluton, Inner Hebrides, Scotland, Great Britain. This type of occurrence is comparable to lunar mare basalt. In spite of the low total number of grains recorded, the relative proportion of 5:4:3 for zirconolite, baddeleyite, and zircon, respectively, is considered to be a relatively accurate reflection of their abundance within the Rhum intrusion [2]. For one of the Rhum zirconolite grains 0.48% ThO_2 is determined [5].

Sharp, prismatic crystals of zirkelite, 1 to 5 mm in size, are attached to phlogopite on the wall of a small pocket in coarse and vuggy jacupirangite ($\sim$ nepheline gabbro) of the Jacupirango mine, São Paulo, Brazil [6].

Zirconolite, together with pyrochlore, thorianite, and fluorocarbonates, phosphates, and silicates of RE, occurs in zones of metasomatic-hydrothermal alteration in nepheline and microcline syenites, Ełk massif, northeastern Poland [7]. Grains of compositionally zoned (refer to pp. 112/3), partially metamict zirconolite, varying in size from <30 up to 300 µm, occur in fine-grained nepheline syenite of the Chilwa Alkaline Province, Malawi. The host syenite has an unusual composition. It contains neither amphibole nor pyroxene, it does contain subhedral alkali feldspar, ragged patches of turbid nepheline, which is sometimes included in the feldspar and partly altered to sericite, and abundant brown biotite; an opaque oxide phase is plentiful, as is apatite [8]. Deep orange-brown (in thin section) zirconolite forms small (< 10 µm) euhedral inclusions in alkali feldspar phenocrysts of syenite, Glen Dessary, Inverness-shire, Scotland, Great Britain. Other associated minerals include Fe-Ti oxides, titanite, allanite, apatite, and zircon [5].

Small zirconolite grains (30 to 100 µm) with contents of 2.0 to 8.38% Th and 0.59 to 2.27% U occur in rocks of the Jimberlana mafic complex, Western Australia [9].

Zirkelite is a typical accessory mineral of alkalic-ultrabasic rocks of the Kola Peninsula, northwestern European USSR. It is formed as an early crystallization product of the postmagmatic stage or, sometimes, metasomatically in intensely altered parts of the ultrabasites. Thus, zirkelite especially is found: (1) In calcite-amphibole-diopside rock of the Afrikanda massif, where it occurs erratically in druses or voids in close association with titanite, rarely with perovskite and calcite. Zirkelite forms irregular segregations up to 5 mm or more in size, but sometimes occurs as poorly developed platy and octahedral crystals and its intergrowths. (2) In apatite-forsterite-magnetite rock and other rocks of the Vuoriyarvi and Kovdor massifs, where zirkelite occurs locally as small (up to 1 mm) hexagonal platelets or crystal intergrowths contained within calcite and apatite. It is associated with baddeleyite, zircon, pyrochlore, and some other minerals. (3) In dolomite-phlogopite rock of the Sebl'yavr massif, where zirkelite occurs as a rare impregnation in the rock, or in the part enriched in apatite, closely associated with phlogopite, titanomagnetite, apatite, and baddeleyite. Zirkelite forms well-developed platy crystals and crystal intergrowths, predominantly in the interstices between apatite grains and aggregates, often intersecting one or the other apatite grain, indicating a very late formation. And (4) in calcite and calcite-dolomite carbonatites of the Kovdor, Vuoriyarvi, and Sebl'yavr massifs, where zirkelite occurs more seldom than in the foregoing types (1) to (3). For instance, at Sebl'yavr, platy crystals and its intergrowths of zirkelite, associated with pyrochlore, always are found in crushed calcite-dolomite carbonatite. Megascopic zirkelite is found with relicts of dolomite-phlogopite rock, but also with chloritized and carbonatized olivinites, enclosed in the carbonatite veins. At Kovdor (Vatsu-Vaara region), zirkelite in close association with pyrochlore, partly forming mutually intergrowths, is found in the calcite carbonatite [10, pp. 261/3], see also [11]. Zirkelite of the Vuoriyarvi massif occurs in higher amounts in altered carbonatite, containing richterite and tetraferriphlogopite, and in richteritized carbonatite. In the latter rock, zirkelite is about 15% of the summary content of zirkelite, pyrrhotite, tetraferriphlogopite, zircon, pyrochlor, and richterite [27]. Metamict pseudocubic niobozirconolite, composed of an aggregate of tabular crystals, occurs in apatite-magnetite rock of the Vuoriyarvi carbonatite [12]. Anisotropic (metamict) and isotropic varieties of zirkelite occur, together with hatchettolite, pyrochlore-I, and baddeleyite-II, as a characteristic accessory mineral in complex apatite-magnetite rocks (ores) formed metasomatically during a late postmagmatic stage of an undisclosed complex of alkalic-ultrabasic rocks and carbonatites, Baltic Shield, northwestern European USSR. The zirkelite has a widely varying composition (see p. 111) and forms small trigonal platelets of 0.3 to 2 mm in size, simple and stellate intergrowths of platy crystals, and dark brown or black grains [13, pp. 147/9, 152/3]; for the intergrowths and inclusions, see pp. 107/8.

Note further that metamict zirkelite is a widespread accessory mineral of the carbonatites from the southern part of an undisclosed massif of alkalic-ultrabasic rocks of the Kola Peninsula, where it occurs as three morphologically different varieties (platy, cubic, and cubo-octahedric) with different chemistry (see p. 114) and optical properties (see below). The three varieties never have been found together in one sample, and in cases with two varieties, always one predominates [14, pp. 29, 31/2]: (1) Platy zirkelite, partly forming star-shaped or rosette-like figures, is dark red, nearly nontransparent, has a refractive index of 2.20, and occurs as scattered inclusion; or is attached to diopside-rich parts of phlogopite-diopside-calcite carbonatite. The zirkelite is found in the interstices between diopside crystals, or is enclosed in calcite. A part of its small crystals adheres to corroded baddeleyite grains and, sometimes, to calzirtite and altered minerals of the pyrochlore group. An interrelationship with dysanalyte or zircon is not observed. Platy zirkelite is also found in only crushed samples of biotite-calcite carbonatite (together with biotite, magnetite, apatite, calzirtite, and sulfides) sporadically in forsterite-calcite carbonatite, and in druses in dolomite-calcite carbonatite. In these rocks, it occurs as an aggregation of tiny crystals adhered to grains, or is oriented with distinct directions in crystals, of strongly altered minerals of the pyrochlore group. (2) Cubic zirkelite is blackish to dark brown; strongly altered varieties are light brown. It has a refractive index of 2.16 to 2.02, and is found in the schistose body of phlogopite-diopside-calcite carbonatite containing single relicts of aegirine-diopside-calcite carbonatite. The greatest concentration of zirkelite is with the apatite-rich parts of the carbonatite, where it occurs in the interstices between apatite grains, at times enclosing the latter. Distribution of color for zirkelite is irregular and patchy. Sometimes parts of the mineral are anomalously birefringent. (3) Cubo-octahedric crystals, and isometric grains, of zirkelite are black, dark brown with a reddish hue, have a refractive index of 2.13 to 2.09, and are found in crushed samples from the schistose and actinolitized part of phlogopite-calcite carbonatite. Associated minerals are phlogopite, magnetite, apatite, actinolite, altered pyrochlor, and diopside. It often encloses apatite.

Dark brown to pitch-black platy crystals of Nb-zirconolite up to 1 mm in size, sometimes forming polysynthetic twins of star-shaped, rosette-like, or pseudocubic aggregates, occur, especially, in biotite-rich parts of amphibole-containing aegirine-biotite-calcite carbonatite of Cape Turii, Kola Peninsula, northwestern European USSR. Occasionally, Nb-zirconolite is found in apatite-magnetite rock and in associated tetraferriphlogopite carbonatite. In the former, it is accompanied by apatite, magnetite, zircon, and pyrochlor; in the latter by apatite, magnetite, ilmenite, hatchettolite, and zircon [15]. Note also that zirconolite, as a characteristic rare-metal mineral, is observed in 2 out of 13 investigated non-altered phlogopite-calcite carbonatites of the Turii Peninsula [16].

Largely metamict zirkelite crystals of different shape (see pp. 121/2) are widespread in forsterite-calcite and diopside-calcite carbonatites of stage II from an undisclosed locality, Eastern Siberia, USSR. In the latter carbonatite type, zirkelite is concentrated in the interstices between individual crystals of diopside schlieren, partly adhered to diopside. And in the forsterite-calcite carbonatite, its distribution is subordinated to the general lamination, and its content becomes 10 to 15%. In pure calcitic zones, zirkelite has suffered recrystallization to (polycrystalline) "block crystals", ranging in size between 0.5 and 0.8 cm. Crystalline zirkelite of the diopside-calcite carbonatite is closely associated with red phlogopite, diopside, and pyrochlor [17, p. 101]; refer also to [3, pp. 109, 112]. Based on analytical data for zirkelite (see pp. 111/2) and associated baddeleyite, dysanalyte-II, and hatchettolite from carbonatites, Eastern Siberia, a trigonal Zr-Ti-(Nb + Ta) diagram is given, which displays nine fields of different stable mineral parageneses. Calculated from the at% content, the (Nb + Ta):(Zr + Ti) ratio is a suitable measure to discriminate between the stable parageneses. For the detailed discussion, also including change of paragenesis in relation to the different source rocks for single carbonate types, see original paper [3, pp. 113/7].

Non-metamict Nb-rich zirconolite, as isolated euhedral crystals and/or aggregates of parallel crystals, occurs as an accessory mineral locally in the main body of magnesian biotite-apatite sövite, in some dikes of pyroxene-apatite-magnetite-magnesian biotite sövite and magnesian biotite-apatite beforsite, as well as in small concentrates in dikes of calcite-amphibole-pyroxene-apatite glimmerite from the Canafistula carbonatitic plug near the mouth of "Ribeira da Barka", Santiago Island, Cape Verde Republic. Zirconolite with average dimensions between 0.1 and 0.2 mm, and grain sizes often up to 1 and 2 mm, seems to be one of the earlier minerals to crystallize within the host rocks [18].

Accessory rounded grains of translucent reddish brown anisotropic zirconolite with a maximum size of ~0.2 mm (the smaller ones are opaque) occur together with titanite, pyrite, and chalcopyrite as well in calcite lenses as in the dark groundmass (consisting of anorthite + phlogopite + relict calcite), both of which are generated during a metasomatic alteration of spinel-calcite marble (skarn) by the Bergell calc-alkaline granite intrusion, Switzerland-Italy [1, pp. 461, 464, 467].

In several samples of metacarbonate rocks of the Oetztal-Stubai region, Tyrol, Austria, anhedral grains of zirconolite occur in association (direct contact) with baddeleyite, chlorite ± ilmenite ± apatite, phlogopite + clinopyroxene, ilmenite + calcite ± phlogopite, and ilmenite + clinopyroxene + phlogopite ± spinel ± chlorite ± apatite ± olivine ± calcite ± dolomite. Additionally, but not in direct contact, all samples contain titanite and/or rutile. Zirconolite may have been formed by a metamorphic reaction involving baddeleyite + calcite + a Ti-bearing phase (rutile, ilmenite, titanite), but the complete process seems to be more complicated [19].

Tiny grains (5 to 20 µm) of possibly zirkelite were found (as a local curiosity) in biotite and in fissures of red porphyroblastic microcline-bearing biotite granodiorite, Erdömecske, Mecsek Mountains, Hungary [20].

Zirconolite is a subordinate component of alluvial/deluvial pyrochlore-rich placer deposits derived from carbonatites [21].

Partly occurring as an oriented (epitactic) **intergrowth** with pyrochlore, zirkelite is found in calcite carbonatite, Kovdor massif (Votsu-Vaara region), Kola Peninsula, northwestern European USSR. Intergrowth with baddeleyite is not observed [10, p. 263], whereas zirkelite of carbonatites, Eastern Siberia, adheres, partly as small brushes of platy crystals, to grains of baddeleyite and calzirtite [17, p. 104]. Platy zirkelite is intergrown with calzirtite in samples from phlogopite-diopside-calcite carbonatite, Kola Peninsula [14, p. 31]; and platy zirkelite crystals, filled with minute grains of apatite and forsterite, are reported for apatite-magnetite rock (ore), Baltic Shield, northwestern European USSR [13, pp. 152/3]. Partly overgrown with hatchettolite crystals are zirkelite lamellas of metasomatic apatite-forsterite/tetraferri-phlogopite-magnetite rocks and of tetraferriphlogopite-calcite carbonatite of the Sebl'yavr massif, Kola Peninsula [22]. Apparently intergrown with Mn-rich ilmenite is a single zirconolite grain of an olivine crescumulate from the Rhum layered ultrabasic pluton, Inner Hebrides, Scotland [23]; for intergrowths with baddeleyite and apatite from Rhum, see p. 104.

Due to overlapping time of formation, there is a difference in type of intergrowth between zirkelite and pyrochlore-group minerals of carbonatites, Eastern Siberia. Early formed platelets of zirkelite lie, partly, with their basis plane close to octahedral pyrochlore grains. Rosette-like polysynthetic zirkelite contains pyrochlore within the re-entring angles, in such way that the latter would heal (continue to form) pseudooctahedral skelets of the former. Very late-formed pyrochlore, in the beginning, heals twin sutures and re-entring angles of simple twins of zirkelite, and later on, the zirkelite platelets are overgrown by small octahedral crystals of pyrochlore [17, p. 104].

Platy crystals with roughly hexagonal outline of sanidine appear as an oriented intergrowth of three crystalline phases of the sanidinite, Campi Flegrei, Campania, Italy; namely with subordinate amounts of pyrochlore and of a monoclinic zirconolite phase and, dominantly, with a trigonal phase showing an axial ratio a:c = 1:2.317, fairly similar to zirkelite from Ceylon [24].

For an interpretation of zirconolite-zirkelite intergrowths, as well as other mimetically twinned intergrowths, in terms of a modular structure arrangement, see [25].

Inclusions. Usually enclosed in apatite or forsterite are the grains and crystals of zirkelite of apatite-magnetite rocks (ores), Baltic Shield, northwestern European USSR. Inclusions of zirkelite in mica platelets and in hatchettolite are frequently surrounded by pleochroic halos, whereas in apatite one finds characteristic radioactive cracking [13, pp. 152/4]. For zirkelite inclusions in calcite and apatite of alkalic-ultrabasic rocks, Kola Peninsula, see p. 105. Inclusions of zirconolite crystals in phlogopite, which is partially replaced by chlorite, are observed in metasomatically altered spinel-calcite marble (skarn) from the contact aureole of the Bergell granite intrusion, Switzerland-Italy. These zirconolites are always surrounded by a broad pleochroic halo owing to its high Th and U concentration [1, pp. 464, 466]. For zirconolite inclusion in alkali feldspar of the Glen Dessary syenite, Scotland, see p. 104.

Inclusions of apatite occur in the cubic and cubo-octahedric varieties of zirkelite from phlogopite-diopside-calcite and phlogopite-calcite carbonatites, respectively, of an undisclosed locality, Kola Peninsula [14, pp. 31/2]. Microinclusions of baddeleyite and other nondefinable minerals in zirkelite from the locus typicus, Brazil, account for shortcomings of the original chemical analysis (see analysis No. 9 of Table 6, p. 115) [26].

Alteration. Replacement of zirkelite by perovskite (knopite), besides overgrowth of zirkelite by titanite, is observed for samples from calcite-amphibole-diopside rock, Afrikanda massif, Kola Peninsula, northwestern European USSR [10, p. 262]. Note also that metamict zirkelite of alkalic-ultrabasic rocks, Kola Peninsula, displays selvages composed of white and brownish red substances [14, p. 33]. For changes in composition during alteration of zirkelite, see pp. 113/4.

References for 2.2.1.3.1:

[1] Gieré, R. (Contrib. Mineral. Petrol. [Berlin] **93** [1986] 459/70).
[2] Williams, C. T. (Contrib. Mineral. Petrol. [Berlin] **66** [1978] 29/39, 33/4).
[3] Pozharitskaya, L. K.; Pavlinskii, G. V.; Razvozzhaeva, E. A.; Samoilov, V. S. (in: Shmakin, B. M., Osobennosti Petrologii, Mineralogii i Geokhimii Karbonatitov Vostochnoi Sibiri, Nauka, Moscow 1966, pp. 109/20).
[4] Kochemasov, G. G. (Mineral Zh. [L'vov] **2** No. 6 [1980] 30/9).
[5] Fowler, M. B.; Williams, C. T. (Mineral. Mag. **50** [1986] 326/8).
[6] Menezes, L. A. D., Jr.; Martins, J. M. (Mineral. Record **15** [1984] 261/70, 267, 269).
[7] Kubicki, S. (Biul. Inst. Geol. No. 347 [1984] 49/54, 53).
[8] Platt, R. G.; Wall, F.; Williams, C. T.; Wolley, A. R. (Mineral. Mag. **51** [1987] 253/63, 254/5).
[9] Ringwood, A. E.; Oversby, V. M.; Sinclair, W. (Sci. Basis Nucl. Waste Management **2** [1980] 273/80, 278; C.A. **94** [1981] No. 213978).
[10] Bulakh, A. G.; Il'inskii, G. A.; Kukharenko, A. A. (Zap. Vses. Mineral. Obshch. **89** [1960] 261/73).
[11] Kukharenko, A. A.; Orlova, M. P.; Bulakh, A. G.; Bagdasarov, E. A.; Rimskaya-Korsakova, O. M.; Nefedov, E. I.; Il'inskii, G. A.; Sergeev, A. S.; Abakumova, N. B. (Kaledonskii Kompleks Ul'traosnovnykh, Shchelochnykh Porod i Karbonatitov Kol'skogo Poluostrova i Severnoi Karelii, Nedra, Moscow 1965, pp. 1/772, 321/2).
[12] Borodin, L. S.; Bykova, A. V.; Kapitonova, T. A.; Pyatenko, Yu. A. (Dokl. Akad. Nauk SSSR **134** [1960] 1188/91; Dokl. Acad. Sci. USSR Earth Sci. Sect. **130/135** [1960] 1022/4).
[13] Zhuravleva, L. N.; Berezina, L. A.; Gulin, E. A. (Geokhimiya **1976** 1512/30; Geochem. Intern. **13** No. 5 [1976] 147/66).
[14] Osokin, A. S. (Mineral. Geokhim. **1979** No. 6, pp. 27/38).
[15] Lapin, A. V. (Mineralog. Issled. **3** [1973] 108/21, 109/10, 117).
[16] Anastasenko, G. F.; Bulakh, A. G.; Vaganov, P. A.; Bulnaev, A. I. (in: Tatarinov, P. M., Mineraly i Paragenezisy Mineralov, Nauka, Leningrad 1978, pp. 38/46, 39).
[17] Zdorik, T. B. (Tr. Mineral. Muzeya Akad. Nauk SSSR No. 20 [1971] 96/106).

[18] Silva, S. C.; Figueiredo, M. D. (Garcia de Orta Ser. Geol. **4** [1980] 1/6, 2/3).

[19] Purtscheller, F.; Tessadri, R. (Mineral. Mag. **49** [1985] 523/9, 524, 528).

[20] Pantó, G. (Acta Geol. Acad. Sci. Hung. **19** [1975] 59/93, 62, 69).

[21] Ginzburg, A. I. (Geol. Mestorozhd. Redk. Elem. No. 14 [1961] 26/86, 76).

[22] Subbotin, V. V.; Kirnarskii, Yu. M.; Kurbatova, G. S.; Strel'nikova, L. A.; Subbotina, G. F. (in: Ivanova, T. N., Petrologiya i Minerageniya Shchelochnykh, Shchelochno-Ul'tra-osnovnykh i Karbonatitovykh Kompleksov Karelo-Kol'skogo Regiona, Akad. Nauk SSSR, Kol'sk. Filial, Apatity 1985, pp. 61/9, 65).

[23] Williams, C. T. (Unpubl. Diss. Univ. London 1976 from [2]).

[24] Mazzi, F.; Munno, R. (Am. Mineralogist **68** [1983] 262/76, 262/3).

[25] White, T. J. (Am. Mineralogist **69** [1984] 1156/72, 1163, 1165/7).

[26] Pudovkina, Z. V.; Dubakina, L. S.; Lebedeva, S. I.; Pyatenko, Yu. A. (Zap. Vses. Mineral. Obshch. **103** [1974] 368/72, 371/2).

[27] Kapustin, Yu. L.; Chernitsova, N. M.; Pudovkina, Z. V. (in: Tatarinov, P. M., Mineraly i Paragenezisy Mineralov Gornykh Porod, Nauka, Leningrad 1973, pp. 17/25, 18).

2.2.1.3.2 Chemistry

The chemical composition of zirkelite varies widely. Besides Nb-rich varieties, there are samples enriched in U, Th, or RE, see Table 6, pp. 115/20, and refer to a trigonal diagram $(Th, U)_2-(Nb, Ta)_2O_5-RE_2O_3$ given in [1, p. 37] and the following paragraphs. For the content (variation and mean) of Y_2O_3 and/or $RE_2O_3 + Y_2O_3$, see "Seltenerdelemente" A4, 1979, pp. 9, 36. The chemical composition is largely dependent on the paragenesis of zirkelite, mainly its associatation with other minerals containing Nb, Ti, and Zr, see pp. 111/2.

Crystallochemically, zirkelite is characterized by a wide variation of isomorphous elements [2, p. 109]; it is one of the few complex minerals which may contain up to 30 chemical elements in concentrations <0.1% [1, p. 30]. The structural behavior of a few elements can be assessed with certain confidence on crystallochemical reasons, but it is not so for Th, U, and Fe^{2+}, which may enter at least two structural positions (refer to the next paragraph and to pp. 124/6) [3]. Experiments investigating the binary systems between $CaZrTi_2O_7$ and various real or hypothetical end-members show that zirconolite solid solutions can accept 27% UO_2, 20% ThO_2, at least 29% RE oxides, 14% Nb_2O_5, 7% Al_2O_3, 4% MgO, 4% FeO, or 5% MnO, but only limited amounts of Na_2O, SrO, and NiO [4]. The content of water in zirkelite/zirconolite (see Table 6, pp. 115/20) seems to be a constitutional part of metamict samples, see pp. 129/30. For an influence of diadochic substitution on the crystal structure ("hexagonality") of zirkelite, see p. 123.

Because there is some dispute over nomenclature (see pp. 102/3), the formulas given for zirkelite/zirconolite in the literature also vary. Thus, zirkelite is viewed as a Ta-bearing zircononiobotitanate $(Ca, U, Th)(Zr, Nb, Ti)_2O_7$ [5, pp. 150/3]. Two possible types of formula are proposed as AB_3O_7 for zirconolite, and as AB_2O_5 for zirkelite [6, p. 33], with the former type considered to more closely represent the observed chemical compositions [7]; refer also to $CaZrTi_2O_7$ assumed as the simplest formula for the polymorphs zirkelite, zirconolite, and polymignite given in [8, p. 459], see also [9], or to $CaZr(Ti, Nb, Fe)_2O_7$ given for zirconolite/zirkelite in [10, pp. 158/62]. The general formula of zirkelite may be given as $AZrB_2O_7$ [11], with group A containing the divalent cations Ca, Fe, and, subordinately, Mg, Mn, Co, etc.; group B contains Ti^{4+} which is replaced isomorphously by Nb^{5+}, Ta^{5+}, Al^{3+}, Cr^{3+}, etc.; and Zr^{4+} may be replaced by U^{4+}, Th^{4+}, or Rb^{2+}. The RE are distributed between the positions of group A and B [12]. Note further, that for the formula type $AZrB_2O_7$, is given $A = Ca$, RE, U, Th, Na, Sr, and Ba; $B = Ti$, Nb, Ta, Fe^{3+} [32, p. 18]. Refer also to other general formulas given as: $(M_I)(M_{II})(M_{III})_2O_7$, where $M_I = Fe$, Ca, and other minor divalent cations; $M_{II} = Zr$ with minor U^{4+}, Th^{4+}, and Pb^{2+}; and $M_{III} = Ti$ with minor Nb^{5+}, Ta^{5+}, Al^{3+}, and Cr^{3+} [6, p. 34], [8, p. 465]. Or as $AB_xC_{3-x}O_7$, where $A = Ca$, Y, Sr, Ba, RE, Th, and U occurring in 8-coordination; $B = Zr$ and Hf in 7-coordination; and $C = Ti$, Mg, Al, Si, Mn, Fe, Nb, and Ta in three distinct sites, two of which are in 6- and one is in 5-coordination [13, p. 255]. For a general formula of the type $(Ca, Na, RE, Th)_2Zr_2(Ti, Nb, Ta)_3(Fe, Ti)O_{14}$ for zirkelite, zirconolite, and polymignite, see [14].

Taking into account the isomorphous replacements for an individual analysis, the conversion into special types of crystallochemical formulas is given for a great number of chemical analyses of zirkelite/zirconolite in [3, 7, 11, 15], [6, pp. 32/5], [16, pp. 29/33], and [17, pp. 101/4]. The composition of natural zirkelites (as from Vuoriyarvi and Eastern Siberia) may be expressed as a mixture of the compounds $CaZrTi_2O_7$, $(Ce, Ca)Zr(Ti, Fe)_2O_7$, $(Ca, U, Th)Zr-(Ti, Fe)_2O_7$, and $CaZr(Ti, Nb, Fe^{3+})_2O_7$. Other samples may additionally contain components such as $CaZrTaFe^{3+}O_7$, $Ca_{0.5}(U^{4+}, Th)_{0.5}ZrTiFe^{3+}O_7$, and also their Hf analogues [32, p. 23].

A recent compilation summarizes the chemical analyses of lunar and terrestrial zirkelites (for the data of samples with $ThO_2 > 1\%$ refer to Table 6, pp. 115/20) and gives the following statements [1, pp. 31, 36/8]:

(1) Contents of Na, K, Fe^{3+} and H_2O are characteristic only for terrestrial samples, which may be related to lower activities of alkali elements, of water, and of oxygen during the late stage of crystallization of lunar rocks.

(2) As follows from plots of $\Sigma(TiO_2, ZrO_2)$ vs. $\Sigma(RE_2O_3, ThO_2, UO_2, Nb_2O_5, Ta_2O_5)$, the relative portion of TiO_2 and the summary content of the main elements TiO_2 and ZrO_2 (adding up from 43 to 75 wt%) increases from mafic to salic host rocks; from mare basalts to breccias (of the Terra regions) of the moon, and from alkalic-ultrabasic to acidic-alkalic rocks of the earth. This relationship is more pronounced for terrestrial zirkelites, whereas for lunar ones only the mean value for $\Sigma(TiO_2, ZrO_2)$ and $\Sigma(RE_2O_3, ThO_2, UO_2, Nb_2O_5, Ta_2O_5)$ gives better evidence.

(3) As follows from plots of FeO vs. "ferruginosity" (zhelezistost', i.e., the ratio $FeO:\Sigma(FeO + MnO + MgO)$), there is a direct correlation between both factors, which is more distinct for lunar samples. And the ferruginosity reaches a limiting value of ~ 1 for samples from mafic environment. The relative portion of MgO in the ferruginosity ratio increases for zirkelites of light-colored (acidic) host rocks from both moon and earth.

(4) Correlations between some individual elements, or element groups, give hints to possible types of diadochic replacement occurring in zirkelite of different genetic types. Thus, for both lunar and terrestrial samples, an inverse correlation between Ca and RE contents corroborates their heterovalent isomorphism. The inverse correlation between Ti and $\Sigma(Nb + Fe)$, more distinct for terrestrial samples, indicates isomorphism of the type Nb + Fe for 2 Ti. But missing correlation between $\Sigma(Th + U)$ and Zr, and between Th and Ca, suggests doubt about isomorphism of this type (as partly is assumed for the zirkelite/zirconolite formulas, see above).

(5) It is assumed that the Ti:Zr ratio diminishes with increasing temperature of formation (as is found for synthetic material, see p. 111). It is also assumed that the terrestrial zirkelites of alkalic-ultrabasic rocks formed at higher temperature than those of acidic rocks, and the lunar zirkelites of mare basalts, altogether, are higher-temperature formations than those of breccia samples from the Terra regions. Comparing the mean value for the Ti:Zr ratio, the resulting sequence, which apparently reflects the temperature of formation, is: lunar mare basalt, terrestrial alkalic-ultrabasic rocks, lunar breccias of Terra regions, and terrestrial acidic rocks.

(6) There are differences in composition of RE and in content of Y_2O_3 for lunar and terrestrial zirkelites. The former have a distinctly yttric composition (with 51 to 78% RE(Y) of the

total RE content) and display a poorly fractionated chondrite-normalized RE pattern, except for negative anomalies for La and Eu; Terra samples are more yttric than mare basalt samples. Terrestrial zirkelites are enriched in light RE (such as Ce and Nd) and, on a large scale, Pr, La, and Sm; their portion of RE(Y) is between 12 and 30%. But note that the general statement given under (6) cannot be confirmed by the data given for zirconolites from the Bergell massif (see p. 114), from the Oetztal-Stubai region (see p. 113), or from one of three localities of Scotland [18].

There is a good constancy in the sum of Nb_2O_5 and TiO_2 contents for typical zirconolites [3].

Compared to the terrestrial type zirkelite, lunar zirkelite (6 grains from Apollo 12 and 14 missions) is lower in Zr, U, and Th but higher in RE and Ti [19]; refer also to Table 6, pp. 115/20. As is shown by phase-equilibria studies (see [12]), the Ti:Zr ratio of zirconolite may be temperature sensitive, for it is 1.30 at 1500°C, 1.53 at 1450°C, and 2.51 at 1300°C. Some samples of terrestrial (Brazil, Ceylon, Kola Peninsula, and Aldan) and lunar zirconolite fit well in this trend, as can be seen from a diagram Ti:Zr ratio vs. T given in [20].

Terrestrial zirconolites have variable, and often non-chondritic, element ratios between Nb and Ta, Zr and Hf, and U and Th. Because of a close similarity of the respective ionic radii and stable valence states, these element pairs are generally considered to be geochemically coherent during magmatic processes, but may alter systematically during post-magmatic processes [18] thus giving hints to formation conditions; refer to the examples given in the following paragraphs.

Zirkelite of apatite-magnetite rocks (ores), genetically related to carbonatites, from an undisclosed complex of alkalic-ultrabasic rocks, Baltic Shield, northwestern European USSR, has a widely varying composition as is given in the following, and as additionally regards the contents of Fe, Ca, and RE; the number of analyses is given in parentheses [5, pp. 150/3]:

ThO_2 0.02 to 3.70, mean 2.03% (29); ZrO_2 18.3 to 41.0, mean 25.99% (21); HfO_2 0.29 and 0.46, mean 0.38%; TiO_2 5.10 to 18.30, mean 11.80% (21); Nb_2O_5 14.63 to 30.20, mean 22.11% (50); Ta_2O_5 0.43 to 2.90, mean 1.67% (63); U_3O_8 0 to 3.37, mean 1.66% (29).

Split up into different host rocks, the composition of the above zirkelites varies as follows [5, p. 156]:

	of apatite-magnetite rock (13)	of phlogopite-calcite carbonatite (7)
ThO_2	0.82 to 3.76, mean 2.08%	0.02 to 3.76, mean 1.72%
ZrO_2	16.0 to 38.20, mean 23.46% [a]	24.90 to 41.00, mean 29.94% [c]
TiO_2	5.10 to 18.15, mean 8.34% [a]	5.90 to 16.43, mean 8.84% [c]
Nb_2O_5	14.75 to 29.80, mean 20.93%	14.63 to 30.20, mean 23.82%
Ta_2O_5	0.62 to 2.68, mean 1.55%	0.75 to 2.90, mean 1.85%
$UO_2 + UO_3$	0.58 to 2.63, mean 1.39% [b]	0.50 to 3.54, mean 1.73% [d]

[a] Only for 7 analyses. — [b] Only for 12 analyses, one analysis gives 0.0% UO_2 and 1.26% UO_3. — [c] Only for 5 analyses. — [d] Only for 6 analyses, one analysis gives 0.75% UO_2 and 2.45% UO_3.

Dependent on its association with other Nb-, Zr-, and Ti-containing minerals, the investigated 42 samples of zirkelite of carbonatites from Eastern Siberia, USSR, display differences in chemical composition, as partly can be seen from the table below (which only contains the samples with > 1% ThO_2). Thus, zirkelite associated with dysanalyte-II has a raised content in TiO_2 in relation to ZrO_2, whereas zirkelite associated with baddeleyite clearly contains more ZrO_2 and minor TiO_2. Zirkelite associated with hatchettolite is very rich in Nb_2O_5. Although there is a relationship between the relative Nb content, as is expressed by the Nb:(Ti + Zr) ratio and the value of Nb:Ta and Th:U ratios, the distribution of Th and U is complicated for zirkelites of different parageneses (see below). Despite a greater variation, there is an obvious tendency for diminishing values of the Th:U ratio with increasing relative Nb content. And at the same relative Nb content, the variation in Th content is stronger than for U. Thus, for the paragenesis zirkelite−dysanalyte-II−baddeleyite with low relative Nb content, the variance in Th content for zirkelite is greatest (between 3 and 30 at%); for the monomineralic zirkelite paragenesis it is something smaller (3 and up to 26 at%). For the paragenesis zirkelite-baddeleyite with higher relative Nb content, the variance in Th content for zirkelite is increased (14 and up to 26 at%). And for the paragenesis zirkelite−dysanalyte-II also with low relative Nb content, the Th content is diminished to 6 at%. The following table gives the content (variance and mean, under the fraction stroke) of the main elements for zirkelite from different parageneses; for the individual data, see original paper [2, pp. 110/1, 113, 118/20]:

zirkelite from paragenesis	content in %					
	ThO_2	ZrO_2	TiO_2	Nb_2O_5	Ta_2O_5	U_3O_8
zirkelite only, 14 analyses						
	1.08 to 6.9	19.4 to 46.0	13.1 to 22.6	11.0 to 24.6	0.7 to 2.73	0.4 to 3.78
	2.99	31.81	18.41[a]	14.04	1.65	1.22
zirkelite−baddeleyite, 8 analyses						
	5.3 to 8.4	20.3 to 57.9	11.4 to 17.3	13.25 to 30.6	0.54 to 7.85	0.52 to 2.78
	6.68	37.36	14.20[b]	20.27	4.13	1.49
zirkelite−baddeleyite−dysanalyte-II, 5 analyses						
	1.23 to 10.06	30.3 to 38.6	21.0 to 23.25	10.2 to 11.6	0.61 to 1.75	0.59 to 1.84
	5.73	33.68	22.77	11.06	0.97	1.36
zirkelite−hatchettolite, 1 analysis						
	6.7	14.6	10.0	30.4	5.26	7.9
zirkelite−dysanalyte-II, 1 analysis						
	1.62	22.2	19.8	19.5	1.31	7.9

[a] Only 12 analyses. − [b] Only 4 analyses.

Two distinct compositional zoning patterns are present in optically zoned zirconolite samples of nepheline syenite, Malawi [13, pp. 255/8]: (1) There is a clear correlation between color variation and the concentrations of several elements, including SiO_2 and CaO. Where in the core region of the mineral the color variation is patchy, the components SrO, ThO_2, and UO_2 are enriched in light brown areas, whereas FeO and Nb_2O_5 are depleted; but the RE content varies little or not at all. The highest concentrations of U and Th in the light colored areas would cause α radiation damage to the structure and produce a partially disordered, metamict state. In addition, microprobe analyses of the light areas reveal water to be present at concentration levels of several percent, which could easily be accomodated in the partially disordered structure (as it was shown as well for zirconolites with < 1% ThO_2 from Afrikanda, Kola Peninsula, and from Aldan, Eastern Siberia, see [7]). (2) By comparing the average rim analysis with that of the dark colored part of the core (as the latter is unaffected by hydration), it can be seen that Y_2O_3 and ThO_2 are enriched, CaO and Ce_2O_3 are depleted in the core relative to the rim. The other elements do not show significant variations. For a detailed discussion based on a ternary diagram Ce-Y-Th and on a plot of chondrite-normalized Ce:Y ratios vs. wt% ThO_2, see original paper. Taking into account the two possible mechanisms of zirconolite formation and zoning given by [21, pp. 527/8], the Malawi zirconolites undoubtedly are of magmatic origin and not produced by metamorphic reactions. The two zoning types described above resulted from separate events. Namely, initial crystallization incorporated high concentrations of Th and U in the core of the mineral, which partially became metamict. Fractionation of the RE accompanied the crystallization of the rims (perhaps in a locally restricted environment) and resulted in an outward increase of enrichment in light RE, accompanied by lower contents of Th and U in the rim region. At some stage there was a gain in SiO_2 and H_2O in the core, and a subsequent loss of other elements, including CaO and FeO, from the metamict core producing the pale brown rim. It is unknown whether this stage occurred during an intermediate hydration stage, or as a late-stage event during which these elements moved by way of microcracks in the crystalline rim [13, pp. 258/9].

One zirconolite grain (intergrown with clinopyroxene and phlogopite) of metacarbonate rock, Oetztal-Stubai region, Austria, shows extreme compositional zoning. The core (= analysis No. 36 of Table 6, p. 118) is enriched in Nb and Ca relative to the rim (= analysis No. 39), whereas Ti and RE are enriched in the rim in accord with a Ti-Nb and Ca-RE substitution, respectively. However, Y does not follow exactly this core-rim relationship. As is evident from the X-ray scanning profile, there is an enrichment of Y near the rim (maximum 3.08% Y_2O_3) with a further decrease at the outer rim. And the Ti:Zr ratio increases from core (1.15) to rim (1.37) which, according to the zirconolite thermometer (see, e.g., [20]), implies a decrease in temperature of formation in this direction [21, pp. 525/7].

Alteration effects (due to mainly primary, hydrothermal alteration) along the rims and fractures of complex Nb-Ta-Ti oxides are usually characterized by increased Ca content and a decrease in U, Th, and RE [22]. Two high-Th zirconolite samples from Sri Lanka, in which fracturing was not observed, show a slight decrease in ThO_2 content. For all other elements the variation between analyses in a given area is greater than the differences between core and rim analyses (see below). The only other indication of alteration is the slightly lower analytical total for the rims. This may be due to increased molecular water content, a common characteristic of metamict minerals. The analyses for the two zirconolite samples are as follows [23]:

	content in % in			
	sample No. B20392		sample No. 111.35	
	core	rim	core	rim
ThO_2	20.50	19.23	19.52	18.00
CaO	8.05	8.17	8.31	8.60
MgO	2.75	2.72	2.32	2.43
U_3O_8	1.71	1.90	1.58	1.90
Al_2O_3	0.60	0.71	0.67	0.81
Fe_2O_3	3.59	2.84	3.50	2.25
ΣRE_2O_3	0.38	0.51	0.84	1.23
ZrO_2	27.02	27.42	28.29	28.10
HfO_2	0.64	0.65	0.67	0.67

Table (continued)

| | content in % in | | | |
| | sample No. B20392 | | sample No. 111.35 | |
	core	rim	core	rim
TiO_2	27.63	28.30	29.04	30.19
Nb_2O_5	3.41	3.19	2.61	2.24
others	0.88 [a]	0.86 [b]	0.77 [c]	0.84 [d]
sum	97.16	96.50	98.12	97.26

[a] Comprises 0.02% Cr_2O_3, 0.11% MnO_2, 0.27% Y_2O_3, and 0.48% PbO. —
[b] Comprises 0.01% Cr_2O_3, 0.10% MnO_2, 0.32% Y_2O_3, and 0.43% PbO. —
[c] Comprises 0.07% MnO_2, 0.26% Y_2O_3, and 0.44% PbO. — [d] Comprises 0.03% Cr_2O_3, 0.05% MnO_2, 0.01% SiO_2, 0.32% Y_2O_3, and 0.43% PbO.

The composition of zirconolites (= analysis Nos. 41 and 42 of Table 6, p. 119, and others with <1% ThO_2) of metasomatically altered marble, Bergell massif, Switzerland-Italy, corresponds well to the generalized formula $(M_I)(M_{II})(M_{III})_2O_7$ (refer to p. 110) and shows high Ti:Zr ratios. This could be due to a minor excess of Zr^{4+} (which affects the ratio as a function of formation temperature), absence of Ta^{5+}, and very small contents of Nb^{5+}. The RE content of Bergell zirconolites may reflect the availability of RE during crystallization; Ce and Nd are always the most abundant. The chondrite-normalized RE patterns rise steeply from La to Ce and then remain virtually constant [8, pp. 465/7].

Non-metamict Nb-rich zirconolite of carbonatite, Canafistula, Cape Verde Republic, shows a slight excess of cations (8.1 instead of 8.0 for 14 oxygens) and has a similar total content of Th + U as is found for other niobozirconolites, but the U content alone is relatively higher. The low Ti:Zr ratio (1.24) suggests a relatively high temperature of formation for the mineral if it is compared to data for synthetic material (see p. 111) [3].

Whereas the chemical composition of platy zirkelite of carbonatite (refer to pp. 105/6) from an undisclosed massif, Kola Peninsula, is similar to data for zirkelites of other Kola carbonatites, the cubic and cubo-octahedric zirkelites show some peculiar features. They are noticeably enriched in Ta, U, Th, and Nb, whereas Ti and Zr are reduced (see analysis Nos. 3, 6, and 13 of Table 6, pp. 115/6) [16, pp. 32/3].

As is exemplified by zirconolite samples from the Afrikanda massif, Kola Peninsula, and from Aldan, Eastern Siberia, variable contents of H_2O and SiO_2 in the mineral are supposed to be caused by secondary alteration [24].

As follows from synthetic preparations of zirconolite, the elements Al, Ce, Nd, Sm, Gd, Yb, Th, and U promote twinning. Stoichiometric and nonstoichiometric zirconolite preparations, which (nominally) only contained Ca, Ti, and Zr oxides, never exhibit the twinning phenomenon [25].

Table 6

Chemical Composition and Density of Zirkelite/Zirconolite from Various Types of Occurrence.
n.d. = not determined; tr. = traces. For the footnotes, see p. 120.

	content in % in sample number								
	1	2	3	4	5	6	7	8	9
ThO_2	20.44	18.78	9.42	9.29	8.51	8.37	8.33	8.22	8.0
CaO	6.87	6.78	13.26	11.89	9.35	14.56	8.55	9.05	14.4
MgO	2.34	3.04	0.54	0.32	1.08	0.44	1.33	0.07	n.d.
UO_2	–	–	n.d.	2.10	–	1.13	–	2.86	–
UO_3	1.04[a)]	0.65[a)]	5.44	5.39	2.08[a)]	3.43	4.66[a)]	–	n.d.
SiO_2	–	–	–	–	–	–	–	0.29	–
FeO	4.07	4.42	–	–	4.65	–	4.72	7.70[h)]	–
Fe_2O_3	–	–	2.08	2.93	–	2.95	–	–	7.9[j)]
Al_2O_3	–	–	1.55	0.19	2.26	1.23	–	0.27	–
ΣRE_2O_3	3.76	1.80	3.76	2.89	0.83	5.22	0.32	7.23	2.0[k)]
ZrO_2	30.73	32.56	15.06[c)]	28.64[c)]	32.65	9.88	34.19	31.52	32.8
HfO_2	–	–	–	–	–	–	–	0.31	–
TiO_2	29.50	30.95	13.55	11.80	36.06	10.15	36.26	28.80	25.3
Nb_2O_5	–	–	23.46	13.36	–	33.35	–	2.71	10.5
Ta_2O_5	–	1.05	5.55	5.81	–	3.63	–	0.26	n.d.
H_2O	0.46	–	2.71[d)]	2.89	1.74	2.34[d)]	1.70	–	n.d.
others	0.41[b)]	–	3.34[e)]	2.30[f)]	–	3.68[g)]	–	0.56[i)]	–
sum	99.64	100.03	99.71	99.80	99.20	100.37	100.06	99.56	100.9
D in g/cm³	5.0	4.72	4.59	–	4.32	4.52	4.47	–	–

Nos. 1 and 2 = Thorian zirkelite from placers (gem gravels), Sri Lanka [26], see also [1, pp. 34/5]. – No. 3 = Zirkelite (cubic) of schistose phlogopite-diopside-calcite carbonatite, Kola Peninsula, USSR [16, p. 30]. – No. 4 = Zirkelite from Kovdor, Kola Peninsula, USSR [27]. – No. 5 = U-Th zirkelite of placers (gem gravels), Sri Lanka [26], see also [1, pp. 34/5]. – No. 6 = Zirkelite (cubo-octahedral) of phlogopite-calcite carbonatite, Kola Peninsula, USSR [16, p. 30]. – No. 7 = U-Th zirkelite of placers (gem gravels), Sri Lanka [26], see also [1, pp. 34/5]. – No. 8 = Zirconolite of syenite, Glen Dessary, Scotland, Great Britain [18]. – No. 9 = Zirkelite from Brazil (mean for 2 samples, one of which contains micro-inclusions of baddeleyite and other minerals) [11, p. 371].

Table 6 (continued)

	content in % in sample number								
	10	11	12	13	14	15	16	17	18
ThO_2	7.37	7.31	5.80	5.68	5.57	5.57	5.17	4.36	4.32
CaO	1.83	10.79	12.10	10.36	12.04	12.01	2.13	9.16	7.22
MgO	0.05	0.22	–	0.19	–	–	–	0.42	0.18
UO_2	1.82	–	0.93[m]	–	0.77[m]	0.76	0.28[m]	–	0.98
UO_3	–	1.40[a]	–	3.72	–	–	–	1.56[a]	–
SiO_2	3.25	–	–	–	–	2.9	8.80	0.09	<0.07
FeO	4.56[h]	7.72	–	–	–	–	–	–	8.32[h]
Fe_2O_3	–	–	8.74	5.68	7.86	5.83	2.01	8.72	–
Al_2O_3	0.43	–	–	0.55	–	–	–	0.02	0.75
ΣRE_2O_3	14.20	2.73	1.86	2.46	1.89	3.77	0.68	1.86	14.32
ZrO_2	27.30	52.89	32.86	34.51	30.64	35.26[o]	24.46	32.54	27.30
HfO_2	0.40	–	–	–	–	0.25[p]	–	0.56	0.43
TiO_2	24.80	14.95	23.37	19.17	25.59	27.50	19.28	33.21	26.00
Nb_2O_5	6.9	–	10.94	14.00	12.77	2.50	20.30	0.41	8.1
Ta_2O_5	0.77	–	0.55	2.43	1.03	–	4.46	–	0.69
H_2O	–	1.02	–	0.50	–	3.15[q]	–	–	–
others	0.79[l]	–	–	0.64[n]	–	0.34[r]	9.52[s]	1.97[t]	0.61[u]
sum	94.41	99.03	97.35	99.89	98.15	99.84	97.10	94.88	99.17
D in g/cm^3	–	4.74	–	4.74	–	4.34	–	–	–

No. 10 = Zirconolite (light core) of nepheline syenite, Malawi; average of 3 analyses [13, p. 254]. – No. 11 = Zirkelite of weathered jacupirangite, São Paulo, Brazil [28]. – No. 12 = Zirkelite of apatite-phlogopite-diopside rock, Sebl'yavr, Kola Peninsula, USSR [29]. – No. 13 = Zirkelite (platy) of phlogopite-diopside-calcite carbonatite, Kola Peninsula, USSR [16, p. 30]. – No. 14 = Zirkelite of apatite-tetraferriphlogopite-magnetite rock, Sebl'yavr, Kola Peninsula, USSR [29]. – No. 15 = Zirkelite of calcite-amphibole-diopside rock, Afrikanda, Kola Peninsula, USSR [15, p. 269]. – No. 16 = Zirkelite of apatite-forsterite-magnetite rock with tetraferriphlogopite, Sebl'yavr, Kola Peninsula, USSR [29]. – No. 17 = Zirconolite from Palabora, Transvaal, R.S.A. [23]. – No. 18 = Zirconolite (dark core) of nepheline syenite, Malawi [13, p. 254].

116

Table 6 (continued)

	content in % in sample number								
	19	20	21	22	23	24	25	26	27
ThO_2	4.25	3.93	3.50	3.26	3.21	2.90	2.90	2.79	2.73
CaO	8.51	11.81	10.00	9.42	5.44	9.70	11.00	10.78	10.22
MgO	0.63	—	0.21	0.40	1.07	tr.	—	—	tr.
UO_2	0.56	1.44[m]	—	—	—	—	—	—	0.96
UO_3	—	—	—	0.56	1.82[a]	1.0[a]	0.40[a]	—	1.23
SiO_2	0.30	—	—	1.33	0.15	0.37	—	—	1.23
FeO	2.51	—	5.24	3.71	4.00	5.33	6.00	5.16	3.19
Fe_2O_3	6.48	8.50	3.48	4.00	2.88	5.07	1.11	1.48	4.81
Al_2O_3	—	—	—	1.76	3.47	tr.	—	—	—
ΣRE_2O_3	4.58	1.69	2.84	5.19	3.06	1.29	4.00	3.97	6.10
ZrO_2	27.68	31.61	28.73	32.20	32.94	30.18	25.00	22.82	33.42
HfO_2	0.35[p]	—	0.28	—	—	—	—	—	0.64[p]
TiO_2	24.30	25.29	16.00	23.45	14.86	22.27	18.19	22.00	20.00
Nb_2O_5	16.17	11.06	24.40	11.91	21.45	21.59	24.84	27.00	11.25
Ta_2O_5	0.87	0.35	1.25	0.82	2.08	0.72	2.00	0.41	1.50
H_2O	2.16	—	2.88	0.85	1.72[x]	3.53	2.48	2.42	3.59[ab]
others	0.97[r]	—	1.26[v]	0.70[w]	1.50[y]	0.43[z]	2.38[aa]	0.90[v]	—
sum	100.32	95.48	100.06	99.56	99.65	104.38	100.30	99.73	99.64
D in g/cm^3	4.52	—	4.60	4.85	4.40	4.33	4.51	4.27	4.27

No. 19 = Nb-zirkelite of apatite-magnetite-olivine rock, Vuoriyarvi, Kola Peninsula, USSR [15, p. 269], see also [1, pp. 34/5]. – No. 20 = Zirkelite of apatite-forsterite-magnetite rock with tetraferriphlogopite, Sebl'yavr, Kola Peninsula, USSR [29]. – No. 21 = Zirkelite of carbonatite, Vuoriyarvi, Kola Peninsula, USSR [10, pp. 160/1]. – No. 22 = Zirkelite of carbonatite, Aldan, Eastern Siberia, USSR [17, p. 100]. – No. 23 = Zirkelite of carbonatite, Kovdor, Kola Peninsula, USSR [10, pp. 160/1]. – No. 24 = Nb-zirconolite of biotite carbonatite, Cape Turii, Kola Peninsula, USSR [31]. – No. 25 = Nb-zirconolite (of apatite-magnetite-olivine rock [1, pp. 34/5]) from Kola Peninsula, USSR [7]. – No. 26 = Zirkelite of carbonatite, Vuoriyarvi, Kola Peninsula, USSR [10, pp. 160/1]. – No. 27 = Zirkelite of calcite-dolomite carbonatite, Sebl'yavr, Kola Peninsula, USSR [15, p. 269].

Table 6 (continued)

	content in % in sample number								
	28	29	30	31	32	33	34	35	36
ThO$_2$	2.72	2.67	2.51	2.51	2.50	2.34	2.05	1.70	1.65
CaO	9.59	13.00	11.44	10.12	11.28	6.15	10.71	12.61	12.24
MgO	0.36	–	0.10	–	0.43	1.15	0.70	0.14	0.56
UO$_2$	–	0.73$^{m)}$	1.00	–	0.86	1.16	–	0.70	0.86
UO$_3$	–	–	–	3.20$^{a)}$	–	–	2.51$^{a)}$	–	–
SiO$_2$	–	2.67	–	–	0.76	0.26	–	–	0.10
FeO	5.37	–	2.84	2.02	4.53	4.23	5.56	2.30	4.93$^{h)}$
Fe$_2$O$_3$	2.72	6.54	3.15	3.46	3.39	–	1.08	2.78	–
Al$_2$O$_3$	0.70	–	0.09	–	–	0.67	0.04	0.10	1.12
ΣRE$_2$O$_3$	3.71	–	–	2.79	4.62	15.23	1.40	5.16	1.08
ZrO$_2$	27.35	35.02	32.96	28.40	32.32	30.06	34.39	33.54	34.72
HfO$_2$	0.29	0.38	0.39	0.46	–	0.86	–	0.55	0.68
TiO$_2$	18.30	30.68	30.43	16.32	22.04	29.62	14.08	31.57	25.86
Nb$_2$O$_5$	22.47	6.31	4.71	24.81	13.65	4.34	24.11	4.80	16.47
Ta$_2$O$_5$	0.43	0.97	<0.01	2.74	1.75	0.40	–	0.75	0.95
H$_2$O	3.30	–	–	1.13	1.38$^{ae)}$	–	2.62	–	–
others	2.46$^{ac)}$	0.04$^{ad)}$	–	2.27$^{v)}$	0.61$^{af)}$	2.69$^{ag)}$	1.22$^{ah)}$	0.15$^{ai)}$	–
sum	98.77	95.96	97.38	100.23	100.12	99.25	100.47	98.55	101.22
D in g/cm^3	4.31	–	–	4.32	4.87	–	4.96	–	–

No. 28 = Zirkelite of carbonatite, Kovdor, Kola Peninsula, USSR [10, pp. 160/1]. – No. 29 = Zirkelite of apatite-forsterite-magnetite rock, Sebl'yavr, Kola Peninsula, USSR [29]. – No. 30 = Zirconolite of calcite-amphibole-diopside rock, Afrikanda, Kola Peninsula, USSR [1, pp. 34/5]. – No. 31 = Zirkelite of carbonatite, Kovdor, Kola Peninsula, USSR [10, pp. 160/1]. – No. 32 = Zirkelite of carbonatite, Aldan, Eastern Siberia, USSR [17, p. 100]. – No. 33 = Zirconolite of recrystallized breccia, Apollo 14 mission, Moon [12]. – No. 34 = Zirkelite of carbonatite, Sayan, Eastern Siberia, USSR [10, pp. 160/1]. – No. 35 = Zirconolite (of carbonatite [1, pp. 34/5]) from Aldan, Eastern Siberia, USSR [12]. – No. 36 = Zirconolite (core) of metacarbonate rock, Oetztal-Stubai region, Austria; sample is intergrown with clinopyroxene and phlogopite [21, p. 525].

Table 6 (continued)

	content in % in sample number					
	37	38	39	40	41	42
ThO_2	1.62	1.44	1.38	1.19	1.03	1.02
CaO	8.08	12.10	10.97	7.31	12.75	12.04
MgO	0.13	0.36	0.63	0.65	0.04	0.01
UO_2	0.57	2.07	0.81	0.30	–	–
UO_3	–	–	–	–	–	–
SiO_2	<0.07	0.06	0.08	1.74	0.17	0.13
FeO	7.96[h]	6.53[h]	5.36[h]	5.95	–	–
Fe_2O_3	–	–	–	–	5.20[j]	5.90[j]
Al_2O_3	0.89	0.14	1.14	1.07	0.82	0.83
ΣRE_2O_3	14.21	5.01	5.26	7.90	4.25	5.24
ZrO_2	28.80	31.07	35.15	37.21	35.00	35.20
HfO_2	0.44	0.42	0.71	0.72	–	–
TiO_2	27.20	24.95	31.12	36.21	37.30	36.00
Nb_2O_5	8.30	12.39	5.71	1.85	–	–
Ta_2O_5	0.59	1.81	0.51	0.12	–	–
H_2O	–	–	–	–	–	–
others	0.39[aj]	0.40[ah]	–	1.19[am]	<0.04[an]	0.07[an]
sum	99.07	99.75	98.78	99.81	99.60	96.40
D in g/cm^3	–	5.0[al]	–	–	–	–

No. 37 = Zirconolite (rim) of nepheline syenite, Malawi; average of 3 analyses [13, p. 254]. – No. 38 = Zirconolite (non-metamict) of carbonatite, Santiago Island, Cape Verde Republic [3, p. 4]. – No. 39 = Zirconolite (rim) of metacarbonate rock, Oetztal-Stubai region, Austria; sample is intergrown with clinopyroxene and phlogopite [21, p. 525]. – No. 40 = Zirconolite of mare basalt, Apollo 16 mission, Moon [12]. – Nos. 41 and 42 = Zirconolite of metasomatically altered marble (skarn), Bergell massif, Switzerland-Italy [8, p. 465].

120

Footnotes to Table 6:

[a] Given as U_3O_8. — [b] Comprises 0.38% PbO and 0.03% MnO. — [c] Includes HfO_2. — [d] H_2O^- only. — [e] Comprises 0.14% SrO, 0.51% MnO, 0.97% PbO, 1.59% Na_2O, and 0.12% K_2O. [f] Comprises 0.16% SrO, 1.65% Na_2O, and 0.49% K_2O. — [g] Comprises 0.11% SrO, 0.32% MnO, 0.45% PbO, 2.43% Na_2O, and 0.37% K_2O. — [h] Total Fe as FeO. — [i] Comprises 0.15% MnO, 0.29% PbO, and 0.12% BaO. — [j] Total Fe as Fe_2O_3.
[k] Comprises only 1.5% Ce_2O_3 and 0.5% Nd_2O_3. — [l] Comprises 0.19% SrO, 0.54% MnO, and <0.06% BaO. — [m] Comprises UO_2 + UO_3. — [n] Comprises 0.03% SrO, 0.24% Na_2O, and 0.37% K_2O. — [o] 35.21% ZrO_2 is given in [30].
[p] Calculated from ZrO_2:HfO_2 ratio obtained by X-ray spectral analysis. — [q] Comprises 2.98% H_2O^+ and 0.17% H_2O^-. — [r] SrO only. — [s] BaO only. — [t] Comprises 0.49% MnO, 1.47% PbO, and 0.01% Cr_2O_3.
[u] Comprises <0.05% SrO, 0.50% MnO, and 0.06% BaO. — [v] Na_2O + K_2O. — [w] Comprises 0.38% MnO, 0.26% Na_2O, and 0.06% K_2O. — [x] Erroneously given as 2.72% H_2O in the original paper, see [30]. — [y] Comprises 0.20% SrO, 0.89% Na_2O + K_2O, and 0.41% others.
[z] Comprises 0.08% SrO, 0.01% MnO, and 0.34% Na_2O. — [aa] Comprises 0.38% MnO, 1.40% Na_2O, and 0.6% F. — [ab] Comprises 3.26% H_2O^+ and 0.33% H_2O^-. — [ac] Comprises 1.0% SrO and 1.46% Na_2O + K_2O. — [ad] Na_2O only.
[ae] Comprises 1.26% H_2O^+ and 0.12% H_2O^-. — [af] Comprises 0.32% MnO, 0.19% Na_2O, and 0.10% K_2O. — [ag] Comprises 2.19% PbO_2 and 0.5% Cr_2O_3. — [ah] Comprises 0.06% SrO, 0.52% Na_2O + K_2O, and 0.64% others. — [ai] PbO only.
[aj] Comprises <0.05% SrO, 0.28% MnO, and <0.06% BaO. — [ak] Comprises 0.28% MnO, 0.10% PbO, and 0.02% P_2O_5. — [al] Calculated value. — [am] Comprises 0.11% MnO, 0.52% PbO_2, and 0.56% Cr_2O_3. — [an] ZnO only.

References for 2.2.1.3.2:

[1] Kochemasov, G. G. (Mineral. Zh. [L'vov] **2** No. 6 [1980] 30/9).
[2] Pozharitskaya, L. K.; Pavlinskii, G. V.; Razvozzhaeva, E. A.; Samoilov, V. S. (in: Shmakin, B. M.; Osobennosti Petrologii, Mineralogii i Geokhimii Korbonatitov Vostochnoi Sibiri, Nauka, Moscow 1966, pp. 109/20).
[3] Silva, S. C.; Figueiredo, M. D. (Garcia de Orta Ser. Geol. **4** [1980] 1/6).
[4] Kesson, S. E.; Sinclair, W. J.; Ringwood, A. E. (Nucl. Chem. Waste Management **4** [1983] 259/65; C.A. **101** [1983] No. 179650).
[5] Zhuravleva, L. N.; Berezina, L. A.; Gulin, E. A. (Geokhimiya **1976** 1512/30; Geochem. Intern. **13** No. 5 [1976] 147/66).
[6] Williams, C. T. (Contrib. Mineral. Petrol. [Berlin] **66** [1978] 29/39).
[7] Borodin, L. S.; Bykova, A. V.; Kapitonova, T. A.; Pyatenko, Yu. A. (Dokl. Akad. Nauk SSSR **134** [1960] 1188/91; Dokl. Acad. Sci. USSR Earth Sci. Sect. **130/135** [1960] 1022/4).
[8] Gieré, R. (Contrib. Mineral. Petrol. [Berlin] **93** [1986] 459/70).
[9] Chukhrov, F. V.; Bonshtedt-Kupletskaya, E. M. (Mineraly Spravochnik, Vol. 2, Pt. 3, Nauka, Moscow 1967, pp. 1/676, 182).
[10] Kapustin, Yu. L. (Mineralogiya Karbonatitov, Nauka, Moscow 1971, pp. 1/286).

[11] Pudovkina, Z. V.; Dubakina, L. S.; Lebedeva, S. I.; Pyatenko, Yu. A. (Zap. Vses. Mineral. Obshch. **103** [1974] 368/72).
[12] Wark, D. A.; Reid, A. F.; Lovering, J. F.; El Goresy, A. (Lunar Sci.-IV, The Lunar Sci. Inst., Houston 1973, pp. 764/6 from [1, p. 30] and [18, p. 330]).
[13] Platt, R. G.; Wall, F.; Williams, C. T.; Woolley, A. R. (Mineral. Mag. **51** [1987] 253/63).
[14] Mazzi, F.; Munno, R. (Am. Mineralogist **68** [1983] 262/76, 265).

[15] Bulakh, A. G.; Il'inskii, G. A.; Kukharenko, A. A. (Zap. Vses. Mineral. Obshch. **89** [1960] 261/73, 270/2).

[16] Osokin, A. S. (Mineral. Geokhim. **6** [1979] 27/38).

[17] Zdorik, T. B. (Tr. Mineral. Muzeya Akad. Nauk SSSR No. 20 [1971] 96/106).

[18] Fowler, M. B.; Williams, C. T. (Mineral. Mag. **50** [1986] 326/8).

[19] Busche, F. D.; Prinz, M.; Keil, K.; Kurat, G. (Earth Planet. Sci. Letters **14** [1972] 313/21, 318).

[20] Raber, E.; Haggerty, S. E. (Proc. 2nd Intern. Kimberlite Conf., Santa Fe, N.Mex., 1977 [1979], Vol. 1, pp. 229/40, 232/3, 237).

[21] Purtscheller, F.; Tessadri, R. (Mineral. Mag. **49** [1985] 523/9).

[22] Ewing, R. C. (Geochim. Cosmochim. Acta **39** [1975] 521/30, 524/6).

[23] Ewing, R. C.; Haaker, R. F.; Headley, T. J.; Hlava, P. F. (Sci. Basis Nucl. Waste Management **1982**, 249/56, 251/2, 254; C.A. **98** [1983] No. 129379).

[24] Borodin, L. S. (in: Vlasov, K. A., Geochemistry and Mineralogy of Rare Elements and Genetic Types of Their Deposits, Vol. 2., Israel Program Sci. Transl., Jerusalem 1966, pp. 336/40, 337/8 [Russian original, Moscow 1964]).

[25] White, T. J. (Am. Mineralogist **69** [1984] 1156/72, 1163, 1169/70).

[26] Blake, G. S.; Smith, G. F. (Mineral. Mag. **16** [1913] 309/16, 313).

[27] Burova, T. A. (in: Borneman-Starynkevich, I. D.; Metody Khim. Analiza Gorn. Porod i Mineralov, Nauka, Moscow 1973, pp. 23/6, 25).

[28] Prior, G. I. (Mineral. Mag. **11** [1895] 180/3).

[29] Subbotin, V. V.; Kirnarskii, Yu. M.; Kurbatova, G. S.; Strel'nikova, L. A.; Subbotina, G. F. (in: Ivanova, T. N.; Petrologiya i Minerageniya Shchelochnykh, Shchelochno-Ul'traosnovnykh i Karbonatitovykh Kompleksov Karelo-Kol'skogo Regiona, Akad. Nauk SSSR, Kol'sk. Filial, Apatity 1985, pp. 61/9, 66).

[30] Kukharenko, A. A.; Orlova, M. P.; Bulakh, A. G.; Bagdasarov, E. A.; Rimskaya-Korsakova, O. M.; Nefedov, E. I.; Il'inskii, G. A.; Sergeev, A. S.; Abakumova, N. B. (Kaledonskii Kompleks Ul'traosnovnykh, Shchelochnykh Porod i Karbonatitov Kol'skogo Poluostrova i Severnoi Karelii, Nedra, Moscow 1965, pp. 1/772, 325).

[31] Lapin, A. V. (Mineralog. Issled. **3** [1973] 108/21, 118).

[32] Kapustin, Yu. L.; Chernitsova, N. M.; Pudovkina, Z. V. (in: Tatarinov, P. M.; Mineraly i Paragenezisy Mineralov Gornykh Porod, Nauka, Leningrad 1973, pp. 17/25, 18).

2.2.1.3.3 Crystal Form and Structure

Zirkelite is trigonal and zirconolite is monoclinic [1, p. 263]; for the latter, refer also to [2].

Non-metamict Nb-rich zirconolite of carbonatite, Santiago Island, Cape Verde Republic, shows well developed **crystal forms**, most frequently combinations of {001}, {h0l}, and {hkl}. The crystals are tabular, flattened parallel to (001) and slightly elongated along [010]. Aggregates of parallel crystals with (001) as composition face are common [3]. Prismatic zirkelite crystals of jacupirangite, São Paulo, Brazil, display typical striations due to polysynthetic twinning [4]. Besides simple twins with re-entering angles, small monocrystals of 0.1 to 0.2 mm in size displaying the rhombohedron and two pinacoids are mentioned for zirkelite of carbonatite, Eastern Siberia, USSR [5]. But note that a similar looking combination of pinacoid and rhombohedron, observed for zirkelite of calcite-dolomite carbonatite, Sebl'yavr massif, Kola Peninsula, was identified (by two-circle goniometer measurement) to be of cubic symmetry [6].

The main habit of zirkelite crystals of carbonatites and associated rocks from Eastern Siberia is given as trigonal or hexagonal platelets with a smooth, lustrous surface, or with a comb-like or parallel ribbed surface due to polysynthetic twinning (with the basis as twinning plane) [5]. Zirkelites from the Kola Peninsula are also smooth, lustrous, and characteristically trigonal (with truncated edges) or hexagonal platelets of octahedrons flattened parallel to one pair of opposite faces. Additionally, there are: dull platelets with nearly always slightly distorted and coarsely striated rips on their "basal" faces; platelets intergrown as twins after (111) forming rosette-like, stellate, or pseudocubic aggregates; and few platelets intergrown as a cubo-octahedron (rarely octahedron) with re-entering angles (concave faces) along its fourfold and threefold axis. For a detailed discussion and for figures, see original paper [7]; refer also to [8]. Polysynthetic twins partly intergrown to rosette-like aggregates and stellate or pseudocubic block crystals for zirkelite from Eastern Siberia, see also [5]. Note further that niobozirconolite of carbonatite and, partly, of apatite-magnetite rock, Cape Turii, Kola Peninsula, occurs as trigonal or hexagonal platelets, sometimes forming polysynthetic twins of stellate, rosette-like, or pseudocubic habit. For the latter, especially, a diagonal striation is mentioned as a characteristic [9]. For an interpretation of zirconolite twins by means of its modular structure, see [10, p. 1163]. For twins refer also to p. 114.

The immediate interest in detailed investigations of the **crystal structure** of zirkelite/zirconolite stems from its inclusion in SYNROC, where it plays a leading role in connection with immobilization of high-level nuclear waste [10, p. 1167]; refer also, for instance, to [11, 12].

The space group for zirkelite is $P3_12$ (No. 152) and for zirconolite C2/c (No. 15) [1, p. 266], or Cc (No. 9) [3].

Number of formula units, based on a formula $(Ca, Na, RE, Th)_2 Zr_2 (Ti, Nb, Ta)_3 (Fe, Ti) O_{14}$, is $= 3$ for zirkelite and $= 4$ for zirconolite, both from Campi Flegrei, Italy [1, p. 266]; and $= 8$ based on a formula with seven oxygens for zirkelite from Kaiserstuhl, southwest Germany [13], and for synthetic zirconolite [14].

The lattice parameters of zirkelite (zirk.) and zirconolite (zirc.), as well as of some corresponding synthetic compounds, are given as follows:

mineral and location, or chemical composition	lattice constants in Å and angle β				Ref.
	a	b	c	β	
zirk.; Campi Flegrei, Italy[a]	7.287(2)	7.287(2)	16.886(9)	90°	[1, p. 266]
Nb-zirk.; Cape Verde Republic	12.33(6)	7.27(4)	11.44(7)	100°	[3]
zirc.; Palabora, R.S.A.[b]	12.39(2)	7.27(1)	11.45(2)	100.4(2)°	[15]
zirc.; Kaiserstuhl, Germany[c]	12.431(1)	7.224(1)	11.483(3)	100.33(1)°	[13]
zirc. (No. 111.35); Sri Lanka[b]	12.438(9)	7.309(6)	11.441(9)	100.4(1)°	[15]
zirc. (No. B20393); Sri Lanka[b]	12.44(1)	7.298(7)	11.45(1)	100.4(2)°	[15]
zirk. (No. 4); Brazil[d]	12.55±0.02	7.23±0.01	11.39±0.02	100°30′	[16]
zirk.; Lovozero, USSR	12.57	7.26	11.38	100°33′	[17, p. 20]
zirk. (No. 5); Brazil[d]	12.58±0.02	7.24±0.01	11.40±0.02	100°22′	[16]
zirc.; Aldan, Asian USSR[d]	12.58	7.27	11.44	100.6°	[16]
zirk.; Campi Flegrei, Italy[a]	12.611(5)	7.311(2)	11.444(8)	100.52(5)°	[1, p. 266]
$CaZrTi_2O_7$[e]	12.43	7.26	11.37	100.5°	[16]
$CaZrTi_2O_7$, synthetic zirc.[f]	12.4458(7)	7.2734(4)	11.3942(9)	—	[14]
$Ca_{0.961}Zr_{0.850}Ti_{2.169}O_7$[g]	12.444(2)	7.266(1)	11.341(2)	100.59(1)°	[18]
$Ca_{0.993}Zr_{1.304}Ti_{1.700}O_7$[h]	12.445(4)	7.288(2)	11.487(3)	100.39(1)°	[18]
$Ca_{0.5}Ce_{0.5}ZrTi_{1.5}Fe_{0.5}O_7$[i]	12.57	7.26	11.39	100°34′	[17, p. 20]
$CaZrNbFeO_7$[i]	12.73	7.36	11.57	100°33′	[17, p. 20]

[a] Derived from a zirkelite-zirconolite-pyrochlore intergrowth. — [b] After heating at 1130°C for 5 h. — [c] Unheated sample, having undergone only minor structural changes owing to the action of natural α-recoil. — [d] After heating at 1200°C. — [e] Nearly identical values are given for synthetic $CaZrTi_2O_7$ obtained by sintering stoichiometric quantities of the corresponding oxides, followed by annealing at 1450°C for 45 h [19]. — [f] Obtained by heating stoichiometric quantities of the corresponding oxides at 1450°C for 4 d. — [g, h] Obtained by heating of a 1:1:2 and 1:0.8:2.2 molar mixture, respectively, of $CaCO_3$, ZrO_2, and TiO_2 at 1773 and 1573 K; for duration of heating and for the specialized individual preparation procedure, see original paper. — [i] Obtained by heating stoichiometric quantities of the corresponding oxides at 1200°C for 3 to 5 h.

Despite its varied behavior during heating (see pp. 129/30), X-ray diffraction patterns (d values) for partly metamict zirconolite (with 0.37% ThO_2) from the Arbarastakh massif, Aldan, and for metamict zirconolite (with about 0.5% ThO_2) from the Afrikanda massif, Kola Peninsula, can be taken, after heating at 1100 to 1200°C, as the standard when identifying minerals of the zirconolite type. Their strongest d values given below [20], see also [21], correspond well to the data given for Nb-rich non-metamict zirconolite from the Cape Verde Republic (= analysis No. 38 in Table 6, p. 119) and for synthetic $CaZrTi_2O_7$ [3]:

I	Aldan d in kX	Afrikanda d in kX	Cape Verde I	d in Å	$CaZrTi_2O_7$ [*] I	d in Å	hkl
10	2.97	2.94	10	2.94	100	2.929	221
2	2.83	2.80	3	2.83	55	2.789	004
3	2.52	2.52	3	2.52	55	2.508	$\overline{2}23$
5	1.82	1.81	5	1.82	25	1.815	040
—	—	—	—	—	65	1.798	$\overline{6}21$
3	1.75	1.75	3	1.75	65	1.740	225
—	—	—	—	—	60	1.737	$\overline{4}06$

[*] Data are for ASTM sample 17-495.

Other diffraction data obtained after heat treatment are given for two zirkelites from Brazil in [16], and for nine zirkelite/zirconolite samples of different types of occurrence from the Kola Peninsula, USSR, in [7], see also [8].

Applying the criterion $-\cos\beta = a/6c$ and $a = \sqrt{3}$ or $a/b = \sqrt{3}$ on the monoclinic lattice constants (a, b, c, and β), some natural zirkelites (e.g., from Aldan, see p. 129) display, on their X-ray diffraction pattern, single reflexes corresponding to a hexagonal lattice. For synthetic compounds, this degree of "hexagonality" increases with changing composition (substitution of Ca by Ce, or of Ti by Nb + Fe) in the series $CaZrTi_2O_7$-$Ca_{0.5}Ce_{0.5}ZrTi_2O_7$ and $CaZrTi_2O_7$-$CaZrNbFeO_7$. Thus, with increasing content of Ce or Nb, the lattice parameter a increases faster than does the parameter b (i.e., the a:b ratio rises and approaches to $\sqrt{3}$) [17, pp. 21/5].

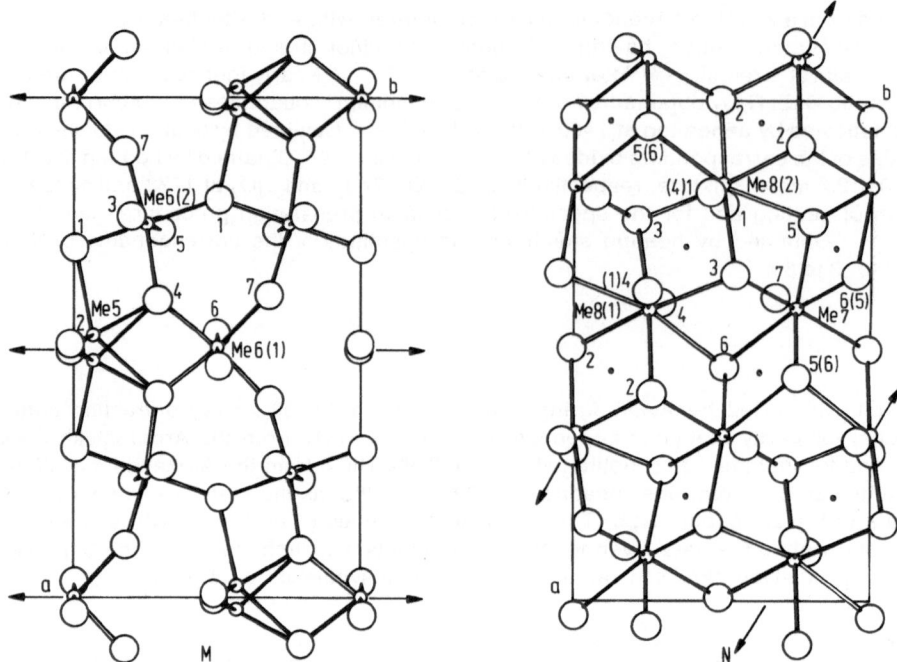

Fig. 5. Zirkelite and zirconolite structure from [1, p. 274]. M- and N-layers of the crystal structure projected along [001] (zirkelite) or [106] (zirconolite); a-edge is [201] of zirkelite or [100] of zirconolite, b-edge is [010] of zirkelite or [0$\bar{1}$0] of zirconolite. Twofold axes (real in the M-layer of both minerals and in the N-layer of zirkelite) and symmetry centers (real only in zirconolite) determined by the arrangement of the atoms in each layer have been traced. Numbering of oxygen atoms corresponds to that reported in the original paper on p. 268 (zirkelite) or by [18] (zirconolite); the same figures generally hold for both minerals: when different, those in parentheses refer to zirconolite. Me8(1) and Me8(2) are equivalent in zirconolite. Bonds in P7 and P64/5 chains are identified by solid lines.

The proposal given at first by [19] that zirkelite/zirconolite, ideally $CaZrTi_2O_7$, is closely related to a structure of the CaF_2-CeO_2 type (fluorite-cerianite) and exhibits a superstructure based on a C-centered monoclinic lattice, is proved by more recent crystal structure refinements using both powder [14] and single-crystal X-ray techniques, as is summarized in [22] as follows: Synthetic zirconolite, best expressed by $CaZr_xTi_{3-x}O_7$ with x between 0.8 and 1.37, has an anion-deficient fluorite-related superstructure derived from the parent fluorite-type lattice by cation-ordering on planes normal (perpendicular) to the superstructure c axis. Alternate plains contain Ti^{4+} only or Ca^{2+} and Zr^{4+}, the latter two elements are ordered within the plane. Calcium is coordinated by eight oxygens positioned at the corners of a distorted cube, whereas zirconium is in 7-fold coordination to oxygen at 7 of the 8 vertices of a distorted cube (but refer also to the next paragraph). Titanium occupies three distinct lattice sites. Whereas Ti(I) and Ti(III) are octahedrally coordinated by six oxygens, forming corner-linked 6- or 3-membered rings of TiO_6 octahedra, respectively, Ti(II) is in trigonal bipyramidal (5-fold) coordination to oxygen, forming a pair of closely-positioned sites, only one of which is randomly occupied by Ti^{4+}. Each unit cell contains 8 Ti(I) and 4 each of Ti(II) and Ti(III).

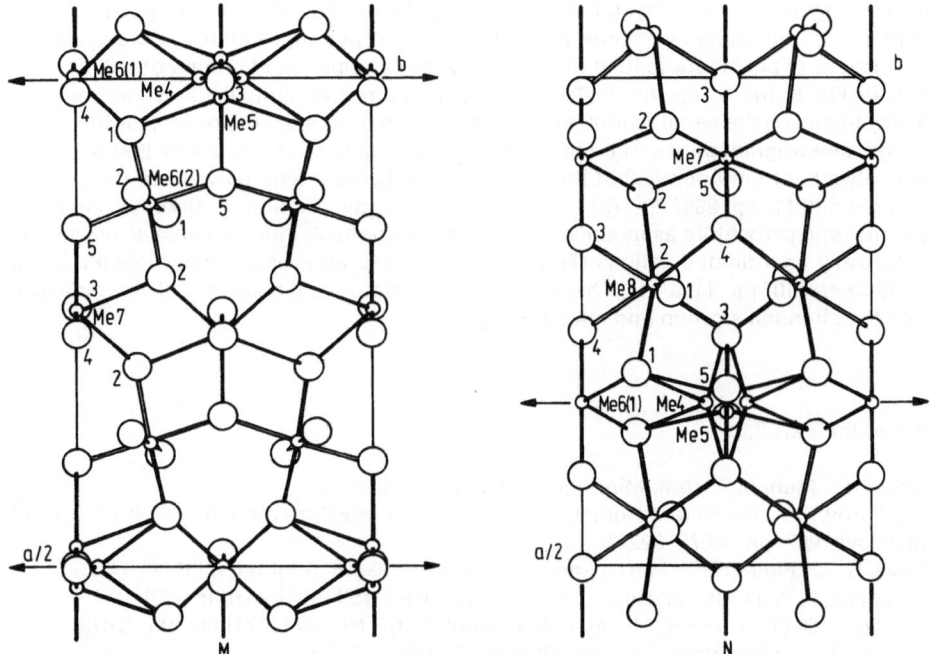

Fig. 6. Polymignite structure from [1, p. 275]. M- and N-layers of the crystal structure projected along [$\overline{1}$10]; a-edge corresponds to [210] and b-edge to [001]. Twofold axes and mirror planes determined by the arrangement of the atoms in each layer have been traced. Numbering of oxygen atoms corresponds to that reported in the original paper on p. 267. Bonds in P7 and P64/5 chains are identified by solid lines.

The Ti(I) and Ti(III) sites are more suitable for smaller cations in isomorphous replacement, whereas the very smallest cations should be preferentially contained in the Ti(II) sites.

Zirconolite has a distorted pyrochlore structure, which itself can be derived from the (fluorite-type [23]) UO_2 structure by systematically deleting one-eight of the oxygen atoms. This converts one-half of the occupied sites to octahedrons, doubles the unit cell edge, lowers the symmetry (from F3m to Fd3m), and leads to a 6-fold coordination for both Ca and Zr [24]. The molar volume of zirconolite is 10% smaller than that of average pyrochlore. Since both have similar cell contents, zirconolite is a more closely packed structure [25].

A most recent reinvestigation shows that the crystal structures of zirkelite, zirconolite, and polymignite (refer also to pp. 232/4) are related to the pyrochlore structure, and are based on the same kind of polyhedra: distorted cubes (P8), octahedra (P6), as well as polyhedra with seven (P7), five (P5), or four (P4) vertices. The cubes are always centered by Me8 cations (Ca, RE, Th, etc.) and the octahedra by Me6 cations (Ti, Nb, Ta). Me7 cations (Zr) are at the centers of P7 polyhedras, and Me5 or Me4 cations (Fe in natural, Ti in synthetic zirconolites) at those of P5 or P4 polyhedra. The bond distances and angles (see original paper) show that the shape of each kind of polyhedra is practically the same in all these above minerals. Based on the dimensions of C-centered "orthorhombic" multiple cells (for details, see original paper), a comparison of the crystal structures of zirkelite, zirconolite, and polymignite can easily be made. All these structures are obtained by repeating, along the c axis, pairs of adjacent layers of polyhedra (called M- and N-layers), composed of chains of polyhedras as is given in **Fig. 5** and **6**, pp. 124/5. The crystal structures of zirkelite and zirconolite are practically based on the same pair of layers, but differ in their stacking order; for the detailed discussion, see original paper. The Figures 5 and 6 are given in such a way that one Me6(1) cation is always at the point 000 of the M-layer on a twofold axis parallel to the b edge of the multiple cell [1, pp. 266, 268/75]. For an alternative description of zirkelite, zirconolite, polymignite, and pyrochlore as modular structures, where each module consists of two layers of octahedrally coordinated cations (Ti, Nb, Ta) arranged as a hexagonal tungsten bronze-type motif, see [10, pp. 1156/61]. This paper also contains a discussion of the similarities between the zirconolite group and mica polytypes.

References for 2.2.1.3.3:

[1] Mazzi, F.; Munno, R. (Am. Mineralogist **68** [1983] 262/76).

[2] Chukhrov, F. V.; Bonshtedt-Kupletskaya, E. M. (Mineraly Spravochnik, Vol. 2, Pt. 3, Nauka, Moscow 1967, pp. 1/676, 322/3).

[3] Silva, S. C.; Figueiredo, M. D. (Garcia de Orta Ser. Geol. [Lisboa] **4** [1908] 1/6, 5).

[4] Menezes, L. A. D., Jr.; Martins, J. M. (Mineral. Rec. **15** [1984] 261/70, 269).

[5] Zdorik, T. B. (Tr. Mineral. Muzeya Akad. Nauk SSSR No. 20 [1971] 96/106, 101/3).

[6] Bulakh, A. G. (Zap. Vses. Mineral. Obshch. **92** [1963] 746/8).

[7] Bulakh, A. G.; Il'inskii, G. A.; Kukharenko, A. A. (Zap. Vses. Mineral. Obshch. **89** [1960] 261/73, 263/6).

[8] Kukharenko, A. A.; Orlova, M. P.; Bulakh, A. G.; Bagdasarov, E. A.; Rimskaya-Korsakova, O. M.; Nefedov, E. I.; Il'inskii, G. A.; Sergeev, A. S.; Abakumova, N. B. (Kaledonskii Kompleks Ul'traosnovnykh, Shchelochnykh Porod i Karbonatitov Kol'skogo Poluostrova i Severnoi Karelii, Nedra, Moscow 1965, pp. 1/772, 322/4).

[9] Lapin, A. V. (Mineralog. Issled. No. 3 [1973] 108/21, 117).

[10] White, T. J. (Am. Mineralogist **69** [1984] 1156/72).

[11] Ringwood, A. E.; Kesson, S. E.; Ware, N. G.; Hibberson, W. O.; Major, A. (Geochem. J. [Tokyo] **13** [1979] 141/65).

[12] Ringwood, A. E.; Oversby, V. M.; Kesson, S. E.; Sinclair, W.; Ware, N.; Hibberson, W.; Major, A. (Nucl. Chem. Waste Management **2** [1981] 287/305).

[13] Sinclair, W.; Eggleton, R. A. (Am. Mineralogist **67** [1982] 615/20).

[14] Rossell, H. J. (Nature **283** [1980] 282/3).

[15] Ewing, R. C.; Haaker, R. F.; Headley, T. J.; Hlava, P. F. (Mater. Res. Soc. Symp. Proc. **6** [1982] 249/56, 251).

[16] Pudovkina, Z. V.; Dubakina, L. S.; Lebedeva, S. I.; Pyatenko, Yu. A. (Zap. Vses. Mineral. Obshch. **103** [1974] 368/72, 370).

[17] Kapustin, Yu. L.; Chernitsova, N. M.; Pudovkina, Z. V. (in: Tatarinov, P. M.; Mineraly i Paragenezisy Mineralov Gornykh Porod, Nauka, Leningrad 1973, pp. 17/25).

[18] Gatehouse, B. M.; Grey, I. E.; Hill, R. J.; Rossell, H. J. (Acta Cryst. B **37** [1981] 306/12, 307/8).

[19] Pyatenko, Yu. A.; Pudovkina, Z. V. (Kristallografiya **9** [1964] 98/100; Soviet Phys.-Cryst. **9** [1964] 76/8).

[20] Borodin, L. S.; Bykova, A. V.; Kapitonova, T. A.; Pyatenko, Yu. A. (Dokl. Akad. Nauk SSSR **134** [1960] 1188/91; Dokl. Acad. Sci. USSR Earth Sci. Sect. **130/135** [1960] 1022/4).

[21] Borodin, L. S. (in: Vlasov, K. A., Geochemistry and Mineralogy of Rare Elements and Genetic Types of Their Deposits, Vol. 2, Israel Program Sci. Transl., Jerusalem 1966, pp. 336/40, 337/8 [Russian original, Moscow 1964]).

[22] Kesson, S. E.; Sinclair, W. J.; Ringwood, A. E. (Nucl. Chem. Waste Management **4** [1983] 259/65, 260; C.A. **101** [1983] 179650).

[23] Strunz, H.; Tennyson, C. (Mineralogische Tabellen, 6th Ed., Geest and Portig, Leipzig 1977, pp. 1/621, 42).

[24] Haaker, R. F.; Ewing, R. C. (PNL-3505 [1981] 1/284, 244).

2.2.1.3.4 Optical and Other Physical Properties. Chemical and Thermal Behavior

Optical Properties. The color of zirkelite/zirconolite is black, pitchy black, brown, dark brown, sometimes light brown, and in very thin splinters it is transparent cinnamon brown [1, p. 184]. Zirkelite from Eastern Siberia is steel gray on its surface, the fracture surface is black, pitchy black, or rarely dark brown. Altered varieties are colored with a light brownish hue [2]. In transmitted light (thin sections), zirkelite/zirconolite is red-brown, intense red, sometimes yellowish or brownish. The color is unevenly distributed [1, p. 184]. Zirconolite from the Bergell massif, Switzerland-Italy, is translucent reddish brown for larger rounded grains of 0.2 mm maximum size, or is opaque for smaller grains [3].

The luster of zirkelite/zirconolite is metallic, pitchy, sometimes adamantine to greasy [1, p. 184], or resinous [4]. Zirkelite from Eastern Siberia is dull (for metamict varieties) or strong semimetallic (for crystalline varieties) on faces, the fracture surface is fatty [2].

The streak of zirkelite/zirconolite is brownish yellow, chocolate brown, or is darker for U-rich varieties [1, p. 184].

The refractive index of zirkelite from the Kola Peninsula, USSR, varies considerably: n is 2.28 ± 0.02 for dark colored, optically isotropic grains from the Sebl'yavr massif and 2.25 ± 0.02 for anisotropic parts of the same crystal: n is 2.18 ± 0.02 for dark brown samples from the Afrikanda massif and between 2.06 and 1.87 for peripheral, light-colored zones of the crystals [5, pp. 265/6], see also [1, p. 184]; n varies between 2.05 and 2.29 for crystalline varieties of zirkelite from the Baltic Shield [6] and is 2.5 for not completely metamict, iron black niobozirconolite from Aldan, both USSR [7]. The refractive index of zirkelite/zirconolite increases with increasing Nb content, but decreases for hydrated samples and with increasing grade of metamictization [8]; note also that hydration accompanies metamictization [9, p. 244]. Synthetic $CaZrTi_2O_7$ is birefringent with $n_g \sim 2.30$, $n_m \sim 2.27$, $n_p \sim 2.23$, and $2V = 80$ to $85°$; optically biaxial negative [1, p. 184].

In reflected light, zirkelite from the Kola Peninsula is isotropic with light gray color (something darker than for magnetite and without a yellowish hue); reflectivity is 12.5% [5, p. 266], see also [1, p. 184]. In its central part, the above zirkelite is sometimes anomalously birefringent [5, p. 265].

For color and refractive index of zirkelite forming morphologically three different varieties, see pp. 105/6.

For an IR spectrum of zirkelite, see [29].

Other Physical Properties. Zirkelite/zirconolite is brittle with conchoidal fracture [4], or is brittle without cleavage; parting along (001) is noticed [1, p. 184]. It has a Mohs hardness of 5½ [4] or 5½ to 6 [1, p. 184]. Microhardness of 760 to 880 kg/mm^2, corresponding to $H_M = 6$, is given for zirkelite from Kola Peninsula [5, p. 265] and of 705 kg/mm^2 for zirkelite from Eastern Siberia [2].

Zirkelite/zirconolite has a calculated density of 4.46, measured values are between 4.02 and 5.2 g/cm^3 [1, p. 184]; refer also to Table 6, pp. 115/20. The density changes with composition as is given above for the refractive index. A difference in density of at least 8% is reported [9, p. 248] between partially and total metamict zirconolite, see [10]. Measured density for synthetic zirconolite, CaZrTi$_2$O$_7$, is 4.45(5) g/cm^3 [11].

Although otherwise mechanicaly durable, zirconolite is susceptible to metamictization [9, p. 248]. Varieties with as little as 1.58% U$_3$O$_8$ and 0.58% ThO$_2$ are metamict [12].

The natural isotopes [238]U, [235]U, [232]Th contained in zirkelite cause damage to the crystal structure predominantly by the recoil of the radioactive nucleus associated with α decay. Most of the damage occurs near the end of the path of high-energy α particles producing 100 to 500 displacements in the lattice per event, resulting in changes of the X-ray diffraction data [13]; for a more detailed discussion of this subject, see [14]. Theoretical models of radiation-damage mechanisms, which assume true diadochy must be regarded with suspicion. Although they predict the correct bulk properties of the metamict material, they will provide an unrealistic description of its microstructural characteristics [15]. Some observation on crystal structures of zirconolite and perovskite show that these materials become metamict by heavy ion bombardment (K$^+$ or Ar$^+$ with energies up to 4 MeV) at relatively small fluence [16, 17].

Related to effects of their total α-recoil dose, natural zirconolites display a range of microstructural features, as can be revealed by transmission electron microscopy. These investigations were carried out especially in relation to the important role synthetic zirconolite plays as a phase in SYNROC for nuclear waste deposition (refer to [18]) and are mentioned only briefly in the following; for the detailed discussion, refer to the cited more recent literature. Thus, the monoclinic crystal structure of zirconolite samples from Sri Lanka, from Palabora in South Africa, from Jacupiranga in Brazil, and from Kaiserstuhl in Germany is retained up to doses of 4.4×10^{24} α/m^3. At intermediate doses of 3×10^{25} α/m^3, the material consists of a mixture of amorphous and crystalline regions, the latter (note also the next paragraph) with an average face-centered cubic fluorite structure. And at doses $> 10^{26}$ α/m^3, the zirconolite is highly disordered and amorphous to electron diffraction [19]; refer also to [20]. Additional data for zirconolite from Sri Lanka annealed at 1000 and 1100°C as well, see [21]. As can be deduced from data given for natural zirconolites (see, e. g., [22, 23]), the recoil dose for complete metamictization is 10^{16} α/mg, similar to the value for zircon [24].

There are two additional important observations concerning the microstructure of natural zirconolites which should be noted: (1) An unusual feature of three metamict samples from Sri Lanka, and of one sample each from Palabora and Jacupiranga, is the presence of predominantly spherical microvoids from a few tens to thousands of nm in size. They may have been caused by nucleation and growth of He bubbles resulting from internal U and Th decay. A zirconolite from the Kaiserstuhl, Germany, has no microvoids [19]. Thus, presence or absence of voids from sample to sample may indicate rather a unique (natural) thermal history for each sample, as the diffusion and accumulation of He in these complex oxides depends on temperature [25, p. 255]. (2) It is relatively easy to recrystallize loosely attached (to the carbon substrate) chips of metamict zirconolite by electron beam-heating during normal examination. This may be responsible for a "partial crystallinity" reported for some samples [19].

Chemical Behavior. Zirkelite/zirconolite is decomposed by HF and by fusion in KHSO₄ [4], or by warm HCl, H₂SO₄, and caustic alkali [1, p. 186].

Although natural zirkelite/zirconolite can have raised contents of radioactive elements, and has suffered strong irradiation therefrom over, partly, long periods of time, there is no mobilization of chemical elements from the mineral during its complex geological history (including elevated temperature and pressure, contact with ground water, weathering, and erosion), see, for instance, [22, 23].

Thermal Behavior. When metamict zirkelite/zirconolite is annealed, partially or completely, at temperatures between 650 and 800 °C, a metastable fluorite-type phase occurs (with a = 5.02 Å [7] or between 5.04 and 5.06 kX [5, p. 266]), and above 1100 °C the original structure is regained (see also below) [9, p. 248]. The formation of such phases, corresponding to different heating temperatures for samples of the same metamict mineral, is apparently not an exactly defined function of heating temperature, but depends substantially on the degree of metamict decomposition [7]. The diffraction patterns of the fluorite-type phase are similar, especially, to those of the cubic modification of ZrO₂ and to polymignite [5, p. 266]. But note that, although being completely metamict, two zirconolite samples from Sri Lanka and one sample from Palabora, South Africa, recrystallized to the zirconolite structure (other phases are absent) after heating at 1130 °C for 5 h. Because of the different orientations of the zirconolite crystallites and their erratic distribution through the sample, they may have been created by recrystallization of the metamict structure, rather than being a preserved primary structure [25, pp. 250, 252, 254/5]; for the lattice parameters of the newly formed zirconolites, see p. 122.

The heating curve of metamict zirkelite of calcite-dolomite carbonatite, Sebl'yavr massif, Kola Peninsula, has endothermic effects clearly expressed at around 200 °C and, less so, at ∼450 and ∼850 °C. The strong exothermic effect at ∼750 °C corresponds to the transition from metamict to crystalline. This is corroborated by the X-ray diffraction patterns which prove the mineral to be X-ray amorphous up to 600 °C, whereas between 740 and 760 °C, a crystalline structure becomes visible. Synchronously, the curve of weight loss of the same sample indicates a loss of about 3.5% water, corresponding well to 3.59% H₂O of the analysis, during heating up at 1000 °C. Loss of water occurs stepwise near 200 °C (1.5%), near 450 °C (1.5%), and minor at 850 °C (0.5%), which gives rise to the assumption that water is a constitutional part (bound as OH⁻) of the mineral and may be segregated after reconstruction of its crystalline structure [5, p. 268]; refer also to [26] where the curve of weight loss is somewhat different and turn out as reversed. For both above curves, see also [1, p. 186]. The heating curve of zirconolite shows a distinct exothermic high-temperature effect between 750 and 800 °C, common for metamict minerals. The corresponding dehydration curve indicates a gradual loss of water up to the stage of recrystallization of the mineral. For the curves, see original paper [10]. Whereas the heating curve of totally metamict zirconolites (e.g., from Afrikanda massif, Kola Peninsula) display the high-temperature exothermic effect, there is no such effect for zirconolite from Aldan, Eastern Siberia, which shows weak and diffuse lines on the X-ray diffraction pattern and thus is not completely metamict [7].

The heating (DTA) curve of metamict zirconolite, Eastern Siberia, shows a strong exothermic effect between 730 and 850 °C, corresponding to recrystallization and formation (at 800 °C) of a metastable cubic phase (a = 5.07 Å) of the fluorite-cerianite structure type (similar in its dimension to cubic ZrO₂ with a = 5.10 Å). A small peak between 1150 and 1200 °C is ascribed to the final reconstruction of the zirconolite structure showing a pseudo-hexagonal symmetry with a = 7.25 Å and c = 8.36 Å [2]. Reconstruction of the crystalline structure after heating at 1200 °C also occurs for two zirkelites (sample No. 4 and 5) from Brazil; for sample No. 5 a cubic phase is obtained at 800 °C. For the lattice parameters, see p. 122 [27].

Whereas the heating curve of platy zirkelite of carbonatite, Kola Peninsula, has a strong exothermic peak at $\sim 750\,°C$ and broad endothermic ones at ~ 200 and $\sim 450\,°C$, corresponding to the curve for a standard sample (see [26]), the heating curve of cubic zirkelite is somewhat different in that all peaks are shifted to lower temperatures: The exothermic effect of recrystallization occurs at about $760\,°C$, endothermic effects are at 180, 400, and $750\,°C$. Two additional exothermic effects, whose nature remains unclear, occur at $770\,°C$ (distinct) and about $450\,°C$ (indistinct and small) [28].

References for 2.2.1.3.4:

[1] Chukhrov, F. V.; Bonshtedt-Kupletskaya, E. M. (Mineraly Spravochnik, Vol. 2, Pt. 3, Nauka, Moscow 1967, pp. 1/676).

[2] Zdorik, T. B. (Tr. Mineral. Muzeya Akad. Nauk SSSR No. 20 [1971] 96/106, 103/4).

[3] Gieré, R. (Contrib. Mineral. Petrol. [Berlin] **93** [1986] 459/70, 464).

[4] Palache, C.; Berman, H.; Frondel, C. (Dana's System of Mineralogy, 7th Ed., Vol. 1, Wiley, New York 1946, pp. 1/834, 741/2).

[5] Bulakh, A. G.; Il'inskii, G. A.; Kukharenko, A. A.ʼ (Zap. Vses. Mineral. Obshch. **89** [1960] 261/73).

[6] Zhuravleva, L. N.; Berezina, L. A.; Gulin, E. A. (Geokhimiya **1976** 1512/30; Geochem. Intern. **13** No. 5 [1976] 147/66, 150/1).

[7] Borodin, L. S.; Bykova, A. V.; Kapitonova, T. A.; Pyatenko, Yu. A. (Dokl. Akad. Nauk SSSR **134** [1960] 1188/91; Dokl. Acad. Sci. USSR Earth Sci. Sect. **130/135** [1960] 1022/4).

[8] Kapustin, Yu. L. (Mineralogiya Karbonatitov, Nauka, Moscow 1971, pp. 1/286, 158/62).

[9] Haaker, R. F.; Ewing, R. C. (PNL-3505 [1981] 1/284).

[10] Borodin, L. S. (in: Vlasov, K. A.; Geochemistry and Mineralogy of Rare Elements and Genetic Types of Their Deposits, Vol. 2, Israel Program Sci. Transl., Jerusalem 1966, pp. 336/40, 337/8 [Russian original, Moscow 1964]).

[11] Rossell, H. J. (Nature **283** [1980] 282/3).

[12] Borodin, L. S.; Nazarenko, I. I.; Rikhter, T. L. (Dokl. Akad. Nauk SSSR **110** [1956] 845/8).

[13] Sinclair, W.; Eggleton, R. A. (Am. Mineralogist **67** [1982] 615/20, 618/9).

[14] Sinclair, W.; Ringwood, A. E. (Geochem. J. [Tokyo] **15** [1981] 229/43).

[15] White, T. J. (Am. Mineralogist **69** [1984] 1156/72, 1167).

[16] Cartz, L.; Karioris, F. G.; Fournelle, R. A.; Gowda, K. A.; Ramasami, K.; Sarkar, G.; Billy, M. (Sci. Basis Nucl. Waste Management **3** [1980/81] 421/7, 421, 425).

[17] Cartz, L.; Karioris, F. G.; Fournelle, R. A. (Radiat. Eff. **54** [1981] 57/64).

[18] Ringwood, A. E.; Kesson, S. E.; Ware, N. G.; Hibberson, W. O.; Major, A. (Geochem. J. [Tokyo] **13** [1979] 141/65).

[19] Ewing, R. C.; Headley, T. J. (J. Nucl. Mater. **119** [1983] 102/9).

[20] Headley, T. J.; Ewing, R. C. (Microbeam Anal. **21** [1986] 141/4; C.A. **105** [1986] No. 118290).

[21] Lumpkin, G. R.; Ewing, R. C.; Chakoumakos, B. C.; Greegor, R. B.; Lyttle, F. W.; Foltyn, E. M.; Clinard, F. W., Jr.; Boatner, L. A.; Abraham, M. M. (J. Mater. Res. **1** [1986] 564/76).

[22] Ringwood, A. E.; Oversby, V. M.; Sinclair, W. (Proc. Australian Ceram. Conf. **9** [1980] 252/6; C.A. **94** [1981] No. 162265).

[23] Ringwood, A. E.; Oversby, V. M.; Sinclair, W. (Sci. Basis Nucl. Waste Management **2** [1980] 273/80; C.A. **94** [1981] No. 213978).

[24] Vance, E. R.; Pillay, K. K. S. (DOE-ET-41900-5 [1981] 1/39; C.A. **96** [1982] No. 112068).

[25] Ewing, R. C.; Haaker, R. F.; Headley, T. J.; Hlava, P. F. (Mater. Res. Soc. Symp. Proc. **6** [1982] 249/56).

[26] Kukharenko, A. A.; Orlova, M. P.; Bulakh, A. G.; Bagdasarov, E. A.; Rimskaya-Korsakova, O. M.; Nefedov, E. I.; Il'inskii, G. A.; Sergeev, A. S.; Abakumova, N. B. (Kaledonskii Kompleks Ul'traosnovnykh, Shchelochnykh Porod i Karbonatitov Kol'skogo Poluostrova i Severnoi Karelii, Nedra, Moscow 1965, pp. 1/772, 23/5).
[27] Pudovkina, Z. V.; Dubakina, L. S.; Lebedeva, S. I.; Pyatenko, Yu. A. (Zap. Vses. Mineral. Obshch. **103** [1974] 368/72).
[28] Osokin, A. S. (Mineral. Geokhim. **1979** No. 6, pp. 27/38, 32/4).
[29] Povarennykh, A. S. (Konst. Svoistva Mineral. No. 13 [1979] 53/78, 62/4).

2.2.1.4 Betafite $\sim (Ca, Na, Ce, U, Th)_2 Zr_2 (Ti, Nb, Fe)_4 O_{14}$[*]

Betafite from granite pegmatite of Ambolotara, Madagascar, was described in 1912 and is named after the nearby place Betafo [1]. Many compositional varieties have been found in nature, see [2 to 5], of which higher amounts of Th carry thorian or Th-betafite, yttrobetafite, and calciobetafite, as can be seen from the following paragraphs. There are also other varieties with a special composition such as, for example, tantal-, titan-, zirconium-, or rare earth-betafite [4].

Synonymous with betafite are mendeleevite and blomstrandite (of Lindström, 1874), both rich in Th, but not the blomstrandinite (of Brögger, 1906), refer to p. 185, [2, 4, 5]. Ellsworthite is synonymously used for mendeleevite, both rich in Ca [3].

References for 2.2.1.4:

[1] Lacroix, A. (Bull. Soc. Franc. Mineral. **33** [1910] 321/3).
[2] Palache, C.; Berman, H.; Frondel, C. (Dana's System of Mineralogy, 7th Ed., Vol. 1, Wiley, New York 1946, pp. 1/834, 803/5).
[3] Kalita, A. P.; Mukhitdinov, G. N. (in: Vlasov, K. A., Geochemistry and Mineralogy of Rare Elements and Genetic Types of Their Deposits, Vol. 2, Israel Program Sci. Transl., Jerusalem 1966, pp. 511/6 [Russian original, Moscow 1964]).
[4] Kalita, A. P. (in: Vlasov, K. A., Geochemistry and Mineralogy of Rare Elements and Genetic Types of Their Deposits, Vol. 2, Israel Program Sci. Transl., Jerusalem 1966, pp. 516/9 [Russian original, Moscow 1964]).
[5] Chukhrov, F. V.; Bonshtedt-Kupletskaya, E. M. (Mineraly Spravochnik, Vol. 2, Pt. 3, Nauka, Moscow 1967, pp. 1/676, 172).

2.2.1.4.1 Occurrence, Paragenesis, Intergrowths, and Inclusions

As can be seen from the following examples for **occurrence and paragenesis**, betafite and its varieties occur mainly in granite pegmatites and in metasomatically altered parts of alkalic rocks and cogenetic carbonatites.

Small crystals of betafite, enclosd in euxenite, occur together with other RE minerals in granite pegmatite of Ippaiyama, Abukuma massif, Fukushima prefecture, Japan [1]. Granite pegmatite dikes contain betafite near Herschel township, Hastings county, and near Cavendish

---

[*] Refer to formula discussion on pp. 134/5.

township, Peterborough county, both Ontario, Canada. In the latter locality, abundant black betafite and magnetite are connected with red feldspar [2, pp. 135, 162]. Acid pegmatite cutting the Hamar gabbro, Hargeisa district, Somalia, contains betafite crystals, reaching 1.25 inches across, together with Th-poor allanite (0.81% ThO_2) and Th-free columbite [3]. A quartz-feldspar vein forming clear contacts in granite gneiss of the Urals, USSR, contains metamict betafite, intimately associated with mendeleevite, as rounded or irregularly shaped inclusions of 3 to 5 mm in size within rose feldspar or gray quartz. Betafite rarely forms octahedral crystals, and its grains in heavy-mineral concentrates often reach 1 to 2 cm in diameter [4, pp. 32/3]. Accessory betafite occurs in early (pre-miaskite) quartz-oligoclase (granite) pegmatite (mine No. 60) and, together with pyrochlore, in post-miaskite granite pegmatite without amazonite, both Il'men Mountains, Southern Urals, USSR [19].

Blomstrandite occurs in syenitic pegmatite veins with low quartz content in the Urals (Il'men Mountains?), USSR. The metamict (X-ray amorphous) blomstrandite occurs in feldspar (microcline perthite) of the central part of the vein, in more or less narrow association with zircon and biotite. It mainly forms rounded or octahedral grains of 2 to 5 mm in size; octahedra of 1 to 2 cm are found in heavy-mineral concentrates [4, pp. 39, 41], [5].

Metamict yttrobetafite, genetically closely related to obruchevite, occurs near the contact between coarse-grained albite (core) and the outer zone of a quartz-albite-oligoclase pegmatite from vein No. 1, Alakurtti, Kola Peninsula, northwestern European USSR. The mineral forms yellowish seams on, or contains irregularly shaped relicts of, obruchevite and occurs as somewhat elongated clusters within the albite, which shows radial fissures around the clusters filled with a yellowish brown or yellowish ochre [6, pp. 52/3]. Yttrobetafite (resembling yellow wiikite-Le of Lokka [7]) fills, together with biotite, fissures in blocky microcline perthite or in quartz of a pegmatite, Nuolainen, Karelia, northwestern European USSR [6, pp. 95/6]. Pegmatites of anatectic plagio-microcline granite from the eastern part of the Baltic Shield, northwestern European USSR, contain yttrobetafite together with allanite, yttrohatchettolite, obruchevite, columbite, monazite, and euxenite. Most of the accompanying minerals contain RE and Th, the latter element reaching, in part, contents of $> 1\%$ ThO_2 [8]. The formation of yttrobetafite is connected with processes of alteration (pyrochloritization) of samarskite [9].

Single grains of brown to cinnamon isotropic betafite (containing $\sim 3\%$ Th) occur in nepheline monzonite and albitite of the Pkhrut complex, Megri pluton, southern Armenian SSR; it is also found in albitized parts of the leucocratic nepheline-bearing variety of alkalic syenite and, relatively often, in pegmatite. Betafite forms octahedral crystals (0.04 to 0.08 mm in size) and is associated with monazite, zircon (cyrtolite), baddeleyite, and albite. According to its mode of occurrence, betafite forms during late metasomatic processes from primary pyrochlore [20].

Biotitized rocks of the ijolite and micro-ijolite series, Oka complex, Quebec, Canada, contain, besides pyrochlore, in places, betafite grains (mostly < 0.5 mm in size), which are invariably strongly radioactive, and crystals (sometimes zoned). Many of the grains and crystals are rimmed, or entirely replaced, by a greenish yellow material [10], refer also to [11].

Good cubo-octahedral crystals of betafite occur in lenticular sill-like calcite bodies in hornblende-plagioclase gneiss near Faraday township, Hastings county, Ontario, Canada [2, pp. 22, 125].

Accessory metamict Th-betafite, rarely forming greater idiomorphic crystals (5 to 8 mm in size, or even larger), is distributed irregularly in carbonatite of the 2nd stage (occurring as phlogopite-diopside-calcite veins dissecting ijolite) and its exocontact glimmerites from an undisclosed alkalic-ultrabasic complex, Kola Peninsula, northwestern European USSR. The distribution and occurrence of Th-betafite as clusters or chain-like segregations of grains, which locally are connected with zones of apatitization and micatization, are suggestive of a metasomatic formation [12]. Tiny (0.15 to 1.5 mm) octahedral or cubo-octahedral crystals of X-ray amorphous Th-betafite occur in altered zones within ijolite of the Tsentral'nye massif, Cape Turii region, Kola Peninsula. Associated minerals are garnet (schorlomite), hornblende, diopside, cancrinite, natrolite, biotite, apatite, calcite, and perovskite. An enrichment of the latter mineral in the unaltered source rock serves as a source for Ti, Nb, and Th fixed in Th-betafite during metasomatic alteration processes preceding the intrusion of vein-like carbonatite bodies [13]. Arfvedsonite metasomatites, forming lens-shaped layers (of several m in size) at the contact between picritic porphyrite and coarse-banded lujavrite, from an undisclosed alkaline rock massif, Kola Peninsula, contain grains of thorian betafite of 1 to 50 μm in size. Accompanying minerals are a metamict hydrous silicate of Th, RE, and Mn (chemically similar to thorosteenstrupine, see "Thorium" Suppl. Vol. A 1b) and uranian betafite (with ThO_2 contents < 1%) [14].

An „aplite" vein (containing pyroxene, titanite, black garnet, magnetite, zircon, and orange magmatic calcite) cutting crystalline limestone (with pyroxene and salite) in mica gneiss near Slyudyanka, Lake Baikal area, Siberia, contains black betafite crystals forming octahedra and, sometimes, dodecahedra. Additionally, the "aplite" contains very small amounts of mendeleevite and allanite [15].

Sanidinite ejecta (phreatomagmatic explosion breccia) from Monte di Procida of the Campi Flegrei, Campania, Italy, contain, besides polymignite and zirkelite, reddish brown and isotropic crystalline calciobetafite as small octahedral crystals of 0.1 to 0.2 mm in size. All three minerals, sometimes, are intergrown mutually [16]. Note further that small (< 1 mm) pyrope-red betafite crystals, forming a well-developed icositetrahedron (trapezohedron), occur together with asbecasite, uranothorite, cryptomelane, and a new radioactive borosilicate (= hellandite-Th) in "sanidinitic" ejecta from Tre Croci near Viterbo, Latium, Italy [17].

Intergrowths. Partly intergrown with apatite grains are Th-betafite crystals from carbonatite of the 2nd stage and its exocontact glimmerites, Kola Peninsula [12]. For betafite-polymignite-zirkelite intergrowths, see above.

Inclusions of opaque rutile, making up some 15% of the sample, occur in betafite of the Hargeisa district, Somalia [3]. Small inclusions of apatite occur in Th-betafite of carbonatite of the 2nd stage and its exocontact glimmerites, Kola Peninsula [12]. Also found in betafite are parallel inclusions of magnetite [18].

References for 2.2.1.4.1:

[1] Omori, K.; Hasegawa, S.; Konno, H. (Sci. Rept. Tohoku Univ. III 6 [1960] 389/96, 392).

[2] Satterly, J. (Ann. Rept. Ontario Dept. Mines **65** VI [1956] 1/181).

[3] Daniels, J. L. (Somaliland Ministry Nat. Resources Mineral Resources Pamphl. No. 3 [1960] 1/25, 12, 14).

[4] Makarochkin, B. A.; Frank-Kamenetskii, V. A.; Gonibesova, K. A. (Geol. Geofiz. **1963** No. 9, pp. 32/51).

[5] Makarochkin, B. A.; Gonibesova, K. A.; Makarochkina, M. S. (Zap. Vses. Mineral. Obshch. **93** [1964] 54/9, 54/5).

[6] Kalita, A. P. (Redkozemel'nye Pegmatity Alakurtti i Priladozh'ya, Akad. Nauk SSSR, Moscow 1961, pp. 1/120).

[7] Lokka, L. (Bull. Comm. Geol. Finlande No. 82 [1928] 1/68, 21 and No. 149 [1950] 1/76, 33).

[8] Kalita, A. P. (in: Kuz'menko, M. V., Osobennosti Raspredeleniya Redkikh Elementov Pegmatitakh, Nauka, Moscow 1969, pp. 79/100, 84/5).

[9] Kalita, A. P. (Nov. Dannye Geol. Geokhim. Genezisu Pegmatitov 1965 266/304, 287).

[10] Rowe, R. B. (Econ. Geol. Rept. Geol. Surv. Can. No. 18 [1958] 1/108, 79).

[11] Hogarth, D. D. (Can. Mineralogist 6 [1957/61] 610/33, 611, 614).

[12] Kirnarskii, Yu. M.; Afanas'ev, B. V.; Men'shikov, Yu. P. (Mater. Mineral. Kol'sk. Polu-ostrova 6 [1968] 258/64, 259).

[13] Lapin, A. V. (in: Lebedeva, S. I.; Kuz'min, V. I.; Matias, V. V., Novoe v Mineralogicheskikh Issledovaniyakh, Moscow 1976, pp. 44/5).

[14] Kalinkin, M. M.; Men'shikov, Yu. P.; Polezhaeva, L. I. (in: Goryainov, P. M., Strukturnyi Kontrol' Orudeneniya v Magmaticheskikh i Metamorficheskikh Kompleksakh Kol'skogo Poluostrova, Kol'sk. Filial. Akad. Nauk SSSR, Apatity 1985, pp. 50/6, 52, 54/6).

[15] Tschernik [Chernik], G. (Bull. Soc. Franc. Mineral. 50 [1927] 485/9).

[16] Mazzi, F.; Munno, R. (Am. Mineralogist 68 [1983] 262/76, 262).

[17] De Casa, C. G.; Della Ventura, G. C.; Parodi, G. C.; Stoppani, F. S. (Riv. Mineral. Ital. 1987 No. 2, pp. 97/104, 99/100, 103/4).

[18] Palache, C.; Berman, H.; Frondel, C. (Dana's System of Mineralogy, 7th Ed., Vol. 1, Wiley, New York 1946, pp. 1/834, 803).

[19] Popov, V. I.; Bazhenova, L. F. (in: Materialy k Topomineralogii Urala, Sverdlovsk 1986, pp. 62/70, 64, 68).

[20] Meliksetyan, B. M. (Zap. Arm. Otd. Mineral. Obshch. No. 2 [1963] 57/80, 69).

2.2.1.4.2 Chemistry

Occurring in nature predominantly in the metamict state, betafites are minerals with a complicated and highly variable composition (see also Table 8, pp. 138/41) containing Ti, Nb(Ta), U, and Ca as the main components, and Na, Pb, Fe, RE, and H_2O are, partly, isomorphous admixtures. The water content seems to be secondary due to the metamict nature of the mineral (see also below) [1, p. 117]. For a compilation of 145 complete analyses, mostly containing <1% ThO_2, of minerals of the pyrochlore group, to which betafite belongs, see [2]. A review of the current (1985) understanding of the chemical and structural properties for minerals of the pyrochlore-microlite-betafite group, besides other Nb-Ta minerals, is given in [3].

According to Bonshtedt-Kupletskaya [2], the distinctive feature for individual mineral species within the pyrochlore group is the proportion between the main cations Nb, Ta, and Ti of group B (see below). Additional improvement is due to the predominating group-B cation. Thus, betafites proper are defined by Ti ≈ Nb > Ta with Ca and U. Another definition given in [4], see also [5], is 2 Ti ≧ Nb + Ta for group-B cations of betafite. But note that the analyses given in Table 8, pp. 138/41, with its mineral name as used in the original paper, do not always correspond to these definitions!

The main difficulty in classifying betafite with pyrochlore is chemical composition as can be seen from the following paragraphs: If both minerals are isostructural, the general formula could be written as $A_{16-x}B_{16}(O,OH)_{48}(F,OH)_8$, where A represents ions with radii 0.19 to 1.1Å and B with radii 0.65 to 0.75 Å. The ions Al^{3+}, P^{5+}, and Si^{4+} appear to be too small for the B position. This formula is in good agreement with about 70 analyses (published and new ones) of betafite and pyrochlore. Writing the formula as $8(A_{2/3}B_2O_6H_2O) \cdot n\ H_2O$, one considers betafite as a hydrous, uranian pyrochlore with 67% of the A ions missing. The role of water, there are two types as is suggested by DTA data (see pp. 147/8), in betafite and pyrochlore is unclear. There appears to be little relation between water determined by analysis and water necessary to balance the positive charge (for the details, see original paper). As follows from the frequency histogram showing the U content of betafite and pyrochlore, it seems logical to restrict the name betafite to a mineral with >15% U metal [7, pp. 617/20, 629/31].

Betafite and its varieties are hydrated titano-tantaloniobates of U and Ca with a pyrochlore structure and a general formula $A_{2-x}B_2(O,OH)_7$. Group-A cations are Ca, U, RE, and to a smaller extent Na, K, Mg, Pb, Mn, Fe^{2+}, and Th. The deficiency of group-A cations varies from 1.5 to 0.5. Group-B cations are Nb, Ti, Ta, Zr, Sn, Fe^{3+}, Al, and W. Betafites differ from pyrochlores by a higher content of Ti and U; their (Nb + Ta) : Ti ratio varies between 2.5 and 0.3 [8, p. 511]. For primary deficiencies in complex titano-niobo-tantalates, refer also to p. 75.

The analysis of Th-betafite (from Kola Peninsula) can be expressed by a formula of the type $A_{2-x}B_2(O,OH,F)_7$, corresponding to the pyrochlore group, characterized by high contents of TiO_2 and ThO_2, low Ta_2O_5, and negligible U_3O_8. A small surplus for the anionic part of this Th-betafite permits the assumption that a part of the water occurs as oxonium ion in group A. In a trigonal diagram (Ca + Na)-(U + Th)-(total RE), Th-betafite plots within the betafite field, and, although its U:Th ratio is very low (0.018) compared with other betafites ($\geqq 10:1$), it is viewed as an individual mineral species within the pyrochlore-microlite-betafite group [9, pp. 261/2].

Note also that betafite is viewed as the Ti-rich member of the pyrochlore group with a general formula $A_{1\ to\ 2}B_2X_6Y_{0\ to\ 1}$, where A = Ca, Na, U, Th, RE, Y, Ba, Sr, Bi, and Pb; B = Nb, Ti, Ta, Zr, Sn, Fe; X = O; and Y = O, OH, and F [10]. Or betafite is a complex Ca-U-Ti-Nb oxide of the pyrochlore group with a formula corresponding to $A_{2-m}B_2X_6Y_{0\ to\ 1} \cdot H_2O$ [11]. Other betafite formulas are given as $(U,Ca)_{2-x}(Nb,Ti,Ta)_2O_{6-x}(OH)_{1+x}$ [8, p. 511], (U,Ca)-$(Nb,Ta,Ti)_3O_9 \cdot n\ H_2O$ [12], or $(Ca,U,RE)_{2-m}(Ti,Nb,Ta)_2O_6(O,F)_{1-x} \cdot n\ H_2O$ [13]. Refer also to p. 143 for a formula type corresponding to the crystal structure.

Usually, betafite and pyrochlore varieties occurring in pegmatites are characterized by low contents of Na and K in the case of higher contents of high-valency cations, such as Y, U, and Th. In addition, there is a straight correlation between the contents of Y + Ce and U + Th, whereby diminished contents of Y and U characterize the strongly hydrated mineral species [14, p. 208].

Partial X-ray spectrographic analysis of two betafite specimens from the Oka complex, Canada, gives 8.2 and 8.0% ThO_2, 1.1 and 0.9% U_3O_8, and 35.5 and 32.8% Nb_2O_5, respectively [15]. Quantitative spectral analysis of a betafite of (post-miaskite) granite pegmatite without amazonite from Il'men Mountains, Southern Urals, gives 4.0% ThO_2, 14.2% U, 40% Nb_2O_5, 3.0% Ta_2O_5, and 14% TiO_2 [6, p. 66].

For the pH of a suspension of betafite, partly indicative of its chemical composition, chiefly of the predominant group-A cations, see p. 76.

136

In connection with precise measurements of the wave-length dependence of the refractive index, see pp. 146/7, X-ray fluorescence analyses with totally reflecting sample support (TRFA), see Table 7, were carried out for betafite samples from the Silver Crater mine, Bancroft area, Ontario, Canada, and from various locations of Madagascar. The optical and microscopical properties of the analyzed samples are as follows [16]:

sample number	location	color and microscopical features of betafite of Table 7
1955	Ontario	brownish black; inhomogeneous
3351	Ontario	dirty coffee brown; numerous crevices and holes, small inclusions
10719	Ontario	dark coffee brown with lighter intercalations
447	Betafo	yellow-brown-olive; inhomogeneous
2243	Ambalahazo	milky coffee brown to yellow-olive; inhomogeneous
10353	Betafo	olive, a few yellow to black phases, thin yellow outer zone; very inhomogeneous
14368	Betafo	tip black, vitreous otherwise gray-yellow to light olive-brown with a yellow outer zone; very inhomogeneous
1960	Ambolotara	nearly black with yellow inclusions; very inhomogeneous
5363	Betafo	inner part (a) yellow-olive with darker intercalations, outer part (b) brown
1995	Ambolotara	milky coffee brown; various phase microscopically visible
1887	Miarinarivo	dark olive to yellow-green, outer part gray-brown-olive; inhomogeneous
5362	Betafo	coffee brown with brown outer zone; fairly well inhomogeneous

Table 7
Chemical Composition of Betafites from Canada and Madagascar after [16].
For a description of the samples, see above.

	content in % in sample number						
	1955	3351	10719	477	2243	10353	14368
Ca	8.87	7.06	7.0	0.1	0.11	0.30	0.17
Ti	10.31	9.42	9.3	12.0	13.20	14.16	12.69
Mn	0.41	0.45	0.39	0.05	0.06	0.07	0.01
Fe	1.54	2.69	2.5	1.9	1.64	2.05	2.4
Sr	0.26	0.23	0.22	0.02	0.02	0.03	0.02
Y	0.09	0.10	0.14	0.03	0.055	0.03	0.30
Zr	0.02	0.02	0.02	0.05	0.03	–	–
Nb	23.39	20.52	23.2	19.7	18.71	19.17	19.70
Sn	0.08	0.06	0.04	0.09	0.11	0.09	0.08
Ba	0.1	0.52	0.5	0.7	0.36	0.52	0.5
Ta	2.50	3.01	2.2	3.0	2.70	2.22	3.32
W	0.07	0.03	0.07	0.62	0.59	0.22	0.35
Pb	1.36	1.56	1.41	1.91	1.99	1.83	1.785
Th	0.35	0.28	0.34	1.24	1.275	1.53	1.60
U	19.63	18.57	18.9	24.7	22.86	21.94	21.27
others	–	0.09[a]	–	–	–	–	–
l.o.i	8.8	10.4	11.5	10.6	10.9	9.6	10.8

Table 7 (continued)

| | content in % in sample number | | | | | |
	1960	5363a	5363b	1995	1887	5362
Ca	0.1	0.19	0.30	0.44	0.255	1.1
Ti	14.1	14.21	13.29	12.5	9.66	13.3
Mn	0.2	0.23	0.034	0.07	—	0.12
Fe	4.3	4.93	2.18	8.7	1.175	1.1
Sr	0.03	0.05	0.02	0.02	0.065	0.04
Y	0.02	0.06	0.13	0.19	0.47	9.3
Zr	0.06	0.03	—	1.60	—	0.18
Nb	22.5	20.09	20.86	21.1	23.79	15.4
Sn	0.09	0.09	0.10	0.12	0.33	0.05
Ba	0.6	—	0.30	1.0	0.60	0.4
Ta	3.3	3.74	3.6	3.7	3.065	2.7
W	0.59	0.23	0.37	0.34	0.35	0.31
Pb	1.43	1.75	1.74	2.76	2.39	1.09
Th	1.32	1.54	1.61	1.75	1.665	3.41
U	18.7	15.34	19.95	7.69	18.325	12.8
others	—	0.04[b]	—	0.14[c]	7.075[d]	—
l.o.i	10.1	11.0	12.6	9.50	13.3	5.3

[a] Comprises 0.07% Cu and 0.09% Ga. — [b] Comprises 0.02% Cu and 0.022% Ga. — [c] Comprises Bi only. — [d] Comprises 0.02% Cu, 0.01% Ga, 0.695% Ce, and 0.06% Bi.

Table 8
Chemical Composition and Density of Betafites from Various Types of Occurrence.
For the footnotes, see p. 141; tr. = traces.

	content in % in sample number									
	1	2	3	4	5	6	7	8	9	10
ThO_2	25.75	21.210	12.84	11.19	7.23	5.12	4.5	3.8	3.58	2.81
U_3O_8	–	1.015	6.03 c)	0.20	1.83	30.07 g)	4.8 c)	16.8	12.70	13.18
CaO	12.42	4.388	–	14.14	17.11	–	16.0	3.5	7.09	8.24
Na_2O	0.87	–	0.20	3.88	2.88	–	1.7	–	0.46	–
K_2O	0.52	–	–	0.02	–	–	–	–	0.32	–
ΣRE_2O_3	3.92	4.350	10.69	5.32	2.40	–	7.3	18.3	3.75	1.92
ZrO_2	1.30	–	0.86	1.72	1.45	–	1.0	0.4	–	–
TiO_2	15.86	31.260	0.11	12.52	10.42	20.95	15.0	27.7	8.34	11.89
Nb_2O_5	29.28	9.447	15.20	45.57	40.53	22.45	32.9	19.7 i)	48.40 i)	48.82
Ta_2O_5	0.35	0.237	27.87	1.30	3.08	13.20	2.4	–	–	–
Fe_2O_3	0.78	0.710	3.40	0.56	–	–	–	1.3	3.73	2.28
FeO	–	–	0.75	0.54	1.80	–	2.0	–	–	–
SiO_2	1.23	–	1.20	0.59	–	1.13	–	–	–	–
H_2O^+	4.37 a)	–	4.47	1.35	7.50	7.0 a)	–	3.7	5.31	7.21
H_2O^-	–	–	1.20	0.15	–	–	–	0.2	3.43	1.03
others	3.43	–	15.11 d)	4.32 e)	4.09 f)	0.13 h)	–	2.3 j)	1.68 k)	3.38 l)
sum	100.08	73.621	99.93	100.34	100.30	100.3	87.6	96.7	99.49	100.76
D in g/cm³	4.36	–	–	4.46	–	–	–	–	–	–

No. 1 = Th-betafite from a metasomatized zone, Turii carbonatite complex, Kola Peninsula, USSR [17]. – No. 2 = Thorian betafite from arfvedsonite metasomatite, Kola Peninsula, USSR [18]. – No. 3 = Yttrobetafite from granite pegmatite, eastern Baltic Shield, USSR [19]. – No. 4 = Th-betafite from phlogopite-diopside-calcite vein cutting ijolite, Kola Peninsula, USSR [9, p. 260]. – No. 5 = "Betafite" (thorian pyrochlore) from carbonate rock (biotitized rock [15]), Oka, Quebec, Canada [7, pp. 611, 614/5]. – No. 6 = Betafite (No. 3 from collection of V.P. Ivanov) [1, p. 119]. – No. 7 = Ca-betafite from ejected sanidinitic material, Campi Flegrei, Campania, Italy [20]. – No. 8 = Betafite from Crocker's Well, Australia [21]. – Nos. 9 and 10 = Blomstrandite from syenitic vein, Urals, USSR [22, pp. 39/41].

Table 8 (continued)

	content in % in sample number								
	11	12	13	14	15	16	17	18	19
ThO_2	2.80	2.7	2.5	1.8	1.71	1.60	1.56[s]	1.30	1.30
U_3O_8	17.69	10.0	17.0	19.0	13.60	23.60	21.69[t]	26.60[g]	26.37[g]
CaO	3.58	4.2	2.3	5.2	8.60	0.22	4.35	3.45	3.93
Na_2O	0.21	2.7	0.3	0.2	0.23	0.06	–	–	–
K_2O	0.33	–	–	–	0.08	0.18	–	–	–
ΣRE_2O_3	4.54	19.8	7.5	3.1	1.41	2.30	–	1.50	4.72
ZrO_2	–	0.2	0.03	0.1	–	0.14	–	–	–
TiO_2	16.65	22.0	11.0	12.0	9.90	18.40	14.88	18.30	16.51
Nb_2O_5	24.00	30.0	46.0	34.0	42.75	33.30	11.54[i]	34.80	37.36
Ta_2O_5	6.62	2.1	3.4	3.3	8.65	2.50	–	–	1.46
Fe_2O_3	4.70	2.0	3.1	2.6	–	1.70	2.76	2.87	2.25
FeO	–	–	–	–	2.19	0.20	–	–	1.20
SiO_2	4.52	1.9	5.0	2.6	–	0.90	3.01	–	0.59
H_2O^+	8.49	–	–	–	7.16	6.90	13.80[a]	7.60[a]	2.47[a]
H_2O^-	2.66	–	–	–	2.52	3.31	–	–	–
others	4.10[m]	2.7[n]	2.4[o]	4.4[p]	1.99[q]	4.52[r]	26.37[u]	2.80[v]	0.99[w]
sum	100.94	100.4	100.5	88.3	100.79	99.83	99.96	99.22	99.15
D in g/cm^3	3.395	–	–	–	–	–	3.9	4.17	–

No. 11 = Betafite (purified material) from acid pegmatite cutting gabbro, Hargeisa district, Somalia [23]. – No. 12 = Betafite from granite pegmatite dike, Bancroft area, Ontario, Canada [24, pp. 20, 162]. – No. 13 = Betafite from granite pegmatite dike, Hastings county, Ontario, Canada [24, pp. 20, 135]. – No. 14 = Betafite from carbonate body in gneiss, Hastings county, Ontario, Canada [24, pp. 20, 125]. – No. 15 = Blomstrandite from syenitic pegmatite vein, Urals, USSR [25]. – No. 16 = Betafite (altered zone with mineral impurities of kaolinite and bauxite) from Betafo, Madagascar [26]. – No. 17 = Betafite from Kaijo near Mukden, Manchuria, China [27]. – No. 18 = Betafite from granite pegmatite, Ambolotara, Madagascar [28]. – No. 19 = Betafite from aplite vein, Slyudyanka, Lake Baikal area, Siberia, USSR [29]; refer also to [12].

Table 8 (continued)

| | content in % in sample number | | | | | | | | |
	20	21	22	23	24	25	26	27	28
ThO_2	1.25	1.20	1.19	1.17	1.14	1.12	1.10	1.07	1.04
U_3O_8	20.60[g]	12.84	21.11	21.80	27.28[ad]	27.15[g]	9.93[c]	19.95[g]	27.12[g]
CaO	11.61	2.23	2.81	4.37	3.06	3.12	5.02	14.09	4.82
Na_2O	–	0.20[y]	–	–	0.10	–	–	1.25	–
K_2O	–	–	–	–	0.20	0.38	–	–	–
ΣRE_2O_3	1.20	11.63	2.04	0.566[ab]	4.21	1.00	8.12	1.10	1.56
ZrO_2	–	0.11	–	0.243	–	–	–	–	–
TiO_2	17.30	15.20	15.07	18.5	12.60	16.20	11.52	12.60	23.20
Nb_2O_5	32.10	27.87	34.11	24.3	25.07	34.80	29.34	40.74	21.40
Ta_2O_5	tr.	7.73	4.94	15.4	6.30	1.00	14.25	2.24	7.12
Fe_2O_3	1.38	4.30	3.23	–	1.07	0.50	3.25	1.90	1.94
FeO	–	–	–	1.76	–	–	–	–	–
SiO_2	–	3.40	–	–	1.83	–	4.49	0.56	–
H_2O^+	5.20[a]	4.47	5.35	–	15.99[a]	12.50[a]	5.97	2.95	8.10[a]
H_2O^-	–	1.20	7.10	4.51	–	–	2.86	–	–
others	0.75[x]	2.55[z]	3.40[aa]	4.05[ac]	0.58[ae]	1.87[af]	2.06[ag]	0.58[ah]	0.32[ai]
sum	99.39	99.93	100.35	96.4	100.43	99.64	100.20	99.39	100.62
D in g/cm³	4.475	4.9	–	–	–	3.75	3.70	–	–

No. 20 = Betafite from Ambatolampikely, Madagascar [30]. – No. 21 = Yttrobetafite from granite pegmatite (Alakurtti [8, p. 513]), eastern Baltic Shield, USSR [14, pp. 202/3]; refer also to [19]. – No. 22 = Betafite from corundum pegmatite, Urals, USSR [22, pp. 32, 36]. – No. 23 = Betafite from Haichun, China [31]. – No. 24 = Betafite (No. 1 from collection of V. K. L'vov) [1, p. 119]. – No. 25 = Betafite from granite pegmatite, Ambolahazo, Madagascar [30]; refer also to [12], [8, p. 512]. – No. 26 = Yttrobetafite (corresponds to wiikite-Le after Lokka) from pegmatite, Priladozh'ya, Karelia, USSR [32]. – No. 27 = Betafite from early (pre-miaskite) granite pegmatite, Il'men Mountains, Southern Urals, USSR [6, pp. 64/5]. – No. 28 = Betafite (No. 4 from collection of A. A. Kukharenko) [1, p. 119].

Footnotes for Table 8:

[a] Total water. — [b] Comprises 2.06% Al_2O_3, 0.50% MnO, 0.67% VO_3, and 0.2% F. — [c] Given as UO_2. — [d] Comprises 4.30% Al_2O_3, 2.23% MnO, 7.73% PbO, 0.55% SnO, and 0.3% loss on ignition. — [e] Comprises 0.45% Al_2O_3, 0.16% MnO, 0.26% MgO, 0.17% SrO, 0.13% BaO + PbO, 2.96% F, and 0.19% P_2O_5.

[f] Comprises 1.17% MnO, 0.1% PbO, 0.65% SrO, and 2.17% F. — [g] Given as UO_3. — [h] Only Al_2O_3. — [i] Includes Ta_2O_5. — [j] Comprises 1.6% Al_2O_3 and 0.7% SnO_2.

[k] Comprises 0.89% Al_2O_3, 0.1% MnO, 0.46% PbO, and 0.23% SnO_2. — [l] Comprises 1.58% MgO, 0.31% MnO, 0.31% PbO, and 1.18% BaO. — [m] Comprises 1.5% Al_2O_3, 0.18% MnO, 0.44% MgO, 0.93% PbO, and 1.05% F. — [n] Comprises 0.2% Al_2O_3, 0.3% MnO, 0.2% MgO, 1.8% PbO, and 0.2% SnO_2. — [o] Comprises 0.4% Al_2O_3, 0.7% MnO, 0.52% MgO, 0.4% PbO, and 0.4% SnO_2.

[p] Comprises 0.3% Al_2O_3, 0.6% MnO, 0.4% MgO, 2.9% PbO, and 0.2% SnO_2. — [q] Comprises 1.61% PbO, 0.26% MgO, and 0.12% MnO. — [r] Comprises 1.3% Al_2O_3, 2.35% PbO, 0.03% SrO, 0.5% BaO, 0.04% MnO, and 0.3% SnO_2. — [s] Comprises $ThO_2 + \Sigma Y_2O_3$. — [t] Comprises 18.06% UO_3 and 3.63% UO_2.

[u] Comprises 15.68% Al_2O_3, 1.83% MnO, 0.92% MgO, and 7.94% SnO_2, etc. — [v] Comprises 2.1% Al_2O_3, 0.4% MgO, and 0.3% SnO_2. — [w] Comprises 0.24% Al_2O_3, 0.15% MnO, 0.13% MgO, 0.1% PbO, and 0.37% SnO_2. — [x] Comprises 0.5% Al_2O_3 and 0.25% MnO. — [y] Includes K_2O.

[z] Comprises 0.86% Al_2O_3, 0.75% MnO, 0.55% PbO, 0.09% MgO, and 0.3% F. — [aa] Comprises 0.55% MgO, 0.16% MnO, 0.21% SnO_2, and 2.48% PbO. — [ab] Only Y_2O_3. — [ac] Comprises 1.65% Al_2O_3, 0.358% MnO, 0.0742% MgO, 0.834% PbO, 1.07% Sb_2O_3, and 0.0632% WO_3. — [ad] Comprises 25.2% UO_3 and 2.08% UO_2. — [ae] Only MgO.

[af] Comprises 1.5% Al_2O_3 and 0.37% SnO_2. — [ag] Comprises 0.17% MnO and 0.89% MgO. — [ah] Only MnO. — [ai] Only Al_2O_3.

References for 2.2.1.4.2:

[1] Komkov, A. I.; Dubik, O. Yu. (Rentgenogr. Mineral'n. Syr'ya **1979** 117/29).

[2] Bonshtedt-Kupletskaya, E. M. (Zap. Vses. Mineral. Obshch. **95** [1966] 134/44).

[3] Černý, P.; Ercit, T. S. (Bull. Mineral. **108** [1985] 499/532).

[4] Greegor, R. B.; Lytle, F. W.; Chakoumakos, B. C.; Lumpkin, G. R.; Ewing, R. C. (Mater. Res. Soc. Symp. Proc. **50** [1986] 387/92 from C.A. **105** [1986] No. 87 167).

[5] Fleischer, M. (Report of the Subcommittee on Pyrochlore Group. Memorandum to the Commission on New Minerals and Mineral Names, U.S. Dept. Interior, Geol. Surv. **1974**, August 7, pp. 1/150 from [6, p. 62]).

[6] Popov, V. I.; Bazhenova, L. F. (in: Materialy k Topomineralogii Urala, Sverdlovsk 1986, pp. 62/70).

[7] Hogarth, D. D. (Can. Mineralogist **6** [1957/61] 610/33).

[8] Kalita, A. P.; Mukhitdinov, G. N. (in: Vlasov, K. A., Geochemistry and Mineralogy of Rare Elements and Genetic Types of Their Deposits, Vol. 2, Israel Program Sci. Transl., Jerusalem 1966, pp. 511/6 [Russian original, Moscow 1964]).

[9] Kirnarskii, Yu. M.; Afanas'ev, B. V.; Men'shikov, Yu. P. (Mater. Mineral. Kol'sk. Poluostrova **6** [1968] 258/64).

[10] Eyal, Y.; Lumpkin, G. R.; Ewing, R. C. (Mater. Res. Soc. Symp. Proc. **50** [1986] 379/86 from C.A. **105** [1986] No. 87 166).

[11] Lumpkin, G. R.; Ewing, R. C.; Eyal, Y. (J. Mater. Res. **3** [1982] 357/68 from C.A. **108** [1988] No. 207 819).

[12] Palache, C.; Berman, H.; Frondel, C. (Dana's System of Mineralogy, 7th Ed., Vol. 1, Wiley, New York 1946, pp. 1/834, 803/5).

[13] Chukhrov, F. V.; Bonshtedt-Kupletskaya, E. M. (Mineraly Spravochnik, Vol. 2, Pt. 3, Nauka, Moscow 1967, pp. 1/676, 172).

[14] Bykova, A. K.; Kukharchik, M. V. (Tr. Inst. Mineral. Geokhim. Kristallokhim. Redk. Elem. No. 8 [1962] 201/11).

[15] Rowe, R. B. (Geol. Surv. Can. Econ. Geol. Rept. No. 18 [1958] 1/108, 79/80).

[16] Junge, W.; Knoth, J.; Rath, R. (Neues Jahrb. Mineral. Abhandl. 147 [1983] 169/83, 170/1, 175).

[17] Lapin, A. V. (in: Lebedeva, S. I.; Kuz'min, V. I.; Matias, V. V., Novoe v Mineralogicheskikh Issledovaniyakh, Moscow 1976, pp. 44/5).

[18] Kalinkin, M. M.; Men'shikov, Yu. P.; Polezhaeva, L. I. (in: Goryainov, P. M., Strukturnyi Kontrol' Orudeneniya v Magmaticheskikh i Metamorficheskikh Kompleksakh Kol'skogo Poluostrova, Kol'sk. Filial. Akad. Nauk SSSR, Apatity 1985, pp. 50/6, 56).

[19] Kalita, P. (Nov. Dannye Geol. Geokhim. Genezisu Pegmatitov 1965 266/304, 286).

[20] Mazzi, F.; Munno, R. (Am. Mineralogist 68 [1983] 262/76, 264).

[21] Whittle, A. W. G. (J. Geol. Soc. Australia 2 [1955] 21/45, 34).

[22] Makarochkin, B. A.; Frank-Kamenetskii, V. A.; Gonibesova, K. A. (Geol. Geofiz. 1963 No. 9, pp. 32/51).

[23] Daniels, J. L. (Somaliland Ministry Nat. Resources Mineral Resources Pamphl. No. 3 [1960] 1/25, 12).

[24] Satterly, J. (Ann. Rept. Ontario Dept. Mines 65 VI [1956] 1/181).

[25] Makarochkin, B. A.; Gonibesova, K. A.; Makarochkina, M. S. (Zap. Vses. Mineral. Obshch. 93 [1964] 54/9, 56).

[26] Van Wambeke, L. (Neues Jahrb. Mineral. Abhandl. 112 [1969/70] 117/49, 135).

[27] Kawai, T. (Nippon Kagaku Zasshi 81 [1960] 1219/20; C.A. 1961 7173).

[28] Lacroix, A. (Bull. Soc. Franc. Mineral. 33 [1910] 321/3 and 35 [1912] 84/92, 88/9).

[29] Tschernik [Chernik], G. (Bull. Soc. Franc. Mineral. 50 [1927] 485/9, 486/7).

[30] Lacroix, A. (Mineralogie de Madagascar, Vol. 1, Challamel, Paris 1922, pp. 1/624, 384).

[31] Nagashima, K.; Yano, Y.; Hamada, M. (Bunseki Kagaku 33 [1984] T91/T94, T94).

[32] Kalita, A. P. (Redkozemel'nye Pegmatity Alakurtti i Priladozh'ya, Akad. Nauk SSSR, Moscow 1961, pp. 1/120, 94).

2.2.1.4.3 Crystal Form and Structure

Crystal Form. Betafite is cubic, $O_h - m3m$, X-ray amorphous, and after heating reconstitutes its original structure [1] or forms a pyrochlore phase [2, p. 173]; refer also to pp. 143/5 and 147/8.

The morphology of betafite crystals is dominated by the octahedron or the dodecahedron (refer also to [3, pp. 170/1]); additional forms observed are: cube, tetrahexahedron {230}, trisoctahedron {233} [2, p. 173], and the trapezohedrons {112}, {113}, {223}, and {332} [1]. Crystals of metamict Th-betafite with {111} and {100} occur in carbonatite of the 2nd stage and its exocontact glimmerites, Kola Peninsula, northwestern European USSR [4, p. 259]. Betafite crystals from Somalia show dominant octahedron and, occasionally, small faces of the rhombic dodecahedron [5].

The habit of natural betafite crystals is octahedral, rarely it is dodecahedral. The crystals often are flattened parallel to (110), or stretched after one of the 4- or 3-fold axis; rarely they are flattened parallel to (100). Mutual intergrowth of crystals is common [2, p. 173].

Twins of betafite are not observed, and cleavage is absent [2, p. 173].

Crystal structure. The space group for betafite is O_h^7—Fd3m (No. 227); Z = 8, and lattice parameters of recrystallized, annealed material (see also the following paragraphs) vary between 10.20 and 10.42 Å [6, p. 511], or between 10.28 and 10.36 Å, which is noticeable lower than for pyrochlores with 10.37 to 10.45 Å (see also next paragraph) [4, p. 262]. For individual betafite varieties are given: 10.2978 Å for calciobetafite (Campi Flegrei, Italy) [7, p. 266]; 10.32 [8] or 10.352 Å for Th-betafite (Cape Turii, Kola Peninsula) [4, p. 262]; and 10.34 or 10.42 Å for blomstrandite (Urals) according to two different investigators [9]. In accordance with observations given in [10], the cell parameters for betafite and pyrochlore from several types of Canadian occurrences appear to decrease as the number of Ti and Fe ions increase in the B position of the structure. Spectrographic analyses do not indicate a relation between cell size and content of Al and Fe [11, pp. 623/4].

The crystal structure of betafite, as restored after ignition, corresponds to that of pyrochlore (see below), which was described at first by [12] using X-ray powder diffraction data. Structure refinement by data collected from a single crystal of calciobetafite, Campi Flegrei, Italy, corresponds well to a formula of the type $(Ca, RE, Th, U, etc.)_2^{VIII}(Nb, Ti, etc.)_2^{VI}O_6(O, F)$ (and displays a relationship with the structures of zirkelite, zirconolite, and polymignite) [7, pp. 266/7].

As is partly summarized in [2, pp. 177/8], heating of metamict betafites leads to several crystalline phases, but pyrochlore is predominant: Thus, two betafites from Madagascar and one sample from Slyudyanka, Siberia, (chemically similar, see [13], and corresponding to sample Nos. 18, 19, and 25 of the Table 8, pp. 139/40) heated in air up to 1300°C yield a cubic pyrochlore phase (with 10.29 to 10.30 kX), a hexagonal UTa_2O_8 phase (with a = 6.41 and c = 3.95 kX) and a tetragonal rutile phase (with greater lattice constants of a = 4.58 and c = 2.95 kX). Heating at 1600°C for 30 min changes the UTa_2O_8 phase to UO_2, and the rutile phase to tetragonal strueverite (with a = 4.60 to 4.63, and c = 2.97 to 3.02 kX) [14]. For the detailed discussion of the X-ray powder diffraction data, and of the recrystallization process (considering also conditions of synthesis), see [15]. Recrystallization between 700 and 1000°C of betafite from Madagascar gives two types of X-ray pattern: A mixture of pyrochlore and rutile in various proportions; or a mixture of two hexagonal, pseudocubic phases identical to those obtained as synthesis products on heating oxides of Nb and U [16]. But note that [2, p. 177] erroneously gives for the latter type, two hexagonal and one pseudocubic phases. Heating at 1000°C for 4 h of a betafite from the Urals (sample No. 22 of Table 8, p. 140) results in two cubic phases, a face-centered (A) with a = 5.187 Å and a predominating body-centered (B) with a = 6.401 Å. Mendeleevite from the Urals (with 1.25% ThO_2) gives predominantly phase A with a = 5.142 Å and small amounts of phase B with a = 6.356 Å [17]. Annealing in air and in argon up to a maximum of ~1000°C (with a rate of 12°C/min) of a number of pyrochlore and betafite samples from Canada (including a few samples with >1% ThO_2) yields mainly a cubic phase with a = 10.27 to 10.39 ± 0.02 Å for betafite; for pyrochlores, a is 10.347 to 10.435 ± 0.005 Å. The size of the unit cell may depend on temperature of ignition; at higher temperatures a shrinkage occurs owing to, perhaps, the reorganization of crystallites, or a further expulsion of water [11, pp. 621/3].

X-ray diffraction (Guinier method) investigation of selected betafite samples from Madagascar and Canada (for its description and chemistry, see pp. 136/7) gives, for the unheated state, the following results: (1) Only very weak betafite reflexes, corresponding to the strongest ones of the standard sample used (= Powder Diffraction File No. 17—746, see [18]), occur for the three Canadian samples. (2) Besides the betafite reflexes for planes (222), (400), (440), and (622), there are additional weak, strongly broadened, and diffuse lines for the sample Nos. 477, 2243, 10353, and 14368 from Madagascar. (3) None of the above betafite reflexes, but a number of other ones, occur for sample Nos. 1960 and 5363a,b (inner and outer part, respectively). (4) Transitional to a completely metamict state (= sample No. 5362) are sample Nos. 1987 and 1995, the former displaying only four reflexes. Heating in air at 1000°C for 4 h improves the degree of crystallization and enlarges both number and intensity of reflexes; some new phases also occur. Exemplified by one sample each from the above groups (1) and (2), sample No. 1935 displays stronger reflexes for a betafite phase with a = 10.27 Å together with reflexes for a rutile-like phase with negligibly expanded lattice; sample No. 2243 does not show reflexes of the cubic betafite phase, instead of which reflexes appear similar to so-called Nb-rutile and to pyrochlore. Whereas very small deviations (negligible shift of lines, small changes in intensity) occur for the group (3) samples, stronger differences are observed for sample Nos. 1987 and 1995; differences are strongest for sample No. 5362, which most largely corresponds with euxenite from Nipissing district, Ontario. These results correspond to data given in [13], who also yields a pyrochlore and a rutile phase, but his UTa_3O_8[*] is not proved [3, pp. 171/2].

With increasing deficiencies in group-A cations (see p. 135) of betafites, the following mixtures of phases may be obtained after heating in air to 1000°C: pyrochlore; pyrochlore + rutile; pyrochlore + UNb_2O_8 + rutile; or rutile + UNb_2O_8. In highly altered betafites Nb_2O_5 may also appear. Primary and especially secondary deficiencies (for definition, see p. 75) are responsible mainly for the complex X-ray diffraction patterns observed [19]. Differences in the chemical composition of the betafite starting material effect the various products (phases) and their relative proportion, obtained during heat treatment at different conditions of temperature and time (including sample Nos. 6, 24, and 28 of the Table 8 on pp. 138 and 140). Thus, the following phases may be observed after heating in air at 1000°C for 3 h: a hexagonal phase U, representing an oxidic compound of U, Nb(Ta), and Ti, with varying lattice constants for the individual samples investigated; a phase U_1 representing on oxidic compound of U:Nb(Ta):Ti with 1:2:1; and phases R_U or R_Y of the rutile or pyrochlore type, respectively. For the details, see original paper [20].

Besides data for betafites and mendeleevite from the literature, the following indexed d values (with arbitrary intensities ≧5; for the others, see original paper) are given for X-ray amorphous thorian betafite from the Kola Peninsula, USSR, after heating at 900°C for 1 h [4, pp. 262/3]; the data for calciobetafite from Campi Flegrei, Italy, are calculated from the experimental lattice parameters [7, p. 272]:

	thorian betafite			calciobetafite	
hkl	d in Å	I	hkl	d in Å	I
(222)	2.975	10	(222)	2.973	100
(400)	2.58	5	(400)	2.574	27
(440)	1.826	10	(331)	2.363	5
(622)	1.558	10	(440)	1.820	53
(444)[*]	1.490	5	(622)	1.552	45
(800)	1.291	6	(444)	1.486	10
(662)	1.185	8	(800)	1.287	7
(840)	1.555	7	(662)	1.181	16
(844)	1.0549	9, broad	(840)	1.151	13
			(844)	1.051	12

[*] In the original paper only given as (44).

[*] Erroneously for UTa_2O_8, see p. 143.

Other examples for d values with arbitrary intensities (≥ 3; for the others, see original paper) of recrystallized betafite (bet.) and its varieties are given below for betafite (= sample No. 13-197 of the ASTM file) from Basin Property, Bancroft, Ontario, Canada [1]; for betafite from Antanifotsy, Madagascar [6, p. 514]; for blomstrandite from the Urals [9]; and for yttrobetafite (with 0.9% ThO_2) from Alakurtti, Kola Peninsula, USSR [21]:

bet., Ontario		bet., Madagascar		blomstrandite		yttrobetafite	
d in Å	I	d in Å	I	d in Å	I	d in Å	I
2.96	100	2.98	10	3.152	4	3.258	8
2.56	20	2.49	3	2.955	5	2.985	7
1.814	45	1.82	8	1.822	10	2.501	6
1.546	40	1.55	6	1.555	8	1.694	10
1.482	15	1.154	3	1.185	4	1.631	3
1.284	10			1.154	5	1.180	6
1.179	20			1.055	7	1.104	3
1.148	15			0.9963	7		

For d values inspect also the papers given above (on pp. 143/4) for recrystallization of betafite during heating.

References for 2.2.1.4.3:

[1] Weiner, K. L.; Hochleitner, R. (Lapis **9** No. 4 [1984] 5/6).
[2] Chukhrov, F. V.; Bonshtedt-Kupletskaya, E. M. (Mineraly Spravochnik, Vol. 2, Pt. 3, Nauka, Moscow 1967, pp. 1/676).
[3] Junge, W.; Knoth, J.; Rath, R. (Neues Jahrb. Mineral. Abhandl. **147** [1983] 169/83).
[4] Kirnarskii, Yu. M.; Afanas'ev, B. V.; Men'shikov, Yu. P. (Mater. Mineral. Kol'sk. Poluostrova **6** [1968] 258/64).
[5] Daniels, J. L. (Somaliland Ministry Nat. Resources Mineral Resources Pamphl. No. 3 [1960] 1/25, 14).
[6] Kalita, A. P.; Mukhitdinov, G. N. (in: Vlasov, K. A., Geochemistry and Mineralogy of Rare Elements and Genetic Types of Their Deposits, Vol. 2, Israel Program Sci. Transl., Jerusalem 1966, pp. 511/6 [Russian original, Moscow 1964]).
[7] Mazzi, F.; Munno, R. (Am. Mineralogist **68** [1983] 262/76).
[8] Lapin, A. V. (in: Lebedeva, S. I.; Kuz'min, V. I.; Matias, V. V., Novoe v Mineralogicheskikh Issledovaniyakh, Moscow 1976, pp. 44/5).
[9] Makarochkin, B. A.; Gonibesova, K. A.; Makarochkina, M. S. (Zap. Vses. Mineral. Obshch. **93** [1964] 54/9).
[10] Ginsburg, A. I.; Gorzhevskaya, S. A.; Erofeeva, E. A.; Sidorenko, G. A. (Geokhimiya **1958** 486/500; Geochem. [USSR] **1958** 615/36, 621).

[11] Hogarth, D. D. (Can. Mineralogist **6** [1957/61] 610/33).
[12] von Gärtner, H. R. (Neues Jahrb. Mineral. Geol. Paläontol. Abhandl. A **61** [1930] 1/30).
[13] Gasperin, M. (Bull. Soc. Franc. Mineral. Crist. **81** [1958] 116/20, 117/8).
[14] Gasperin, M. (Bull. Soc. Franc. Mineral. Crist. **80** [1957] 232/4).
[15] Gasperin, M. (Bull. Soc. Franc. Mineral. Crist. **83** [1960] 1/21, 8/14).

[16] Orcel, J.; Fauquier, D. (Chemistry Earth Crust, Transactions Geochem. Conf. Celebrating 100th Anniv. Birth Acad. A. I. Vernadskii, Moscow 1964, Vol. 2, pp. 340/9, 342 [Russian original, Moscow 1964]).

[17] Makarochkin, B. A.; Frank-Kamenetskii, V. A.; Gonibesova, K. A. (Geol. Geofiz. **1963** No. 9, pp. 32/51, 33).

[18] Berry, L. G. (Powder Diffraction File, Sets 11 — 15 (revised) and 16 to 18. Inorganic Vol. No. PDIs — 15iRB, Joint Committee on Powder Diffraction Standards, Swarthmore, Pennsylvania, 1972, pp. 1/1081 from [3, p. 182).

[19] Van Wambeke, L. (Neues Jahrb. Mineral. Abhandl. **112** [1969/70] 117/49, 141/2).

[20] Komkov, A. J.; Dubik, O. Yu. (Rentgenogr. Mineral. Syr'ya Voronezh **1979** 117/29).

[21] Kalita, A. P. (Redkozemel'nye Pegmatity Alakurtti i Priladozh'ya, Akad. Nauk SSSR, Moscow 1961, pp. 1/120, 50).

2.2.1.4.4 Optical and Other Physical Properties. Chemical and Thermal Behavior

Optical Properties. The color of betafite, often irregularly distributed, is cinnamon, greenish brown, yellowish, green, dark brown, grayish black to black [1, p. 174], refer also to [2, pp. 170/1]; or black to reddish brown for samples from Ontario, Canada, changing to yellow-brown or greenish brown by alteration (hydration) [3], and olive green for a sample from Somalia [4]. Betafite from the Oka complex, Canada, is reddish brown on smooth surfaces, dark purplish gray on rough surfaces [5]. Metamict Th-betafite from carbonatite of the 2nd stage and its exocontact glimmerites, Kola Peninsula, is characteristically red-brown, crystal faces are brown or brownish cinnamon [6, p. 259]. Irregularly distributed greenish dark (core more densely, rim lighter colored), partly with a reddish brown tint, is blomstrandite from the Urals [7], refer also to [8, p. 55].

In transmitted light (under the microscope), betafite is: light yellowish to nearly colorless, sometimes also brown [1, p. 174]; yellowish green to colorless (sample from Somalia) [4]; or yellowish to cinnamon (blomstrandite from the Urals) [7], refer also to [8, p. 55]. In reflected light, betafite is: light gray with a brownish tint [1, p. 174]; light yellowish in the core and cinnamon in the peripheral part (samples from the Urals) [7]; or cinnamon in the core and light yellowish in the rim (blomstrandite from the Urals) [7], refer also to [8, p. 55].

Betafite is transparent to opaque (in dark colored specimens) [1, p. 174]; or translucent yellowish in thin splinters (samples from the Urals) [7].

Luster of betafite is fatty to vitreous, also waxy [1, p. 174]. Th-betafite from carbonatites of the 2nd stage and its exocontact glimmerites, Kola Peninsula, is vitreous on crystal faces, and lustrous resinous on the fracture [6, p. 59].

Refractive indices of betafite, as measured for thermally nontreated samples, are given as: between <1.89 and 2.10, mainly dependent on the degree of hydration [1, p. 174]; between 1.915 and 1.925 [9]; between 1.89 and 1.92 (Somalia) [4]; or between 1.91 and 1.915 (Urals) [7]. Betafite has a refractive index between 1.895 and 1.925, after heating it raises to 2.02 [10]. Th-betafite (Kola Peninsula) has $n = 2.02$ [11] or 2.025 [6, p. 259], and for blomstrandite (Urals) it varies between 1.95 and 1.99 [7]. Based on precise measurements of the refractive index in relation to wavelength (λ between 316 and 2505 nm), and regarding the crystal-chemical formula given by [12, p. 617], it is shown for selected betafite samples (for its description and chemistry, see pp. 136/7) that the refractive index for $\lambda = 337$ nm (a region with most pregnant changes) increases with increasing content of Fe^{2+}, whereas U and Ti are of minor influence, and Nb has practically none. For the detailed discussion, see original paper [2, pp. 176/82].

Reflectivity is about 13% for Th-free betafite (with n = 2.10) from Vishnevye Gory, Southern Urals [13], refer also to [1, p. 174].

The IR spectrum of betafite displays a clear line at 1250 cm^{-1} and a broad region between 2380 and 3570 cm^{-1} [1, p. 174].

In contrast to many other uranium minerals, betafite is nonfluorescent [14].

Other Physical Properties. Mohs hardness for betafite is between 4 and 5 [1, p. 173] or 5½ [3]. X-ray amorphous Th-betafite (Cape Turii, Kola Peninsula) has a microhardness between 400 and 450 kg/mm^2, corresponding to H_M = 4.5 [6, p. 259]. Microhardness is between 313 and 676 kg/mm^2 (at 100 g load) or 350 and 490 kg/mm^2 (at 50 g load) for betafites, displaying differences between individual parts of the same crystal [1, pp. 173/4]; and between 512 and 657 kg/mm^2 for Th-betafite (Kola Peninsula) [11].

Betafite is brittle, its fracture is irregular to conchoidal [1, pp. 173]; or is uneven conchoidal for Th-betafite (Cape Turii, Kola Peninsula) [6, p. 259].

Measured density for betafite varies between 3.7 and 5.0 g/cm^3, due to differences in chemical composition, and depends on the degree of metamictization or hydration [1, pp. 173/4]; refer also to the data given in Table 8, pp. 138/40. Further, D is between 3.75 and 5.25 g/cm^3, depending on the Nb$_2$O$_5$ content and the degree of alteration, for betafites from Ontario, Canada [3]; or varies between 3.65 and 4.11 g/cm^3, depending on the degree of hydration, for betafites from the Urals, and increases after heating at about 10 to 18%. Blomstrandite from the Urals has a density between 3.83 and 4.51, increasing to 4.78 g/cm^3 after heating [3]; for the latter, refer also to [8, p. 55].

Chemical Behavior. Data in the literature for chemical behavior of betafite/blomstrandite against mineral acids are nonuniform: Betafite dissolves slightly in hot HCl and HNO$_3$, in hot concentrated H$_2$SO$_4$ it dissolves slowly leaving an abundant yellowish greenish residue [15]. Or it dissolves completely in strong sulfuric acid on heating; in other mineral acids, in contrast to most niobates, it dissolves partially [16, p. 345], Powdered betafite dissolves less in HCl, somewhat better in H$_2$SO$_4$ (when concentrated and heated [1, p. 177]), and distinctly in a melt of KHSO$_4$ [14]. Blomstrandite is weakly soluble in hot aqua regia [8, p. 56], and dissolves easily in HCl [16, p. 348].

Naturally occurring minerals with the pyrochlore structure (as, e.g., betafite) often occur in the metamict state, owing to α-recoil damage on the structure, which influences the dissolution behavior of radio-isotopes of U and Th in a bicarbonate-carbonate solution, a fact being of utmost importance in nuclear waste deposition, see, for example, [17, 18].

Thermal Behavior. The differential heating curves of minerals of the pyrochlore structural type have, in general, two endothermic and one exothermic peaks. The former is related to liberation of water, the latter to the recrystallization process. The first endothermic effect occurs between 100 and 300°C, with a maximum at 170 to 200°C. The second endothermic effect is between 400 and 600°C, with a maximum at 422 to 523°C. The exothermic effect has its peak between 630 and 650°C [19]. Note further that titano-tantalo-niobates of varying composition (as obruchevite, ellsworthite, hatchettolite, betafite, and titanobetafite) have two endothermic effects at about 200 and 450°C, an exothermic one is between 655 and 730°C [20]. Their intensities may vary for various localities according to the different degrees of metamictization or hydration of the mineral [12, pp. 619/21]; refer also to [1, pp. 177/8]. Note also that betafites may be recrystallized well below the exothermic peak temperature [12, pp. 622/3]. The differential heating curve for Th-betafite (Kola Peninsula) displays a broad endothermic depression between 100 and 270 °C, a small endothermic effect between 330 and 350 °C, and a strong exothermic effect between 600 and 680 °C with a clear maximum at 636 °C [6, pp. 262/3]. The heating curve for yttrobetafite has a small endothermic effect at about 250 °C and an exothermic one of the same order at about 700 °C [21]. For figures of the heating curves, see the relevant references and refer also for betafite and blomstrandite to [7, p. 33] and [8, p. 59]. A heating curve together with the curve of weight loss is given for betafite in [1, pp. 177/8].

For the crystalline phases formed during heat treatment of metamict betafite and its varieties, see pp. 143/4. For a betafite phase formed during recrystallization of euxenite or priorite, see pp. 175/6.

References for 2.2.1.4.4:

[1] Chukhrov, F. V.; Bonshtedt-Kupletskaya, E. M. (Mineraly Spravochnik, Vol. 2, Pt. 3, Nauka, Moscow 1967, pp. 1/676).
[2] Junge, W.; Knoth, J.; Rath, R. (Neues Jahrb. Mineral. Abhandl. **147** [1983] 169/83).
[3] Satterly, J. (Ann. Rept. Ontario Dept. Mines **65** VI [1956] 1/181, 21).
[4] Daniels, J. L. (Somaliland Ministry Nat. Resources Mineral Resources Pamphl. No. 3 [1960] 1/25, 14).
[5] Rowe, R. B. (Geol. Surv. Can. Econ. Geol. Rept. No. 18 [1958] 1/108, 79).
[6] Kirnarskii, Yu. M.; Afanas'ev, B. V.; Men'shikov, Yu. P. (Mater. Mineral. Kol'sk. Polu-ostrova **6** [1968] 258/64).
[7] Makarochkin, B. A.; Frank-Kamenetskii, V. A.; Gonibesova, K. A. (Geol. Geofiz. **1963** No. 9, pp. 32/51, 33, 41).
[8] Makarochkin, B. A.; Gonibesova, K. A.; Makarochkina, M. S. (Zap. Vses. Mineral. Obshch. **93** [1964] 54/9).
[9] Kalita, A. P.; Mukhitdinov, G. N. (in: Vlasov, K. A., Geochemistry and Mineralogy of Rare Elements and Genetic Types of Their Deposits, Vol. 2, Israel Program Sci. Transl., Jerusalem 1966, pp. 511/6, 514 [Russian original, Moscow 1964]).
[10] Getseva, R. V.; Savel'eva, K. T. (Rukovodstvo po Opredeliniyu Uranovykh Mineralov, Gosgeoltekhizdat, Moscow 1956, pp. 1/260, 145).

[11] Lapin, A. V. (in: Lebedeva, S. I.; Kuz'min, V. I.; Matias, V. V., Novoe v Mineralogicheskikh Issledovaniyakh, Moscow 1976, pp. 44/5).
[12] Hogarth, D. D. (Can. Mineralogist **6** [1957/61] 610/33).
[13] Chesnokov, B. V. (Zap. Vses. Mineral. Obshch. **93** [1964] 73/4).
[14] Weiner, K. L.; Hochleitner, R. (Lapis **9** No. 4 [1984] 5/6).
[15] Tschernik [Chernik], G. (Bull. Soc. Franc. Mineral. **50** [1927] 485/9).
[16] Soboleva, M. V.; Pudovkina, I. L. (AEC-tr-4487 [1961] 1/455; N.S.A. **15** [1961] No. 23739).
[17] Eyal, Y.; Lumpkin, G. R.; Ewing, R. C. (Mater. Res. Soc. Symp. Proc. **50** [1986] 379/86; C.A. **105** [1986] No. 87166).
[18] Lumpkin, G. R.; Ewing, R. C.; Eyal, Y. (J. Mater. Res. **3** [1988] 357/68 from C.A. **108** [1988] No. 207819).
[19] Gorzhevskaya, S. A. (Geol. Mestorozhd. Redk. Elem. No. 23 [1964] 26/7).
[20] Gorzhevskaya, S. A. (Geol. Mestorozhd. Redk. Elem. No. 10 [1960] 32/41).

[21] Bykova, A. K.; Kukharchik, M. V. (Tr. Inst. Mineral. Geokhim. Kristallokhim. Redk. Elem. No. 8 [1962] 201/11, 205).

2.2.2 Complex Oxides with Nb, Ta, and Ti

Since their initial description in the early 1800's, the AB_2O_6-type Nb-Ta-Ti oxides have been a "mineralogical headache" owing to their complex and variable compositions (see statements given below) often combined with a metamict state and pervasive alteration (refer to p. 150). These facts have led to a contradictory nomenclature and to inconsistent values for physical and structural parameters (refer to the corresponding paragraphs given for individual minerals in this section) [1]. For a comprehensive compilation of present knowledge on titano-tantalo-niobates refer to the following papers discussing the topics as mentioned: Characteristics [2, pp. 5/10], metamict state [2, pp. 10/21], structure types [2, pp. 22/32], thermal studies [2, pp. 32/41], chemical composition [2, pp. 41/7], physical properties [2, pp. 47/60], and classification [2, pp. 60/4] of titano-tantalo-niobates in general. The detailed descriptions (comprising the afore-mentioned topics) for the individual minerals, arranged by their different structure types, are given for cubic and tetragonal titano-tantalo-niobates in [2, pp. 64/131] and [2, pp. 132/57], respectively, and for rhombic titano-tantalo-niobates in [3].

By convention, the composition of complex multiple oxides of Nb and Ta, commonly containing U and Th, may be expressed by the general formula $A_xB_yO_z$, where A = RE, U, Ca, Th, Fe^{2+}, and Na; B = Nb, Ta, Ti, and Fe^{3+}; and O = O, OH, and F. Approximately 25 such minerals are known, of which about half belong to well-defined isomorphous series such as the pyrochlore group, the euxenite-polycrase series, and the fergusonite-formanite series [4]. The orthorhombic, often metamict, Ti-Nb-Ta oxides of the formula type AB_2O_6 (where A = RE, Fe, Mn, Ca, U, Th and B = Nb, Ta, Ti) include euxenite, polycrase, priorite-blomstrandine, and aeschynite. Present mineralogic systematics are contradictory, but euxenite and aeschynite are ideally defined as $YNbTiO_6$ and $CeNbTiO_6$, respectively. Polycrase and blomstrandine are considered as Ti-rich varieties [5]. Priorite (= aeschynite-(Y) [6]) is the low-temperature dimorph of euxenite [7] and isostructural with aeschynite [8]. For a compilation of dimorphism and isomorphism occurring in the system $CeNbTiO_6$-$YNbTiO_6$, see [12].

The complex, metamict Nb-Ta-Ti oxides of the formula type AB_2O_6 (where A = RE, Fe^{2+}, Mn, Ca, Th, U, Pb and B = Nb, Ta, Ti, Fe^{3+}, W) include three structure types: orthorhombic euxenite structure, orthorhombic aeschynite structure, and monoclinic (pseudo-orthorhombic) samarskite structure [9]. Structural and chemical relationships between the first and second structure type are summarized in [10, 11]. For euxenite-aeschynite polymorphism, refer to p. 187.

Titano-niobo-tantalates with a complex chemical composition very often show deficiencies only in A-site cations, which are capable of being leached out during alteration processes, a fact not taken widely into account when calculating the chemical formula [13]. The deficiencies may be primary or secondary, see p. 75.

Using 91 analyses from the literature for orthorhombic AB_2O_6-type Nb-Ta-Ti oxides, means and standard deviations were calculated for the principal chemical components, considering the relative effects of A- and B-site cations on the matrix absorption of low-energy γ rays (47 and 74 keV). From these data, seven idealized but plausible compositions (see original paper) were calculated that represent typical compositions for euxenite, polycrase, and aeschynite. Data recorded from natural γ-ray spectra for 10 euxenites, 12 polycrases, and 1 aeschynite correspond well with the theoretical assumptions (for the details, see original paper), but may be used only as a qualitative tool in the identification of Nb-Ta-Ti oxides. Whereas for extreme compositions (e. g., with Nb + Ta ⪢ Ti) polycrase and euxenite may be distinguished from one another, for less extreme compositions the A-site cationic variations may make this difficult or impossible (i.e., it is not possible to relate the absorption data to unique compositions) [15].

Alteration is a common feature of metamict, rare earth, AB_2O_6-type Nb-Ta-Ti oxides and is mostly investigated for the cubic pyrochlores or the orthorhombic samarskite group. Minerals of the euxenite and aeschynite groups are less well studied. But alteration (of the primary and secondary type, at first defined by [16]) has a bearing on nomenclature and identification procedures, and affects composition, physical properties, and the results of an annealing study of the Nb-Ta-Ti oxides. As follows from an investigation of eight specimens (euxenite, aeschynite, and blomstrandine) in which the unaltered, dark, nearly opaque areas are surrounded by a characteristically honey yellow seam, primary alteration (e.g., hydrothermal) causes: a consistent increase in Ca content; generally a decrease in U and Th contents; a decrease in the absolute RE concentration (except for one case, the relative RE composition remains constant); a slight decrease in B-site cations; and an increase in structural and absorbed water. Secondary alteration (e.g., weathering) acts similarly, but produces a decrease in Ca content, an increased leaching of A-site cations, and a relative increase in B-site cations. In all eight samples, the alteration, whether primary or secondary, results in a decrease in specific gravity, refractive index, and reflectance. The Vickers hardness, although variable, does not differ by more than one standard deviation from altered to unaltered zones. And, in general, annealed rims of altered zones are more complicated because of non-stoichiometric mixtures of A- and B-site cations, and result in the formation of several new phases, together with variations in temperature for the priorite-aeschynite to euxenite transition. For the details, see original paper [1]; refer also to [13]. Alteration of niobate tantalates may be accompanied by selective leaching of A ions with corresponding replacement of the non-metallic elements by OH groups and H_2O molecules, a feature that affects both the physical properties and X-ray diffraction patterns [4].

The reason for the metamict state of the complex Nb-Ta oxides is controversial (for the details, refer to the original papers mentioned): Whereas Graham, Thornber [17], following Pyatenko [18], assume as the main cause complex chemical composition (via effects of diadochic substitution) and segregation into domains of microregions of different structure, Ewing [19] reviews the literature in support of radiation damage as a necessary condition for metamictization. He suggests that the structural stability of the metamict "glass" relative to that of the crystalline pre-metamict material is an important consideration in predicting whether a substance will occur in the metamict state or not. If the minerals of the euxenite-polycrase and priorite-blomstrandine series have not been completely decomposed by metamict processes into a mixture of oxides, they regain their pre-metamict structures on heating, and they may be uniquely determined by X-ray methods. If the metamict processes completely destroy the lattice, it is impossible to distinguish the minerals of these series by X-ray data, except for aeschynite, which regains its pre-metamict structure when heated above 1000°C regardless of the disturbance to its lattice [8].

An annealing study at 50°C intervals over the temperature range 400 to 1100°C was performed on seven selected natural metamict members of the orthorhombic, rare earth, AB_2O_6-type Nb-Ta-Ti oxides comprising euxenite (analysis Nos. R17 and R25 in the table on p. 163), blomstrandine (analysis Nos. R12, R13, and R24 in the table on p. 181), and aeschynite (analysis Nos. R23 and R26 in the table on p. 204). Recrystallization to a priorite-aeschynite phase occurs at lower temperatures (~400°C), followed by the euxenite phase at higher temperatures (>700°C). The transition of priorite-aeschynite to euxenite occurs over the range 550 to 750°C for Y-rich compositions, and between 900 and 1000°C for Ce-rich compositions. Both reactions are clearly expressed by their different types of DTA pattern. Other phases identified are a cubic pyrochlore (present throughout the temperature range 400 to 1000°C), brannerite, Nb_2TiO_7, rutile, and anatase. Despite general similarities, each annealing run is unique in detail (see original paper) and may have been influenced by the degree of metamictness, by compositional variations, and/or by experimental conditions. Even for standardized experimental techniques, recrystallization temperatures of metamict complex Nb-Ta-Ti oxides vary over a wide range (refer to [14]), and additional data (such as morphology) are required for positive identification of the pre-metamict material [10].

Complex multiple oxides of Nb and Ta commonly contain U and Th, and typically occur in granite and syenite bodies and their pegmatitic facies, and in carbonatite. Compositions of the minerals vary greatly because of complex substitution and locally present difficulties in assignment of a precise formula. The niobate tantalates are virtually impossible to distinguish from one another in hand specimen. Positive identification of the minerals normally requires X-ray diffraction, but this is complicated by their metamict nature because ignition may yield one or more phases (refer also to the preceding paragraphs). Generally, niobate tantalates vary in color from black through brown to yellow, with black to deep brown predominating, and they show a high luster, have a subconchoidal fracture, and are hard but brittle [4]. Limited Vickers hardness and reflectance data are given in the literature for orthorhombic, often metamict Ti-Nb-Ta oxides of the formula type AB_2O_6 (refer to the relevant paragraphs given for individual minerals in this section), but are not accompanied by the determination of weight percentages for TiO_2, Nb_2O_5, and Ta_2O_5 [5]. According to a regular relationship between refractive index n and density D, a coefficient $k = n - 1/D$ may be used to discriminate individual members of the metamict titano-tantalo-niobates [2, pp. 55/8].

For a compilation of properties utilizable for the identification of individual members of the complex Nb-Ta-Ti oxides, see [20 to 22]; refer also to [23].

References for 2.2.2:

[1] Ewing, R. C. (Geochim. Cosmochim. Acta **39** [1975] 521/30).

[2] Ginzburg, A. I.; Gorzhevskaya, S. A.; Erofeeva, E. A.; Sidorenko, G. A. (Geol. Mestorozhd. Redk. Elem. No. 10: Titano-Tantalo-Niobaty I [1960] 1/168).

[3] Gorzhevskaya, S. A.; Sidorenko, G. A. (Geol. Mestorozhd. Redk. Elem. No. 23: Titano-Tantalo-Niobaty II [1964] 1/116).

[4] Steacy, H. R.; Kaiman, S. (Short Course Handbook — Mineral. Assoc. Canada **3** [1978] 107/40, 113/4).

[5] Ewing, R. C. (Am. Mineralogist **58** [1973] 942/4).

[6] Levinson, A. A. (Am. Mineralogist **51** [1966] 152/8, 153).

[7] Mitchell, R. S. (Southeast. Geol. **14** No. 1 [1972] 59/72, 62/5).

[8] Komkov, A. I. (Dokl. Akad. Nauk SSSR **126** [1959] 641/4; Dokl. Acad. Sci. USSR Earth Sci. Sect. **124/129** [1959] 584/6).

[9] Ewing, R. C.; Snetsinger, K. G.; Bunch, T. E. (Can. Mineralogist **15** [1977] 92/6).

[10] Ewing, R. C.; Ehlmann, A. J. (Can. Mineralogist **13** [1975] 1/7).

[11] Ewing, R. C. (Can. Mineralogist **14** [1976] 111/9).

[12] Chzhan, Pei-Shan' [Chang, P'ei-Shan] (Sci. Sinica [Peking] **12** [1963] 237/43; C.A. **58** [1963] 9962).

[13] Van Wambeke, L. (Neues Jahrb. Mineral. Abhandl. **112** [1969/70] 117/49; 118/20).

[14] Mitchell, R. S. (Mineral. Rec. **4** [1973] 214/23).

[15] Ewing, R. C.; Krumhansl, J. L. (Can. Mineralogist **12** [1974] 357/9).

[16] Borodin, L. S.; Nazarenko, I. I. (Geokhimiya **1957** 278/95, 293).

[17] Graham, J.; Thornber, M. R. (Am. Mineralogist **59** [1974] 1047/50).

[18] Pyatenko, Yu. A. (Geokhimiya **1970** 1077/83; Geochem. Intern. **7** [1970] 758/63).

[19] Ewing, R. C. (Am. Mineralogist **60** [1975] 728/33).

[20] George, D'Arcy (RMO-563 [1949] 1/198, 52/8; N.S.A. **5** [1951] No. 833).

[21] Getseva, R. V.; Savel'eva, K. T. (Rukovodstvo po Opredeleniyu Uranovykh Mineralov, Gosgeoltekhizdat, Moscow 1956, pp. 1/260, 30/3, 52, 145/59).

[22] Soboleva, M. V.; Pudovkina, I. L. (AEC-tr-4487 [1961] 1/455, 315/94; N.S.A. **15** [1961] No. 23739).

[23] Chukhrov, F. V.; Bonshtedt-Kupletskaya, E. M. (Mineraly Spravochnik, Vol. 2, Pt. 3, Nauka, Moscow 1967, pp. 1/676, 342/4, 348/56, 359/71).

2.2.2.1 Euxenite-Polycrase Series

Euxenite and polycrase are the end members of an isomorphous series of minerals that are multiple oxides of Nb, Ta, Ti, and RE(Y), and usually contain U, Th, and RE(Ce) [1].

Naturally occurring minerals of the euxenite/priorite-polycrase/blomstrandine isomorphous series display a pronounced Ti-Nb isomorphism. All members in which the content (in wt%) of Nb + Ta exceeds that of Ti are classified as euxenite, whereas the Ti-rich members of the series are regarded as polycrase. But the complexity of composition (refer to Table 9, pp. 164/8) and the great variety of isomorphous substitution (refer to pp. 161/3) complicate the classification of minerals of this group. Although minerals of the euxenite-polycrase series have largely identical chemical compositions, there are differences in their crystal habit (refer to pp. 170/1). It is still not clear whether this difference is indicative of specific features of the crystal structure or is merely fortuitous. And X-ray analysis after ignition of the commonly metamict specimens does not permit the distinction between minerals of the euxenite-polycrase series [2, pp. 470/1].

Minerals of the euxenite-polycrase series are orthorhombic and rarely form stout prismatic crystals. More often they occur as jet black to brown or yellow irregular patches in zoned granite pegmatite dikes (as, e.g., in the Bancroft area, Ontario, Canada) [1]. They are widespread in granite pegmatites, but as a rule do not form large accumulations, and are generally confined to the feldspathic or quartz-feldspathic (block) zones, sometimes also to the albitized zones. In certain cases they occur in fine-grained contact zones (between pegmatite and host rock, or within the pegmatites?). As a rule they penetrate feldspar, occasionally biotite or muscovite, sometimes quartz of the pegmatites. Minerals of the euxenite-polycrase series further occur (at times in considerable amounts) in albitized zones of quartz or quartz-free alkali syenites and related pegmatites, where they are found in albite, rarely in quartz, potash feldspar, or potash hornblende. In small amounts, euxenite-polycrase minerals occur in placers [2, pp. 478, 482]. For examples of occurrence, see pp. 154/61 and 178/9, respectively. As follows from a "typical" mineral suite associated with tin-bearing placer deposits compiled from literature (and being capable for prospecting purposes), the Th-bearing minerals euxenite, huttonite, polycrase, and aeschynite rarely occur (in 10% or less of the mineral assemblages) in tin-bearing placers [3].

A short compilation of types of occurrence, of composition, and of properties is given for the minerals of the euxenite-polycrase series in [4] and, more extensive, in [5]; the compilations include characteristics for identification.

For a multivariant analysis of the consistency of nomenclature, and for a classification improvement of euxenite-polycrase minerals, see [6].

References for 2.2.2.1:

[1] Satterly, J. (Ann. Rept. Ontario Dept. Mines **65** VI [1956/57] 1/181, 23).
[2] Aleksandrov, V. B. (in: Vlasov, K. A., Geochemistry and Mineralogy of Rare Elements and Genetic Types of Their Deposits, Vol. 2, Israel Program. Sci. Transl., Jerusalem 1966, pp. 470/82 [Russian original, Moscow 1964]).
[3] Callaghan, C. C. (Ann. Geol. Surv. South Africa **17** [1983/84] 35/41, 36).
[4] George, D'Arcy (RMO-563 [1949] 1/198, 52/5; N.S.A. **5** [1951] No. 833).
[5] Getseva, R. V.; Savel'eva, K. T. (Rukovodstvo po Opredeleniyu Uranovykh Mineralov, Gosgeoltekhizdat, Moscow 1956, pp. 1/260, 151/9).
[6] Ewing, R. C. (Can. Mineralogist **14** [1976] 111/9).

2.2.2.1.1 Euxenite $(Y, Ca, Ce, U, Th) (Nb, Ta, Ti)_2O_6$

Euxenite from Jölster, Norway, occurring as coarse masses (and probably identical with blomstrandine [10]) was described in 1840, and its name is derived from the Greek word "euxenos" = hospitable or friendly to strangers in allusion to the rare elements it contains [1]; refer also to [2]. Euxenite crystals from near Tvedestrand, southern Norway, were described in 1847 [11] and are taken for the type material; the correct type locality is Alve on the island Tromö, near Arendal, see [10]. Synonymous are eschwegeite and priorite [3], but note that it is possible only for part of the latter mineral, namely aeschynite-(Y) [4], refer also to p. 185. Note further that priorite is also viewed as a dimorph of euxenite, see [5] and below. Loranskite [6, p. 348] and a part of the wiikites (black wiikite) are euxenite [7].

As varieties of euxenite are given: tanteuxenite, to which delorenzite is synonymous [3], [6, p. 354]; Ce-euxenite, Th-euxenite (with 8.4% ThO_2), U-euxenite, Ti-euxenite (= polycrase), Ca-Nb-euxenite, and, possibly, hydro-euxenite (with 6% H_2O) [8].

For an attempt to define experimentally an equilibrium curve for conditions of formation (temperature and water vapor pressure) of euxenite and its low-temperature polymorph priorite, see [9]. For nomenclature of the polymorphs, refer to p. 162.

References for 2.2.2.1.1:

[1] Scheerer, T. (Ann. Physik Chem. [2] **50** [1840] 149/53).
[2] Palache, C.; Berman, H.; Frondel, C. (Dana's System of Mineralogy, 7th Ed., Vol. 1, Wiley, New York 1946, pp. 1/834, 787/91).
[3] Aleksandrov, V. B. (in: Vlasov, K. A., Geochemistry and Mineralogy of Rare Elements and Genetic Types of Their Deposits, Vol. 2, Israel Program. Sci. Transl., Jerusalem 1966, pp. 470/82, 470/1 [Russian original, Moscow 1964]).
[4] Ewing, R. C. (Can. Mineralogist **14** [1976] 111/9, 117).
[5] Komkov, A. I. (Dokl. Akad. Nauk SSSR **148** [1963] 679/80; Dokl. Acad. Sci. USSR Earth Sci. Sect. **148** [1963] 103/4).
[6] Chukhrov, F. V.; Bonshtedt-Kupletskaya, E. M. (Mineraly Spravochnik, Vol. 2, Pt. 3, Nauka, Moscow 1967, pp. 1/676).
[7] Beus, A. A.; Kalita, A. P. (Dokl. Akad. Nauk SSSR **141** [1961] 705/8; Dokl. Acad. USSR Earth Sci. Sect. **139/141** [1961] 1278/80).

154

[8] Semenov, E. I. (Tipochimism Mineralov Shchelochnykh Massivov, Nedra, Moscow 1977, pp. 1/119, 54).

[9] Komkov, A. I. (Dokl. Akad. Nauk SSSR **171** [1966] 181/2; Dokl. Acad. USSR Earth Sci. Sect. **171** [1966] 145/6).

[10] Brøgger, W. C. (Skr. Nor. Vidensk. Akad. I Mat.-Naturvidensk. Klasse No. 6 [1906] 1/162, 82/3).

[11] Scheerer, T. (Ann. Physik Chem. [2] **72** [1847] 566/71).

2.2.2.1.1.1 Occurrence, Paragenesis, Intergrowths, Inclusions, and Alteration

Occurrence and Paragenesis. The main part of euxenite occurs with granite pegmatites of the RE type. Fairly often and usually in small amounts, on average 1 to 1.5 g/t, euxenite is an accessory mineral of granites, predominantly of the alkalic and subalkalic series; less frequently it is found in granites of the acid and ultra-acid series [1, p. 205]. Mean contents of euxenite, given as $\bar{x}$ and $(\bar{x} + s)$, for special rock types are as follows; s is the mean standard deviation [2]:

	content in g/t of euxenite	
	$\bar{x}$	$(\bar{x} + s)$
107 biotite granites	0.87	6.38
16 leucocratic granites	1.57	6.50
20 alaskites	0.017	0.09

For other contents of euxenite in granites, see pp. 156/7.

A crude statistical study of mineral associations in granite pegmatite reveals the following for euxenite from one Canadian source and a number of Swedish occurrences: There is a high degree of association with biotite and primary feldspar; there is a low degree of association with quartz and, most remarkably, with muscovite. This suggests that euxenite most definitely is confined to the pegmatoid zone of first-group pegmatites (composed essentially of quartz, primary feldspar, and biotite). The Canadian specimen (from Lyndoch, Ontario) shows that euxenite may also be found in a different mineral association, which is otherwise more typical of columbite [3].

Euxenite-polycrase, next to allanite, is one of the most common radioactive minerals found in zoned granite pegmatite of the Grenville subprovince, Bancroft area, Ontario. So far, it has not been found in the unzoned granite and syenite pegmatites of this area [4].

Rounded subhedral metamict euxenite crystals (1 inch in diameter and composed of three phases differing somewhat in color, see p. 159) are restricted to a zone of pink albite with some interstitial microcline near the center of a pegmatite body, St. Pierre de Wakefield, Quebec, Canada. Locally, deep purple fluorite is associated [5]; or purple-white fluorite, zircon (cyrtolite), and thorite occur near the euxenite-feldspar contact. These minerals do not seem to be alteration products of the euxenite [6, p. 225].

Black and brown X-ray amorphous euxenite from the intermediate zone III (of about 10 cm thickness) of the Gloserheia granite pegmatite, Froland, southern Norway, occurs as euhedral to subhedral crystals and aggregates (sometimes with a radiating structure) of a size up to

1 cm, or forms anhedral masses up to a few cm in diameter. Euxenite is generally intergrown with small amounts of zircon and, to a variable degree, with other minor or trace minerals (as, e. g., allanite, biotite, apatite, etc.). It is intimately associated with secondary fourmarierite. Closely associated, more frequently with the brown rather than the black variety of euxenite, are the two dimorphs β-uranophane and uranophane (for the formation mechanism, see p. 158) [7, pp. 246, 252, 257]. Large crystals of black metamict euxenite, surrounded by samarskite, occur in the quartz-feldspar zone and in the mineralized zone of a pegmatite vein in hornblende-bearing quartz monzonite (or farsundite) at Rømteland, Vest-Agder, southern Norway. The contact between host rock and pegmatite is completely gradational; and euxenite disappears when samarskite disappears [8, pp. 125, 130/3, 159].

Yellowish brown euxenite of rock-crystal-bearing granite pegmatite, central Kazakhstan, is present usually as massive granular concretions of 10 to 12 cm in diameter within the zones of pink microcline-perthite or white albite. Small (1 to 5 mm) thick-prismatic plates of euxenite with well-developed faces (010) and (100) occur in the yellowish sandy-clayey masses filling chambers and nests containing the rock-crystal [9], [10, p. 44]. However, this euxenite is obviously composed of at least two phases, see pp. 162/3. Euxenite of quartz-fluorite pegmatites (of the second group), Zarenda, Kazakhstan, forms, together with zircon (cyrtolite) and biotite, roundish dark brown segregations up to 0.3 cm in size. The zoned pegmatite occurs as schlieren in the host rock, and is genetically related to a late facies of alkali granites [11].

Euxenite is found only in the outer zones of several complex pegmatites of the Abukuma massif, Fukushima prefecture, Japan. In the Ippaiyama pegmatite, it occurs together with black ilmenite in perthite and quartz (refer also to [12]). In the Uzumine pegmatite, it occurs (without crystal faces) in association with monazite, perthite, and quartz. And in the Nekonaki pegmatite of Ishikawa town, it is associated with quartz, perthite, lepidomelan, and garnet and occurs either as prismatic crystals (prismatic or tabular crystals of polycrase, 5 mm long [13]) or as massive aggregate (of 3 to 8 mm in size [13]) in quartz without other minerals [14, pp. 389, 392]; the Nekonaki euxenite is black, in small flakes reddish black [13].

Black euxenite occurs as small accumulations of poorly developed prismatic crystals (often reaching 1 to 3 cm in length, with sometimes 0.2 to 0.5 cm in cross-cuts) and, rarely, as irregular segregations mainly within the zone of blocky oligoclase, microcline-perthite, and quartz of the (Lower Proterozoic [15]) pegmatite vein Kapraevo, Kola Peninsula, northwestern European USSR. In this zone, euxenite is connected with relics of quartz and feldspar enclosed by plates of biotite and muscovite. Associated minerals are columbite, zircon, altered allanite, monazite, etc. Considerably lesser and as small segregations, euxenite occurs in the graphic pegmatite; and associated with lathes of muscovitized biotite (occurring in fissures) it occurs in quartz-microcline-plagioclase pegmatite, where it sometimes is accompanied by allanite [16]. The Kapraevo pegmatite belongs to an anatectic plagioclase-microcline granite, and is characterized by a Nb-Y-RE mineralization. Together with other Th-containing accessory minerals (fergusonite, thorite, abukumalite, and gadolinite), euxenite occurs in pegmatites of an alkali arfvedsonite-microcline granite, eastern part of the Baltic Shield, European USSR. These pegmatites are characterized by a Zr-Th-Nb-Y mineralization, sometimes with Be [17]. Greenish, sometimes yellowish, cinnamon brown euxenite occurs in quartz-albite-microcline pegmatite within aegirin-bearing gneiss-granite of an undisclosed location of the USSR (Keivy, Kola Peninsula, after [24, pp. 54/5]). Euxenite forms crystals 1 to 1.5 cm in length and 0.5 to 1 cm in diameter, which are sometimes considerably deformed, and it is in narrow association with fergusonite, zircon, and ferrithorite. It is practically absent in areas of the pegmatite that are enriched in britholite-(Y) (abukumalite). The longer axis of euxenite crystals often is parallel to the longer axis of the pegmatite body [41].

Black euxenite forming scattered impregnations or small veinlets (differing somewhat in their lattice parameters, see p. 171) occurs in a pegmatite vein from an undisclosed locality, USSR. The impregnations occur as crystals with a prismatic outline (see p. 170) of 0.01 to, rarely, 15 mm in size; or as xenomorphic grains sometimes reaching 40 to 50 mm in diameter. In the veinlets (2 to 7 mm thick and 5 to 55 mm long) euxenite forms monomineralic aggregates or is intergrown with magnetite. Both types of occurrence (impregnations or veinlets), intimately associated with tourmaline, zircon, magnetite, and apatite, are mainly connected with the albitized parts of the pegmatite [18, p. 408].

Light chocolate brown metamict euxenite is a characteristic accessory component of the albitic part of pegmatite veins from the exocontact aureole of the Transangara (or Tatarskii) nepheline syenite massif, western part of the Enisei Ridge, Siberia. It forms stretched crystals (of 0.5 by 2 mm) with a ragged surface and is associated with columbite, and also with zircon and monazite [19].

Mainly metamict, but also partially anisotropic euxenite, together with fergusonite, occurs in a pegmatite vein of the Central Urals, USSR. Euxenite is brownish and greenish black to velvet black, forms short prismatic-to-platy crystals of up to 1 mm in size, or is found as small grains. It is enclosed in fergusonite and columbite, but also in albite, muscovite, and quartz adjacent to the fergusonite [20].

Chemically variable, highly metamict euxenite occurs as olive to dark brown irregular pods (as much as 3 inches long) along the core-margin replacement zone of a pegmatite dike intruding hornblende gneiss near Sappington, Montana. It replaces albite and microcline, which have been stained red. In many places the replacement was controlled by the feldspar cleavage [21]. Together with xenotime, monazite, strüverite, and tantalian anatase, two types of elongated euxenite crystals (see p. 170) occur in a microcline pegmatite vein containing albite within gneiss from D'ortovoto near Chepelare, central Rhodopes, Bulgaria. The complex Ti-Nb-Ta oxides are formed by interaction of pegmatite fluids with the country rock (gneiss and amphibolite) rich in titanite, ilmenite, and rutile [22]. Small amounts of euxenite fragments are found in quartz-feldspar veins, probably of pegmatitic origin, cutting granite gneiss, Kanuku Mountains, British Guiana [23].

Other occurrences of euxenite in pegmatites, whithout specification for site of deposition, are given as follows: Chocolate brown metamict euxenite occurs in pegmatite of biotite-hornblende granite, Kaveripatnam, India. It forms irregular segregations up to 1 cm in size and is associated with allanite and magnetite [24, p. 90]. Besides a great number of other accessory minerals, especially (arranged in order of its crystallization) allanite, zircon, euxenite, and polycrase are abundant in pegmatite of the biotite type connected with Precambrian granites of Inner Mongolia, P.R.C. [25]. Small quantities of euxenite, samarskite, zircon, etc. occur in few of the allanite pegmatites (forming small irregular bodies in porphyritic rapakivi-like granite, see [26]) of the Gold Butte district, Clark county, Nevada [27]. Tanteuxenite is found in a pegmatite at Woodstock, Western Australia [28].

Tiny crystals of prismatic or platy euxenite (0.04 to 1 mm long) were found in all varieties of rock suffering postmagmatic alteration from a (ultra-acid) biotite granite massif, eastern Transbaikalia region, Siberia. In connection with the degree of muscovitization (which is most intense in the endocontact zone of the massif), the euxenite content increases from 7.4 g/t in predominantly biotite granite, over 9.95 g/t in two-mica granite, to 14.8 g/t in muscovite granite. Simultaneously, the composition of accompanying accessory minerals changes in the same direction: ilmenite diminishes nearly completely; the contents of zircon, monazite, and apatite are strongly reduced and release Ti, Ta, Nb, and RE [1, pp. 205, 207/8].

Partly metamict black euxenite(?) (identification is hitherto not confirmed by X-ray studies) seems to be widely but sparsely distributed through many of the hydrothermal veins and lodes of the Lemhi Pass thorium (mainly thorite) mineralization, Idaho. Euxenite(?) is formed at the late stages of mineralization and seems to occur with somewhat greater concentrations near Agency Creek and particularly in the Black Bear deposit on the high slope south of Agency Creek. In intimate association with hematite (specularite), euxenite(?) appears in thin sections as short, rather broad prismatic crystals forming parallel or slightly radial aggregates (and displaying parallel extinction) enclosed in hematite (see p. 160); in addition, isolated free aggregates occur. Whereas elsewhere the euxenite(?) is subordinate to thorite, no thorite was observed with it at the Black Bear deposit, but sparse grains of monazite and xenotime were present in areas outside the hematite masses [29, pp. 180/1, 187/8, 191].

Euxenite is a characteristic accessory mineral of pegmatitic bodies of the Zolotaya Mountain connected with a fissure system within gray porphyric biotite granite of the Borshchevskii pluton, eastern Transbaikal region, Siberia. The pegmatitic bodies have transitions into rose vein granite cutting the gray granite with clear contacts. Within the microcline (block) zone of the pegmatitic body, euxenite forms accumulations of small elongated flat crystals displaying step-like intergrowths with fan-type endings; or it forms radiating aggregates up to 1 to 1.5 cm^2 in size. Individual lanceolate crystals also occur. Associated minerals are monazite, zircon (cyrtolite), magnetite octahedrons, and ilmenite (altered to brookite and anatase) [30, pp. 197/8]. For inclusions and alteration of the Zolotaya Mountain euxenite, see pp. 158 and 160, respectively.

An enrichment of a metamict mineral, transitional in composition between euxenite and polycrase, is reported for late, intrusive granitic veins of the Yan Shan complex, composed mainly of medium-grained biotite granite, People's Republic of China [31].

A little wax-brown euxenite occurs both as small granular masses (ranging in size from 1 to 4 mm in its greatest dimension) within brannerite nodules (see p. 84) and as small grains dispersed in the biotite-rich parts of probable Precambrian light-gray granitic gneiss, San Bernardino Mountains, California. The quantity of euxenite seems to be no more than several percent that of brannerite [32].

Euxenite, uranothorite, and monazite occur in black sand placers of the Victor and McCalla placers, Montana, and of the Bear Valley area, Idaho [33]; note also euxenite (along with brannerite, uranothorite, fergusonite, etc.) as a rare component in black sand placers of Idaho [34]. Dependent on the mineral content of the source rock in the Idaho batholite (namely granitic rock and associated pegmatite dikes with euxenite content up to 0.03 kg/m^3), the heavy-mineral composition of alluvial fills in the Bear Valley area contain in kg/m^3: euxenite 0.6, monazite 0.3, columbite 0.1, zircon 0.03, garnet 3, ilmenite 1.2, and magnetite 3.0. Well-washed gravel in favourably situated streamlets may contain as much as 2 kg/m^3 euxenite. Worthy of notice is the fact that euxenite and thorite seem to be mutually exclusive and that placers being worked for monazite contain only traces of euxenite and thorite [35]. Well-shaped crystals (see p. 170) and irregular grains of black euxenite, of pegmatitic origin, were found in heavy-mineral concentrates from the youngest alluvial depositions of streams cutting across injection biotite gneiss and pegmatites west of the village Monastir, central Rhodopes, Bulgaria [36]. Tanteuxenite, together with columbite, euxenite, and uranium microlite, is found as rough flat prisms in the alluvials of river Liha, North Kivu, Congo [37, p. 131]; see also [38].

Different types of **intergrowths** are typical for euxenite from a pegmatite vein of an undisclosed locality, USSR. Sometimes, they occur parallel to the plane of the 2nd pinacoid [18,

p. 410]. Around its margins, lath-like euxenite is intergrown with albite in a pegmatite body near St. Pierre de Wakefield, Quebec [5]. Euxenite is intergrown with, generally, zircon in the Gloserheia granite pegmatite, Norway, see pp. 154/5, or, sometimes, with magnetite in a pegmatite vein, USSR, see p. 156.

Inclusions. Small crystals of betafite occur in euxenite from the Ippaiyama pegmatite, Fukushima prefecture, Japan [14, pp. 392/3]; and titanite and zircon (alvite) were found within one large crystal of euxenite from a pegmatite dike at Rømteland, Vest-Agder, southern Norway [8, pp. 125, 159].

Metamict euxenite is included in microcline of a graphic pegmatite, Harrar, Ethiopia [39]. A small subhedral piece of brownish black euxenite (3 mm across) is embedded in a muscovite crystal from the Rutherford pegmatite, Amelia county, Virginia [40]. And small euxenite crystals, completely free of secondary products, occur in monazite of pegmatitic bodies, Zolotaya Mountain, eastern Transbaikal region. In contrast, small euxenite crystals in microcline are substituted by dirty olive products of alteration, which also fill the radial crevisses starting from euxenite grains in microcline [30, p. 200].

For an inclusion of euxenite in fergusonite, columbite, and other minerals of a pegmatite vein, Central Urals, see p. 156; inclusions (spherulites) of an intermediate member of the euxenite-fersmite isomorphous series in polycrase from Hitterö, Norway, see pp. 179/80.

Alteration of both crystals and segregations of black euxenite from the pegmatite vein Kapraevo, Kola Peninsula, sometimes leads to cinnamon-brown minerals of the pyrochlore group (with $D = 4.5$ g/cm^3 and $a = 10.20$ Å) [16]; especially when the euxenite crystals show crevices or spots with a more yellow coloration [15].

Euxenite from the Gloserheia granite pegmatite, Froland, southern Norway, is partially or completely altered to fourmarierite. The generally supposed mechanism of selective leaching of A-site cations (refer to p. 163) does not hold for the Gloserheia euxenites, for U is leached in some samples to form β-uranophane and is concentrated in others to form fourmarierite. A possible mechanism would be that small amounts of ions in both A- and B-site positions are leached out, the latter somewhat more. Thus, there would be A-site ions present to form β-uranophane and, as alteration proceeds, the residue will be enriched in A-site ions to form fourmarierite. Most probably as an earthy pseudomorph after euxenite, kasolite occurs in one specimen only [7, pp. 259/61].

Euxenite crystals from muscovite granite, Transbaikal region, USSR, are covered with yellowish cinnamon-brown crusts of secondary minerals with a jarosite structure (shown by X-ray analysis) [1, p. 205].

Owing to differential leaching by chemical weathering (in a tropical climate) of euxenite from a graphic pegmatite, Harrar, Ethiopia, an alteration margin was produced which is relatively depleted in Nb and U and, to a lesser extent, in Th and Zr; whereas there is more Y and Ti. The alteration is caused by intergranular (low-temperature) water forming solution paths that subsequently developed into channels of element migration. The oxides and hydrated oxides of the above elements will be the neocrystallizations of the altered margin. Ore-microscopic observations reveal a progressive change from the dark metamict euxenite → euxenite with weak internal reflections → translucent euxenite with strong internal reflec-tions → fine aggregate of anatase + translucent isotropic grains (= transition phase between euxenite and the alteration product) + ampangabeite. The changes for some element ratios for the uneffected euxenite and its alteration margin are as follows [39]:

	U:Th	Y:Nb	Y:Zr	Y:U
uneffected euxenite	4:1	1:1	9:1	3:1
alteration margin	3:2	4:1	14:1	17:1

Euxenite crystals from pegmatite near St. Pierre de Wakefield, Quebec, exhibit a mottled appearance, partly as a result of variation in chemical composition during crystallization, and partly the result of alteration [6, p. 223]. As can be seen in reflected light (polished sections), it consists of light gray (a), medium gray (b), and dark gray phases (c). In transmitted light, phase (a) appears red, phase (b) is yellow, and phase (c) is colorless and always present along fractures and margins of the crystals. Phase (c) is thought to be a product of incipient supergene alteration [5]. For the variance in content of the main elements Ti, Nb, Y, U, Pb, and Fe in the above phases, see [6, pp. 216/7, 223/7].

Alteration (hydration) on the surface and along crevisses of dark brown euxenite from rock-crystal-bearing granite pegmatite, central Kazakhstan, leads to a similar-looking yellowish material strongly depleted in ThO_2 and RE_2O_3, slightly depleted in Nb_2O_5 and TiO_2, and enriched, besides water, in CaO, Fe_2O_3, and U_3O_8 (for the detailed composition, see pp. 162/3) [10, pp. 45/6]. Whereas alteration of light chocolate brown metamict euxenite from pegmatite of the Transagara (or Tatarskii) nepheline syenite massif, Siberia, leads to a cream-colored, porcelain-like or powdry, anisotropic crust containing noticeable amounts of Nb, Ti, and mainly cerian REE [19].

Tanteuxenite from alluvials of the river Liha, North Kivu, Congo, has a fairly high content of OH groups owing mainly to crystallization from hydrothermal residual solutions or mineralizers. Relative to the slightly altered brown-black to black inner zone A, the leaching out of

	content in % in	
	zone A	zone B
ThO_2	4.1	4.8
U_3O_8	2.75	2.5
CaO	3.70	2.6
SrO	0.03	0.02
FeO	0.25	0.31
MuO	0.16	0.14
PbO	0.50	0.58
ΣRE_2O_3	13.76	11.31
Nb_2O_5	9.5	9.7
Ta_2O_5	38.4	40.2
TiO_2	20.0	20.6
Fe_2O_3	2.75	2.80
Al_2O_3	0.50	0.42
ZrO_2	0.11	0.17
SnO_2	0.10	0.12
H_2O^-	0.13	0.29
H_2O^+	3.47	4.03
sum	100.21	100.59

about 15% of the A-site cations in the outer yellow-brown zone B affects mainly the RE, Ca, and to some extent U as can be seen from the analyses given on p. 159. After heating in air at 1000 °C, the slightly altered part gives a complex X-ray diffraction pattern characteristic of the euxenite-polycrase series (refer to p. 176) [37, pp. 128/9, 131].

Euxenite(?) completely enclosed within veinlets and masses of hematite (specularite) from the Lemhi Pass thorium mineralization, Idaho, generally escaped replacement by the aggregate feldspar [29, 184/5].

Due to radioactive effects, minerals enclosing euxenite display alterations: Thus, albite from a pegmatite vein, USSR, has a pink color and displays a system of radial crevisses, which become visible megascopically only around big euxenite inclusions [18, p. 410]; feldspar from the pegmatite vein Kapraevo, Kola Peninsula, has a brick-red color and fine radial crevisses, reaching 2 to 3 cm in length [16]; and microcline from pegmatitic bodies, Zolotaya Mountain, eastern Transbaikal region, is usually red and is impregnated on crevisses with powdery, yellow, secondary minerals of unknown composition [30, p. 198]. Feldspar from a pegmatite near St. Pierre de Wakefield, Quebec, that surrounds euxenite contains numerous fractures coated with red crusts. At the euxenite contact, the feldspar is altered to micaceous clay [6, p. 223].

References for 2.2.2.1.1.1:

[1] Zalashkova, N. E.; Distler, V. V.; Smirnova, N. V. (Tr. Mineral. Muzeya Akad. Nauk SSSR No. 20 [1971] 205/8).
[2] Lyakhovich, V. V. (Sov. Geol. **1963** No. 9, pp. 30/50, 34).
[3] Brotzen, O. (Geol. Fören. Stockholm Förh. **81** [1959] 231/96, 249, 282/3).
[4] Satterly, J. (Ann. Rept. Ontario Dept. Mines **65** VI [1956/57] 1/181, 23).
[5] Robinson, S. C.; Loveridge, W. D.; Rimsaite, J.; van Peteghem, J. (Can. Mineralogist **7** [1962/63] 533/46, 534/5).
[6] Rimsaite, J.; Lachance, G. R. (Intern. Mineral. Assoc. Papers Proc. 4th Gen. Meeting, New Delhi 1964 [1966], pp. 209/29).
[7] Amli, R. (Norsk Geol. Tidsskr. **57** [1977] 243/62, 246, 252, 257, 259/60).
[8] Sverdrup, T. L. (Norg. Geol. Unders. No. 211 [1959] 124/96).
[9] Erdzhanov, K. N. (Tr. Kazakh. Nauchn.-Issled. Inst. Mineral'n. Syr'ya No. 2 [1960] 92/5, 92/3).
[10] Mineev, D. A.; Kataeva, Z. T.; Malyshev, A. G. (in: Lebedeva, S. I., Fizicheskie Svoistva Redkometal'nykh Mineralov i Metody Ikh Issled., Nauka, Moscow 1968, pp. 43/7).

[11] Semenov, E. I.; Kostyunina, L. P.; Kulakov, M. P. (in: Mineralogiya Pegmatitov i Gidrotermalitov Shchelochnykh Massivov, Nauka, Moscow 1967, pp. 137/49, 137, 141, 148).
[12] Omori, K.; Hasegawa, S. (Ganseki Kobutsu Kosho Gakkaishi **42** [1958] 280/4 from Ref. Zh. Geol. **1960** No. 14777).
[13] Omori, K.; Hasegawa, S. (Kobutsugaku Zasshi **2** [1955] 268/74 from C.A. **51** [1957] 5641).
[14] Omori, K.; Hasegawa, S.; Konno, H. (Sci. Rept. Tohoku Univ. [3] **6** [1960] 389/96).
[15] Kalita, A. P. (in: Kuz'menko, M. V.; Nedumov, J. B., Novye Dannye po Geologii, Geokhimii i Genezisu Pegmatitov, Akad. Nauk. SSSR, Inst. Mineral Geokhim. i Kristallokhim. Redk. Elem., Moscow 1965, pp. 266/304, 284; C.A. **65** [1966] 5237).
[16] Kalita, A. P. (Redkozemel'nye Pegmatity Alakurtti i Priladozh'ya, Akad. Nauk SSSR, Moscow 1961, pp. 1/120, 46/7).

[17] Kalita, A. P. (in: Kuzmenko, M. V., Osobennosti Raspredeleniya Redkikh Elementov Pegmatitakh, Nauka, Moscow 1969, pp. 79/100, 83/5).
[18] Sokolova, E. P. (Zap. Vses. Mineral. Obshch. **88** [1959] 408/18).
[19] Sveshnikova, E. V.; Semenov, E. I.; Khomyakov, A. P. (Zaangarskii Shchelochnoi Massiv, Ego Porody i Mineraly, Nauka, Moscow 1976, pp. 1/80, 31).
[20] Nefedov, E. I. (Vses. Nauchn. Issled. Geol. Inst. No. 3 [1956] 82/5, 84/5).

[21] Heinrich, E. W. (Am. Mineralogist **50** [1965] 2083/8).
[22] Arnaudov, A.; Petrusenko, S. (Izv. Geol. Inst. Bulg. Akad. Nauk. Ser. Geokhim. Mineral. Petrol. **21** [1972] 77/84, 77, 79/81, 84).
[23] Henderson, G. (Brit. Guiana Geol. Surv. Depart. Rept. **1950/51** 31/9, 33).
[24] Semenov, E. I. (Tipokhimizm Mineralov Shchelochnykh Massivov, Nedra, Moscow 1977, pp. 1/119).
[25] Go, Chen-Tszi (Sci. Record [Peking] [2] **2** [1958] 400/13, 400, 404/5, 408, 410).
[26] Volborth, A. (Econ. Geol. **57** [1962] 209/16, 214).
[27] Garside, L. J. (Bull. Nevada Bur. Mines Geol. No. 81 [1973] 1/121, 20).
[28] Greenhalgh, D.; Jeffery, P. M. (Geochim. Cosmochim. Acta **16** [1959] 39/57, 52/3).
[29] Anderson, A. L. (Econ. Geol. **56** [1961] 117/97).
[30] Kornetova, V. A.; Aleksandrov, V. B.; Kazakova, M. E. (Tr. Mineral. Muzeya Akad. Nauk SSSR No. 18 [1968] 197/202).

[31] Shen, Gan-Fu (Diqiu Huaxue **1976** No. 4, pp. 279/84 from C.A. **87** [1977] No. 26128).
[32] Hewett, D. F.; Stone, J.; Levine, H. (Am. Mineralogist **42** [1957] 30/8, 33/4).
[33] Eilertsen, D. E.; Lamb, F. D. (RME-3140 [1956] 1/46 from C.A. **1957** 7968).
[34] Savage, C. N. (Bull. Idaho Bur. Mines Geol. No. 17 [1961] 1/160).
[35] Mackin, J. H.; Schmidt, D. L. (Geol. Surv. Profess. Paper [U.S.] No. 300 [1956] 375/80).
[36] Breskovska, V.; Marinova, R. (Spisanie Bulg. Geol. Druzh. **36** [1975] 187/90).
[37] Van Wambeke, L. (Neues Jahrb. Mineral. Abhandl. **112** [1969/70] 117/49, 131).
[38] Van Wambeke, L. (Bull. Soc. Belge Geol. Paleontol. Hydrol. **67** [1958] 121/7 from C.A. **1959** 9908).
[39] Augustithis, S. S.; Mposkos, E.; Vgenopoulos, A. (Form. Uranium Ore Deposits, Proc. Symp., Athens 1974, pp. 61/71; C.A. **83** [1975] No. 196304).
[40] Mitchell, R. S. (Southeast. Geol. **14** [1972] 59/72, 68).

[41] Kalita, A. P. (Tr. Inst. Mineral. Geokhim. Kristallokhim. Redk. Elem. No. 16 [1963] 107/25, 107/8, 117/8).

2.2.2.1.1.2 Chemistry

Minerals of the euxenite-polycrase series, including tanteuxenite, may be expressed by a formula of the type AB_2O_6, where A = Y, Ce, Ca, U, Th and B = Ti, Nb, Ta, and Fe^{3+} [1]; see also [2, pp. 412/4]. Or minerals of the euxenite-polycrase and priorite-blomstrandine series are titano-niobates of Y and RE. A part of the Y may have been substituted by U, Th, Ca, or Fe^{3+} [3]. Note also that members of the euxenite-polycrase series have a wide tolerance to RE ions of different size [4]. The introduction of higher charged U and Th into group-A cations results in the formation of defect structures with deficiencies in this group. Owing to large-scale isomorphism (for the types, see below), the calculation of formulas of minerals of the euxenite-polycrase series is impeded, especially in cases in which altered material is analyzed, or the initial valence of some elements (as U and Fe) is not clear [5]. For an influence of alteration processes (of the primary or secondary type) on calculation of formulas, see [6].

Minerals of the euxenite group, $(Y, RE)(Nb, Ti)_2O_6$, are characterized by wide variations in composition [7]; refer also to Table 9, pp. 164/8. For a compilation of 40 older (pre-1950) analyses of euxenite/polycrase, partly with contents of ThO_2 between 0.44 and 6.41%, see [8]; refer also to [9, 10] and [2, pp. 351, 412/3, 472]. But note that, in a few cases, the latter reference gives values for the analyses deviating in composition, or total sum, or density from the values given in the mentioned literature source.

It is not clear in many cases what element isomorphously entered the crystal structure of euxenite-group minerals, or is adsorbed owing to the colloid-like metamict state of the minerals, or simply forms mechanical admixtures. As follows from experimental investigations of corresponding compounds with various RE, there are discontinuieties in the isomorphous miscibility (for the details, see original paper). Especially for Th, a heterovalent substitution $2 RE^{3+} \rightleftharpoons Ca^{2+} + Th^{4+}$ is possible. For example, the solid solution of composition $(Y_{0.5}Ca_{0.25}Th_{0.25})(Nb, Ti)_2O_6$ is a homogeneous euxenite phase. The experimental data further showed that there is a limited substitution of 25 to 30% for the "molecule" $ThTi_2O_6$ (refer also to solid solutions of this molecule in minerals of the thorutite-brannerite series on p. 77). The above statements are valid only for the synthesis temperature of 1200°C. At 1000°C, the region of immiscibility is appreciably greater for systems $CaNb_2O_6$-$RE(Nb, Ti)_2O_6$. Probably this is also valid for other cases of isomorphism [7]. Assuming $YNbTiO_6$ as the more correct basic formula of the euxenite-polycrase series, there are possible heterovalent isomorphic substitutions of the type: $Nb^{5+} + RE^{3+} \rightleftharpoons Ti^{4+} + Th^{4+}$ [5]; or Th + Ti for RE + Nb and Th + $O_{0.5}$ for RE in Th-euxenite [11, p. 54].

As follows from a discriminant analysis of chemical analyses given in the literature, $\Sigma U (= U_3O_8, UO_2, UO_3)$ and ThO_2 content may be critical to the chemical distinction between euxenite and priorite (= aeschynite-(Y)), which both should be considered as chemically equivalent. Euxenite should be used in preference to priorite in the absence of structural or morphological data, or if $\Sigma U > ThO_2$. If priorite and aeschynite are the endmembers of a solid-solution series, the Th content of priorite may be particularly important [12].

Based on the content of Ti and on the RE composition, minerals corresponding to either the euxenite-polycrase series or priorite may have $>15\%$ TiO_2 and Y as a predominant element over RE; additionally, the contents of heavier RE exceed that of the lighter RE [13].

Euxenite, the high-temperature phase of $YTi(Nb, Ta)O_6$, tends to an extensive mixed-crystal formation. Not only with isotypic components containing RE or $CaNb_2O_6$, but also with heterotypic components. Hereby, vacancies and disorder of position are formed in the crystal structure. Thus, recalculation of analyses given in [9] into components leads, for example, for the components $(U, Th)Ti_2O_6$ and $(U, Th)O_2$, respectively, to portions of 1.3 and 12.8% for analysis No. 4, or 9.6 and 3.2% for analysis No. 6. Incorporation of, for example, the $(U, Th)O_2$ component must lead to a disorder in the oxygen partial-lattice of euxenite if U^{4+}, with a coordination number of 8, is also incorporated [14].

An investigation of the chemical composition of unaltered brown euxenite from rock-crystal-bearing pegmatite, central Kazakhstan, shows that incorporation of higher-valency cations (Th^{4+}, U^{4+}) in group-A positions is compensated by divalent Fe, Ca, and Ba. Simultaneously, incorporation of lower-valency cations (Al^{3+} and Fe^{3+}) in group-B positions causes a stoichiometric compensation of oxygen by hydroxyl. And a small surplus of Ti^{4+} and Zr^{4+} is balanced by corresponding amounts of F. Thus, the crystal-chemical formula may be given as $(RE_{0.75}Th_{0.11}Ca_{0.07}Fe^{2+}_{0.04}Mn_{0.01}Ba_{0.01})_{\Sigma0.99}$ $(Ti_{1.61}Nb_{0.77}Al_{0.12}Fe^{3+}_{0.07}Zr_{0.02}Ta_{0.01})_{\Sigma2.00}$ $(O_{5.56}(OH)_{0.38}F_{0.05})_{\Sigma5.99}$ for the dark brown sample, corresponding to a thorian Ce-Y-euxenite.

Alteration, due to a raised alkalinity of the mineralizing solution, changed this euxenite to a more yellowish product with a composition $(Ca_{0.30}RE_{0.27}U_{0.15}Th_{0.01})_{\Sigma 0.73}$ $(Ti_{0.61}Nb_{0.51}Al_{0.37}Si_{0.27}$ $Fe^{3+}_{0.19}Mg_{0.04}Ta_{0.02})_{\Sigma 2.01}$ $(O_{4.28}(OH)_{1.78})_{\Sigma 6.06} \cdot 0.3\,H_2O$. Obviously, this alteration, besides hydration, changed in group-A positions: RE and Th by removal, Ca and U (or UO^{2+}?) by addition; and in group-B positions: Ti and B by removal, Al, Si, Fe, Mg, and Ta by addition. Simultaneously, the RE composition became relatively higher in light (cerian) RE. The crystal-chemical formula for the original composition (as given in [15]) of this euxenite from Kazakhstan is nearly intermediate between both the above given formulas [16].

Leaching experiments with euxenite from Canon City, Fremont county, Colorado, reveal a preferential leaching of [228]Th, considered to be situated at a site around which atomic arrangements are disturbed by the first α-recoil displacement. The [232]Th atoms are less mobile, for they are incorporated in the lattice after the time of mineral formation [17].

Euxenite samples from the Betsiboka valley, Madagascar (Nos. R18 and R21 [18]) and from Iveland, Eitland, and Sjaen, all Norway (Nos. R17, R25 [19], and R20 [18], respectively) have the following chemical compositions and physical properties (VHN$_{50}$ = Vickers hardness in kg/mm^2 at 50 g load; r.i. = refractive index; and R (546 nm) = reflectivity in % for the given wavelength):

	content in % and physical properties for sample				
	R21	R18	R17	R25	R20
ThO_2	4.60	4.16	3.10	2.85	2.65
U_3O_8 [a]	13.90	10.41	13.20	12.12	12.25
ΣRE_2O_3	11.91	24.61	23.60	22.70 [b]	16.78
CaO	0.48	0.42	1.60	2.40	2.57
MnO	>0.10	>0.10	—	0.46	>0.10
PbO	0.73	~0.60	1.00	0.36	~0.60
Fe_2O_3 [a]	1.85	1.36	1.50	3.89	3.90
Nb_2O_5	31.40	27.36	24.20	23.19	27.20
Ta_2O_5	1.80	2.48	3.00	0.49	3.40
TiO_2	28.45	24.51	24.50	25.45	26.00
H_2O^-	0.10	0.20	0.20	—	0.10
H_2O^+	2.00	2.50	3.20	3.63	3.10
sum	97.32	98.71	99.10	97.54 [c]	98.65
D in g/cm^3	5.14	4.85	5.11	4.826	5.22
VHN$_{50}$	633	626	690	641	679
r.i.	2.22	2.14	2.21	2.18	2.13
R (546 nm)	15.62	14.55	14.79	13.68	13.55

[a] Oxidation state not determined. — [b] 21.50% as calculated from the given contents for individual RE. — [c] Given as 99.26% in the cited paper. But note that [20] gives for this sample: 7.80% UO_2 and 4.32% UO_3; additionally occur 0.83% FeO, 0.51% SiO_2, 0.07% Al_2O_3, 0.06% Sc_2O_3, 0.05% MgO, and 0.2% SnO_2, which together gives the correct sum 99.26%.

Table 9

Chemical Composition and Density of Euxenite and Polycrase Samples from Various Types of Occurrence.

tr. = traces; n.d. = not determined; n.f. = not found. For the footnotes, see p. 168.

	content in % in sample number								
	1	2	3	4	5	6	7	8	9
ThO_2	8.40	8.2	6.90	4.80	4.77	4.75	4.67	4.5	4.30
UO_2	—	—	—	—	2.18	7.75	7.97	11.5	—
UO_3	0.15[a]	3.9[a]	5.05[a]	4.03[a]	9.20	tr.	0.08	—	5.04[a]
ΣRE_2O_3	30.31	14.5	26.56	26.90	24.58	27.53	23.50	24.65	28.82
CaO	1.65	4.7	2.72	1.82	1.16	1.22	1.65	—	2.44
MgO	0.02	0.5	0.15	0.13	0.09	—	0.01	—	0.15
PbO	n.d.	0.2	—	0.30[e]	0.82	—	0.37	—	—
Nb_2O_5	28.75	52.0	18.39	21.05	25.83	29.69	28.74	18.0	18.45
Ta_2O_5	0.78	1.4	3.84	—	2.10	6.56	2.48	22.5	6.40
TiO_2	22.74	1.5	28.82	31.20	21.13	17.85	26.06	20.0	29.24
SiO_2	n.f.	3.1	1.56	0.55	—	0.50	0.28	—	0.66
Al_2O_3	1.72	0.2	—	0.20	0.50	—	0.16	—	—
Fe_2O_3	1.46	3.2	2.30	3.54	1.11[g]	2.62	1.08	1.4	1.22
FeO	0.71	—	0.10	—	—	—	0.14	—	0.07
H_2O^-	0.26	—	0.06	—	—	—	0.40	—	0.04
H_2O^+	1.49	—	2.66	—	3.21	—	2.89	—	3.14
others	1.40[b]	1.9[c]	0.60[d]	3.48[f]	0.27[h]	1.89[i]	0.37[j]	0.15[k]	0.22[l]
sum	100.33	95.3	100.25	98.00	99.95	100.36	100.85	102.7	100.21
D in g/cm^3	4.34	—	—	5.00	—	5.46	4.92 ± 0.2	~5.38	—

No. 1 = Eux. from pegmatite containing rock-crystal (Akzhailyaru [11, p. 55]), central Kazakhstan [16]. – No. 2 = Eux. from granite pegmatite dike, Bancroft area, Ontario, Canada [21]. – No. 3 = Polycr. (eux.) from pegmatite, Kenhardt district, Cape Province, R.S.A. [22]. – No. 4 = Eux. from Trout Creek Pass, Colorado [23]. – No. 5 = Eux.-polycr. from Sahamandrevo, southwest of Ampangabé, Madagascar; corrected values after deduction of silica and hygroscopic moisture from the totals [24]. – No. 6 = Eux. from pegmatite, USSR [2, p. 412]. – No. 7 = Eux. ("ampangabeite") from pegmatite, Ampangabé, Madagascar [25]. – No. 8 = Tanteux. from pegmatite near Chepelare, Bulgaria [26]. – No. 9 = Polycr. (eux.) from pegmatite, Kenhardt district, Cape Province, R.S.A. [22].

Table 9 (continued)

	content in % in sample number								
	10	11	12	13	14	15	16	17	18
ThO_2	3.80	3.74	3.30	3.20	3.12	3.0	2.86	2.85	2.82
UO_2	—	—	7.22	11.48	17.2	—	7.37	—	—
UO_3	—	5.30[a]	tr.	0.67	—	11.0[a]	4.50	12.12[v]	11.70[a]
ΣRE_2O_3	38.07	7.31	28.85	25.64	19.4	26.0	23.79	22.70	23.82
CaO	1.01	0.21	1.26	0.64	2.5	2.0	2.34	2.40	0.16
MgO	—	0.17	—	0.22	0.1	—	0.15	0.05	0.15
PbO	0.35	2.09	—	—	0.7	1.0	—	0.36	1.11
Nb_2O_5	20.06	27.64	35.85[e]	30.76[e]	21.7	29.0	21.37[t]	23.19	22.20
Ta_2O_5	—	7.90	—	—	2.90	1.0	5.49	0.49	5.86
TiO_2	29.89	36.15	17.03	24.74	26.7	22.0	24.72	25.45	28.45
SiO_2	2.16	—	1.62	0.33	—	—	1.03	0.57	—
Al_2O_3	—	—	—	—	0.4	—	0.16	0.07	—
Fe_2O_3	2.40	1.14	2.62	1.07	1.8	5.4	2.89	3.89	0.80
FeO	—	0.29	—	—	—	0.1	—	0.83	0.20
H_2O^-	n.f.	3.53	—	—	—	—	0.31	—	0.50
H_2O^+	1.46	4.15	—	0.23	—	2.18[r]	1.80	3.63[r]	1.92
others	1.06[m]	0.38[n]	1.89[o]	1.16[q]	—	0.63[s]	0.58[u]	0.66[w]	0.31[x]
sum	100.26	100.0	99.64[p]	99.14	96.52	103.31	99.36	99.26	100.00
D in g/cm³	—	—	5.32	~5.5	5.28	—	5.10	4.826	—

No. 10 = Polycr. (red) from albitite, Eastern Siberia [27]. – No. 11 = Polycr. (primarily altered outer zone) from the Pomba pegmatite, Brazil; analysis is corrected for impurities (3.71% kaolinized feldspar) and recalculated to 100% [34]. – No. 12 = Eux. from pegmatite vein, USSR [2, p. 412]. – No. 13 = Eux./polycr. from granite pegmatite, Showa Base, Antarctica [28]; refer also to [29, 44]. – No. 14 = Polycr. (black) from alaskite, Zenkovskii massif, Urals, USSR [30]. – No. 15 = Eux. from pegmatite near St. Pierre de Wakefield, Quebec, Canada [31]. – No. 16 = Polycr. from granite pegmatite Nekonaki, Fukushima prefecture, Japan [32]; refer also to [33]. – No. 17 = Eux. from Farsund, Norway [20]; refer also to analysis No. R25 on p. 163. Note that this sample is named as polycr. in [8]. – No. 18 = Polycr. (unaltered inner zone) from the Pomba pegmatite, Brazil; analysis is corrected for impurities (0.88% kaolinized feldspar) and recalculated to 100% [34].

Table 9 (continued)

	19	20	21	22	23	24	25	26	27
					content in % in sample number				
ThO_2	2.76	2.40	2.35	2.32	2.30	2.16	2.15	2.15	1.96
UO_2	13.7	7.95	—	4.71	4.18	—	—	—	10.76
UO_3	—	—	0.23	3.50	—	9.62[a]	9.90[a]	—	—
ΣRE_2O_3	22.96	25.77	25.23	29.42[ab]	15.19	25.57	25.15	27.10	23.36
CaO	8.5	2.8	2.84	1.05	0.03	0.97	2.00	1.55	0.92
MgO	—	0.08	—	0.10[ab]	0.18	—	—	—	tr.
PbO	1.1	0.12	—	—	—	—	—	—	—
Nb_2O_5	17.3	23.49	37.00	27.75	26.92	24.48	20.55	38.60	29.01
Ta_2O_5	2.4	2.55	0.05	7.05	23.43	5.74	2.18	2.23	2.88
TiO_2	25.0	25.64	19.80	18.12	19.80	25.54	25.41	20.25	22.01
SiO_2	0.1	0.47	2.60	0.86	0.04	—	1.80	—	tr.
Al_2O_3	0.3	0.07	0.46	0.45	0.10	0.50	1.33	—	—
Fe_2O_3	2.3	4.66	1.46	2.23	—	2.59	3.68	—	3.98
FeO	—	0.40	—	0.16	5.13	—	—	1.90	2.23[af]
H_2O^-	—	—	—	0.12	—	—	—	—	—
H_2O^+	—	3.26	5.76[r]	1.56	—	3.78[r]	6.00[r]	1.75[r]	—
others	—	0.33[x]	0.55[z]	0.33[ac]	2.98[ad]	—	—	2.91[ae]	3.07[ag]
sum	96.42	99.99	98.33[aa]	99.73	100.31	99.95	100.15	100.05	100.18
D in g/cm³	4.86	4.797	4.355	5.23[ab]	—	—	4.47	5.40	5.24[af]

No. 19 = Polycr. (brown) from alaskite, Zenkovskii massif, Urals, USSR [30]. – No. 20 = Eux. from Kalstadgangen, Kragerø, Norway [20]. – No. 21 = Eux. from pegmatite of the exocontact aureole of a nepheline syenite massiv, Transangara (or Tatarka [11, p. 55]), Enisei Ridge, Siberia [35]. – No. 22 = Eux. from granite pegmatite Nekonaki, Fukushima prefecture, Japan [33]. – No. 23 = Tanteux. from Korea [36]. – No. 24 = Eux. from muscovite granite, eastern Transbaikalia, USSR [37]. – No. 25 = Eux. from quartz-albite-microcline pegmatite, USSR [45]. Note that [11, pp. 54/5] gives 2.25% ThO_2 for the same analysis and as location Keivy, Kola Peninsula. Note further that this mineral may be blomstrandine, see p. 189. – No. 26 = Eux. (non-metamict) from alkali granite, Pamir, USSR [11, pp. 54/5]. – No. 27 = Eux. from a pegmatitic body related to rose vein granite in gray porphyric biotite granite, Zolotaya Mountain, eastern Transbaikalia, USSR [38].

Table 9 (continued)

	content in % in sample number							
	28	29	30	31	32	33	34	35
ThO_2	1.95	1.9	1.77	1.69	1.33	1.10	1.06	1.04
UO_2	—	8.0	—[ai]	1.80	—	—	11.18	1.78
UO_3	11.50	3.5	14.42[a]	6.45	—	7.45[a]	0.32	10.05
ΣRE_2O_3	21.53	24.4	9.27	24.91	39.06	24.84	20.06	22.08
CaO	1.91	—	3.08	0.94	0.73	1.64	0.65	1.45
MgO	0.10	—	—	0.05	0.30	0.39	tr.	0.38
PbO	—	1.9	—	—	—	—	—	—
Nb_2O_5	21.97[ah]	22.2	39.76	25.98	18.95	16.89	26.68	15.02
Ta_2O_5	4.74	2.5	2.69	1.42	0.65	20.48	16.14	14.78
TiO_2	24.84	25.0	n.d.	27.50	32.24	20.25	13.73	25.60
SiO_2	0.89	0.2	4.05	1.15	3.43	1.28	0.48	1.18
Al_2O_3	0.34	4.4	2.54	1.46	—	—	0.16	0.21
Fe_2O_3	4.50	4.4	6.00	2.66	2.37	2.65	4.80	0.96
FeO	—	—	—	—	n.f.	0.09	2.60	2.34
H_2O^-	0.15	—	—	—	—	0.06	0.21	0.33
H_2O^+	5.32	2.2[r]	12.25[r]	2.67[r]	1.13	2.44	0.88	1.67
others	0.22[k]	—	4.94[ai]	0.45[ak]	—	0.46[al]	0.86[am]	0.57[an]
sum	99.96[ah]	100.8	100.77	99.13	100.19	100.02	99.81	94.44
D in g/cm³	4.0	—	—	5.19	5.05	—	5.47	5.27

No. 28 = Eux. from pegmatite vein Kapraevo, near Alakurtti, Kola Peninsula, USSR [39]. – No. 29 = Eux. from (pegmatitic) quartz-feldspar vein in granite gneiss, Kakanuku Mountains, British Guiana [41]. – No. 30 = Eux. from a pegmatite dike in hornblende gneiss, Sappington, Montana [42]. – No. 31 = Eux. from pegmatite of biotite-hornblende granite, Kaveripatnam, India [11, pp. 89/90]. – No. 32 = Polycr. (yellow-brown) from albite, Eastern Siberia [27]. – No. 33 = Tantalian polycr. (eux.) from pegmatite, Gordonia district, Cape Province, R.S.A. [22]. – No. 34 = Eux. from pegmatite, Uzumine, Fukushima prefecture, Japan [33]. – No. 35 = Eux. from pegmatite, Ipaiyama, Fukushima prefecture, Japan [33]; refer also to [43].

Footnotes for Table 9:

a) Given as U_3O_8. — b) Comprises 0.13% MnO and 1.27% l.o.i. — c) Comprises 0.5% MnO, 1.2% Na_2O, and 0.2% ZrO_2. — d) Comprises 0.4% MnO, 0.12% Na_2O, and 0.08% K_2O. — e) Includes Ta_2O_5.
f) Comprises 0.04% MnO, 0.29% Na_2O, 0.12% K_2O, 0.03% SO_3, and 3.0% l.o.i. — g) Total iron given as Fe_2O_3. — h) Comprises 0.09% MnO, 0.11% Na_2O, 0.03% K_2O, 0.02% CuO, and 0.02% ZnO. — i) Comprises 0.23% MnO, 0.29% SnO_2, and 1.37% l.o.i. — j) Comprises 0.18% MnO, 0.11% SnO_2, and 0.08% WO_3.
k) Only MnO. — l) Comprises 0.02% MnO, 0.10% Na_2O, and 0.10% K_2O. — m) Comprises 0.38% MnO and 0.68% Na_2O. — n) Comprises 0.02% MnO and 0.36% SnO_2. — o) Comprises 0.14% MnO, 0.2% SnO_2, 0.7% l.o.i., and 0.85% insoluble.
p) In the original paper given as 100.53%. — q) Comprises 0.05% MnO and 0.11% SnO_2. — r) Total water. — s) Comprises 0.5% MnO and 0.13% F. — t) Erroneously given as 21.75% Nb_2O_5 in [33].
u) Comprises 0.43% MnO and 0.15% SnO_2. — v) Comprises $UO_3 + UO_2$. — w) Comprises 0.46% MnO and 0.20% SnO_2. — x) Comprises 0.03% MnO and 0.28% SnO_2. — y) Comprises 0.18% MnO, 0.12% SnO_2, and 0.03% Sc_2O_3.
z) Comprises 0.3% Na_2O and 0.25% F. — aa) In the original paper given as 99.11%. — ab) Given as 21.42% ΣRE_2O_3, 0.10% MgO, and 5.25 g/cm³ in [32]. — ac) Comprises 0.24% MnO and 0.09% SnO_2. — ad) Comprises 2.66% MnO, 0.3% ZrO_2, and 0.02% SnO_2.
ae) Comprises 0.55% MnO, 0.63% ZrO_2, 0.25% F, 0.21% BaO, and 1.27% l.o.i. — af) Calculated from the crystal-chemical formula; total iron determined is 6.46% Fe_2O_3. — ai) Comprises 0.32% MnO and 2.75% l.o.i. — ag) Calculated value; mean for observed density is 5.4 g/cm³. — ah) Given as 21.77% Nb_2O_5 and 99.76% in [40]. — ai) No UO_2 present.
aj) Comprises 0.55% BaO and 4.39% WO_3. — ak) Comprises 0.13% MnO, 0.16% Na_2O, 0.09% K_2O, and 0.07% SnO_2. — al) Comprises 0.24% MnO, 0.12% Na_2O, and 0.10% K_2O. — am) Comprises 0.54% MnO and 0.32% SnO_2. — an) Comprises 0.18% MnO and 0.39% SnO_2.

References for 2.2.2.1.1.2:

[1] Arnott, R. J. (Am. Mineralogist **35** [1950] 386/400, 390/9).

[2] Sokolova, E. P. (Zap. Vses. Mineral. Obshch. **88** [1959] 408/18).

[3] Komkov, A. I. (Dokl. Akad. Nauk SSSR **126** [1959] 641/4; Dokl. Acad. Sci. USSR Earth Sci. Sect. **124/129** [1959] 584/6).

[4] Butler, J. R. (Econ. Geol. **56** [1961] 442/4).

[5] Aleksandrov, V. B. (in: Vlasov, K. A., Geochemistry and Mineralogy of Rare Elements and Genetic Types of Their Deposits, Vol. 2, Israel Program Sci. Transl., Jerusalem 1966, pp. 470/82, 471 [Russian original, Moscow 1964]).

[6] Van Wambeke, L. (Neues Jahrb. Mineral. Abhandl. **112** [1969/70] 117/49).

[7] Aleksandrov, V. B. (Dokl. Akad. Nauk SSSR **153** [1963] 672/5; Dokl. Acad. Sci. USSR Earth Sci. Sect. **153** [1963] 129/31).

[8] Gorzhevskaya, S. A.; Sidorenko, G. A. (Geol. Mestorozhd. Redk. Elem. No. 23 [1964] 1/116, table between pp. 20/1).

[9] Palache, C.; Berman, H.; Frondel, C. (Dana's System of Mineralogy, 7th Ed., Vol. 1, Wiley, New York 1946, pp. 1/834, 789/90).

[10] George, D'Arcy (RMO-563 [1949] 1/198, 53/4; N.S.A. **5** [1951] No. 833).

[11] Semenov, E. I. (Tipokhimizm Mineralov Shchelochnykh Massivov, Nedra, Moscow 1977, pp. 1/119).

[12] Ewing, R. C. (Can. Mineralogist **14** [1976] 111/9, 117).

[13] Butler, J. R. (Mineral. Mag. **31** [1958] 763/80, 764/6).

[14] Schröcke, H. (Neues Jahrb. Mineral. Abhandl. **106** [1967] 1/54, 10/1).

[15] Erdzhanov, K. N. (Tr. Kaz. Nauchn. Issled. Inst. Mineral'n. Syr'ya No. 2 [1960] 92/5; C.A. **60** [1964] 3876).

[16] Mineev, D. A.; Kataeva, Z. T.; Malyshev, A. G. (in: Lebedeva, S. I., Fizicheskie Svoistva Redkometal'nykh Mineralov i Metody Ikh Issled., Nauka, Moscow 1968, pp. 43/7).

[17] Kobashi, A.; Sato, J.; Saito, N. (Radiochim. Acta **26** [1979] 107/11, 109).

[18] Ewing, R. C. (Geochim. Cosmochim. Acta **39** [1975] 521/30, 522/3).

[19] Ewing, R. C.; Ehlmann, A. J. (Can. Mineralogist **13** [1975] 1/7, 2).

[20] Hongslo, T.; Langmyhr, F. J. (Norsk Geol. Tidsskr. **40** [1960] 157/64, 160).

[21] Satterly, J. (Ann. Rept. Ontario Dept. Mines **65** VI [1957] 1/181, 20).

[22] Hugo, P. J. (Mem. South Africa Geol. Surv. No. 58 [1970] 1/94, 54/6).

[23] Muench, O. B. (Bull. Geol. Soc. Am. **61** [1950] 129/32).

[24] Smellie, J. A. T.; Cogger, N.; Herrington, J. (Chem. Geol. **22** [1978] 1/10, 2/3, 6).

[25] Ewing, R. C.; Snetsinger, K. G.; Bunch, T. E. (Can. Mineralogist **15** [1977] 92/6, 92/3).

[26] Arnaudov, A.; Petrusenko, S. (Izv. Geol. Inst. Bulg. Akad. Nauk. Ser. Geokhim. Mineral. Petrol. **21** [1972] 77/84, 81).

[27] Proshchenko, E. G. (in: Mineralogiya Pegmatitov i Gidrotermalitov Shchelochnykh Massivov, Nauka, Moscow 1967, pp. 103/36, 132).

[28] Hayashi, S.; Nagashima, K. (Nankyoku Shiryo No. 11 [1961] 27/30).

[29] Watanuki, K.; Torii, T.; Nishiyama, T.; Hideki, M. (Mem. Natl. Inst. Polar Res. Spec. Issue [Japan] No. 14 [1979] 52/6, 53).

[30] Chashchukhina, V. A. (Sb. Nauchn. Tr. Ural. Nauchn. Tsentr Akad. Nauk SSSR, Sverdlovsk 1987, pp. 65/7).

[31] Rimsaite, J.; Lachance, G. R. (Papers Proc. 4th General Meeting Intern. Mineral. Assoc., New Delhi, India, 1964 [1966], pp. 209/29, 226).

[32] Omori, K.; Hasegawa, S. (Kobutsugaku Zasshi **2** [1955] 268/74 from C.A. **1957** 5641).

[33] Omori, K.; Hasegawa, S.; Konno, H. (Sci. Rept. Tohoku Univ. III **6** [1958/60] 389/96, 389, 396).

[34] Van Wambeke, L. (Neues Jahrb. Mineral. Abhandl. **112** [1969/70] 117/49, 129).

[35] Sveshnikova, E. V.; Semenov, E. I.; Khomyakov, A. P. (Zaangarskii Shchelochnoi Massiv, Ego Porody i Mineraly, Nauka, Moscow 1976, pp. 1/80, 29, 31).

[36] Hata, S.; Imori, T. (Bull. Inst. Phys. Chem. Res. [Tokyo] **21** [1942] 1160/2).

[37] Zalashkova, N. E.; Distler, V. V.; Smirnova, N. V. (Tr. Mineral. Muzeya Akad. Nauk SSSR No. 20 [1971] 205/8).

[38] Kornetova, V. A.; Aleksandrov, V. B.; Kazakova, M. E. (Tr. Mineral. Muzeya Akad. Nauk SSSR No. 18 [1968] 197/202, 198/200).

[39] Kalita, A. P. (Redkozemel'nye Pegmatity Alakurtti i Priladozh'ya, Akad. Nauk SSSR, Moscow 1961, pp. 1/120, 46/7).

[40] Kalita, A. P. (in: Kuz'menko, M. V.; Nedumov, J. B., Novye Dannye po Geologii, Geokhimii i Genezisu Pegmatitov, Akad. Nauk SSSR, Inst. Mineralog. Geokhim. i Kristallokhim. Redk. Elem., Moscow 1965, pp. 266/304, 282).

[41] Henderson, G. (Brit. Guiana Geol. Surv. Dept. Rept. **1950/51** 31/9, 39).

[42] Heinrich, E. W. (Am. Mineralogist **50** [1965] 2083/8, 2085).

[43] Omori, K.; Hasegawa, S. (Ganseki Kobutsu Kosho Gakkaishi **42** [1958] 280/4 from Ref. Zh. Geol. **1960** No. 14777).

[44] Saito, N.; Tatsumi, T.; Sato, K. (Nankyoku Shiryo No. 12 [1961] 31/6 from Mineral. Abstr. **16** [1963/64] 63).

[45] Kalita, A. P. (Tr. Inst. Mineral. Geokhim. Kristallokhim. Redk. Elem. No. 16 [1963] 107/205, 117/8).

2.2.2.1.1.3 Crystal Form and Structure

Crystal Form. Euxenite/polycrase is orthorhombic dipyramidal [1], but usually it is partially or totally metamict and X-ray amorphous; non-metamict euxenite is rare in nature [2, p. 348]. Morphological axial ratio is a:b:c = 0.3789:1:0.3527 [1], [2, p. 348] and corresponds well to the ratios which can be derived from lattice constants, refer to pp. 171/2.

Crystal forms, with their symbols, observed for euxenite/polycrase are given as follows: a{100}, b{010}, c{001}, m{110}, w{011}, d{101}, e{201}, γ{131}, q{121}, and p{111} [1]; additionally observed are h{160}, k{150}, l{140}, g{130}, t{251} [2, p. 349], and {120}? [3]. Commonest forms are bdmpa [1], bamdp, {102}, and e [4, pp. 475/7], or bamep [5, pp. 31/5] sometimes with c. Crystals are imperfect and their faces often dull; goniometric measurements can be done only roughly [2, p. 349]. For the table of polar coordinates (φ and ϱ) for the above forms, see [1], [2, p. 349]; or [3], which gives the crystal forms separate for euxenite and polycrase. Additionally, the latter reference gives hkl in the corresponding crystallographic arrangement for aeschynite. For a compilation of figures of crystal morphology of euxenite/polycrase from different occurrences, see [3].

Examples for observed combinations of crystal forms from individual occurrences of euxenite are the following: Crystals exhibiting the forms bmp and partly, a and e occur with three different habits (see **Fig. 7**) in a pegmatite vein of the USSR [6, p. 409]. Tiny prismatic or platy crystals elongated along the c axis and with forms bamdpe occur in (ultra-acid) muscovite granite, Transbaikalia [7]. A large crystal (23 by 12 by 8 mm) from the Rutherford pegmatite, Amelia county, Virginia, has the forms mha and {120}? [3]. Thin or thick platelets from alluvial deposits near Monastir, central Rhodopes, Bulgaria, show three combinations of forms: abp, mbp, and abmep. Whereas the first and the second combination is flattened parallel to (010) and elongated along the c axis, the last combination looks somewhat "stocky" owing to some elongation along the b axis (for the figures, see original paper) [8]. Long-prismatic crystals of tanteuxenite from a pegmatite near Chepelare, central Rhodopes, are up to 2 cm long and up to 2 mm wide (the length:width ratio varies between 5 and 100) and have the forms ab with subordinate mep, occurring with two different habits. One type is thin tabular along the b axis, the other is almost equally developed parallel to the a and b axis [9].

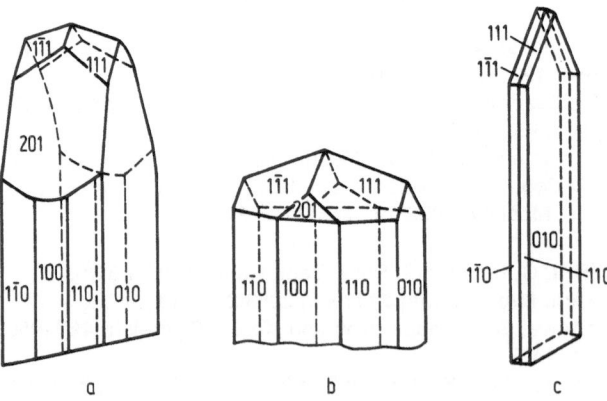

Fig. 7. Morphological types (habits) of euxenite crystals from a pegmatite vein, USSR, after [6, p. 409]. a = columnar chisel-like, b = prismatic, and c = tabular habit.

Generally, the habit of euxenite/polycrase is stout prismatic along the c axis, sometimes flattened parallel to the b axis (polycrase) [1] forming thin- or thick-tabular crystals (polycrase is found only in the latter form) [5]. On vertical (prism) faces [2, p. 349], polycrase and, to a lesser extent, euxenite are striated parallel to their intersection with {010}. Euxenite/polycrase also occurs in parallel or subparallel and slightly radial aggregates of crystals (for examples, refer to pp. 154, 157, and to [7, 10]) [1].

Twins are common on {201} (which are flattened parallel to (010) and striated parallel to [001]) and, possibly, on {100} and {013} [1], or {111} [2, pp. 349/50]; for figures of twins, see the latter reference.

Cleavage of euxenite/polycrase is not observed [2, p. 350].

Crystal Structure. Euxenite has space group $D_{2h}^{14}-Pcan$ (No. 60); $Z = 4$ [2, p. 348]. But minerals are generally metamict, crystallizing on ignition to $>1000\,°C$ (refer to pp. 175/6), and considerable fluctuations in lattice constants (see below) may be due to variation in chemical composition and to differences in the conditions of ignition [4, p. 471].

Lattice constants for Th-containing euxenites, heated at 1000 to 1100°C, are given as follows:

locality	lattice constants in Å			Ref.
	a	b	c	
Central Urals, USSR[a]	5.19[b]	14.72[b]	5.73[b]	[11]
Mattawan township, Ontario	5.520	14.57	5.166	[12, p. 397]
USSR[c]	5.54	14.68	5.18	[6, p. 417]
Sabine township, Ontario	5.552	14.16	5.194	[12, p. 397]
eastern Transbaikalia, USSR	5.558	14.54	5.193	[13]
eastern Transbaikalia, USSR	5.57	14.67	5.18	[7]
USSR[d]	5.56	14.66	5.18	[6, p. 417]
Ampangabé, Madagascar[e]	5.567	14.64	5.198	[14]
USSR[f]	5.75	14.65	5.18	[6, p. 417]

[a] Anisotropic part of the unheated sample. — [b] Recalculated from kX given in the original paper. — [c] Xenomorphic sample. — [d] Crystal sample. — [e] Originally named ampangabeite. — [f] Partially anisotropic sample with hour-glass structure; data by rotating crystal method.

Euxenite samples from Eitland (No. R25) and Iveland (No. R17), both Norway, heated at 900°C, give the following lattice constants and axial ratios [15]; for the chemical analyses, see p. 163:

sample	lattice constants in Å			axial ratio
	a	b	c	a:b:c
No. R25	5.53	14.61	5.17	0.379:1:0.354
No. R17	5.55	14.70	5.16	0.378:1:0.351

Synthetic pure $YNbTiO_6$, which may be used as a standard, has lattice constants $a = 5.550 \pm 0.005$ Å, $b = 14.61 \pm 0.01$ Å, $c = 5.185 \pm 0.005$ Å for the euxenite structure, and $a = 5.185 \pm 0.005$ Å, $b = 10.96 \pm 0.01$ Å, $c = 7.410 \pm 0.005$ Å for the priorite structure [19].

The axial ratio 0.388:1:0.352 derived from Laue patterns for anisotropic euxenite from the Central Urals, is very similar to that derived from morphological data (see p. 170) [11]. Axial ratios derived from Debye-Scherrer patterns are 0.3789 : 1 : 0.3546 and 0.3800 : 1 : 0.3556 for euxenite samples heated at 1000°C from Mattawan and Sabine townships, respectively [12]. For samples from Norway, see above.

As euxenite usually is metamict (see p. 170), it is normally also X-ray amorphous and diffraction data can be obtained only after heat treatment, see pp. 175/6. For an "euxenite analyzer" to index X-ray powder patterns, giving the variations of indexed d values with b (based on a morphological axial ratio 0.3789 : 1 : 0.3527), see [12, pp. 393/4]. For indexed d values, observed and calculated, of synthetic pure $YNbTiO_6$, which may be used as standards for euxenite and priorite, see [19].

Euxenite is structurally analogous to fersmite. Relative to oxygen, group-A cations are 8-coordinated, group-B cations are 6-coordinated [4, p. 475] in the high-temperature euxenite structure (Pcan). The B-site octahedra are slightly distorted and connected along two edges into zig-zag chains parallel to the c axis. Chains are arranged into two-dimensional layers that are connected by single layers of A-site cations [20]. For the low-temperature aeschynite/priorite structure (Pbnm), distinguished by its three-dimensional framework, see pp. 222/3. A description of the fersmite structure including a figure is given in [16]; refer also to [2, pp. 342/4], [5, pp. 9/10]. Thus, euxenite has similar cell parameters, a nearly identical density, and the same space group as fersmite [17]. For a crystal structure refinement of euxenite, $Y(Nb_{0.5}Ti_{0.5})_2O_6$, see [18].

References for 2.2.2.1.1.3:

[1] Palache, C.; Berman, H.; Frondel, C. (Dana's System of Mineralogy, 7th Ed., Vol. 1, Wiley, New York 1946, pp. 1/834, 787/8).
[2] Chukhrov, F. V.; Bonshtedt-Kupletskaya, E. M. (Mineraly Spravochnik, Vol. 2, Pt. 3, Nauka, Moscow 1967, pp. 1/676).
[3] Mitchell, R. S. (Southeast. Geol. **14** No. 1 [1972] 59/72, 68).
[4] Aleksandrov, V. B. (in: Vlasov, K. A., Geochemistry and Mineralogy of Rare Elements and Genetic Types of Their Deposits, Vol. 2, Israel Program Sci. Trans., Jerusalem 1966, pp. 470/82 [Russian original, Moscow 1964]).
[5] Gorzhevskaya, S. A.; Sidorenko, G. A. (Geol. Mestorozhd. Redk. Elem. No. 23 [1964] 1/116).
[6] Sokolova, E. P. (Zap. Vses. Mineral. Obshch. **88** [1959] 408/18, 409).
[7] Zalashkova, N. E.; Distler, V. V.; Smirnova, N. V. (Tr. Mineral. Muzeya Akad. Nauk SSSR No. 20 [1971] 205/8).
[8] Breskovskaya, V.; Marinova, R. (Spisanie Bulg. Geol. Druzh. **36** [1975] 187/90).
[9] Arnaudov, A.; Petrusenko, S. (Izv. Geol. Inst. Bulg. Akad. Nauk. Ser. Geokhim. Mineral. Petrol. **21** [1972] 77/84, 79, 84).
[10] Roubault, M. (Geologie de l'Uranium, Masson, Paris 1958, pp. 1/462, 33).
[11] Nefedov, E. I. (Vses. Nauchn.-Issled. Geol. Inst. **1956** No. 3, pp. 82/5).
[12] Arnott, R. J. (Am. Mineralogist **35** [1950] 386/400).
[13] Kornetova, V. A.; Aleksandrov, V. B.; Kazakova, M. E. (Tr. Mineral. Muzeya Akad. Nauk SSSR No. 18 [1968] 197/202, 198).
[14] Ewing, R. C.; Snetsinger, K. G.; Bunch, T. E. (Can. Mineralogist **15** [1977] 92/6, 94).
[15] Ewing, R. C.; Ehlmann, A. J. (Can. Mineralogist **13** [1975] 1/7, 3/4).
[16] Aleksandrov, V. B. (Dokl. Akad. Nauk SSSR **132** [1960] 669/72; Dokl. Acad. Sci. USSR Earth Sci. Sect. **130/135** [1960] 597/600).
[17] Komkov, A. I. (Dokl. Akad. Nauk SSSR **126** [1959] 641/4; Dokl. Acad. Sci. USSR Earth Sci. Sect. **124/129** [1959] 584/6).

[18] Weitzel, H.; Schröcke, H. (Z. Krist. **152** [1980] 69/82).
[19] Komkov, A. I. (Rentgenogr. Mineral'n. Syr'ya **1965** No. 5, pp. 38/44, 42/4).
[20] Ewing, R. C.; Chakoumakos, B. C. (Short Course Handb. Mineral. Assoc. Can. **8** [1982] 239/65, 249/50).

2.2.2.1.1.4 Optical and Other Physical Properties

Optical Properties. Minerals of the euxenite-polycrase group are usually black or brownish black, dark brown, and rarely green. Streak is yellowish gray or light brown. Surface luster is from sub-metallic to waxy, on fractures it is resinous. In transmitted light, these minerals are isotropic, seldom partly or completely anisotropic, and have light brown, yellow-brown, occasionally red-brown color [1].

For color and luster of individual Th-containing euxenites is stated: Euxenite from muscovite granite, Transbaikalia, is dark brown to brownish black, in thin splinters translucent yellowish cinnamon to brownish, and in immersion media reddish brown. Luster is semi-metallic to waxy on the surface, and resinous to fatty on fracture [2]. Euxenite from pegmatitic bodies in porphyric biotite granite, Zolotaya Mountain, eastern Transbaikalia, is dark cinnamon on fresh fracture, later becomes black, and is nonuniformly colored in thin splinters. In thin sections, it is transparent dark cinnamon and, sometimes, shows a zonal color distribution: black or dark cinnamon (with a reddish hue for the marginal part) in the center of a grain, followed outwards by lighter colored zones (fresh yellow or tobacco yellowish and olive), possibly caused by a change in composition [3, pp. 198/200]. Euxenite from pegmatite veins, USSR, is black, on the margin of thin splinters translucent red-brown, and becomes yellowish brown after heating at ~1000 °C. Luster is variable between clear and dull resinous, sometimes changing to fatty. Usually, it is isotropic, although a few crystals have anisotropic parts. With crossed Nicols, an hour-glass structure becomes visible with anisotropic marginal parts of the crystal adjacent to the prism zone; the inner parts are isotropic [4, pp. 410/1, 415]. Metamict euxenite with anisotropic parts from a pegmatite vein, Central Urals, is brownish to greenish black or velvet black, streak is brownish gray, and its luster is glassy to semi-metallic. Heating of the anisotropic part at 1000 °C, partly decolorizes the mineral to a golden yellowish tint without noticeable changes in the optical properties and structure [5].

In reflected light, minerals of the euxenite-polycrase group are cream gray [1], or grayish white [4, p. 411]. Reflectivity is 11.7 to 13% [1]. For reflectivity at a wavelength of 546 nm for euxenite samples with a given chemical composition from Madagascar and Norway, see p. 163. Depending on wavelength and considering the contents of TiO_2, Nb_2O_5, and Ta_2O_5, the reflectivity of six euxenite samples from Madagascar, Brazil, and Norway vary as follows (for the individual data, see original paper): 13.7 to 15.6% for 470 nm; 13.0 to 15.6% for 546 nm; 12.5 to 15.0% for 589 nm; and 12.4 to 15.0% for 650 nm. The variance in element content is 24.5 to 34.5% TiO_2, 15.2 to 31.4% Nb_2O_5, and 1.1 to 7.7% Ta_2O_5 [6]. The reflectivity of an unoriented euxenite (data on its chemistry are lacking) is 13.5% for 700 μm; 14.4% for 600 μm, 14.3% for 575 μm, 18.9% for 520 μm, 19.9% for 470 μm, and 13.5% for white light (610 μm) [7].

Refractive indices of minerals of the fersmite-euxenite structural group are high, values of 2.23 to 2.375 are normal for euxenite-polycrase and the lower values of 2.035 to 2.175 are typical for more hydrated specimens [8, p. 34]. Refractive indices increase after ignition [1]. Thus, for Th-containing euxenites n is 1.83 to 1.84 for a sample very rich in water (= analysis No. 30 in Table 9, p.167) from pegmatite, Sappington, Montana [17]; or is >2.07 for samples from the Transangara nepheline syenite massif, eastern Enisei Ridge, Siberia [9]; or is 2.18

for an isotropic sample from muscovite granite, Transbaikalia [2]; or is between 2.23 and 2.28 (white light) for samples from pegmatite veins, USSR. After heating at $\sim 1000\,°C$, these euxenites become anisotropic, optical biaxial with a straight extinction and an intermediate optical angle. Their refractive indices are $N_p = 2.25$ to 2.28 and $N_g = 2.30$ to 2.35 [4, pp. 411, 415]. Euxenite (ampangabeite) from Ampangabé, Madagascar, has prior to heating $n = 2.01 \pm 0.1$, which increases after heating to 2.14 (500 °C) and 2.19 (1200 °C); each value is the mean of five determinations. The dramatic increase at 500 °C is due largely to loss of water [10]. For refractive indices of euxenite samples with a given chemical composition from Madagascar and Norway, see p. 163. Anisotropic euxenite from a pegmatite vein, Central Urals, is optical biaxial positive with $2 V \approx 70°$ and has $N_p = 2.18 \pm 0.02$, $N_m = 2.21$ (calculated), $N_g = 2.26 \pm 0.02$; optical orientation is $N_p = [001]$ and $N_m = [100]$ [5].

The coefficient k (see p. 151) for euxenite varies between 0.24 and 0.26 [15, p. 57].

Other Physical Properties. The Mohs hardness for euxenite-polycrase is 5.5 to 6.5 [1], or 5 to 6 for minerals of the fersmite-euxenite structural type [8, pp. 31/2]. Microhardness (VHN_{50}) for six euxenites from Madagascar, Brazil, and Norway varies between 633 and 692 kg/mm^2; for the corresponding contents of TiO_2, Nb_2O_5, and Ta_2O_5, see p. 173 [6]. For VHN_{50} for euxenite samples with a given chemical composition from Madagascar and Norway, see p. 163. Microhardness (70 to 100 g load) for euxenite from Zolotaya Mountain, eastern Transbaikalia (= analysis No. 27 in Table 9, p. 166), varies between 621 and 699, mean 663 kg/mm^2 [3, p. 198]. Note also microhardness for euxenite (with data on its chemistry lacking) between 530 and 767 kg/mm^2 (100 g load) [11], or between 680 and 735 kg/mm^2 [1].

Fracture of euxenite-polycrase is conchoidal [1] to subconchoidal [8, pp. 31/2].

Density for minerals of the fersmite-euxenite structural type varies between 4.29 and 5.90 g/cm^3. The higher values (5.4 to 5.9) are characteristic of Ta-rich varieties, for euxenite-polycrase they vary between 4.29 and 5.47. Hydrated varieties have the lowest values of 3.8 to 4.14 [8, pp. 32/4]; refer also to Table 9, pp. 164/8. Density of euxenite increases with contents of Th, U, and Ta, and decreases with contents of Ca, Ti, and H [12]. For individual Th-containing euxenites is given: D varies between 5.37 and 5.39 g/cm^3 for tanteuxenite from pegmatite near Chepelare, Bulgaria [13]. Euxenite (poor in Ta, but with considerable amounts of Th and U) from pegmatite veins, USSR, has an increased D varying, depending on the content of the mentioned elements, from 5.31 to 5.33 g/cm^3 (for crystals) to 5.45 to 5.47 g/cm^3 (for xenomorphic grains) [4, pp. 410, 415]. D is 4.85 and 4.90 g/cm^3 for brown, and 5.03 and 5.08 g/cm^3 for black euxenite from the Gloserheia granite pegmatite, southern Norway. The brown variety is a little higher in content of Ca, Fe, and Mn, whereas Pb is a little lower [14]. Euxenite (ampangabeite) from Ampangabé, Madagascar, has $D = 4.92 \pm 0.2$ g/cm^3 prior to heating, which increases after heating to 5.29 (500 °C) and 5.32 g/cm^3 (1200 °C). The dramatic increase at 500 °C is due largely to loss of water [10].

Euxenite-polycrase is electromagnetic [1]. The dielectric permeability ε for 10 euxenites varies between 3.74 and 5.92 [15, pp. 52/4], most typical are values between 4.02 and 4.63 [8, p. 34]. For behavior of dielectric constants during recrystallization of euxenite by heat treatment, see [16].

References for 2.2.2.1.1.4:

[1] Aleksandrov, V. B. (in: Vlasov, K. A., Geochemistry and Mineralogy of Rare Elements and Genetic Types of Their Deposits, Vol. 2, Israel Program Sci. Transl., Jerusalem 1966, pp. 470/82, 477 [Russian original, Moscow 1964]).

175

[2] Zalashkova, N. E.; Distler, V. V.; Smirnova, N. V. (Tr. Mineral. Muzeya Akad. Nauk SSSR No. 20 [1971] 205/8, 205).
[3] Kornetova, V. A.; Aleksandrov, V. B.; Kazakova, M. E. (Tr. Mineral. Muzeya Akad. Nauk SSSR No. 18 [1968] 197/202).
[4] Sokolova, E. P. (Zap. Vses. Mineral. Obshch. **88** [1959] 408/18).
[5] Nefedov, E. I. (Vses. Nauchn. Issled. Geol. Inst. No. 3 [1956] 82/5, 84/5).
[6] Ewing, R. C. (Am. Mineralogist **58** [1973] 942/4, 943).
[7] Gray, J. M.; Millman, A. P. (Econ. Geol. **57** [1962] 325/49, 334).
[8] Gorzhevskaya, S. A.; Sidorenko, G. A. (Geol. Mestorozhd. Redk. Elem. No. 23 [1964] 1/116).
[9] Sveshnikova, E. V.; Semenov, E. I.; Khomyakov, A. P. (Zaangarskii Shchelochnoi Massiv, Ego Porody i Mineraly, Nauka, Moscow 1976, pp. 1/80, 31).
[10] Ewing, R. C.; Snetsinger, K. G.; Bunch, T. E. (Can. Mineralogist **15** [1977] 92/6, 94).

[11] Young, B. B.; Millman, A. P. (Bull. Inst. Mining Met. No. 689 [1964] 437/66, 449).
[12] Semenov, E. I. (Tipokhimizm Mineralov Shchelochnykh Massivov, Nedra, Moscow 1977, pp. 1/119, 54).
[13] Arnaudov, A.; Petrusenko, S. (Izv. Geol. Inst. Bulg. Akad. Nauk. Ser. Geokhim. Mineral. Petrol. **21** [1972] 77/84, 84).
[14] Amli, R. (Norsk Geol. Tidsskr. **57** [1977] 243/62, 257, 259).
[15] Erofeeva, E. A. (Geol. Mestorozhd. Redk. Elem. No. 10 [1960] 47/60).
[16] Ukai, Y.; Korekawa, M.; Mochida, Y. (Kobutsugaku Zasshi **2** [1955] 252/62 from C.A. **1957** 4884).
[17] Heinrich, E. W. (Am. Mineralogist **50** [1965] 2083/8, 2084).

2.2.2.1.1.5 Chemical and Thermal Behavior

Powdered euxenite, usually, is slowly decomposed in concentrated acids (H_2SO_4, HCl, HF); it is soluble in HF + H_2SO_4, and is easily fusible in KOH and $KHSO_4$ [1]. Euxenite is decomposed in a mixture of HF and HCl, or when sintered with an excess (4- to 10-times) of sodium peroxide. Also borax is a suitable and powerful flux for euxenite [2]. Diagnostic etching of euxenite may be done by HF: Within a second, concentrated HF causes a mat whitish coating, sometimes with a bluish tint. Etching by a 50% solution of HF causes, after 10 to 15 s, an iridescent cover. Structural etching by HF is successful for metamict euxenites [3], but note that in no case has it disclosed the internal structure of X-ray amorphous euxenite crystals from the Zolotaya Mountain, eastern Transbaikalia [4, p. 200].

As follows from a literature review of heat treatment studies on euxenite and priorite, the recrystallization histories of both minerals when metamict are similar. Usually priorite forms at lower temperatures and euxenite at higher temperatures. Other phases formed by decomposition of the starting materials are a cubic betafite-like phase (or phases, see below), a brannerite-like phase, and a rutile phase (for the detailed examples, see original paper). The temperatures at which metamict euxenite and priorite begin to recrystallize vary over a wide range, and apparently depend on the chemical composition of the specimen, the degree of metamictization, and whether the original material was euxenite or priorite [5, pp. 62/5]. Thus, minerals of the euxenite-polycrase and priorite-blomstrandine series subjected to the initial stage of metamict decomposition (i.e., containing under the microscope anisotropic sections with distinct extinction) restore their premetamict structure after heating for several seconds at 700 to 750°C, regardless of the rate of heating and the degree of grinding before heating (for the details, see original paper). Minerals, completely disintegrated into mechanical finely-dispersed mixtures of oxides, do not reveal either euxenite or priorite phases under the above conditions, but always yield an X-ray diffraction pattern corresponding to a cubic pyrochlore-like phase with a = 10.10 to 10.20 Å (10.09 to 10.16 kX [6, p. 20] or 10.11 to 10.15 kX [6, p. 37]). Therefore, a distinct identification of the original material is impossible. But note that prolonged heating at 750°C (longer than 1 h) may give an euxenite pattern for the completely disintegrated material. Minerals with partially differing degrees of metamictization within a single grain give, within 45 to 60 min and with a steady rate of heating (about 10 to 20°C/min), X-ray patterns corresponding to a cubic pyrochlore-like phase plus euxenite or priorite [7]. Recrystallization of metamict euxenite during heating to 1000°C begins with the appearance of a cubic pyrochlore-like phase I (with lattice constants as given above) which then is converted into euxenite. This is accompanied by the formation of a new cubic phase with a = 10.16 to 10.24 kX [6, p. 20]; see also [8]. Most of the euxenite specimens ignited at 500 to 900°C give an X-ray powder pattern corresponding to an aeschynite-like orthorhombic phase with a = 5.18 Å, b = 10.89 Å, and c = 7.36 Å, or to a cubic phase with a = 5.08 to 5.15 Å; sometimes both phases are present. At 1110°C, all specimens analyzed gave identical patterns corresponding to an orthorhombic cell with euxenite (and polycrase) constants (see p. 171) [9], refer also to [10].

Results from heat treatment of individual Th-containing euxenite samples are: After heating in air at about 1000°C for 20 h, brown and black X-ray amorphous varieties from the Gloserheia granite pegmatite, southern Norway, give slightly different X-ray diffraction patterns corresponding to a composite of euxenite, pyrochlore-like phase, and rutile (for details, see original paper) [11]. After heating at about 900°C, a metamict sample (with partially crystalline parts) from muscovite granite, eastern Transbaikalia, gives a diffraction pattern between euxenite and polycrase [12]; for the lattice constants after heating at 1100°C, see p. 171. After heating at 800 to 850°C for 1.5 h, an X-ray amorphous crystal from pegmatitic bodies of the Zolotaya Mountain, eastern Transbaikalia, gives a diffraction pattern of nearly pure euxenite. A cubic phase with a = 5.12 Å is present in negligible amounts [4, p. 198]. After heating in air at 930°C, metamict tanteuxenite from alluvials of the river Liha, Congo, gives a diffraction pattern corresponding to a cubic member of the microlite-pyrochlore series with a = 10.25 ± 0.03 Å and some rutile [13]. Annealing of euxenite (ampangabeite) from Ampangabé, Madagascar, over the range 300 to 1200°C yield the following phases during the recrystallization sequence: an isometric pyrochlore-type phase with a ~ 10 Å, monoclinic $YNbO_4$, monoclinic $FeNbO_4$ (?), Nb_2TiO_7, and $YNbTiO_6$ (with euxenite structure). For the indexed X-ray diffraction data (d values), observed and calculated, after heating at 1100°C for 24 h, see original paper [14]. As follows from X-ray quantitative analysis, approximately 70% of the Norwegian euxenite samples investigated undergo a polymorphic transition from an aeschynite structure to the euxenite structure at 700 to 750°C. At 770°C, a residual glassy phase crystallizes directly to the euxenite structure, producing a sharp relief recrystallization heat previously absorbed by the aeschynite-euxenite structural transition. This transition is accompanied by a decreased rate of crystallite formation. A decreased euxenite crystallinity observed at 500 to 600°C is attributable to condensation of atom groupings in the amorphous phase. Heating time does effect considerably the rate of ordering of the structure [21].

As is characteristic for minerals with a pyrochlore-type structure, the DTA curves of minerals of the euxenite-polycrase series in general have 2 endothermic and 1 exothermic maxima, but the latter (related to the crystallization of the substance) is shifted against pyrochlore to somewhat higher temperatures (735 to 775°C, occasionally 710°C) [6, pp. 26/7], refer also to [15], and displays a wide range [5, p. 66] as is corroborated by the following examples (the DTA curves are given in the references cited):

euxenite from	exothermic peak at	Ref.
rock-crystal bearing pegmatite, Kazakhstan	700°C	[16, 17]
alluvial stream deposits, Monastir, Bulgaria	750°C	[18]
Ampangabé, Madagascar[a]	760°C	[14]
pegmatite, Harrar, Ethiopia	770°C[b]	[19]
pegmatite, USSR	720 to 785°C	[20]

[a] Ampangabeite. — [b] Additionally, an exothermic peak occurs at 580°C.

The dehydration curves of euxenites (which, as a rule, are hydrated to a large degree and contain $\leqq 7.9\%$ H_2O; refer also to Table 9, pp. 164/8) show no liberation of water during heating to 200 to 300°C. The maximum amount of water (loss in weight of 45%) is liberated at 300 to 400°C, and at 700°C all water present is completely removed [6, pp. 26/7].

Heating at 1000°C of partially metamict euxenite crystals from a pegmatite vein, Central Urals, USSR, restores the optical properties (birefringence and refractive indices; refer also to p. 174) without a disturbance of their monocrystalline character; that means, the original crystalline structure is reconstructed [22].

References for 2.2.2.1.1.5:

[1] Chukhrov, F. V.; Bonshtedt-Kupletskaya, E. M. (Mineraly Spravochnik, Vol. 2, Pt. 3, Nauka, Moscow 1967, pp. 1/676, 352).
[2] Doležal, J.; Povondra, P.; Šulcek, Z. (Decomposition Techniques in Inorganic Analysis, Iliffe Books Ltd., London 1968, pp. 1/224, 45, 121, 163 [Czechish original, Prague 1966]).
[3] Sergeev, V. N. (Izv. Sibirsk. Otd. Akad. Nauk SSSR Geol. Geofiz. No. 1 [1958] 103/5; C.A. **1959** 17776).
[4] Kornetova, V. A.; Aleksandrov, V. B.; Kazakova, M. E. (Tr. Mineral. Muzeya Akad. Nauk SSSR No. 18 [1968] 197/202).
[5] Mitchell, R. S. (Southeast. Geol. **14** [1972] 59/72).
[6] Gorzhevskaya, S. A.; Sidorenko, G. A. (Geol. Mestorozhd. Redk. Elem. No. 23 [1964] 1/116).
[7] Komkov, A. I. (Rentgenogr. Mineral'n. Syr'ya **1966** No. 5, pp. 38/44; C.A. **65** [1966] 13411).
[8] Gorzhevskaya, S. A.; Sidorenko, G. A. (Geol. Mestorozhd. Redk. Elem. No. 30 [1966] 43/51, 46; C.A. **68** [1968] No. 71391).
[9] Aleksandrov, V. B. (in: Vlasov, K. A., Geochemistry and Mineralogy of Rare Elements and Genetic Types of Their Deposits, Vol. 2, Israel Program Sci. Transl., Jerusalem 1966, pp. 470/82, 475 [Russian original, Moscow 1964]).
[10] Aleksandrov, V. B.; Pyatenko, Yu. A. (Dokl. Akad. Nauk SSSR **124** [1959] 179/82; Dokl. Acad. Sci. USSR Earth Sci. Sect. **124/129** [1959] 124/7).

[11] Amli, R. (Norsk Geol. Tidsskr. **57** [1977] 243/62, 257/8).
[12] Zalashkova, N. E.; Distler, V. V.; Smirnova, N. V. (Tr. Mineral. Muzeya Akad. Nauk SSSR No. 20 [1971] 205/8, 206).
[13] Van Wambeke, L. (Bull. Soc. Belge Geol. Paleontol. Hydrol. **67** [1958] 121/7, 124).
[14] Ewing, R. C.; Snetsinger, K. G.; Bunch, T. E. (Can. Mineralogist **15** [1977] 92/6).

178

[15] Gorzhevskaya, S. A. (Geol. Mestorozhd. Redk. Elem. No. 10 [1960] 32/41; C.A. **58** [1963] 3203).
[16] Erdzhanov, K. N. (Tr. Kaz. Nauchn. Issled. Inst. Mineral'n. Syr'ya No. 2 [1960] 92/5; C.A. **60** [1964] 3876).
[17] Mineev, D. A.; Kataeva, Z. T.; Malyshev, A. G. (in: Lebedeva, S. I., Fiz. Svoistva Red-kometal'nykh Mineralov i Metody Ikh Issled., Nauka, Moscow 1968, pp. 43/7).
[18] Breskovskaya, V.; Marinova, R. (Spisanie Bulg. Geol. Druzh. **36** [1975] 187/90).
[19] Augustithis, S. S.; Mposkos, E.; Vgenopoulos, A. (Form. Uranium Ore Deposits Proc. Symp., Athens 1974, pp. 61/71, 65; C.A. **83** [1975] No. 196304).
[20] Sokolova, E. P. (Zap. Vses. Mineral. Obshch. **88** [1959] 408/18, 414/5).

[21] Janeczek, J. (Mineral. Polonica **9** [1978] 41/9 from C.A. **92** [1980] No. 44818 and Ref. Zh. Geol. **1980** 1V210 and Mineral. Abstr. **31** [1980] No. 80-1577).
[22] Nefedov, E. I. (Vses. Nauchn. Issled. Geol. Inst. No. 3 [1956] 82/5, 84/5).

2.2.2.1.2 Polycrase $(Y,Ca,Ce,U,Th)(Ti,Nb,Ta)_2O_6$

Polycrase from a granite vein of Hitterö (Hidra), Vest-Agder, southern Norway, was described in 1844. Its name is derived from the Greek words "poly" = many and "krasis" = mixture in allusion to the complex composition [1]; refer also to [2]. Synonymous is blomstrandine/blomstrandinite (of Brögger, 1906); a variety is tantalpolycrase [3], to which tantpolycrase and tantal-polycrase are differing spellings [4].

Occurrence and Paragenesis. The type polycrase, together with xenotime ("Ytterspath") from granite veins of Hitterö accompanies allanite, and fairly often is enclosed in the latter [1]. For spherulitic polycrase from Hitterö, formed by alteration processes, see p. 179. A tiny polycrase rich in Th, identified by a fission track method, is enclosed in quartz of the Mittagsfluh granite, central Aare massif, Switzerland [5]. Black or red-brown to brown (partly spotted) polycrase occurs in leucocratic two-mica granites and pegmatites and alaskites of several massifs of the Urals, USSR. It forms prismatic crystals, flattened along [100], with a size of hundredths of mm up to 0.5 to 0.8 mm (for samples from pegmatite), or it forms granular masses. Genetically, polycrase is connected with pneumatolytic, more rarely with postmagmatic mineral associations. Especially for alaskites of the Zenkovskii massif, the absolute contents of polycrase and associated RE minerals vary with the grain size of the rock as follows [6]:

grain size of the host alaskite	content in g/t			
	polycrase	columbite	xenotime	monazite
fine	—	78	80	12.8
fine to medium	traces	75	100	107
medium	62	40	20	107

In amounts of tenths of wt% from the total accessory minerals of the nonmagnetic fraction, brown-colored (in various shades) polycrase occurs in aplite and pegmatite of the Ashchikol', and in aplitoid granite of the Borovskii plutons, northern Kazakhstan [23].

As a rare accessory mineral, euxenite (actually polycrase), forming subhedral to anhedral masses of half an inch to 2 inches in diameter, occurs together with other RE minerals (gadolinite, allanite, monazite, fergusonite, xenotime, and zircon) along the outer part of the intermediate zone of simple (by its mineral composition) inhomogeneous pegmatites of the Kenhardt and Gordonia districts, Cape Province, R.S.A. Polycrase and the other RE minerals, in most cases, are found to be attached to biotite flakes. The pegmatites formed from a segregated fluid injected into pre-existing tensional openings in the granitic-gneissic country rock [7, pp. 10/1, 40, 54/5, 71/2]. For eluvial polycrase of this area, see below. Polycrase, together with allanite, euxenite, and zircon, often is found in the intermediate zone of pegmatites of the biotite type connected with Precambrian granites of Inner Mongolia, P.R.C. [8]. Black massive polycrase (by its chemical composition, whereas its physical properties correspond better to euxenite) occurs in a (granite [9]) pegmatite vein cutting hornblende-pyroxene dioritic gneiss, Showa Base, Antarctica [10].

Yellow-cinnamon polycrase (without secondary alteration, and displaying some anisotropic blocks in single grains) occurs as irregularly shaped depositions up to 1.5 cm in size, rarely as prismatic crystals up to 3 mm in length and up to 1 mm thick (refer to p. 181), in metasomatic albitite, which is genetically related to alkali syenite of an undisclosed locality of the USSR (doubtless Tuva, see [11]). Besides the rock-forming minerals albite, riebeckite, and aegirine, associated minerals are zircon (cyrtolite), ilmenite, xenotime; sometimes astrophyllite is observed. Polycrase and zircon crystallized prior to albite, and in a number of cases polycrase is replaced by ilmenite [12, pp. 70, 76]. And note that [24] identified the afore-mentioned polycrase as a pseudomorph of blomstrandine after polycrase. Except for a difference in the RE composition (see p. 180), entirely similar polycrase occurs in albitites of Eastern Siberia. It forms flattened crystals, which are <1 cm along its axis, have red or yellow-brown to black color (depending on inclusions, see below), and mostly are metamict; small anisotropic blocks may be visible in thin sections [11].

Polycrase, along with other minerals (as, e.g., uranothorite, brannerite, euxenite, fergusonite, etc.), is a rare component of black sand placers of central Idaho [13]. Eluvial polycrase (euxenite) of the Kenhardt and Gordonia districts, Cape Province, R.S.A., are usually covered by a thin veneer of a yellowish alteration product [7, p. 55].

Whereas the red polycrase from albitite, Eastern Siberia, is fresh, clear, and transparent, the yellow-brown crystals, usually, are opaque and crowded with ilmenite **inclusions**. In some cases, the color changes gradually from red (central part) to yellow-brown towards the periphery of individual crystals [11].

Black, completely metamict polycrase from the Pomba pegmatite, Brazil, has a brown crust, which gives, without heating, a relict of the main X-ray diffraction lines of the polycrase series. This crust is supposed to be a product of **primary alteration** caused by solutions acting on the mineral prior to its change to a metamict state. The effects of primary alteration (leaching of about 60% of the A-group ions, see p. 180) are the same as those of weathering [14, pp. 131/2]. The formation of spherulites during primary alteration and recrystallization of metamict Nb-Ta-Ti oxides may be a temporary feature which becomes obscured as the recrystallization progresses and the spherulites coalesce and themselves recrystallize (be obliterated). Thus, honey yellow metamict polycrase (originally viewed to be aeschynite) from Hitterö, Norway, contains moderately birefringent pale yellow spherulites (30 to 150 µm in diameter) of Ti-rich RE fersmite (or polycrase [15]). Probably initiated by a thermal (e.g., hydrothermal) event postdating metamictization, the spherulites occur throughout the specimen, but are more concentrated along numerous microfractures, where they coalesce, forming alteration zones parallel to the microfracture. Except for a fivefold higher content of CaO for the spherulite, the chemical composition of both components is nearly identical [16], as can be seen from electron microprobe analyses along a traverse across a spherulite and its matrix [15]:

	ThO_2	U_3O_8 [*]	RE_2O_3	content in wt% CaO	Nb_2O_5	Ta_2O_5	TiO_2	Fe_2O_3 [*]
matrix	7.40	3.40	25.49	0.27	13.76	4.15	36.60	0.75
edge	7.00	3.10	24.19	1.88	13.14	3.85	35.21	0.67
center	7.35	3.20	25.10	1.87	12.92	4.05	35.82	0.75
edge	7.10	3.15	24.74	1.86	13.71	4.00	37.02	0.76
matrix	7.75	3.50	25.44	0.31	13.57	4.15	37.27	0.71

[*] Oxidation state not determined.

For a pseudomorph of blomstrandine after polycrase from albitite connected with alkali syenite of Siberia, see [24].

Chemistry. For chemical analysis of polycrase, see Table 9, pp. 164/8, and table above; for its general formula type, see p. 161, refer also to [17].

The formula, based on a unit cell content of 12 oxygen, for polycrase (euxenite) samples from the Cape Province (= sample Nos. 3, 9, and 33 in Table 9, pp. 164 and 167) are in good agreement with the formula type AB_2O_6, and correspond to $A_{2.25}B_{4.03}O_{12}$, $A_{2.10}B_{4.01}O_{12}$, and $A_{2.00}B_{4.08}O_{12}$, respectively, where A = RE, U, Th, Si, Ca, Fe^{2+}, Na, K, Mn and B = Ti, Nb, Ta, Fe^{3+}, Mg. The presence of water in the above analyses is attributed to metamictization [7, pp. 55/6]. The crystal-chemical formula for polycrase samples of different colors from albitites is similar: $(Y,Dy,Gd,Na,Ca,Th)_{1.13}(Ti_{1.3}Nb_{0.52}Fe,Mn)_{1.95}O_6$ for the red, or $(Y,Dy,Gd,Ca,ThMg)_{1.06}(Ti_{1.4}Nb_{0.49}Fe)_{2.1}O_6$ for the yellow-brown sample from Eastern Siberia (= sample Nos. 10 and 32 in Table 9, pp. 165 and 167), and $(Y,Dy,Er)_{0.9}(Ti_{1.3}Nb_{0.7})_{2.0}O_6$ for the yellow-cinnamon sample from Tuva (= sample No. 45 in Table 11, p. 211). The sole difference is a predominance of Er over Gd [11], or an unusually strong predominance of RE(Y) in the last sample. Further, it displays a nearly total absence of Ta, and a deficit for group-A cations points to vacant lattice sites, compensated by a statistical substitution of a part of the oxygens by hydroxyl groups [12, pp. 74/5]. For euxenite-polycrase from Madagascar (= sample No. 5 in Table 9, p. 164), taking an average value of 166 for the atomic weight of RE, the formula is $A_{1.00}B_{1.93}(O,OH)_{6.50}$, where A = Y, RE, Ca, U, Th, Pb, Mg, Mn, Na, K and B = Nb, Ta, Ti, Fe^{3+}, Al. Taking oxygen as equal to 6, the formula reads as $A_{0.92}B_{1.78}O_{6.00}$ [18]. Neglecting H_2O (which is attributed to alteration) and adopting average atomic weights of 127 and 140 for the RE(Y) and RE(Ce), respectively, the crystal-chemical formula for euxenite/polycrase (= sample No. 13 in Table 9, p. 165) from the Showa Base, Antarctica, becomes approximately $A_{1.00}B_{1.96}O_{5.83}$ [9].

Primary alteration of polycrase from the Pomba pegmatite, Brazil, affected mainly the RE and to a lesser degree U contents; the Ca content is insignificant in this mineral. Derived from the chemical analyses (= sample Nos. 11 and 18 in Table 9, p. 165), the crystal-chemical formula is $[RE(Y)_{0.65}U_{0.15}Th_{0.04}(Pb, Mg, Ca, Fe^{2+}, RE(Ce), Mn)_{0.5}]$ $[Ti_{1.27}Nb_{0.595}Ta_{0.095}(Fe^{3+}, Sn)_{0.04}]O_{5.38}(OH)_{0.38} \cdot 0.1\ H_2O$ for the inner zone and $[RE(Y)_{0.145}Th_{0.04}Pb_{0.025}(Mg, Fe^{2+}, Ca, Mn, RE(Ce))_{0.035}]$ $[Ti_{1.27}Nb_{0.585}Ta_{0.10}(Fe^{3+}, Sn)_{0.045}]O_{4.77}(OH)_{0.65} \cdot 0.55\ H_2O$ for the outer crust [14, p. 129, 132]. For an attempt to recalculate an approximate formula of the unaltered material from a given chemical analysis of highly altered polycrase (having lost about 65% of its original cations) from Nauseke, Kenya, see [14, pp. 135/6].

Poorly developed isomorphous substitution of RE by Th and U (as otherwise is common for polycrase) in the Tuva polycrase sample may be the cause for its optically weakly anisotropic (crystalline) domains [12, p. 74].

Polycrase (blomstrandine) samples from Iveland (Nos. R12 and R24) and Arendal (No. R13), Norway, have the following chemical compositions and physical properties (VHN_{50} = Vickers hardness in kg/mm^2 at 50 g load; r.i. = refractive index; and R (546 nm) = reflectivity in % for the given wavelength) [19, p. 2]:

	content in % and physical properties for sample		
	R12	R24	R13
ThO_2	7.70	5.80	5.40
U_3O_8	5.40	4.40[a]	3.80
ΣRE_2O_3	27.40	24.90[b]	32.40
CaO	1.00	1.70	0.70
MnO	—	0.20	—
PbO	0.90	0.05	0.80
Fe_2O_3	1.30	3.33	1.50
Nb_2O_5	16.40	17.72	19.00
Ta_2O_5	4.70	4.29	2.00
TiO_2	29.50	31.71	31.90
H_2O^-	0.20	—	0.40
H_2O^+	4.20	3.99	1.30
sum	98.70	98.09	99.20
D in g/cm^3	4.86	4.838	4.49
VHN_{50}	730	673	692
r.i.	2.16	2.13	2.19
R (546 nm)	13.11	14.61	13.57

[a] Given as 4.32% UO_2 and 0.08% UO_3 in [20]. — [b] 21.42% when calculated from the given contents for individual RE.

The main **crystal forms** of prismatic crystals of polycrase, 0.1 to 0.5 mm in size, from pegmatite of the Ashchikol' massif, northern Kazakhstan, are (010), (110), and (201); rarely occur (111) and (100), see **Fig. 8**a, b, c on p. 182. The crystal surface is always dull, covered by furrows, and striated on the plaques (010) and (110). There are also zoned and parallel intergrown crystals, see Fig. 8d, e, f. The former remember on "puff pastry", for distinct zones (furrows along the plane (010), see Fig. 8g) of the crystal are replaced by canary or straw yellow secondary products [23]. For polycrase crystals from albitite (of Tuva) are abundant the rhombic prism (110) and dipyramid (111), the pinacoid (010) is small for most of the crystals; and the prism (130) sometimes is lacking. A single crystal displays an extremely small prism (150). The resulting morphological axial ratio a:b:c = 0.381:1:0.350 is similar to that for euxenite (see p. 171) and may be correlated with the diadochic substitution of Nb by Ti in polycrase [12, p. 71].

The **crystal structure** of polycrase is partly restored after heat treatment, refer to pp. 183/4. For three polycrase (blomstrandine) samples with differing composition (see above) from

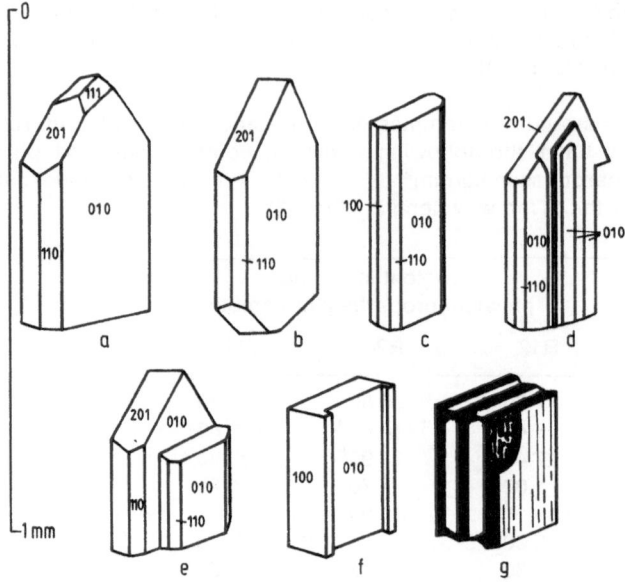

Fig. 8. Crystal forms of polycrase from pegmatite, Ashchikol' massif, northern Kazakhstan, after [23].

Norway, heated at 900 °C, the following lattice constants and axial ratios are given, corresponding to a newly formed euxenite structure (refer to p. 183; see also p. 171) [19, p. 4]:

| sample | lattice constants in Å | | | axial ratio |
	a	b	c	a:b:c
No. R12	5.56	14.69	5.17	0.378:1:0.351
No. R24	5.53	14.63	5.16	0.378:1:0.353
No. R13	5.54	14.48	5.15	0.383:1:0.356

Optical Properties. The color for individual Th-containing polycrase samples is black (from pegmatites) [9, 14]; black or red-brown to brown (from alaskite) [6]; brown in various shades (from aplite, pegmatite, and aplitoid granite), for instance, brownish orange and reddish cinnamon, dark cinnamon, or chestnut brown [23]; usually yellow-cinnamon and rarely black (from albitite, Tuva) [12, p. 71] or yellow-brown to black (from albitite, Eastern Siberia) [11]; and honey yellow and pale yellow for a metamict matrix and spherulitic inclusions, respectively [16].

The luster of polycrase is vitreous [10], fatty or resinous [23], strong fatty to metallic [11], or pitchy to adamantine [12, p. 71].

Polycrase from albitite (Tuva), in thin section, mainly is isotropic with an intense yellow color. Sporadically, single grains display parts with a weak anisotropism (interference color of low-order gray) and straight extinction along the length of the crystal. In reflected light, it has a low reflectivity (smaller than for sphalerite) and strong yellow-brown internal reflections [12, p. 71]. Anisotropic polycrase from albitite (Eastern Siberia) has a refractive index of 2.28 to 2.30. After heating, it becomes uniaxial birefringent with $N_o = 2.36$ to 2.37 and $N_e = 2.28$ to 2.30 [11].

For reflectivity at 546 nm of 3 polycrase (blomstrandine) samples from Norway, see p. 181. A somewhat narrower range, 14.2 to 14.6% (at 546 nm), is given for 3 polycrase samples from Brazil with similar TiO_2 content, but higher Nb_2O_5 and lower Ta_2O_5 contents than for the Norwegian samples. Reflectivity of the Brazilian samples varies between 15.0 and 15.5% for 470 nm, 13.4 and 14.1% for 589 nm, and 13.5 and 14.4% for 650 nm [21].

Other Physical Properties. Mohs hardness for polycrase from albitites is ~ 5 (Eastern Siberia) [11] or ~ 5.5 (Tuva) [12, p. 71]. For the Vickers hardness, VHN_{50}, for 3 polycrase (blomstrandine) samples from Norway, see p. 181. A somewhat higher VHN_{50} between 713 and 746 kg/mm^2 is given for 3 polycrase samples from Brazil (with somewhat deviating composition, see above) [21].

Fracture of polycrase is uneven conchoidal [11] or conchoidal [10], [12, p. 71].

For density of polycrase, refer to Table 9, pp. 164/8.

Chemical and Thermal Behavior. Polycrase from albitite is insoluble in acids (Tuva) [12, p. 71] as, for example, HCl (Eastern Siberia) [11]. Polycrase (euxenite) from a pegmatite vein, Showa Base, Antarctica, is readily attacked by mineral acids such as hot concentrated HF and H_2SO_4, and by fusion with $KHSO_4$ [9].

The X-ray diffraction patterns obtained after heating at different temperatures for the partially anisotropic polycrase from albitite (Tuva) are identical according to their position of diffraction lines, the intensity of which increases at about 400°C. This indicates that only recrystallization of the original structure occurs, leading to a textured (see [22]) paramorph of fine-grained polycrase with some coarser parts (representing the originally anisotropic blocks); for the detailed discussion, see original paper [12, pp. 72/4]. Identical patterns are also obtained after heating at 900 to 1100°C for the polycrase samples from albitite of Eastern Siberia [11].

After heating in air at 1000°C of zoned polycrase from the Pomba pegmatite, Brazil, its completely metamict inner black zone gives the typical X-ray diffraction pattern of polycrase with the following phases in decreasing importance: euxenite, a cubic phase, and rutile. For the outer, primarily altered zone (refer to p. 179) the lines of rutile become predominant over those of euxenite and the cubic phase. Additionally, the principle lines of the compound UNb_3O_8 appear [14, p. 132]. Three polycrase (blomstrandine) samples from Norway recrystallize between 400 and 450°C to a priorite-aeschynite phase, which itself becomes unstable between 750 and 800°C. At about 650°C, euxenite begins to form, and is stable up to 1100°C. Additionally, a cubic pyrochlore-type phase (with a = 10.22 to 10.33 Å for the 900°C runs) is formed over the temperature range investigated [21, pp. 3/4]. X-ray diffraction patterns comparable with those of typical euxenite (for the details, see original papers cited) are obtained for polycrase (euxenite) from Showa Base, Antarctica, after heating in air at 1000°C for 7 h [9]; from northern Kazakhstan, after heating at 1000°C for 4 h [23]; and for black and brown polycrase from the Zenkovskii massif, Urals [6].

184

The DTA curve for polycrase (euxenite) from Showa Base, Antarctica, shows one significant exothermic peak at about 760°C, typical for euxenite (refer to pp. 176/7); two endothermic peaks are slightly visible on the curve [9]. As is atypical for metamict tantalo-niobates (refer to pp. 150/1), heating and cooling curves for the partially anisotropic polycrase from albitite (Tuva) indicate a main exothermic effect starting at as low as 390°C, and reaching its maximum at 450°C, which corresponds to the restoration of the original structure. Two considerably weaker exothermic peaks at about 600/680°C and 740/770°C may be connected with oxidation processes of some chemical components [12, pp. 71/2]. Note that [24] identified this polycrase as a pseudomorph of blomstrandine after polycrase.

References for 2.2.2.1.2:

[1] Scheerer, T. (Ann. Physik Chem. [2] **62** [1844] 429/34, 429/30).
[2] Palache, C.; Berman, H.; Frondel, C. (Dana's System of Mineralogy, 7th Ed., Vol. 1, Wiley, New York 1946, pp. 1/834, 787/91).
[3] Aleksandrov, V. B. (in: Vlasov, K. A., Geochemistry and Mineralogy of Rare Elements and Genetic Types of Their Deposits, Vol. 2, Israel Program Sci. Transl., Jerusalem 1966, pp. 470/82, 470/1 [Russian original, Moscow 1964]).
[4] Chukhrov, F. V.; Bonshtedt-Kupletskaya, E. M. (Mineraly Spravochnik, Vol. 2, Pt. 3, Nauka, Moscow 1967, pp. 1/676, 355).
[5] Bajo, C.; Rybach, L.; Weibel, M. (Chem. Geol. **39** [1983] 299/318, 300, 303/5).
[6] Chashchukhina, V. A. (Mineraly Mestorozhdenii Urala, Sverdlovsk 1987, pp. 65/7).
[7] Hugo, P. J. (Mem. South Africa Geol. Surv. No. 58 [1970] 1/94).
[8] Go, Chen-Tsi (Sci. Record [Peking] [2] **2** [1958] 400/13, 400, 405, 410).
[9] Hayashi, S.; Nagashima, K. (Nankyoku Shiryo No. 11 [1961] 27/30).
[10] Saito, N.; Tatsumi, T.; Sato, K. (Nankyoku Shiryo No. 12 [1961] 31/6 from Mineral. Abstr. **16** [1963/64] 63).

[11] Proshchenko, E. G. (in: Mineralogiya Pegmatitov i Gidrotermalitov Shchelochnykh Massivov, Nauka, Moscow 1967, pp. 103/36, 130/3).
[12] Aleksandrov, V. B. (Tr. Inst. Mineral. Geokhim. Kristallokhim. Redk. Elem. No. 1 [1957] 70/6).
[13] Savage, C. N. (Idaho Bur. Mines Geol. Bull. No. 17 [1961] 1/160, 8/10).
[14] Van Wambeke, L. (Neues Jahrb. Mineral. Abhandl. **112** [1969/70] 117/49).
[15] Ewing, R. C. (Ceram. Glass Radioact. Waste Forms **1977** 139/46, 141/3; C.A. **90** [1979] No. 30882).
[16] Ewing, R. C. (Science **184** [1974] 561/2).
[17] Gorzhevskaya, S. A.; Sidorenko, G. A. (Geol. Mestorozhd. Redk. Elem. No. 23 [1964] 1/116, 28/31).
[18] Smellie, J. A. T.; Cogger, N.; Herrington, J. (Chem. Geol. **22** [1978] 1/10, 7).
[19] Ewing, R. C.; Ehlmann, A. J. (Can. Mineralogist **13** [1975] 1/7).
[20] Hongslo, T.; Langmyhr, F. J. (Norsk Geol. Tidsskr. **40** [1960] 157/64, 160).

[21] Ewing, R. C. (Am. Mineralogist **58** [1973] 942/4).
[22] Pyatenko, Yu. A.; Aleksandrov, V. B. (Kristallografiya **4** [1959] 248/9; Soviet Phys.-Cryst. **4** [1959] 227/8).
[23] Platonov, A. N. (Sb. Nauchn. Rab. Nauchn. Issled. Sekt. Kiev. Gos. Univ. No. 2 [1964] 110/5, 113/5).
[24] Lunts, A. (Izv. Akad. Nauk Latv. SSR **1965** No. 2, pp. 51/67, 62).

2.2.2.2 Aeschynite $\sim (RE, Ca, Fe, Th, U, Na)(Ti, Nb, Ta)_2(O, OH)_6$ *)

Owing to its variable chemical composition (see pp. 196/215) and nearly exclusively metamict state (see, e.g., [1 to 3]), there were given a number of names (see the following paragraphs) to mineral samples found in nature belonging in reality to aeschynite, a name which should be solely used in the future (refer to p. 187).

Aeschynite (or correctly aeschynite-(Ce), refer to p. 187) from the Ilmen Mountains, northern part of the Southern Urals, USSR, was first found in 1825 [4, p. 161] and, using material from a coarse-grained feldspar-albite-mica rock from Miass (or, as usually named in western literature, Miask), was described in 1828. Its name is derived from the Greek word "aischyne" = shame, in allusion to the inability of contemporary chemists to separate some of its chemical constituents, see [5, pp. 2/5], [4, p. 161], [6].

Blomstrandine (a mineral similar to the foregoing by its morphological axial ratio, but diverging by its predominating yttrian RE) from a pegmatite near Urstad, Hitterö, southwest Norway, was described in 1879 [7] and later (1906) named in honor of the analyst Christian Wilhelm Blomstrand (1826–1897), Professor of Chemistry at the University of Lund, Sweden [8, pp. 98/9]; see also [5, pp. 80/4], [6], [9, p. 41]. Synonymous to blomstrandine are blomstrandinite, but not blomstrandite of [10] (which possibly is uranopyrochlore [11]), see [5, pp. 80/4] and [6], and titanopriorite [12], see also [4, p. 376]. Refer also to priorite below.

Lyndochite (although metamict, displaying a strong tendency to form almost perfect crystals, refer to p. 218) from a beryl-containing granite pegmatite dike at Lyndoch township, Ontario, Canada, was known since 1897 and was described in 1927. The name is in allusion to the locality. On the basis of morphological data, the mineral was considered probably to be a variety of euxenite, but its chemical composition resembles aeschynite rather than euxenite [13]. Owing to a narrow intergrowth with columbite (see [14]), lyndochite is viewed as a pseudomorph after columbite by [15]. But crystalline samples from the Southern Urals, USSR, without any sign of pseudomorphism and displaying crystal forms analogous to that of the Canadian type material, indicate for lyndochite its proper nature [16]. For lyndochite classified as a variety of aeschynite rich in Nb and poor in Th, see [17]; or tentatively viewed as a thorian neodymian aeschynite, see [18].

Priorite (containing 0.61% ThO_2 and similar by its crystal morphology and physical properties to the above aeschynite) from stanniferous gravels of the Embabaan district, Swaziland, South Africa, was described in 1899 as an euxenite-like mineral by [19] and later (1906) was named in honor of the analyst George Thurland Prior (1862–1936), Keeper of Minerals in the British Museum, London [8, pp. 110/1]; see also [5, pp. 496/9], [6], [9, pp. 287/8]. Although there are differences in the composition of group-A cations (see pp. 196/7), priorite and blomstrandine are taken as synonymous to each other, refer to [4, pp. 371, 376], [6], or both as synonymous to aeschynite-(Y), refer to [20, pp. 20, 148]. For yttrium aeschynite refer to p. 186.

According to variances in chemical composition, the following **varieties of aeschynite** containing >1% ThO_2 are named as a mineral species in the literature:

Thoroaeschynite from an exocontact zone of a miaskite intrusion, Vishnevye Gory, Middle Urals, USSR, was found in 1958 and described in 1964 [21, pp. 150/1]. But note that this name is rejected by the Nomenclature Commission of the International Mineralogical Association [22].

---

*) Refer to formula discussion on pp. 196/9.

Yttrium aeschynite (containing at about equal amounts of RE(Ce) and RE(Y)) from pegmatite of the Ilmen Mountains, Southern Urals, USSR, was described in 1959 [23]; similar in composition is sinicite, see pp. 234/6. Aeschynite-(Y) rich in Gd and Dy from a U-Th ore-bearing coarse-grained biotite granite, People's Republic of China, was described in 1987 [24].

Aeschynite-(Nd) with an empirical formula (Nd, Ce, Ca, Th)(Ti, Nb)$_2$(O, OH)$_6$ from rocks of Beiyun-Obo, Inner Mongolia, P.R.C., was described in 1982 [25] and is approved as a valid mineral species, see [20, p. 1].

Nb-containing aeschynites were described as metamict nioboaeschynite containing 2.52% ThO$_2$ from fenite of the Vishnevye Gory, Middle Urals, USSR, in 1960 [26]; as slightly crystalline niobian aeschynite containing 2.15% ThO$_2$ from the northern People's Republic of China in 1962 [27]; and as non-metamict niobian aeschynite from a placer deposit of Alaska in 1975 [28].

Ta-rich aeschynite from granitic pegmatites of Siberia, USSR, was discovered in 1958 and described in 1963 [29], and has been named erroneously Ta aeschynite (see p. 187). True Ta aeschynite-(Y) from pegmatite of the Borborema region, northeastern Brazil, was described in 1974 [30] and is approved as a valid mineral species, see [20, p. 179].

Alumoaeschynite containing 1.83% ThO$_2$ from a quartz veinlet within fenite of the Vishnevye Gory, Middle Urals, USSR, was found in 1959 and described in 1964 [21, p. 154].

The distribution in special host rock types is given for compositional varieties of aeschynite as follows: Aeschynite and its Th-Ti variety are typical particularly for pegmatites and albitites related to miaskite and syenite (as, e.g., from Ilmen Mountains and Vishnevye Gory, Tatarka, or Korea); CaNb aeschynite (partly rich in Nd) is concentrated in (especially calcitic-dolomitic) hydrothermalites in connection with alkali syenite or granosyenite (as, e.g., from Madras, Mongolia, the Urals, or Montana); and aeschynite occurs also in carbonatite (as, e.g., from Kovdor and Aldan). Y aeschynite (enriched in U and displaying a Dy, Yb maximum for its RE composition) is characteristic for aegirine- and lepidomelane-containing granite (e.g., Pamir) and the related pegmatites and albitites (as, e.g., from Keivy and Eastern Sayan) [31]; refer also to Section 2.2.2.2.1, pp. 188/96.

The **nomenclature of aeschynite and** its **related minerals**, based on chemical and X-ray diffraction data, may be represented as follows: An aeschynite-blomstrandine series, originally proposed from chemical considerations in 1906 by Brögger [8, pp. 102/4], was later adopted to a concept of an aeschynite-priorite series, in which the Ce-dominant[*] mineral is aeschynite and the Y-dominant[*] (or Y-Nb-dominant [32]) mineral is priorite, blomstrandine and lyndochite are only varietal names for this series (refer also to the following paragraphs), see [18, 33, 34]. Blomstrandine may be characterized by its higher contents of Th and Ti [32].

As is proved by X-ray diffraction studies reported up to 1961 in the literature, there is a solid-solution series in the system CeNbTi$_2$O$_6$-YNbTi$_2$O$_6$ forming the low-temperature orthorhombic (D$_{2h}^{16}$ − Pbnm) aeschynite-priorite series, which, according to the Nb$_2$O$_5$:TiO$_2$ ratio (given in parenthesis), may comprise as varieties: aeschynite (1:1 to 2:1) with titanoaeschynite ($<$1:1) and nioboaeschynite ($>$2:1), all containing predominantly RE(Ce); and priorite (1:1 to 2:1) with titanopriorite or blomstrandine ($<$1:1) and niobopriorite ($>$2:1), all containing predominantly Y and/or RE(Y). Yttrian aeschynite is the variety containing at about equal amounts RE(Ce) and RE(Y). Hitherto, titanoaeschynite and niobopriorite are

---

[*] Implies the cerium-group elements La to Eu, or the yttrium-group elements Gd to Lu plus Y.

not found in nature, but may exist, however, as unknown minerals (having a crystal from corresponding to β-sinicite; refer also to Section 2.2.2.4, pp. 234/6) [12]; see also [27]. And, as follows from heating experiments (refer to pp. 228/31), the discovery in nature of aeschynite that has the crystal form of euxenite is quite possible [12], and has happened more recently, see p. 191.

According to a proposal given by [35] (which is favored by the International Mineralogical Association, refer to [20, Preface]), the correct nomenclature for aeschynite and its related minerals is aeschynite-(Ce) and aeschynite-(Y) for formerly aeschynite and priorite, respectively. These names can be modified for the suffix by other main elements in group-A positions; additional modification may be given by praefixes (e. g., niobian aeschynite-(Ce)). Thus, the Ta-rich aeschynite, described and named Ta aeschynite by [29], with an atomic ratio Ti:Nb:Ta = 0.89:0.565:0.525 should better have been called tantalian aeschynite-(Y) as Y is the main RE and Ta is predominant in the group-B elements, whereas real Ta aeschynite-(Y) has 0.62:0.75:0.595 for that ratio [30].

As the chemistry (especially the Nb and Ta concentrations together with high Th and U contents) of zonal aeschynites from the Habachtal, Austria, can only limitedly be assigned to a classification system (cluster types) given by [36], it is concluded that the prevailing data for nomenclature of the aeschynite-group minerals are insufficient. And the strong variations in element distribution within a single grain (see p. 202) may call in question a consistent, precise denomination, or even may render it impossible [37].

For complex oxides of the formula type AB_2O_6, according to [33, 34, 38], there is an euxenite series and an aeschynite series, displaying polymorphism to each other. Whereas minerals/compounds of the aeschynite series (showing a space group $D_{2h}^{16} - Pbnm$ and axial ratios of about 0.382:1:0.676) may represent the "low-temperature modifications", that of the euxenite series (showing a space group $D_{2h}^{14} - Pbcn$ and 0.382:1:0.385) may be viewed as the "high-temperature modifications", for they can be generated from the former by annealing at temperatures above 800 °C (refer to pp. 228/31). The reverse change was not observed hitherto [39].

References for 2.2.2.2:

[1] Pabst, A. (Am. Mineralogist **37** [1952] 137/57).

[2] Bouska, V. (Acta Univ. Carolinae Geol. **3** [1970] 143/69).

[3] Mitchell, R. S. (Mineral. Rec. **4** [1973] 177/82, 214/23).

[4] Chukhrov, F. V.; Bonshtedt-Kupletskaya, E. M. (Mineraly Spravochnik, Vol. 2, Pt. 3, Nauka, Moscow 1967, pp. 1/676).

[5] Hintze, C. (Handbuch der Mineralogie, Ergänzungsband 1, Walter de Gruyter, Berlin–Leipzig 1938, pp. 1/760).

[6] Palache, C.; Berman, H.; Frondel, C. (Dana's System of Mineralogy, 7th Ed., Vol. 1, Wiley, New York 1946, pp. 1/834, 793/6).

[7] Brögger, W. C. (Z. Krist. **3** [1879] 481/6).

[8] Brögger, W. C. (Skrifter Norske Videnskaps Akad. I Mat. Naturvidensk. Klasse No. 6 [1906] 1/162).

[9] Embrey, P. G.; Fuller, J. P. (A Manual of New Mineral Names 1892–1978, Oxford Univ. Press, Oxford 1980, pp. 1/467).

[10] Lindström, G. (Geol. Fören. Stockholm Förhandl. **2** [1874] 162/4).

[11] Hogarth, D. D. (Am. Mineralogist **62** [1977] 403/10, 406).

[12] Chzhan, Pei-Shan' [Chang, P'ei-Shan] (Sci. Sinica [Peking] **12** [1963] 237/43; C.A. **58** [1963] 9962).

188

[13] Ellsworth, H. V. (Am. Mineralogist 12 [1927] 212/8).
[14] Ellsworth, H. V. (Geol. Surv. Can. Econ. Geol. Rept. No. 11 [1932] 1/272, 229).
[15] Butler, J. R. (Am. Mineralogist 42 [1957] 671/6).
[16] Svyashin, N. V.; Gaidukova, M. K. (Zap. Vses. Mineral. Obshch. 96 [1967] 206/10, 207).
[17] Gorzhevskaya, S. A.; Sidorenko, G. A. (Dokl. Akad. Nauk SSSR 146 [1962] 1176/8; Dokl. Acad. Sci. USSR Earth Sci. Sect. 146 [1962] 131/3).
[18] Fleischer, M. (Mineral. Mag. 35 [1966] 801/9).
[19] Prior, G. T. (Mineral. Mag. 12 [1899] 96/101).
[20] Fleischer, M. (Glossary of Mineral Species, 5th Ed., Mineralogical Record Inc., Tucson 1987, pp. 1/227).

[21] Es'kova, E. M.; Zhabin, A. G.; Mukhitdinov, G. N. (Mineralogiya i Geokhimiya Redkikh Elementov Vishnevykh Gor, Nauka, Moscow 1964, pp. 1/318).
[22] International Mineralogical Association (Mineral. Mag. 36 [1968] 1143/5).
[23] Makarochkin, B. A.; Es'kova, E. M.; Gonibesova, K. A. (Tr. Inst. Mineral. Geokhim. Kristallokhim. Redk. Elem. No. 3 [1959] 145/50).
[24] Cai, Genqing (Kuangwu Xuebao 7 [1987] 66/73 from C.A. 108 [1988] No. 97990).
[25] Zhang, Peishan; Tao, Kejie (Dizhi Kexue 1982 424/8 from C.A. 98 [1983] No. 164080).
[26] Zhabin, A. G.; Mukhitdinov, G. N.; Kazakova, M. E. (Tr. Inst. Mineral. Geokhim. Kristallokhim. Redk. Elem. No. 4 [1960] 51/72, 67/8).
[27] Chang, Pei-Shan (Sci. Sinica [Peking] 11 [1962] 969/76 from C.A. 58 [1963] 2274).
[28] Rosenblum, S.; Moiser, E. L. (Am. Mineralogist 60 [1975] 309/15).
[29] Kornetova, V. A.; Aleksandrov, V. B.; Kazakova, M. E. (Tr. Mineral. Muzeya Akad. Nauk SSSR No. 14 [1963] 108/21).
[30] Adusimilli, M. S.; Kieft, C.; Burke, E. A. J. (Mineral. Mag. 39 [1974] 571/6).

[31] Semenov, E. I. (Tipochimizm Mineralov Shchelochnykh Massivov, Nedra, Moscow 1977, pp. 1/119, 49, 54).
[32] Kostin, N. E.; Petrova, E. A.; Sidorenko, G. A.; Skorobogatova, N. V. (Mineral'n. Syr'e No. 1 [1960] 121/8, 126).
[33] Komkov, A. I. (Dokl. Akad. Nauk SSSR 126 [1959] 641/4; Dokl. Acad. Sci. USSR Earth Sci. Sect. 124/129 [1959] 584/6).
[34] Komkov, A. I. (Dokl. Akad. Nauk SSSR 171 [1966] 181/2; Dokl. Acad. Sci. USSR Earth Sci. Sect. 171 [1966] 145/6).
[35] Levinson, A. A. (Am. Mineralogist 51 [1966] 152/8).
[36] Ewing, R. C.; Ehlmann, A. J. (Can. Mineralogist 14 [1976] 111/9).
[37] Grundmann, G.; Koller, F. (Neues Jahrb. Mineral. Abhandl. 135 [1979] 36/47, 45).
[38] Seifert, H.; Beck, B. (Neues Jahrb. Mineral. Abhandl. 103 [1965] 1/20).
[39] Sommerauer, J.; Weber, L. (Schweiz. Mineral. Petrogr. Mitt. 52 [1972] 75/91, 84/5).

2.2.2.2.1 Occurrence, Paragenesis, Intergrowths, Inclusions, and Alteration

As can be seen from the following paragraphs, **occurrence and paragenesis** of aeschynite and its varieties is connected mainly with granitic pegmatites, especially its metasomatically altered parts, and with alkali syenites and alkali metasomatites. Subordinately, it occurs in carbonatites and, merely as a mineralogical speciality, in other types of occurrence.

Aeschynite is a relatively widespread mineral, but nowhere it occurs in greater amounts [1]. For instance, its variety blomstrandine is found within rocks of the Aktau granitoid massif, Uzbekistan, in contents of 0.00009% in biotite granite of the 2nd intrusion phase; of 0.00299% in alaskite and leucocratic granite of the 3rd intrusion phase; of 0.00092% in a granite aplite vein; and of 0.0075% in pegmatite [2]. The mean content of blomstrandine for the big individual alkali granite massifs of the Kola Peninsula, northwestern European USSR, is given as follows: 32 g/t for 33 samples from western Keivy; 31 g/t for 23 samples from Purkachskii; 18 g/t for one sample from Puchinskii; 8 g/t for 4 samples from Strel'ninskii; and 2 g/t for 11 samples from Lavrent'evskii [3].

Aeschynite, replacing brannerite, is reported from granite of northern Kazakhstan, see p. 81. Note also that grains or tabular fragments of blomstrandine (0.1 to 1.0 mm, rarely greater) occur in 25 out of 150 samples investigated from several granite massifs of central Kazakhstan [3].

Idiomorphic prismatic crystals (see p. 218) of metamict Ta aeschynite-(Y) occur together with small crystals of titanomagnetite (?), columbite, and beryl in the homogeneous Raposa pegmatite (composed of quartz, strongly red-colored microcline, and biotite) within mica schist of the Borborema region, near São José do Sabugí, Paraiba, northeastern Brazil [4].

Blomstrandine is a component (predominantly in the altered parts) of RE pegmatites of the zircon-blomstrandine (= vein No. 5) and the abukumalite-fergusonite type (= veins No. 13 and 14) of a pegmatite area (Keivy) in the northwestern European USSR. These veins are genetically related to alkali granites, have suffered albitization and silicification, and are located in the contact-near zone of metasomatically altered gneissic rocks. Blomstrandine is associated with yttrotitanite, zircon, an allanite-like mineral, P-free abukumalite, and, rarely, magnetite in vein No. 5; and with magnetite, zircon, and, rarely, fergusonite and Th-containing minerals in veins No. 13 and 14. Blomstrandine with red and brown to cinnamon-brown colors from vein No. 5 displays four types with different crystal morphologies, which have a different spatial distribution within the vein (see pp. 218/9) and, according to literature data, are characteristic of minerals of the priorite-blomstrandine series. Blomstrandine with cinnamon-brown color from veins No. 13 and 14 clearly differs from that of vein No. 5 and displays two types of crystal morphologies, which, according to literature data, are characteristic of minerals of the euxenite-polycrase series (and altogether resembles more polycrase). As thermal and X-ray diffraction data indicate, the blomstrandine from veins No. 13 and 14 is a pseudomorph (paramorph) preserving chemical composition, external appearance, and crystal form (see p. 219) of the earlier polycrase. Euxenite previously described from these veins (see p. 155 and analysis No. 25 of Table 9, p. 166) may also be blomstrandine [5], [6, pp. 52/4, 61]. Forming, besides well-developed crystals, also metacrystals with a complex internal structure (for details, see original paper), the above-mentioned blomstrandines are not syngenetic to the processes of silicification or albitization. Its formation is connected rather with hydrothermal processes affecting not only the zones of silicification and albitization but also non-altered parts of the pegmatite and its host rocks [6, pp. 65/7]. For a detailed discussion (blomstrandine included) of the paragenesis of rare-element minerals and of the subsequent stages of their formation in RE pegmatites of the northwestern European USSR, see [6, pp. 145/53].

Aeschynite is a widespread accessory niobate found in pegmatite veins of different types (for details, see original paper) at about 80 locations of the Ilmen Mountains, northern part of the Southern Urals, USSR. Usually, it forms incomplete columnar crystals (see pp. 216/7), mostly stretched along the axis [001]. Dependent on the type of host pegmatite, there is a distinct change in composition for the aeschynite (see pp. 200/2) [7, pp. 18/9, 22]. As a rare mineral of corundum-bearing syenite pegmatite from the northern part of the Ilmen Mountains, Y aeschynite occurs nearly exclusively within the feldspar parts and forms imperfect prismatic crystals some mm to 2 cm long and 0.3 cm wide. A brown halo is always present around the mineral [8]. But a more recent investigation of the same area does not confirm this mineral, and gives only samarskite and fergusonite as Y-rich accessories. And blomstrandine (forming a platy crystal up to 1 cm in size, see p. 217) from pegmatite of the southern part of the Ilmen

Mountains (pit No. 205, originally described by [9]) does not differ from the normal aeschynites of this area by its chemical analysis and thermal behavior [7, p. 28]. Black prismatic aeschynite, forming crystals up to 6 cm long and up to 2 cm wide, occurs in a nepheline-feldspar vein within granite gneiss near Lake Ilmen. It is grown together with mica and feldspar. Associated minerals are gray and red zeolite, sodalite, zircon, magnetite, and pyrochlore [43]. And accessory black aeschynite occurs as single crystals or groups of crystals in nepheline-feldspar and quartz-feldspar veins (genetically related to miaskitic pegmatites?) of the Southern Urals. The prismatic crystals, stretched parallel to the c axis, have a length from some mm up to 6 to 7 cm (sometimes up to 10 cm) and reach 0.5 to 2.0 cm in diameter. They are mainly bent, and usually sprout within feldspar. Owing to the various degree of acidity and alkalinity of the vein composition, the chemical composition of the aeschynite differs, expecially for the content of Th and Ce (see analysis Nos. 7 and 18 in Table 11, pp. 207/8) [44].

Minerals of the aeschynite group (e.g., normal aeschynite, thoroaeschynite, alumo-aeschynite, and nioboaeschynite, the latter with a ThO_2 content $<1\%$) occur in alkali peg-matites of the Vishnevye Gory, southern part of the Middle Urals, USSR. Thus, black normal aeschynite is a characteristic accessory mineral (associated with zircon, biotite/muscovite, and corundum) in nepheline-feldspar (miaskitic) pegmatite veins located within the country rocks (fenite and metamorphites) surrounding complexes of alkali rocks. Aeschynite is formed as a late deposition mainly in the endocontact zone of the vein, that is depleted in, or free of, nepheline by processes of silicification (interaction between country rock and the pegmatite). Normal aeschynite also occurs (associated with betafite-group minerals, allanite, zircon, muscovite/biotite, and corundum) in feldspar and quartz-feldspar (granitic) pegmatite veins, that may be, partially, only border zones of the above silicified veins. In quartz-feldspar (granitic) pegmatite, suffering alteration at different degrees during fenitization of the country rock and as a result of postmagmatic processes, the normal aeschynite is associated with pyrochlore, fersmite (+ columbite), allanite, and ilmenorutile [10, pp. 146/8]. For normal aeschynite in albitic rocks of the Vishnevye Gory, see p. 192. Brown to dark brown metamict thoroaeschynite occurs as an irregular deposition of $10 \times 20 \times 3$ mm in a microcline veinlet within the contact zone between biotite syenite and gneiss near the pegmatite vein No. 35, which is located in the exocontact zone of a Vishnevye Gory miaskite intrusion [10, p. 150]. And alumoaeschynite, forming irregular depositions of up to 0.5×1 cm and fine veinlets within titanian columbite, occurs in quartz veinlets within fenitized rocks [10, p. 154]. In quartz-arfvedsonite veinlets cutting fenites of the exocontact aureole of a miaskite (nepheline syenite) intrusion of the Vishnevye Gory [11], orthorhombic crystals of black Nb aeschynite (varying in size from 0.2×0.5 to 0.5×2 cm and containing 2.52% ThO_2) form, together with arfvedsonite, quartz, and other minerals, drusy aggregates growing on the walls of fractures [12]. For the paragenesis of other aeschynite generations in rocks of the Vishnevye Gory and Ilmen Mountains, see [13].

Irregular segregations up to 1 mm in size of X-ray amorphous cinnamon-brown normal aeschynite (of which, sometimes, zoned crystals are recognized under the microscope) and metamict light brown to greenish brown Th aeschynite crystals (of prismatic and tabular habit and up to 1 cm long) occur in a number of pegmatite veins of the exocontact area near the southwestern margin of the Transangara (or Tatarskii) nepheline-syenite massif, western part of the Enisei Ridge, Eastern Siberia. The aeschynites are located, together with microcline-perthite and arfvedsonite, in the peripheral, more melanocratic part of the veins [14]. As a rare accessory mineral, black isotropic blomstrandine occurs in some small (muscovite) pegmatite veins of the northern slope of the Stanovoi Range, southern Yakutia. It forms irregular grains up to 2 mm in size or, as a rarity, well-developed crystals 5 to 7 mm long.

Within the pegmatite, blomstrandine is connected with the rose-to-red microcline-perthite zone with blocky structure; connection with magnetite-rich parts is also observed [15].

As is atypical for granite pegmatites (which usually contain euxenite), Ta aeschynite (according to its X-ray data and thermal behavior, but chemically being similar to euxenite) forms during the initial phase of the albitization process in pegmatite bodies of the Na-Li type, forming thick lenses with a sharp boundary in slightly cataclastic biotite granite of an undisclosed locality of Siberia. The exocontact zone of pegmatite consists of tourmalinized and albitized granite. The Nb aeschynite, reaching 6×1 cm in size, occurs as two chemically identical types: Well-developed crystals (see p. 218), usually coated surficially by a brownish crust, which are associated with greenish beryl, rose apatite, and albite; or columnar aggregates of fresh crystals with lustrous faces enclosed in light-colored smoky quartz. The formation of this complicated mineral (by its structural and chemical data, see above) may have occurred, perhaps, in two stages. An early-formed euxenite structure suffers an internal modification of its lattice to an aeschynite structure with decreasing temperature, comparable to a way that otherwise forms paramorphs (refer also to p. 189) [16]. In a paragenetical association with tabular albite-oligoclase, beryl, ilmenorutile, and quartz, priorite occurs in nearly all veins of albitized amazonite pegmatite of an undisclosed pegmatite area, Eastern Siberia. Priorite forms small (0.1 to 0.2×1 to 2 mm in size) tabular crystals (flattened parallel to (001) and stretched along [100]), isometric short-prismatic crystals, or rounded "droplike" depositions of 1 to 1.5 cm in size. Sometimes, in bulge parts of thicker pegmatite veins, priorite occurs, together with beryl, in nest-like segregates, where aggregates of poorly-circumscribed priorite crystals, together with ilmenorutile, adhere as "crusts" to prismatic beryl. These crusts are 1 to 3 cm thick and cover an area of 10 to 20 cm. At times, priorite also occurs on fissures within the beryl [17].

Priorite, together with biotite, occurs in thin veins (up to 8 mm thick) in jointed quartz of the pegmatoid zone and within the quartz core of complex pegmatites, western Priazov'e, Ukraine. Individual grains of priorite are also scattered in the late fine-grained albite, where they are associated with priazovite, bismuthinite, and samarskite [18]; see also [19].

Small (0.1 to 1.0 mm in size) flattened crystals of aeschynite occur in veins of ankerite-dolomite carbonatite cutting nepheline-bearing fenite, Kola Peninsula, northwestern European USSR. Associated minerals are pyrite, strontiobarite, orthoclase, chlorite, and quartz [20]. Dark brown to orange-red crystalline lyndochite, associated with tetraphlogopite and actinolite, occurs in amphibolitized parts of medium-grained forsterite-calcite carbonatite of an undisclosed massif, Eastern Siberia. It forms either radiate and sheaf-like intergrowths of tiny thin-prismatic crystals (1 to 2 mm long, 0.1 to 0.2 mm wide) elongated, and partly bent, along its a axis; or it is peripherically grown on dark brown grains of Ta- and Ti-bearing pyrochlore. In amphibolitized parts of huge-grained forsterite-calcite carbonatite (that differs by some peculiarities from the above medium-grained), a grayish yellow mineral with a structure identical to lyndochite (but possibly being natroniobite) occurs as a fine aggregation pseudomorphous after cubic crystals of lueshite [21]. Dark brown grains (up to 2 cm in size) of strongly pleochroic aeschynite (and giving a clear X-ray diffraction pattern) occur, together with barite and allanite, in carbonatite of Pakkanadu, Tamil Nadu, India [22].

Dark brown, microscopically partly metamict aeschynite (originally suggested to be euxenite) occurs as aggregates of parallel fibers, as much as 6 inches long and 4 to 5 inches across, in carbonatic veins in gneisses of the upper part of Sheep Creek, Ravalli county, Montana. Clearly defined textural relationships indicative of age differences lead to a generally consistent paragenetic sequence. Together with allanite, monazite, barite, columbite, and Nb rutile, the aeschynite seems to belong to the RE-Nb stage of metasomatic replacements [23].

Aeschynite, intimately intergrown with ilmenite and ilmenorutile, occurs in all parts of a zoned greisen (which developed after alaskite and along steep clefts) from an undisclosed alaskite-biotite granite massif of the USSR (Kazakhstan?). Particularly, big accumulations of aeschynite (up to 3×7 cm) are found in the quartz zone and in quartz fillings within muscovite, where also well-developed tabular crystals (see p. 218) with a maximum size of 3×1 cm occur. The crystallization of aeschynite (evidently at temperatures of 250 to 300°C) is related to processes of pneumatolysis, as is corroborated by outer zones associated with the greisen containing tourmaline and high amounts of fluorite [24].

Lyndochite, displaying three varieties of different color and degree of crystallinity, occurs as a characteristic mineral in metasomatic amphibole-feldspar rock and, to a lesser degree, in aegirine and in hematite-feldspar rocks of a rare-metal mineralization, which is genetically related to a granite intrusion into slightly metamorphic rocks of the Bakal'(skaya) suite, Southern Urals, USSR. The highly crystalline dark red variety (containing 31.2% ΣRE_2O_3 + ThO_2) is hosted in an arfvedsonite-biotite-albite rock and forms flattened-prismatic platy crystals or parallel, sometimes fan-like, intergrowths of tiny lamellar crystals of 0.02 to 0.3 mm, predominantly 0.08 to 0.05 mm, in size. A slightly pleochroic brownish black to black variety is scarce and sometimes occurs together with the dark red in a single grain; their verge is very distinct (growing zone?). Crystals and fragments of the black lyndochite usually are tabular. And a slightly pleochroic to isotropic yellowish brown variety, mainly associated with albite, is 2-times rarer than the dark red and rarely forms crystals showing single faces of the 2nd pinacoid. Partly, compact depositions of the yellowish brown lyndochite are surrounded by many much smaller grains of the same variety. The mutual relationship of the above color varieties of lyndochite is not entirely clear. Partly, the black variety forms the inner part of the dark red; and the dark red, sometimes, is peripherically carpeted with the yellowish brown variety [25].

Normal aeschynite occurs in albitite bodies within fenite of the Vishnevye Gory, southern part of the Middle Urals. It also occurs as single prismatic crystals (with a rhombic section and striated parallel to planes of the rhombic prism) and is a component of concentric-zonal "ovoids" in albitic stringers within miaskite of the Western Laminated Body, northern part of the Vishnevye Gory. The "ovoids" represent a metacryst that has a core of monazite surrounded by apatite and furthermore by allanite. The latter, on its periphery, is seamed by aeschynite [10, p. 148]. Together with fergusonite, pyrochlore, and some other ore minerals, priorite occurs within the lying albitic part of an asymmetrical riebeckite-zircon (malacon)-albite vein cutting albitites of an undisclosed granite dome, Eastern Siberia. Priorite forms 1.5 to 2 cm-sized irregular accumulations of grains; more rarely, crystals 0.2 to 0.5 mm in length. There are two mineral varieties with different chemical composition and micro-hardness, but identical refractive index: A widespread red-brown metamict, and a yellow slightly crystalline species (see pp. 199, 228 and 226, respectively). The latter sometimes occurs within the former [26].

An accessory yellow niobate of Y (containing 1.02% Th by spectral analysis), probably priorite, occurs as short prismatic crystals in albitized biotite-bearing nepheline syenite and its pegmatites from the Borsuksaiskii massif, Mugodzhary, Kazakhstan. These crystals form chains along fissures (vestiges) of albitization within microcline; their main axis is oriented perpendicular to the wall of the fissure. Accompanying minerals within the fissures are zircon, colorless thorite, red-brown pyrochlore, and (additionally in pegmatites only) gadolinite and xenotime [27],

Blomstrandine of pneumatolytic-hydrothermal veins (composed of albite, quartz, fluorite, and calcite), that are genetically related to Paleozoic riebeckite-aegirine granite of Eastern Siberia, has different colors (see p. 225) and forms grains and aggregates of irregular shape.

Sometimes, prismatic crystals occur that are striated parallel to (010) and reach 0.5 to 0.7 cm along their longer axis. Usually, the mineral is related to the endocontact zone of the vein, where it is closely attached to fluorite and calcite. Regarding its spatial relationships with the other minerals, blomstrandine was formed at a relatively late stage of vein formation, during a phase of acid leaching at low temperature [28]. For changes in chemical composition of the above blomstrandines, see p. 203.

Completely metamict but fresh lyndochite, associated with smoky quartz, tends to become dominant to the exclusion of allanite (otherwise present) in massive quartz reefs transecting as veins the country-rock gneisses at Tura dukas, 35 miles north of Nanyuki, Kenya [29].

Accessory blomstrandine, 15 μm in size, was found in a quartz fissure in kaolinitic granite (rich in accessories) from a quarry at Nadap, Velence Mountains, Hungary [30].

Yttrocrasite (containing $7.3 \pm 0.1\%$ Th and $3.3 \pm 0.1\%$ U) from stanniferous veins (composed of quartz, muscovite, feldspar, and tourmaline) of the periphery of a granite batholith at Mitwaba, Kibara Mountains, Katanga [31], turned out to be priorite [32], whose occurrence in high-temperature hydrothermal cassiterite-quartz veins is exceptional because this mineral normally crystallizes in high-temperature pegmatites [33]. For the changes in chemical composition during weathering processes, see p. 204.

Hypidiomorphic to idiomorphic (see p. 218), dark brown to black metamict aeschynite, varying in size between 0.1 and 20 mm, occurs in three phenakite-bearing biotite-chlorite lenses, only a few meter apart, of the talc deposit near the abandoned Habachtal emerald mine, Salzburg, Austria. Aeschynite, probably of (poly-phase) metamorphic origin, is found as inclusions in phenakite and, with a distinct variation in chemical composition (see p. 202), also in the biotite-chlorite groundmass. The associated accessory minerals, somewhat differing for the individual lenses, are allanite, epidote, zircon, apatite, monazite (only in lenses 1 and 2), scheelite, cassiterite, and an unidentified RE silicate (only in lens 3) [34].

Tiny (0.1 to 0.5, rarely up to 2 mm in length) prismatic reddish brown crystals of aeschynite-(Y, Gd, Dy, Er), displaying various morphological habits (see p. 218), are mainly grown on (010) planes of a twinned adular crystal from an open cleft ("Zerrkluft") in the Fibbia gneiss near the contact to Rotondo granite, Gotthard massif, Switzerland [35]. Red or black aeschynite-(Y) also occurs in four samples of cavernous Rotondo granite from the Safety Tunnel South, Furka Basis Tunnel [36]. Note further that totally metamict crystals of aeschynite, strongly resembling and therefore erroneously named prior rutile, are reported from open clefts in gneiss of the Val Nalps, Graubünden, Switzerland [37], see also [38].

Non-metamict nioboaeschynite-(Ce) of unknown origin occurs as brittle, red-brown to dark brown, striated, slender prismatic grains in a placer deposit of the Tofty tin belt, Tanana quadrangle, central Alaska. The grains are 0.2 to 1.5 mm in size, mainly tabular, but often stubby with a prominent cross-fracture [39].

Anisotropic Ca-rich aeschynite, associated with allanite (orthite), zoisite, albite, microcline, monazite, calcite, apatite, and titanite (sphene), is found in an undisclosed deposit of Northern Korea [40].

Small prismatic crystals of crystalline lyndochite are reported from an undisclosed host rock and locality of the People's Republic of China. It is associated with xenotime, ilmenorutile, zoisite, an unidentified carbonate, albite-oligoclase, and phlogopite [41].

Parallel **intergrowths** along the 2nd pinacoid were observed for blomstrandine from pneumatolytic-hydrothermal veins genetically related to riebeckite-aegirine granite of Eastern Siberia [28, p. 44]; refer also to intergrowths between aeschynite, Eastern Siberia, on p. 191.

Ovoid intergrowths of crystalline aeschynite and metamict allanite occur in the Pakkanadu carbonatite, Tamil Nadu, India [22]. Skeleton-like intergrowths of aeschynite with titanite, rutile, and ilmenite are found in samples from biotite-chlorite lenses, Habachtal, Salzburg, Austria [34]. For intergrowths of aeschynite with ilmenite and ilmenorutile, USSR, see p. 12. Intergrowths of aeschynite with magnetite and zircon, occurring in crevisses within feldspar, are typical for samples from a number of pegmatite veins of the Ilmen Mountains, northern part of the Southern Urals. Samples from late granite pegmatites of the northern part of this area are intergrown with quartz, feldspar, and, sometimes, betafite; that from amazonite pegmatite (pit No. 248) are intergrown with quartz and feldspar [7, p. 22]. Dark red crystalline lyndochite from a metasomatic arfvedsonite-biotite-albite rock, Southern Urals, is intergrown with albite, hematite, and apatite (dehrnite), but also with quartz, arfvedsonite, and anatase (a paramorph after rutile). The lyndochite was formed later than the above first three minerals. And anatase (forming well-developed crystals 0.1 to 0.2 mm in size) is intergrown with dark red lyndochite that was formed prior [25]. Closely intergrown with Mn clinozoisite is crystalline lyndochite from an undisclosed host rock and locality of the People's Republic of China [41]. For intergrowths, refer also to alteration below.

Inclusions of priorite in quartz and feldspar of albitized amazonite pegmatite, Eastern Siberia, are surrounded, as a rule, by diverging cracks. And around inclusions of priorite, the blue or green beryl becomes yellowish [17].

Small inclusions of magnetite occur in priorite from muscovite pegmatite, Priazov'e, Ukraine [19]. Possibly titanite (according to its optical properties, and as a secondary product) occurs as single grains within the area of altered aeschynite of pegmatite (pit No. 205), Ilmen Mountains [42].

Besides changes in composition of group-A and/or group-B cations (examples are given on pp. 204/5), **alteration** of aeschynite may lead to newly formed mineral phases, often with a size too small to be identified with certainty. Thus, beside replacement by zircon (cyrtolite) and allanite (orthite), reported also for other localities, the megacrystals of blomstrandine from RE pegmatites, northwestern European USSR, show replacement by yttrotitanite. This replacement, often occurring to a degree of 50 to 70%, leads to several types of mutual intergrowths (for details, see original paper). Subsequent replacement of blomstrandine and/or yttrotitanite by ilmenite occurs, too [6, pp. 67/73]. Occurring as newly formed minerals or as included minerals, magnetite, ilmenite, albite, microcline, quartz, biotite, and thorite were found in aeschynite from pegmatites of the Ilmen Mountains, Southern Urals. Also presumably occurring were hematite, pyrochlore, columbite and Mn columbite, fersmite, titanite, rutile, and euxenite. Some of the minerals within the area of aeschynite crystallized prior to, sometimes simultaneously with aeschynite. But in all cases investigated, there are seeds of altered aeschynite at the contact with magnetite and ilmenite [42]. For lyndochite pseudomorphous after lueshite, Eastern Siberia, see p. 191. For aeschynite replacing brannerite of the Borovskii pluton, northern Kazakhstan, see p. 81.

References for 2.2.2.2.1:

[1] Azimov, P. T. (Dokl. Akad. Nauk Uzbek. SSR **27** [1970] 45/7).
[2] Batieva, I. D.; Bel'kov, I. V. (in: Vinogradov, A. N., Zakonomernosti Kontsentratsii Rudnykh Elementov v Granitoidnykh Formatsiyakh Karelo-Kol'skogo Regiona, Geol. Inst. Akad. Nauk SSSR, Apatity 1985, pp. 62/73, 64).
[3] Gogel', G. N. (Izv. Akad. Nauk Kaz. SSR Ser. Geol. **1958** No. 4, pp. 99/105).
[4] Adusimili, M. S.; Kieft, C.; Burke, E. A. J. (Mineral. Mag. **39** [1973/74] 571/6).
[5] Lunts, A. (Izv. Akad. Nauk Latv. SSR **1965** No. 2, pp. 51/67, 52/4, 61).

[6] Lunts, A. Ya. (Mineralogiya Geokhimiya i Genezis Redkozemel'nykh Pegmatitov Shchelochnykh Granitov Severo-Zapada SSSR, Nedra, Moscow 1972, pp. 1/176).

[7] Popova, V. I.; Bazhenova, L. F. (in: Popov, V. A., Mineralogicheskie Issledovaniya v Il'menskom Zapovednike, Akad. Nauk SSSR, Sverdlovsk 1981, pp. 18/29; C.A. **98** [1983] No. 146714).

[8] Makarochkin, B. A.; Es'kova, E. M.; Gonibesova, K. A. (Tr. Inst. Mineral. Geokhim. Kristallokhim. Redk. Elem. No. 3 [1959] 145/50, 145, 150).

[9] Simonov, A. I. (in: Mineraly Il'menskogo Zapovednika, Akad. Nauk SSSR, Moscow 1949, pp. 476/80 from [7, p. 29]).

[10] Es'kova, E. M.; Zhabin, A. G.; Mukhitdinov, G. N. (Mineralogiya i Geokhimiya Redkikh Elementov Vishnevykh Gor, Nauka, Moscow 1964, pp. 1/318).

[11] Zhabin, A. G.; Mukhitdinov, G. N.; Kazakova, M. E. (Tr. Inst. Mineral. Geokhim. Kristallokhim. Redk. Elem. No. 4 [1960] 51/72, 51, 57, 66/8).

[12] Zhabin, A. G.; Aleksandrov, V. B.; Kazakova, M. E. (Tr. Inst. Mineral. Geokhim. Kristallokhim. Redk. Elem. No. 7 [1961] 108/12, 108/9).

[13] Zhabin, A. G.; Mukhitdinov, G. N. (Dokl. Akad. Nauk SSSR **126** [1959] 1055/7; Dokl. Acad. Sci. USSR Earth Sci. Sect. **124/129** [1959] 490/1).

[14] Sveshnikova, E. V.; Semenov, E. I.; Khomyakov, A. P. (Zaangarskii Shchelochnoi Massiv, Ego Porody i Mineraly, Nauka, Moscow 1976, pp. 1/80, 31).

[15] Shaposhnikov, G. N. (Zap. Vses. Mineral. Obshch. **89** [1960] 709/10).

[16] Kornetova, V. A.; Aleksandrov, V. B.; Kazakova, M. E. (Tr. Mineral. Muzeya Akad. Nauk SSSR No. 14 [1963] 108/21, 115, 120/1).

[17] Fel'dman, L. G.; Konopleva, I. B. (Tr. Mineral. Muzeya Akad. Nauk SSSR No. 19 [1969] 86/112, 86/9).

[18] Litovchenko, E. I.; Kuts, V. P. (Mineral. Sb. [Lvov] **29** No. 4 [1975] 32/41, 32/3, 38/9).

[19] Litovchenko, E. I.; Osadchii, V. K. (in: Mitskevich, B. F.; Shcherbak, N. P., Aktsessornye Mineraly Ukrainskogo Shchita, Naukova Dumka, Kiev 1976, pp. 112/3).

[20] Kapustin, Yu. L. (Mineralogiya Karbonatitov, Nauka, Moscow 1971, pp. 1/286, 144, 147).

[21] Zdorik, T. B.; Finyagina, K. K. (Tr. Mineral. Muzeya Akad. Nauk SSSR No. 17 [1966] 209/15, 209/10).

[22] Semenov, E. I.; Upendran, R.; Subramanian, V. (J. Geol. Soc. India **19** [1978] 550/7, 551).

[23] Heinrich, E. W.; Levinson, A. A. (Am. Mineralogist **46** [1961] 1424/47, 1436, 1438/9).

[24] Podol'skii, A. M.; Golikov, I. V.; Burova, T. A. (Tr. Mineral. Muzeya Akad. Nauk SSSR No. 16 [1965] 175/86, 175/7, 185).

[25] Svyazhin, N. V.; Gaidukova, M. K. (Zap. Vses. Mineral. Obshch. **96** [1967] 206/10).

[26] Kostin, N. E.; Petrova, E. A.; Sidorenko, G. A.; Skorobogatova, N. V. (Mineral'n. Syr'e No. 1 [1960] 121/8, 121/2; C.A. **56** [1962] 3143).

[27] Voloshin, A. V. (Mater. Geol. Polezn. Iskop. Zapadn. Kaz. **3** [1966] 169/75, 169, 174/5).

[28] Romanov, I. A. (In: Mineraly i Paragenezisy Mineralov Endogennykh Mestorozhdenii, Nauka, Leningrad 1975, pp. 43/9, 44, 46, 48).

[29] Horne, J. E. T.; Butler, J. R. (Mineral. Mag. **34** [1965] 237/48, 237/8).

[30] Pantó, G. (Acta Geol. [Budapest] **19** [1975] 59/93, 61, 69).

[31] Ledent, D.; Picciotto, E.; Poulaert, G. (Bull. Soc. Belge Geol. Paleontol. Hydrol. **65** [1956] 233/50, 234, 249).

[32] Van Wambeke, L. (Bull. Soc. Belge Geol. Paleontol. Hydrol. **66** [1957] 35/53, 35, 38/40).

[33] Van Wambeke, L. (Neues Jahrb. Mineral. Abhandl. **112** [1969/70] 117/49, 127).

[34] Grundmann, G.; Koller, F. (Neues Jahrb. Mineral. Abhandl. **135** [1979] 36/47, 38/9, 45/6).

[35] Sommerauer, J.; Weber, L. (Schweiz. Mineral. Petrogr. Mitt. **52** [1972] 75/91, 75/6).

[36] Kipfer, A.; Stuker, P. (Schweizer Strahler **5** [1979] 45/91, 65/6, 69).

[37] Kipfer, A. (Schweizer Strahler **3** [1973] 133/69, 158/9).

[38] Kipfer, A. (Schweizer Strahler **5** [1979] 6/10).

[39] Rosenblum, S.; Moiser, E. L. (Am. Mineralogist **60** [1975] 309/15, 309/10).

[40] Komkov, A. I. (Tr. Vses. Nauchn.-Issled. Geol. Inst. **96** [1963] 213/25, 213).

[41] Gorzhevskaya, S. A.; Sidorenko, G.A. (Dokl. Akad. Nauk SSSR **146** [1962] 1176/8; Dokl. Acad. Sci. USSR Earth Sci. Sect. **146** [1962] 131/3).

[42] Popova, V. I. (in: Materialy po Mineralogii Mestorozhdenii Urala, Akad. Nauk SSSR, Sverdlovsk 1984, pp. 12/8; Ref. Zh. Geol. **1985** 3V213).

[43] Vladimirova, M. E. (Tr. Radievogo Inst. Akad. Nauk SSSR **6** [1957] 139/66, 146).

[44] Makarochkin, B. A. (Geol. Geofiz. **1979** No. 1, pp. 150/2).

2.2.2.2.2 Chemistry

The **general formula** of aeschynite and related varieties (refer to pp. 185/6) is AB_2O_6 with A = RE, Th, Ca, Fe^{2+}, Mn, Mg, and other divalent ions, B = Nb and Ti, and O may include OH and F [1], see also [2]; or A = Y, RE, U, Th, Ca, Fe and B = Ti, Nb, Ta, W [3, p. 84]; or A = Y, RE, Th, U, Ca and B = Ti, Nb, Ta, Fe^{3+}, Al, Si [4, p. 62]; or A = Ca, Ce^{3+}, Th and B = Nb, Ta, Ti, Fe^{3+}, Fe^{2+} [5]; or A = Ca, Mn, Na, Y, Ce, La, Th, U, Pb, and others and B = Ti, Nb, Fe, Zr, Ta, Mg, and others [6].

Against the theoretical composition AB_2O_6, a deficiency of cations and/or of the anionic part may occur for aeschynite and its varieties, owing to the oxidation state (valency) of Fe refer to p. 197) and/or the composition of RE [7, pp. 47/8, 67/9]. A small deficit in the total content of group-A and -B cations may also be explained by a raised content of water (as is supposed, e.g., for Y aeschynite of the Ilmen Mountains, Southern Urals) [6]. For examples of cation/anion deficiencies, see pp. 198/9.

For a compilation of chemical analyses of aeschynite and priorite/blomstrandine, see Tables 10 and 11, pp. 206 and 207/13; and for additional analyses, see also pp. 199/204. Older, pre-1910 analyses are given in [8]. For contents (variation and mean) of Y_2O_3 and/or RE_2O_3 + Y_2O_3 in aeschynite and blomstrandine, see "Seltenerdelemente" A4, 1979, pp. 7, 10, 14.

The distinction between aeschynite and priorite is not in their contents of CaO, RE_2O_3, TiO_2, and $(Nb,Ta)_2O_5$ but in the relative importance of their light RE on the one hand and of Y and heavy RE on the other [9, p. 244]. The cerian RE (comprising 55 to 92% of the total RE content) predominate in aeschynite and lyndochite, the yttrian RE (comprising 44 to 75% of the total RE content) in priorite/blomstrandine [7, pp. 13/6]. It is probable that there is a tendency for Th to be lower and U higher in priorite compared with aeschynite, but this is a measure of the geochemical association of Y, Dy, and Er with U^{4+}, which is closer than that of La and Ce with Th [9, p. 245].

Aeschynite is a titano-niobate of RE and Th fitting the general formula AB_2X_6. The principal group-A components are RE(Ce), generally with Ce (that apparently enters as Ce^{3+}) and occasionally with Nd predominating; RE(Y) sometimes account for more than half of the total RE content (Y aeschynite). In some aeschynites Th may constitute the principal group-A cation. The main group-B components are Nb and Ti, with Ta and Fe^{3+} in subordinate amounts; Si and Al may also be present in small quantities. The Nb:Ti ratio may vary within wide limits, although it is lower than in euxenite and polycrase. Group X generally contains O, a small part of which may be replaced by OH and, sometimes, by F [10]. In correspondence with the geochemical peculiarities of alkaline rocks, aeschynite from this source is totally devoid of

Ta, or contains it only in negligible amounts [11, p. 108]. Using plots of $\Sigma(La + Ce + Pr)$ vs. the ratios $100\,Y : \Sigma(Y + RE)$ or La : Nd, it is shown that minerals of the aeschynite-priorite series from granitic pegmatites are enriched in Y and those from alkaline rocks are enriched in light RE [12].

According to its chemical composition, aeschynite/lyndochite is a titano-niobate of cerian RE (for the details of RE composition, see [7, pp. 62/5]) containing Th and, characteristically, Ca (0.75 to 5.46% CaO, the higher values are typical for lyndochites) and Fe (1.14 to 5.45% Fe oxides). By its content of Th, there can be separated two groups of minerals: rich in ThO_2 (9.51 to 20.34%) and poor in ThO_2 (0.72 to 4.95%), corresponding to normal aeschynite and lyndochite, respectively. Further, normal aeschynite is poor in Ta; and contents of Si, Al, Mn, Mg, etc. often are found in both minerals [7, pp. 58/62]; refer also to [13, pp. 49/51], where the ranges for ThO_2 are given as 10.8 to 20.3% for normal aeschynite and 2.5 to 4.95% for lyndochite. Considering the above mentioned chemical components, the formula may be expressed as $(RE, Th, Ca)(Ti, Nb)_2O_6$ for aeschynite and as $(RE, Ca, Th)(Nb, Ti)_2O_6$ for lyndochite, where the composition of RE is predominantly Ce + Nd, Y for the former and Nd + Ce, Y for the latter. For a detailed calculation of the formula, one has to make allowance not only for the ratios Y : RE and RE(Ce) : RE(Y), but also for the degree of oxidation (valency) of Fe. Thus, for 23 analyses of aeschynite/lyndochite (see original paper) regarding total Fe as Fe^{3+}, the general formula becomes $A_{1+n}B_2O_{6+m}$, with $n = 0.01$ to 0.27 and $m = 0.07$ to 0.53; and for Fe^{2+} it reads as $A_{1-n}B_2O_{6-m}$, with $n = 0.21$ to 0.03 and $m = 0.65$ to 0.05, where $m > n$ [7, pp. 67/9]; for a special example, see p. 198.

Diadochic Replacement. Priorite/blomstrandine is a complex oxide of Nb-Ti with yttrian RE (for the details of RE composition, see [7, pp. 43/5]) and has much less ThO_2 than aeschynite but always contains U and Fe (0.50 to 6.01% Fe oxides, mainly of Fe^{2+}); small amounts of Ta (up to 2.04% Ta_2O_5) and Ca (up to ~4% CaO) occur constantly. Priorite is the variety rich in Nb, and blomstrandine is rich in Ti. The ThO_2 contents vary between 0.61 and 4.66% (characteristic are 1.3 to 2.75%) for priorite, and between 1.21 and 8.48 for blomstrandine. As a characteristic admixture, Al (traces to 2.8% Al_2O_3) and Si (0.02 to 2.49% SiO_2) occur in priorite, whereas blomstrandine contains SnO_2 (0.08 to 0.2%) [7, pp. 40/4]. For 8 analyses of priorite/blomstrandine (see original paper), the general formula can be expressed as $(Y, RE(Y), Ca, Th)_{1-n}(Ti, Nb)_2O_{6-m}$, where sometimes $m > n$ [7, pp. 47/8].

In aeschynite, the RE are largely replaced by Th; they are also replaced by U, Ca, Mg, and Pb and, possibly, by Fe^{2+}, Mn, and alkalis [10]. Further, there are widespread heterovalent replacements as, for instance, Ca^{2+} by Th^{4+} [13, p. 51] and Nb + RE by Ti + Th and/or Nb + Ca by Ti + RE [7, p. 61]. It seems likely that in aeschynite the appreciable Th content is to be correlated with the content of light RE, whereas in priorite (where U exceeds Th) substitution of Y and heavy RE is more normal [14]. But in many cases it is not clear what element isomorphically entered the crystal structure of aeschynite, or is adsorbed owing to the colloid-like metamict state of that mineral, or simply forms mechanical admixtures. As follows from experiments, studying at 1200°C the corresponding compounds with various individual RE, there are discontinuities in the isomorphous miscibility (for the details, see original paper). Especially for Th, a heterovalent substitution $2\,RE^{3+}$ by $Ca^{2+} + Th^{4+}$ is possible. For example, the compound of composition $Ca_{0.5}Th_{0.5}(Nb, Ti)_2O_6$ has the structure of aeschynite (with a disordered distribution of Ca and Th) and is isomorphously and unlimitedly miscible with titaniferous niobates of the aeschynite type. Probably, compounds with the aeschynite structure should have a limited miscibility with $Ca_{0.5}Th_{0.5}(Nb, Ti)_2O_6$, however, the region of homogeneity of the euxenite phase (refer to p. 162) in this case is apparently great. The synthesis further showed that there is a limited substitution of 25 to 30% for the "molecule" $ThTi_2O_6$.

The preceding statements are valid only for the temperature at which synthesis was done. At 1000°C, the region of immiscibility is appreciably greater for systems $CaNb_2O_6$-$RE(Nb,Ti)_2O_6$. Probably this is also valid for other cases of isomorphism [15].

Minerals of the aeschynite-blomstrandine series, which may be viewed as the low-temperature phase of $YTi(Nb,Ta)O_6$, tend to an extensive mixed crystal formation. Thus, recalculation into components of analyses (given in [8]) leads, for example, for the components $(U,Th)Ti_2O_6$ and $(U,Th)O_2$ to portions of 1.8 and 18.4% (for analysis No. 1) or 12.5 and 9.5% (for analysis No. 2), respectively. For the other possible components, summing up to 100%, see original paper [16].

As can be seen from examples of crystal-chemical formulas (see next paragraphs and the examples on p. 200), which reflect the wide range and type of diadochic replacement supposed in the original paper to occur, there are variances (deficit or surplus) against the general AB_2O_6-type formula not only for the cationic part (especially the group-A elements; partly owing to primary alterations, refer to p. 75), but also for the anionic part. Variances in the crystal-chemical formula also appear if one makes special assumptions for group-A and -B cations. Thus, Ta aeschynite from a granite pegmatite, Siberia, may have the composition $(RE_{0.537}Ca_{0.267}Sr_{0.11}Th_{0.019}Fe^{2+}_{0.03})_{\Sigma0.867}(Ti_{0.857}Ta_{0.549}Nb_{0.504}Al_{0.075}Fe^{3+}_{0.019})_{\Sigma2.00}(O_{5.63}(OH)_{0.439})_{\Sigma6.070}$ or $(RE_{0.557}Ca_{0.279}Sr_{0.011}Fe^{2+}_{0.055}Th_{0.019}Al_{0.079})_{\Sigma1.00}(Ti_{0.897}Ta_{0.575}Nb_{0.528})_{\Sigma2.00}(O_{5.889}(OH)_{0.460})_{\Sigma6.349}$. The first formula displays a big deficit in group-A elements, as is usual for highly altered metamict minerals. The second formula makes allowance for only Fe^{2+}, regards small amounts of Fe^{3+} as caused by oxidation, and includes Al (possibly an adsorbed colloidal component) in group A. The resulting surplus for the anionic part may be attributable, partly, to changes the mineral suffered during metamict disintegration [11, pp. 116/7]. As is characteristic for most of the metamict minerals, there is a considerable deficit for group-A cations and also for oxygen. Irrespective of the degree of metamictization, the calculation of a water-free composition gives nearly identical results for cinnamon-brown and black priorite from amazonite pegmatite, Eastern Siberia: $A_{0.82}B_2O_{5.63}$ and $A_{0.83}B_2O_{5.67}$, respectively [17, p. 97].

The following examples are arranged by the type of their host rock, and the number given in parenthesis corresponds to the analysis number given in Table 11, pp. 207/12. Sometimes, the content of water in the analyses is not included in the crystal-chemical formula (e.g., of blomstrandine from Keivy) as there are statements that minerals of the titano-tantalo-niobate group contain only adsorbed and zeolitic water [4, p. 64].

Aeschynite (No. 30) from pegmatite, Inner Mongolia, P.R.C., is $(Ce_{0.55}Y_{0.22}Th_{0.06}Na_{0.06}Ca_{0.05}Fe^{2+}_{0.04})_{\Sigma0.98}(Ti_{0.95}Nb_{0.92}Fe^{3+}_{0.13})_{\Sigma2.00}O_{5.76}$ [18].

Red (No. 34) and brownish cinnamon blomstrandine (No. 24) from RE pegmatite in alkali granite (Keivy), northwestern European USSR, is $(RE_{0.81}Ca_{0.24}Th_{0.05}U^{4+}_{0.03}U^{6+}_{0.03})_{\Sigma1.16}(Ti_{1.42}Nb_{0.37}Ta_{0.02}Fe^{3+}_{0.08}Al_{0.05}Si_{0.03}Mg_{0.03})_{\Sigma2.00}O_{5.85} \cdot 0.74 H_2O$ and $(RE_{0.72}Ca_{0.012}Th_{0.07}U^{4+}_{0.01}U^{6+}_{0.01}Pb_{0.01})_{\Sigma0.94}(Ti_{1.48}Nb_{0.37}Ta_{0.02}Fe^{3+}_{0.11}Si_{0.01}Zr_{0.01})_{\Sigma2.00}O_{5.59} \cdot 0.65 H_2O$, respectively [4, p. 64].

Blomstrandine from a complex pegmatite, northern Priazov'e, Ukraine, is $(Y_{0.74}Ca_{0.14}Th_{0.11}Fe^{2+}_{0.09}U_{0.04}Ce_{0.04}Na_{0.02}Mn_{0.02}K_{0.01}Mg_{0.01})_{\Sigma1.23}(Ti_{1.50}Nb_{0.49}Ta_{0.01})_{\Sigma2.00}O_6$ [19]. Priorite (No. 52) from pegmatite, Priazov'e, Ukraine, is $(Y_{0.8}Ce_{0.02}Ca_{0.19}Mg_{0.04}U_{0.06}Th_{0.02}Pb_{0.03})_{\Sigma1.16}(Nb_{0.79}Ta_{1.1}Ti_{0.92}Fe^{3+}_{0.05}Al_{0.14})_{\Sigma3.00}O_{5.39} \cdot 1.64 H_2O$ [20, p. 39], see also [19].

Ta aeschynite-(Y) (No. 20) from the Raposa pegmatite, northeastern Brazil, is $(Ca_{0.47}(Y,RE)_{0.425}(Th,U)_{0.11})_{\Sigma1.005}(Ta_{0.75}Ti_{0.62}Nb_{0.595}Fe_{0.05})_{\Sigma2.015}O_6$ [21, p. 574].

Aeschynite samples from a quartz-feldspar vein (No. 7) and a nepheline-feldspar vein (No. 18), both Southern Urals, are $(Th_{0.22}Ce_{0.27}RE_{0.32}Ca_{0.18}(Pb,Mn)_{0.01})_{\Sigma1.00}(Nb_{0.82}Ta_{0.01}Ti_{1.04}(Sn,Fe^{2+})_{0.14}Mg_{0.01})_{\Sigma2.02}(O_{5.41}(OH)_{0.58})_{\Sigma5.99}$ and $(Th_{0.14}Ce_{0.32}RE_{0.32}Ca_{0.15}(Pb,Mn)_{0.02})_{\Sigma0.95}(Nb_{0.90}Ta_{0.01}Ti_{0.88}(Sn,Fe^{2+})_{0.13}Mg_{0.02})_{\Sigma1.94}(O_{5.50}(OH)_{0.49})_{\Sigma5.99}$, respectively [22].

Lyndochite (No. 25) from carbonatite, Eastern Siberia, has the composition $(RE_{0.65}Ca_{0.25}$-$Th_{0.07}U_{0.02})_{\Sigma 0.99}(Nb_{0.79}Ti_{1.06}Fe_{0.08}^{3+}Si_{0.04}Al_{0.03})_{\Sigma 2.00}O_{3.76}$ [23] and aeschynite (No. 21) from an ankerite-dolomite carbonatite vein in fenite, Kola Peninsula, both USSR, is $(RE_{0.83}Ca_{0.09}$-$Th_{0.08})_{\Sigma 1.00}$ $(Nb_{0.67}Ti_{1.29}Fe_{3.004})_{\Sigma 2.00}O_{5.81}$ [24].

Aeschynite (No. 26) from apo-alaskite greisen, USSR, is $(RE_{0.58}Ca_{0.14}Na_{0.10}Th_{0.07}U_{0.06}Mn_{0.04}$-$K_{0.01})_{\Sigma 1.00}(Ti_{0.91}Nb_{0.76}Ta_{0.10}Fe_{0.12}^{3+}Al_{0.11})_{\Sigma 2.00}(O_{5.74}F_{0.10}(OH)_{0.16})_{\Sigma 6.00}$ [25]. Red-brown (No. 40) and yellow priorite (No. 46) from an albite-zircon-riebeckite vein in albitite, Eastern Siberia, are $((Y_{0.31}Dy_{0.18}Gd_{0.06}Er_{0.06})_{\Sigma 0.73}Th_{0.03}U_{0.01})_{\Sigma 0.77}(Ti_{1.03}Nb_{0.66}Fe_{0.08}^{3+}Si_{0.11}Al_{0.10}Zr_{0.02})_{\Sigma 2.00}(O_{4.82}$-$(OH)_{1.18})_{\Sigma 6.00}$ and $((Y_{0.31}Dy_{0.18}Er_{0.11}Gd_{0.08})_{\Sigma 0.84}Th_{0.02}U_{0.01}Ca_{0.02}Na_{0.02}K_{0.02}Mg_{0.01})_{\Sigma 0.94}(Ti_{0.97}Nb_{0.92}$-$Al_{0.04}Si_{0.04}Fe_{0.02}^{3+}Ta_{0.01})_{\Sigma 2.00}(O_{5.72}(OH)_{0.20}F_{0.08})_{\Sigma 6.00}$, respectively [26, pp. 125/6].

Nioboaeschynite (No. 43) from fenite, Vishnevye Gory, Middle Urals, is $(RE_{0.58}Ca_{0.29}Th_{0.03}$-$(OH)_{0.15})_{\Sigma 1.05}(Nb_{1.05}Ti_{0.79}Fe_{0.12}Si_{0.02}Al_{0.02})_{\Sigma 2.00}O_6$ [27] or $(RE_{0.58}Ca_{0.30}Th_{0.03})_{\Sigma 0.91}(Nb_{1.07}Ti_{0.81}Fe_{0.12}^{3+}$-$Si_{0.02}Al_{0.02})_{\Sigma 2.04}(O_{5.70}(OH)_{0.16})_{\Sigma 5.86}$ [28].

Corresponding with a wide diadochic replacement that occurs in aeschynite and related minerals (refer to the foregoing paragraphs), there are **variations in composition for samples from an individual deposit, or between** samples from **different locations**, as is exemplified in the following paragraphs. Partly, these variations (expressed by indicative ratios between group-A and -B elements) were correlated with the physico-chemical conditions of formation, see p. 203.

As follows from electron microprobe profiles, the element distribution within single crystals of aeschynite-(Y, Gd, Dy, Er) from the central Swiss Alpes is very inhomogeneous, but shows no zonal arrangement. There is a tendency for RE to be separated in regions enriched predominantly in light or heavy RE. Also visible is a regularity for isomorphous substitution within the group-B elements and, partly, within group-A elements. Microprobe analysis for three crystals gives (as a mean for 15 to 20 measurements) the following contents [3, pp. 82/3, 87/8]:

crystal	content in %											
	Th	Y	RE	Ca	U^{4+}	Fe^{3+}	Nb	Ta	Ti	W	O*)	sum
1	2.2	14.4	14.4	0.29	4.3	0.23	14.0	1.5	20.5	2.2	27.7	101.7
2	2.7	13.1	14.08	0.26	7.3	0.28	12.3	2.5	19.8	2.3	26.8	101.4
3	1.6	14.2	15.51	0.16	3.6	0.16	11.6	1.3	21.8	0.79	27.1	97.7

*) Calculated value.

The variance in composition of aeschynite-(Nd) from Beiyun-Obo, Inner Mongolia, P.R.C., is given as follows; sample numbers are as given in the original paper [59]:

	content in % in sample number					
	G-1	1	L-e-22	3	2	C-6
ThO_2	7.72	5.79	3.65	1.33	1.28	1.08
ΣRE_2O_3	36.54	35.49	34.60	37.99	38.32	37.45
CaO	0.73	2.29	0.65	2.96	4.58	2.17
MgO	0.35	1.03	0.41	0.44	0.47	—
MnO	—	0.07	—	0.06	0.08	0.37

Table (continued)

| | content in % in sample number | | | | | |
	G-1	1	L-e-22	3	2	C-6
Fe_2O_3	1.36	1.61	2.21	1.99	1.51	1.94
Na_2O	0.07	0.14	—	0.42	0.35	—
Nb_2O_5	21.40	24.72	33.45	30.16	29.66	33.69
Ta_2O_5	1.24	0.06	0.81	0.04	0.07	0.026
TiO_2	28.65	27.18	22.23	22.46	21.73	18.97
SiO_2	0.04	0.19	0.01	0.65	0.41	—
Al_2O_3	0.64	0.15	0.59	0.54	0.19	2.42[d]
H_2O^-	0.10	—	0.16	—	—	—
H_2O^+	1.27	1.28	0.32	0.96	0.92	—
sum	100.17[a]	100.00	99.30[b]	100.00	100.00[c]	98.116

[a] Includes 0.06% U_3O_8. — [b] Includes 0.21% FeO. — [c] Includes 0.43% BaO. — [d] Includes U_3O_8.

The corresponding crystal-chemical formulas, depicting at the same time the variance in RE composition, for the above aeschynite-(Nd) samples are given as follows; note that ΣY^* comprises Y, Gd, Tb, and the remaining heavy RE (if present) [59]:

$(Ce_{0.16}Pr_{0.05}Nd_{0.40}Sm_{0.10}Eu_{0.01}Gd_{0.03}\Sigma Y^*_{0.04}Th_{0.11}Ca_{0.05}Mg_{0.03}Na_{0.01})_{\Sigma 1.00}(Ti_{1.29}Nb_{0.58}Fe_{0.06}Al_{0.05}Ta_{0.02})_{\Sigma 2.00}(O_{5.45}(OH)_{0.55})_{\Sigma 6.00}$ for G-1;

$(La_{0.02}Ce_{0.19}Pr_{0.06}Nd_{0.38}Sm_{0.08}(Eu, Gd)_{0.03}Dy_{0.02}Y_{0.02}Ca_{0.15}Th_{0.09}Mg_{0.09}Na_{0.02})_{\Sigma 1.13}(Ti_{1.23}Nb_{0.67}Fe^{3+}_{0.07}Al_{0.01}Si_{0.01})_{\Sigma 1.99}(O_{5.49}(OH)_{0.51})_{\Sigma 6.00}$ for 1;

$(La_{0.01}Ce_{0.11}Pr_{0.04}Nd_{0.35}Sm_{0.11}Eu_{0.02}Gd_{0.04}\Sigma Y^*_{0.05}Th_{0.05}Ca_{0.04}Mg_{0.04}Fe^{2+}_{0.01})_{\Sigma 0.88}(Ti_{0.97}Nb_{0.88}Fe^{3+}_{0.10}Al_{0.04}Ta_{0.01})_{\Sigma 2.00}(O_{5.81}(OH)_{0.19})_{\Sigma 6.00}$ for L-e-22;

$(La_{0.05}Ce_{0.29}Pr_{0.06}Nd_{0.30}Sm_{0.05}(Eu, Gd)_{0.04}Dy_{0.02}Y_{0.03}Ca_{0.19}Na_{0.05}Mg_{0.04}Th_{0.02})_{\Sigma 1.14}(Ti_{1.00}Nb_{0.82}Fe^{3+}_{0.09}Al_{0.04}Si_{0.04})_{\Sigma 1.99}(O_{5.62}(OH)_{0.38})_{\Sigma 6.00}$ for 3;

$(La_{0.06}Ce_{0.30}Pr_{0.07}Nd_{0.33}Sm_{0.06}(Eu, Gd)_{0.04}Dy_{0.01}Y_{0.02}Ca_{0.31}Th_{0.02}Mg_{0.04}Na_{0.04})_{\Sigma 1.30}(Ti_{1.04}Nb_{0.85}Fe^{3+}_{0.07}Si_{0.03}Al_{0.01})_{\Sigma 2.00}(O_{5.61}(OH)_{0.39})_{\Sigma 6.00}$ for 2;

and $(La_{0.04}Ce_{0.26}Pr_{0.06}Nd_{0.30}Sm_{0.07}Eu_{0.02}Gd_{0.04}\Sigma Y^*_{0.02}Ca_{0.14}Th_{0.02}Mn_{0.02})_{\Sigma 0.99}(Nb_{0.90}Ti_{0.84}Al_{0.17}Fe_{0.09})_{\Sigma 2.00}O_{6.00}$ for C-6.

Blomstrandine from the Ilmen Mountains, Southern Urals, has a much lower ThO_2 content (7.93 to 8.48%) than Y aeschynite from this region (17.42 to 20.34%) and the content of RE is much higher in blomstrandine, which leads to a uniform total sum of group-A cations. Whereas for group-B cations the sum of TiO_2 and $(Nb, Ta)_2O_5$ contents is the same in both minerals, there are differences for the ratio $Nb_2O_5:TiO_2$, that is 1:1 in Y and Ce aeschynites and varies between 1:1.4 and 1:2.2 in blomstrandine [6].

According to the relationship between the main chemical components, aeschynite from various types of pegmatite veins of the Ilmen Mountains forms two groups: (1) samples from old granitic, from feldspar and corundum-feldspar pegmatites, and from syenite; and (2) samples from late granitic non-amazonitic pegmatites. The samples from the latter group are relatively rich in Nb and Ta, and are diminished in Ti, Th, and RE. For both groups, there occurs a straight correlation between the contents of $(Nb, Ta)_2O_5$ and $\Sigma RE(RE_2O_3?)$, whereas for TiO_2 the correlation is inverse. But for the contents of ThO_2 and TiO_2, the correlation is straight for samples of group 1 and inverse for group 2. The opposite behavior occurs for ThO_2 and ΣRE (group 1 inverse, group 2 straight). For the RE composition, the differences for samples of both groups are not so distinct (for details, see original paper). A variation of the content of some elements in aeschynite samples from veins displaying the same mineralogical composition, possibly, can be related with the inhomogeneity in chemical composition observed (e.g., for Th) in the aeschynite crystals. The chemical composition of individual aeschynite samples from different types of host pegmatite are as follows; n.d. = not determined [29, pp. 20/4]:

host rock type and sample No. [a]	content in % in aeschynite						
	ThO_2	ΣRE	TiO_2	$(Nb, Ta)_2O_5$	Ta_2O_5 [b]	Fe_2O_3 [c]	CaO
old granite pegmatite							
223 L	10.06	26.25	19.82	32.53	1.6	3.55	n.d.
syenite							
11006	14.04	26.91	19.82	31.80	1.4	3.21	n.d.
feldspar pegmatite							
49	14.01	22.2	20.39	30.77	1.2	n.d.	2.37
51	15.84	24.59	18.77	32.78	n.d.	4.20	n.d.
60	13.17	23.00	17.00	36.20	1.6	n.d.	n.d.
75	13.02	26.61	19.15	32.00	1.6	4.01	n.d.
96	12.68	20.70	19.29	39.05	1.8	4.37	5.39
108	12.75	25.0	20.25	31.60	1.4	5.87	3.93
119	13.19	24.10	20.59	35.69	1.5	n.d.	3.09
127	17.80	~26	20.15	28.20	0.9	5.41	2.35
128	14.73	19.66	16.67	37.40	n.d.	4.68	n.d.
150	20.66	~20	20.17	31.86	1.4	4.97	2.27
181	12.33	28.55	22.95	32.44	1.6	2.09	n.d.
205	16.32	27.77	21.05	28.50	1.1	3.00	n.d.
205-b	17.65	25.0	24.24	27.10	2.2	n.d.	2.44
11-98	12.02	27.88	17.84	35.78	1.4	2.96	n.d.
miaskitic pegmatite							
76	9.52	26.21	20.72	33.00	1.6	2.29	n.d.
78-II	11.21	23.5	20.75	33.20	1.6	n.d.	3.12
92	16.25	26.3	20.0	30.60	0.7	n.d.	2.83
120	9.45	28.41	18.47	34.00	1.8	3.39	4.30
corundum-feldspar pegmatite							
31	25.25	19.20	20.83	27.48	1.6	n.d.	3.02
32	18.28	24.02	23.55	29.35	1.3	n.d.	2.80
362	10.09	23.45	n.d.	n.d.	1.8	n.d.	3.18
386	10.86	25.21	16.64	35.50	1.2	n.d.	4.01
late granite pegmatite							
350 l	7.29	19.00	15.25	44.00	3.6	4.22	5.51
354	5.36	22.25	15.71	n.d.	3.3	n.d.	6.91
356-II	6.79	20.60	13.97	45.00	n.d.	n.d.	3.85

Table (continued)

host rock type and sample No. [a]	content in % in aeschynite						
	ThO_2	ΣRE	TiO_2	$(Nb,Ta)_2O_5$	Ta_2O_5 [b]	Fe_2O_3 [c]	CaO
371	4.52	~20	15.16	48.39	3.7	n.d.	8.16
385-I	4.20	16.40	16.00	41.80	3.0	n.d.	5.06
399	6.79	18.87	15.50	45.32	3.3	4.48	5.60

[a] Corresponds to the vein No. Note that all numbers over 300 occur within an area of about 2×2 km and are from the northern part of the Vishnevye Gory, about 200 to 220 km apart from the southern part (covering an area of about 2×10 km); sample No. 11006 is located in between, see [29, p. 19]. — [b] Determined by quantitative spectral analysis. — [c] Total iron.

For the composition of aeschynites from different types of occurrence in phenakite-bearing biotite-chlorite lenses of the Habachtal, Austria, is stated: Aeschynite occurring as inclusions in phenakite porphyroblasts (see, e.g., analysis Nos. A4 and A5 below) is largely homogeneous and always rich in Y, the ratio U:Th is 2.15 (standard deviation for 15 analyzed points is 0.59); unimportant variations in group-A elements may indicate a weak zoning. As opposed to this, aeschynite from the biotite-chlorite groundmass (see, e.g., analyses A2 and A9) has a strong zonal structure comprising a Y-rich core with U:Th = 0.66 (standard deviation for 30 analyzed points is 0.13), followed by a zone rich in U and Th, and leading to an outer part rich in Ce. The ratio U:Th for the outer part is variable and is 0.60 for aeschynites from lens 1 and 2, and is 1.60 for that from lens 3 (for the lenses, see p. 193). Generally, the Ta content increases from core to rim. The microprobe analyses for the main elements are as follows; for the RE are given here only those that have been determined for all samples (for the others, see original paper) [30]:

	content in % in sample number and part [*]							
	A4	A5		A2		A9		
	—	c	r	c	r	c	i.r.	o.r.
ThO_2	5.13	3.63	5.71	7.01	6.74	14.6	18.4	9.75
Y_2O_3	16.0	15.0	13.4	10.6	8.24	12.4	8.6	4.27
Ce_2O_3	0.17	1.84	2.89	2.39	5.78	0.52	1.10	9.21
Gd_2O_3	1.34	1.21	1.24	1.41	1.19	1.07	1.38	1.26
CaO	0.63	1.00	1.15	2.00	2.09	1.34	0.81	2.25
U_3O_8	9.98	9.03	8.33	5.33	10.6	7.77	11.4	4.16
Fe_2O_3	0.40	0.20	0.23	0.44	0.73	0.31	0.24	0.62
Nb_2O_5	22.7	17.0	17.0	20.7	18.9	13.0	7.32	20.6
Ta_2O_5	5.55	5.95	5.52	7.53	10.5	6.75	5.74	9.95
TiO_2	25.9	27.5	27.6	26.0	26.5	29.2	31.2	22.2

[*] Core = c; rim = r; inner and outer rim = i.r. and o.r.

Whereas blomstrandines from Norway and from the Ilmen Mountains, Southern Urals, are similar in composition, there occurs a much lower content of RE and a higher content of Th and Nb in the Y aeschynite from the Ilmen Mountains [6]. By its content of ThO_2, ΣRE_2O_3, TiO_2, and CaO the Kenya lyndochite is very similar to that from the type locality, but there is an appreciable difference for the $(Nb, Ta)_2O_5$ and U_3O_8 contents [9, pp. 243/4]:

	ThO$_2$	ΣRE$_2$O$_3$	CaO	(Nb,Ta)$_2$O$_5$	TiO$_2$	U$_3$O$_8$	l.o.i.	sum
	\multicolumn{8}{c}{content in % in lyndochite from}							
Kenya	10.13	21.35	4.4	32.50	22.0	2.8	3.3	96.48
Lyndoch	10.77	21.05	3 ± 0.2	42.50	19.1	0.4	2.1	98.92

Changes in composition in group-A and -B cations (expressed by the ratios U:Th and Nb:Ta, respectively) for differently colored blomstrandine samples from pneumatolytic-hydrothermal veins, Eastern Siberia, were correlated with the physico-chemical conditions prevailing in the mineral-forming solutions. Thus, during the course of crystallization the content of Ta and Nb increases, that of Ti decreases, in the latest formed blomstrandine. But the relationship between these elements remains that characteristic of blomstrandine $((Nb, Ta)_2O_5 < TiO_2)$ and does not reach a priorite relationship of $(Nb, Ta)_2O_5 > TiO_2$, whereas the ratio Nb:Ta decreases. Simultaneously, the change in group-A cations leads to a high value for the ratio U:Th, corresponding with an increased acidity of the mineral-forming solution. The content of the main components and the indicative ratios change with color as follows for the above mentioned blomstrandines [31, pp. 45/6]:

color of blomstrandine	content in %					ratio	
	Ta$_2$O$_5$	Nb$_2$O$_5$	TiO$_2$	U	Th	Nb:Ta	U:Th
yellow with a reddish hue	1.30	19.8	37.2	4.24	3.90	13.0	1.09
reddish yellow	1.70	22.9	35.0	4.8	3.92	11.5	1.23
cinnamon	1.72	21.0	37.0	4.92	3.60	10.3	1.36
black, slightly transparent reddish	2.40	25.0	33.6	5.5	3.60	8.9	1.52

The content of Th and Ce in aeschynites from the Southern Urals is indicative of the acidity and alkalinity of the solutions that formed the host pegmatite veins: High Th content (15.21 to 24.50% ThO_2, 8.74 to 12.37% CeO_2 for 5 samples) is characteristic of acidic (quartz-feldspar) pegmatite; and high Ce content (9.96 to 15.19% CeO_2, 5.36 to 11.91% ThO_2 for 6 samples) of alkalic (feldspar and nepheline-feldspar) pegmatite [32]. For crystal-chemical formulas of two aeschynite samples (Nos. 7 and 18) from the above veins, see p. 198.

A few examples are given in Table 10, p. 206, and in the following table from [33] for the relationship between chemical composition and some physical properties of aeschynite and blomstrandine from different occurrences. The physical properties are D = density in g/cm^3;

VHN_{50} = Vickers hardness in kg/mm^2 at 50 g load; r.i. = refractive index; and R(546 nm) = reflectivity in % for the given wavelength; n.a. = not analyzed:

| | content in % and physical properties for sample | | | | |
	R26	R12	R24	R13	R23
ThO_2	16.98	7.70	5.80	5.40	2.26
U_3O_8[a]	0.62	5.40	4.40	3.80	0.25
ΣRE_2O_3	13.33[b]	27.40	24.90[c]	32.40	14.84[e]
CaO	2.55	1.00	1.70	0.70	0.99
MnO	0.20	—	0.20	—	1.26
PbO	0.40	0.90	0.05	0.80	0.55
Fe_2O_3[a]	2.56	1.30	3.33	1.50	1.24
Nb_2O_5	25.21	16.40	17.72	19.00	35.72
Ta_2O_5	1.21	4.70	4.29	2.00	—
TiO_2	22.65	29.50	31.71	31.90	25.09
H_2O^-	0.20	0.20	—	0.40	n.a.
H_2O^+	3.20	4.20	3.99	1.30	n.a.
sum	89.11	98.70	98.09[d]	99.20	82.20
D in g/cm^3	5.339	4.86	4.838	4.94	5.18
VHN_{50}	690	730	673	692	—
r.i.	2.21	2.16	2.13	2.19	2.24
R(546 nm)	13.57	13.11	14.61	13.57	14.83

Nos. 12 and 13 = Blomstrandine from Kåbuland and Morefjaer, respectively, Norway. — No. 23 = Aeschynite from Luanga Bridge, Zambia. — No. 24 = Blomstrandine from Kåbuland, Norway. — No. 26 = Aeschynite from Ilmen Mountains, Southern Urals, USSR.

[a] Oxidation state not determined. — [b] Comprises only 10.82% Ce_2O_3 and 2.51% Y_2O_3, other RE not analyzed. — [c] 21.42% as calculated from the given contents for the individual RE. — [d] Given as 99.53% in the original paper. — [e] Comprises only 13.21% Ce_2O_3 and 1.63% Y_2O_3, other RE not analyzed.

Changes in Chemical Composition During Alteration. During alteration by weathering, the fresh black priorite (aeschynite-(Y)) from the Kibara Mountains, northern Katanga, Zaire (Congo-Kinshasa), becomes an outer brown to yellow-brown zone (for the analyses, see No. P1 in Table 10, p. 206). In the fresh part of the crystal more than 85% of the group-A sites are occupied and the (primary; refer to p. 75) deficiencies are mainly due to complex substitution during crystallization at hydrothermal conditions (as is indicated by a high content of OH). The outer zone has lost, by selective leaching, about 40% of the group-A ions initially present; especially lost are the RE, Ca, and U. The apparent loss of Th is related mainly to a small change of the chemical composition from the nucleus to the external zone of the fresh part. The crystal-chemical formulas change from $(RE(Y)_{0.365}Th_{0.13}Ca_{0.115}RE(Ce)_{0.075}(Mg, Pb, Fe^{2+}, Mn)_{0.07})_{\Sigma 0.692}(Ti_{1.47}Nb_{0.385}Ta_{0.095}(Fe, Sn, Zr)_{0.05})_{\Sigma 2.00}(O_{5.58}(OH)_{0.76})_{\Sigma 6.24} \cdot 0.3 H_2O$ for the fresh part to $(RE(Y)_{0.155}Th_{0.115}Ca_{0.055}U_{0.05}RE(Ce)_{0.03}(Mg, Pb, Fe^{2+}, Mn)_{0.06})_{\Sigma 0.465}(Ti_{1.49}Nb_{0.36}Ta_{0.10}(Fe, Sn, Zr)_{0.05})_{\Sigma 2.00}(O_{4.94}(OH)_{0.90})_{\Sigma 5.84} \cdot 0.8 H_2O$ for the outer zone [34].

Primary and secondary alteration (refer to p. 75) of aeschynite and related minerals from Norway and the Urals leads to changes in composition, that, partly, behave differently for group-A and for group-B cations. Primary alteration causes for group-A elements a constant increase in Ca content; generally a decrease in the U and Th content; and a decrease in the total RE content (with the exception of sample R16, its RE composition remains constant). Group-B elements display a slight decrease. Secondary alteration produces a decrease in Ca content; an increased leaching of group-A elements; and a relative increase in group-B elements. Both types of alteration lead to an increase in structural and adsorbed water content. The behavior of U and Th is variable: In samples R1, R2, and R15, there occur slight decreases in these elements; in sample R3 they increase; and in sample R16 there is a reversal of the ratio U:Th in the altered area. Whereas previous work used group-B elements (that are assumed to be nearly unchanged during primary alteration) to calculate chemical formulas for altered material (see above), their use may be questionable, as for samples R1, R2, R3, and R15 a slight but consistent decrease in Nb_2O_5 and TiO_2 content occurs during alteration. The chemical composition of the dark, nearly opaque, unaltered areas and of characteristically honey yellow altered areas of five specimens of aeschynite/blomstrandine are given in Table 10, p. 206. Additionally given are some physical properties [35, pp. 521/7].

Relative to the unaltered aeschynite from pegmatite (pit No. 205) of the Ilmen Mountains, Southern Urals, some newly formed phases (see also p. 194) in the alteration crust are enriched in Nb, Th, Ti, Fe, Ca, Mn, Al, Si, and K. As follows from pictures of element distribution (obtained by qualitative X-ray spectral microanalysis), there are (e.g., for Th) besides un-altered aeschynite (containing 17.65% ThO_2): slightly altered aeschynite, enriched in Ca and depleted in Th; thorite, strongly enriched in Th; and fersmite, strongly enriched in Nb, Ca and depleted in Th [58].

Table 10

Chemical Composition and Physical Properties of Aeschynite and Related Minerals (Fresh and Altered) from Different Occurrences.
VHN$_{50}$ = Vickers hardness in kg/mm^2 at 50 g load; r.i. = refractive index; and R (546 nm) = reflectivity in % for the given wavelength. Altered samples are denoted by an asterisk.

sample No.	R2	R2*	R15	R15*	R1	R1*	P1	P1*	R3	R3*	R16	R16*
ThO$_2$	12.40	12.05	10.40	9.20	7.50	7.50	9.98	9.74	4.70	4.90	3.00	10.10
U$_3$O$_8$ [a]	>0.10	>0.10	4.82	4.50	3.30	3.23	7.65	4.67	6.20	6.62	10.90	8.30
ΣRE$_2$O$_3$	22.00	21.12	19.92	18.62	25.49	23.88	17.70	8.86	22.13	20.71	17.75	18.11
CaO	2.93	3.10	0.84	1.92	0.39	1.80	1.84	1.01	1.03	2.41	1.66	2.65
MnO	0.12	0.11	>0.10	>0.10	>0.03	>0.03	0.14	0.06	0.18	0.16	>0.10	>0.10
PbO	0.13	0.08	0.33	0.35	0.27	0.21	1.12	1.72	0.20	0.10	0.70	0.75
Fe$_2$O$_3$ [a]	2.40	2.31	1.81	0.75	0.71	0.76	1.31 [b]	1.39 [d]	1.20	1.61	1.40	0.50
Nb$_2$O$_5$	40.10	39.81	14.80	13.40	12.80	12.53	14.68	15.93	28.00	27.20	25.00	13.40
Ta$_2$O$_5$	0.80	0.80	1.90	1.70	4.20	4.30	5.81	7.41	5.00	5.08	4.50	3.20
TiO$_2$	15.31	14.46	37.30	34.45	37.40	37.00	33.64	39.37	27.20	26.70	28.20	33.60
H$_2$O$^-$	0.30	0.50	1.30	1.70	0.70	0.90	1.62	3.88	0.20	0.30	0.60	0.70
H$_2$O$^+$	2.10	3.50	3.30	3.70	4.70	5.10	3.90	5.36	3.20	3.60	3.40	3.80
sum	98.69	97.94	96.82	90.39	97.69	97.24	100.00 [c]	100.00 [e]	99.24	99.39	97.81	95.21
D in g/cm^3	5.06	4.61	5.01	4.17	4.82	4.79	5.06	3.60	5.13	4.61	4.97	4.46
VHN$_{50}$	645	621	727	694	665	675	—	—	683	699	672	631
r.i.	2.20	2.14	2.19	2.12	2.14	2.11	—	—	2.22	2.19	2.20	2.15
R (546 nm)	14.55	12.29	14.03	12.68	12.74	11.62	—	—	13.89	12.39	14.12	14.01

No. P1 = Priorite from Kibara Mountains, Zaire [34]. – Nos. R1 and R3 = Aeschynite from Hitterö, Norway [35]. – No. R2 = Aeschynite from Miask, Southern Urals, USSR [35, pp. 522/3]. – Nos. R15 and R16 = Blomstrandine from Hitterö and Morefjaer, respectively, Norway [35, pp. 522/3].

[a] Oxidation state not determined. – [b] Comprises 1.09% Fe$_2$O$_3$ and 0.22% FeO. – [c] Includes 0.41% MgO, 0.03% ZrO$_2$, and 0.17% SnO$_2$. Recalculated after deduction of 0.60% arsenopyrite, 0.84% kaolinite, and 0.37% bauxite. – [d] Comprises 1.14% Fe$_2$O$_3$ and 0.25% FeO. – [e] Includes 0.35% MgO, 0.05% ZrO$_2$, and 0.20% SnO$_2$. Recalculated after deduction of 0.59% arsenopyrite, 0.86% kaolinite, and 0.39% bauxite.

Table 11

Chemical Composition and Density of Aeschynite, Priorite, and Blomstrandine from Various Types of Occurrence.

n.d. = not determined; n.f. = not found; tr. = traces. For the footnotes, see p. 213.

	content in % in sample number								
	1	2	3	4	5	6	7	8	9
ThO_2	29.56	22.56	20.34	20.27	18.47	17.42	15.21	14.94	14.90
ΣRE_2O_3	15.94	19.00	18.98	19.38	26.73	14.32[i]	25.97[l]	19.84	22.40
CaO	2.73	0.95	2.53	2.56	1.64	2.65	2.60	1.84	2.20
UO_2	–	–	2.94	n.d.	–	–	–	–	–
UO_3	0.95[a]	0.87	–	–	0.40	–	–	–	3.60
FeO	–	–	1.89	–	–	1.14	–	–	–
Fe_2O_3	1.50	1.4	–	2.92	2.47	1.56	3.04	2.04	1.14
Nb_2O_5	16.15	7.4	24.23	24.97[f]	26.25[f]	25.35	28.20	6.70	26.45
Ta_2O_5	0.55	–	0.58	–	–	0.94	0.87	–	0.35
TiO_2	29.55	33.0	25.71	24.63	24.04	23.79	21.83	39.50	23.50
SiO_2	0.29	1.66	–	–	–	0.50	–	2.75	0.88
Al_2O_3	0.60	0.8	–	–	–	–	–	4.90	2.35
H_2O^-	n.f.	–	tr.	0.18	–	–	0.10	–	–
H_2O^+	2.28	–	2.78	0.54	0.55[h]	1.08[j]	–	4.32[h]	1.65[h]
others	0.30[b]	0.94[c]	0.52[e]	2.91[g]	0.15[i]	0.22[k]	2.13[m]	3.10[c]	–
sum	100.40	88.58[d]	100.50	98.36	100.70	88.97[i]	99.95	99.93	99.42
D in g/cm³	5.25	4.6	5.19	5.19	–	5.339	5.27	4.29	–

No. 1 = Thoroaesch. from microcline veinlet within the contact zone between biotite syenite and gneiss, region of pegmatite vein No. 35, Ilmen Mountains, Southern Urals, USSR [36, p. 150]; see also [37, pp. 365/6]. – No. 2 = Thoroaesch. from pegmatite vein of the exocontact area of the Transangara nepheline syenite massif, Enisei Ridge, Eastern Siberia, USSR [38]. – Nos. 3 and 4 = Y aesch. from pegmatite, Ilmen Mountains, Southern Urals, USSR [6]. – No. 5 = Aesch. from the USSR [7, table 13 between pp. 58/9]. – No. 6 = Aesch. from pegmatite vein (with slightly albitized microcline) cutting granite gneiss, Aesch. Hillock, Ilmen Mountains, Southern Urals, USSR [5]; see also [36, p. 149]. [37, pp. 365/6]. – No. 7 = Aesch. from quartz-feldspar vein (No. 181 [32]), Southern Urals, USSR [22]. – No. 8 = Thoroaesch. from a pegmatite vein of the exocontact area of the Transangara nepheline syenite massif, Enisei Ridge, Eastern Siberia, USSR [38]. – No. 9 = Aesch. from the USSR [7, table 13 between pp. 58/9]; see also [37, pp. 365/6].

Table 11 (continued)

	content in % in sample number								
	10	11	12	13	14	15	16	17	18
ThO_2	14.45	13.06	12.65	12.35	12.00	11.91	10.81	10.22	9.90
ΣRE_2O_3	22.93	29.30	33.99	28.91	19.12	29.12	26.18	29.01	29.82[l]
CaO	4.25	2.73	–	–[p]	4.30	2.81	2.32	2.04	2.37
UO_2	–	–	–	–	–	–	–	–	–
UO_3	–	0.06[a]	–	–	2.11	–	0.03	–	–
FeO	–	–	–	n.f.	–	1.62	1.60	3.29	–
Fe_2O_3	1.44	1.74	1.90	1.05	2.50	0.13	1.83	–	2.96
Nb_2O_5	32.29	23.59	26.51	29.60	29.50	31.03[l]	30.00	31.14[f]	33.53
Ta_2O_5	0.15	0.26	0.01	1.93	5.40	–	0.28	–	0.75
TiO_2	19.48	26.65	21.50	23.18	18.47	23.14	20.96	21.84	19.28
SiO_2	0.70	1.05	0.80	0.18	1.33	–	0.48	–	–
Al_2O_3	2.15	0.65	1.72	0.10	1.20	–	3.54	–	–
H_2O^-	–	n.f.	–	–	–	–	–	–	–
H_2O^+	1.05[h]	1.40	–	1.00	2.25[h]	0.63[h]	0.80[h]	1.02	2.01[u]
others	0.53[n]	–	0.49[o]	0.71[q]	0.43[r]	–	0.65[s]	1.20[t]	–
sum	99.42	100.49	99.57	99.01[l]	98.61	100.39	99.48	99.76	100.62[v]
D in g/cm³	–	5.096	5.17	5.30	4.88	–	5.239	5.23	5.33

No. 10 = Aesch. from the USSR [7, table 13 between pp. 58/9]; see also [37, pp. 365/6]. – No. 11 = Aesch. from a small nepheline-feldspar pegmatite veinlet in miaskite, Ilmen Mountains, Southern Urals, USSR [36, p. 149]; see also [37, pp. 365/6], but note that there Vishnevye Gory is given as the location. – No. 12 = Aesch. from the USSR [7, table 13 between pp. 58/9]; see also [36, p. 155], where as host rock and location is given a feldspar vein (No. 133) of the exocontact of a miaskite intrusion, Vishnevye Gory. – No. 13 = Aesch. from microcline pegmatite vein No. 133 in the exocontact of an alkali rock complex, Ilmen Mountains, Southern Urals, USSR [36, p. 149]; see also [37, pp. 365/6] and [39, pp. 52/3], but note that Vishnevye Gory is given as the locality. – Nos. 14 and 15 = Aesch. from the USSR [7, table 13 between pp. 58/9]. – No. 16 = Aesch. from the USSR [40]. – No. 17 = Aesch. from a nepheline-feldspar vein in granite gneiss, near Lake Ilmen, Southern Urals, USSR; mean values from 4 determinations, H_2O, MgO, and MnO are from 2 determinations [43]. PbO is from [37, pp. 365/6]. – No. 18 = Aesch. from nepheline-feldspar vein (No. 78 [32]), Southern Urals, USSR [22].

Table 11 (continued)

					content in % in sample number				
	19	20	21	22	23	24	25	26	27
ThO_2	8.48	6.5	6.20	5.80	5.78	5.27	5.17	4.96	4.95
ΣRE_2O_3	31.35	14.8	36.00	24.90	27.71	30.5	30.13	26.95	22.56
CaO	1.35	6.2	1.44	1.70	1.0	2.0	3.97	2.24	4.86
UO_2	–	0.4	–	4.32	1.27	0.98	–	–	0.67
UO_3	3.52[a]	–	–	0.08	2.40	1.20	1.44	4.87	0.04
FeO	–	–	–	0.34	0.53	–	–	–	0.77
Fe_2O_3	–	0.9	0.90	3.33	3.96	2.80	1.65	2.49	1.32
Nb_2O_5	21.01	18.7	24.05	17.72	14.26	14.90	30.88	27.62	41.43
Ta_2O_5	1.05	39.3	1.45	4.29	2.59	1.22	n.f.	6.20	3.84
TiO_2	29.01	11.7	28.12	31.71	35.20	36.15	23.86	20.09	16.39
SiO_2	0.01	–	–	0.28	0.24	0.24	1.45	0.59	0.07
Al_2O_3	–	–	–	0.60	0.60	–	0.50	2.04	0.13
H_2O^-	–	–	–	–	–	0.07	–	–	0.06
H_2O^+	1.08	–	1.86[h]	3.99	3.53	3.60	–	0.57	1.90
others	2.49[w]	–	–	0.47[y]	0.45[z]	1.00[aa]	–	1.71[ab]	1.22[ad]
sum	99.35	98.5	100.02	99.53	99.52	99.93	99.05	100.33[ac]	100.24
D in g/cm³	–	6.39[x]	5.16	4.838	4.786	4.70	4.81	5.08	4.909

No. 19 = Blomstr. from the Urals, USSR [41]. – No. 20 = Ta aesch.-(Y) from the Raposa pegmatite, São José do Sabugi, Paraiba, northeastern Brazil; average of five analyses [21, p. 573]. – No. 21 = Aesch. from ankerite-dolomite carbonatite vein in fenite (Kovdor [39, pp. 52/3]), Kola Peninsula, USSR [24]. – No. 22 = Blomstr. from Kåbuland, Iveland, Norway [42]; see also [37, p. 374]. – No. 23 = Blomstr. from Rasvåg, Hitterøy, Norway [42]; see also [37, p. 374]. – No. 24 = Blomstr. (brownish cinnamon) from RE pegmatite in alkali granite (Keivy), northwestern European USSR [4, pp. 55, 62]; see also [37, p. 374]. – No. 25 = Lyndochite from carbonatite (Aldan [39, pp. 52/3]), Eastern Siberia, USSR [23]. – No. 26 = Aesch. from apo-alaskite greisen, USSR [25]. – No. 27 = Lyndochite from a pegmatite vein, Lyndoch township, Renfrew county, Ontario, Canada [44].

Table 11 (continued)

	content in % in sample number								
	28	29	30	31	32	33	34	35	36
ThO_2	4.80	4.66	4.33	4.29	4.02	4.08	3.9	3.75	3.70
ΣRE_2O_3	21.90[ae]	30.54	27.87	35.74	27.32	24.02	31.93	27.03	32.36
CaO	3.95	–	–	0.75	–	5.55	3.90	5.34	2.50
UO_2	4.50	0.60[ag]	–	–	5.40[a]	–	2.30	–	–
UO_3	0.92	–	5.64	–	–	0.38	2.15	0.08	–
FeO	–	–	–	2.83	–	–	–	0.26	–
Fe_2O_3	5.00	0.87	1.10	–	1.11	2.36	1.72	1.28	1.20
Nb_2O_5	20.61	27.82	21.0	32.71	20.5	48.00	13.75	35.90	17.64
Ta_2O_5	2.04	0.09	1.40	0.37	1.39	0.45	1.55	–	3.30
TiO_2	32.48	26.20	36.40	20.57	36.80	13.78	31.95	17.10	30.10
SiO_2	0.55	2.49	0.95	0.13	1.05	–	0.60	3.84	0.13
Al_2O_3	0.60	2.34	–	0.06	–	–	0.7	3.80	4.16
H_2O^-	–	–	–	0.28	–	–	0.50	0.40	–
H_2O^+	2.40[h]	3.10[h]	2.32	0.57	2.44	1.11[h]	3.75	0.57	–
others	1.12[af]	0.06[ah]	0.05[ai]	1.08[aj]	0.05[ak]	–	0.86[am]	1.646[an]	5.75[ao]
sum	100.87	98.77	101.06	100.38	100.26	99.73[al]	99.56	100.99	100.74
D in g/cm³	–	–	–	4.90	–	5.00	4.72	4.54 to 4.82	4.9

No. 28 = Blomstr. from an undisclosed host rock and locality [45]; see also [37, p. 374]. Probably a pegmatite of the Middle Urals, refer to [46]. – No. 29 = Priorite from the USSR [7, table 7 between pp. 40/1]. – No. 30 = Blomstr. from (pneumatolytic-hydrothermal vein?) Eastern Siberia [31, p. 47]. – No. 31 = Aesch. from pegmatite, Inner Mongolia, P.R.C. [18]. – No. 32 = Blomstr. from (pneumatolytic-hydrothermal vein?) Eastern Siberia, USSR [31, p. 47]. – No. 33 = Aesch. from carbonatite, Pakkanadu, Tamil Nadu, India [47]. – No. 34 = Blomstr. (red) from RE pegmatite in alkali granite (Keivy), northwestern European USSR [4, pp. 55, 62]. – No. 35 = Lyndochite (crystalline) from the People's Republic of China [48]; see also [40]. – No. 36 = Aesch. from Bayun Obo, P.R.C. [49]; see also [37, pp. 365/6].

Table 11 (continued)

	content in % in sample number								
	37	38	39	40	41	42	43	44	45
ThO_2	3.3	3.28	3.14	2.75	2.75	2.73	2.52	2.15	1.97
ΣRE_2O_3	25.4	36.77	25.79	33.20	25.72	30.09	28.17	31.87	33.43
CaO	4.2	2.88	5.46	–	0.90	5.75	4.82	3.54	0.86
UO_2	–	–	–	0.68	3.44	0.08	–	–	n.d.
UO_3	–	0.31	0.40[a]	–	–	0.20	–	0.83	–[av]
FeO	–	–	–	–	–	–	–	6.12	–
Fe_2O_3	2.0	2.43	2.10	2.0	0.73	2.00	2.75	–	1.17
Nb_2O_5	49.4	20.17	36.83	27.62	17.62	29.23	41.41	41.13	28.56
Ta_2O_5	–	1.27	–	0.22	8.25	0.65	n.f.	0.51	–
TiO_2	15.7	15.24	18.17	26.10	33.73	21.75	18.73	12.13	39.42[av]
SiO_2	–	6.13	3.81	2.04	2.48	4.20	0.35	0.55	tr.
Al_2O_3	–	2.41	2.83	1.60	–	0.10	0.35	0.015[at]	–
H_2O^-	–	–	–	2.40	–	0.53	0.04	–	–
H_2O^+	–	6.34[h]	0.83[h]	–	2.75[h]	0.90	0.41	1.39[at]	0.08[aw]
others	–	3.04[aq]	0.47[aq]	0.60[ah]	1.10[ar]	0.83[as]	–	1.23[au]	1.34
sum	100.0	100.27	99.83	99.21	99.47	99.04	99.95	(101.465)[at]	99.85[aw]
D in g/cm³	4.20	4.378	4.9	5.03	4.95	5.04	5.1323	5.056	5.003[av]

No. 37 = Aesch. from a carbonate vein in gneiss, Ravalli county, Montana; analysis is recalculated after subtraction of impurities [50]. – No. 38 = Aesch. from pegmatite vein of the exocontact of the Transangara nepheline syenite massif, Enisei Ridge, Eastern Siberia, USSR [38]. – No. 39 = Aesch. from Korea [39, pp. 52/3]. – No. 40 = Priorite (red-brown) from albite-zircon (malacon)-riebeckite vein in albite, Eastern Siberia, USSR [26, pp. 123/4]; see also [37, p. 374]. – No. 41 = Blomstr. from pegmatite, Stanovoi Range, southern Yakutia, USSR [51]; see also [37, p. 374]. – No. 42 = Aesch.-(Ce) from northern Korea [52]. – No. 43 = Nb aesch. from (a quartz-arfvedsonite veinlet [36, p. 155]) in fenite, Vishnevye Gory, Middle Urals, USSR [27]; see also [28]. – No. 44 = Nb aesch. from the (northern [37, pp. 365/6]) People's Republic of China [53]. – No. 45 = Polycrase (= blomstr. after [4, p. 62]) from albitite (Tuva [54]), USSR [55]. Note that the same analysis is cited as for euxenite from albitite, western Sayan, in [39, pp. 54/5].

Table 11 (continued)

	content in % in sample number								
	46	47	48	49	50	51	52	53	54
ThO$_2$	1.85	1.83	1.82	1.71 [ba]	1.66	1.62	1.58	1.33	1.31
ΣRE$_2$O$_3$	36.55	18.89	25.09	27.13	26.04	26.08	26.07	28.38	22.66
CaO	0.39	7.15	0.34	0.40	3.07	3.00	3.01	4.52	4.01
UO$_2$	0.51	–	–	–	4.40 [ag]	–	–	–	n.f.
UO$_3$	0.57	–	6.06 [a]	3.09 [a]	–	4.30 [a]	5.01 [a]	0.01 [a]	0.77
FeO	–	–	–	–	–	–	–	1.09	0.77
Fe$_2$O$_3$	0.50	7.68	2.18	1.62	1.30	–	1.28	–	0.28
Nb$_2$O$_5$	34.37	45.48	22.81	24.00	29.62	28.91	28.89	47.0	17.87
Ta$_2$O$_5$	0.41	–	15.74	15.74	5.68	5.54	6.21	<0.1	32.13
TiO$_2$	21.77	10.76	19.96	20.75	21.92	21.41	20.46	20.2	18.22
SiO$_2$	0.64	–	0.64	1.18	0.55	0.54	0.38	3.51	0.25
Al$_2$O$_3$	0.64	7.37	1.45	1.28	–	2.80	1.98	n.d.	1.03
H$_2$O$^-$	–	0.07	–	–	–	–	2.50	n.d.	0.09
H$_2$O$^+$	0.52	0.89	–	–	2.46 [h]	2.40 [h]	–	n.d.	1.05
others	1.53 [ax]	0.36 [ay]	4.08 [az]	2.925 [bb]	3.27 [bc]	2.56 [bd]	2.52 [bc]	–	0.30 [bg]
sum	100.25	100.48	100.16	98.85	99.97	99.16	99.89	(106.14)	99.97
D in g/cm^3	5.08	–	–	–	–	4.957	–	5.04 [bf]	5.87

No. 46 = Priorite (yellow) from albite-zircon(malacon)-riebeckite vein in albitite, Eastern Siberia, USSR [26, pp. 123/4], see also [37, p. 374]. – No. 47 = Alumoaesch. from quartz veinlet within fenitized rocks, region of pegmatite vein No. 35, Ilmen Mountains, Southern Urals, USSR [36, p. 154]; see also [37, pp. 365/6], but note that there Vishnevye Gory is given as the location. – No. 48 = Priorite (black, opaque) from pegmatite, Eastern Siberia, USSR [17, p. 94]. – No. 49 = Priorite (cinnamon brown, semitransparent) from pegmatite, Eastern Siberia, USSR [17, p. 94]. – No. 50 = Priorite from the USSR [7, table 7 between pp. 40/1]. – No. 51 = Priorite from the USSR [56]; see also [37, p. 374]. – No. 52 = Priorite (blomstr.) from a complex pegmatite, northern Priazov'e, Ukraine, USSR [20, p. 36]; see also [19]. – No. 53 = Nioboaesch.-(Ce) (non-metamict) from a tin placer, central Alaska [1]. – No. 54 = Ta aesch. from granite pegmatite, Siberia, USSR [11, p. 116].

Footnotes for Table 11:

a) Given as U_3O_8. — b) Comprises 0.10% MgO and 0.20% F. — c) Comprises 0.08% Na_2O and 0.86% l.o.i. — d) In the original paper given as 88.68. — e) Comprises 0.21% MgO, 0.20% ZrO_2, and 0.11% PbO.

f) Includes Ta_2O_5. — g) Comprises 2.56% Na_2O, 0.22% MnO, and 0.13% MgO. — h) Total water. — i) Comprises 0.08% MgO and 0.07% PbO. — j) Note that for this analysis is given in [37, pp. 365/6]: 10.84% CeO_2, 13.43% ΣCe_2O_3, 0.89% ΣY_2O_3, and 0.94% H_2O^- and 0.14% H_2O^+. This leads to a sum of 99.81 against 99.86 given in the original paper.

k) Comprises 0.10% MgO and 0.12% PbO. — l) Includes CeO_2. — m) Comprises 0.17% MgO, 0.21% MnO, 0.24% PbO, 0.12% SnO_2, and 1.39% l.o.i. — n) Comprises 0.26% MgO and 0.27% ZrO_2. — o) Comprises 0.30% ZrO_2, 0.11% SnO_2, and 0.08% WO_3.

p) 1.77% CaO is given in [37, pp. 365/6] and [39, pp. 52/3], which leads to a sum of 100.78 against 100.07 given in the original paper. — q) Comprises 0.22% PbO, 0.30% ZrO_2, 0.11% SnO_2, and 0.08% WO_3. — r) Comprises 0.15% MgO and 0.28% PbO. — s) Comprises 0.12% MnO, 0.14% Na_2O, 0.14% K_2O, 0.07% PbO, and 0.18% F. — t) Comprises 0.13% MnO, 0.08% MgO, and 0.99% PbO.

u) Comprises 0.18% MgO, 0.31% MnO, 0.15% PbO, 0.13% SnO_2, and 1.24% l.o.i. — v) In the original paper given as 100.77. — w) Comprises 0.21% MnO, 0.10% MgO, 0.03% PbO, 0.07% SnO_2, and 0.04% P_2O_5. — x) As is calculated from the chemical analysis and the unit cell data. Measured values range from 5.73 to 6.13 g/cm^3 and may reasonably be attributed to an unidentified mineral, forming veinlets in the Ta aeschynite, and to the metamict state of the Ta aeschynite [21, p. 572]. — y) Comprises 0.20% MnO, 0.06% MgO, 0.05% PbO, 0.005% SnO_2, and 0.11% P_2O_5.

z) Comprises 0.21% MnO, 0.10% MgO, 0.03% PbO, 0.07% SnO_2, and 0.04% P_2O_5. — aa) Comprises 0.73% PbO and 0.27% ZrO_2. — ab) Comprises 0.68% MnO, 0.89% Na_2O, 0.14% K_2O, and (not contained in the sum) 0.50% F. — ac) In the original paper given as 100.05. — ad) Comprises 0.59% MnO, 0.34% PbO, 0.13% MgO, 0.12% SnO_2, and 0.04% ZrO_2.

ae) Only Y_2O_3. — af) Comprises 0.14% MgO and 0.98% PbO. — ag) Comprises $UO_2 + UO_3$. — ah) Only ZrO_2. — ai) Comprises 0.02% PbO and 0.03% ZrO_2.

aj) Comprises 0.58% MgO, 0.42% Na_2O, and 0.08% K_2O. — ak) Comprises 0.03% PbO and 0.02% ZrO_2. — al) In the original paper given as 100.10. — am) Comprises 0.39% PbO, 0.32% MgO, and 0.15% MnO. — an) Comprises 0.56% MgO, 0.063% Na_2O, 0.063% K_2O, 0.39% P_2O_5, 0.15% F, and 0.42% l.o.i.

ao) Comprises 0.26% Na_2O, 0.23% K_2O, 1.60% F, and 3.66% l.o.i. — ap) Comprises 0.42% MnO and 2.62% ZrO_2. — aq) Only MgO. — ar) Comprises 0.69% PbO, 0.10% MgO, 0.15% SrO_2, 0.08% SnO_2, and 0.08% BeO. — as) Comprises 0.73% MgO and 0.10% PbO.

at) Note that in [37, pp. 365/6] is given for this analysis 0.13% Al_2O_3, 0.10% H_2O^-, and 0.64% H_2O^+; sum is given as 99.77. — au) Comprises 0.05% MgO, 0.39% P_2O_5, 0.15% F, and 0.64% l.o.i. — av) Given as 0.08% U_3O_8, 34.24% TiO_2, 0.02% MnO, and 5.05 g/cm^3 in [57]. — aw) Given as 0.10% H_2O^- and 99.87% in [54]. — ax) Comprises 0.15% MgO, 0.27% K_2O, 0.21% Na_2O, and 0.90% F.

ay) Only MnO. — az) Comprises 0.41% MnO, 0.65% WO_3, 0.19% SnO_2, and 2.83% l.o.i. — ba) Two other cinnamon brown samples contain 1.97 and 2.06% ThO_2 [17, p. 95]. — bb) Comprises 0.35% MnO, 0.04% MgO, 0.45% WO_3, 0.095% SnO_2, and 1.99% l.o.i. — bc) Comprises 0.56% MgO, 2.06% PbO, and 0.65% K_2O.

bd) Comprises 0.55% MgO and 2.01% PbO. — be) Comprises 0.53% MnO and 1.99% PbO. — bf) Average of seven determinations ranging from 4.99 to 5.11 g/cm^3. — bg) Only SrO.

214

References for 2.2.2.2.2:

[1] Rosenblum, S.; Moiser, E. L. (Am. Mineralogist **60** [1975] 309/15, 311/2).
[2] Ewing, R. C.; Ehlmann, A. J. (Can. Mineralogist **14** [1976] 111/9, 111).
[3] Sommerauer, J.; Weber, L. (Schweiz. Mineral. Petrogr. Mitt. **52** [1972] 75/91).
[4] Lunts, A. (Izv. Akad. Nauk Latv. SSR **1965** No. 2, pp. 51/67).
[5] Borneman-Starinkevitch (Compt. Rend. Acad. Sci. URSS [2] **30** [1941] 234/7).
[6] Makarochkin, B. A.; Es'kova, E. M.; Gonibesova, K. T. (Tr. Inst. Mineral. Geokhim. Kristallokhim. Redk. Elem. No. 3 [1959] 145/50, 148/9).
[7] Gorzhevskaya, S. A.; Sidorenko, G. A. (Geol. Mestorozhd. Redk. Elem. No. 23 [1964] 1/116).
[8] Palache, C.; Berman, H.; Frondel, C. (Dana's System of Mineralogy, 7th Ed., Vol. 1, Wiley, New York 1946, pp. 1/834, 795).
[9] Horne, J. E. T.; Butler, J. R. (Mineral. Mag. **34** [1965] 237/48).
[10] Aleksandrov, V. B.; Es'kova, E. M.; Zhabin, A. G. (in: Vlasov, K. A., Geochemistry and Mineralogy of Rare Elements and Genetic Types of Their Deposits, Vol. 2, Israel Program Sci. Transl., Jerusalem 1966, pp. 489/96, 490/1 [Russian original, Moscow 1964]).

[11] Kornetova, V. A.; Aleksandrov, V. B.; Kazakova, M. E. (Tr. Mineral. Muzeya Akad. Nauk SSSR No. 14 [1963] 108/21).
[12] Fleischer, M. (Mineral. Mag. **35** [1966] 801/9, 804/7).
[13] Gorzhevskaya, S. A.; Sidorenko, G. A. (Mineral'n. Syr'e Vses. Nauchn. Issled. Inst. Mineral'n. Syr'ya No. 8 [1963] 48/57).
[14] Butler, J. R. (Econ. Geol. **56** [1961] 442/4).
[15] Aleksandrov, V. B. (Dokl. Akad. Nauk SSSR **153** [1963] 672/5; Dokl. Acad. Sci. USSR Earth Sci. Sect. **153** [1963] 129/31).
[16] Schröcke, H. (Neues Jahrb. Mineral. Abhandl. **106** [1966] 1/54, 11).
[17] Fel'dman, L. G.; Konopleva, I. B. (Tr. Mineral. Muzeya Akad. Nauk SSSR No. 19 [1969] 86/112).
[18] Chzhan, Pei-Shan' [Chang, P. S.] (Ti Chih Lun P'ing **18** [1958] 360/4, 383 from Ref. Zh. Geol. **1960** No. 12751).
[19] Litovchenko, E. I.; Osadochii, V. K. (in: Mitskevich, B. F.; Shcherbak, N. P.; Aktsessornye Mineraly Ukrainskogo Shchita, Naukova Dumka, Kiev 1976, pp. 112/3; C.A. **88** [1978] No. 64156).
[20] Litovchenko, E. I.; Kuts, V. P. (Mineral. Sb. [Lvov] **29** No. 4 [1975] 32/41).

[21] Adusimili, M. S.; Kieft, C.; Burke, E. A. J. (Mineral. Mag. **39** [1973/74] 571/6).
[22] Makarochkin, B. A. (Geol. Geofiz. **1979** No. 1, pp. 150/2).
[23] Zdorik, T. B.; Finyagina, N. K. (Tr. Mineral. Muzeya Akad. Nauk SSSR No. 17 [1966] 209/15, 210/1).
[24] Kapustin, Yu. L. (Mineralogiya Karbonatitov, Nauka, Moscow 1971, pp. 1/286, 147/8).
[25] Podol'skii, A. M.; Golikov, I. V.; Burova, T. A. (Tr. Mineral. Muzeya Akad. Nauk SSSR No. 16 [1965] 175/86, 178/9).
[26] Kostin, N. E.; Petrova, E. A.; Sidorenko, G. A.; Skorobogatova, N. V. (Mineral'n. Syr'e No. 1 [1960] 121/8; C.A. **56** [1962] 3143).
[27] Zhabin, A. G.; Mukhitdinov, G. N.; Kazakova, M. E. (Tr. Inst. Mineral. Geokhim. Kristallokhim. Redk. Elem. No. 4 [1960] 51/72, 67).
[28] Zhabin, A. G.; Aleksandrov, V. B.; Kazakova, M. E. (Tr. Inst. Mineral. Geokhim. Kristallokhim. Redk. Elem. No. 7 [1961] 108/12; C.A. **56** [1962] 5656).

[29] Popova, V. I.; Bazhenova, L. F. (in: Popov, V. A.; Chesnokov, B. V., Mineralogiches-kie Issledovaniya v Il'menskom Zapovednike, Ural. Nauchn. Tsentr, Sverdlovsk 1981, pp. 18/29; C.A. **98** [1983] No. 146714).

[30] Grundmann, G.; Koller, F. (Neues Jahrb. Mineral. Abhandl. **135** [1979] 36/47, 41/4).

[31] Romanov, I. A. (in: Tatarinov, P. M.; Rundkvist, D. V., Mineraly i Paragenezisy Mine-ralov Endogennykh Mestorozhdenii, Nauka, Leningrad 1975, pp. 43/9; C.A. **85** [1976] No. 80929).

[32] Makarochkin, B. A. (Geol. Geofiz. No. 6 [1981] 150/1; C.A. **95** [1981] No. 100651).

[33] Ewing, R. C.; Ehlmann, A. J. (Can. Mineralogist **13** [1975] 1/7, 2).

[34] Van Wambeke, L. (Neues Jahrb. Mineral. Abhandl. **112** [1969/70] 117/49, 127/8, 140).

[35] Ewing, R. C. (Geochim. Cosmochim. Acta **39** [1975] 521/30).

[36] Es'kova, E. M.; Zhabin, A. G.; Mukhitdinov, G. N. (Mineralogiya i Geokhimiya Redkikh Elementov Vishnevykh Gor, Nauka, Moscow 1964, pp. 1/318).

[37] Chukhrov, F. V.; Bonshtedt-Kupletskaya, E. M. (Mineraly Spravochnik, Vol. 2, Pt. 3, Nauka, Moscow 1967, pp. 1/676).

[38] Sveshnikova, E. V.; Semenov, E. I.; Khomyakov, A. P. (Zaangarskii Shchelochnoi Massiv, Ego Porody i Mineraly, Nauka, Moscow 1976, pp. 1/80, 29).

[39] Semenov, E. I. (Tipochimizm Mineralov Shchelochnykh Massivov, Nedra, Moscow 1977, pp. 1/119, 52/3).

[40] Gorzhevskaya, S. A.; Sidorenko, G. A. (Mineral'n. Syr'e No. 8 [1963] 48/57, 50/1).

[41] Chernik, G. P. (Izv. Akad. Nauk Ser. 6 **1912** 6 from [6]).

[42] Hongslo, T.; Langmyhr, F. J. (Norsk Geol. Tidsskr. **40** [1960] 157/64, 160).

[43] Vladimirova, M. E. (Tr. Radievogo Inst. Akad. Nauk SSSR **6** [1957] 139/66, 149).

[44] Ellsworth, H. V. (Am. Mineralogist **12** [1927] 212/8, 215).

[45] Komkov, A. I. (Dokl. Akad. Nauk SSSR **126** [1959] 641/4; Dokl. Acad. Sci. USSR Earth Sci. Sect. **124/129** [1959] 584/6).

[46] Nefedov, E. I. (Vses. Nauchn. Issled. Geol. Inst. Inform. Sb. No. 3 [1956] 82/5).

[47] Semenov, E. I.; Upendran, R.; Subramanian, V. (J. Geol. Soc. India **19** [1978] 550/7, 551).

[48] Gorzhevskaya, S. A.; Sidorenko, G. A. (Dokl. Akad. Nauk SSSR **146** [1962] 1176/8; Dokl. Acad. Sci. USSR Earth Sci. Sect. **146** [1962] 131/3).

[49] Semenov, E. I. (Mineralogiya Redkikh Zemel', Moscow 1963, pp. 1/412, 81).

[50] Heinrich, E. W.; Levinson, A. A. (Am. Mineralogist **46** [1961] 1424/47, 1436).

[51] Shaposhnikov, G. N. (Zap. Vses. Mineral. Obshch. **89** [1960] 709/10).

[52] Komkov, A. I. (Tr. Vses. Nauchn. Issled. Geol. Inst. [2] **96** [1963] 213/25, 214, 217).

[53] Chang, Pei-Shan (Sci. Sinica [Peking] **11** [1962] 969/76 from C.A. **58** [1963] 2274).

[54] Saltykova, V. S. (Tr. Inst. Mineral. Geokhim. Kristallokhim. Redk. Elem. No. 2 [1959] 189/208, 195).

[55] Aleksandrov, V. B. (Tr. Inst. Mineral. Geokhim. Kristallokhim. Redk. Elem. No. 1 [1957] 70/6, 71, 75).

[56] Soboleva, M. V.; Pudovkina, I. E. (AEC-tr-4487 [1957] 1/455, 363; N.S.A. **15** [1961] No. 23739).

[57] Proshchenko, E. K. (Mineral. Pegmatitov Gidroterm. Shchelochnykh Massivov, Akad. Nauk SSSR, Inst. Mineral. Geokhim. Kristallokhim. Redk. Elem. **1967** 103/36, 132).

[58] Popova, V. I. (Materialy po Mineralogii Mestorozhdenii Urala, Sverdlovsk 1984, pp. 12/8).

[59] Zhang, Peishan; Tao, Kejie (Dizhi Kexue **1982** No. 4, pp. 424/8; C.A. **98** [1983] No. 164080; Ref. Zh. Geol. **1983** No. 8V240).

2.2.2.2.3 Crystal Form and Structure

Aeschynite/priorite is orthorhombic dipyramidal, axial ratio a:b:c = 0.4867:1:0.6737 [1], [2, pp. 362/3], or 0.48665:1:0.67366 [3]. The morphological axial ratio 0.4873:1:0.6734 of Y aeschynite (from syenite pegmatite, Ilmen Mountains, Southern Urals) is similar to that for normal aeschynite of this area (0.48665:1:0.67366), but differs clearly from the axial ratio of blomstrandine (or priorite [2, pp. 371/2]) with 0.4746:1:0.6673 [4]. Aeschynite has an axial ratio 0.486:1:0.682 [38]. For additional axial ratios, refer to pp. 119/20.

Observed **crystal forms**, with their symbols, for aeschynite are c {001}, b {010}, n {130}, r {120}, m {110}, x {021}, d {101}, p {111} [1], and l {160} for Ta aeschynite. The characteristic angles between individual faces of aeschynite are 51°54' for (110)/(1$\bar{1}$0), 73°10' for (021)/(02$\bar{1}$), and 69°26' for (021)/(110) [2, pp. 362/3]. Additionally to the forms given for aeschynite, priorite shows a {100}, t {140}, u {023}, v {045}, z {658} (doubtful after [2, pp. 371/2]), π {121} [1], l {210} *), w {011}, e {120} **), and f {502}. The characteristic angles between individual faces of priorite are 50°54' for (110)/(1$\bar{1}$0), 26°42' for (210)/(2$\bar{1}$0); 73°41' for (021)/(02$\bar{1}$), 70°10' for (130)/ ($\bar{1}$30), 67°26' for (011)/(0$\bar{1}$1), and 70°51' for (101)/(10$\bar{1}$) [2, pp. 371/2].

The commonest forms of aeschynite/priorite are bcmnx (the latter is prominent on aeschynite) [1], or bamx [3], [5, pp. 70/2]. Crystal habit is prismatic to short prismatic along the c axis; less often prismatic along the a axis (priorite); sometimes tabular parallel to (010) (especially Ta aeschynite [2, pp. 362/3]). And there is a striation parallel to [100] on {010} [1]; see also, aeschynite mostly displays a vertical striation on the face (110) [2, pp. 362/3] and priorite a horizontal one on (010) [2, pp. 371/2]. Prismatic lyndochite crystals show the faces (010), (110), (120), and occasionally the narrow face (140) is present. As a rule, these crystals are bent in the direction of the elongation, and a distinct striation occurs parallel to the elongation [5, pp. 70/2].

There are distinct differences in crystal morphology and habit for aeschynite and priorite (refer also to the figures given in the original paper), that both mostly form imperfect crystals with dull faces: Aeschynite usually displays the forms {110} and {021}, sometimes also {010}, and there are subparallel intergrowths between individual crystals [2, pp. 362/3]. Priorite displays {001} and {010} in more or less equal size, the former may also occur as a narrow face or is absent. Usually, the crystals are stretched parallel to the a axis; shortening parallel to the c axis occurs rarely; and sometimes crystals are shortened parallel to the b axis and somewhat stretched parallel to the c axis [2, pp. 371/2].

For individual occurrences of aeschynite and related minerals, the following crystal morphologies and habits were observed:

Aeschynite crystals, pseudomorphous after brannerite (refer to p. 87), from granite of the Borovskii pluton, northern Kazakhstan, clearly display the forms {100}, {110}, {010}, {021}, and {111} [6].

Variously colored aeschynite crystals from different pegmatite types of the Ilmen Mountains, Southern Urals, display a columnar crystal habit (mostly stretched parallel to the c axis) that is dominated only by the form {110}, or by {110} and {010}, sometimes together with {130}. Subordinate forms are {120}, {021}, and {111}; rarely {100}, {101}, and {350} occur. The ratio between length and cross section for aeschynite is different for the type of host pegmatite and is about 2:1 for samples from old granite pegmatite; between 3:1 and 4:1 for samples from feldspar, miaskitic, and corundum-feldspar pegmatites; or between 2:1 and

---

*) Note that the same letter is given above for another form of aeschynite. — **) Note that another letter is given above for the same form of aeschynite.

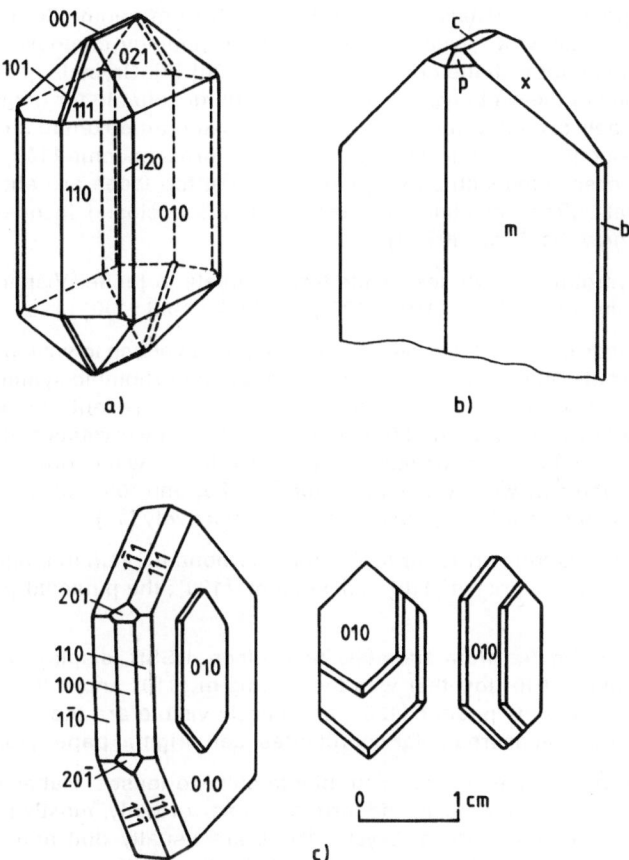

Fig. 9. Crystal forms of aeschynite and blomstrandine: a = Aeschynite from apo-alaskite greisen, USSR, after [15, p. 117]; b = Y aeschynite from syenite pegmatite, Ilmen Mountains, Southern Urals, after [4, p. 146]; c = Blomstrandine pseudomorphous after polycrase from RE pegmatite (veins No. 13 and 14 of Keivy), northwestern European USSR, after [21, p. 54].

1.5:1 for samples from late granite (inclusive amazonitic) pegmatite [7, p. 19]. A crystal of blomstrandine, platy parallel to (010), from pegmatite pit No. 205 of the Ilmen Mountains has {010} as the main form; less developed are {110} and {120}; and {021} and {111} are only very small [7, p. 28]. For blomstrandine from pegmatite of the northwestern European USSR, see pp. 218/9.

The imperfect prismatic crystals of Y aeschynite (see **Fig. 9**b) from syenite pegmatite, Ilmen Mountains, usually are flattened parallel to the a axis and, mostly, are composed of the prisms m and x, a second pinacoid b, and the dipyramid p; the latter both forms are only weakly developed. Very rarely the base c occurs. Out of five crystals measured by goniometer, only two display top faces. The measured polar coordinates (see original paper) for the above forms are identical to those for normal aeschynite (given in [1, pp. 793/4]) [4].

Idiomorphic prismatic crystals of Ta aeschynite-(Y) from pegmatite, Borborema region, northeastern Brazil, show a typical orthorhombic morphology composed of the pinacoids {010} and {001} together with the prisms {110} and {130} [8]. Crystals of Ta aeschynite from albitized pegmatite of the Na-Li type, Siberia, although unsuitable for goniometric measurement, reveal as their crystal forms {010} and {160}. Using an euxenite setting of the crystal (that better corresponds to chemical composition), the forms become {100} and {110}. Except one case (when associated with beryl, refer to p. 191), top faces are absent for the above mentioned crystals. Fresh crystals with lustrous faces enclosed in quartz also occur as subparallel intergrowths [9, pp. 109/11].

Crystals of hydrothermal Nb aeschynite from veinlets in fenite, Vishnevye Gory, Middle Urals [10], show the forms {010}, {110}, {120}, and only small {140} [11].

Excellent crystals of lyndochite from the type locality (Lyndoch township, Ontario, Canada) display rough and distorted faces, may have only a pseudo-rhombic symmetry, and owing to parallel intergrowths and twinning, are composite. The forms present were identified as {100}, {010}, {110}, {310} or {410} or both, {201}, and {111} [12]. Tiny crystals of dark red lyndochite (refer to p. 192) from a metasomatic rock, Southern Urals, show the forms {010}, {100}, {111}, and {110}; rarely, the only weakly developed forms {120} and {001} occur. The ratio between thickness, width, and length for crystals is 1:2:5, respectively [13].

Aeschynite crystals from an ankerite-dolomite carbonatite vein in fenite, Kola Peninsula, display small faces of the prisms {110} and, seldom, {120}; the pinacoid {010} is only poorly developed [14].

Aeschynite (see Fig. 9a) from apo-alaskitic greisen, USSR, displays an especially well-developed pinacoid b {010} together with the prisms m {110} and x {021}; smaller are the bipyramide p {111} and the prism r {120}; and hardly visible are the basis c {001} and the prism d {101}. For the measured polar coordinates, see original paper [15, pp. 177/8].

Orthorhombic aeschynite crystals from biotite-chlorite lenses, Habachtal, Austria, often are stretched parallel to their c axis and form columnar or platy, mostly parallel-intergrown aggregates with a striated surface. Crystal faces are usually dull and often furrowed by intergrown biotite-chlorite leaflets. Isolated crystals display the planes (100), (110), (010), (001), and (250); rarely (130) and (021) occur [16].

For aeschynite-(Y) rich in Gd and Dy from an U-Th ore-bearing biotite granite, People's Republic of China, as crystal forms were identified {100}, {010}, {110}, {310}, {001}, {021}, {111}, {211}, {311}, and {702} [17]. Orthorhombic crystals of aeschynite-(Y, Gd, Dy, Er) from the central Swiss Alps display the forms {010}, {001}, {110}, {210}, and {111}, which do not always occur together. Some crystals are platy parallel to (010); others are prismatic parallel to [001], often with a strongly developed (021) face [18, pp. 77/9].

Some crystals of a niobate of Y, probably priorite, from albitized nepheline syenite, Mugodzhary, Kazakhstan, clearly display rhombic prism and dipyramide together with a pinacoid face [19]. A priorite crystal, platy parallel to (010), from stanniferous veins, Kibara Mountains, Katanga, shows the forms {010}, {110}, {210}, and {011} [20].

The crystal morphology of blomstrandine from RE pegmatites of the northwestern European USSR (Keivy) varies with spatial distribution within a vein, but also between veins of different type (refer also to p. 189). Thus, blomstrandine from the zircon-blomstrandine-type vein No. 5 displays four crystal morphologies: (1) Short prismatic habit, 0.7 to 1.5 cm long and 0.8 to 1 cm in diameter. (2) Prismatic, or stretched-prismatic habit, 2.5 to 3 cm long and 1.8 to 4 cm in diameter; rarely, a columnar-chisel type habit occurs. (3) Thick- and thin-tabular habit, rarely thin-platy, and flattened parallel to the 2nd pinacoid (010); the plates are

1 to 4 mm thick and have an areal extension between 1.5×1 cm and 2.5×3 cm. (4) Needle-like habit, 0.4 to 1.5 cm long and 0.1 to 0.4 cm in diameter; sometimes, the needles form radiate and radial-radiating aggregates ("blomstrandine suns"). Within vein No. 5, the crystal morphologies (1) and (2) occur, in the main, in non-altered and in albitized parts, whereas (4) is restricted to the southern, strongly silicified part (where quartz extensively is broken in pieces). The country rock of vein No. 5, an alkali granite, contains skeletal, dendritic blomstrandine in its albitized part (with sugary albite), and the crystal morphology (3) is restricted exclusively to parts with a pegmatoid structure. Blomstrandine (containing 0.6 to 2.10% ThO_2) from the abukumalite-fergusonite-type veins No. 13 and 14 is distinctly different and forms in the main thick- and thin-tabular crystals (see Fig. 9c, right part, p. 217) and prismatic crystals stretched parallel to the c axis (see Fig. 9c, left part) that display the forms {010}, {100}, {110}, and {111}; {201} rarely occurs. Sometimes, minute tables and lamellas are grown upon the face (010). Thermal and X-ray diffraction data (see original paper) show this blomstrandine to be a pseudomorph (paramorph) after earlier polycrase [21, p. 52/4, 61/2].

Cleavage was observed for aeschynite in traces parallel to {100} [1], or is imperfect parallel to (100) for Ta aeschynite [9, p. 114]. Aeschynite (lyndochite from Korea) displays, under the microscope, a cleavage along the length of the crystal [22]. Crystalline dark red lyndochite (from alkaline metasomatite of the Urals, USSR) has a perfect cleavage parallel to (010); parallel to (100) it is less distinct and irregular [13]. No cleavage is observed for priorite [2, pp. 371/2].

Crystal Structure. Minerals of the aeschynite-priorite series have space group D_{2h}^{16} – Pbnm (No. 62), $Z = 4$ [23]; see also [2, pp. 361/2, 372], [3, p. 491], [24]. Aeschynite has space group D_{2h} – mmm [38]. Mostly, these minerals are metamict and X-ray amorphous; cell parameters can be obtained only after heat treatment [2, pp. 361/2, 372], [24], and are given as a = 5.42, b = 10.97, c = 7.55 kX for (synthetic, see [25]) aeschynite and as a = 5.18, b = 10.84 to 10.91, c = 7.34 to 7.38 kX for priorite/blomstrandine [23]. But note that [26], based on synthetic material, gives for minerals of the aeschynite-priorite series (the second decimal place may vary) a = ~5.0x, b = 10.0x, and c = 7.0x Å as characteristic lattice constants. The lattice parameters of aeschynite/lyndochite are a = 5.26 to 5.41, b = 10.86 to 11.03, c = 7.41 to 7.51 kX [5, pp. 57, 70], see also [27], and those of priorite/blomstrandine are similar with a = 5.14 to 5.18, b = 10.85 to 11.04, and c = 7.32 to 7.39 kX [5, p. 49].

In addition to the lattice parameters and their corresponding axial ratios given (partly also for Th-poor samples) in [2, pp. 361/2, 372], the following data for Th-containing aeschynite minerals are given in the literature:

Unheated, partly metamict aeschynite-(Y, Gd, Dy, Er) from the central Swiss Alps has a = 5.18 ± 0.04, b = 11.04 ± 0.04, and c = 7.44 ± 0.04 Å, that correspond well with data for synthetic material. The axial ratio is 0.469:1:0.674 [18, pp. 82, 86/7]. Without heat treatment, crystalline lyndochite from carbonatite, Eastern Siberia, has a = 5.26, b = 10.74, and c = 7.36 kX. Its noticeably lowered lattice constants, especially for b, are caused by the raised contents of Ti (23.56% TiO_2) and Th (5.17% ThO_2). The former influences the dimension of the B polyhedron, the latter that of the A polyhedron (refer to pp. 222/3) [28]. Non-metamict nioboaeschynite-(Ce) from Alaska has a = 5.396 ± 0.001, b = 11.085 ± 0.02, c = 7.585 ± 0.003 Å and the cell volume is 453.7 ± 0.1 Å^3. After heating at 800°C for 1 h, these values changed to a = 5.342 ± 0.02, b = 11.007 ± 0.003, c = 7.508 ± 0.002 Å and 441.4 ± 0.2 Å^3 [29].

After heat treatment at 600°C, anisotropic aeschynite from Northern Korea has a = 5.32 ± 0.02, b = 10.94 ± 0.03, and c = 7.48 ± 0.02 kX [30, pp. 214, 218, 221]. Heated at 580°C for 30 min, blomstrandine (probably from the Middle Urals) has a = 5.16, b = 10.93, and c = 7.38 kX; axial ratio is 0.472:1:0.674 [25]. And partially metamict blomstrandines (with different color) from pneumatolytic-hydrothermal veins, Eastern Siberia, recrystallize at 720°C uniformly to a mineral that gives X-ray diffraction data (see original paper) corresponding to priorite-blomstrandine with a = 5.16 ± 0.01, b = 10.93 ± 0.01, and c = 7.37 ± 0.01 Å [31]. After heating at 1100°C, Ta aeschynite from the Ilmen Mountains, Southern Urals, has a = 5.308, b = 10.983, c = 7.463 Å and the axial ratio is 0.4833:1:0.6795. These values correspond to an intermediate member between normal aeschynite (Th-poor) and Y aeschynite (containing ~20% ThO₂, see [4]) from the area in discussion. The Y aeschynite has a = 5.285, b = 10.975, c = 7.443 Å and 0.4815:1:0.6782 [9, pp. 112/4]. After heating at 600 to 700°C, crystalline lyndochite from the People's Republic of China has a = 5.33 ± 0.02, b = 10.97 ± 0.05, and c = 7.50 ± 0.02 Å [22]. Heated in air at 1300°C, Ta aeschynite-(Y) from the Borborema region, Brazil, has a = 5.34 ± 0.02, b = 10.97 ± 0.02, c = 7.38 ± 0.02 Å and the axial ratio is 0.487:1:0.673 [8].

Compared with normal aeschynite, the lattice parameters are something greater for samples enriched in Nb (see also table of the following paragraph), whereas for Th aeschynite the parameters a and c are lower (the former to a noticeable degree, the latter only insignificantly) [2, p. 362]. For example, with increasing content of Nb₂O₅, the parameter a increases from about 5.30 to 5.41 kX, and c from about 7.40 to 7.47 kX for aeschynites [5, p. 56]. And compared with synthetic (ideal) aeschynite devoid of Th and Ca (given in [25]), replacement of a part of Ce by Th and Ca in aeschynite from the Urals (with Th ≫ Ca) and from Northern Korea (with Ca > Th) caused a decrease in the lattice parameters a and c from 5.42 to 5.34 and 5.32 kX and from 7.55 to 7.50 and 7.48 kX, respectively. The parameter b varies only slightly (10.97, 10.98 and 10.94 kX) [30, pp. 217/20]. As apposed to this, variations in content of Th and Ca apparently do not seem to influence lattice parameters [5, p. 56].

The relationship between lattice parameters and content of main elements and Th for aeschynites from the Vishnevye Gory, Middle Urals, is given as follows; for the host rocks, see below [34]:

	aeschynite (aesch.) sample number						
	1	2	3	4	5	6	7
a in Å	5.405	5.352	5.355	5.38	5.316	5.32	5.317
b in Å	11.145	11.01	11.01	11.08	10.98	10.99	10.98
c in Å	7.581	7.051	7.516	7.56	7.477	7.49	7.478
% Nb₂O₅	52.22	45.42	41.41	38.70	25.61	25.35	23.59
% TiO₂	15.61	10.76	18.73	22.23	21.50	23.79	26.65
% ΣRE₂O₃	23.58	18.89	28.17	31.93	33.99	13.43	29.30
% ThO₂	0.86	1.83	2.52	0.78[*]	12.65	17.42	13.06

[*] Or 0.72% as is given on p. 157 of the original paper.

No. 1 = Aesch. (anisotropic) from albitite, region of vein No. 125. − No. 2 = Alumoaesch. from columbite-quartz vein in fenite, region of vein No. 35. − No. 3 = Nb aesch. (see [10]) from quartz-arfvedsonite veinlet in fenite. − No. 4 = Aesch. (anisotropic) from quartz-potassium feldspar veinlet in fenite, region of vein No. 125. − No. 5 = Aesch. from feldspar vein No. 133 of the exocontact of a miaskite intrusion. − No. 6 = Aesch. from feldspar vein within granite gneiss at Aeschynite Hillock. − No. 7 = Aesch. from nepheline-feldspar vein in miaskite, region of vein No. 140.

As follows from X-ray diffraction data for unaltered minerals of the aeschynite-priorite series, the most important compound obtained after heating in air at 1000°C is aeschynite;

sometimes a cubic phase (a = 5.1 Å) is associated, see, for instance, [35]. Further, recrystallization (restoration of the original crystal structure) of aeschynite and priorite proceeds somewhat differently, and normally results in a polycrystalline aggregate (for an exception, see below). Thus, recrystallization of aeschynite/lyndochite starts (only slightly visible in the diffraction pattern) at about 700°C, but is better at 800 to 850°C, and is accompanied by formation of a cubic phase with pyrochlore structure (a = 10.28 to 10.32 ± 0.02 kX). At higher temperatures (≧1000°C), the cubic phase diminishes and the aeschynite structure usually remains unchanged. The recrystallization proceeds at a faster rate for poorly metamict samples, and occurs at higher temperatures for higher-grade metamict samples. After the formation of the above normal cubic phase, hydrated varieties of lyndochite form a second cubic phase with smaller lattice parameters (a = 10.19 to 10.20 kX), which also diminishes at higher temperatures [5, pp. 53/8]; see also [36, pp. 45/6]. As an exception, after heat treatment at 550 to 700°C, distinct X-ray amorphous grains of Ta aeschynite from a Siberian pegmatite recrystallize as a monocrystal, and do not form a polycrystalline aggregate, as follows from X-ray diffraction studies (oscillating crystal method) that yield lattice constants of ~7.44 ± 0.003 Å parallel to the length direction; of 11.00 ± 0.04 Å perpendicular to the especially well-developed pinacoid face; and of 5.33 ± 0.02 Å perpendicular to both aforementioned directions. These values correspond well with the lattice parameters given for non-metamict aeschynite from Vishnevye Gory, Middle Urals (see [32]). Heating up to 1200°C reveals no phase change [9, pp. 112/4].

Recrystallization of priorite/blomstrandine starts at about 600°C (rarely at 500°C) and gives only a priorite structure up to 800°C. Between 800 and 900°C, this structure is replaced by a cubic phase with pyrochlore structure (a = 10.11 to 11.15 kX, corresponding to an oxide of RE(Y)). Heating at 900 to 1200°C gives (for all priorites) euxenite coexisting with a cubic phase of usually a greater lattice parameter (a = 10.16 to 10.22 kX); the formation of euxenite alone is very rare. The temperature of this polymorphic transformation, priorite → euxenite, depends on the degree of metamictization and varies between 800 and 1200°C (the latter is valid for crystalline samples). The transformation itself occurs in a discontinuous manner (for details, see original paper) [5, pp. 36/40]; see also [36, pp. 44/5]. For example, after heat treatment at 700 to 900°C for 30 min, most of the originally X-ray amorphous 17 samples of variously colored priorite from albitized amazonite pegmatite, Eastern Siberia, display a more or less distinct diffraction pattern of the priorite/blomstrandine type. The most metamict of the samples, after heating at 800°C, contain a cubic phase with pyrochlore structure (a = 10.17 Å). Other samples preserve the priorite structure even after heating at 900°C, and pass over to euxenite at higher temperatures (1000 to 1200°C); note that a cubic phase (see above) sometimes occurs along with priorite (this is especially characteristic for samples composed of black and cinnamon priorite, representing different stages of metamict destruction). Some samples out of the group with preserved priorite structure contain two phases after heating at 1200°C: mainly euxenite and a cubic pyrochlore (a = 10.18 to 10.22 Å). Another part of the investigated samples, giving the priorite structure at 700 to 750°C, displays a pattern of slightly recrystallized euxenite at 800°C [33].

Indexed d values for non-metamict aeschynites (that may be used as a reference) are given for a sample (containing 0.72% ThO_2) from the Vishnevye Gory, Middle Urals, as follows [32]:

hkl	I	d in Å	hkl	I	d in Å	hkl	I	d in Å
130	6	3.04	230	1	2.165	062	2	1.661
112	10	2.98	222	2	2.036	134, 312	5	1.604
131	2	2.83	151	1	1.971	252, 322	3	1.550

Table (continued)

hkl	I	d in Å	hkl	I	d in Å	hkl	I	d in Å
200	3	2.69	004	3	1.891	170	1	1.509
140	1	2.47	043	1	1.866	171, 224	2	1.483
221	2	2.29	061	1	1.785			
042	2	2.23	242	3	1.713			

And nioboaeschynite-(Ce) from Alaska gives [29]:

hkl	I	d in Å		hkl	I	d in Å	
		observ.	calc.			observ.	calc.
020	2	5.539	5.543	140	1	2.464	2.465
021	1	4.479	4.475	221	1	2.309	2.311
111	1	4.082	4.087	103	1	2.289	2.290
002	1	3.789	3.793	231	1	2.091	2.094
121	2	3.445	3.445	222	2	2.044	2.044
022	4	3.132	3.130	151	1	1.980	1.980
130	7	3.048	3.049	004	1	1.896	1.896
112	10	2.983	2.988	242	2	1.722	1.722
131	2	2.827	2.829	134	2	1.611	1.610
040	1	2.773	2.771	322	1	1.560	1.559
200	2	2.699	2.698	170	1	1.521	1.520
041	1	2.602	2.603				

After heating at 800°C for 1 h, the above pattern of nioboaeschynite-(Ce) is shifted to smaller d values only to a small degree (0.01 to 0.03 Å) for all peaks; three peaks increased and eleven decreased in relative height (intensity), but eight remained unchanged, and one small peak disappeared [29].

After heating in air or nitrogen at 700°C for 3 h, at 1000°C for 1 h, and finally at 1300°C for 1 h, the X-ray powder diffraction patterns of completely metamict Ta-aeschynite-(Y) from the Borborema region, Brazil, are essentially the same, and are similar to the aeschynite pattern of the Powder Diffraction File (card No. 11-627). The five strongest d values (out of 30 observed, see original paper) are for the sample heated in air at 1300°C as follows; given is hkl, intensity, and d value observed and calculated (the latter is given in parentheses): 130, 10, 3.00 (3.02); 112, 10, 2.94 (2.93); 200, 5, 2.65 (2.67); 242, 5, 1.70 (1.70); and (330, 312, 134), 7, 1.581 (1.602, 1.588, 1.574) [8]. Synthetic standards ("etalon") are often used for an evaluation of X-ray diffraction patterns in the Russian literature. For example, $CeNbTiO_6$ is used for aeschynite and $YNbTiO_6$ for priorite; for the corresponding complete data (indexed d values and lattice parameters), see [25] and [33], respectively. Numerous X-ray diffraction data are also given in the papers cited for lattice constants and for X-ray data after heat treatment of Th-containing aeschynite minerals, see pp. 219/21; refer also to [2, pp. 370, 376], [4], [5, pp. 38, 54], [13, 37], [15, pp. 184/5], and [21, pp. 57/60].

Unlike other AB_2O_6-type titano-niobates that are characterized by a typical hexagonal closest packing arrangement of oxygen atoms (as, e. g., euxenite and fersmite), the aeschynite structure is characterized by a three-dimensional framework of slightly distorted (Nb, Ti)

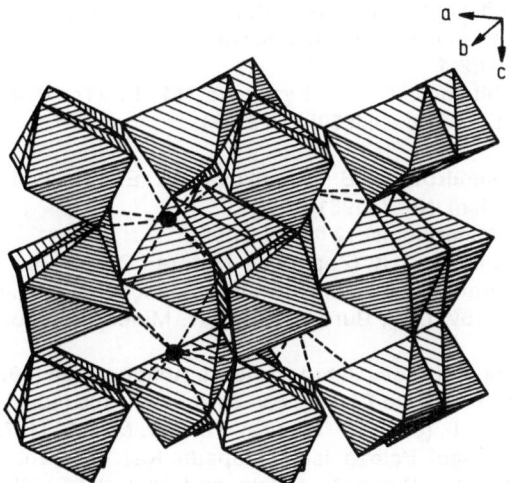

Fig. 10. Crystal structure of aeschynite from [24]. The (Nb,Ti) octahedra are hatchured; the black point denotes the position of the RE atom.

octahedra, that build up the basic structure, see **Fig. 10**. The (Nb,Ti) octahedra (or B polyhedron) are joined in pairs along their edges and are interconnected at the apices of the octahedra. Thus, within the framework zig-zag bands of octahedra pairs are formed (leaving channels between them) parallel to the c axis. The RE polyhedra (an irregular figure with eight apices) are connected at their edges into isolated chains that also extend along [001]; and each RE polyhedron has a common edge with six adjoining (Nb,Ti) octahedra. For the interatomic distances in the octahedron and the polyhedron, see original paper [24]; see also [2, p. 362], [3]. Isostructural with aeschynite/lyndochite is priorite/blomstrandine [25]; see also [5, pp. 11/2].

References for 2.2.2.2.3:

[1] Palache, C.; Berman, H.; Frondel, C. (Dana's System of Mineralogy, 7th Ed., Vol. 1, Wiley, New York 1946, pp. 1/834, 793/4).
[2] Chukhrov, F. V.; Bonshtedt-Kupletskaya, E. M. (Mineraly Spravochnik, Vol. 2, Pt. 3, Nauka, Moscow 1967, pp. 1/676).
[3] Aleksandrov, V. B.; Es'kova, E. M.; Zhabin, A. G. (in: Vlasov, K. A., Geochemistry and Mineralogy of Rare Elements and Genetic Types of Their Deposits, Vol. 2, Israel Program Sci. Transl., Jerusalem 1966, pp. 489/96, 491/3 [Russian original, Moscow 1964]).
[4] Makarochkin, B. A.; Es'kova, E. M.; Gonibesova, K. A. (Tr. Inst. Mineral. Geokhim. Kristallokhim. Redk. Elem. No. 3 [1959] 145/50).
[5] Gorzhevskaya, S. A.; Sidorenko, G. A. (Geol. Mestorozhd. Redk. Elem. No. 23 [1964] 1/116).
[6] Platonov, A. N. (Sb. Nauchn. Rab. Nauchn. Issled. Sekt. Kiev. Gos. Univ. No. 2 [1964] 110/5, 111).
[7] Popova, V. I.; Bazhenova, L. F. (in: Popov, V. A.; Chesnokov, B. V., Mineralogicheskie Issledovaniya v Il'menskom Zapovednike, Akad. Nauk SSSR Ural. Nauchn. Tsentr, Sverdlovsk 1981, pp. 18/29).

224

[8] Adusumilli, M. S.; Kieft, C.; Burke, E. A. J. (Mineral. Mag. **39** [1974] 571/6).

[9] Kornetova, V. A.; Aleksandrov, V. B.; Kazakova, M. E. (Tr. Mineral. Muzeya Akad. Nauk SSSR No. 14 [1963] 108/21).

[10] Zhabin, A. G.; Mukhitdinov, G. N.; Kazakova, M. E. (Tr. Inst. Mineral. Geokhim. Kristallokhim. Redk. Elem. No. 4 [1960] 51/72, 67).

[11] Zhabin, A. G.; Aleksandrov, V. B.; Kazakova, M. E. (Tr. Inst. Mineral. Geokhim. Kristallokhim. Redk. Elem. No. 7 [1961] 108/12).

[12] Ellsworth, H. V. (Am. Mineralogist **12** [1927] 212/8).

[13] Svyazhin, N. V.; Gaidukova, M. K. (Zap. Vses. Mineral. Obshch. **96** [1967] 206/10, 207).

[14] Kapustin, Yu. L. (Mineralogiya Karbonatitov, Nauka, Moscow 1971, pp.1/286, 147).

[15] Podol'skii, A. M.; Golikov, I. V.; Burova, T. A. (Tr. Mineral. Muzeya Akad. Nauk SSSR No. 16 [1965] 175/86).

[16] Grundmann, G.; Koller, F. (Neues Jahrb. Mineral. Abhandl. **135** [1979] 36/47, 39, 41).

[17] Cai, Genqing (Kuangwu Xuebao **7** [1987] 63/73 from C.A. **108** [1988] No. 97990).

[18] Sommerauer, J.; Weber, L. (Schweiz. Mineral. Petrogr. Mitt. **52** [1972] 75/91).

[19] Voloshin, A. V. (Mater. Geol. Polezn. Iskop. Zapadn. Kaz. **1966** Vol. 3, pp. 169/75, 174).

[20] Van Wambeke, L. (Bull. Soc. Belge Geol. Paleontol. Hydrol. **66** [1957] 35/53, 36).

[21] Lunts, A. (Izv. Akad. Nauk Latv. SSR **1965** No. 2, pp. 51/67, 52/4, 61/2).

[22] Gorzhevskaya, S. A.; Sidorenko, G. A. (Dokl. Akad. Nauk SSSR **146** [1962] 1176/8; Dokl. Acad. Sci. USSR Earth Sci. Sect. **146** [1962] 131/3).

[23] Sidorenko, G. A. (Geol. Mestorozhd. Redk. Elem. No. 10 [1960] 22/32, 30).

[24] Aleksandrov, V. B. (Dokl. Akad. Nauk SSSR **142** [1962] 181/4; Dokl. Acad. Sci. USSR Earth Sci. Sect. **142** [1962] 107/9).

[25] Komkov, A. I. (Dokl. Akad. Nauk SSSR **126** [1959] 641/4; Dokl. Acad. Sci. USSR Earth Sci. Sect. **124/129** [1959] 584/6).

[26] Chzhan, Pei-Shan' [Chang, P'ei-Shan] (Sci. Sinica [Peking] **12** No. 2 [1963] 237/43; C.A. **58** [1963] 9962).

[27] Gorzhevskaya, S. A.; Sidorenko, G. A. (Mineral'n. Syr'e No. 8 [1963] 48/57, 52).

[28] Zdorik, T. B.; Finyagina, N. K. (Tr. Mineral. Muzeya Akad. Nauk SSSR No. 17 [1966] 209/15, 210).

[29] Rosenblum, S.; Moiser, E. L. (Am. Mineralogist **60** [1975] 309/15).

[30] Komkov, A. I. (Tr. Vses. Nauchn. Issled. Geol. Inst. **96** [1963] 213/25).

[31] Romanov, I. A. (in: Tatarinov, P. M.; Rundkvist, D. V., Mineraly i Paragenezisy Mineralov Endogennykh Mestorozhdenii, Nauka, Leningrad 1975, pp. 43/9, 44/5).

[32] Zhabin, A. G.; Aleksandrov, V. B.; Kazakova, M. E.; Feklichev, V. G. (Dokl. Akad. Nauk SSSR **143** [1962] 686/9; Dokl. Acad. Sci. USSR Earth Sci. Sect. **143** [1962] 86/9).

[33] Fel'dman, Konopleva, I. B. (Tr. Mineral. Muzeya Akad. Nauk SSSR No. 19 [1969] 86/112, 92/3).

[34] Es'kova, E. M.; Zhabin, A. G.; Mukhitdinov, G. N. (Mineralogiya i Geokhimiya Redkikh Elementov Vishnevykh Gor, Nauka, Moscow 1964, pp. 1/318, 155).

[35] Van Wambeke, L. (Neues Jahrb. Mineral. Abhandl. **112** [1969/70] 117/49, 143).

[36] Gorzhevskaya, S. A.; Sidorenko, G. A. (Geol. Mestorozhd. Redk. Elem. No. 30 [1966] 43/51).

[37] Kostin, N. E.; Petrova, E. A.; Sidorenko, G. A.; Skorobogatova, N. V. (Mineral'n. Syr'e No. 1 [1960] 121/8, 127/8).

[38] Anonymous (Lapis **10** No. 3 [1985] 8/10).

2.2.2.2.4 Optical and Other Physical Properties. Chemical and Thermal Behavior

Optical Properties. Generally, aeschynite-priorite series minerals are black to various shades of brown and yellow, transmit light in only the thinnest splinters and are reddish brown to light brown; and become anisotropic on ignition and the refractive index increases somewhat (see below and p. 226) [1, p. 794]. The color of aeschynite/lyndochite varies between black and brown in different shades. Aeschynite, usually, is black, sometimes brown-black. Lyndochite is characteristically brown, varying from dark brown to red-brown and light brown; sometimes, it is black. In transmitted light (under the microscope), aeschynite is dark cherry red and sometimes red-brown; lyndochite is red-brown to golden brown, metamict samples are dark brown (clouded) [2, pp. 70/2]. Priorite/blomstrandine is brownish black to dark brown and brown, sometimes with a reddish hue. Under the microscope, it is golden and golden-brown with red-brown and dark red spots. After heating, the mineral gets a porcelain-like appearance and color, usually, brightens. Above 900°C, a greenish shade occurs [2, pp. 49/50].

The following statements are given in the literature for variation and distribution of color visible on megascopic scale and/or in transmitted light (for reflected light, see p. 227) in individual Th-containing samples of aeschynite and related minerals:

Anisotropic aeschynite is translucent ruby red or light cinnamon [3, p. 156], normal aeschynite is pitch-black or brownish black, and thoroaeschynite (all from alkaline pegmatites, Vishnevye Gory, Middle Urals) is brown to dark brown [3, pp. 149/50]. Aeschynite (from apo-alaskitic greisen, USSR) is brown, yellowish cinnamon in thin splinters, or cinnamon yellowish with a faint greenish tint in thin section [4, p. 178].

Lyndochite (from alkaline metasomatite, Urals) is (partly mottled) dark red, brownish red, yellowish red, and even black [5]. Crystalline lyndochite (from China) is brown in various shades (pale golden brown to reddish brown) in transmitted light. The color is often in-homogeneous but is not affected by heating to 800°C; between 900 and 1100°C it becomes reddish brown, and at 1200°C (or at 1100°C in vacuum) brownish black [6].

Aeschynite-(Nd) (from Baiyun-Obo, Inner Mongolia, P.R.C.) is brown to brownish black, yellowish orange to reddish orange in thin section [7]. Y aeschynite (from syenite pegmatite, Ilmen Mountains, Southern Urals) is black, greenish or reddish brown in thin section [8]. Non-metamict nioboaeschynite-(Ce) (from Alaska) is in thin fragments (0.01 to 0.1 mm) red and red-brown, when viewed against a glazed white-porcelain background [9]. Ta aeschynite (from granite pegmatite, Siberia) is transparent olive-brown in transmitted light, or is cinnamon in various shades depending on the degree of alteration; the fresh material enclosed in quartz is dark cinnamon to red-brown [10, p. 114]. Isotropic Ta aeschynite-(Y) (from the Borborema region, Brazil) is brownish black to black, transparent yellowish orange in thin section [11].

Priorite/blomstrandine (from a complex pegmatite, northern Priazov'e, Ukraine) is dark cinnamon, red-brown [12], or (from albitized amazonite pegmatite, Eastern Siberia) red-orange to dark cinnamon [13, pp. 88/9]. Blomstrandine (from RE pegmatite vein No. 5, Keivy, northwestern European USSR) is red, reddish yellow, and reddish cinnamon in its early-formed varieties; later formed varieties are brownish cinnamon and dark cinnamon. In transmitted light, both varieties are yellow and grayish yellow, the early-formed variety is also brownish yellow [14]. And blomstrandine (from pneumatolytic-hydrothermal vein, Eastern Siberia) is reddish yellow to cinnamon brown and black; the first and the last color is subordinate in frequency. The mean proportion between dark- (black to cinnamon) and light-colored crystals is 3:1 [15]. A niobate of Y, probably priorite (from albitized nepheline syenite, Mugodzhary, Kazakhstan), is yellow [16].

The distribution of color within single grains or crystals of aeschynite and related minerals may vary. Thus, blomstrandine from a pneumatolytic-hydrothermal vein, Eastern Siberia, sometimes has a central yellowish zone, followed by a cinnamon brown and an outer black part. But substitutions of cinnamon by yellowish, or black by cinnamon also occur [15]. For an inverse color sequence (central black, outside yellowish brown) in lyndochite of metasomatic rocks, Southern Urals, see p. 192. The distribution of color within priorite from albitized amazonite pegmatite, Eastern Siberia, is very complicated and displays no visible regularity; it seems that lighter colored areas more often occur at the periphery of a grain [13, p. 89]. In some cases, red blomstrandine of the RE pegmative vein No. 5 (Keivy), northwestern European USSR, is overgrown by brown to cinnamon material, pointing to different stages of formation (or generations?) [14, p. 53]. And light-colored blomstrandine occurs as rather tiny, more irregularly shaped grains within the dark (black to cinnamon) blomstrandine from a pneumatolytic-hydrothermal vein, Eastern Siberia [15].

The luster of aeschynite, usually, is vitreous and in certain cases resinous. Lyndochite, characteristically, has a resinous luster, that sometimes turns into adamantine [2, p. 71]. See also Ta aeschynite from granite pegmatite, Siberia, has a strong luster, unaltered crystals are vitreous [10, p. 114]; and luster is glassy on surface and pitchy on fracture for crystalline lyndochite from Eastern Siberia [17]. The luster of priorite/blomstrandine is strongly vitreous, sometimes up to adamantine; not infrequently, on the surface fatty (resinous) [2, p. 50]. Greasy (fatty) to adamantine is the luster on fresh fracture of blomstrandine from RE pegmatite, northwestern European USSR [14, p. 55], and strong semi-adamantine for a niobate of Y, probably priorite, from albitized nepheline syenite, Mugodzhary, Kazakhstan [16].

Streak is almost black, to brown (aeschynite) and reddish yellow (priorite) [1, p. 794]; or orange to light reddish brown (non-metamict nioboaeschynite-(Ce)) [9]; or faint yellowish and (for darker colored samples) light cinnamon to yellowish (Ta aeschynite) [10, p. 114].

The refractive index of metamict aeschynite and lyndochite is similar and varies between 2.23 and 2.375, or 2.20 and 2.33, respectively [2, pp. 70/2]. Isotropic thoroaeschynite (from the Transangara massif, Enisei Ridge) has $n > 2.07$ and displays anisotropic areas with 2.15 to 2.20 [18]. Thoroaeschynite has $n = 2.16$ and normal aeschynite (both from Vishnevye Gory, Middle Urals) 2.21 to 2.27 (pitch-black) and 2.30 (brownish black) [3, pp. 149/50]. Ta aeschynite (from granite pegmatite, Siberia) has n between 2.21 and 2.23 [10, p. 115]. Aeschynite-(Nd) (from Beiyun-Obo, Inner Mongolia) gives average values of $n \doteq 2.1$ to 2.4 [7]. And Y aeschynite (from syenite pegmatite, Ilmen Mountains, Southern Urals) has $n = 2.27$, that increases to 2.45 after annealing [8]. The refractive index varies with color (and degree of metamictization) for lyndochites (from alkaline metasomatites, Urals) and is $n = 2.10$ (crystalline, dark red), 2.147 (brownish black and black), and 2.00 or 2.06 after heat treatment (metamict, yellowish brown) [5]. After heating at 800°C, aeschynite becomes anisotropic, biaxially positive, with $N'_m = 2.20$ to 2.25 and $2V = \sim 60°$ [2, pp. 70/2].

The refractive index of priorite/blomstrandine is between 2.15 and 2.33; sometimes, it may be diminished to 2.075 or is raised up to 2.375 [2, pp. 49/50]. See also, for example, $n = 2.115 \pm 0.06$ for clear red-brown priorite (from albitized amazonite pegmatite, Eastern Siberia) [13, p. 89]; 2.15 ± 0.01 for blomstrandine (from RE pegmatite, north-western European USSR) [14]; 2.16 to 2.24 for priorite/blomstrandine (from a complex pegmatite, northern Priazov'e, Ukraine) [12]; or 2.18 ± 0.05 for red-brown and yellow priorite (from albite-zircon-riebeckite vein in albitite, Eastern Siberia) [19, p. 123].

The following data are given in the literature for refractive index, optical axial angle, pleochroism, optical orientation, and some other optical properties of non-metamict (crystalline) unheated samples of aeschynite/lyndochite:

Aeschynite (containing 0.72% ThO_2) from the Vishnevye Gory, Middle Urals, has $N_g = 2.34 \pm 0.01$, $N_p = 2.28 \pm 0.01$ and is biaxially positive, $2 V = 75 \pm 2°$. Pleochroism is γ = darkish brown, β = brown, α = yellowish brown; N_g parallel to c and N_p parallel to a axis [20]; see also [3, p. 156].

Nioboaeschynite-(Ce) from Alaska has $\alpha = 2.27$, $\beta = 2.32$, and $\gamma = 2.36$; optically biaxial, $2 V_\alpha$ varies between 75 and 97° (average 84.6°, median value 85°). Pleochroism with reddish brown colors for all directions is weak to nil despite a strong anisotropy. Optical orientation is γ = a, β = c, and α = b; optical axial plane is (001). Note that α and γ are reversed, probably because of dominant Y, for aeschynite-(Y) from Switzerland (see [21]) [9].

Lyndochite has $N_g = 2.34$ to 2.50, $N_p = 2.28$ to 2.39 and $2 V = 75 \pm 2°$; pleochroism is N_g = brown to brown-red, N_m = cherry red, and N_p = yellow-cinnamon to cinnamon-yellow [2, pp. 70/2]. Lyndochite from China has $\beta = 2.375 \pm 0.015$, $\gamma = 2.36$ to 2.50 (2.43 $\pm$ 0.07, but closer to 2.50) and 2 V is large and positive. Pleochroism is weak from lighter to darker brown tones. Elongation is positive or negative, β is parallel to the length of the prismatic crystal [6]. Dark red lyndochite from alkaline metasomatite, Urals, is optically biaxial positive, $2 V = 60$ to 70°; pleochroism from light yellowish to orange-red [5]. And lyndochite from carbonatite, Eastern Siberia, has $N_e = 2.3$; optical orientation is N_m parallel to [001] [17].

Partly metamict aeschynite-(Y, Gd, Dy, Er) from the central Swiss Alpes has $\alpha = 2.280 \pm 0.005$, $\beta = 2.305 \pm 0.005$, $\gamma = 2.34 \pm 0.01$ (as is calculated from α, β, and $2 V_\gamma$) and is optically biaxial positive, $2 V_\gamma \approx 80°$ ($\pm 5°$). Optical orientation is α = a, β = c, and γ = b. Dispersion is strong r $\ll$ v [21].

In reflected light, aeschynite is cream-colored gray [22, p. 364]. Aeschynite (from apo-alaskitic greisen, USSR) is light gray [4, p. 178]; crystalline dark red lyndochite (from alkaline metasomatite, Urals) is gray [5]; and aeschynite-(Nd) (from Beiyun-Obo, Inner Mongolia) is dark gray [7]. Ta aeschynite (from granite pegmatite, Siberia) is nearly white, yellowish brown when viewed under crossed Nicol prisms [10, p. 115]. Crystalline lyndochite (from China) is colorless, clearly anisotropic [6]. In reflected light, priorite is golden brown or reddish brown [22, p. 372].

In polished section, isotropic (metamict) Ta aeschynite-(Y) (from the Borborema region, Brazil) is dark gray in oil immersion with abundant internal reflections (white, yellowish orange, dark red) [11]. Internal reflections are (appreciable to strong) red and pale brown for crystalline lyndochite (from China) [6]; red-brown for priorite (from albitized amazonite pegmatite, Eastern Siberia) [13, pp. 88/9]; or brown, brownish red for priorite/blomstrandine (from complex pegmatite, northern Priazov'e, Ukraine) [12]. Internal reflections are faint brownish yellow and reddish for aeschynite, and red (at strong illumination) for priorite [22, pp. 364, 373].

Reflectivity of aeschynite at different wavelength is 15.6% (white), 22.5% (blue), 19.3% (green), 14.5% (yellow), and 14.0% (orange and red) [22, p. 364]. Against a standard carborundum, R is 14.5% at 470 nm, 14.2% at 546 nm, and 14% at both 589 and 650 nm for metamict Ta aeschynite-(Y) (from the Borborema region, Brazil) [11]. Reflectivity is 15% for isotropic thoroaeschynite (Transangara massiv, Enisei Ridge) [18]; about 16.9% (in daylight) for aeschynite (apo-alaskitic greisen, USSR) [4, p. 178]; 16 to 18% for crystalline dark red lyndochite (alkaline metasomatite, Urals) [5]; and 17.1% for priorite/blomstrandine (complex pegmatite, northern Priazov'e, Ukraine) [12].

The IR spectrum of aeschynite and priorite (see original paper) is similar in configuration; its differences are caused by the differing influence of Ce and Y on the strength of inter-atomic connection [23].

Crystalline lyndochite (from China) does not fluoresce in UV light [6].

Other Physical Properties. Microhardness of aeschynite/lyndochite varies between 473 and 710 kg/mm^2, corresponding to Mohs hardness $H_M = 4.9$ to 5.8 [2, p. 71], or is between 593 and 734; mean 690 kg/mm^2 for aeschynite and between 724 and 834 kg/mm^2 for priorite [22, pp. 364, 373]. Microhardness and corresponding H_M for individual Th-containing samples is given as follows: 473 kg/mm^2 (as a mean of eight measurements), $H_M = 4.9$ for Y aeschynite (from syenite pegmatite, Ilmen Mountains, Southern Urals) [8]; 657 to 673 kg/mm^2 (mean maximum and minimum values VHN_{100} of fifteen indentations), $H_M = 5.5$ to 6 for Ta aeschynite-(Y) (from the Borborema region, Brazil) [11]; between 670 and 690 kg/mm^2, $H_M \approx 5.6$ for aeschynite (from apo-alaskitic greisen, USSR) [4, p. 178]; 712 kg/mm^2, $H_M \approx 5.75$ for normal aeschynite and 714 kg/mm^2 (both Vishnevye Gory, Middle Urals) [3, pp. 149/50]; 734 and 706 kg/mm^2 for early- and late-formed blomstrandine, respectively (from RE pegmatite, north-western European USSR) [14]; 750 kg/mm^2 for priorite/blomstrandine (from complex pegmatite, northern Priazov'e, Ukraine) [12]; 764 kg/mm^2, $H_M \approx 5.9$ for Ta aeschynite (from granite pegmatite, Siberia) [10, p. 114]; and 724 and 834 kg/mm^2 for red-brown and yellow priorite, respectively (from albite-zircon-riebeckite vein in albitite, Eastern Siberia) [19, p. 123].

Fracture is conchoidal for aeschynite and irregular (splintery) for lyndochite, that, sometimes, also displays a conchoidal fracture [2, p. 71]; see also, uneven (ragged) fracture for crystalline lyndochite (from China) [6]. Fracture is conchoidal, sometimes smooth conchoidal, for priorite/blomstrandine [2, p. 50].

Parting parallel to (100) was observed for crystalline lyndochite (from carbonatite, Eastern Siberia) [17].

Density, corresponding to the ThO_2 content, is between 5.10 and 5.38 g/cm^3 for aeschynite (containing 9.51 to 20.34% ThO_2) and between 4.55 and 5.132 g/cm^3 for lyndochite (containing 0.72 to 4.95% ThO_2) [2, p. 71]; D is 5.19 g/cm^3 for Y aeschynite (from syenite pegmatite, Ilmen Mountains, Southern Urals) [8], and is calculated as 5.5 ± 0.1 g/cm^3 for partly metamict aeschynite-(Y, Gd, Dy, Er) (from the central Swiss Alps) [21]. Density of priorite/blomstrandine varies between 4.50 and 5.34 g/cm^3 [2, p. 50]. For variances of D related to chemical composition, see also Tables 10 and 11, pp. 206 and 208/13, respectively.

Chemical Behavior. The data given for solubility are somewhat inconsistent: Aeschynite is strongly soluble in H_2SO_4, HF, and in melts of KOH and $KHSO_4$; it quickly dissolves in phosphoric acid [22, p. 368]. Fine powder of aeschynite is partially decomposed by H_2SO_4, completely by HF and in a fusion with $KHSO_4$. Fine powder of priorite is more or less completely decomposed by hot HCl or H_2SO_4 [1, p. 796]. Priorite/blomstrandine is insoluble in HCl and HNO_3, boiling H_2SO_4 dissolves it with difficulty [2, p. 51]. Niobian aeschynite (from China) is insoluble in HCl and HNO_3, soluble in H_3PO_4, and slightly soluble in H_2SO_4 [24]. And crystalline lyndochite (from China) does not dissolve in cold HCl, HNO_3, and H_2SO_4 [6].

Polished surfaces of aeschynite are etched by $H_2SO_4 + KMnO_4$, by boiling H_2SO_4, and very intensely by HF [22, p. 368]; see also, polished Ta aeschynite (from granite pegmatite, Siberia) is corroded immediately by HF [10, p. 115]. Diagnostic etching of aeschynite may be done by a 50% solution of HF that causes, after 1 to 2 s, an irridescent cover [25]. Whereas red-brown priorite (from albite-zircon-riebeckite vein in albitite, Eastern Siberia) is corroded immediately by HF, revealing its internal structure. The internal structure of the lesser metamict yellow variety cannot be seen immediately after acid treatment [19, p. 123].

Thermal Behavior. For a short literature review on heat treatment studies performed up to 1970 for priorite and euxenite, see [26]; refer also to pp. 175/7.

As can be seen from the curve of loss of water for aeschynite/lyndochite (see original paper), both minerals only contain small amounts of water (<3.5%) that is completely set free during heating to 600°C; the main portion is lost between 300 and 400°C [2, pp. 66/7]; see also, the loss of weight is only small (4.6 to 6.8 mg) for aeschynite (from apo-alaskite greisen, USSR) and is related to loss of water [4, pp. 181, 184]. For loss of weight in aeschynite from pegmatites, Ilmen Mountains, see below and p. 230. The curve of loss of water for priorite/blomstrandine (see original paper) shows that water is set free within the interval of temperature between 200 to 300°C and 600 to 800°C; the main portion is lost between 300 to 400°C and up to 600°C [2, pp. 45/6].

The DTA curve of aeschynite/lyndochite (see original paper) is characterized by a rather flat endothermic effect between 100 and 300°C (maximum is at 200 to 210°C) caused by loss of water, the intensity depends on the degree of hydration; sometimes the endothermic effect is absent. Recrystallization gives an acute exothermic peak with a maximum between 763 and 782°C for lyndochite and between 780 and 815°C for aeschynite; the low recrystallization temperature of about 735°C given by [27] could not be verified in this study, for the precision of the thermal reaction is ±50°C. For aeschynite, there is usually an additional exothermic effect of unknown origin at 720 to 750°C, in single cases raised up to 760 to 785°C, that forms a step (kink) near the main exothermic effect and seldom occurs as a separate peak (see also the examples given in the following paragraphs). Thus, recrystallization occurs for aeschynite during a greater interval of temperature (60 to 80°C and even up to 120°C) than for lyndochite (30 to 50°C), and the process itself proceeds with two steps for the former. Sometimes, the DTA curve of aeschynite/lyndochite displays a small second exothermic effect between 880 and 950°C that, apparently, is connected with the entire formation of euxenite and the disappearance of a cubic phase (see X-ray diffraction data, pp. 220/1) [2, pp. 65/7].

For the thermal behavior of individual Th-containing samples of aeschynite, the following statements are given in the literature:

The heating curve of aeschynite from apo-alaskitic greisen, USSR, displays weak exo- and endothermic effects between 100 and 200°C (apparently caused by loss of water); is clearly curved at 350 and 460°C; and has its maximum exothermic peak at 740 to 746°C corresponding to crystallization (and comparable with the exothermic peak at 735°C (for Nb aeschynite) [4, pp. 181, 184].

The heating curve (see original paper) of thoroaeschynite from the Vishnevye Gory, Middle Urals, displays a very strong exothermic effect at 790°C; at 750°C, a small break (kink, see above) is visible in the smooth course of the heating curve [3, pp. 150/1]. Alumoaeschynite, from the same area, always displays an exothermic effect at 750 to 800°C corresponding to recrystallization [3, p. 155].

The differential heating curve (see original paper) of Ta aeschynite from granite pegmatite, Siberia, displays an endo-exothermic effect at about 200°C. The former is, probably, connected with liberation of water, the latter may be explained by a change in valency of elements (perhaps of the cationic part). The maximum exothermic peak at 800°C, corresponding to recrystallization, has in its smooth course, a break of unknown nature at about 775°C that is typical for aeschynite group minerals. A small exothermic peak at 950°C is, probably, connected with a phase change not detected by the X-ray diffraction data [10, pp. 118/20]. And the DTA curve of metamict Ta aeschynite-(Y) from the Borborema region, Brazil, gives a sharp exothermic effect at 740°C that is characteristic for aeschynite and is caused by crystallization [11].

The thermal behavior of aeschynite differs somewhat for samples from different types of host pegmatite, Ilmen Mountains, Southern Urals. Thus, corresponding to recrystallization, the DTA curve displays an exothermic effect at 780 to 800°C for samples from feldspar pegmatite, whereas this effect occurs at 770°C and between 750 and 760°C for samples from corundum-feldspar and from late granite pegmatites, respectively. Also the loss of weight differs, namely 1 to 1.5% for aeschynite from feldspar pegmatite and about 3 to 3.5% for aeschynite from late granite pegmatite. In the latter case, there is a small increase (1.2%) in weight within the interval of temperature from 0 to 250°C [28].

The DTA curve of priorite/blomstrandine (see original paper) is very similar to that of metamict pyrochlore, and displays an endothermic effect between 20 and 300°C (maximum is at 200 to 222°C) caused by loss of water. Recrystallization gives an exothermic effect between 400 and 600°C, the maximum is between 473 and 532°C and depends on the degree of metamictization. Two other exothermic peaks at higher temperatures belong to the disintegration of the priorite/blomstrandine structure to a cubic phase (between 680 and 780°C) and a subsequent formation of euxenite (between 900 and 920°C) [2, pp. 45/6]. The heating curve of priorite (from stanniferous veins, Kibara Mountains, Katanga) has two exothermic peaks at 320 and 720°C [29]. Another form displays the DTA curve (see original paper) of priorite (from Inner Mongolia) that is characterized by a split-up main exothermic peak composed of a main peak at 473 to 491°C and a subsidiary peak at 533 to 576°C. The relation between both peaks is variable; very rarely, the former or the latter peak occurs alone, sometimes both are of equal size [2, pp. 45/6].

For the thermal behavior of (further) individual Th-containing samples of priorite/ blomstrandine, the following statements are given in the literature:

The exothermic effect between 400 and 500°C, corresponding to recrystallization, on the heating curve (see original paper) is much stronger and more distinct for red-brown than for yellow priorite from albite-zircon-riebeckite vein in albitite, Eastern Siberia. This difference may be explained by the different degree of metamictization; the yellow variety is more crystalline (as follows from X-ray diffraction data, see original paper). A second exothermic effect between 600 and 700°C is equal in intensity and may be caused by admixed fergusonite [19, pp. 123, 126/8].

The DTA curve (see original paper) is identical for red and brown-cinnamon blomstrandine from RE pegmatite (vein No. 5, Keivy), northwestern European USSR, and displays: an endothermic effect between 50 and 200°C (maximum is at about 140 to 150°C, connected with release of water that is not bound to the lattice but adsorbed during metamictization); a flat exothermic effect at 340°C (connected with the beginning of recrystallization or, possibly, with oxidation of Fe^{2+}); the main exothermic peak at 500 to 505°C (recrystallization); and a weak exothermic effect between 730 and 770°C (crystallization of a cubic phase or transition from blomstrandine to euxenite structure). Very similar is the DTA curve (see original paper) for four samples of cinnamon blomstrandine (pseudomorphous after polycrase) from pegmatite veins No. 13 and 14 of the same area: endothermic effect between 20 and 300°C (maximum at 180 to 200°C, release of water); a small exothermic effect between 360 and 380°C (oxidation of Fe^{2+} or start of recrystallization); a distinct exothermic peak at 400 to 460°C (recrystallization); and a broad, flat exothermic effect between 680 and 720°C [14].

Besides a flat endothermic effect between 200 and 400°C (release of water), the DTA curve (see original paper) of both dark- and light-colored priorite from albitized amazonite pegmatite, Eastern Siberia, displays (for six samples) a small, distinct exothermic effect with a maximum between 705 and 770°C. Such a high temperature of recrystallization, usually, is not characteristic of priorite, but is typical of euxenite (refer to [30]). The peak characteristically swells at its beginning, and the peak height varies in the main with the sample's degree of metamictization (dark samples have a greater height). Two other, substantially weaker exothermic peaks occur at 900 to 1000°C (maximum at 960 to 970°C, connected with the formation of euxenite) and at 550 to 650°C (maximum at 612°C, most probably connected with oxidation of U); the latter peak is much weaker for dark samples. A smooth exothermic effect at 300 to 400°C may be caused by oxidation of Fe (and of a part of U). If the sample has a high content of water (2 to 3%), there is also an endothermic effect at 600 to 700°C [13, pp. 90/2].

For results of hydrothermal resynthesis (that may act, possibly, as an identification means) of metamict titano-tantalo-niobates at 350 to 500°C and 760 to 800 atm, corresponding to p-T conditions as are indicated by liquid-gas inclusions in minerals of rare-metal greisens, see [31].

References for 2.2.2.2.4:

[1] Palache, C.; Berman, H.; Frondel, C. (Dana's System of Mineralogy, 7th Ed., Vol. 1, Wiley, New York 1946, pp. 1/834).
[2] Gorzhevskaya, S. A.; Sidorenko, G. A. (Geol. Mestorozhd. Redk. Elem. No. 23 [1964] 1/116).
[3] Es'kova, E. M.; Zhabin, A. G.; Mukhitdinov, G. N. (Mineralogiya i Geokhimiya Redkikh Elementov Vishnevykh Gor, Nauka, Moscow 1964, pp. 1/318).
[4] Podol'skii, A. M.; Golikov, I. V.; Burova, T. A. (Tr. Mineral. Muzeya Akad. Nauk SSSR No. 16 [1965] 175/86).
[5] Svyazhin, N. V.; Gaidukova, M. K. (Zap. Vses. Mineral. Obshch. **96** [1967] 206/10).
[6] Gorzhevskaya, S. A.; Sidorenko, G. A. (Dokl. Akad. Nauk SSSR **146** [1962] 1176/8; Dokl. Acad. Sci. USSR Earth Sci. Sect. **146** [1962] 131/3).
[7] Zhang, Peishan; Tao, Kejie (Dizhi Kexue **1982** No. 4, pp. 424/8).
[8] Makarochkin, B. A.; Es'kova, E. M.; Gonibesova, K. A. (Tr. Inst. Mineral. Geokhim. Kristallokhim. Redk. Elem. No. 3 [1959] 145/50, 146).
[9] Rosenblum, S.; Moiser, E. L. (Am. Mineralogist **60** [1975] 309/15).
[10] Kornetova, V. A.; Aleksandrov, V. B.; Kazakova, M. E. (Tr. Mineral. Muzeya Akad. Nauk SSSR No. 14 [1963] 108/21).

[11] Adusumilli, M. S.; Kieft, C.; Burke, E. A. J. (Mineral. Mag. **39** [1974] 571/6).
[12] Litovchenko, E. I.; Kuts, V. P. (Mineral Sb. [Lvov] **29** No. 4 [1975] 32/41, 34/5).
[13] Fel'dman, L. G.; Konopleva, I. B. (Tr. Mineral. Muzeya Akad. Nauk SSSR No. 19 [1969] 86/112).
[14] Lunts, A. (Izv. Akad. Nauk Latv. SSR **1965** No. 2, pp. 51/67, 55/7).
[15] Romanov, I. A. (in: Mineraly i Paragenezisy Mineralov Endogennykh Mestorozhdenii, Nauka, Leningrad 1975, pp. 43/9, 44/5).
[16] Voloshin, A. V. (Mater. Geol. Polezn. Iskop. Zapadn. Kaz. **1966** Vol. 3, pp. 169/75, 174).
[17] Zdorik, T. B.; Finyagina, N. K. (Tr. Mineral. Muzeya Akad. Nauk SSSR No. 17 [1966] 209/15, 210).
[18] Sveshnikova, E. V.; Semenov, E. I.; Khomyakov, A. P. (Zaangarskii Shchelochnoi Massiv Ego Porody i Mineraly, Nauka, Moscow 1976, pp. 1/80, 31).
[19] Kostin, N. E.; Petrova, E. A.; Sidorenko, G. A.; Skorobogatova, N. V. (Mineral'n. Syr'e No. 1 [1960] 121/8; C.A. **56** [1962] 3143).
[20] Zhabin, A. G.; Aleksandrov, V. B.; Kazakova, M. E.; Feklichev, V. G. (Dokl. Akad. Nauk SSSR **143** [1962] 686/9; Dokl. Acad. Sci. USSR Earth Sci. Sect. **143** [1962] 86/9).

232

[21] Sommerauer, J.; Weber, L. (Schweiz. Mineral. Petrogr. Mitt. **52** [1972] 75/91, 80).

[22] Chukhrov, F. V.; Bonshtedt-Kupletskaya, E. M. (Mineraly Spravochnik, Vol. 2, Pt. 3, Nauka, Moscow 1967, pp. 1/676).

[23] Povarennykh, A. S. (Konst. Svoistva Mineral. No. 13 [1979] 53/78, 69/70).

[24] Chang, Pei-Shan (Sci. Sinica [Peking] **11** [1962] 969/76 from C.A. **58** [1963] 2274).

[25] Sergeev, V. A. (Izv. Sibirsk. Otd. Akad. Nauk SSSR Geol. Geofiz. No. 1 [1958] 103/5; C.A. **1959** 17776).

[26] Mitchell, R. S. (Southeast. Geol. **14** [1972] 59/72, 62/6).

[27] Zhabin, A. G.; Aleksandrov, V. B.; Kazakova, M. E. (Tr. Inst. Mineral. Geokhim. Kristallokhim. Redk. Elem. No. 7 [1961] 108/12).

[28] Popova, V. I.; Bazhenova, L. F. (in: Popov, V. A.; Chesnokov, B. V., Mineralogicheskie Issledovaniya v Il'menskom Zapovednike, Ural. Nauchn. Tsentr, Sverdlovsk 1981, pp. 18/29, 24/5).

[29] Van Wambeke, L. (Bull. Soc. Belge Geol. Paleontol. Hydrol. **66** [1957] 35/53, 39).

[30] Gorzhevskaya, S. A.; Sidorenko, G. A. (Geol. Mestorozhd. Redk. Elem. No. 30 [1966] 43/51).

[31] Golikov-Zavolzhenskii, I. V.; Trufanov, V. N. (Geokhimiya **1971** 974/9; Geochem. Intern. **8** [1971] 630).

2.2.2.3 Polymignite *)

Metamict polymignite (or polymignyte) from a pegmatite vein at Fredriksvärn (Stavern) near Larvik, southeast Norway, was described in 1824 by Berzelius and later (1890) studied in greater detail by Brögger (see [1]); it also occurs in the country rock (augite syenite or larvikite); and associated minerals are Na orthoclase, barkevikite, nepheline, magnetite, zircon, and pyrochlore. The name is derived from the Greek words "polys" = many and "migninai" = to mix in allusion to the complex composition [2 to 4].

Dark red elongated prisms (maximum length 0.4 mm) of polymignite are enclosed in sanidine and sometimes in the interstitial glass bordering K feldspar of a piece of ejected sanidinite, Campi Flegrei, Campania, Italy [5, p. 262].

Polymignite is an oxide or niobate-titanate-zirconate of uncertain formula, possibly ABO_4, with A = Ca, Fe^{2+}, (Y,Er,Ce), Zr, Th and B = Nb, Ti, Ta, Fe^{3+} [2]; see also a formula $(Ca,Fe,Y,Th)(Nb,Ti,Ta,Zr)O_4$ in [6]. Or it is a complex oxide of variable composition that fits the formula AB_2X_6, with A = Ca, Fe^{2+}, Ce and B = Zr, Ti, Nb, Ta, Fe^{3+}; X is oxygen [3]. See also a formula $(Ce,La,Y,Th,Mn,Ca)(Ti,Zr,Nb,Ta)_2O_6$ given in [7]. For a formula type of $AZrB_2O_7$, see [8]. A most recent reinvestigation suggests a crystal-chemical formula $(Ca,Na,RE,Th)_2Zr_2(Ti,Nb,Ta)_3(Fe,Ti)O_{14}$, that is the same as for zirkelite/zirconolite (refer to p. 110) and is confirmed by crystal structure investigations (refer to pp. 124/6).

The chemical analysis of polymignite from Norway (No. 1) [1], see also [2, 4, 8], and (incomplete) from Italy (No. 2) [5, p. 264] is given as follows:

---

*) Formula uncertain, see text.

	content in % in sample	
	No. 1	No. 2
ThO_2	3.92	3.5
ΣRE_2O_3	13.30 [a]	9.9 [b]
CaO	6.98	8.6
UO_2	—	0.7
FeO	2.08	8.0
Fe_2O_3	7.66	—
Nb_2O_5	11.99	12.7
Ta_2O_5	1.35	0.6
TiO_2	18.90	19.5
ZrO_2	29.71	25.8
Na_2O	0.59	0.4
others	3.71 [c]	—
sum	100.19	89.7

[a] Note that [5, p. 264] gives 6.20% CeO_2 instead of 5.91% Ce_2O_3 as in the original paper. — [b] Ce is given as CeO_2. — [c] Comprises 1.32% MnO, 0.77% K_2O, 0.45% SiO_2, 0.39% PbO, 0.19% Al_2O_3, 0.16% MgO, 0.15% SnO, and 0.28% H_2O.

The space group for polymignite is Acam (No. 64); lattice constants are a = 10.148(4), b = 14.147(5), c = 7.278(3) Å; and Z = 2, based on the above crystal-chemical formula [5, p. 266]; based on another formula type and Z = 8, one gets a = 10.31, b = 14.48, and c = 7.41 Å [8]. Using a non-standard c-centered unit cell (for details, see original paper), the lattice constants become a = 2 × 12.37, b = 7.28, and c = 17.41 Å [9, pp. 1157/8]. For a detailed crystallographic investigation considering the relationship between orthorhombic and monoclinic setting for polymignite and zirconolite, see [8].

Indexed d values for polymignite from Italy are similar to that of zirkelite/zirconolite and synthetic $CaZrTi_2O_7$ (see p. 123), and therefore cannot be used as a method of identification or discrimination. The strongest indexed d values (out of 22 given in the original paper) are as follows; I = intensity [5, pp. 272, 276]:

d in Å	I	hkl	d in Å	I	hkl	d in Å	I	hkl
3.687	12	131	1.819	15	004	1.534	13	602
2.957	93	202	1.794	68	442	1.526	15	640
2.901	100	240	1.768	14	080	1.518	22	282
2.536	50	042, 400	1.541	24	244			

Another set of 39 d values obtained after heating at 1200°C (see original paper) gives the following strongest values; intensity is given in parenthesis: 2.96(100), 2.80(30), 2.52(30), 1.819(60), 1.749(30), and 1.550(30) [8].

For a description of crystal forms, physical properties, and chemical behavior of polymignite, see [2 to 4]; data on thermal behavior are given in [4].

For the energy dispersive X-ray spectrum and the $[100]_0$ selected area diffraction pattern of synthetic perfectly crystalline polymignite $Ca_{0.7}Th_{0.3}ZrTi_{0.7}Fe_{0.3}O_7$, see [9, pp. 1165, 1171].

References for 2.2.2.3:

[1] Brögger, W. C. (Z. Krist. **16** Pt. II [1890] 1/663, 387/96).
[2] Palache, C.; Berman, H.; Frondel, C. (Dana's System of Mineralogy, 7th Ed., Vol. 1, Wiley, New York 1946, pp. 1/834, 764/6).
[3] Noneshnikova, V. I. (in: Vlasov, K. A., Geochemistry and Mineralogy of Rare Elements and Genetic Types of Their Deposits, Vol. 2, Israel Program Sci. Transl., Jerusalem 1966, pp. 496/8 [Russian original, Moscow 1964]).
[4] Chukhrov, F. V.; Bonshtedt-Kupletskaya, E. M. (Mineraly Spravochnik, Vol. 2, Pt. 3, Nauka, Moscow 1967, pp. 1/676, 188/90).
[5] Mazzi, F.; Munno, R. (Am. Mineralogist **68** [1983] 262/76).
[6] Fleischer, M. (Glossary of Mineral Species, 5th Ed., Mineral Record Inc., Tucson 1987, pp. 1/227, 147).
[7] Strunz, H.; Tennyson, C. (Mineralogische Tabellen, 6th Ed., Geest and Portig, Leipzig 1977, pp. 1/621, 208).
[8] Pudovkina, Z. V.; Chernitsova, N. M.; Pyatenko, Yu. A. (Zap. Vses. Mineral. Obshch. **98** [1969] 193/9).
[9] White, T. J. (Am. Mineralogist **69** [1984] 1156/72).

3.2.2.4 Sinicite *)

Sinicite from granite pegmatite (of Inner Mongolia [1, p. 400]), People's Republic of China, was described in 1957 [2]. It was named also as sinisit, sinitsit, or sinitit [3].

For the type locality, the occurrence of sinicite is given as follows: It occurs as a blackish brown to reddish brown mineral (forming long columnar, sometimes irregular, deposits and aggregates [4, pp. 49/50]) in the zone of coarse feldspar and mica of the pegmatite, where it is associated with quartz, biotite, muscovite, topaze, fluorite, tourmaline, and beryl [2]; see also [5, 6]. It occurs in the feldspar zone of granite pegmatite in association with fluorite, tourmaline, beryl, monazite, ferrithorite, uranothorite, ilmenorutile, ilmenite, magnetite, and hematite [4, p. 52]. Or, together with fergusonite, zircon, thorite (uranothorite), it is found in the intermediate zones mainly in pegmatites of the biotite type, more subordinately in those of the two-mica type [1, pp. 400, 405, 410]. Sinicite also occurs in the quartz-mica(-feldspar) cement of Middle Cambrian quartz-pebble conglomerates of the Urals, USSR. Together with ilmenorutile, Mn columbite, allanite (orthite), RE uranothorite, rhabdophane, xenotime, thorite, and other ore minerals, it forms minute impregnations (shows of ores) in the conglomerate (metamorphosed at greenschist-facies condition) within the exocontact of a granitoid massif [7].

---

*) Formula uncertain, see text.

Table 12
Chemical Composition of Sinicite from China after [4, table between pp. 40/1].

| | content in % in sample number | | | | |
	1	2	3	4 [d]	5 [f]
ThO_2	5.64	6.46	7.24	7.93	8.29
$UO_3 + UO_2$	6.30	5.54	6.92 [a]	6.83 [a]	5.98 [a]
RE_2O_3	23.39	23.86	25.50	24.41	25.57
CaO	2.55	2.57	1.19	1.36	1.20
MgO	0.04	0.17	0.33	0.55	—
MnO	0.02	0.08	0.09	0.12	0.15
TiO_2	35.07	30.39	49.56 [b]	47.48 [b]	24.96
Nb_2O_5	16.46	16.87	—	—	20.85
Ta_2O_5	1.09	1.06	—	—	2.75
Al_2O_3	2.38	1.74	1.27	2.08	1.81
Fe_2O_3	2.19	3.05	2.82	3.62	4.44
SiO_2	0.04	0.60	0.68	0.31	0.59
H_2O	3.17	6.16	3.15 [c]	4.22 [e]	2.74 [g]
sum	98.34	98.55	98.75	98.91	99.33

[a] Given as UO_3 in [2]. — [b] Comprises TiO_2 + Nb_2O_5 + Ta_2O_5. — [c] Comprises 0.42% H_2O^- and 2.73% H_2O^+ [2]. — [d] Reddish brown after [2], see also [6]. — [e] Comprises 0.85% H_2O^- and 3.37% H_2O^+ [2], see also [6]. — [f] Blackish brown after [2], see also [6]. — [g] Comprises 0.21% H_2O^- and 2.53% H_2O^+ [2], see also [6].

Chemically (refer to Table 12), sinicite is a niobotitanate of RE (containing U, Th, and some Ca), which displays some relationship to priorite-blomstrandine (see below), but differs from aeschynite by the presence of U oxides and of enhanced contents of water. For group-A cations, the summary content of radioactive elements in sinicite is fairly stable (11.94 to 14.76%) and similar to that of euxenite-polycrase or something lower than for priorite-blomstrandine. Usually, Th predominates over U, and the ratios $ThO_2:UO_2$ vary between 1.04 and 1.16, or are even up to 1.38. The RE content is high, with mainly Ce and RE(Ce). The content of CaO is small. For group-B cations, the content of TiO_2 predominates over that of Nb_2O_5; Ta_2O_5 is only subordinate. Constantly present in this group are also Al, Fe, and Si. Thus, according to its composition of group-A cations, sinicite is closer to blomstrandine than priorite. Its group-B cations also are similar to blomstrandine. Therefore and owing to its X-ray diffraction data (see below), true sinicite may be viewed as cerian blomstrandine, but not as aeschynite [4, pp. 36, 39/43] as is given in [3].

The formula of sinicite may be expressed by 2 RO · 4 R_2O_3 · ThO_2 · 10 TiO_2 · 3 $(Nb,Ta)_2O_5$ [2], see also [5], or by $(Nd,Y,Th,U,Ca)_{1-n}(Ti,Nb)_2O_{6-m}$; for a detailed calculation of this formula regarding the valence state of Fe, see original paper [4, pp. 46/7].

Similarities in X-ray diffraction patterns obtained after heating for sinicite with priorite-blomstrandine indicate structural relationships leading to lattice parameters for sinicite of a = 5.15, b = 10.97, and c = 7.35 kX [4, pp. 37/9]. Heated at 1000°C for 2 h, its d values (in Å) with intensities given in parenthesis are as follows: 3.04(4), 2.94(10), 2.56(2), 2.45(2), 2.04(2), 1.92(2), 1.81(2), 1.74(1), 1.63(8), and 1.58(6) [2], see also [5, 6].

Sinicite is very brittle, fracture is conchoidal. Mohs hardness about 6.5. Density varies between 4.92 and 5.0 g/cm^3 (or 4.919 and 4.976 g/cm^3 [2], see also [5, 6]). Luster is vitreous to resinous. The brown color remains unchanged up to 800°C, lightens at 900°C (and the mineral looks like porcelain). In thin section, it is golden, optically isotropic, and has n = 2.23 to 2.33 [4, pp. 49/50].

As is corroborated by the DTA curve (see original paper), sinicite starts to recrystallize at a much higher temperature (724 to 730°C, or nearly 800°C from X-ray data) than priorite-blomstrandine. Between 800 and 900°C, a cubic phase with a = 10.28 to 10.32 kX occurs besides sinicite and a rhombic phase. The lattice constant of this cubic phase is much higher than for the cubic phase obtained from priorite-blomstrandine, and corresponds to oxides of RE(Ce). Heating at >1000°C produces a phase with a sinicite-type*) structure similar to the behavior of priorite-blomstrandine [4, pp. 37/9, 45/6].

Sinicite dissolves with difficulty in H$_2$SO$_4$, but quickly (after 15 min) in phosphoric acid (and colors the solution light yellowish, as is typical for U^{6+}) [4, p. 51].

References for 2.2.2.4:

[1] Go, Chen-Di (Sci. Record [Peking] New Ser. Vol. 2 [1958] 400/13).
[2] Go, Chen-Tszi; Chzhun, Chzhi-Chen (Kesyue Tunbao Nauchn. Vestn. Scientia **1957** No. 12, p. 378 from Ref. Zh. Geol. **1958** No. 19339).
[3] Chukhrov, F. V.; Bonshtedt-Kupletskaya, E. M. (Mineraly Spravochnik, Vol. 2, Pt. 3, Nauka, Moscow 1967, pp. 1/676, 361).
[4] Gorzhevskaya, S. A.; Sidorenko, G. A. (Geol. Mestorozhd. Redk. Elem. No. 23 [1964] 1/116).
[5] Bonshtedt-Kupletskaya, E. M. (Zap. Vses. Mineral. Obshch. **87** [1958] 479).
[6] Fleischer, M. (Am. Mineralogist **44** [1959] 467).
[7] Koptyaev, A. F. (Geokhimiya **1969** 894/9).

2.3 Carbonates

2.3.1 Thorbastnaesite Th(Ca, Ce)(CO$_3$)$_2$F$_2$ · 3 H$_2$O

Thorbastnaesite from alkalic rocks of the Sangilen Plateau, eastern Tuva, Siberia, USSR, was described in 1961 [3, pp. 582, 585]. It is named for its composition as a fluorocarbonate with contents of Th > Ca and RE [1, p. 111], refer also to [2].

In the following paragraphs only examples with >5% ThO$_2$ will be given.

Occurrence, Paragenesis, and Alteration. Thorbastnaesite in alkalic rocks of the Balyktygkhem massif, Sangilen Plateau, occurs in a tectonic zone of post-ore character on the periphery of the principal mineralized area; it is formed as a later mineral during hydrothermal alteration of an earlier mineral association [3, pp. 582, 589]. Further, this thorbastnaesite is mentioned as an accessory mineral occurring as segregations (of various size, see p. 239) in "one of the nepheline syenite intrusions of Eastern Siberia", where it is found in alkaline metasomatic rocks (iron-rich saccharoidal microcline albitites) of the exocontact, and was formed by pseudomorphic replacement of ferrithorite under the influence of fluorocarbonate solutions with high oxygen activity [1, pp. 105, 111], refer also to [2].

---

*) But note that priorite-blomstrandine gives euxenite after heating at about 1000°C, see pp. 221 and 230.

Thorbastnaesite is a rare mineral in dawsonite-rich silicocarbonatite sills of the Francon quarry near Montreal, Quebec, Canada, where it is associated with baddeleyite and zircon (cyrtolite) in the assemblage B (from the lower level sill) [4, pp. 115, 118], and it occurs as a rare mineral in white heterogeneous pegmatite of the Pond Lake Prospect, north of Mont-Laurier area, Grenville Province, Quebec, mainly associated with chloritized biotite and uranophane. Other associated minerals are hydroxylbastnaesite, serpentine, sulfides, allan-ite, zircon, uraninite, uranothorite, and complex U-, Th-, RE-, and Zr-bearing compounds. These minerals are products of alteration of primary radioactive minerals by introduced hydrothermal mineralizing solutions, and they crystallize in rims on the radioactive minerals or in fractures [6]. Both thorbastnaesite (identified by X-ray data) and chlorite replace thorite in the fractured rocks [7]. Note further, that in the fractured radioactive granitic rocks of the above-mentioned area, Th-bearing allanite has altered to a mixture of Th-free hydroxyl-bastnaesite and allanite, accompanied by a loss of Th [5].

In the Th-RE-fluorite-barite deposit of Beylikahir, Sivrihisar, western Anatolia, Turkey, bastnaesite with 22.5% Th (an intermediate member of the series bastnaesite-thorbastnaesite) is deposited as the last mineral in the final stage of a multiphase mineralization process (after barite, fluorite, calcite, limonite, and psilomelane) in the calcite cement of veins and fissure zones in brecciated and silicified volcanic rocks [8, pp. 59, 62/3].

A tendency for Th enrichment in secondary bastnaesites, formed pseudomorphously after allanite and, more rarely, other primary RE minerals, is corroborated by the following examples: Secondary Th-bearing bastnaesite, thorogummite, and cerianite occur as alter-ation products of primary allanite in a lateritized near-surface ore body at Morro do Ferro, Poços do Caldas, Minas Gerais, Brazil, where Th and RE are enriched in alkalic rocks and are closely associated with a magnetite-vein network cutting these rocks [9]. Microprobe investigations of concentrated material from this alteration zone (clay lenses) show that the most abundant secondary mineral is thorbastnaesite (or another RE-Th mineral), containing Th and RE in varying proportions (1 to 40% Th). A conclusive identification of thorbastnaesite, however, requires additional analyses [10]. In quartz-feldspar pegmatite lenses from the center of biotite granite of the Akzhailyau alkali massif, northwestern Tarbagatai, Kazakhstan, USSR, Th-rich bastnaesite constitutes about 40% of the pseudomorphs after allanite, which further contain illite ($\sim$53%) and Fe hydroxide ($\sim$5%) [17, table 2 after p. 140], or only illite ($\sim$59%) [18, p. 88]. For these pseudomorphs, the product of a complete metasomatic transformation of allanite, was calculated initially an addition of K, RE, Th, Al, C, O, and F and a loss of Na, Ca, Mg, Mn, Fe, Si, and numerous other elements [17, p. 142]. But this addition-loss process may be appreciably simplified to an addition of K, F, and CO_3^{2-} and a loss of Ca and O, because the original allanite and the pseudomorphs have essentially the same chemi-cal composition [18, p. 89]. A bastnaesite with 5.36% ThO_2, formed as a late-stage mineral in ankerite-barite veins of an Eastern Siberian carbonatite massif, replaces an earlier formed RE mineral, probably burbankite [11].

Chemistry. The formula for thorbastnaesite from the type locality, eastern Tuva, is calcu-lated from the analysis of contaminated material (No. 2, see p. 238) as $ThCe_{0.3}Ca_{0.7}(CO_3)_2F_2$ $\cdot$ 2 H_2O, similar to $ThCe_{0.2}Ca_{0.8}(CO_3)_2F_2$ $\cdot$ 3 H_2O according to X-ray data. With the surplus of Al_2O_3 (left after recalculation of its portion ascribed to albite inclusions) and the PbO content (due to radioactive decay of Th) other formulas can be calculated as $(Th_{0.96}Ca_{0.77}Ce_{0.24}Pb_{0.01}Al_{0.02})$ $\cdot$ $(C_{1.83}Al_{0.17})_2O_6$ $\cdot$ F_2 $\cdot$ 3 H_2O or, if a small surplus of positive charges is compensated by the entry of hydroxyl groups, as $(Th_{0.98}Ca_{0.65}RE_{0.28}Pb_{0.02}Al_{0.02})$ $\cdot$ $(C_{1.74}Al_{0.26})_2O_6[F_{1.77}(OH)_{0.18} \cdot O_{0.05}]_2$ $\cdot$ n H_2O.

Two analyses of the thorbastnaesite are as follows [1, pp. 110/1]:

sample	content in %							
	ThO_2	$\Sigma\,(RE,Y)_2O_3$	CaO	Al_2O_3	Fe_2O_3	PbO	SiO_2	CO_2
No. 1	46.79	7.46	7.97	1.76	3.62	0.33	2.01	14.78
No. 2	42.09	7.37	5.88	2.76	11.52	0.56	4.59	12.38

sample	content in %					
	F	H_2O^+	H_2O^-	others	sum (total)	sum (-O=F_2)
No. 1	6.87	9.04	2.14	0.16[a]	102.93	100.04
No. 2	5.44	4.98	3.63	1.53[b]	102.72	100.43

[a] Comprises 0.11% U_3O_8 and 0.05% TiO_2. — [b] Comprises 0.14% U_3O_8, 0.43% $(Nb,Ta)_2O_5$, and 0.96% TiO_2.

An older, tentative analysis of the Eastern Tuva thorbastnaesite gives 34.3% Th, 1.7% U, and 0.1648% ^{206}Pb [3, p. 585].

From the analysis of relatively pure thorbastnaesite (No. 1, see above), ThO_2, RE_2O_3, CO_2, F, and H_2O have been considered as the main components of the mineral, whereas Fe_2O_3, Al_2O_3, and SiO_2 may be ascribed to unseparated admixtures of iron hydroxide, albite, and quartz (which have been found in the analysis of less pure material). Total Pb is of radiogenic origin (refer also to [3, p. 585]) and the RE composition is dominated by the light RE, Ce itself being anomalously low. Whereas the content of CO_2 and F in analyses No. 1 and No. 2 is similar to that of supergene bastnaesite (from other localities), the H_2O content is higher. This surplus, together with a surplus of positive charges, suggests for both samples, an existence of hydroxyl groups (whose position in the IR spectra may be masked by the broader bands for molecular water, see p. 240) [1, pp. 109/11]. For RE content and composition in the eastern Tuva thorbastnaesite (and other Th-containing bastnaesites), see "Rare Earth Elements" A7, 1984, pp. 109/16.

Bastnaesite from an Eastern Siberian ankerite-baryte carbonatite vein contains 5.36% ThO_2, 63.70% $\Sigma\,RE_2O_3$, 20.4% CO_2, 0.98% F, and $\sim$1% CaO (+MgO) [11]. Assuming the total content of 2.09% ThO_2 for an allanite pseudomorph from Kazakhstan (composed of 40.65% bastnaesite and 59.27% illite) to be solely with the bastnaesite, the latter contains 5.14% ThO_2 [18, p. 88].

The replacement schemes given for Th in hydroxyl-bastnaesite in "Rare Earth Elements" A7, 1984, p. 108, are also relevant to thorbastnaesite. Additionally are given the schemes: 3 Th for 4 Ce, Th + F for Ce, and Ca + Th for 2 Ce [13]. The latter scheme, especially, is suggested by X-ray and chemical data [14, p. 444]. Refer also to reduced lattice parameters due to substitution of RE by Ca + Th, see pp. 239/40. The replacement of RE(Ce) by only $\leq$3% Th suggested by [16, p. 614] must be extended, probably, to the whole series bastnaesite — thorbastnaesite, for an intermediate member with 22.5% Th was found in Turkey [8, pp. 63/4].

During bastnaesitization of allanite, the Th shows a different behavior: It either enters the lattice of bastnaesite (Kazakhstan) [17, table 2 after p. 140], especially in the case of a total pseudomorphic replacement of allanite [18, p. 88], or, less commonly, forms thorite [19], for instance, as a result of special conditions in the solutions which cause a fractionation of Th and RE (Th combines with SiO_4, and RE with F and CO_2), and lead to formation of a second generation of allanite with only a very minor Th content (Canadian Shield) [20]. For bastnaesite pseudomorphous after thorite refer also to "Thorium" Suppl. Vol. A 1 b.

Crystal Form and Structure. Thorbastnaesite (Siberia) is hexagonal (ditrigonal-dipyramidal according to its space group, see below) [1, p. 106].

Siberian thorbastnaesite occurs as irregular or rounded segregations of microscopic grains, and reaches up to 3 cm in diameter; more rarely it forms little chains of quadratic to rectangular grains near relicts of microcline. Under the microscope, it displays a cryptocrystal-line structure with the microindividuals arranged subparallel. Under the electron microscope, it shows little globules of 0.2 to 0.5 µm in diameter and, sometimes, shows fine lamellae with a hexagonal outline [1, pp. 105/6]. The thorbastnaesite from Quebec occurs as spheres of radiating fibres up to 750 µm in diameter, or as aggregates of fine fibres forming coatings [4, p. 118]; that from Brazil is found as small botryoidal or mammillary masses, readily disintegrating into a fine powdery material [9].

The space group of thorbastnaesite is $D_{3h}^4 - P\bar{6}2c$ (No. 190) [1, p. 106], [2]. The lattice constants are a = 6.99 ± 0.02 and c = 9.71 ± 0.02 Å [1, p. 107], refer also to [2]. In contrast to normal bastnaesite (with Z = 6, refer also to "Rare Earth Elements" A7, 1984, p. 123), thorbastnaesite has Z = 3 [1, p. 110], [16, p. 618].

The X-ray powder diffraction data for thorbastnaesite from Siberia lead to the following indexed d values (only strongest lines given), intensity is given in parenthesis [1, p. 107], [2]:

hkl	d in Å	hkl	d in Å
11$\bar{2}$2	2.886(10)	14$\bar{5}$2	1.275(6)
30$\bar{3}$0	2.017(10)	22$\bar{4}$2	1.644(5)
30$\bar{3}$1	1.976(10)	41$\bar{5}$4	1.160(5)
11$\bar{2}$0	3.499(8)	50$\bar{5}$3	1.134(5)
30$\bar{3}$2	1.863(7) *)	33$\bar{6}$2	1.133(5)
21$\bar{3}$3	1.868(7)		

*) For this line a misprint is supposed because it precedes 1.868 [2].

The values for the sample from Turkey deviate somewhat with 2.86(10), 3.53(8), 2.054(6), 4.85(5), 1.886(5), and 1.667(4) [8, p. 63].

The X-ray data confirm that thorbastnaesite is isostructural with bastnaesite, see "Rare Earth Elements" A7, 1984, pp. 123/4, although the lattice parameters calculated for thorbastnaesite are somewhat less precise owing to diffuse lines. Compared with normal bastnaesite, the lattice parameters are somewhat reduced owing to the smaller ionic radius of Th^{4+} (1.02 Å) compared to that of Ce^{3+} (1.07 Å) [1, p. 106]. See also reduced lattice parameters for bastnaesites with appreciable quantities of Th, formed by alteration of allanite, from Nigeria (a = 7.056, c = 9.730 Å) [21, pp. 324, 327] and from Virginia (a ≅ 7.07, c ≅ 9.72 Å) [15]. Marked deviations occur also in the lattice parameters of bastnaesite with 5.36% ThO_2, formed by replacement of burbankite, in Siberian carbonatite (a = 7.11, c = 9.90 Å; recalculated from kX given in the original paper) [11]. Further, reduced lattice parameters are mentioned for bastnaesite, accompanied by thorite, from a radioactive pegmatite zone at Jabal Sayid, Saudi Arabia [12], and also for bastnaesites containing RE of higher atomic number or Y, see, for instance, bastnaesite-(Nd) [22] and bastnaesite-(Y) [21, pp. 325/6]; refer also to "Rare Earth Elements" A7, 1984, pp. 122/3.

Optical Properties. Thorbastnaesite (Siberia) in megascopic segregations is brown (similar to iron-stained albite) with refractive index n = 1.670 to 1.678, and shows aggregate polarization [1, p. 105]; that from Quebec is white, with silky to dull luster [4, p. 118]; and that from Turkey is white and shows a green-yellow color under the microscope [8, p. 63]. Thorian bastnaesite from a Siberian carbonatite is pinkish brown with greasy luster, and has n_o = 1.740, n_e = 1.800 [11]. Subparallel arranged micro-individuals of thorbastnaesite from Siberia give a very unclear conoscopic picture of an uniaxial crystal [1, p. 105].

The IR spectrum of unheated thorbastnaesite from Siberia shows absorption bands with wave numbers 755, 875, 1090, and 1400 to 1500 cm^{-1}, corresponding to valence and deformation oscillation of the CO_3 groups. A splitting of the very strong band at 1400 to 1500 cm^{-1} apparently is due to different grouping of CO_3 ions around cations (e.g., Ce^{3+} and Th^{4+}) and further may be influenced by α decay of Th. Diffuse bands at 3000 to 3600 cm^{-1}, with a maximum at 3450 cm^{-1}, are characteristic of molecular water and disappear after heating to 200°C (according to the text) or to 300°C (according to the figure given in the original paper), whereas the existence of hydroxyl groups with bands between 3000 and 3700 cm^{-1}, maxima at 3470 and 3620 cm^{-1}, may be masked by the broad bands of the molecular water [1, pp.107/9, 111], refer also to [2]. An additional investigation of this thorbastnaesite shows unusual wavenumbers for the CO_3 groups (750, 860 to 880, 1050, and 1420 to 1520 cm^{-1}), but shows the normal range for molecular water (3000 to 3600 cm^{-1}). The analogous spectra of unheated thorbastnaesite on the one hand, and of fluorbastnaesite and parisite, both heated to 400°C, on the other hand, are probably due to α decay of Th atoms, which cause local disarrangements of the structure along the path of the α particles and, mainly, of the nuclear recoil. In contrast to Th silicates (e.g., thorite), the absence of annealing effects in the case of Th carbonates is explained by the substantially covalent character of the bonds within the CO_3 group, compared with the SiO_4 tetrahedron, and by the dissociation process of CO_3 groups, which causes loss of volatile CO_2 and local irreversible reconstruction of the mineral structure due to the effect of α decay [14, pp. 445, 447/51]. The IR spectrum of Th-rich bastnaesite from Nigeria reveals the presence of CO_3 groups by absorption bands at 870 and 1450 cm^{-1} [21, pp. 325/7].

Other Physical Properties. Thorbastnaesite (Siberia) has a Mohs hardness of 4.5 to 5 [16, p. 618].

The measured density (by micropycnometer) for thorbastnaesite from eastern Tuva of ~ 4.04 g/cm^3 is markedly lower than 5.70 g/cm^3 calculated from unit cell data; the discrepancy is ascribed to the presence of molecular water [1, p. 106], refer also to [2].

Thermal Behavior. The DTA curve of thorbastnaesite (Siberia) shows a diffuse endothermic effect in the range up to 350°C owing to loss of water (and in accordance with the IR spectrum) and a sharp endothermic peak at 410°C caused by the dissociation of the mineral [1, pp. 106, 109]; refer also to [2] and to "Rare Earth Elements" A7, 1984, p. 135.

Thermic investigation of the mixture of thorian bastnaesite with illite in pseudomorphs after allanite from Kazakhstan shows weight losses of 2% between 0 and 250°C, owing to loss of water from illite; of 8% between 250 and 600°C by loss of CO_2; and of 3% between 600 and 950°C by loss of F, both from bastnaesite [18, p. 140].

References for 2.3.1:

[1] Pavlenko, A. S.; Orlova, L. P.; Akhmanova, M. V.; Tobelko, K. I. (Zap. Vses. Mineral. Obshch. **94** [1965] 105/13).
[2] Fleischer, M. (Am. Mineralogist **50** [1965] 1504/8, 1505).
[3] Zykov, S. I.; Stupnikova, N. I.; Pavlenko, A. S.; Tugarinov, A. I.; Orlova L. P. (Geokhimiya **1961** 547/60; Geochem. [USSR] **1961** 581/95).
[4] Sabina, A. P. (Paper Geol. Surv. Can. No. 79-1A [1979] 115/20).
[5] Rimsaite, J. (Paper Geol. Surv. Can. No. 82-1B [1982] 253/66, 256, 259).
[6] Rimsaite, J. (Paper Geol. Surv. Can. No. 85-1A [1985] 47/64, 53, 55).
[7] Rimsaite, J. (Paper Geol. Surv. Can. No. 83-1A [1983] 23/37, 31).
[8] Çagatay, N. (Turk. Jeol. Kurum Bul. **24** [1981] 59/65).
[9] Wedow, H., Jr. (Geol. Surv. Bull. [U.S.] No. 1185-D [1967] 1/34, 25/6).
[10] Barretto, P. M. C.; Fujimori, K. (Chem. Geol. **55** [1986] 297/312, 310/1).

[11] Gaidukova, V. S.; Zdorik, T. B. (Geol. Mestorozhd. Redk. Elem. No. 17 [1962] 86/117, 112/3).
[12] Turkistany, A. R. A.; Koyama, K. (I. A. G. [Inst. Appl. Geol.] Bull. No. 3 [1980] 99/104, 100, 103).
[13] Semenov, E. I. (Mineralogiya Redkikh Zemel', Akad. Nauk SSSR, Moscow 1963, pp. 1/412, 97).
[14] Akhmanova, M. V.; Orlova, L. P. (Geokhimiya **1966** 571/8; Geochem. Intern. **3** No. 3 [1966] 444/51).
[15] Riesmeyer, W. D. (Virginia J. Sci. [2] **18** [1967] 188).
[16] Povarennykh, A. S. (Crystal Chemical Classification of Minerals, Vol. 2, translated from Russian by Bradley, J. E. S., Plenum, New York – London 1972, pp. 459/766 [Russian original, Kiev 1966]).
[17] Chernov, V. I.; Krol', O. F.; Stepanov, A. V. (Geol. Geokhim. Mineral. Mestorozhd. Redk. Elem. Kaz. **1966** 138/44).
[18] Mineev, D. A. (Geokhimiya Apogranitov i Redkometal'nykh Metasomatitov Severo-Zapadnogo Tarbagataya, Nauka, Moscow 1968, pp. 1/183).
[19] Mineev, D. A.; Rozanov, K. I.; Smirnova, N. V.; Matrosova, T. I. (Dokl. Akad. Nauk SSSR **210** [1973] 925/8; Dokl. Acad. Sci. USSR Earth Sci. Sect. **210** [1973] 149/52).
[20] Littlejohn, A. L. (Paper Geol. Surv. Can. No. 81-1B [1981] 95/104, 100, 103).

[21] Hughes, J. C.; Brown, G. (Clay Minerals **12** [1977] 319/29).
[22] Podporina, E. K.; Niyazov, A. P.; Brylin, M. D. (in: Burkov, V. V., Ekzogennye Mestorozhdeniya Redkikh Elementov Sbornik Nauchnykh Statei, Akad. Nauk SSSR, Moscow 1980, pp. 21/6, 25).

2.3.2 Unnamed (Fe-)Y-Th Fluorocarbonate

A new (Fe-)Y-Th fluorocarbonate was found in 1982 as an accessory mineral in the radioactive granitic rocks from the Lawler Peak pluton, Bagdad area, and in the Dells granite near Prescott, both west-central Arizona, USA [1], refer also to [2, p. 408]. In the latter location, this unusual Y-Th fluorocarbonate phase is intergrown with apatite and monazite, or occurs as rims on the latter; it contains thorite as an inclusion [2, pp. 198, 323, 327]. In two monazite grains, the Fe-Y-Th fluorocarbonate phase is suggested to be the major phase in the outer rim. In one grain, the Th content is in the same level of about 3.5% in outer and inner zones. In the other grain with more intimate intergrowth of the monazite with other species, the polycrystalline rim also contains inclusions of thorite and probably hematite. The Th content of the outer fluorocarbonate zone is two- to three times higher compared with the monazite core containing about 6% Th [2, pp. 323, 327/34].

The chondrite-normalized RE pattern of the Y-Th fluorocarbonate from the Lawler Peak muscovite granite shows enrichment in heavy RE and in Ce (positive Ce anomaly), that of the sample from the Dells granite shows enrichment in the intermediate RE (compared with the light and heavy RE and Y) and has Ce < La [2, pp. 207/9, 324/6].

References for 2.3.2:

[1] Silver, L. T.; Woodhead, J. A.; Williams, I. S. (Uranium Explor. Methods Rev. NEA/IAEA R and D Programme Proc. Symp., Paris 1982, pp. 355/67, 360).
[2] Silver, L. T.; Woodhead, J. A.; Williams, I. S.; Chappell, B. W. (GJBX-7-84 [1984] 1/412).

2.4 Phosphates, Silicophosphates

According to [1], binary Th phosphates belong to three structural groups that are used as a principle of classification for the sections of this chapter: (1) Cheralite and the unnamed mineral $Fe^{2+}Th[PO_4]_2$ (as well as brabantite [2], unnamed Th-Y phosphate, and a Th-Y-Ca-P phase) are assigned to the monoclinic monazite group; see Section 2.4.1. (2) Brockite and the unnamed mineral $Fe^{2+}Th[PO_4]_2 \cdot H_2O$ belong to the hexagonal rhabdophane group; see Sections 2.4.2 and 2.4.3. (3) The unnamed mineral $(Fe^{2+}, Ca)_{1-x(?)}Th_{1-x(?)}(Fe^{3+}, RE)_{2x(?)}[PO_4]_2 \cdot 1-3H_2O$, see Section 2.4.8 (as well as the monophosphate grayite [3], see Section 2.4.5, and the unnamed mineral $Ca_{1-x}Th_{1-x}RE_{2x}[PO_4]_2 \cdot 2H_2O$ [4], see Section 2.4.7) belong to the orthorhombic ningyoite group. Note also that brockite and grayite are viewed as members of the rhabdophanite group [5].

Hexagonal eylettersite belongs to the crandallite series; see Section 2.4.4. As a questionable orthorhombic Th phosphate, kivuïte is mentioned; see Section 2.4.6. Quite recently, the triclinic uranyl phosphate althupite was described; see Section 2.4.9.

A comparison of the chemical composition of minerals of the monazite, rhabdophane, and ningyoite groups suggests the existence of a series of compounds with isomorphous replacement for cations (erroneously designated as anions in the original paper) as well as possibly a substitution of PO_4 by OH. The general formulas for the members of the above groups are given as follows (where M^{2+} and M^{3+} denote divalent and trivalent metals, respectively, and M^{4+} is Th^{4+} or U^{4+}): monazite group $= M^{2+}M^{4+}[PO_4]_2$; rhabdophane group $= M^{2+}M^{4+}[PO_4]_2 \cdot H_2O$; and ningyoite group $= M^{2+}_{1-x}M^{4+}_{1-x}M^{3+}_{2x}[PO_4]_2 \cdot 1-3\,H_2O$ [6, p. 23].

For the minerals of the above groups, hydration causes a transformation series from monoclinic cheralite through hexagonal brockite to an orthorhombic phase of the ningyoite group. Introduction of H_2O into the mineral structure is a main factor for the modification of the structure, whereas the partial substitution of PO_4 groups by CO_3 (leading to the segregation of Fe as $Fe_3^{2+}H_2[PO_4]_2$) is less important. Thus, starting with a monoclinic mineral (intermediate in composition between cheralite $(CaTh[PO_4]_2)$ and $Fe^{2+}Th[PO_4]_2$) displaying an extension of the X-ray diffraction spots (probably due to a substitution of the monoclinic structure by a hexagonal structure developing parallel to (1$\bar{1}$1)), addition of H_2O leads to the development of a diffraction pattern of orthorhombic $Ca_{1-x}Th_{1-x}RE_{2x}[PO_4]_2 \cdot 2\,H_2O$ "in statu nascendi" besides a pattern of hexagonal symmetry. The final stage of transformation is orthorhombic $(Ca, Fe)_{1-x}Th_{1-x}RE_{2x}[PO_4]_2 \cdot 2\,H_2O$ [1]. Note also that the H_2O content increases progressively during the transformation of monoclinic $Fe^{2+}Th[PO_4]_2$ (with $\leqslant 0.5\,H_2O$) through hexagonal $Fe^{2+}Th[PO_4]_2 \cdot H_2O$ into orthorhombic $(Fe^{2+}, Ca)_{1-x}Th_{1-x}(Fe^{3+}, RE)_{2x}[PO_4]_2 \cdot 1-3\,H_2O$ [6, pp. 17, 20].

Hydrous phosphates related to rhabdophane contain Ca and U or Th in place of RE and are either isostructural with rhabdophane (i.e., hexagonal) or are orthorhombic (pseudo-hexagonal) [3].

For SiO_2-rich mineral phases transitional in composition between Th phosphates and silicates, refer to the discussion of P-rich varieties of thorite ($ThSiO_4$) in "Thorium" Suppl. Vol. A1b.

References for 2.4:

[1] Kucha, H.; Wieczorek, A. (Mater. Konf. 6th Konf. Mikrosk. Elektron. Ciala Stalego, Krynica, Pol., 1981, pp. 449/52; C.A. **99** [1983] No. 143271).
[2] Rose, D. (Neues Jahrb. Mineral. Monatsh. **1980** 247/57, 249).
[3] Fisher, F. G.; Meyrowitz, R. (Am. Mineralogist **47** [1962] 1346/55, 1346).
[4] Kucha, H.; Wieczorek, A. (Mineral. Polonica **11** No. 1 [1979] 123/36 from C.A. **96** [1982] No. 146295).
[5] Floran, R. J.; Abraham, M. M.; Boatner, L. A.; Rappaz, M. (Sci. Basis Nucl. Waste Management **3** [1981] 507/14, 508).
[6] Kucha, H. (Mineral. Polonica **10** No. 1 [1980] 3/28).

2.4.1 Monazite Group

The monoclinic monazite group includes the Th phosphates monazite (see pp. 245/360), cheralite (see pp. 360/8), and brabantite (see pp. 368/70). As brabantite was discovered just recently, in the previous literature synthetic $CaTh[PO_4]_2$ had been used instead as the theoretical end member for this group.

The nomenclature of the monazite-group minerals can be connected with the system $2\,Ce[PO_4] - CaTh[PO_4]_2 - 2\,Th[SiO_4]$, as is given in **Fig. 11**. The monazite group comprises minerals with the general formula $A_2B_2X_8$ (where A = Ca, Pb, Ce, Th; B = P, Si, S; and X =

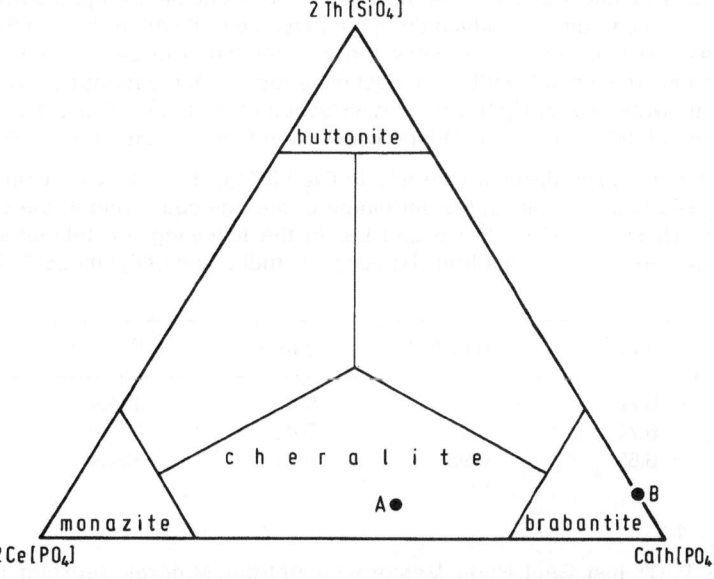

Fig. 11. Nomenclature scheme for the system $2\,Ce[PO_4] - CaTh[PO_4]_2 - 2Th[SiO_4]$ from [3]. Contents of U are added to Th; A is type cheralite and B is type brabantite.

O, OH) displaying an isomorphous replacement in all positions. End members of this group are RE-free huttonite $Th_2Si_2O_8$ (for its description, see "Thorium" Suppl. Vol. A1b), Th-free monazite $Ce_2P_2O_8$ [1], and brabantite containing between 75 and 100 mol% $CaTh[PO_4]_2$ [3]; see also [5]. Intermediate members comprise cheralite $ThCaP_2O_8$ and monazites with the general formula $(Ce, Th, Ca)_2(P, Si, S)_2(O, OH)_8$ [1]. In cheralite, Th generally predominates over RE [4].

In minerals with monazite structure, two main types of isomorphous replacement may occur when tetravalent cations (e.g., Th^{4+}) are incorporated instead of trivalent RE: Either PO_4^{3-} is substituted by SiO_4^{4-} leading to the formation of huttonite, or the compensation is achieved by equivalent amounts of divalent cations, mostly Ca^{2+}, leading to the formation of cheralite or brabantite. A substitution by U^{4+}, instead of Th^{4+}, may lead to the formation of a U analog of cheralite [3]. Evidence is provided for a substitution of the type $Th^{4+} + Si^{4+} \rightleftharpoons RE^{3+} + P^{5+}$ in the monoclinic system $(RE, Th, M^{2+}, U)[PO_4, SiO_4, OH, F]$ [6].

The monazite group apparently includes a complete solid-solution series from $(La, Ce)PO_4$ to $CaTh[PO_4]_2$ [7]; note also that there seems to be no significant miscibility gap in the series between monazite and brabantite [3]. It is not clear whether minerals of an isomorphous cheralite–huttonite series exist in the monazite group [8]. A solid solution is proposed between monazite, huttonite, cheralite, and $Fe^{2+}Th[PO_4]_2$, which may also include hypothetical monoclinic $ThPO_4OH$ or $Th(OH)_4$ [9]. The existence of an undisturbed mixture between monazite, brabantite, and $CaU[PO_4]_2$ is very likely, as is indicated when comparing the lattice constants for both synthetic $CaTh[PO_4]_2$ and $CaU[PO_4]_2$ [3]. However, a possible gap in Th and U contents between minerals with monazite structure is due, probably, to a partial miscibility or to a limited variation in the conditions of formation [10].

Monazite, cheralite, and $CaTh[PO_4]_2$ are isostructural with huttonite $(ThSiO_4)$ [2]; see also [8, 11]. In the structure of monazite and cheralite, Ce^{3+} ions are surrounded by eight oxygens (one half of the O^{2-} ions occurring nearer to Ce^{3+} than the other) and are connected with PO_4^{3-} groups [12]. But note that in natural minerals of the monazite group, the trivalent metal cations have a 9-fold coordination which does not place severe constraints on the symmetry of these cations and thus renders possibly the accomodation of such chemically diverse cations such as described above [13]. A monoclinic structure of the monazite type is capable of attaching some water (OH or H_2O) without destruction of the lattice. But at a content as low as <0.5 H_2O per unit $(M^{2+}, Th)PO_4$, H_2O-type water would destroy the monoclinic lattice [9].

On going from monazite through cheralite to $CaTh[PO_4]_2$, the lattice constants decrease systematically (see below), indicating a shrinkage of the unit cell owing to the smaller ionic radii of Ca and Th as compared to Ce and La. In the following are tabulated the lattice constants of monazite and cheralite from Travancore, India, and of synthetic $CaTh[PO_4]_2$ [2]:

mineral	a in Å	b in Å	c in Å	β	V in Å³
monazite	6.79	7.04	6.47	104.4°	299
cheralite	6.74	7.00	6.43	104.6°	294
$CaTh[PO_4]_2$	6.69	6.93	6.38	104.4°	286

References for 2.4.1:

[1] Sokova, K. P. (Tr. Inst. Geol. Rudn. Mestorozhd. Petrogr. Mineral. Geokhim. No. 81 [1962] 3/22, 21/2).
[2] Bowie, S. H. U.; Horne, J. E. T. (Mineral. Mag. **30** [1953/55] 93/9).

[3] Rose, D. (Neues Jahrb. Mineral. Monatsh. **1980** 247/57, 252/5).

[4] Mannucci, G.; Diella, V.; Gramaccioli, C. M. (Can. Mineralogist **24** [1986] 469/74).

[5] Speer, J. A. (Rev. Mineral. **1980** No. 5, pp. 113/35, 132).

[6] Kucha, H. (Mineral. Mag. **43** [1980] 1031/4).

[7] Jensen, B. B. (Norsk Geol. Tidsskr. **47** [1966] 9/19, 15).

[8] Nekrasov, I. Ya. (Dokl. Akad. Nauk SSSR **204** [1972] 941/3; Dokl. Acad. Sci. USSR Earth Sci. Sect. **204** [1972] 134/6).

[9] Kucha, H. (Mineral. Polonica **10** No. 1 [1980] 3/28, 7).

[10] Gramaccioli, C. M.; Segalstad, T. V. (Am. Mineralogist **63** [1978] 757/61).

[11] Deer, W. A.; Howie, R. A.; Zussman, J. (Rock-Forming Minerals, Vol. 5, Non-Silicates, Longmans, London 1962, pp. 1/371, 340).

[12] Sidorenko, G. A. (Geol. Mestorozhd. Redk. Elem. No. 26 [1966] 5/23, 12/3).

[13] Beall, G. W.; Boatner, L. A.; Mullica, D. F.; Milligan, W. O. (J. Inorg. Nucl. Chem. **43** [1981] 101/5).

2.4.1.1 Th-Containing Monazite (Ce, La, Nd, Th)PO$_4$

Although monazite, strictly speaking, is an RE mineral, it is included in this volume since it is the principal ore mineral of Th, see Chapter Deposits in "Thorium" Suppl. Vol. A1b and, for example, [1 to 5], and since its chemical composition seldom lacks Th, see Section 2.4.1.1.2, pp. 283/331. The following sections deal only with those monazite specimens containing $\geqq$ 1% ThO$_2$. Only such references are considered that combine mineralogical, paragenetical, structural, or physical data with chemical analyses; or that put emphasis on the occurrence of Th in monazite (not necessarily including chemical analyses).

Monazite from zircon granites of the Ilmen Mountains, Southern Urals, USSR, was described in 1829. The name is from the Greek "monazein" = to be solitary, in allusion to its (at that time) rather rare occurrence [6].

Originally applied to independent minerals, the following mineral names later on turned out to be synonymous with monazite (for the detailed discussion, see [7]): cryptolite (kryptolith), edwardsite, eremite, korarfveite (kårarfveit), monazitoid, phosphocerite, turnerite, and urdite (compilations of these mineral names are given also in [8 to 12]). Also synonymous with monazite are kularite (see below) and llalagualite [11].

A monazite variety with high U and Th content (see analysis No. 16 in Table 14, p. 306) but a low (Th + U + Ca):(Ce + La + Pr + Nd) ratio of 0.91 differs significantly from normal monazite in chemical composition, whereas its physical properties are similar [13]. Green U- and Th-rich monazite from Travancore, India, provisionally was called travancorite [14]. Monazite containing > 0.95% U$_3$O$_8$ is defined as U-rich monazite [15]. Monazite-(Ce, Nd) is a distinct variety characterized by a low ThO$_2$ content [16]. Dark Eu-rich monazite has a low ThO$_2$ content (rarely exceeding 1%) [17]; see also [18]. Europium monazite from the USSR is characterized by a low ThO$_2$ content between 0.64 and 0.94, mean ~0.7% ThO$_2$ [19].

Silicomonazite is a monazite variety intermediate in composition between monazite and huttonite (see, e.g., analysis No. 209 in Table 14, p. 328), and close to cheralite in its optical properties, X-ray powder diffraction pattern, and unit-cell data, but containing as much as 13% SiO$_2$ and only ~1.5% (ThO$_2$ + U$_3$O$_8$). It is suggested that minerals with a PO$_4$:SiO$_4$ ratio of 3:1 to 1:1 should be called silicomonazite to distinguish them from monazite, which sometimes may be rich in SiO$_2$ ($\leqq$5%) [20]. More recently it was shown that a sizeable SiO$_2$ content is caused by inclusions of amorphous SiO$_2$ in the gray or black monazite (with 0.48

to 1.06% ThO$_2$) named kularite (in allusion to its type occurrence Kular Ridge, Yakutia, USSR) [21]. It is identical with monazite [22].

Erikite from lujavrites of the Ilimaussaq massif, southern West Greenland, shows an X-ray powder pattern identical with that of monazite [23] and probably is a pseudomorph of monazite after eudialyte [24]. Erikite, occurring in pegmatites of the Lovozero alkaline massif, Kola Peninsula, USSR [25], perhaps is related to rhabdophane [26]. It is a pseudomorph of at least two components after an orthorhombic or pseudohexagonal mineral, possibly rhabdophane, grayite, or RE silico-apatite [12, p. 381].

Guadarramite in a feldspar matrix was described in 1906 [27]. It is an intergrowth of radioactive monazite and some ilmenite [28], and the name should be dropped [29], [30, pp. 549/50].

Pisekite originally was considered either as a new mineral isomorphous with monazite, or as a pseudomorph after monazite [31]. It is similar to monazite with respect to crystallographical properties and probably is a metamict Nb-U-RE monazite [12, p. 313], or it is an isotropized monazite containing Nb, Ta, Ti, U, Y, Yb, and Th [32]. But note that pisekite is also regarded as a mineral closely related to minerals of the orthorhombic samarskite [33] or samarskite-yttrotantalite group, with an affinity to the fergusonite-formanite group in certain properties [34]. Possibly, it is a urano-tantalo-niobate and -titanate of RE(Ce) and RE(Y) related to ampangabeite (samarskite) [30, pp. 453, 600/1]; see also "Rare Earth Elements" A7, 1984, p. 64.

References for 2.4.1.1:

[1] Cooper, P. F.; MacNevin, A. A.; Winward, K. (Mineral Ind. New South Wales Nos. 38, 40, 44 [1979] 1/80, 18).
[2] Davidson, C. F. (Mining Mag. [London] **94** [1956] 197/208, 197).
[3] Jarrard, L. D. (Bull. Montana Bur. Mines Geol. No. 15 [1957] 1/90, 41).
[4] Kelly, F. J. (U.S. Bur. Mines Inform. Circ. No. 8124 [1962] 1/38, 5).
[5] Twenhofel, W. S.; Buck, K. L. (Proc. Intern. Conf. Peaceful Uses At. Energy, Geneva 1955 [1956], Vol. 6, pp. 562/7).
[6] Breithaupt, A. (J. Chem. Phys. **55** [1829] 301/3).
[7] Hintze, C.; Linck, G. (Handbuch der Mineralogie, Vol. 1, Pt. 4, De Gruyter, Berlin 1921 [1931], pp. 1/632, 315/6).
[8] Palache, C.; Berman, H.; Frondel, C. (Dana's System of Mineralogy, 7th Ed., Vol. 2, Wiley, New York 1963, pp. 1/1124, 691).
[9] Frondel, C. (Geol. Surv. Bull. [U.S.] No. 1064 [1958] 1/400, 150/1).
[10] Hey, M. H. (An Index of Mineral Species and Varieties Arranged Chemically, 2nd Ed., Jarrold and Sons, Norwich, U.K., 1962, pp. 1/728, 238).

[11] Semenov, E. I. (Mineralogiya Redkikh Zemel', Akad. Nauk SSSR, Moscow 1963, pp. 1/412, 115).
[12] Strunz, H. (Mineralogische Tabellen, 5th Ed., Geest & Portig, Leipzig 1970, pp. 1/621).
[13] Gramaccioli, C. M.; Segalstad, T. V. (Am. Mineralogist **63** [1978] 757/61).
[14] Brown, J. C.; Dey, A. K. (India's Mineral Wealth, Oxford Univ. Press, London 1955, pp. 1/761, 278).
[15] Overstreet, W. C.; White, A. M.; Warr, J. J., Jr. (Geol. Surv. Profess. Paper [U.S.] No. 700-D [1970] D169/D175).
[16] Andersen, T. (Mineral. Mag. **50** [1986] 503/9).
[17] Rosenblum, S.; Mosier, E. L. (Geol. Surv. Profess. Paper [U.S.] No. 1181 [1983] 1/67, 39, 41).

[18] Kosterin, A. V.; Alekhina, K. N.; Kizyura, V. E. (Soobshch. Dal'nevost. Filiala Sibirsk. Otd. Akad. Nauk SSSR **1962** No. 15, pp. 23/6).

[19] Semenov, E. I.; Khomyakov, A. P. (Redk. Elem. No. 6 [1971/72] 77/81).

[20] Nekrasov, I. Ya. (Dokl. Akad. Nauk SSSR **204** [1972] 941/3; Dokl. Acad. Sci. USSR Earth Sci. Sect. **204** [1972] 134/6).

[21] Nekrasova, R. A.; Nekrasov, I. Ya. (Dokl. Akad. Nauk SSSR **268** [1983] 688/93; Dokl. Acad. Sci. USSR Earth Sci. Sect. **268** [1983] 151/5).

[22] Dunn, P. J.; Grice, J. D.; Fleischer, M.; Pabst, A. (Am. Mineralogist **69** [1984] 210/5).

[23] Danø, M.; Sørensen, H. (Medd. Groenland **162** No. 5 [1959] 1/35, 13).

[24] Fleischer, M. (Am. Mineralogist **44** [1959] 1329).

[25] Vlasov, K. A.; Kuz'menko, M. V.; Es'kova, E. M. (Lovozerskii Shchelochnoi Massiv (Porody, Pegmatity, Mineralogiya, Geokhimiya i Genezis), Akad. Nauk SSSR, Moscow 1959, pp. 1/623, 425).

[26] Fleischer, M. (Am. Mineralogist **45** [1960] 1135).

[27] Munoz del Castillo, J. (Bol. Real. Soc. Espan. Hist. Nat. **4** [1906] 479/81 from [29]).

[28] Frondel, J. W.; Fleischer, M. (Geol. Surv. Bull. [U.S.] No. 1009-F [1955] 169/209, 185/6).

[29] Switzer, G. (Am. Mineralogist **37** [1952] 1061).

[30] Chudoba, K. F. (Handbuch der Mineralogie, Neue Mineralien und Mineralnamen, Erg.-Bd. 2, De Gruyter, Berlin 1960, pp. 1/958).

[31] Krejči, A. (Casopis Mineral. Geol. **1** [1923] 2/5 from C.A. **1925** 952).

[32] Ramdohr, P.; Strunz, H. (Klockmanns Lehrbuch der Mineralogie, 16th Ed., Enke, Stuttgart 1978, pp. 1/876, 626).

[33] Bouška, V. (Univ. Carolina Geol. **3** [1957] 139/45).

[34] Bouška, V. (Rozpr. Cesk. Akad. Ved Rada Mat. Prir. Ved **70** No. 3 [1960] 1/35, 24, 33).

2.4.1.1.1 Occurrence, Paragenesis, Intergrowths, Inclusions, and Alteration. Synthesis

Introductory Note. Owing to the vast number of papers dealing, partly only in a cursory manner, with the occurrence and/or paragenesis of monazite, the following paragraphs will give only locations not mentioned by W. C. Overstreet, 1967, in a special paper on the geological occurrence of monazite (Geol. Surv. Profess. Paper [U.S.] No. 530 [1967] 1/327). Further, not included here are statements concerning only host rock, locality, and/or chemical analysis without additional information. For data on host rock and/or locality of chemically analyzed monazites, the reader is referred also to Table 14, pp. 305/28.

2.4.1.1.1.1 Occurrence and Paragenesis

As shown by examples in the following paragraphs, Th-containing monazite occurs mainly in granites (see pp. 250/3), pegmatites (see pp. 254/7), and metasomatically altered rocks and metasomatites (see pp. 266/70). Economically utilizable occurrences of monazite are found in placer deposits (see pp. 259/63). Moderately abundant are monazite occurrences in alkalic igneous rocks (see pp. 249/50), hydrothermal veins (see pp. 257/8), and metamorphic schists and gneisses (see p. 265). Only a few occurrences of monazite containing $\geq 1\%$ ThO_2 are known in carbonatites (see p. 259); in igneous rocks ranging in composition from monzonite to granodiorite (see p. 250); in alkali granites and alaskites (see pp. 253/4); in aplites and pneumatolytic to hypothermal deposits (see p. 257); in volcanic and metavolcanic porphyritic rocks (see pp. 250 and 266, respectively); in consolidated sedimentary rocks comprising conglomerates, sandstones, and black shales (see pp. 263/4); in residual clays, soils and sands, and bauxites (see p. 264); in migmatites and anatectic rocks (see pp. 265/6); in metamorphic carbonaceous rocks (see p. 265); and in quartzites (see p. 266).

General Statements on Occurrence, Paragenesis, and Abundance

Accessory monazite occurs in igneous rocks ranging in composition from diorite to muscovite granite, and also in associated pegmatites, greisens, and quartz veins. It is not observed in mafic plutonic and volcanic rocks and does not occur as a primary constituent in silicic lavas. Commonly it is observed in biotite-quartz monzonites, two-mica granites, muscovite granites, and cassiterite-bearing granites. Monazite is found very rarely in syenites, but does occur locally in nepheline syenites and syenite pegmatites. It is abundant in carbonatites and in related alkalic volcanic rocks and dikes as well as in a wide variety of veins (ranging from quartz-chlorite veins and epithermal deposits through mesothermal quartz veins to hypothermal W- and Sn-bearing quartz veins and alteration zones). The most effective agents in the sedimentary cycle of monazite are those related to rock weathering, fluvial transport, and accumulation of the weathered material in beach placers which represent the economically most important monazite deposits. Monazite is an abundant minor accessory mineral in intermediate- and high-grade metamorphic rocks derived from argillaceous sediments, whereas it is less common and less abundant in metamorphic rocks derived from arenaceous, calcareous, or carbonaceous sediments. It is especially common in pelitic schists, gneisses, and migmatites of the upper subfacies of the amphibolite facies and of the granulite facies. As an accessory mineral, monazite rarely makes up more than a few hundredths of a percent. Extraordinary concentrations are observed for Th-rich monazite in veins and for Th-poor monazite in marbles and carbonatites [1, pp. 11/2]; see also [2]. Rarely, monazite occurs as a massive vein deposit. Detrital monazite occurs in ancient placers (consolidated and cemented conglomerates) as well as in recent beach and river sands and placer deposits [3].

Accessory monazite is associated with zircon, xenotime, and titanite in granitic igneous rocks and gneisses or schists. In pegmatites derived from granitic or syenitic igneous rocks, coarse monazite occurs together with zircon, xenotime, allanite, apatite, gadolinite, magnetite, and niobate-tantalates. Well-crystallized monazite is associated with anatase, rutile, zircon, epidote, titanite, ilmenite, and albite in pegmatitic and quartz-rich veins in metamorphic rocks [3]. Usually, monazite forms in pegmatites, sometimes in granites or gneisses in paragenesis with feldspars, zircon, and sometimes magnetite or ilmenite and in hydrothermal dolomite veins together with magnetite. During weathering of these primary deposits, monazite accumulates in fluvial and marine placers [4]. The paragenesis of monazite with other RE- and Ti-containing minerals in granitoid rocks (comprising biotite granite, alaskite, plagiomicrocline granite, alkali granite, and metasomatic rocks) is determined by the physico-chemical parameters (primarily by temperature and redox potential) of the mineral-forming medium. The most frequent associations are as follows: monazite + ilmenite (without titanite and allanite) + zircon; monazite + ilmenite + titanite (without allanite) + zircon; monazite + ilmenite + titanite + allanite; and monazite + titanite + allanite (without ilmenite) + zircon [5, pp. 135, 139]. For additional summaries of the occurrence and paragenesis of monazite, see, for example, [6 to 10], [11, p. 198], [12, p. 339].

Temperature (T) and pressure (p) conditions existing at the time of monazite crystallization control the abundance of monazite in igneous and metamorphic rocks. Thus, monazite is rare in rocks of the epizone (with lowest T and p), as, for example, in epizonal granite, slate, phyllite, or schist. In the intermediate mesozone, accessory monazite is moderately common in pelitic schist and gneiss and is more abundant in syn- or posttectonic granitic rocks (particularly including quartz monzonite, two-mica granite, muscovite granite, and pegmatite). In the katazone (with high T and p), monazite is about as abundant in metasedimentary rocks as in granitic rocks (e. g., in sillimanite schist and gneiss, granulite, silicic charnockite, and synkinematic quartz monzonite) [1, pp. 12, 16]. But note that T and p are unimportant factors in controlling the occurrence of primary accessory monazite because it is stable in the domain of magmatic activity [13, p. 451].

The abundance of monazite may be controlled also by the age of the host rock. Thus, monazite occurrences are more abundant in Precambrian than in younger rocks because regions occupied by Precambrian rocks contain greater proportions of high-grade metamorphic rocks and of plutonic igneous rocks, which are the preferred host rocks of monazite [1, pp. 25/6]. But see p. 251 for higher contents of monazite in younger granites as compared to older granodiorites.

Igneous Rocks

General. Differentiation under plutonic conditions yields granitic batholiths in which monazite is a minor accessory mineral, but large volumes of monazite-rich rocks are not formed. Locally, differentiation produces monazite-rich veins in the meso- and epizone. Extreme differentiation of alkalic rocks leads to the formation of Th-poor monazite in carbonatites. Fractionation during crystallization produces Th-rich monazite in pegmatites [1, p. 21]. All acid igneous rocks, especially granites, syenites, and foyaites, as well as their pegmatites, contain monazite [14]. Thus, accessory monazite is widespread in granites, granite aplites, quartz monzonites, granodiorites, and syenites [15, p. 75]. Mafic to intermediate igneous rocks appear to lack monazite [16]. Among 478 samples of monazite from igneous rocks (described for the most part in literature published prior to 1955 and including monazite-bearing rocks formed by partial or complete anatexis of sedimentary rocks and veins and alteration zones related to these rocks), none is from a mafic igneous rock. Only one sample is from diorite and two are from granodiorite, whereas 254 samples are from granite and 204 from pegmatite [1, p. 21].

Monazite-bearing varieties of **alkalic igneous rocks** are rare, except for syenites [1, p. 23] and lujavrites (see p. 250). Th-containing monazite occurs as an accessory mineral in syenite-diorites, alkali syenites, and nepheline monzonites of the Megri pluton, Little Caucasus, southern Armenian SSR [17, p. 103], where it amounts to 0.1 vol% in leucocratic pegmatoid syenite, especially in the albitized and fluoritized parts, and to 0.03 to 0.05 vol% in medium-grained syenite and nepheline monzonite [266]. Th-containing monazite is also found in two small stocks of medium-grained quartz syenite in the southern contact-near part of the Tommot massif, Verkhoyansk-Chukotka Fold Region, Eastern Siberia [18, p. 80]. Monazite is one of the most typical accessory minerals in brecciated granosyenites of a sedimentary-metamorphic complex in Siberia [19, pp. 56/7, 72]. Rarely, monazite occurs, as single grains, in the Khodzha-Achkan(skii) and Dzhilisui(skii) nepheline-syenite massifs (comprising mainly acmite (aegirine)-augite and biotite-nepheline syenites) on the upper Khodzha-Achkan River, Alai Mountain Range, USSR [20]. Radioactive accessory monazite is associated with thorite, an unidentified isotropic mineral, and Fe-oxide stain in syenites near Samalpatti and Kumandapatti, Salem district, Madras, India [21, pp. 112/3]. Yellow-brown and subordinately green monazite occur in mineralized coarse-grained radioactive quartz-syenite bodies intruded into granite gneisses of Maturigiri and Dhurakantagiri, western Garo Hills district, Meghalaya, India. Monazite formed after the associated uraninite [22]. Possibly as a relict mineral of primary source rocks, rare accessory monazite occurs as oval grains and tabular crystals in fenitized nepheline-syenite xenoliths in the contact zone between trachytoid khibinites and rischorrites of the Khibiny massif, Kola Peninsula, USSR. Monazite is associated with chrysoberyl as well as zircon, corundum, cordierite, ilmenite, ilmenorutile, hercynite, apatite, celestine, pyrrhotite, and sphalerite [23].

Several clusters of monazite grains, which most probably formed at the expense of eudialyte (see p. 274), occur in lujavrites (nepheline syenites) from several parts of the Ilímaussaq massif, southern West Greenland [24], as in syenite and analcime-rich lujavrite that often contain monazite as an important constituent of pseudomorphs after eudialyte [25]; and in green medium- to coarse-grained lujavrite and in black laminated arfvedsonite lujavrite where dark-colored monazite occurs in large cloudy aggregates of small elongate crystals [26, pp. 20, 35/6]; see also [27]. Clusters of small angular grains of variably radioactive monazite occur in lujavrite as well as in naujaite (sodalite-nepheline syenite) and kakortokite (eudialyte-arfvedsonite-nepheline syenite) from the lower part of the Ilímaussaq massif [28]. As part of a Th-U mineralization, monazite occurs in medium- to coarse-grained lujavrites from the southernmost part of the Kvanefjeld plateau. It is associated with sodalite, acmite (aegirine), blue apatite, neptunite, sphalerite, britholite, Li mica, sorensenite, pectolite, and steenstrupine. In addition, it occurs (with amounts up to 20 vol%) in thin veins enriched in analcime or ussingite-albite in the eastern part of the plateau. Associated are pyrochlore, "igdloite", neptunite, blue apatite, chkalovite, thorite, and steenstrupine. Acmite(aegirine)-rich greenish to greenish brown lujavrite in the northernmost part, near the contact against volcanic rocks, contains monazite, together with steenstrupine, sphalerite, Li mica, fluorite, calcite, and pseudomorphs after eudialyte and steenstrupine (?). In the radioactive volcanic rocks of the Kvanefjeld, monazite is associated with natrolite, analcime, arfvedsonite, acmite (aegirine), and steenstrupine [29].

Accessory monazite occurs in porphyritic **monzonite** of the Kirovograd complex [30, p. 141] and in black monzonite of the Novo-Ukrainkii massif, both Ukraine, USSR [31, pp. 161/2].

In granitic rocks ranging in composition from **quartz monzonite to granodiorite** of the Mount Wheeler mine area, White Pine county, Nevada, monazite is the only accessory mineral in samples containing $\lesssim 0.7\%$ CaO. Both monazite and allanite are present in rocks with intermediate CaO content, and monazite is absent at a CaO content $\gtrsim 2\%$ [32]. But note that this relationship between the relative distribution of monazite and allanite and the CaO content of their host rock may be of only local application [267]. Primary monazite is a major accessory mineral (besides apatite and rutile) in red porphyroblastic microcline-bearing biotite granodiorite of Erdösmecske, Mecsek Mountains, Hungary, and varies in size between 5 and 100 μ. Rarely, finer-grained monazite occurs as an alteration product of feldspar and, associated with sphalerite and chalcopyrite, in the granodioritic part of the boundary between heavily biotitized amphibolite and red microcline granodiorite [33, pp. 62/3, 79/81]. Honey yellow monazite occurs in granodiorites of the Megri pluton, Little Caucasus, southern Armenian SSR [17, p. 103], and of the Ilin-Emnekenskii massif, Verkhoyansk-Chukotka Fold Region, Eastern Siberia, reaching a content of 3×10^{-4} vol% [18, pp. 182, 193, 195]. Monazite occurs in coarse-grained pegmatitic hornblende granodiorite (sobite) of Kamenka and in coarse-grained biotite-hornblende granodiorite of Skakunki near Lipovets, Ukraine [34, p. 570]; refer also to monazite in sobites of the Kirovograd region, Ukraine [31, pp. 161/2].

Monazite is very common in **granites**, especially in the most felsic and aluminous varieties. As follows from experimental studies, the solubility of monazite is low in peraluminous magmas and increases as melt composition becomes more meta-aluminous (for details, see original paper) [16]. Early crystallization of monazite occurs only in SiO_2-rich ($>70\%$ SiO_2), peraluminous granites (see p. 251) [35]. Due to its limited solubility, crystallization of monazite leads to a progressive depletion in Th and light RE for certain felsic differentiation suites.

This is consistent with a decreasing monazite solubility with proceeding differentiation (i.e., increasing SiO_2 content, decreasing temperature and CaO, MgO, and FeO contents) [36]. Thus, fractionation of monazite can cause a depletion of Th and light RE in leucocratic two-mica granites [37], and monazite, in contrast to allanite, is concentrated, for example, in granitoid rocks that are unsaturated in CaO and oversaturated in Al_2O_3 and SiO_2 of the Megri pluton, Little Caucasus, southern Armenian SSR [17, p. 103].

Together with allanite, monazite is a common accessory mineral in granites of the interior parts of North and South Carolina, northern Georgia, and Alabama [38] and of the Idaho batholith [1, p. 142]. In biotite granite, in coarse porphyritic two-mica granite, and in muscovite granite of the Lawler Peak pluton, Yavapai county, west-central Arizona, as well as in the Dells granite, Arizona, primary monazite occurs as grains of varying form, color, and clarity with highest degree of solid solution. It was at least metastable and suffered exsolution (see p. 270) [39, pp. 39, 105/6, 109/10, 123, 159, 197, 323]. Straw yellow through orange to turbid orange-brown equant grains of primary monazite (some as large as 150 μm, but mostly <75 μm) are associated with zircon, thorite, xenotime, an unidentified isotropic mineral, apatite, and aggregates of magnetite and Mn ilmenite in coarse porphyritic biotite-muscovite granite of the Lawler Peak pluton [40, pp. 14, 80, 85, 193, 206/7, 345].

Monazite is a main Th mineral in the Cacciola granite, central Gotthard massif, Switzerland [41]. It is associated with biotite, zircon, apatite, ThO_2-poor uraninite, and varying amounts of xenotime and Ti oxides in U-rich biotite-muscovite leucogranites of southeastern England, of the Massif Central and of Vendée, both France, and of Portugal [13, p. 448]. Monazite is very abundant in biotite-muscovite granites hosting U deposits at Urgeiriça, Bica, and Cunha Baixa, all Beira region, Portugal [42], and occurs as minute crystals (up to 0.18 mm long in younger and 0.05 mm in older rocks) in granites of the Visen district, northern Portugal. Younger granites show an increase in the amounts of monazite, zircon, and xenotime relative to the older suite which comprises largely potash-feldspathized granodiorites [43]. Monazite is the most abundant transparent mineral with important radioactivity in syntectonic alkalic two-mica granites of the Morvan region, eastern Massif Central, France [44]. Primary magmatic monazite is associated with uraninite, zircon, and apatite in granites of the Bourbonnaise Mountains, northeastern Massif Central. A reddish mineral, similar to monazite in composition and surrounded by zircon and apatite, occurs in two-mica granite of Pragniot-Vimont and in epigranular microgranite of Bois-Noirs, Bourbonnaise Mountains [45]. Accessory monazite is more abundant and has a higher radioactivity than zircon in unweathered two-mica granites of the uraniferous Saint-Sylvestre massif, Limousin region [46], and forms microscopically small inclusions in biotite in a granite near Vannes, Morbihan [47]. In the albitic topaze-lepidolite leucogranite of Echassières, Beauvoir, Brittany, monazite and brockite are the only Th phases observed. The extreme depletion of Th in the granite may result from the strong fractionation of monazite, either in the parental magma or in the source rock, during partial melting [48]. Early crystallized monazite is responsible for the depletion of Th in more evolved granites of the United Kingdom. It is associated: with apatite, zircon, xenotime, thorite, uraninite, and complex Nb-Ta-U oxides and silicates in the slightly peraluminous part of the Cairngorm granite containing 75 to 77% SiO_2 (whereas it is absent in the less evolved meta-aluminous part containing 70 to 72% SiO_2); with xenotime, apatite, zircon, titanite, allanite, and very rare thorite in the Loch Doon granite; and with apatite, zircon, and uraninite in the Skiddaw and Carnmenellis granites [35]. In coarse-grained porphyritic biotite granite of the Carnmenellis pluton, monazite and zircon are the main accessory minerals and occupy similar textural positions (mostly encrusted upon the surface of ilmenite or apatite crystals), suggesting that both minerals nucleated after crystallization of ilmenite or apatite had ceased. In medium-grained porphyritic biotite granite from the central part of the pluton, the abundance of the monazite-zircon-apatite-ilmenite association is lower [49]; refer also to the occurrence of rounded grains of monazite, 40 to 70 µm in size, in monzogranite of the Carnmenellis pluton [50].

Monazite and zircon, often contiguous or directly intergrown with equal portions, occur in the core of the Malsburg granite, Black Forest, F.R.G., and, with a predominance of zircon, in the border zone. An association or intergrowth of monazite with biotite indicates a common growth phase in a medium of low viscosity in which adhesive forces between already precipitated crystallites were effective [51]. As a frequent accessory mineral associated with zircon, monazite occurs in kaolinitic granite from Nadap, Velence Mountains, Hungary [33, pp. 61, 79/80]. Pale yellow to orange monazite (mostly 0.25 to 0.1 mm in size, rarely coarser) amounts to 17 g/t in coarse-grained and to 4.6 to 6.6 g/t in fine-grained biotite granites of the Rila-Planina region, Bulgaria. The amounts of monazite and zircon change nearly parallel, and only in two samples their behavior is reverse [52, pp. 6/7, 12/5].

The content of monazite in granitic rocks from different regions of the USSR varies: 113 g/t for the Talitskii massif, Altai Mountains; 398 g/t for eastern Sayan; 175 g/t for the Murzhinskii massif, Urals; 20 g/t for the El'dzhurtinskii massif, northern Caucasus [53]; and up to 30 g/t for the Kuramin zone, Central Asia [54].

Monazite is a typical accessory mineral in various granitoid rocks from some complexes of the Ukrainian Crystalline Shield: In the Chudnov-Berdichev complex, the content of yellow or clear yellow monazite ($\sim$ 120 g/t) is the same in leuco- and melanocratic medium-grained to pegmatoid granite types. Dark gray monazite in plagioclase granites is distributed more irregularly [55]; see also [56, p. 119]. Note also that in this complex, accessory monazite occurs as inclusions in rock-forming minerals of granites of the 1st intrusive phase; as separate crystals between biotites [30, p. 130], [57], [58, p. 19]; and as coarser-grained later formed mineral at the junctions of rock-forming minerals [58, p. 152]. Monazite formation was closely connected with the formation of these granites and was the result of "mobilization" of RE, Th, and P from the enclosing rock [56, pp. 130/1]; refer also to [30, p. 130], [57], [58, p. 19]. Granites of the Kirovograd-Zhitomir complex are extremely low in accessory monazite, which occurs as small irregularly shaped crystals only in aplite-pegmatitic varieties and is formed simultaneously with the rock-forming minerals. Fine-grained granites are practically free of monazite [56, pp. 120, 131]. Imperfect crystals of dark gray, yellow, or brown to black monazite occur in trachytoid granites of the Kirovograd-Zhitomir complex [55]. Fine-grained light yellow, yellow, or greenish yellow monazite occurs in porphyritic and trachytoid granites that formed by intense granitization of the surrounding rocks [31, pp. 159, 161/2, 168/70]. Monazite is associated with almandine, magnetite, ilmenite, apatite, zircon, rutile, and sometimes tourmaline in porphyritic K-Na granites (including unaltered biotite granites and muscovitized granites) of the Kirovograd-Zhitomir complex. The more acid K-rich granites have the highest monazite content [59, pp. 539/40]. But note that the content of monazite is lower in granites of the Kirovograd-Zhitomir complex than in those of the Chudnov-Berdichev complex [30, pp. 130, 141], [57]. In red porphyritic granites of the Verblyuzhka and Bokovyan massifs, monazite is associated with apatite, zircon, ilmenite, magnetite, and/or rutile [58, p. 16]. Monazite, displaying heterogeneous color, form, size, and degree of alteration, occurs in rose-colored banded medium-grained and pegmatoid granites of the Rusava suite, basin of the Ol'shanka, Markovka, and Rusava rivers, southern Ukraine [60, pp. 32, 65, 67], and in two-mica and biotite granites along the lower and middle reaches of the Berda, Kal'chik, and Kal'mius rivers, southeastern Ukraine. It is associated with apatite, zircon and its variety malacon, ilmenite, garnet, and tourmaline. Rarely associated are columbite-tantalite, topaze, fluorite, xenotime, and cassiterite [61]. Note also that monazite and ilmenite are the major accessory minerals in the medium- to coarse-grained biotite-containing Anatoliiskyi granites, basin of Kal'chik River [62]. Monazite occurs in gray plagiomicrocline granite of the Mokromoskovsky massif and in palingenetic plagiogranite [63] as well as in red porphyritic granite from the Novo-Danilovsk region, Dnepropetrovsk-Zaporozh'e region, Ukraine [64, 65].

Monazite (ranging in average content from <1 to 21×10^{-3}%) occurs only in microcline-containing biotite, two-mica, porphyritic, and alaskitic granites of the Dzhabyk-Sanar complex, Southern Urals, USSR. It crystallized simultaneously with the rock-forming minerals and zircon [66, pp. 196/7, 202/3]. Note also that monazite is the main carrier of Th (containing 50 to 90% of the total-rock Th content) in microcline-rich biotite and two-mica granites of Sanar, Kochkar region (as is inferred from balance calculations) [67], as well as in so-called anchi-eutectic plagioclase-microcline granites of the Dzhabyk complex [68].

In porphyritic granites of the Megri pluton, Little Caucasus, southern Armenian SSR, single grains of accessory monazite formed during the pneumatolytic stage [17]. Monazite occurs as brown translucent prismatic crystals in coarse-grained early granite and in lower amounts as smaller isometric yellow transparent crystals in most of the medium-grained late granites of the Kuu massif, central Kazakhstan, USSR. It generally predominates over other P-containing minerals (as, e.g., apatite) [69]. In unaltered rare-metal granite of the Orlinogor complex, northern Kazakhstan, isomorphous brownish yellow monazite is associated with biotite, plagioclase, and rarely with quartz [70]. Monazite occurs in biotite granite from the intermedi-ate zone of the Kalbinskii complex, Altai Mountains, where it is associated with magnetite, zircon, polycrase, tanteuxenite-(Y) (delorenzite), and ilmenite [71]. In small bodies of leuco-cratic granites and aplites of the Beloiyussk pluton, eastern Kuznetskii-Alatau, Kazakhstan, the monazite content is on average ~3.14 g/t and may even reach 50.21 g/t [72]. Accessory monazite is often covered by a pink film and is partly spotted by reddish patches in both two-mica and biotite granites of Siberia [73]. It occurs as an early formed phase in biotite and acmite (aegirine) granites of the Tommot massif, Verkhoyansk-Chukotka Fold Region, Eastern Siberia [18, pp. 80, 101]. Brown and yellow monazite in granites of the Tarakska intrusion, Enisei Ridge, Central Siberia, probably are derived from the enclosing rocks during a process of assimilation [74].

Monazite occurs in medium-grained porphyritic biotite granite of Ranong and in horn-blende-containing biotite granite of Phuket, both southern Thailand, and in coarse-grained porphyric biotite granite of Chiang Mai, northern Thailand [75, p. 32]. In minor amounts, monazite occurs in unaltered biotite granite of the Ririwai ring complex, Nigeria (together with zircon, columbite, TiO_2 minerals, thorite, ilmenite, marcasite, pyrite, sphalerite, galena, chalcopyrite, and cassiterite) [76], as well as in granites of the Mananjeba massif, Ambilobe district, northern Madagascar, and of the Vohimena Mountain, Androy district, southern Madagascar [77].

See pp. 267/8 for the occurrence of monazite in both metasomatically formed muscovite granites and tourmaline granite.

Light honey yellow lens-shaped grains of monazite with varying morphology occur in fine-to coarse-grained, partly pegmatoid **alkali granites** of the Kola Peninsula, USSR [5, pp. 127, 131]. Monazite is associated with fluocerite and bastnaesite in a peralkaline granite of Mulanje, Chilwa Alkaline Province, southern Malawi [78].

As a main host of Th, monazite occurs in black-quartz **alaskites** of the Ingulets River and its right tributaries, southern Ukraine, USSR. It is associated with magnetite, apatite, ilmenite, pyrite, zircon, garnet, rutile, and molybdenite [59, pp. 539, 542, 544/5, 547/8]. Its content is 0.0081% in alaskites and leucocratic granites of the Aktau massif, western Uzbekistan [79]; see also [80]. Orange and honey yellow monazite crystals up to 0.5 mm in size occur in alaskitic granites of the Or-Tau massif, Karaganda region, central Kazakhstan [81, pp. 205, 212/4].

Monazite occurs in granitic (see the following paragraphs) and, less commonly, in syenitic (see p. 257) or dioritic **pegmatites** [15, p. 75]. For the occurrence of monazite formed metasomatically in pegmatites, see pp. 266/7 and 268/9. In pegmatites, monazite often is associated with zircon, xenotime, cassiterite, columbite, apatite, wolframite, or magnetite [12, p. 344]. In many pegmatites, massive monazite crystals occur that often are secondary and replace feldspars [82]. In granitic pegmatites of the RE-niobate type, such coarse-grained monazite is associated with allanite, samarskite, and zircon [83].

Monazite is associated with xenotime, uraninite, uranothorite, zircon, and apatite in granite pegmatites of an area between Baie-Johan-Beetz, Costebelle Lake, and Baie-Pontbriand, Quebec, Canada. These radioactive and RE minerals crystallized either within the paleosome-plagioclase grains (at the boundary between plagioclase and quartz, adjacent to biotite), or in neosome-microcline portions of the rock [84, pp. 129, 133, 138]. Monazite and allanite are the most common Th- and RE-bearing minerals in pegmatites of the Piedmont region in North and South Carolina, northern Georgia, and Alabama [38]. Microscopically small crystals of monazite are associated with polycrase in a pegmatite at Day, New York [85]. Monazite of clearly magmatic origin is associated with zinnwaldite, spessartine, gahnite, blue tourmaline, and columbite, with muscovite and schorl, or with traces of beryl in the wall-rock zone of pegmatites of Brown Derby, Colorado. Monazite of probably magmatic origin is associated with euxenite, allanite, and radioactive ilmenite in the wall-rock zone of the Baumer pegmatite, Guffey area, and with euxenite and fluorite in albite-rich parts of zoned pegmatites of the Trout Creek Pass area, both Colorado [86, pp. 225, 227, 229/30].

Monazite occurs in considerable amounts in pegmatites of Argentina (e.g., in the Valle Fértil region, San Juan, and near Cosquín, Córdoba) [87], and it forms prismatic crystals some cm in size (some of which have small intercalations of mica) in samples of the Sierra de la Huerta [88]. The respective average contents of monazite and xenotime in two zoned pegmatites of Brazil are 225 and 123 (Estevão Pinto) and 49 and 164 mg/5 L of rock (Ibitiguaia). The contents of monazite and xenotime vary greatly across both pegmatites and, in general, the two minerals increase or decrease together. From the wall of the pegmatite inward there appear to be two zones of enrichment (one is in a mica-rich part near the quartz core). A kaolinite-rich zone in contact with the quartz core is low in both minerals. The ratio xenotime:monazite increases as the crystallization proceeds from the wall towards the core [89].

Red-brown crystals (1 to 5 mm in diameter) of monazite, or of a mineral quite close to monazite in the monazite—cheralite series, are associated with lepidolite in tourmaline-rich dark bands (composed of quartz, schorl, and albite) in the Haapaluoma granitic pegmatite, Peräseinäjoki-Alavus region, southern Pohjanmaa, Finland [90]. Yellowish brown monazite crystals are most abundant (up to 10 to 15 vol%) in biotite-rich parts of radioactive vein-like bodies with pegmatitic affinities which consist of plagioclase, quartz, biotite, and microcline and occur in cordierite-garnet gneiss, or between gneiss and pegmatite, at Puumala, eastern Finland. In biotite-poor zones, the content of monazite is much lower (≤ 2 to 6 vol%). Monazite is found partly as inclusions in biotite, partly on the borders of the biotite crystals, and partly entirely outside the biotite plates. It is surrounded by strongly altered zircon [91].

Green Th- and U-rich monazite occurs, together with uraninite and zircon, in a pegmatite at Piona, southern Alps [92, 93], as well as at Rio Graia and as large yellow crystals, together with euxenite and gadolinite, at Arvogno, both Val Vigezzo, Italy. Monazite with normal Th and U contents in pegmatites of Italy occurs as greenish crystals in the Val Dombastone, Sondalo, and as yellowish crystals in the Val Masino, Val Codera, and at Mondei, Valle Antrona [93]. Red monazite, zircon, and fergusonite are the main sources of radioactivity in pegmatites of the Szklarska Poręba region, Sudetes, Poland, and are concentrated in fractures and at contact surfaces of mineral grains [94].

In the European USSR, monazite is a prevailing accessory mineral in granitic pegmatites of the northeastern Lake Ladoga region, northern Karelia, and of the southwestern and central Kola Peninsula. It is associated with allanite, RE-enriched zircon (cyrtolite), and both rare uraninite and xenotime in oligoclase-muscovite-microcline pegmatites and with zircon in muscovite-albite pegmatites [95]. Monazite is the most abundant accessory mineral in zoned pegmatites of northern Karelia. In plagioclase-microcline pegmatites, it occurs in plagioclase and is associated with uraninite, secondary U minerals, zircon (cyrtolite), and xenotime [96, pp. 243/4]. In three plagioclase-microcline pegmatites of Chkalova and Chernaya Salma, and in three microcline-plagioclase pegmatites of the Kheto Lambino deposit, Chupa region, monazite and associated garnet, uraninite, zircon (cyrtolite), and xenotime are connected with block plagioclase near the selvages. Monazite crystallized earlier than, partly simultaneously with plagioclase, garnet, biotite, and quartz [97]. Monazite, occurring as single crystals or radial intergrowths of crystals, is associated with garnet, columbite, biotite, and xenotime in amazonitic pegmatites of the Ilmen Mountains, Southern Urals [98].

In the Ukrainian Crystalline Shield, coarse-grained monazite is found in syngenetic pegmatite bodies enclosed in garnet migmatites and Chudnov-Berdichev granites [99, p. 235]. Pale yellow idiomorphic monazite is distributed irregularly in syngenetic aplitic-pegmatitic granites and pegmatites of the Dnepr and Berdichev complexes. It forms lens- or schlieren-like segregations with coarse-grained or pegmatoid structure. Its content in these pegmatites reaches 100 g/t [56, pp. 119/21]. Flattened grains of yellow or black-brown monazite (with contents $\geq$ 100 g/t) are associated with allanite and magnetite in the silicic parts of rose-colored aplite-pegmatites from the Dnepropetrovsk-Zaporozh'e region [63]. Monazite occurs in rose-colored and light gray aplite-pegmatoid granites of the Kirovograd-Zhitomir complex [31, pp. 161/2] and, as rather coarse (0.25 to 1 mm in size) flattened brownish crystals, associated with dark brown short-prismatic zircon crystals, in mineralized zones in dark charnockitic and red aplitic-pegmatitic granites of the charnockite formation [100]. Note also that relatively coarse-grained (up to 3 mm) brown monazite is associated with black zircon in the pegmatoid parts of charnockitic rocks and in rose-colored pegmatitic segregations in charnockite at several places, Vinnitsa district, southeastern Ukraine [34, pp. 567/8]. Brown and subordinately yellowish brown, honey yellow, or yellow monazite crystals (between 0.1 and 0.75 mm, mostly between 0.2 and 0.4 mm in size) precipitated during the formation of rose-colored pegmatitic granites and pegmatites of the 2nd intrusive phase of the Bug-Podolian complex. Monazite occurs in biotite or between biotite and feldspar [30, pp. 130, 136/7], [57]. In these postorogenic granites and pegmatites, monazite is associated with magnetite, zircon, apatite, and pyrite. Coarser-grained monazite is formed later at the junctions of rock-forming minerals [58, pp. 15, 152]. In pegmatites occurring along the lower and middle reaches of the Berda, Kal'chik, and Kal'mius rivers, brownish red or lemon yellow monazite is associated with apatite, zircon and its variety malacon, ilmenite, garnet, and tourmaline. Rarely, it is associated with columbite-tantalite, topaze, fluorite, xenotime, and cassiterite [61]. Very coarse-grained monazite, partly forming well developed crystals, is the main accessory mineral in pegmatites of the central Azov massif (the southeastern part of the Ukrainian Crystalline Shield), whereas pegmatites from the western part are poor in monazite [62]. In relatively big pegmatitic bodies of the Azov massif, monazite is concentrated predominantly in marginal biotite-rich zones [63].

Honey yellow monazite formed mainly during the final phase of crystallization of the residual melt in pegmatites and lamprophyres of the Megri pluton, Little Caucasus, southern Armenian SSR [17, pp. 103, 106/7].

In the Asian USSR, monazite occurs in a coarse-grained porphyritic granite pegmatite of the Akchatau granite intrusion, central Kazakhstan [101] (forming 0.00994% of the rock's mineral content [79]); see also [81, p. 212]. It is associated with xenotime in pegmatites of the Aktau massif, western Uzbekistan [80]. Brown and yellow monazite occur in pegmatites of the Tarakskii intrusion [74] and of the exocontact aureole of the Transangara alkaline massif, both Enisei Ridge, Central Siberia. The Transangara monazite is associated with arfvedsonite, aeschynite, zircon, ferrithorite, and apatite [102]. As a rare accessory mineral, brownish rose monazite (as grains up to 4 mm in diameter) occurs in the pegmatite veins near the mouth of the Slyudyanka River, Lake Baikal, Siberia. It is associated with zircon, betafite, fergusonite, thorite, polymignite, and ferri-orangite [103]. Magmatic and postmagmatic monazite are closely associated with albite in pegmatites of the Tommot massif and in aplite-pegmatites of the Ninkatskii massif, Verkhoyansk-Chukotka Fold Region, Eastern Siberia [18, pp. 101, 193, 195].

Monazite crystallized during the entire process of pegmatite formation but is concentrated during the first stages in biotite granopegmatites of Inner Mongolia, People's Republic of China. It is associated with allanite, xenotime, euxenite, thorite, and uranothorite [104]. In pegmatites of India, monazite occurs as secondary alteration product of complex U minerals [105]. In high-temperature pegmatite of muscovite granite at Kurumbapatti, southern India, brown monazite with a size up to 2 cm is associated with beryl and columbite [106]. Monazite occurs in situ in weathered pegmatites at Nuwara Eliya, at Nugatenne near Teldeniya, and at Denegama, Balangoda, all Sri Lanka [107]. In Sn pegmatites of Thailand, the most common detrital minerals include monazite, magnetite, ilmenite, tourmaline, garnet, and zircon [108]. Monazite occurs in Thailand in Nb-Ta-Sn pegmatites of the Sutut Mine, Phuket, and of the Thung Kamin Mine, Takua Pa, and in Sn and Li-Sn pegmatites of the Tanti Ko Vit Mine [75, pp. 36/7]. Monazite as a major Th host is associated with fergusonite, samarskite, thorite, columbite-tantalite, xenotime, allanite, and zircon in granite pegmatites of Japan [109], as, for example, in the Ushiroda pegmatite, Ishikawa, Fukushima prefecture [110].

Reddish brown thorian monazite occurs in complex pegmatites of the Bundali Hills, Tukuyu district, Southern Highlands, Tanzania [111]. Primary monazite is associated with samarskite, beryl, and ilmenorutile in a pegmatite of Mount Darwin, Zimbabwe [112]. Monazite is associated with uraninite, tantalite-columbite, RE minerals, and some secondary U minerals in pegmatites of the Republic of South Africa [113, pp. PEL 112-8/9]. It is closely associated with apatite in the pegmatitic intrusions near Bandolierkop Station, Northern Transvaal [114], occurring as numerous yellow prisms in bands of massive apatite and as reddish brown grains distributed throughout a narrow dike but concentrated especially around apatite [112]. Monazite may be connected with the pegmatitic phase in the fluorite mineralization of leptite inclusions of the Bushveld granite at Buffelsfontein [114]. Brown monazite occurs in a pegmatite at Kakamas, Bushman Land [112], and as small single crystals in vein- or lens-like bodies of quartz-microcline perthite-oligoclase-albite pegmatites between Onseepkans and Putsonderwater, Kenhardt and Gordonia districts, Cape Province. In larger irregularly shaped pegmatites, it forms clusters ranging from 2 to 6 inches in diameter and is associated with muscovite [115]. Monazite occurs in pegmatoid masses or veins and sometimes is associated with apatite within the Anosyenne Chains and their eastern border, Madagascar [116]. At Marovato near Ampasimena, a rock consisting almost completely of small grains of monazite and apatite is enclosed in a pegmatite with smoky quartz, kalifeldspar, and biotite [117].

Lenses of monazite and apatite occur in a pegmatitic vein cutting leptynites north of Fort-Dauphin, southeastern Madagascar [77]. Monazite occurs in uraniferous pegmatites from the pegmatite field of Ankazobe, between Berere and Vohambohitra, and in an orthoclase-biotite pegmatite at the Manangotry pass [118].

Monazite occurs as a minor constituent in pegmatite lodes of the New England and Broken Hill districts, Australia [119, p. 18].

Within an area between Baie-Johan-Beetz, Costebelle Lake, and Baie-Pontbriand, Quebec, Canada, **syenite pegmatites**, consisting mainly of paleosome remnants rich in oligoclase, contain monazite commonly associated with apatite, allanite, and galena [84, p. 139]. Monazite is associated with aeschynite, zircon, and black mica in syenitic pegmatites connected with a fenitic aureole around a miaskitic intrusion of the Ilmen Mountains, Southern Urals, USSR [98]. In pegmatitic derivatives of syenites at Samalpatti and Pakkanadu, India, monazite forms brown-yellow crystals with a size up to 10 cm [106].

Monazite is less common in **aplites** (i. e., in granitic differentiation products low in volatiles) than in the more volatile fractions of the magma [1, p. 22]. It occurs in an aplite vein associated with the Romele granite at Stenberget on Romelåsen, southern Sweden [120], and as a rare primary accessory mineral in the grain-size fraction 10 μ in an aplite at Székesfehérvár, Velence Mountains, Hungary [33, pp. 61, 83]. The content of monazite is 0.0075% in granite aplite of the Aktau massif, western Uzbekistan, USSR [79]. For monazite in aplite-pegmatites of the Ukrainian Crystalline Shield and of Eastern Siberia, see pp. 255/6.

In **pneumatolytic to hypothermal deposits** of southwestern Japan (including Mo and W deposits), radioactive minerals include monazite, uraninite, coffinite, and Bi minerals [109].

Occasionally, monazite occurs in **hydrothermal veins** [14], forming over a considerable temperature-pressure range [15, p. 77]. It occurs in quartz veins containing fluorite and molybdenite; or hematite, carbonate, and thorite (see next paragraph); and in carbonate veins containing other Th minerals and Ti and Nb minerals (see p. 258) [1, p. 23]. In Alpine-type veins, monazite is associated with titanite, albite, rutile, and other minerals [12, p. 344]. In low-temperature Alpine-type veins, especially those cutting metasedimentary rocks, monazite is associated with anatase [15, p. 77]. Monazite occurs as a "classic" fissure mineral in the Alps, as, for example, at Kollergraben, Binn, Switzerland, and at the Rauris Plattenbrüche, Salzburg, Austria [93].

Monazite occurs in veins composed chiefly of quartz, potassium feldspar, and Fe oxide which often are adjacent to alkalic complexes and carbonatites [121]. Monazite occurs in some of the quartz-hematite-thorite veins and in replacement bodies of the Lemhi Pass area, Idaho-Montana [122]. Thus, in quartz veins cutting impure quartzitic and phyllitic rocks, it is subordinate to thorite and occurs as small well-shaped crystals or as irregular grains if engulfed within grains or aggregates of barite, feldspar, and quartz. Monazite partly mantles thorite and is penetrated by hematite (specularite), barite, and other minerals, indicating that it is one of the early formed minerals (and perhaps only a little younger than thorite) and has resulted from the combination of Th and RE, which do not enter thorite and allanite, with the phosphate radical [123, pp. 178, 185, 195]. Monazite is a minor mineral in quartz-microcline-Fe oxide-Th veins which also contain barite, calcite, siderite, rutile, pyrite, muscovite, and apatite. Brownish orange to tan granular monazite and thorite are the principal Th minerals in the Contact vein, North Frying Pan Creek, and coarse monazite occurs with microcline, barite, apatite, thorite, chrysocolla, brockite, rutile, magnetite, and hematite (specularite) in irregularly banded quartz-limonite-Mn oxide veins. These low-temperature Th veins probably were precipitated from fluids derived from an alkalic magma and following major through-going fractures [124, pp. A34, A40, A42/A43, A53, A62/A63, A68]. Monazite is the second most important Th mineral (following thorite) in roughly one third of the quartz-microcline-Fe oxide veins studied in the Lemhi Pass area. In a few samples, it is the most abundant or only Th mineral observed. It occurs together with thorite along fractures in early quartz [125, pp. 57, 60/4, 70/1]; see also [126]. Fine-grained monazite occurs in disseminated Th-RE deposits and in Th- and RE-bearing quartz-potassium feldspar-cristobalite (or calcite)-Fe oxide veins near the center of a trachyte and phonolite intrusion, southern Bear Lodge Mountains, Crook county, Wyoming [127]. The formation of monazite is connected with residual solutions in hydrothermal veins of the syenitic ridge of Crni Kamen near Alinci, southeastern Yugoslavia, which are composed of microcline, albite, quartz, two distinct varieties of titanite, ilmenite, arfvedsonite, and davidite [128, pp. 140, 147]. Hydrothermal solutions enriched in RE and volatile components led to the formation of monazite in quartz veins, in quartz crystals and druses, and in enclosing rocks of the central Pamirs and Urals, USSR. In simple and metasomatically reworked quartz veins of the central Pamirs, honey yellow monazite associated with apatite, rutile, zircon, ankerite, dolomite, pyrite, barite, and hematite occurs in two generations (as early strongly altered and as younger transparent light yellow crystals). Its content in the veins, druses, and hydrothermally altered quartzites is 2.78, 5.90, and 15.56 g/t, respectively. In quartz veins of the Southern Urals, pinkish yellow monazite occurs together with apatite, titanite, rutile, xenotime, and pyrite. Its content in rock crystal-containing and rock crystal-free veins and enclosing schists is 0.018, 0.005, and 0.226 g/t, respectively. In quartz veins of the Polar Urals, monazite is associated with titanite, rutile, pyrite, ilmenite, tourmaline, and anatase. Its content in the veins and adjacent crystalline schists is 0.881 and 0.014 g/t, respectively. The content of monazite is lower in veins containing xenotime or titanite because these earlier separated minerals apparently take up large amounts of P and RE [129, pp. 87/90, 93]. Monazite crystallized simultaneously with magnetite, ferrithorite, and columbite in RE-Th-Fe quartz veins located at the endocontact of a leucocratic biotite-granite massif in Kazakhstan, USSR. The content of monazite ranges from 1 to 8% [130]. Monazite occurs in quartz veins of the Akchatau granite intrusion [101] and as a high-temperature hydrothermal phase together with xenotime, fluorite, and fluocerite in variably mineralized pegmatites of Kentau, both central Kazakhstan [131]. Note also that monazite, xenotime, and allanite crystallized during a second hydrothermal stage simultaneously with smoky quartz, colorless quartz, and fluorite in druses in granitic pegmatites of Kazakhstan [132].

Monazite occurs in monazite-apatite veins of South Africa [121]. Thus, monazite is associated with apatite, zircon, pyrite, chalcopyrite, galena, magnetite, ilmenite, and quartz in monazite-bearing veins of Steenkampskraal [133]; see also [134, p. 710]. Monazite and apatite account for ~80% of the ore, and monazite occurs much in excess of apatite [135].

In the northern Wet Mountains of Colorado, in RE-Th carbonate veins, which are composed mainly of dolomite and small amounts of calcite, siderite, rhodochrosite as well as Fe oxides, the minerals monazite and bastnaesite never occur together. The reason may be that monazite forms preferentially if sufficient P is available [136]. Light yellow or pale rose-colored monazite formed together with calcite, ankerite, parisite, thorite, rutile, pyrite, and apatite during the hydrothermal calcitic stage in dense medium-grained calcite ores arranged in a vertical zone in brecciated granosyenites of an RE-fluorocarbonate deposit of Siberia [19, pp. 56/7, 72, 76/7]. Monazite occurs in hydrothermal veins in carbonate rocks of the Shinkolobwe U deposit, Congo [119, p. 20].

Star-shaped monazite (10 to 15 μm in size) occurs in calcite veinlets in granodiorite of Erdösmecske, Mecsek Mountains, Hungary. It is situated in the veinlets next to the rock-vein boundary, displays a distinct chemical composition (see original paper), and probably is the result of metasomatic carbonate mobilization [33, pp. 80/2].

Monazite occurs in **carbonatites** and associated calcite veins [15, p. 77]. In the former, it is moderately widespread and occurs in two distinctly different modifications: As early formed brown anhedral moderately to strongly radioactive grains (e. g., in the San Bernardino county, California, and at Ravalli, Montana) and as lately formed green, spherulitic aggregates with low or no radioactivity (e. g., at Kangankunde, Malawi; at Nkombwa Hill, Zambia; and at Wigu Hill, Tanzania) [137]. Monazite, Ca-bearing members of the bastnaesite group, and apatite are minor or moderately abundant accessory minerals in both ferrocarbonatite and magnetite ore as well as in hematite-calcite(-dolomite) carbonatite (rødbergite) of the Fen complex, Telemark, southeastern Norway. These minerals remain stable throughout the crystallization and alteration history of the carbonatites [138]. Sulfate monazite is disseminated in quartz, and rarely in carbonates, in siderite-dolomite and ankerite-calcite carbonatites as well as in quartz-carbonate veins located in fenitized gneisses and migmatites of the Namo-Vara region, Kola Peninsula, USSR. It is associated with pyrite, barite, celestine, and some other minerals and crystallized (instead of a normal RE phosphate) from low-temperature RE-deficient aqueous solutions saturated with Ca^{2+} and SO_4^{2-} and rich in P [139, pp. 374/5, 380]; see also [268]. Monazite is widely distributed in riebeckite-barite-calcite carbonatites of Samalpatti and Pakkanadu, Tamil Nadu, India [140], and in intrusive carbonatites of Central and East Africa [119, p. 20].

Pyrochlore-bearing carbonatites always contain monazite, which, however, is practically devoid of Th [14].

Sedimentary Rocks

Because monazite resists weathering, it tends to become concentrated in placer deposits [143] together with other heavy minerals (such as zircon, ilmenite, magnetite, rutile, garnet, cassiterite, and gold) [113, p. PEL 112-3] and is derived mainly from nearby source rocks, as can be seen from the following examples. Monazite is very common in fluvial (stream) and beach placers (see below and pp. 260/2, respectively) as well as in "fossil" or consolidated placers (see pp. 262/3) [15, p. 78]. The heavy-mineral fractions in placer deposits are richer in monazite than the igneous or metamorphic source rocks [143].

In **stream placers**, monazite generally is the principal and sometimes only Th mineral [121]. During fluviatile transport, detrital monazite lags behind detrital quartz, feldspar, and other common minerals and is concentrated with resistant minerals (such as ilmenite, rutile, zircon, and sillimanite) in the coarse fraction of stream placers [1, p. 24].

Alluvial dark monazite in a stream placer at Ruby, Yukon River, Alaska, originated from weakly metamorphosed black shales or siltstones [144, pp. 6, 38]. In stream placers of Idaho, the Th-bearing minerals (including monazite) and other heavy minerals were derived by erosion from the Idaho batholith [145]. Note also that the Idaho batholith is the source of detrital monazite (as principal Th mineral) as well as of euxenite, brannerite, samarskite, xenotime, (urano)thorite, ilmenite, zircon, tantalite, columbite, and allanite in the black-sand placers of Idaho [146]. Detrital monazite from stream gravels and sands in the Oconee, Spalding, Pike, Harris, and Troup counties, Georgia, was derived from weathered metamorphic rocks in the drainage basins of streams [147]. Monazite occurs in stream-placer deposits in the Piedmont of North and South Carolina [145], and monazite placers in alluvial sediments covering the floor of shallow narrow valleys underlain by deeply weathered monazite-bearing rocks are the most important type of Th and RE deposits in the Appalachian region of North and South Carolina (for localities, see original paper). The placers are highest in grade near the headwaters of the streams and become larger but less rich downstream [38]. Detrital grains of U-rich monazite occur in the gravel from placers of small streams in North and South Carolina and are derived from granitic rocks and associated pegmatites as well as from schists or gneisses [141]. Low-grade fluvial monazite placers of the western Piedmont (containing on average 0.06 pounds of monazite per cubic yard) occur in the Savannah River — Catawba River area, South and North Carolina, in the Yadkin River — Dan River area, North Carolina and Virginia, and in the Oconee River, Flint River, and Chattahoochee River areas, Georgia. Among a total of 140 concentrates studied, 111 samples contain 1 to 45% monazite, 15 contain <1% monazite, and 13 contain no monazite. The percentage of monazite in concentrates from fluvial placers in the Inner Piedmont belt, between the Savannah and Catawba rivers, rises from the flanks (generally <1 to 10%) to the core of the belt (generally 10 to 30%, locally as much as 60%). There is a concordant trend between the distribution of monazite and that of sillimanite, rutile, and garnet (for details, see original paper) [148, pp. 3/4, 16, 18/21, 63].

In Argentina, monazite occurs in placers of the provinces Jujuy (on Orosmayo and Cincel rivers), Córdoba (on Tercero River), and San Luis (on Quinto and La Carolina rivers) [87]. Note also that the alluvium of the Tercero River consists of fragments of quartz, feldspar, schists, and altered minerals as well as 5% heavy minerals (comprising mainly monazite and garnet) [149]. Monazite occurs in alluvial and eluvial fluviatile placers of Brazil, see Chapter Deposits in "Thorium" Suppl. Vol. A 1 b.

Brown-yellow well-rounded grains (with diameters between 0.15 and 0.30 mm, very rarely >0.50 mm) or crystals of monazite occur in Quaternary alluvial placers (including gravel and sand) of the Otava River and its tributaries, southwestern Bohemia, Czechoslovakia [150]. Dirty green to greenish yellow oval, spherulitic, or isometric grains of silicomonazite occur in alluvial placers of the Kyugyulyur Creek, Kular Ridge [151], and gray to black ellipsoidal to nodular grains of silicomonazite (kularite) are found in alluvial placers of small streams draining black shales in the interfluvial region of the Omoloy and Yana rivers, all Yakutia [152]. Alluvial placers of the Indian peninsula contain monazite along streams and rivers traversing Precambrian granitic, gneissic, or schistose rocks and pegmatites [21, pp. 118/9]. Monazite is associated with ilmenite, garnet, and zircon in stream placers of the Bentota River and its tributaries, southwestern Sri Lanka [153]. Note also that it occurs in gem gravels of Ratnapura, Rakwana, and Elahera, all Sri Lanka [142]. Monazite constitutes 25% of the heavy-mineral fraction (corresponding to ~1% of the total sediment) in placers on the Ezoma River, a tributary of the Mananara River [154, p. 52], and occurs in alluvial deposits at the mouth of the Mananjary River, both southeastern Madagascar [77].

Where monazite-bearing streams flow into oceans (see below) or lakes (see p. 262), **beach placers** tend to form at the mouth of the river and along downdrift shores. Coastal deposits are also formed by the constant sorting action of currents, waves, tides, storms, and wind [1, p. 24]. In beach placers, monazite generally occurs in minor amounts [121].

Monazite occurs in white beach sands at Pacific Grove, Monterey county, and in black sands situated at storm level on a beach immediately south of the mouth of the Año Nuevo Creek between Halfmoon and Monterey bays, California [155]. Monazite in beach-placer and -sand deposits in the western Piedmont, especially in South Carolina, was derived from monazite-bearing igneous and metamorphic rocks and associated pegmatites and hydrothermal veins [119, p. 20]. Thus, the heavy minerals (including 0.64 to 2.3, mean 1.15% monazite as well as ilmenite, zircon, rutile, and others) in composite samples of the coast of Hilton Head Island were eroded from the crystalline rocks of the Piedmont [156]. Monazite occurs in beach placers between Rio Grande do Norte and Estado do Rio de Janeiro, Brazil [157].

Monazite is associated with (urano)thorite, allanite, perrierite, and a Th-RE-Si titanate in radioactive beach placers washed out from tuffs and volcanic rocks south of the delta of the Tevere River (between Lido du Lavinio and the mouth of Garigliano River), Latian coast, Italy [158]. Monazite occurs among the heavy minerals in placer sands along the coast of the Baltic Sea, Poland [159].

In beach placers and dunes of the Malabar and Coromandel coasts, India, monazite was accumulated by the concentrating action of sea waves along with ilmenite, rutile, and zircon [160], and in black sands, its content ranges from 0.5 to 2% [105]. Monazite is associated with ilmenite and other heavy minerals in beach placers along the southwestern Indian coast [161], and its content is ~0.34% in the Chavara deposit, Kerala, and ranges from 0.269 to 2.54% in littoral placers of Kerala, and from 0.097 to 2.34% in sea-bed placers between Puttantura and Colachel Fort, Madras [21, p. 119]. Monazite occurs in beach sands of Manavala, Kurichi, and Quilon, southwestern India [162]. It is enriched together with zircon in the grain-size fraction <90 μm of these placers and of beach placers of Cape Comorin [163]. At Manavala, Kurichi, and Chavara, the productive fraction of the placers contains 0.5 to 1% monazite associated with ilmenite, zircon, rutile, sillimanite, garnet, and quartz [164]. Unlike placers along the western coast, black-sand concentrates along the Indian eastern coast occasionally contain magnetite or garnet as major constituent in addition to monazite, ilmenite, sillimanite, zircon, and rutile. The monazite content varies between 0.14 and 8% [161]. Littoral placers of Mailavani, Chihelanka, and Antarvedi, all Andhra Pradesh, contain 0.02 to 0.24% monazite [21, p. 119]. Monazite is associated with ilmenite, garnet, rutile, and zircon in seasonal beach placers at Kaikawala and Beruwala, Sri Lanka [153]. The source of monazite-rich concentrates at the Induruwa beach of Sri Lanka are pegmatites outcropping in the vicinity [107]. Dark monazite occurs in beach placers of southwestern Taiwan [144, p. 38]. Yellow monazite constitutes 2.51 to 3.06% of heavy-mineral concentrates from littoral placers (comprising beach sands and offshore bars) of three locations of Taiwan. Associated are black (Th-free) monazite, zircon, rutile, leucoxene, garnet, epidote, and Fe minerals. The yellow monazite originated from the plutonic core of the Central Range and from the Fukien granites [165, pp. 16, 19, 27/8]. Monazite is associated mainly with zircon, ilmenite, and magnetite in heavy sands from placers of western Taiwan (occurring along beaches, streams, and offshore bars) and originated from Tertiary sedimentary rocks [166]. In a heavy-mineral suite of a columbite-producing beach sand from Malaya, monazite is associated with xenotime and ilmenite and minor tourmaline, columbite, allanite, Ta-Nb rutile, cassiterite, and zircon [167]. Yellowish brown monazite occurs together with gold in heavy sands from placers of South Korea. Its content is 46% at Sukchon, 28% at Sunan, and 20% at Chonan [168].

About 85% of monazite in black sands at Rosetta, Nile delta, Egypt, occurs in the grain-size fraction between 135 and 73 μ [169]. It is deposited somewhat faster than most of the other heavy minerals and thus is transported for a shorter distance along the beaches east- and westwards of Rashîd (Rosetta) and between Dumyât (Damietta) and Port Said [170]. Well-rounded pale yellow to yellow-brown grains of monazite occur in superficial beach sands between Robertsport and Greenville, Liberia [171]. In beach and dune placers along the southeastern coast of Madagascar, monazite is associated with ilmenite and zircon, as is the case, for example, on the beaches around Fort-Dauphin and on old dunes bordering the southeastern coast of Madagascar [116], between the mouth of the Mandrare River and Vangaindrano [118] (see also [154, p. 78]), and in the beach sands of Vohibarika near Manantenina [172]. Monazite placers occur also on the coast between Vangaindrano and Tamatave, eastern Madagascar [154, p. 73].

Zircon-rutile placers on the beaches and dunes of southeastern Queensland and northeastern New South Wales, Australia, contain on average about 0.3% monazite [134, p. 711]. And it is an important host of Th in heavy-mineral black sands deposited mainly during cyclonic storms along the coast of New South Wales and the adjoining coast of Queensland. Its content ranges between 0.2 and 0.4% of the heavy-mineral concentrate and increases up to 2 to 4% in the area around Newcastle and Port Stephens. Monazite is associated with zircon, rutile, and ilmenite and minor cassiterite, gold, and platinoid metals [173]. About 0.5% monazite is contained in wave-concentrated zircon-rutile-ilmenite beach placers between the mouth of the Shoalhaven River, New South Wales, and near Frazer Island off the coast of Queensland. Biotite-rich Precambrian rocks are suggested as source of most of the monazite [174]. Note also that the monazite content ranges from 0 to 2.5% in heavy beach concentrates on the eastern Australian coast [119, pp. 23, 27].

Rarely, monazite occurs in beach placers of lakes, as, for example, together with magnet-ite-ilmenite, garnet, and zircon in fine-grained consolidated sand on raised beaches of the western shore of Lake Malawi (formerly Lake Nyasa) [134, pp. 713/4]. Its source appears to be monazite-bearing aplite dikes outcropping on the eastern side of Lungu Hill, Fort Johnston district. Further, monazite concentrations occur on the eastern shore of the lake [175].

Eluvial placers of monazite originate during in-situ weathering by removal of the soluble fraction of the source rock, whereas the insoluble residue forms a cover on its surface [1, p. 24]. Thus, extremely fine-grained monazite was derived from gneissose granites in eluvial deposits at Tonj, Mlanje district, Malawi [175]. And monazite and fluocerite are primary minerals in an eluvial pebble collected on albite pegmatite of the Afu Hills, Nigeria [176].

In **Sn-bearing (cassiterite) placers**, monazite commonly is associated with xenotime, al-lanite, and uranothorite and rarely with euxenite, huttonite, polycrase, or aeschynite [177]. Monazite is concentrated mainly in the fractions <0.25 and <0.1 mm in Sn placers of the Rolava River at Nová Role, north of Karlovy Vary (Karlsbad), Czechoslovakia. Its content is highest (and reaches 20.9%) for the electromagnetic fraction between 0.04 and 0.075 mm in size. The source of monazite is the nearby Erzgebirge granite [178]. Monazite occurs in Sn-placer fields of Nigeria [134, p. 714]. Moderately well-rounded crystals of alluvial monazite, derived almost certainly from cassiterite-bearing granite (and rarely from pegmatite), occur in areas actively mined for cassiterite in Malaya. Associated are, besides others, cassiterite, columbite, garnet, tourmaline, zircon, ilmenite, Fe oxides, Ta-Nb rutile, xenotime, allanite, or gahnite [167]. Monazite occurs in Sn placers of the western coastal region of the provinces Phuket, Phang Nga, and Ranong and near Chiang Mai, Thailand [75, p. 9], and is found in many cassiterite concentrates of alluvial and eluvial Sn-mining operations of New England, Australia [179].

Inland-placer deposits of Bihar and West Bengal, India, are similar in mineralogical composition to those of the Indian beach sands (refer to p. 261). But additionally columbite-tantalite, uraniferous allanite, traces of barite, cassiterite, and possibly also celestine occur [161]. Monazite constitutes roughly a quarter of the total heavy-mineral content [134, p. 712]. Inland placers of Purulia, West Bengal, and Ranchi Plateau, Bihar, contain 0.01 and 0.95% monazite, respectively [21, p. 119].

In **fossil placers** (at scattered localities between Canada and Mexico), monazite presumably was deposited by processes similar to those forming placers in the present-day sedimentary cycle [1, p. 25]. In the Devonian Pizhma-Umba Ti placer, Central Timan Ridge, USSR, consider-able quantities of ash to brown-gray monazite are associated with metamorphic metacolloidal leucoxene. Monazite formed during metamorphism and was supplied to the placer from remote greenschist-facies metasedimentary rocks [180]. Monazite occurs in Upper Jurassic thoriferous, ferruginous, consolidated placers (comprising sandstones) on the Kathiawar peninsula and in the Kutch district, India [21, p. 119].

In consolidated sedimentary rocks, monazite generally is a very minor accessory detrital mineral. It mostly occurs in conglomerates and sandstones (see below) [1, pp. 23/4].

Primary Th-containing monazite occurs in the **conglomerates** of the Blind River area, Ontario, Canada [119, p. 20]. It appears to be present in all radioactive clastic sedimentary rocks (including quartz-pebble conglomerates) of this area [181]. In the conglomerates, it is associated with uraninite, brannerite, zircon, and minor uranoan thorite and thucholite [182] and occurs mainly as rounded grains, ranging in size from 0.1 to 0.3 mm, disseminated in the cement [183]. Rounded grains of monazite, about 0.3 mm in diameter, are abundant in most uraniferous conglomerates near Elliot Lake, Ontario, whereas smaller grains are common in less radioactive conglomerate, grit, and subarkose [184]. Monazite occurs as detrital grains in the matrix of basal conglomerates of the Sheridan and Big Horn counties [145], and of the Deadwood Formation near the Bald Mountains, all Wyoming, where it constitutes 8.7% of the heavy-mineral concentrate [185] (see also [119, p. 20]). Detrital monazite grains occur in the cement of oligomictic quartz conglomerates from the USSR (and from abroad) [186]. Accessory monazite varies in content from single grains to a few percent of the heavy fraction of conglomerates, grits, and sandstones from the interfluvial region of the Dnestr and southern Bug rivers, Ukraine, USSR, which are derived from crystalline rocks of the central Ukrainian Crystalline Shield (especially from the Chudnov-Berdichev granites and pegmatites). Samples with abundant monazite are characterized by the assemblage almandine-ilmenite or horn-blende-ilmenite [187]. Note also that monazite, normally occurring as a minor mineral, locally reaches 6 to 7% of the heavy fraction in coarse-grained clastic material of the Baltic suite in the interfluvial region of the Dnestr and southern Bug rivers and is derived from pegmatites of the Chudnov-Berdichev complex [188]. Thoriferous conglomerates in the detritical formation of the Francevillien stratigraphic series, southeastern Gabon, are rich in heavy minerals, which comprise mainly monazite, zircon, and thorite [189]. Monazite occurs as gravels in conglomerates at Denny Dalton, White Mfolozi, northern Zululand [190], and as yellowish brown to brown rounded or oval grains in the fine-grained portions of the matrix of the Dominion Reef conglomerate, western Transvaal, both R.S.A., especially accompanying pri-mary uraninite in a thin sedimentary capping layer [191]. Refer also to oval grains of detrital monazite (mostly < 100 μ in size) comprising up to ~5% of the heavy minerals in the Dominion Reef conglomerate [192]. Monazite is one of the most abundant heavy minerals present in the Upper Reef (together with uraninite), reaching 6 vol% in consolidated black sand, whereas its concentration is much lower in the Lower Reef, both Dominian Reef [113, pp. 11/4]. It is derived from pegmatites. Some contribution from granites and metamorphic rocks of intermediate to low grade is possible [193]. The estimated distribution of radioactive monazite is 3.7% by mass in a heavy-mineral concentrate from a uraniferous conglomerate of the Rietkuil Mine, Upper Reef. Monazite occurs as well-rounded grains with a diameter of 130 to 250 μm and is associated with pyrite, garnet, and leucoxene as major heavy minerals as well as with zircon, nonradioactive oxides, sulfides, and radioactive constituents [194].

In titaniferous black **sandstones** from 14 deposits of Wyoming, fine-grained yellow monazite is a common constituent and one of the main sources of radioactivity. Its content in the heavy fraction is rather erratic and ranges from entirely absent to 4% [195]. Monazite is associated with zircon, anatase, apatite, rutile, and tourmaline in rose-colored medium-grained sand-stone and in gray-green sandy letten of the southwestern Frankenwald, F.R.G. [196]. Silicomonazite forms isometric, oval, or spherulitic grains in crushed Permian sand- and siltstones of the Kyugyulyur Creek, Kular Ridge, Yakutia [151]. Small angular grains or, rarely, prismatic crystals of pale green and yellow-cream monazite in porous sandstone-clay deposits of the Tom'-Kolyvan fold structure, Western Siberian Lowlands, are derived from Hercynian granitoids [197].

In **black shale** of the Kular Ridge, Yakutia, authigenic gray or black silicomonazite (kularite) is associated with pyrite, arsenopyrite, realgar, gold, scheelite, and other ore minerals. Probably, monazite was deposited in a local zone of mixing of submarine hydrothermal alkaline phosphate solutions with chlorides of the seawater. During an initial step, the RE phosphate precipitated as hydrated $REPO_4 \cdot n\,H_2O$ (which is analogous to hexagonal rhabdophane), and silica is coprecipitated as amorphous hydrated gel. Subsequent dehydration of this phosphate converted rhabdophane into anhydrous monoclinic monazite [152].

For the occurrence of monazite in, at least, partly consolidated fossil placers, see pp. 262/3.

Monazite-bearing **residual clay** (saprolite), consisting chiefly of kaolinite, halloysite, and unaltered quartz, occurs on quartzose gneiss and granodiorite exposed in the Freeman clay pit and on sillimanite gneiss south of Saltese Flats, Spokane county, both Washington. In contrast to contents of <0.001 to about 0.003% in the source rocks, the content of monazite reaches about 20% in saprolite on granodiorite and sillimanite gneiss. In the saprolite on granodiorite, monazite is associated with zircon, magnetite, ilmenite, cassiterite, rutile, tourmaline, topaze, lepidolite, and anatase [198]. High contents of globular, ovoid, or sub-rounded grains of monazite (ranging in size from 0.09 to 0.15 mm) are found in saprolite on sillimanite schist in the drainage basin of the Knob Creek, Shelby quadrangle, North Carolina [199]. Monazite is also found in kaoline formed by the decomposition of the Chudnov-Berdichev and Zhitomir granites, Ukraine [200], and it occurs in Tertiary clays of Bahia and Rio de Janeiro, Brazil [157].

In a special accumulation of red **soil and sand** of southern India, known as the Teri deposits, the content of monazite is 0.03 to 0.05% at Sattankulam and Muttam, Madras; 0.002 to 0.85% at Perumalpuram, Pudukkada, and Vattakottai Fort; and 0.045 to 0.077% at Mahadanapuram [21, p. 119].

In **bauxites** at Gánt, Iszkaszentgyörgy, and Nagyharsány, Hungary, well-rounded small grains of monazite probably were derived from the weathering of granitic rocks [201]; see also [202]. In bauxites of the USSR (Urals, Kazakhstan, Siberia, and Timan Ridge) and of Guinea, formed from alkaline ultrabasic rocks, terrigenous sediments, kimberlites, and carbonatites, a high content of Th is due mainly to accessory minerals (e.g., monazite) [203].

Metamorphic Rocks

Monazite is a widespread accessory mineral in various metamorphic rocks including schists, gneisses, granulites, charnockites, metaconglomerates, and sanidinites [15, p. 77]. Accessory monazite is exceedingly rare in greenschist-facies, rare to sparse in epidote amphibolite-facies, sparse to common in amphibolite-facies, and common to abundant in granulite-facies parametamorphic rocks. During early stages of metamorphism, detrital monazite in sediments becomes unstable and breaks down, sharing its components with other minerals. Thus, detrital monazite disappears at the onset of regional metamorphism. As the grade of regional metamorphism increases, monazite becomes stable again and begins to form as a metamorphic mineral at a few centers of crystallization which multiply with increasing grade of metamorphism. The components of the newly formed monazite apparently are derived from trace amounts of Th, RE, and P in some of the other original detrital constituents

of the sediment. The metamorphic formation of monazite in parametamorphic rocks is supported: by a direct relationship between the amount of monazite and the grade of metamorphism;
by an inverse relationship between the amount of monazite and grain size of the original
sediment; by a lack of similarity between the grain size of monazite particles in
parametamorphic and in sedimentary source rock; by a correlation between physical properties of monazite and grade of metamorphism; by the identity of inclusions in monazite
with metamorphic minerals in the host rock; by intergrowths of monazite with metamorphic
minerals; by an inverse relationship between monazite and other Th-bearing minerals (e.g.,
allanite); and by a relationship between the Th content of monazite and metamorphic grade
(see pp. 325/6) [1, p. 17]. The type and grade of metamorphism determine whether monazite
is formed. Monazite is especially widespread in metamorphic rocks of garnet to sillimanite
grade [121].

Monazite occurs in epidote-free mica **schists** and **gneisses** of the Ticino region, Lepontine
Alps, Switzerland [204], and in a paragneiss of the Strona-Ceneri zone, Val Grande, southern
Alps, Italy [205]. Together with florencite, thorite, and thorianite, monazite occurs in various
schists of the Sopron Mountains, Hungary [206]. It occurs as irregular rounded grains in
paragneisses (especially those of the almandine-amphibolite facies) of the southern
Carpathian Mountains, Romania [207]. Accessory monazite is a newly formed or a relict
mineral in amphibolite-facies gneisses and schists. Its average content is: 55.91 and 22.60 g/t
in orthogneisses and 18.48 and 105.82 g/t in paragneisses from the Urals and the Kola
Peninsula, respectively; and 7.25 g/t in schist from Tuva, Siberia [208]. The average content of
accessory monazite varies from <1 to $3.5 \times 10^{-3}\%$ in mica schists, gneisses, and granite
gneisses of the Kochkar and Dzhabyk-Suunduk regions, Southern Urals [66, p. 198]. Monazite
is a major host of Th (accounting for 27 to 96% of the Th in the rock) in gneiss, plagioclase
gneiss, and altered gneiss of the Aldan granulite complex, Siberia, and in plagiogneiss of the
White Sea granulite complex, Kola Peninsula [209]. Deep brownish red crystals of monazite
occur principally in biotite-rich varieties of gneisses and migmatites (e.g., in biotite gneiss,
graphite gneiss, and pyroxene-magnetite gneiss [62]) along the lower and middle reaches of
the Berda, West Kal'chik, East Kal'chik, and Kal'mius rivers, southeastern Ukraine [61].
Monazite also occurs in muscovite schists of this region [62]. It is found in biotite
microgneisses from the basin of the Katugin River, Lake Baikal area, Siberia [210, pp. 20, 22,
41]. The earliest granulite-facies event in a felsic garnet-rich paragneiss of Zircon Point,
Casey Bay area, northeastern Antarctica, is the time of original growth of isometric grains of
monazite (ranging in size from <45 to 300 μm) [211].

Monazite occurs only in high-grade **metamorphic carbonaceous rocks**, and probably the
metamorphic facies determines whether monazite is formed or not in these rocks [1, pp. 18/9].
Thus, anhedral monazite of matrix size and a few euhedral monazite porphyroblasts,
included in other minerals or within the matrix, occur in virtually all graphite-bearing
metashales of staurolite-kyanite grade of the Anakeesta Formation west of Bryson City, North
Carolina, whereas in the garnet zone, monazite occurs only in a few units of this formation
[212].

Migmatites and Anatectic Rocks. Monazite occurs in the St. Malo migmatite dome, Brittany,
France, consisting of a regionally low- to high-grade metamorphosed Al-rich graywacke
sequence displaying all stages of progressive migmatization [213]. Euhedral prismatic or
tabular crystals of monazite occur in migmatitic gneisses and granitoid rocks formed by
granitization in the southern Carpathian Mountains, Romania [207]. Monazite occurs in
migmatites from various regions of the Ukraine, USSR [31, pp. 161/2], as, for example, in
garnet migmatites of the Kotyuzhany suite [60, p. 117]; in migmatites near Taromskoe and

Mishurin Rog, Dnepr zone, where it occurs together with magnetite, apatite, zircon, and pyrrhotite [58, p. 14]; and in gray, rose-gray, or rose-colored migmatites of the Azov massif that contain monazite, together with zircon and apatite, mainly in the biotite-containing migmatites (and rarely in the amphibole-biotite varieties) [62]. For monazite in biotite-rich varieties of migmatites, southeastern Ukraine, see also p. 265. Local high concentrations of monazite in sillimanite or biotite schists around several sills of Toluca quartz monzonite, Shelby quadrangle, North Carolina, appear to be associated with migmatization of these wall-rock schists by lit-par-lit introduction of the monazite-bearing quartz monzonite and of microcline-oligoclase-quartz pegmatite [199]. Monazite, as the principal carrier of Th, has contents of 140 and 249 g/t in migmatites and anatectic granites, respectively, from a polymetamorphic complex of the Nyasyuk-Titovka metallogenetic zone, Kola Peninsula, USSR. During subsequent diaphthoresis, the primary monazite-type mineralization was replaced by a uraninite-zircon-thorite mineralization [214].

Monazite rarely is formed in anatectic rocks. Thus, it is found together with molybdenite and possibly uraninite in an anatectic tonalite of the Karin Lake area [215] and in a massive Fe-Ti oxide-monazite-apatite vein deposit enclosed in pegmatitic, graphic, and aplitic granites near Kulyk Lake, both Saskatchewan, Canada. Both this monazite deposit and its enclosing rocks possibly originated by anatexis during regional metamorphism, and the oxide-monazite-apatite assemblage precipitated from an aqueous phase generated during cooling and emplacement of an H_2O-rich granitic magma [216]; see also [217]. At Kulyk Lake, monazite and associated minerals probably were concentrated in a premetamorphic placer deposit [216], as is also supposed for monazite aggregates in the anatectic granite of Sainte-Anne-d'Auray, Brittany, France [45].

In **quartzites**, monazite is of detrital origin [1, p. 17]. Thus, it occurs in the Precambrian Goodrich quartzite near Palmer, Michigan [119, p. 20], which contains rounded to subrounded detrital grains of interstitial monazite (comprising locally > 50% of the matrix) and several lenses of monazite grains as much as 2 mm in diameter. The monazite is associated with hematite, magnetite, ilmenite, and rutile [218, 219]. Since monazite seems to be restricted to a relatively small area in the Goodrich quartzite, it may be derived from a local source to the southeast, underlain by abnormally radioactive porphyritic granite [219]. Note also that the Goodrich quartzite is considered as a metamorphosed placer (with average monazite contents of 1400 and 2800 g/t in two horizons) by [220].

In acid **metavolcanic porphyritic rocks** from the northwestern termination of the Marmaroshsk massif, middle reaches of Shopurka River, southern Ukraine, USSR, the average content of yellow or reddish yellow accessory monazite is 3 to 5 g/t. It is associated with titanite, apatite, zircon, and anatase [221].

Metasomatically Altered Rocks and Metasomatites

As is shown in the following paragraphs, monazite may originate as a secondary mineral during metasomatic alteration of primary rocks. Thus, well-developed isometric or prismatic crystals of accessory monazite, formed during a final stage of recrystallization and metasomatism, occur in muscovite-plagioclase pegmatites of Perti-varaka, Eki-varaka, and the Lake Loukhskoe region, all northern Karelia, USSR. In pegmatites with apographic structure at Eki-varaka, a striation of monazite faces (often on (100)) adjacent to late quartz indicates a simultaneous growth of these minerals [97]. Note also that in zoned muscovite-microcline pegmatites of the Chupa region, northern Karelia, monazite occurs in block-quartz pegmatites (together with late garnet and muscovite) or in plagioclase pegmatites with apographic structure, but rarely in muscovite-plagioclase pegmatites (together with garnet and sulfides)

[96, pp. 243, 247]. It is associated with uraninite, carburan, xenotime, and/or zircon (cyrtolite) [96, p. 244], [97]. Monazite (and a second generation of allanite) formed simultaneously with muscovite during the metasomatic transformation of quartz-plagioclase-biotite pegmatites of metamorphic origin and of quartz-plagioclase-biotite-microcline pegmatites of magmatic origin of the Mama and Gutar-Biryusa regions, Eastern Siberia. In the resulting muscovite pegmatites, monazite and allanite rarely occur together. Monazite mostly occurs in the muscovite pegmatites of magmatic heritage and is found in quartz-muscovite aggregates grown on microcline [222]. Monazite especially formed during postmagmatic metasomatic reworking of potash feldspar-containing magmatic pegmatites, whereas allanite originated mainly in plagioclase-containing metamorphic pegmatites of Eastern Siberia. In the allanite-containing contact zone of metamorphic pegmatites, monazite possibly was formed metasomatically by alteration of allanite [223]. Minerals of the monazite — huttonite series form grains up to 300 μm in size and belong to an Fe-Nb-Th-RE-PO_4 mineralization in pegmatites, which recrystallized or originated metasomatically along supposed tectonic zones, in the Rumburk granite near Bogatynia, southwestern Poland (see also p. 361) [224].

Magmatic and postmagmatic monazite occur in Sn-containing muscovite and tourmaline granites formed metasomatically from older granites at Ranong, northern Thailand [75, pp. 31/2].

Fine-grained aggregates of monazite originated metasomatically, and partly at the expense of allanite and apatite, during an inflow of pneumatolytic-hydrothermal emanations (which promote the dissolution of allanite and apatite with liberation of RE and P) in a thin layer of a talc-actinolite-carbonate rock in crystalline rocks of the Urushten complex on the left bank of Little Laba River, northern Caucasus, USSR [225, pp. 87, 92]. Note also that Th accumulated in monazite and allanite during an early pneumatolytic stage in the acid intrusion of the Urushten complex [226].

Postmagmatic solutions led to the formation of relatively coarse-grained monazite (ranging from 0.3 to 3.5 cm in size) during the final muscovite stage in muscovite-quartz and, rarely, in albite-quartz veins of an Mo deposit of the European USSR. In muscovite-quartz veins, the monazite is associated with muscovite, fluorite, and sometimes with pyrite; in albite-quartz veins, it occurs together with allanite and pyrochlore [227, pp. 123/4, 133].

In quartz-kalifeldspar-acmite (aegirine) veins in quartzites, Timan Ridge, European USSR, monazite occurs both as radial or isometrically grained aggregates formed during **K-Na metasomatism** and, mainly, as well-rounded detrital yellow grains associated with zircon [228, pp. 105/7, 109/10]. Fibrous radial or rosette-shaped grains of yellowish gray monazite with regeneration rims formed metasomatically by K- and Na-rich solutions in fracture and shatter zones in quartz-chlorite-sericite schists and micaceous quartzites of the central Timan Ridge [180].

During metasomatic **biotitization**, monazite originated in altered gneiss xenoliths in the apical part of a diallage-syenite intrusion of the southern Ukrainian Crystalline Shield, USSR, and is confined together with Cl apatite, chevkinite, and fluorite to the biotite accumulations. An impoverishment of the postmagmatic solutions in F, Fe, and Al during biotite precipitation possibly caused the decomposition of the RE-fluoride complexes in which the RE might well have migrated in these emanations [229].

The formation of small isometric grains and prismatic crystals of monazite is closely connected with postmagmatic **K metasomatism** in muscovite granites and quartz microclinites of the upper Buzurgata River, eastern Kuznetskii-Alatau, Kazakhstan. The average content of monazite for both rock types is 37.8 g/t; the maximum content is 150 g/t. In the central part of muscovite granites, its content is 25- to 30-times higher than in the border zone. Prismatic monazite crystals are associated with muscovite, topaz, hematite pseudomorphous after magnetite (martite), and zircon (cyrtolite); isometric monazite grains (with a size up to 2 mm) are enclosed in the peripheral parts of sericitized plagioclase [72]. Xenomorphic pale yellow grains of monazite formed as a second generation during K metasomatism (microclinization) of rare-metal granites of the Orlinogor complex, northern Kazakhstan [70]. Very small rounded grains of accessory yellow monazite originated during a K-metasomatic stage in biotite-rich parts of cataclastic rose-colored trachytic granites of the Novo-Ukrainkii massif, Ukraine. Monazite formed simultaneously with microcline perthite and somewhat later than biotite [230]. Note also that monazite formed together with xenotime at the beginning of microclinization of biotite at the junctions of cleavage in biotite and between grains of biotite and microcline in migmatites of the Bug (Pobuzh'e) complex, Ukraine [231]. In microclinites (of the USSR?), which formed at high temperature and alkalinity during K metasomatism in a deep fracture zone, yellow to brown monazite occurs as phenocrysts (ranging from 0.01 to 2 mm in size) on fractures in pyroxene and biotite and, rarely, in feldspar and quartz. It is associated with accessory uraninite, zircon and its variety cyrtolite, and xenotime [232].

As part of the mineralization accompanying an extensive **albitization**, hydrothermal monazite formed in pegmatites of the Petaca district (where it is associated with columbite, samarskite, and fluorite) and in a pegmatite near the head of Manzanares Creek, Elk Mountain district, both New Mexico (where it occurs together with spessartine, fluorite, columbite, and gadolinite) [86, pp. 224/5]. Red-brown crystals of monazite, or of a mineral of the monazite — cheralite series quite close to monazite, occur together with columbite, cassiterite, muscovite, spodumen, Mn garnet, apatite, xenotime, brockite, lepidolite, and tourmaline in Na-stage replacement bodies (composed of quartz, albite, and schorl) in the Haapaluoma granitic pegmatite, Peräseinäjoki-Alavus region, southern Pohjanmaa, Finland [90]. Accessory monazite is connected spatially and genetically with the replacement of primary graphic pegmatites, block microcline, and, to a lesser degree, block quartz by an apographic quartz-albite-oligoclase zone in zoned quartz-plagioclase-microcline pegmatites of the Alakurtti region, northern Karelia, USSR. Thus, dark red monazite in the pegmatite vein Alakurtti No. 1 is particularly abundant in intensely fissured parts of coarse-grained albite and is surrounded by coarse lamellar muscovite which causes imprints on the surface of monazite. Monazite is associated with coarsely crystalline garnet and intergrowths of xenotime and zircon (cyrtolite); rarely associated are yttropyrochlore (obruchevite) or allanite. In a coarse pegmatitic vein of the Kaita-Tundra, somewhat smaller crystals of reddish brown monazite ($\leqq 1$ cm in length and 0.5 cm in diameter) occur in intensely biotitized, albitized, and muscovitized microcline perthite or, rarely, in fractured quartz. Associated minerals are allanite, zircon (cyrtolite), uraninite, and apatite. Rarely, yellow monazite with poorly developed crystal forms (and a size between 1.5 and 2 cm in length and 0.3 to 0.5 cm in diameter) is disseminated in the pegmatite vein of Kapraevo. More frequently, it forms segregations in oligoclase blocks. It is associated with allanite, euxenite, columbite, yttropyrochlore (obruchevite), and zircon [233, pp. 56/9, 113]. In albitized pegmatites of Siberia, monazite is associated usually with columbite and samarskite, with ilmenorutile, with xenotime, or with beryl, zircon, ilmenorutile, and "wolframoixiolite" (a heterogeneous mixture of columbite and wolframite) [73]. In two Siberian Na-Li pegmatite fields, situated in gneissic biotite granites (field A) and in plagioclase amphibolites containing titanite (field B), monazite is associated with ilmenorutile and is genetically closely connected with albitization [234, pp. 96, 99, 103, 107]. Also genetically connected with albitization are intergrowths of monazite and xenotime (associated with albite, quartz, and columbian apatite and rarely with spessartine, zircon (cyrtolite), cassiterite and columbite) in a strongly albitized pegmatite vein of Siberia [235]. For monazite closely associated with albite in a Siberian pegmatite, see also p. 256.

Coarse-grained monazite occurs in intensely metasomatized epigenetic pegmatite veins in contact zones of charnockites and garnet migmatites of the southeastern Ukraine [99, p. 235] and formed immediately after albite and biotite [236]. Epigenetic pegmatite veins and pegmatites with superimposed metasomatic mineralization of the Bug (Pobuzh'e) complex, Ukraine, are characterized by a relatively high content of accessory monazite which mainly is bound to the fine-grained parts of the pegmatite bodies and formed during albitization and biotitization subsequent to cataclasis and crushing of coarse feldspar and quartz into fine-grained aggregates [56, pp. 120, 131]. In the Vishnevye Gory, Middle Urals, monazite formed together with thorianite, thorogummite, zircon, and pyrochlore during the pneumatolytic-hydrothermal albitization of fenitized granite pegmatites from the exocontact of miaskitic intrusions [237, pp. 54/6, 69], where it is associated with allanite and aeschynite in albitized corundum-muscovite and alkali-feldspar pegmatites. In albitized biotite-calcite veinlets and in albitized miaskites, it is associated with RE-containing apatite, allanite, aeschynite, and ilmenorutile (forming concentric-zonal aggregates, see p. 270) [238, pp. 164/5]. Monazite and other RE minerals are connected with late albitization (and sometimes with sericitization) in a quartz-fluorite pegmatite of Tarbagatai, Kazakhstan [239].

Dispersed oxide mineralizations (comprising U-rich pyrochlore, Th-rich monazite, cryolite, fluorite, thomsenolite, zircon (malacon), and ilmenite in the cupola zone of peralkaline biotite granites and columbite, Th-rich monazite, cassiterite, thorite, Hf-rich zircon, and xenotime in peraluminous arfvedsonite granites) of Nigerian ring complexes were disseminated during albitization processes [240]. Albitization, as the earliest alteration process caused by fluids enriched in Na, Fe, U, Th, Zr, Nb, and heavy RE, occurred mainly in the apical region of the cupola zone of alkaline biotite granite of the Ririwai complex, northern Nigeria. This process led to the formation of a mica in the compositional range between Li annite and zinnwaldite, the destruction of Fe-Ti oxides, the enrichment of the rock in U and RE, and the introduction of columbite with minor Th-rich monazite, cassiterite, thorite, xenotime, and Hf-rich zircon [241, pp. 197, 200]; see also [242]. Yellow monazite occurs in an albite-quartz rock formed metasomatically from coarse-grained early granites of the Kuu massif, central Kazakhstan [69].

Accessory monazite, zircon, and ilmenite recrystallized during the transformation of crystalline schists and gneisses of the Katugin field, Lake Baikal area, Siberia, into biotite metasomatites. This process was connected with albitization and sericitization of oligoclase and andesine [210, pp. 20, 26, 41/2].

The formation of monazite started simultaneously with albitization, or somewhat later, in albitized quartzites of the central Pamir, USSR [129, pp. 87, 89].

In contrast to the foregoing examples for albitization, monazite was unstable and its content decreased sharply (up to a complete disappearance in locally formed albitites) during albitization of rare-metal granites of the Orlinogor complex, northern Kazakhstan, USSR [70].

Dependent on physico-chemical conditions during **greisenization** of rare-metal granites of the Orlinogor complex, monazite occurs in quartz-topaz greisens owing to a high concentration of F in the greisenizing solution. In contrast, monazite is absent and apatite occurs in quartz-mica greisens, which formed from F-poor solutions [70].

In hydrothermal greisen deposits, monazite is associated with molybdenite, cassiterite, and wolframite [83]. It is associated with zircon, cassiterite, molybdenite, and fluorite within early formed Li siderophyllite and zinnwaldite in gray greisens and their microclinized wall rock in the Ririwai biotite-granite complex, northern Nigeria [241, pp. 205/6, 213]. Monazite is associated with uraninite, thorite, and brannerite in greisens of Kazakhstan, USSR [243]. It may be a relict or a newly formed mineral in biotite-quartz greisens of the Kuu massif, central

Kazakhstan, where it is associated with fluorite and zircon [69]. Magmatic and postmagmatic monazite occur in greisen Sn ores and disseminated cassiterite ores from southern Thailand [75, pp. 36/7].

Monazite is associated with wolframite and xenotime in a greisenized acid volcanic rock of the Mount Pleasant W-Bi-Mo deposit, New Brunswick, Canada [244].

Early formed monazite in high-temperature Mo-quartz veins adjacent to quartz greisens of the eastern Kounrad rare-metal deposit, Kazakhstan, mostly occurs in lumps with sellaite and phlogopite; rarely it forms isomorphous grains (ranging in size from ~0.5 to 1 mm) associated with ilmenorutile [245, pp. 16, 47, 136]. In quartz veins accompanying biotite-quartz greisens of the Kuu massif, central Kazakhstan (see above), monazite displays a different morphological appearance against accompanying minerals. It is xenomorphic against zircon and wolframite and idiomorphic against quartz, mica, and ilmenorutile [69].

2.4.1.1.1.2 Intergrowths, Inclusions, and Alteration. Synthesis

A great diversity of **intergrowths** is observed for monazite, as is shown in the following paragraphs. Thus, it is intergrown, for example, with (arranged by the mineralogical system):

galena [84, p. 133]	fluorocarbonate [39, pp. 331/4]	zircon [51, 167], [84, p. 133]
fluocerite [78]	bastnaesite [78] and an	and its variety cyrtolite [97]
hematite, quartz [151]	unidentified Th-containing	thorite [125, p. 71] and its
ilmenorutile [234, p. 105]	mineral (possibly bastnaesite	varieties uranothorite [84,
tapiolite [151]	or fluocerite) [76]	p. 133] and ferrithorite [180],
polycrase [85]	xenotime [151, 235]	[228, pp. 109/10]
euxenite [86, p. 227]	rhabdophane, churchite [151]	muscovite [97, 151]
uraninite [84, p. 133]	garnet [97]	feldspar [103, 151]

Intergrowths of monazite with thorite and apatite, or with cheralite and apatite, are the most common forms of occurrence of monazite in biotite-muscovite, biotite, and muscovite granites of the Lawler Peak pluton, Yavapai county, and of the Dells granite, both Arizona. The intimate nature and the subhedral form of these intergrowths suggest that they formed by exsolution at lower temperature of apatite and thorite or cheralite from primary monazite originally higher in Ca, Th, and Si. Whether thorite or cheralite is exsolved depends on the original amount of Si substituting for P [39, pp. 105/6, 109/10, 112, 123, 159/62, 197, 327]. In the Lawler Peak biotite-muscovite granite, monazite and apatite commonly are intergrown with volume portions between 6:1 and 1:1 [40, pp. 193, 216/7, 328]. Monazite crystals are intergrown with white grains of apatite and with the marginal parts of ilmenite crystals in granitic pegmatites of Kazakhstan, USSR [132]. Regular intergrowths of monazite and biotite along their (100) and (001) face, respectively, occur in a plagioclase-microcline vein of Karelia [97]. For an intergrowth of monazite with zircon or biotite, see also p. 252.

Ovoid aggregates in miaskites of the Vishnevye Gory, Southern Urals, USSR, have a mono-crystalline core of monazite, followed outwards by concentric layers of RE apatite and allanite and, rarely, of aeschynite and ilmenorutile. The layers formed by successive crystallization of the RE minerals [238, pp. 165, 236/7]. Very fine-grained aggregates of monazite, apatite, and Ca-bearing members of the bastnaesite group are intergrown with chlorite and carbonates in ferrocarbonatite, and with hematite, calcite, and dolomite in rødbergite, both of the Fen complex, southeastern Norway [138].

As is shown in the following examples, **inclusions of monazite in other minerals** occur predominantly in biotite, see, for example, [33, p. 62/3], [42, 45 to 47, 50, 72, 212, 229], [84, p. 139], which implies that monazite is an early crystallizing phase (provided that biotite is also), or that these inclusions are residual monazite enclosed in refractory biotite [36].

Monazite also forms inclusions in (arranged by the mineralogical system):

molybdenite [245, pp. 136/7]	scheelite [69]	beryl [245, pp. 136/7] and its
green fluorite [132]	apatite [225, p. 87]	variety emerald [69]
hematite [123, p. 185]	an unknown opaque, brown	muscovite [40, p. 193], [97, 222]
ilmenite [49]; Mn ilmenite	mineral similar to huttonite	phlogopite [69]
[39, pp. 95/6]	[237, p. 69]	feldspar [33, pp. 61, 63]
quartz [69, 72, 222]; smoky	kyanite [212]	microcline [18, p. 101], [72]
quartz [132]	allanite [225, p. 87]	plagioclase [72, 212], [84, p. 139]

Grains of accessory monazite, enclosed in microcline from granites and pegmatites of the southwestern Ukrainian Crystalline Shield, USSR, have a variable size: 3 to 5 mm in pegmatoid granites and pegmatites and $\leq$ 0.1 to 0.2 mm in medium-grained granites [99, p. 228]. Small chains of monazite crystals between plates of biotite are enclosed in plagioclase from pegmatites of the Chupa region, northern Karelia [97].

Monazite inclusions are surrounded by pleochroic haloes, as, for example, in biotite from granites of the Rila Mountains, Bulgaria [52, pp. 6/7]; in biotite (see also p. 357) and apatite from granites of the Visen district, northern Portugal [43]; and in biotite, garnet, cordierite, and feldspar from granites and pegmatites of the Ukrainian Crystalline Shield, USSR [30, p. 130], [57], [58, pp. 19, 152]. Light yellow monazite grains form the nuclei (ranging from 0.005 to 0.2 mm in diameter) of pleochroic haloes in biotites from mica schists and biotite-plagioclase gneisses of the Ticino region, Lepontine Alps, Switzerland, and constitute more than half of the inclusions (besides apatite, titanite, and allanite) [204]. Monazite inclusions in feldspar are surrounded by Fe-rich dark haloes with elevated Pb, U, and Th contents in biotite-muscovite granite of the Lawler Peak pluton, Yavapai county, west-central Arizona [40, pp. 14, 80, 93]. For a color change of feldspar around inclusions of monazite, see also p. 357.

Data in the literature on **inclusions of other minerals in monazite** are inconsistent: Monazite generally lacks inclusions [1, p. 5], or it may be almost free of inclusions [39, p. 327], [128, p. 142], [246]. Often it contains inclusions of black or dark minerals or particles [19, p. 72], [63, 69, 150, 152, 180, 188, 195, 230], [129, p. 89], [247, p. 44].

Omitting unidentified or poorly defined substances, the following inclusions of minerals, which may affect mainly the optical properties of monazite (for examples, see pp. 349/50 and 352), were reported in the literature (arranged by the mineralogical system):

graphite [212]

sulfides [60, p. 117] including mainly pyrite [31, p. 164], [76, 181, 184, 248], [241, p. 213], [247, p. 44] as well as rarely chalcopyrite [241, p. 213] or galena [190, 192]

gahnite [167]

magnetite [81, p. 214]

hematite [76, 152], [225, pp. 88, 90], [241, p. 213], [247, p. 44]

ilmenite [81, p. 214], [167]

quartz [75, pp. 23/4], [129, p. 89], [180, 212, 249] and amorphous [152] or opaline SiO_2 [144, p. 25]

cassiterite [167]

rutile [144, p. 25], [167, 212]

anatase [60, p. 117]

columbite [167]

Fe hydroxides [30, p. 135], [97], [139, p. 377]

including goethite [247, p. 44] and limonite [76, 152], [247, p. 44]

carbonates [225, p. 90]

xenotime [75, pp. 23/4]

apatite [61, 155, 195], [144, p. 25]

garnet [61]

zircon [30, p. 134], [58, p. 19], [60, pp. 65, 67], [61, 250], [75, pp. 23/4]

thorite [39, pp. 160/2], [40, p. 193], [75, pp. 23/4], [233, p. 58]; uranothorite [181, 248]

titanite [152, 249]

allanite [56, p. 127], [225, p. 90]

tourmaline [144, p. 25], [167]

tremolite [247, p. 44]

micas [144, p. 25]; muscovite [75, pp. 23/4], [247, p. 44]; biotite [30, p. 134], [58, p. 19], [247, p. 44], [61, 230, 249]

chlorite [249]; clinochlore [247, p. 44]
kaolinite [30, p. 135]
amesite [247, p. 44]

feldspars [152, 249]; soda feldspar
[247, p. 44]; albite [129, p. 89], [234, p. 105]

Brick red monazite contains inclusions of orange-red monazite (partly with sprouts of a black ore mineral in the core) in granites of the Rusava suite, Ukrainian Crystalline Shield, USSR [60, pp. 65, 67]. Sometimes inclusions or nuclei of earlier-formed monazite are observed in monazite from pegmatites of the southeastern Ukraine [61].

In gray strongly radioactive monazite from conglomerates of the Blind River region, Ontario, Canada, inclusions of uranothorite are introduced subsequent to the deposition of the conglomerate [184]. In monazite from the Pizhma-Umba Ti placer, Central Timan Ridge, European USSR, mineral admixtures (including Ti-Fe hydroxides) comprise mechanical inclusions and segregations formed during its crystallization [180]. Generally, in monazite from Sn placers, pegmatites, greisens, and granites of Thailand, thorianite inclusions, formed by simple exsolution, do not exceed 10 μm in size. In some large monazite crystals, the formation of idiomorphic thorianite inclusions, which reach a size of 100 μm and are oriented parallel to (100) and (010) of the host mineral, is caused by the incorporation of Th^{4+} with concomitant diadochic replacement of PO_4^{3-} by SiO_4^{4-} [75, pp. 23/4].

A zonal arrangement of films of inclusions of goethite and hydrogoethite causes a banded color of sulfate monazite from carbonatites and quartz-carbonate rocks of the Kola Peninsula, USSR [139, p. 375]. Varying concentrations of minute inclusions of relict organic substances cause a zonality, viewed under crossed nicols, in monazite of Central Asia, Siberia, and the Far East. These spherical and elliptic inclusions often are rimmed by a discontinuous fine-grained aggregate of mica and quartz [251]. Particles of galena usually are concentrated along the periphery or form trains of specks delineating zones within monazite grains from uraniferous conglomerates of the Dominian Reef, R.S.A. [194].

Uniformly distributed microinclusions of dusty opaque substances as well as of quartz and possibly sericite cause a screen-like (poikilitic) structure in monazite from placers of the Verkhoyansk-Chukotka Fold Region and of the Timan Ridge, USSR [252].

Monazite containing inclusions of apatite is more euhedral than monazite alone in biotite granite of the Lawler Peak pluton, Yavapai county, west-central Arizona [39, pp. 160/2].

Monazite is split radially around very small inclusions of elongated crystallites (possibly thorite) in Siberian pegmatites [234, p. 100]. Radial fractures, which often are filled with reddish brown and yellowish ocherous products, are developed around numerous inclusions of rose-colored relict thorite in monazite from a pegmatite of the Kaita Tundra, northern Karelia, USSR [233, p. 58]. Adjacent to dark brown uniformly dispersed dusty inclusions or patches of a decomposed radioactive mineral (possibly thorite), monazite is bright yellowish red, owing to decomposition products, and sometimes fractured radially in metamorphic rocks, granites, and pegmatites of the southeastern Ukraine [61].

Alteration of Monazite. Monazite is moderately resistant to weathering [12, p. 344] and alteration processes. Despite the presence of easily leachable elements, such as Ca and RE, deficiencies in ions at A positions (see pp. 243/4) are almost absent [253]. Despite this durability, monazite rarely is found without superficially altered surfaces and seems to be susceptible to oxidation during weathering [8]. Altered monazite often is covered by reddish brown or white films containing hydroxides and carbonates of Fe, La, Ce, and other RE. These films may penetrate into the monazite crystals along cleavage planes, causing brown colors [254]. Although extremely stable, monazite alters to members of the rhabdophanite group of minerals (including rhabdophane, brockite, and grayite) [7].

Monazite from albitized granites of the Orlinogor complex, northern Kazakhstan, is altered to RE carbonates and fluorocarbonates (including yttrosynchisite, synchisite, fluorbastnaesite, or fluocerite) as well as to zircon (cyrtolite), fluorite, and thorite [70]. Monazite from pegmatites of the Baie-Johan-Beetz area, Quebec, Canada, is altered to amorphous and poorly crystalline aggregates of bastnaesite and hydroxylbastnaesite (which, on their part, are replaced by phyllosilicates and goethite) in samples from granite pegmatites, and to bastnaesite (which, on its part, is replaced by phyllosilicates) in samples from syenite pegmatites. About one-third of the original monazite is replaced by bastnaesite and about one-third of monazite and bastnaesite combined is replaced by phyllosilicates [84, pp. 133, 138/9, 143].

Sometimes, monazite is replaced by rhabdophane, as, for example, in carbonatites from Samalpatti, India [140], and in granites of the southern Timan Ridge, European USSR [180]; or by rhabdophane and, rarely, bastnaesite in granites of the Kola Peninsula [5, p. 131]; or by cheralite, which covers monazite together with pure apatite, in a granite aplite of the Mecsek Mountains, Hungary [33, pp. 81, 83]. Monazite is coated by a very thin Ca- and Th-rich, possibly secondary crust in a bauxite of Kislöd, Hungary [201]. Minerals of the monazite—huttonite series are replaced by a later Fe-Th mineralization in samples of the Bogatynia area, Lower Silesia, Poland [255].

Some grains of monazite, possessing surfaces heavily pitted by chemical attack rather than by mechanical impact, are coated by an Fe crust in beach placers of Taiwan [247, p. 44]. And colorless to yellow-green monazite from offshore bars of Taiwan shows a veined alteration to a reddish brown, probably goethitic material [165, p. 3]. Monazite is stained by Fe oxides in Th veins of the Lemhi Pass area, Idaho-Montana [124, p. A45], and by hematite in the Ririwai biotite granite, Nigeria [76]. Some silicomonazite grains are coated by hematite in alluvial placers of the Kyugyulyur Creek, Kular Ridge, Yakutia [151]. Monazite is altered to limonite-stained masses or cloud-like clusters of minute opaque particles that spread through the crystals, sometimes following cleavage planes, in soil and fractured blocks near Chester, Morris county, north-central New Jersey [256]. It is altered to a greater or lesser degree and is covered by a thin reddish brown or yellowish brown isotropic Fe_2O_3-containing crust in pegmatoid granites of the Novo-Ukrainkii massif, central Ukraine [230].

During hydrothermal processes (for monazite as a hydrothermal alteration product, see p. 274), monazite seems to be unstable and susceptible to intense replacement by secondary products [61]. Thus, as a result of hydrothermal alteration, some monazite grains enclosed in partially chloritized biotite from biotite granite of the Carnmenellis pluton, Cornwall, Great Britain, are marginally dissolved and replaced by apatite and Ca-, Th-, and Si-rich minerals. A decrease in P content during this replacement is insufficiently balanced by an increasing Si content [257]. Later hydrothermal solutions, especially those rich in sulfide, alter monazite in pegmatites of northern Karelia. The monazite is coated by a greenish crust, consisting mainly of pyrite and possibly some pyrrhotite, which also penetrates the mineral along small fissures [258]. Surfaces of monazite are covered by a porous reddish brown rim in microcline pegmatites of the Chupa region, northern Karelia [97], and a secondary mineral occurs as an earthy reddish brown film or rim on, or forms punctuate inclusions in the core of monazite from muscovite-quartz veins of the European USSR [227, pp. 123, 125]. Due to the influence of later hydrothermal solutions, a secondary opaque RE fluorocarbonate may form after monazite in F-rich Siberian pegmatites of field A. This interpretation is suggested by the presence of CO_2 and an excess of CaO in the studied P-deficient monazite (see analysis No. 88 and corresponding footnote in Table 14, pp. 314 and 329, respectively) [234, pp. 101/2]. Under the influence of halide compounds, monazite is partly replaced by secondary RE minerals, deep violet fluorite, and quartz in altered gneiss xenoliths in a diallage-syenite intrusion of the southern Ukrainian Crystalline Shield [229].

Although monazite is considered to be among the most stable, resistant minerals, it is altered under the influence of solutions in a uraniferous environment and liberates phosphoric acid that can react with U to form secondary uraniferous phosphates [84, p. 139]. Thus, monazite is extensively corroded in uraniferous conglomerates of the Pongola Supergroup [190] and of the Dominian Reef, both R.S.A. [194].

For changes of chemical composition, cleavage, optical properties, density, or DTA curve during alteration, see pp. 304 and 331, 339, 349/50, 356, and 359, respectively.

As is shown in the following paragraphs, **monazite** may occur **as an alteration product of other minerals** (e.g., various silicates, phosphates, or carbonates). Thus, monazite often is a constituent of radioactive pseudomorphs after eudialyte*) in syenite and analcime-rich lujavrite of Kvanefjeld, Ilimaussaq massif, southern West Greenland [24], or it occurs as an alteration product of eudialyte in naujaite (sodalite-nepheline syenite), kakortokite (eudialyte-arfvedsonite-nepheline syenite), and lujavrite from the lower part of that massif [28]. In the lujavrite, some fine- or coarse-grained monazite aggregates with crystal outlines may be secondary after steenstrupine [26, p. 35]. Monazite replaces chevkinite in calcite veinlets of the Vishnevye Gory, Southern Urals, USSR [238, p. 165]. Quartz-muscovite aggregates containing monazite and apatite may replace microcline in muscovite pegmatites of the Mama and Gutar-Biryusa regions, Eastern Siberia [222]. For the formation of monazite by alteration of allanite in these pegmatites and for the occurrence of monazite as an alteration product of feldspar, see pp. 267 and 250, 254, respectively. Fine-grained aggregates of monazite corrode the edges or almost completely replace seams of allanite and apatite grains in a talc-actinolite-carbonate rock on the left bank of Little Laba River, northern Caucasus. Apatite is replaced more strongly than allanite [225, pp. 87/8]. Elongated rounded grains of wine yellow or yellow monazite form pseudomorphs after apatite in granites of the Rusava suite, Ukraine [60, pp. 65, 68]. Monazite partly is a conversion product of florencite from an RE mineralization in crystalline schists of the Sopron Mountains, Hungary [206]. Green spherulitic aggregates of monazite, as a late phase, replace dolomite in a carbonatite at Nkombwa Hill, Zambia [137].

Black cryptocrystalline Th-poor monazite from beach placers of Taiwan may be a hydrothermal alteration product related to an autolysis process (i.e., purification of isomorphous impurities) of a primary mineral, which genetically is close to associated yellow monazite. When monazite minerals are affected by a hydrothermal solution, which may be P-bearing or may establish a P concentration through digestion of some phosphoric minerals, monazite partly is dissolved (leading to a cavernous structure) and partly is recrystallized with differential leaching to a cryptocrystalline structure. The Th ions of the primary material are not stable enough to stay in the crystal lattice at lower temperature [247, pp. 56, 58/9].

For monazite as a secondary alteration product of complex U minerals, see p. 256. For the origin of monazite as a secondary mineral in metasomatically altered rocks and metasomatites, refer to pp. 266/70.

For the **synthesis** of pure monazite with the formula $CePO_4$, see, for example, [260 to 263]. Crystals of Th-containing monazite are synthesized hydrothermally by heating at 300°C for 48 h of dried Ce hydroxide gel and an excess of 85% phosphoric acid together with an $Na_2SiO_3 \cdot 9H_2O$ solution and ThO_2 gel [264]; or by fusion of precipitated ceric oxide with KH_2PO_4 in a platinum crucible at $\sim$ 1000°C for several days and mixing of the precipitated Th and Ca oxides with the ceric oxide in the fusion [265].

---

*) Note that the monazite variety erikite (see p. 246) probably is a pseudomorph of monazite after eudialyte [259].

References for 2.4.1.1.1.1 and 2.4.1.1.1.2:

[1] Overstreet, W. C. (Geol. Surv. Profess. Paper [U.S.] No. 530 [1967] 1/327).
[2] Overstreet, W. C. (Geol. Surv. Profess. Paper [U.S.] No. 400-B [1960] B55/B57).
[3] Frondel, C. (Geol. Surv. Bull. [U.S.] No. 1064 [1958] 1/400, 158/60).
[4] Betechtin, A. G. (Lehrbuch der Mineralogie, VEB Verlag Technik, Berlin 1953, pp. 1/681, 410 [Russian original, Moscow 1951]).
[5] Bel'kov, I. V. (Aktsessornye Mineraly Granitoidov Kol'skogo Poluostrova, Nauka, Leningrad 1979, pp. 1/184).
[6] Andresen, L. (Erzmetall **39** [1986] 152/7).
[7] Floran, R. J.; Abraham, M. M.; Boatner, L. A.; Rappaz, M. (Sci. Basis Nucl. Waste Management 3 [1981] 507/14, 508).
[8] Haaker, R. F.; Ewing, R. C. (Sci. Basis Nucl. Waste Management 2 [1980] 281/8, 283).
[9] Kaplan, G. E.; Uspenskaya, T. A.; Zarembo, Yu. I.; Chirkov, I. V. (Torii, Ego Syr'evye Resursy, Khimiya i Tekhnologiya, Atomizdat, Moscow 1960, pp. 1/224, 17).
[10] Perez Mateos, J.; Galvan Garcia, J. (Anales Edafol. Agrobiol. **26** [1967] 1227/44, 1241).

[11] Davidson, C. F. (Mining Mag. [London] **94** [1956] 197/208).
[12] Deer, W. A.; Howie, R. A.; Zussman, J. (Rock-Forming Minerals, Vol. 5, Non-Silicates, Longmans, London 1962, pp. 1/371).
[13] Pagel, M. (Uranium Explor. Methods, Rev. NEA/IAEA R&D Programme Proc. Symp., Paris 1982, pp. 445/56).
[14] Ramdohr, P.; Strunz, H. (Klockmanns Lehrbuch der Mineralogie, 16th Ed., Enke, Stuttgart 1978, pp. 1/876, 626).
[15] Heinrich, E. W. (Mineralogy and Geology of Radioactive Raw Materials, McGraw-Hill, New York 1958, pp. 1/654).
[16] Rapp, R. P.; Ryerson, F. J.; Miller, C. F. (Geophys. Res. Letters **14** [1987] 307/10).
[17] Meliksetyan, B. M. (in: Smorchkov, I. E., Aktsessornye Mineraly Izverzhennykh Porod, Nauka, Moscow 1968, pp. 95/108).
[18] Nekrasov, I. Ya. (Magmatizm i Rudonosnost' Severo-Zapadnoi Chasti Verkhoyano-Chukotskoi Skladchatoi Oblasti, Akad. Nauk SSSR, Moscow 1962, pp. 1/412).
[19] Khomyakov, A. P. (Mineralogiya i Geticheskie Osobennosti Shchelochnykh Massivov, Moscow 1964, pp. 56/82).
[20] Omel'yanenko, B. I.; Sirotinina, N. A. (Mater. Geol. Rudn. Mestorozhd. Petrogr. Mineral. **1959** 414/22, 414, 422).

[21] Dar, K. K. (J. Geol. Soc. India **5** [1964] 112/20).
[22] Basu, A. N.; Sharma, G. S.; Varma, H. M.; Dhana Raju, R.; Dougall, N. K.; Raju, B. N. V. (J. Geol. Soc. India **30** [1987] 267/72).
[23] Tikhonenkova, R. P. (Dokl. Akad. Nauk SSSR **266** [1982] 1236/9).
[24] Danø, M.; Sørensen, H. (Medd. Grønland **162** No. 5 [1959] 1/35, 13, 16).
[25] Bondam, J.; Sörensen, H. (Proc. 2nd U.N. Intern. Conf. Peaceful Uses At. Energy, Geneva 1958, Vol. 2, pp. 555/9).
[26] Wollenberg, H. A. (RISO-228 [1971] 1/40; N.S.A. **26** [1972] No. 28036).
[27] Wollenberg, H. A. (Geochem. Explor. Proc. 4th Intern. Geochem. Explor. Symp., London 1972 [1973], pp. 347/58, 355).
[28] Buchwald, V.; Sørensen, H. (Medd. Grønland **162** No. 11 [1961] 1/35, 11, 14).
[29] Sørensen, H.; Rose-Hansen, J.; Nielsen, B. L.; Løvborg, L.; Sørensen, E.; Lundgaard, T. (Rappt. Grønl. Geol. Unders. No. 60 [1974] 1/54, 15, 22/3, 33/6, 44).
[30] Ivantishin, M. M. (Aktsesorni Ridkisni Minerali ta Rozsiyani Elementi v Granitakh i Pegmatitakh Ukrains'kogo Kristalichnogo Shchita, Akad. Nauk Ukr. SSR, Kiev 1960, pp. 1/243).

276

[31] Komlev, L. V.; Danilevich, S. I.; Ivanova, K. S.; Kuchina, G. N.; Savonenkov, V. G.; Filippov, M. S.; Chupakhin, M. S. (Tr. Sess. Kom. Opred. Absolyut. Vozrasta Geol. Formatsii Akad. Nauk SSSR 5 1956 [1958] 159/75).

[32] Lee, D. E.; Bastron, H. (Geochim. Cosmochim. Acta 31 [1967] 339/56, 340/1).

[33] Pantó, G. (Acta Geol. [Budapest] 19 [1975] 59/93).

[34] Komlev, L. V.; Danilevich, S. I.; Ivanova, K. S.; Mikhalevskaya, A. D.; Savonenkov, V. G.; Filippov, M. S. (Geokhimiya 1957 566/72).

[35] Webb, P. C.; Tindle, A. G.; Barritt, S. D. (Geophys. Res. Letters 14 [1987] 299/302).

[36] Rapp, R. P.; Watson, E. B. (Contrib. Mineral. Petrol. [Berlin] 94 [1986] 304/16, 313/4).

[37] Montel, J. M. (Geology [Boulder] 14 [1986] 659/62).

[38] Adams, J. W.; Fish, G. E., Jr. (Geol. Surv. Profess. Paper [U.S.] No. 580 [1968] 430/5).

[39] Silver, L. T.; Woodhead, J. A.; Williams, I. S.; Chappell, B. W. (GJBX-7-84 [1984] 1/412; C.A. 102 [1985] No. 152440).

[40] Silver, L. T.; Williams, I. S.; Woodhead, J. A. (GJBX-45-81 [1980] 1/380; C.A. 95 [1981] No. 65599).

[41] Oberli, F.; Sommerauer, J.; Steiger, R. H. (Schweiz. Mineral. Petrogr. Mitt. 61 [1981] 323/48, 331).

[42] Pagel, M. (Vein-Type Similar Uranium Deposits Rocks Younger Proterozoic Proc. Tech. Comm. Meeting, Lisbon 1979 [1982], pp. 323/47, 324, 335).

[43] Schermerhorn, L. J. G. (Commun. Serv. Geol. Port. 37 [1956] 1/617, 1, 14, 282).

[44] Carrat, H. G. (Mineralium Deposita 6 [1971] 1/22, 3/4, 12).

[45] Le, V. T.; Stussi, J. M. (Sci. Terre 18 [1973] 353/79, 373/6).

[46] Ranchin, G. (Sci. Terre 13 [1968] 159/205, 163, 183).

[47] Roubault, M.; Coppens, R. (Compt. Rend. 246 [1958] 3086/9).

[48] Cuney, M.; Brouand, M. (Geol. France 1987 247/57, 248, 250, 255).

[49] Jefferies, N. L. (Proc. Ussher. Soc. 6 [1984] 35/41).

[50] Charoy, B. (J. Petrol. 27 [1986] 571/604, 574/6).

[51] Willgallis, A. (Neues Jahrb. Mineral. Abhandl. 114 [1970] 48/60, 48/50, 52/4).

[52] Aleksiev, E.; Tsvetkova, V. (Tr. Vurkhu Geol. Bulg. Ser. Geokhim. Polezni Izkop. 3 [1962] 5/24).

[53] Lyakhovich, V. V. (Tr. Inst. Mineral. Geokhim. Kristallokhim. Redk. Elem. No. 7 [1961] 212/25, 217).

[54] Smorchkov, I. E.; Tseitlin, S. G.; Batyreva, N. N. (Tr. Inst. Geol. Rudn. Mestorozhd. Petrogr. Mineral. Geokhim. No. 99 [1963] 60/7).

[55] Shcherbak, M. P. (Dopov. Akad. Nauk Ukr. SSR 1959 No. 2, pp. 188/91).

[56] Shcherbak, N. P. (Aktsessornye Mineraly Ukr. Shchita 1976 118/32, 244/58).

[57] Ivantishin, M. N. (Metallog. Dokembr. Shchitov Drevnikh Podvizhnykh Zon Dokl. 2nd Vses. Ob'edin. Sess., Kiev 1960, Pt. 1, pp. 116/31, 116/8).

[58] Ivantishin, M. N.; Kul'skaya, O. A.; Gornyi, G. Ya.; Eliseeva, G. D. (Tr. Inst. Geol. Nauk Akad. Nauk Ukr. SSR Ser. Petrogr. Mineralog. Geokhim. No. 21 [1964] 1/168).

[59] Filippov, M. S.; Komlev, L. V. (Geokhimiya 1959 437/48; Geochem. [USSR] 1959 535/49).

[60] Slenzak, O. I. (Charnockity Pridnestrov'ya i Nekotorye Obshchie Voprosy Petrologii, Akad. Nauk Ukr. SSR, Kiev 1960, pp. 1/212).

[61] Marchenko, E. Ya. (Dokl. Akad. Nauk SSSR 176 [1967] 1153/5; Dokl. Acad. Sci. USSR Earth Sci. Sect. 176 [1967] 142/5).

[62] Kuts, V. P. (Mineral. Sb. [Lvov] 20 [1966] 279/84).

[63] Orsa, V. I. (Dopov. Akad. Nauk Ukr. SSR B 31 [1969] 694/6).

[64] Komlev, L. V.; Filippov, M. S.; Danilevich, S. I.; Ivanova, K. S.; Kryukova, N. F.; Kuchina, G. N.; Mikhalevskaya, A. D. (Geokhimiya **1959** 110/5; Geochem. [USSR] **1959** 133/40, 138).

[65] Komlev, L. V.; Filippov, M. S.; Danilevich, S. I.; Ivanova, K. S.; Kryukova, N. F.; Kuchina, G. N.; Mikhalevskaya, A. D. (Tr. Sess. Kom. Opred. Absolyut. Vozrasta Geol. Formatsii Akad. Nauk SSSR **7** 1958 [1960] 123/30, 127).

[66] L'vov, B. K.; Zhangurov, A. A. (in: Smorchkov, I. E., Aktsessornye Mineraly Izverzhennykh Porod, Nauka, Moscow 1968, pp. 196/204).

[67] L'vov, B. K. (Tr. Vses. Nauchn. Issled. Geol. Inst. No. 95 [1963] 13/44, 31, 37/8, 40).

[68] L'vov, B. K.; Polyanskii, E. A. (Vopr. Magmatizma Metamorfizma **2** [1964] 96/114, 96, 101/2).

[69] Povilaitis, M. M.; Varshal, G. M. (Tipomorfizm Miner. **1969** 196/209, 198/201).

[70] Krasil'nikova, A. V. (Geologiya [Alma-Ata] **11** [1976] 104/8).

[71] Amshinskii, N. N. (Geol. Geofiz. **1960** No. 1, pp. 38/50, 38, 47).

[72] Vladimirova, E. V.; Makarochkin, B. A. (Geol. Geofiz. **1977** No. 2, pp. 153/5).

[73] Kornetova, V. A.; Kazakova, M. E. (Nov. Dannye Mineral. No. 30 [1982] 191/4).

[74] Zhirov, K. K.; Artemov, Yu. M.; Volobuev, M. I.; Zhirova, V. V.; Knorre, K. G.; Stupnikova, N. I.; Sten'ko, V. A.; Tikhonov, V. E.; Arakelyan, V. A. (Tr. Sess. Kom. Opred. Absolyut. Vozrasta Geol. Formatsii Akad. Nauk SSSR **7** 1958 [1960] 135/42, 136, 141/2).

[75] Pluhar, E. (Berlin. Geowiss. Abhandl. Reihe A **12** [1979] 1/53).

[76] Ixer, R. A.; Ashworth, J. R.; Pointer, C. M. (Geol. J. **22** [1987] 403/27, 404, 408/9).

[77] Ahrens, L. H.; Besairie, H.; Burger, A. J. (Compt. Rend. **248** [1959] 3088/90).

[78] Platt, R. G.; Wall, F.; Williams, C. T.; Woolley, A. R. (Mineral. Mag. **51** [1987] 253/63, 261).

[79] Azimov, P. T. (Dokl. Akad. Nauk Uzbek. SSR **27** [1970] 45/7).

[80] Khamrabaev, I. K.; Urunbaev, K. (in: Smorchkov, I. E., Aktsessornye Mineraly Izverzhennykh Porod, Nauka, Moscow 1968, pp. 141/9, 147).

[81] Komlev, L. V.; Filippov, M. S.; Danielevich, S. I.; Kryukova, N. F.; Kuchina, G. N.; Mikhalevskaya, A. D. (Tr. Sess. Kom. Opred. Absolyut. Vozrasta Geol. Formatsii Akad. Nauk SSSR **7** 1958 [1960] 205/15).

[82] Kelly, F.-J. (Inform. Circ. U.S. Bur. Mines No. 8124 [1962] 1/38, 6).

[83] Semenov, E. I. (Mineralogiya Redkikh Zemel', Akad. Nauk SSSR, Moscow 1963, pp. 1/412, 126/7).

[84] Rimsaite, J. (Geol. Surv. Can. Paper No. 84-1A [1984] 129/45).

[85] Rowley, E. B. (Rocks Minerals **35** [1960] 328/30).

[86] Heinrich, E. W.; Borup, R. A.; Levinson, A. A. (Geochim. Cosmochim. Acta **19** [1960] 222/31).

[87] Linares, E.; Toubes, R. O. (Actas Jorn. Geol. Argent. **3** 1960 [1962] 191/205, 195).

[88] Kittl, E.; Tosi, A. (Rev. Fac. Ing. Quím. Univ. Nacl. Litoral Santa Fe Arg. **23** No. 36 [1955] 167/72).

[89] Murata, K. J.; Dutra, C. V.; Teixeira Da Costa, M.; Branco, J. J. R. (Geochim. Cosmochim. Acta **16** [1959] 1/14, 5, 7, 9, 12).

[90] Haapala, I. (Bull. Comm. Geol. Finlande No. 224 [1966] 1/98, 36/7, 53).

[91] Haapala, I.; Ervamaa, P.; Löfgren, A.; Ojanperä, P. (Bull. Geol. Soc. Finland No. 41 [1969] 117/24, 117, 119/20).

[92] Gramaccioli, C. M.; Segalstad, T. V. (Am. Mineralogist **63** [1978] 757/61).

[93] Mannucci, G.; Diella, V.; Gramaccioli, C. M. (Can. Mineralogist **24** [1986] 469/74).

[94] Skrzat, Z. (Kwart. Geol. **4** [1960] 874/88, 881, 888).

[95] Kalita, A. P. (in: Kuz'menko, M. V., Osobennosti Raspredeleniya Redkikh Elementov Pegmatitakh, Nauka, Moscow 1969, pp. 79/100, 80, 82/5, 87).

278

[96] Nikitin, Yu. V.; Sal'e, M. E.; Leonova, V. A.; Safronova, G. P.; Shurkin, K. A. (in: Shurkin, K. A.; Gorlov, N. V.; Sal'e, M. E.; Duk, V. L.; Nikitin, Yu. V., Belomorskii Kompleks Severnoi Karelii i Yugo-Zapada Kol'skogo Poluostrova, Akad. Nauk SSSR, Moscow-Leningrad 1962, pp. 223/55).

[97] Leonova, V. A.; Nikitin, Yu. V. (Zap. Vses. Mineral. Obshch. 91 [1962] 136/45, 136/9).

[98] Makarochkin, B. A. (Geol. Geofiz. 1975 No. 2, pp. 155/9).

[99] Shcherbak, M. P.; Lavrov, D. A. (Pitannya Geokhim. Mineral. Petrogr. 1963 223/35).

[100] Komlev, L. V.; Danilevich, S. I.; Ivanova, K. S.; Mikhalevskaya, A. D.; Filippov, M. S.; Chupakhin, M. S. (Tr. Sess. Kom. Opred. Absolyut. Vozrasta Geol. Formatsii Akad. Nauk SSSR 5 1956 [1958] 176/85, 176/8).

[101] Komlev, L. V.; Gerling, E. K.; Zhirov, K. K. (Tr. Sess. Kom. Opred. Absolyut. Vozrasta Geol. Formatsii Akad. Nauk SSSR 6 1957 [1960] 129/32).

[102] Sveshnikova, E. V.; Semenov, E. I.; Khomyakov, A. P. (Zaangarskii Shchelochnoi Massiv, Ego Porody i Mineraly, Nauka, Moscow 1976, pp. 1/80, 48).

[103] Zhirova, V. V.; Zykov, S. I.; Tugarinov, A. I. (Geokhimiya 1957 592/9, 593).

[104] Kuo, C.-C. (Sci. Record [Peking] 2 No. 11 [1958] 400/13, 408).

[105] Wadia, D. N. (Sci. Cult. [Calcutta] 21 [1956] 561/5).

[106] Semenov, E. I. (Typochimizm Mineralov Shchelochnykh Massivov, Nedra, Moscow 1977, pp. 1/119, 94/5).

[107] Herath, J. W. (Econ. Bull. Sri Lanka Geol. Surv. Dept. No. 2 [1975] 1/72, 24).

[108] Aranyakanond, P.; Nilkhuha, C. (Thai Sci. Bull. Rept. Invest. 1955 No. 1, pp. 41/6).

[109] Murakoshi, T.; Koseki, K. (Proc. 2nd U.N. Intern. Conf. Peaceful Uses At. Energy, Geneva 1958, Vol. 2, pp. 720/31, 722, 730).

[110] Hasegawa, S. (Ganseki Kobutsu Kosho Gakkaishi 46 [1961] 57/61 and C.A. 60 [1964] 14259).

[111] McKie, D. (Records Geol. Surv. Tanganyika 6 1956 [1958] 87/93).

[112] Burger, A. J.; Nicolaysen, L. O.; Ahrens, L. H. (J. Geophys. Res. 72 [1967] 3585/94, 3586/8).

[113] von Backström, J. W. (PEL-112 [1966] 1/19; N.S.A. 21 [1967] No. 22051).

[114] Burger, A. J.; Oosthuyzen, E. J.; Van Niekerk, C. B. (Ann. Geol. Surv. South Africa 6 1967 [1969] 85/9).

[115] Hugo, P. J. (Rept. South Africa Dept. Mines Geol. Surv. Mem. No. 58 [1970] 1/94, 1, 10/1, 56).

[116] Murdock, T. G. (Inform. Circ. U.S. Bur. Mines No. 8196 [1963] 1/147, 89, 102/3).

[117] Behier, J. (Ann. Geol. Madagascar No. 29 [1960] 1/78, 50).

[118] Behier, J. (Rappt. Ann. Serv. Geol. [Malagasy] No. 89 [1958] 1/146, 75/6).

[119] Cooper, P. F.; MacNevin, A. A.; Winward, K. (Mineral Ind. New South Wales Nos. 38, 40, 44 [1979] 1/80).

[120] Welin, E.; Blomqvist, G. (Geol. Foeren. Stockholm Foerh. 88 [1966] 3/18, 8, 12).

[121] Staatz, M. H.; Olson, J. C. (Geol. Surv. Profess. Paper [U.S.] No. 820 [1973] 468/76, 471/2).

[122] Austin, S. R.; Hetland, D. L.; Sharp, B. J. (Miner. Resources Rept. Idaho Bur. Mines Geol. No. 11 [1970] 1/10, 2, 4).

[123] Anderson, A. L. (Econ. Geol. 56 [1961] 177/97).

[124] Staatz, M. H.; Sharp, B. J.; Hetland, D. L. (Geol. Surv. Profess. Paper [U.S.] No. 1049-A [1979] A1/A90).

[125] Staatz, M. H. (Geol. Surv. Bull. [U.S.] No. 1351 [1972] 1/94).

[126] Staatz, M. H.; Shaw, V. E.; Wahlberg, J. S. (Econ. Geol. 67 [1972] 72/82, 72).

[127] Staatz, M. H. (Geol. Surv. Profess. Paper [U.S.] No. 1049-D [1983] D1/D52, D2, D32, D34/D36, D43, D46/D48).

[128] Bermanec, V.; Tibljaš, D.; Gessner, M.; Kniewald, G. (Mineral. Petrol. **38** [1988] 139/50).

[129] Komov, I. L.; Mel'nikova, E. M.; Kokarev, G. N. (Tr. Mineral. Muzeya Akad. Nauk SSSR No. 23 [1974] 87/93).

[130] Krol, O. F. (Geol. Geokhim. Mineral. Mestorozhd. Redk. Elem. Kaz. **1966** 60/5).

[131] Chistyakova, M. B. (Tr. Mineral. Muzeya Akad. Nauk SSSR No. 23 [1974] 113/76, 137, 153, 165).

[132] Chistyakova, M. B.; Kazakova, M. E. (Tr. Mineral. Muzeya Akad. Nauk SSSR No. 18 [1968] 245/9).

[133] Coetzee, C. B. (Handbook South Africa Geol. Surv. No. 7 [1976] 199/201).

[134] Bowie, S. H. U. (J. Roy. Soc. Arts **107** [1959] 704/18).

[135] von Backström, J. W. (Handbook South Africa Geol. Surv. No. 7 [1976] 209/12).

[136] Staatz, M. H.; Conklin, N. M. (Geol. Surv. Profess. Paper [U.S.] No. 550-B [1966] B130/B134).

[137] Heinrich, E. W. (The Geology of Carbonatites, Rand McNally, Chicago 1966, pp. 1/555, 178/9).

[138] Andersen, T. (Mineral. Mag. **50** [1986] 503/9).

[139] Kukharenko, A. A.; Bulakh, A. G.; Baklanova, K. A. (Zap. Vses. Mineral. Obshch. **90** [1961] 371/81).

[140] Semenov, E. I.; Upendran, R.; Subramanian, V. (J. Geol. Soc. India **19** [1978] 550/7, 550, 553/4).

[141] Overstreet, W. C.; White, A. M.; Warr, J. J., Jr. (Geol. Surv. Profess. Paper [U.S.] No. 700-D [1970] D169/D175).

[142] Rupasinghe, M. S.; Senaratne, A.; Dissanayake, C. B. (J. Gemmol. **20** [1986] 177/84).

[143] Troxel, B. W. (Bull. Calif. Div. Mines Geol. No. 176 [1957] 635/40).

[144] Rosenblum, S.; Mosier, E. L. (Geol. Surv. Profess. Paper [U.S.] No. 1181 [1983] 1/67).

[145] Paone, J. (U.S. Bur. Mines Bull. No. 585 [1960] 863/72, 865/6).

[146] Savage, C. N. (Idaho Bur. Mines Geol. Bull. No. 17 [1961] 1/160, 12, 14/5).

[147] Overstreet, W. C.; Warr, J. J., Jr.; White, A. M. (Southeast. Geol. **10** [1969] 63/76, 64/5).

[148] Overstreet, W. C.; White, A. M.; Whitlow, J. W.; Theobald, P. K., Jr.; Caldwell, D. W.; Cuppels, N. P. (Geol. Surv. Profess. Paper [U.S.] No. 568 [1968] 1/85).

[149] Angelelli, V.; Chaar, E. (Argent. Repub. Com. Nacl. Energ. At. Informe No. 139 [1965] 1/39 from C.A. **64** [1966] 468).

[150] Kantor, J. (Geol. Prace Zpravy **11** [1957] 5/28, 23 and C.A. **1958** 9894).

[151] Nekrasov, I. Ya. (Dokl. Akad. Nauk SSSR **204** [1972] 941/3; Dokl. Acad. Sci. USSR Earth Sci. Sect. **204** [1972] 134/6).

[152] Nekrasova, R. A.; Nekrasov, I. Ya. (Dokl. Akad. Nauk SSSR **268** [1983] 688/93; Dokl. Acad. Sci. USSR Earth Sci. Sect. **268** [1983] 151/5).

[153] Rupasinghe, M. S.; Gocht, W.; Dissanayake, C. B. (J. Natl. Sci. Counc. Sri Lanka **11** [1983] 99/110, 102).

[154] Boulanger, J. (Territ. Madagascar Trav. Bur. Geol. No. 87 [1958] 1/104).

[155] Hutton, C. O. (Spec. Rept. Calif. Div. Mines Geol. No. 59 [1959] 1/32, 19/20).

[156] McCauley, C. K. (Geol. Notes South Carolina Div. Geol. Surv. No. 26 [1960] 1/13, 3, 5).

[157] de Oliveira, A. I. (Engenharia Miner. Metalurgia **24** [1956] 163/4; Ref. Zh. Geol. **1959** 3328).

[158] Bonatti, S. (Atti 1st Conv. Geol. Nucleare, Rome 1955, pp. 17/21).

[159] Loziński, J. (Pr. Mineral. Pol. Akad. Nauk Oddzial Krakowie Kom. Nauk Mineral. No. 16 [1969] 43/9, 44).

[160] Boissevain, H. (Geol. Mijnbouw [2] 19 [1957] 505/10).

[161] Bhola, K. L.; Chatterji, B. D.; Dar, K. K.; Mahadevan, C.; Mahadevan, V.; Mehta, N. R.; Nagarajarao; Nandi, H.; Narayandas, G. R.; Sahasrabudhe, G. H.; Shirke, V. G.; Udas, G. R. (Proc. 2nd U.N. Intern. Conf. Peaceful Uses At. Energy, Geneva 1958, Vol. 2, pp. 100/2).

[162] Titus, P. O.; Nair, N. R. (Plant Maint. Import Substitution 11 [1978] 86/102, 98).

[163] Chaudhuri, J. N. B. (Berlin. Geowiss. Abhandl. Reihe A 52 [1984] 1/123, 8, 31/3).

[164] Kogan, B. I. (Tr. Inst. Geol. Akad. Nauk Latv. SSR 6 [1960] 183/214, 189).

[165] Matzko, J. J.; Overstreet, W. C. (Chung-Kuo Ti Chih Hsueh Hui Hui K'an No. 20 [1977] 16/35).

[166] Shen, J. T. (Proc. Intern. Conf. Peaceful Uses At. Energy, Geneva 1955 [1956], Vol. 6, pp. 147/51).

[167] Flinter, B. H.; Butler, J. R.; Harral, G. M. (Am. Mineralogist 48 [1963] 1210/26, 1211/3).

[168] Yun, T. S. (Proc. Intern. Conf. Peaceful Uses At. Energy, Geneva 1955 [1956], Vol. 6, pp. 176/7).

[169] El-Hinnawi, E. E. (Beitr. Mineral. Petrogr. [Berlin] 9 [1963/64] 519/32, 528).

[170] El Shazly, E. M. (Tech. Rept. Ser. Intern. At. Energy Agency No. 52 [1966] 186/98, 191).

[171] Rosenblum, S. (J. Res. U.S. Geol. Surv. 2 [1974] 689/92).

[172] Besairie, H. (Ann. Geol. Madagascar No. 30 [1961] 1/116, 39/40).

[173] Rayner, E. O. (Australian At. Energy Symp. Proc., Sydney 1958, pp. 27/34, 29).

[174] Whitworth, H. F. (New South Wales Dept. Mines Tech. Rept. 4 1956 [1959] 7/60, 9, 25, 38).

[175] Bowie, S. H. U. (Chron. Mines d'Outre-Mer Rech. Min. No. 271 [1959] 304/8).

[176] Styles, M. T.; Young, B. R. (Mineral. Mag. 47 [1983] 41/6).

[177] Callaghan, C. C. (Ann. Geol. Surv. [South Africa] 17 [1984] 35/41).

[178] Lomozová, V.; Pácal, Z. (Vestn. Ustred. Ustavu Geol. 43 [1968] 131/4).

[179] Rayner, E. O. (Geol. Surv. New South Wales Rept. GS-099 [1958] [unpubl.] from [119, p. 47]).

[180] Serdyuchenko, D. P.; Kochetkov, O. S. (Dokl. Akad. Nauk SSSR 218 [1974] 1175/7; Dokl. Acad. Sci. USSR Earth Sci. Sect. 218 [1974] 124/5).

[181] Roscoe, S. M.; Steacy, H. R. (Proc. 2nd U.N. Intern. Conf. Peaceful Uses At. Energy, Geneva 1958, Vol. 2, pp. 475/83, 481/2).

[182] Roscoe, S. M. (Can. Mining J. 80 No. 7 [1959] 65/6).

[183] Rudnitskaya, L. S. (Uran Drevnikh Konglomeratakh 1963 129/40, 135).

[184] Roscoe, S. M. (Paper Geol. Surv. Can. No. 68-40 [1969] 1/205, 139).

[185] Osterwald, F. W.; Osterwald, D. B.; Long, J. S., Jr.; Wilson, W. H. (Bull. Geol. Surv. Wyoming No. 50 [1966] 1/287, 220/1).

[186] Bezgubov, A. I.; Dement'ev, P. K.; Kruglova, V. G.; Modnikov, I. S. (Geokhim. Mineral. Radioakt. Elem. Sibiri 1970 156/64, 156, 159/60, 163).

[187] Dishchuk, V. L.; Druzhinin, L. N.; Smirnov, G. I.; Stasiv, T. V.; Yur'eva, A. L. (Mineral. Sb. [Lvov] 24 [1970] 431/4).

[188] Bobrievich, A. P.; Druzhinin, L. N.; Smirnov, G. I.; Yur'eva, A. L. (Mater. Mineral. Petrogr. Geokhim. Osad. Porod Rud No. 2 [1974] 88/99, 88, 95).

[189] Gauthier-Lafaye, F. (Sci. Geol. Mem. No. 78 [1986] 1/206, 3, 5, 179).

[190] Stupp, H. D. (Diss. Univ. Köln 1984, pp. 1/280, 22, 85/6).

[191] Liebenberg, W. R. (J. South African Inst. Mining Met. 57 [1956] 153/208, 162).

[192] Liebenberg, W. R. (Proc. 2nd U.N. Intern. Conf. Peaceful Uses At. Energy, Geneva 1958, Vol. 2, pp. 379/87, 381/2).

[193] Reimer, T. O. (Neues Jahrb. Mineral. Abhandl. **158** [1987] 13/46, 15/6, 22).

[194] Glatthaar, C. W.; Feather, C. E. (Appl. Mineral. Proc. 2nd Intern. Congr. Appl. Mineral. Minerals Ind., Los Angeles 1984 [1985], pp. 617/28, 618/20, 622).

[195] Houston, R. S.; Murphy, J. R. (Bull. Geol. Surv. Wyoming No. 49 [1962] 1/120, 6, 39).

[196] Berger, K.; Salger, M. (Geol. Bavarica No. 53 [1964] 174/93, 187, 189, 192).

[197] Yankovskii, V. A. (Sb. Statei Geol. Sibiri **1975** 30/4).

[198] Hosterman, J. W.; Overstreet, W. C.; Warr, J. J., Jr. (Geol. Surv. Profess. Paper [U.S.] No. 475-D [1964] D128/D130).

[199] Overstreet, W. C.; Yates, R. G.; Griffitts, W. R. (Geol. Surv. Bull. [U.S.] No. 1162-F [1963] F1/F31, F10/F13).

[200] Burkser, E. S.; Eliseeva, G. D.; Lechekhleb, V. R.; Shcherbak, N. P. (Byull. Kom. Opred. Absolyut. Vozrasta Geol. Formatsii Akad. Nauk SSSR No. 5 [1962] 48/52).

[201] Bárdossy, G.; Pantó, G.; Várhegyi, G. (Trav. Com. Intern. Etude Bauxites Alumine Alum. No. 13 [1976] 221/31, 222/3).

[202] Bardoshshi, D'.; Varkhedi, D'.; Panto, D'. [Bárdossy, G.; Várhegyi, G.; Pantó, G.] (Geol. Rudn. Mestorozhd. **19** No. 3 [1977] 72/81, 73/4).

[203] Parshakov, N. S.; Menyailov, A. A.; Laubenbakh, A. I.; Skosyreva, L. N.; Slavyagina, I. I. (Izv. Vysshikh Uchebn. Zavedenii Geol. Razvedka **19** No. 6 [1976] 32/6).

[204] Schwander, H.; Wenk, E. (Schweiz. Mineral. Petrogr. Mitt. **45** [1965] 797/806, 798/9, 805/6).

[205] Köppel, V.; Grünenfelder, M. (Schweiz. Mineral. Petrogr. Mitt. **51** [1971] 385/409, 390, 394, 408).

[206] Fazekas, V.; László, K.; Béla, S. (Foldt. Kozl. **105** [1975] 297/305 from C.A. **85** [1976] No. 49363).

[207] Pavelescu, L.; Pavelescu, M. (TMPM Tschermaks Mineral. Petrogr. Mitt. [3] **17** [1972] 208/14).

[208] Lyakhovich, V. V. (Mineral. Sb. [Lvov] **20** [1966] 199/208, 199, 201, 204/5).

[209] Komarov, A. N.; Berman, I. B.; Kol'tsova, T. V. (Geokhimiya **1985** 979/87; Geochem. Intern. **23** No. 1 [1986] 60/8, 63/4).

[210] Arkhangel'skaya, V. V. (Redkometal'nye Shchelochnye Kompleksy Yuzhnogo Kraya Sibirskoi Platformy, Nedra, Moscow 1974, pp. 1/127).

[211] Black, L. P.; Fitzgerald, J. D.; Harley, S. L. (Contrib. Mineral. Petrol. [Berlin] **85** [1984] 141/8, 141/2, 146).

[212] Mohr, D. W. (Am. Mineralogist **69** [1984] 98/103).

[213] Weber, C.; Barbey, P.; Cuney, M.; Martin, H. (Contrib. Mineral. Petrol. [Berlin] **90** [1985] 52/62, 53/5).

[214] Vinogradov, A. N.; Vinogradova, G. V. (Metamorf. Metamorfog. Rudoobraz. Rannego Dokembr. **1984** 37/47, 38, 40).

[215] Parslow, G. R.; Brandstätter, F.; Kurat, G.; Thomas, D. J. (Can. Mineralogist **23** [1985] 543/51, 545/6, 549).

[216] Watkinson, D. H.; Mainwaring, P. R. (Can. J. Earth Sci. **13** [1976] 470/5).

[217] Watkinson, D. H.; Mainwaring, P. R. (Can. Mineralogist **12** [1973] 148).

[218] Vickers, R. C. (U.S. Geol. Surv. Bull. No. 1030-F [1956] 171/85, 179).

[219] Vickers, R. C. (Proc. Intern. Conf. Peaceful Uses At. Energy, Geneva 1955 [1956], Vol. 6, pp. 597/9).

[220] Burkov, V. V. (in: Ovchinnikov, L. N.; Solodov, N. A., Mestorozhdeniya Litofil'nykh Redkikh Metallov, Nedra, Moscow 1980, pp. 314/22, 320).

[221] Stepanyuk, L. M.; Eliseeva, G. D. (Geokhim. Rudoobraz. No. 13 [1985] 69/74).

[222] Shmakin, B. M.; Shiryaeva, V. A. (Zap. Vses. Mineral. Obshch. **100** [1971] 274/81, 274/5).

282

[223] Shmakin, B. M.; Shiryaeva, V. A. (Geokhimiya **1970** 1263/8; Geochem. Intern. **7** [1970] 908).

[224] Banas, M.; Kucha, H. (Proc. 6th Quadrenn. IAGOD Symp., Tbilisi 1982 [1984], Vol. 1, pp. 241/6).

[225] Ploshko, V. V. (Izv. Akad. Nauk SSSR Ser. Geol. **1961** No. 1, pp. 86/93).

[226] Ploshko, V. V.; Knyazeva, D. N. (Aktsessornye Miner. Elem. Kriter. Komagmat. Metallog. Spets. Magmat. Kompleksov **1965** 146/52, 146/7, 151 and C.A. **64** [1966] 13955).

[227] Kupriyanova, I. I.; Volkova, M. I.; Goroshchenko, Z. I. (Tr. Mineral. Muzeya Akad. Nauk SSSR No. 15 [1964] 123/33).

[228] Kochetkov, O. S. (Redk. Elem. Porodakh Razlichnykh Metamorf. Fatsii **1967** 105/25).

[229] Marchenko, E. Ya. (Dokl. Akad. Nauk SSSR **155** [1964] 349/52; Dokl. Acad. Sci. USSR Earth Sci. Sect. **155** [1964] 118/22).

[230] Yalovenko, I. P.; Yur'eva, A. L. (Mineral. Sb. [Lvov] **21** [1967] 299/304).

[231] Kononov, Yu. V.; Nechaev, S. V. (Pitannya Geokhim. Mineral. Petrogr. **1963** 289/301, 295).

[232] Korneva, N. G.; Gavrusevich, I. B. (Geol. Zh. **47** No. 4 [1987] 59/63).

[233] Kalita, A. P. (Redkozemel'nye Pegmatity Alakurtti i Priladozh'ya, Akad. Nauk SSSR, Moscow 1961, pp. 1/119).

[234] Kornetova, V. A. (Tr. Mineral. Muzeya Akad. Nauk SSSR No. 14 [1963] 96/107).

[235] Kornetova, V. A.; Varshal, G. M. (Khim. Anal. Miner. Ikh Khim. Sostav **1964** 83/9).

[236] Shcherbak, N. P.; Alekseeva, K. N.; Gol'denfel'd, I. V.; Eliseeva, G. D. (Tr. Sess. Kom. Opred. Absolyut. Vozrasta Geol. Formatsii Akad. Nauk SSSR **10** 1961 [1962] 85/93, 87).

[237] Zhabin, A. G.; Mukhitdinov, G. N.; Kazakova, M. E. (Tr. Inst. Mineral. Geokhim. Kristallokhim. Redk. Elem. No. 4 [1960] 51/73).

[238] Es'kova, E. M.; Zhabin, A. G.; Mukhitdinov, G. N. (Mineralogiya i Geokhimiya Redkikh Elementov Vishnevykh Gor, Nauka, Moscow 1964, pp. 1/318).

[239] Semenov, E. I.; Kostyunina, L. P.; Kulakov, M. P. (Mineral. Pegmatitov Gidroterm. Shchelochnykh Massivov Akad. Nauk SSSR Inst. Mineral. Geokhim. Kristallokhim. Redk. Elem. **1967** 137/49, 137, 139).

[240] Kinnaird, J. A. (J. African Earth Sci. **2** [1984] 81/90, 85, 89).

[241] Kinnaird, J. A.; Bowden, P.; Ixer, R. A.; Odling, N. W. A. (J. African Earth Sci. **3** [1985] 185/22).

[242] Kinnaird, J. A. (J. African Earth Sci. **3** [1985] 229/51, 233, 241).

[243] Plyushchev, E. V. (Tr. Inst. Geol. Geofiz. Akad. Nauk SSSR Sibirsk. Otd. No. 286 [1975] 161/6, 272/83, 161, 163/4).

[244] Petruk, W.; Owens, D. (Can. Mineralogist **13** [1975] 298/9).

[245] Chukhrov, F. V. (Tr. Inst. Geol. Rudn. Mestorozhd. Petrogr. Mineral. Geokhim. No. 50 [1960] 3/237).

[246] Jobbins, E. A.; Tresham, A. E.; Young, B. R. (J. Gemmol. **15** [1977] 295/9).

[247] Soong, K.-L. (Acta Oceanogr. Taiwan. No. 8 [1978] 43/62).

[248] Robertson, J. A. (Ontario Dept. Mines Geol. Rept. No. 57 [1968] 1/162, 92).

[249] Sokova, K. P. (Tr. Inst. Geol. Rudn. Mestorozhd. Petrogr. Mineral. Geokhim. No. 81 [1962] 3/22, 18).

[250] Kamel Hassan, K. E. (Diss. Assint Univ. 1964 from [170]).

[251] Li, A. F.; Grebennikova, O. T. (Zap. Vost. Sib. Otd. Vses. Mineral. Obshch. **1962** 155/61, 155).

[252] Semenov, E. I.; Khomyakov, A. P. (Redk. Elem. Syr'e Ekon. No. 6 **1971** [1972] 77/81).

[253] Van Wambeke, L. (Am. Mineralogist **56** [1971] 1366/84, 1366/7).

[254] Komlev, L. V.; Ivanova, K. S. (Radiokhimiya **6** [1964] 585/94; Soviet Radiochem. **6** [1964] 565/73, 565).

[255] Kucha, H. (Mineral. Mag. **43** [1979/80] 1031/4).

[256] Molloy, M. W. (Am. Mineralogist **44** [1959] 510/32, 510/2).

[257] Jefferies, N. L. (Mineral. Mag. **49** [1985] 495/504, 495, 498, 502/4).

[258] Zhirov, K. K.; Baranovskaya, N. V.; Litvina, L. A. (Geokhimiya **1958** 167/70; Geochem. [USSR] **1958** 218/23).

[259] Fleischer, M. (Am. Mineralogist **44** [1959] 1329).

[260] Karkhanavala, M. D. (Current Sci. [India] **25** [1956] 166/7).

[261] Hikichi, Y.; Hukuo, K. (Nippon Kagaku Kaishi **1975** 1311/4).

[262] Hikichi, Y.; Hukuo, K. (Nagoya Kogyo Daigaku Gakuho **29** 1977 [1978] 203/8).

[263] Hikichi, Y.; Nomura, T. (J. Am. Ceram. Soc. **70** [1987] C-252/C-253).

[264] Anthony, J. W. (Am. Mineralogist **42** [1957] 904).

[265] Anthony, J. W. (Am. Mineralogist **50** [1965] 1421/31, 1422).

[266] Meliksetyan, B. M. (Zap. Arm. Otd. Vses. Mineral. Obshch. No. 2 [1963] 57/80, 69).

[267] Overstreet, W. C.; Warr, J. J., Jr.; White, A. M. (Southeast. Geol. **13** [1971] 99/125, 119).

[268] Bulakh, A. G.; Kukharenko, A. A.; Knipovich, Yu. N.; Kondrash'eva, V. V.; Baklanova, K. A.; Baranova, E. N. (Mater. Sess. Uchen. Soveta VSEGEI Rezul'tatam Rabot **1959** [1961] 114/6).

2.4.1.1.2 Chemistry

Analyses

General. The theoretical composition of monazite, if one regards it as a pure RE phosphate, is 69.73% ΣRE_2O_3 and 30.27% P_2O_5 [1, p. 694], [2, p. 116] or 69.81% Ce_2O_3 and 30.19% P_2O_5 [3, p. 1228]. But natural monazite free from Th is very rare [4], and its chemical composition is highly variable (see the following paragraphs) and may depend on its mode of occurrence (see pp. 294/304) and its degree and kind of alteration (see pp. 304 and 331).

For a compilation of chemical analyses of 7 average values for monazites and of chemical analyses and corresponding densities of 215 individual monazites from various source rocks and localities, see Tables 13 and 14, pp. 284 and 305/30, respectively. Compilations of monazite analyses published prior to 1955 are given, for example, by [1, p. 694], [2, p. 116], [4], [5, p. 668]. For a comprehensive survey of older monazite analyses, see [6].

Ranges and average contents of the main constituents of monazite are given as 2 to 31, 8.3% ThO_2; 43 to 73, 59.5% ΣRE_2O_3; 0 to 5, 1.2% Y_2O_3; 0 to 6, 1.0% CaO; 0 to 4, 0.5% U_3O_8; 20 to 30, 27.5% P_2O_5; and 0.3 to 6.1, 2.3% SiO_2 [7, p. 306]. Generally, in monazite, the sum of RE oxides, ThO_2 and U oxide is 68 to 69% [8]. Besides the main components, monazite may contain small to minor amounts of Mg, Mn, Fe^{2+}, Fe^{3+}, Pb, Zr, Be, Sn, and Al [9]; refer also to pp. 288/9. Other compositional ranges given in the literature for monazite are as follows: 2.5 to 12% (mostly 3.5 to 10%) ThO_2, 55 to 68% (rarely up to 74%) ΣRE_2O_3 (comprising mainly oxides of Ce, La, and Nd), 18.4 to 31.5% P_2O_5, and rarely more than1% (generally 0.1 to 0.4%) UO_2 [10]; 0 to 18% ThO_2, 39 to 74% Ce_2O_3, 0 to 7% ZrO_2, and 0 to 5% Y_2O_3 [11], or up to 10% (sometimes up to 28%) ThO_2, 50 to 68% ΣRE_2O_3 (comprising mainly oxides of Ce and La), up to 5% Y_2O_3, 22 to 31.5% P_2O_5, and up to 7% ZrO_2 together with up to 6% SiO_2; sometimes contents of CaO and SO_3 occur [12]. The average composition of 70 samples of

Table 13
Average Chemical Composition of Monazite Samples (with ThO_2 contents > 1%) From Various
Types of Occurrences.
For analysis numbers marked with an asterisk, see original paper for the individual analyses.

	mean content in % in sample number						
	1	2	3*)	4*)	5*)	6*)	7*)
ThO_2	6.8 a)	3.52	3.17	3.14	2.92	1.57	1.16
ΣRE_2O_3	58.3	61.88	56.64	57.90	57.36	58.84	60.15
Y_2O_3	—	3.22 b)	—	—	—	1.80	—
CaO	—	—	—	—	—	0.79	—
Fe_2O_3	—	0.28	—	—	—	0.65	—
Al_2O_3	—	—	—	—	—	0.24	—
U_3O_8	—	0.87	—	—	—	—	—
UO_2	—	—	—	—	—	0.42	—
P_2O_5	27.1	29.63	28.60	25.15	28.30	29.67	28.37
SiO_2	—	—	3.13	4.73	3.07	1.28	2.30
others	—	— b)	—	—	—	—	—
sum	92.2	99.40	91.54	90.92	91.65	95.26	91.98

No. 1 = Monazite from an oxide-apatite-monazite segregation in granitic gneiss, Kulyk Lake
area, northern Saskatchewan, Canada; average of 3 grains [20]. — No. 2 = Monazite from
hydrothermal quartz vein, Alinci, southeastern Yugoslavia; mean of 3 analyses on each of 2
monazite samples; D is 5.10 ± 0.05 g/cm³ and is the average of 10 weighings on each of 2
crystal fragments [21, pp. 140, 142/3]. — No. 3 = Monazite from Mozaan conglomerate, Denny
Dalton, Pongola Supergroup, northern Zululand, R.S.A.; average of 27 grains [22, appendix
A-2]; see also [23]. — No. 4 = Monazite from conglomerate, C-Reef, Quirke Mine No. 2, Elliot
Lake, Ontario, Canada; average of 20 grains [22, appendix A-2]; see also [23]. — No. 5 =
Monazite from conglomerate, Dominion Reef, western Transvaal, R.S.A.; average of 25 grains
[22, appendix A-2]; see also [23]. — No. 6 = Monazite from uraniferous conglomerate,
Dominion Reef, western Transvaal, R.S.A.; mean of 32 grains [24]. — No. 7 = Monazite from
Mozaan conglomerate, Amsterdam, Pongola Supergroup, northern Zululand, R.S.A.; average
of 16 grains [22, appendix A-2]; see also [23].

a) The range of the ThO_2 content is 1.87 to 10.27%. — b) Additionally are given 160 ppm each
of Ho and Tm, 80 ppm Yb, 40 ppm Cr, and 130 ppm Ti.

monazite from different regions of the USSR is given as 7.53% ThO_2, 58.86% ΣRE_2O_3, 1.59%
CaO, 0.34% MgO, 0.98% Fe_2O_3, 1.54% Al_2O_3, 27.18% P_2O_5, and 2.81% SiO_2 [13, pp. 156,
159]. Total analyses of 18 monazite samples (which, however, are badly sorted and contain
admixtures of various minerals [14]) from various sources of the USSR show the following
ranges and average contents (for two partially analyzed samples, see original paper): 1.56
to 10.60, 5.88% ThO_2; 53.38 to 68.30, 60.35% ΣRE_2O_3; 0.22 to 7.14, 1.28% CaO; 0.03 to 0.37,
0.12% MgO (16 samples); traces to 0.14, ~0.03% MnO (5 samples); traces to 1.43, ~0.35%
Fe_2O_3; 0 to 3.34, 0.60% Al_2O_3; 0 to 0.46, 0.16% U_3O_8 (14 samples); 0 to 1.27, ~0.37% PbO
(15 samples); 0.12 to 2.20, 0.73% ZrO_2 (4 samples); 0.44% TiO_2 (1 sample); 24.05 to 29.35,
26.91% P_2O_5; 0.36 to 4.93, 2.32% SiO_2; 0 to 0.79, 0.18% SO_3 (17 samples); 0.09 to 1.60, 0.50%
H_2O; 0 to 1.65, 0.68% CO_2 (6 samples); and 0.13 to 4.28, 1.19% F (5 samples) [15]. Five spot

analyses of monazite from a peralkaline granite of the Chilwa Alkaline Province, Malawi, give as average contents 3.21% ThO_2, ~68.43% ΣRE_2O_3, 0.7% Y_2O_3, <0.05% each of CaO, MgO, MnO, FeO, SrO, Al_2O_3, TiO_2, and Nb_2O_5, 0.08% UO_2, 0.09% ZrO_2, 0.19% BaO, 28.3% P_2O_5, and 1.24% SiO_2 [16]. The respective average compositions corresponding to each 11 U-poor and U-rich monazite particles from stream sediments of the southeastern Piedmont of Georgia and South and North Carolina are 5.0 and 4.8% ThO_2, 65.1 and 62.1% ΣRE_2O_3, 0.7 and 2.1% Y_2O_3, 0.2 and 1.7% UO_2, both 1.2% CaO, 27.9 and 29.0% P_2O_5, and 0.4 and 0.3% F [18, pp. 7, 28, 36].

A mineral similar to monazite or xenotime in a microgranite of Bois-Noirs Nord, Bourbonnaise Mountains, Massif Central, France, contains 6.80% Th, between 6.50 and 13.60% of individual RE, 17.63% Ca, 1 to 5.40% U, 13.70% P, and traces of Si [19].

For an influence of the chemical composition of monazite on its crystal habit, lattice constants, d values, optical properties, and density, see pp. 338/9, 339/40, 341/2, 347/51, and 356, respectively.

Variation in Content of the Main Cations (refer also to pp. 283/5, and Table 14, pp. 305/30). The content of RE ranges from 56 to 68% ΣRE_2O_3 in most pure samples of monazite. Because Th substitutes RE (see p. 291), a high Th content causes a correspondingly lower RE content [8]. Monazite is characterized by a predominance of RE(Ce) [2, p. 117], [25], or principally is composed of the lighter lanthanides with atomic numbers ≤ 63, in contrast to brockite (which is enriched in Gd and other heavy RE; see p. 376) [26] and cheralite (which is enriched in intermediate and heavy RE; see pp. 363/4) [27, pp. 552, 554]. For an enrichment of light RE in monazite, see also, for instance [21, p. 146], [28, pp. 54/6], [29, pp. 135, 137], [30, pp. 81, 83/4]. Yttrian RE and/or Y account for up to ~4% [9], 8% [31], or ~15% of total RE in monazite [32]. The enrichment of heavy RE (excluding Y) and the depletion of La, both relative to the North American Shale composite, in the Th- and Ca-rich core of monazite from graphite-bearing metashales of the Anakeesta Formation, North Carolina, are explained by the increasing stability of the heavier RE in the slightly smaller unit cell resulting from the increased contents of Th and Ca [33]. Note also that Ca- and hence Th-enriched monazite is relatively enriched in RE(Y) in biotite granites of the Carnmenellis pluton, Cornwall, Great Britain [34, p. 498]. The presence of the small-sized U^{4+} ion facilitates the uptake of the heavy RE in alluvial monazite from Malaya [35]. In kularite (see pp. 245/6), intermediate RE, heavier than Sm and including Er, are enriched relative to the average RE distribution in 119 hypogene monazite samples (described by [2, pp. 118/25]). The concentrations of Eu and other intermediate RE are 10- and 1.5- to 3-fold higher, respectively, in the kularite than in normal monazite [36]. The contents and distribution of Y and/or RE in monazite will not be discussed further in detail; for more comprehensive surveys, see "Phosphor" A, 1965, pp. 256/8 and "Seltenerd-elemente" A 4, 1979, pp. 5, 24/6, 53/9, and 72/6.

Monazite invariably contains 2 to 30% Th [7, p. 306]. For the most common monazite varieties, the ranges of Th content are given as 4 to 12% ThO_2 [37], ~4 to 9% ThO_2 (more rarely up to 12%) [4], or 0.5 to 16% ThO_2 [3, p. 1237]; the average is 5% ThO_2 [38]. Note also that in monazite the Th content varies between a few percent and 10.6%, but probably may reach 26.4% [39, 40] or 30% [1, p. 693], [9, 41, 42]. Maximum contents of 33% [43, p. 197], [44] or 35% ThO_2 [38] given as upper limit for the compositional range of monazite actually may refer to the Th phosphate cheralite (see Section 2.4.1.2, pp. 360/8). The highest measured contents of Th in monazite (not included in Table 14, pp. 305/30) are found in a sample from a pegmatite of the Yucca Valley, California (36.6% ThO_2) [45], and from a biotite granite of the Rila Mountains, Bulgaria (27.42% ThO_2) [46, pp. 14, 19].

High-grade commercial monazite commonly carries between 3 and 10% ThO_2 (usually together with 0.1 to 0.4% U oxide) [43, pp. 197/8]. Thus, average commercial monazite of various locations contains 8 to 10.5% ThO_2 [47, 48], 8.9% ThO_2 [49], or 8.5% ThO_2 in India; 4.5% ThO_2 in Florida; 6% ThO_2 [50], 4 to 7% ThO_2 [11], 5 to 6% ThO_2 [48], or 5.1% ThO_2 in Brazil; 2.8 to 3.2% ThO_2 in Indonesia (the Netherlands' East Indies) [49]; and 7.2% ThO_2 in Liberia [51]. Minable monazite from beach placers contains 10.0 to 12.6% ThO_2 in Sierra Leone; 8.5 to 9.5% ThO_2 in Travancore, India; 8.9 to 9.2% ThO_2 on Sri Lanka; 6.8% ThO_2 on Formosa; 6.3 to 7.4, mean $\sim 6.6\%$ ThO_2 in Australia; 5 to 6% ThO_2 in Brazil; 5.6% ThO_2 in Galicia, Spain; $\sim 5\%$ ThO_2 in the Republic of South Africa; and 4.4% ThO_2 in Florida [43, pp. 201/2]. For Th contents of monazite from beach placers of various locations, see also, for example, [52 to 57]. In cassiterite placers of various localities (mainly of Malaya, Australia-Tasmania, and England as well as of Africa, America, Europe, and the Far East), the content of ThO_2 in monazite varies between 2.5 and 20% [58]. In Sn placers, it averages $\sim 6.5\%$ in Nigeria and $\sim 6\%$ in Malaya and is 3.4 and 3.27% ThO_2 in Indonesia [43, p. 203]. Monazite contains between 7.5 and 9.5% ThO_2 in inland placers of Bihar and West Bengal, India [56], and 4.21 to 5.96, mean 5.27% ThO_2 in seven fluviatile placers of South and North Carolina [59]. Moreover, commercial monazite contains 12.30 to 16.10, mean 14.70% ThO_2 in three specimens from an allanite-monazite vein, Rock Corral area, California [60], and the content of ThO_2 ranges from 5.3 to 12.4% in five analyses of monazite from pegmatitic deposits of southeastern Madagascar [61, 62]. It averages 8.8% ThO_2 in the Deadwood conglomerate, Wyoming [63], and $\sim 8\%$ in monazite-apatite veins of Steenkampskraal, R.S.A. [64].

The distribution of Th within individual monazite grains is diverse. Thus, Th is uniformly distributed within grains of monazite from the Cacciola granite, central Gotthard massif, Switzerland, but the variation between different grains is from ~ 1 to 20% Th [65]. It is homogeneously distributed in monazite crystals from Sn placers of Thailand, and sparse strong deviations of the Th content are caused by mineral inclusions (see pp. 271/2) [66, p. 23]. Within the same monazite crystal from two-mica granite of the Massif Saint-Sylvestre, Limousin region, France, the Th content varies from 6.96 to 8.4% [67]. Only a weak chemical zonation is discernible in monazite from biotite granites of the Carnmenellis pluton, Cornwall, Great Britain, which only concerns Sm, Dy, and Y, but not Th, Ca, and Ce [34, pp. 496, 498]. Radioactive elements (including Th, U, and Pb) are relatively evenly distributed in purified monazite from alluvial gravel and sand of the Otava River and its tributaries, southwestern Bohemia, Czechoslovakia, that shows in 69 samples only a weak zonality without strong radioactive centers and contains 3.49 to 8.89, mean 5.00% Th [68]. In a pegmatite of Piona, southern Alps, Italy, monazite shows a faint zoning with generally higher amounts of Th and U and lower amounts of RE in the core relative to the rim (for the composition of the rim, see analysis No. 16 in Table 14, p. 306) [69]. Thorium is distributed irregularly within some grains of monazite from stream sediments of the southeastern United States; its content ranges from 4 to 12.6% ThO_2 in three spot analyses of one grain [70]. Monazite from biotite-muscovite granite of the Lawler Peak pluton, Yavapai county, west-central Arizona, shows an internal zonation, notably an increase from core to rim in the ThO_2 content from ~ 2.5 to $> 8\%$ with a correlated zonation of U, Y, and Pb [71]. Relatively broad cores of essentially constant composition (corresponding to optically visible cores) in monazite from graphite-bearing metashales of the Anakeesta Formation, North Carolina, are surrounded by zones of ~ 10 μm across, showing noticeable decreases in Th and Ca contents, followed outwards by rims with only slight further decreases. Whereas in the cores, Th is slightly in excess relative to Ca, the reverse occurs in the rim [33].

Uranium is common in monazite but rarely reaches contents $> 0.5\%$ [6, p. 6], and monazite averages 1 part U_3O_8 per 50 parts ThO_2 [41]. Generally, the content of U in monazite varies between 0.05 and 0.3% U [3, p. 1240], or between 0.1 and 0.4% UO_2 [10], and may reach 1%

U_3O_8 [4] or 4.5% U [44]. A content of $>1\%$ UO_2 is found in 14.5% of the monazite particles analyzed from stream sediments of the southwestern Piedmont of Georgia and South and North Carolina [18, p. 29]. The maximum measured U content in monazite is 15.64% UO_2 in a sample from a pegmatite, Piona, southern Alps, Italy [69]. For U bound to accessory minerals included in monazite, see p. 289.

The contents of Th and U in monazite may vary with grain size. Thus, monazite in lujavrite of the Kvanefjeld plateau, southern West Greenland, has higher contents of Th and U in samples with an average grain size of 240 μ than in those with an average grain size of 48 μ [72]. Monazite (with 3.76% Th in the total fraction) from a paragneiss of Ponte Casletto, Val Grande, Strona-Ceneri zone, southern Alps, Italy, contains 3.71% Th in the $+75$ μ size fraction and 3.473% Th in the -42 μ size fraction [73]. Coarse detrital grains of monazite tend to be richer in ThO_2 (and, to a lesser degree, also in U_3O_8) than fine grains of the same concentrate from stream placers of North and South Carolina. This difference is inferred to be related to the sources of monazite, and pegmatitic monazite tends to be richer in ThO_2 than is monazite from schists. Thus, both the ThO_2 and the U_3O_8 contents of 33 monazite specimens show a strong tendency to be higher in the fraction coarser than 40 mesh (3.6 to 7.2, mean 5.5% ThO_2 and 0.12 to 1.8, mean 0.56% U_3O_8) than in the fraction finer than 40 mesh (3.2 to 6.9, mean 5.2% ThO_2 and 0.11 to 1.3, mean 0.50% U_3O_8) [74]. In contrast, for detrital monazite from drainage basins of streams of the Piedmont of Georgia, no real difference in Th content is attributable to differences in particle size, because the average ThO_2 content is 4.7% in five monazite samples coarser than, and 4.8% in ten samples finer than 40 mesh; the U_3O_8 content is 0.48 and 0.54%, respectively (for details, see original paper) [75, pp. 64, 67/9].

Correlations Between Contents of the Main Cations. The ThO_2 content in monazite decreases with increasing $\Sigma(La+Ce+Pr)$ (in at% of $\Sigma RE_2O_3 = 100\%$) [76], as, for example, in granitic and pegmatitic monazites from the southwestern United States [77, pp. 106/7, 115], suggesting that end-member monazite (i.e., monazite containing only Ce, La, and Pr on cationic positions) partitions Th less strongly than its less pure equivalents [7, p. 314]. However, alluvial monazite of Malaya lacks a correlation between Th content and RE distribution, because the presence of variable amounts of Th in monazite does not act as a control or as a restraint on the RE composition [35]. No correlation between the contents of ThO_2 (varying between 2.8 and 15.9% with a mean of 11.3%) and of any of the RE is apparent in ten monazite samples from stream sediments, pegmatites, and veins of the United States [78].

Monazite may be rich in Th and poor in U, or vice versa. Monazite originating from granitic magmas, which represents the most frequent type of monazite, normally contains 10- to 60-times more Th than U. The interrelationship between Th and U contents in monazite probably is very complex and is determined by the various geochemical conditions of monazite formation in deposits of different genetic type, or of similar type but of different provinces [10]. For example, U is only enriched in such monazite varieties that show high Th contents. Monazite containing $\sim 5\%$ ThO_2 contains $\sim 0.1\%$ UO_2, whereas monazite with $\sim 10\%$ ThO_2 contains $\sim 0.3\%$ UO_2. Finally, cheralite with $\sim 30\%$ ThO_2 contains a maximum amount of $\sim 4\%$ UO_2 (see also Table 18, p. 363) [79]. But note that, based on literature data, it is shown that a low content of U_3O_8 is associated with a high content of ThO_2 in monazite from saprolite derived from pegmatites of North Carolina and Virginia [77, pp. 106/7, 109, 111]. For monazite from granitoid rocks of the Bug-Podolian group, southeastern Ukraine, USSR, samples with high Th and low U content, and vice versa, are found, as is shown in the following table (the contents are determined radiochemically; for contents determined chemically and for the content of Pb, see original paper) [80]:

source rock	content in % in monazite	
	Th	U
Chudnov-Berdichev granites	4.43 ± 0.18	0.48 ± 0.023
	5.48 ± 0.14	0.55 ± 0.022
	6.12 ± 0.15	0.37 ± 0.026
	6.16 ± 0.15	0.55 ± 0.022
	5.21 ± 0.13	0.41 ± 0.010
rose-colored granites and pegmatites	10.36 ± 0.26	0.07 ± 0.007
	8.90 ± 0.22	0.10 ± 0.012
	8.30 ± 0.20	0.10 ± 0.012
	9.27 ± 0.23	0.08 ± 0.010
charnockite	11.28 ± 0.26	0.059 ± 0.10
granitic veins	2.38 ± 0.07	0.32 ± 0.024
	6.91 ± 0.17	0.39 ± 0.028
metasomatically altered granites and mylonites	8.57	0.19
	10.24	0.20
	9.63	0.19
	11.51	0.27

The content of Pb usually increases with increasing Th and U contents in monazite from granitoid rocks of the Azov region, southeastern Ukraine, USSR [81].

Variation in Content of Subordinate Elements and Water (refer also to pp. 283/5 and Tables 13 and 14, pp. 284 and 305/30, respectively). The content of Fe in monazite is variable and partly depends on the presence of inclusions in monazite (see p. 289).

Small amounts of Si are usually reported in analyses of pure monazite [37], which may reach 6% SiO_2 [4] or even more. Rarely, Si is absent in monazite, as, for example, in monazite from hydrothermal veins near Alinci, southeastern Yugoslavia [21, pp. 142/3]. An increase in the SiO_2 content generally is accompanied by an increased ThO_2 content in monazite from black sands of the Nile delta, Egypt [82]. In monazite, Si may be due to a huttonite component (see pp. 293/4) or to inclusions of silicate minerals (see p. 289).

The content of H_2O in monazite normally ranges from 0.1 to 1% [83], and higher contents are also observed. Water is considered as a constituent part of monazite from various sources by [15] (see p. 284), however, H_2O is lacking in two of three newer analyses of the same but purified samples (see analyses No. 163 and 171 in Table 14) [14]. Although in sulfate-monazite from carbonatite of the Namo-Vara region, Kola Peninsula, USSR, a part of H_2O may be combined with inclusions of Fe hydroxide, probably a part enters into the mineral (see analysis No. 125 in Table 14). The presence of H_2O in this monazite depends on the hydration of the mineral and on its partial conversion into rhabdophane [84]. The content of H_2O is unusually low in monazite from a muscovite pegmatite of the Mama region, Eastern Siberia (see analyses No. 133 and 187 in Table 14) [83]; see also analyses No. 146, 156, and 161 (in Table 14) with in part even lower H_2O contents, of monazite from muscovite pegmatites of the Gutar-Biryusa region, Eastern Siberia.

In handbooks, the presence of SO_3 in monazite is only reported by [12]; for examples of monazites containing SO_3, see footnotes for analyses No. 79, 88, 116, 124, 125, 135, 146, 150, 156, 161, and 174 in Table 14. Metasomatically formed monazite contains $\sim 1.8\%$ S in calcite veinlets of the Mecsek Mountains, Hungary [85].

Elements as Constituents of Mineral Inclusions. As is shown in the following paragraphs, some elements analyzed in monazite may partly, or entirely, be constituents of mineral inclusions and not of monazite itself. Thus, admixtures of Fe, Al, Ti, Zr, Mg, Mn, Be, and Sn in monazite may partly be caused by small inclusions of other minerals [10], as, for example, Fe, Al, Ti, Ca, and Si in the monazite variety kularite [36] and Fe, Ti, Al, Ca, Mg, and Si in black monazite from beach placers of Taiwan [28, pp. 43, 53/4].

The content of Fe_2O_3 is explained by the occurrence of inclusions of Fe hydroxide and does not characterize the chemical composition of sulfate-monazite from carbonatite of the Kola Peninsula (see analysis No. 125 in Table 14) [84] and of monazite from the Chudnov-Berdichev granite, southeastern Ukraine, both USSR (see analyses No. 107, 109, 120, and 141 in Table 14) [86, pp. 133/5]. In monazite from granites of the Novo-Ukrainkii massif, Ukraine, the (very low) content of Fe_2O_3 (see analyses No. 124, 150, and 174 in Table 14) is caused by the entry of Fe into secondary alteration products (see p. 273), because a direct relation is observed between the content of Fe_2O_3 and the degree of alteration of monazite [87].

Aluminium is present as kaolinite inclusions in monazite from the Chudnov-Berdichev granite, southeastern Ukraine (see analyses No. 107, 109, 120, and 141 in Table 14) [86, pp. 133/5], and possibly as admixtures of albite in monazite from pegmatites of Siberia (see analyses No. 57 and 88) [88, pp. 102, 105].

Zirconium dioxide occurs mainly in inclusions of detrital zircon in black monazite from a beach placer of southwestern Taiwan [28, p. 54] and in monazite from an amazonitic pegmatite of the Ilmen Mountains, Southern Urals, USSR [14]; see footnotes for analyses No. 204 and 5, respectively, in Table 14.

In monazite, some Si may be due to thorite [9]. An increasased content of SiO_2 in monazite of Central Asia is explained by the presence of enclosed relict minerals, such as quartz and mica [13, pp. 159/60]. Excess Si and Ca in dark gray monazite from granites of the Chudnov-Berdichev and Kirovograd-Zhitomir complexes may be caused by very fine inclusions of allanite [89, p. 127].

The major part of U is bound to transparent accessory minerals in monazite from two-mica granite of the Massif Saint-Sylvestre, Limousin region [67], and to pyrite inclusions in a mineral similar to monazite or xenotime from a microgranite of the Bourbonnaise Mountains, Massif Central, both France [19]. A higher content of U ($< 2.8\%$ UO_2) in monazite from the Elliot Lake conglomerate, Ontario, Canada, is caused by brannerite occurring in fissures of this monazite [22, p. 135].

Formulas

The generalized formula of monazite is ABO_4, where A = RE, Th, Fe, Al, Ca, and Mg and B = P and Si [130, p. 37], or where A = RE, U, Th, and Ca [32]. Thus, ideal formulas of monazite are given in reference books as: $Ce[PO_4]$ [9, 25, 159]; $(Ce, La, Nd)(PO_4)$ [4]; $(Ce, La, Th)PO_4$ [37, 39]; $(Ce, La, Y, Th)(PO_4)$ [1, p. 691], [42]; $(Ce, Ca, Th)(P, Si)O_4$ [2, p. 115]; or $(Ce, La, Th, Ca) [PO_4, SiO_2, SO_4]$ [12]. Another general formula type for monazite is $A_2B_2X_8$ (see also [14]), where A = Ca, Pb, Ce, and Th; B = P, Si, and S; and X = O and OH [15]. Oxidic formulas for monazites are given as 1.01 $RE_2O_3 \cdot$ 1.00 $P_2O_5 \cdot$ 0.12 $ThO_2 \cdot$ 0.18 SiO_2 (rare-metal deposit Kounrad, Kazakhstan; see analysis No. 115 in Table 14, p. 317) [129, p. 137] and as $n(ThSiO_4)m(RE_2O_3 \cdot P_2O_5)$ for samples from granites of the Chudnov-Berdichev and Zhitomir complexes, southeastern Ukraine, all USSR. The n:m ratio in monazite is 0.09 and 0.10 in the Chudnov-Berdichev and varies between 0.12 and 0.17 in the Zhitomir granites (see analyses No. 151, 168 and 100, 104, 118, and 128, respectively, in Table 14, pp. 316/23) [127].

Taking diadochic substitutions into account (see pp. 291/3), crystal-chemical formulas of monazite are derived. The examples given in the following are arranged by increasing complexity for the A positions, distinguishing between formulas including the total sum of RE and/or Y (see below), formulas distinguishing between Y, Ce, Σlight RE, and/or Σheavy RE, and formulas listing individual RE (see p. 291). The numbers added in parentheses for the monazite samples in question, if not otherwise marked, refer to the analysis number in Table 14 (pp. 305/30).

Monazite is nearly stoichiometric $(RE)PO_4$ in ferrocarbonatites of the Fen complex, Telemark, southeastern Norway (Nos. 197 and 207) [150]. Approximately $RE_{0.99}Th_{0.01}PO_4$ are monazites from placers of the Verkhoyansk-Chukotka Fold Region and of the Timan Ridge, and also Eu monazite of Primor'e, Siberia, and of the Timan Ridge, all USSR [155]. Monazite is $(Ce_{0.95}Th_{0.05})(P_{0.95}Si_{0.05})O_4$ in granosyenite of Siberia (No. 144) [136]; $(RE, Th)_{0.9}P_{1.0}O_4$ in a talc-actinolite-carbonate rock of the Little Laba River, northern Caucasus (No. 191) [145, 146]; $Ca_{0.5}Ce_3Th_{0.5}P_4O_{16}$ in granitic pegmatites of Kurumbapatti, southern India (No. 8) [95, pp. 94/5]; and $[RE_{0.91}Ca_{0.06}Th_{0.03}]_{\Sigma1.01} [(PO_4)_{0.91}(SiO_4)_{0.10}]_{\Sigma1.01}$ (No. 176) in an undisclosed occurrence [142]. Silicomonazite is $4(RE_{0.9}Ca_{0.04}Th_{0.06})(Si_{0.25}P_{0.75})O_4$ in an alluvial placer of the Kyugyulyur Creek, Kular Ridge, Yakutia (No. 209) [32]. Sulfate monazite is $(RE_{0.77}Ca_{0.20}Th_{0.05})_{\Sigma1.02}[(P_{0.89}S_{0.09}Si_{0.03})_{\Sigma1.01}O_{4.00}] \cdot 0.2H_2O$ in carbonate of the Kola Peninsula, USSR (No. 125) [84]; see also [132]. The average formula is $(RE+Y, Th, U, Fe)_{1.001}P_{0.988}O_4$ for two monazite samples (No. 2 in Table 13, p. 284) from a hydrothermal quartz vein of Alinci, southeastern Yugoslavia [21, pp. 140, 143]. Monazite is $(RE, Ca, Fe^{3+}, U, Th)_{4.1}(P, Si)_{3.9}O_{16}$ in simple and complex pegmatites of the Kenhardt district, Cape Province, R.S.A. (Nos. 62 and 164) [115]; $Ce_x^{3+}Ca_x^{2+}(RE, Y, Pb, Th, U)_{1-2x}\{(P_{1-x}Si_x)O_{4-2x}(OH)_{2x}\}$, with $x \leq 0.5$, in a heavy-mineral placer of southern India (No. 55) [30, p. 84]; or $[Ca_{0.057}Fe_{0.005}^{3+}Th_{0.118}$-$(RE+Y)_{0.757}]_{\Sigma0.875}(Si_{0.140}P_{0.924})_{\Sigma1.064}(OH)_{0.168}O_{4.026}$ in a granitic pegmatite of Siberia (No. 7) [94]. For monazite from pegmatites of Siberia (Nos. 57 and 88), a mean formula $[RE_{3.639}Th_{0.324}$-$Mg_{0.026}Fe_{0.001}]_{\Sigma3.99}[P_{3.744}Si_{0.351}]_{\Sigma4.095}O_{16.21}(OH)_{0.04}$ is recalculated on the assumption that a part of Th enters the mineral as $ThSiO_4$ (possibly as thorite inclusions?) and that the remaining SiO_2, combined with Al_2O_3 in the ratio 1:3, probably forms albite [88, pp. 101/2, 105]. For 17 crystal-chemical formulas of monazite without specified sources and containing 1 to 3% foreign admixtures, see [15].

Crystal-chemical formulas, separating Y, Ce, Σlight (lRE), and/or Σheavy RE (hRE), are given as follows for monazites from pegmatites of the Chupa region, northern Karelia, USSR: $(lRE_{0.78}hRE_{0.01}Y_{0.06}Th_{0.08}Ca_{0.07})_{\Sigma1.00}(P_{0.99}Si_{0.01})_{\Sigma1.00}O_4$ (No. 64) and $(lRE_{0.79}(hRE+Y)_{0.08}Th_{0.08}$-$Ca_{0.05})_{\Sigma1.00}(P_{0.94}Si_{0.06})_{\Sigma1.00}O_4$ (based on literature data) in plagioclase-microcline pegmatites; $(lRE_{0.86}hRE_{0.01}Y_{0.02}Th_{0.07}Ca_{0.04})_{\Sigma1.00}(P_{0.98}Si_{0.02})_{\Sigma1.00}O_4$ (No. 89) in a microcline-plagioclase pegmatite; $(lRE_{0.90}Y_{0.03}Th_{0.05}Ca_{0.02})_{\Sigma1.00}(P_{0.98}Si_{0.02})_{\Sigma1.00}O_4$ (No. 130) and $(lRE_{0.92}Y_{0.01}Th_{0.04}$-$Ca_{0.03})_{\Sigma1.00}(P_{0.99}Si_{0.01})_{\Sigma1.00}O_4$ (No. 157) in muscovite-plagioclase pegmatites [116]. Average crystal-chemical formulas of monazite from the Chudnov-Berdichev complex, southeastern Ukraine, USSR, are as follows (for the individual formulas, see original paper): $[Ce_{0.78}$-$lRE_{0.76}hRE_{0.16}Th_{1.06}]_{\Sigma1.814}O_3(P_{1.916}Si_{0.084})_{\Sigma2}O_5$ in samples from four granites and one pegmatite of the 1st intrusive phase (Nos. 107, 109, 120, and 141 and 99, respectively); and $[Ce_{0.762}$-$lRE_{0.955}hRE_{0.056}Th_{0.095}]_{\Sigma1.868}O_3(P_{1.850}Si_{0.150})_{\Sigma2}O_5$ in samples from rose-colored pegmatitic granites and pegmatites of the 2nd intrusive phase (Nos. 19, 25, 28, 40, 56, 75, 95, 105, 111, 113, 143, 148, and 152). The formulas of each one monazite from Zhitomir and Kirovograd granite are given as $[Ce_{1.23}lRE_{0.77}hRE_{0.03}Th_{0.14}]_{\Sigma 2.17}O_3(P_{1.69}Si_{0.11})_{\Sigma 1.80}O_5$ (No. 128) and $[Ce_{1.73}lRE_{0.05}hRE_{0.02}Th_{0.14}]_{\Sigma 1.94}O_3(P_{1.52}Si_{0.47})_{\Sigma 2}O_5$ (No. 106), respectively [86, pp. 135, 140/1]. For average formulas of the same monazite samples, which additionally contain Ca, Si, and/or Zr on A positions, see [160]. In an altered gneiss xenolith, southeastern Ukraine, the composition of monazite fits to the formula $[Ce_{0.585}lRE_{0.371}hRE_{0.020}Y_{0.015}Th_{0.028}]_{\Sigma 1.020}[P_{0.914}Si_{0.084}]_{\Sigma 0.998}O_{4.00}$ (No. 185) [143]. Recalculation of the chemical analysis to $P + Si = 1$ leads to the formula $(lRE_{0.97}hRE_{0.02}Th_{0.01}Ca_{0.01}Mg_{0.01}Fe^{3+}_{0.01}Al_{0.01})_{\Sigma 1.04}[(P_{0.99}Si_{0.01})_{\Sigma 1.00}O_{4.06}] \cdot 0.09\ H_2O$ for monazite in a muscovite-quartz vein of the Kola Peninsula, USSR (No. 199) [151].

Formulas of monazite listing separately Y and individual RE are given as $(Ce_{0.380}La_{0.226}Nd_{0.160}Pr_{0.046}Sm_{0.020}Gd_{0.006}Dy_{0.006}Al_{0.032}Fe^{3+}_{0.012}Th_{0.060}Ca_{0.040})_{\Sigma 0.988}[(P_{0.884}Si_{0.115}(OH)_{0.120})_{\Sigma 1.069}O_4]$ in a sample from a terrigenous sediment of the interfluvial region of Dnestr and southern Bug rivers, Ukraine, USSR (No. 98) [126], and as $(Ce_{0.408}La_{0.19}Nd_{0.142}Pr_{0.042}Gd_{0.017}Sm_{0.012}Eu_{0.007}Yb_{0.005}Y_{0.005}U_{0.003}Dy_{0.002}Ho_{0.002}Tm_{0.001}Th_{0.077}Ca_{0.106}Pb_{0.003})_{\Sigma 1.00}PO_4$ in a sample from a heavy-mineral placer of southern India (No. 55) [30, p. 81]. For other formulas of monazite depicting individual RE, see, for example, [100], [105, pp. 113, 164, 171, 174, 176, 178, 316], [161].

Diadochy

The **diadochic substitution** in monazite is manifold and may concern the cationic and anionic part, as is shown in the following paragraphs that only give examples concerning Th. For additional examples, refer also to pp. 293/4.

Whereas Th was thought in the older literature to be present in monazite as mechanically admixed thorite or other Th compounds, and a few such occurrences were observed more recently (see, e.g., [66, p. 23], [101, 162]), there is nowadays a general agreement that Th is present in solid solution substituting for RE [41]. The upper limit of substitution is not known, but may be near 30% ThO_2 [4]. Most probably, Th enters the crystal lattice of monazite by isomorphous replacement of Ce and other RE, primarily of the cerian subgroup [10]. As Th is tetravalent, any substitution for the trivalent RE in the crystal structure of monazite may be compensated electrostatically by a concomitant substitution of P^{5+} by Si^{4+}, or of Th^{4+} by divalent cations such as Ca^{2+} and Mg^{2+}, or by cation omissions [4]. Note also that Th may substitute in small amounts for Ce and La in monazite [163]. Commonly, monazite contains Th by the following coupled substitutions: $Th^{4+} + Si^{4+}$ for $Ce^{3+} + P^{5+}$, see, for example, [15], [86, p. 135], [118, p. 162], and [141]; or $Th^{4+} + Si^{4+}$ for $RE^{3+} + P^{5+}$ and $Th^{4+} + Ca^{2+}$ for $2\ RE^{3+}$ [7, p. 306], see also [2, p. 115], [28, p. 54], [38, 116]; or ThCa for 2Ce [139, p. 146]. Often, Ca may substitute for the slightly larger Ce ion simultaneously with the substitution of La by Th [37]. Note also that possible heterovalent substitutions in monazite include those of RE by Th and Ca, and of P by Si, Al, and Fe [29, p. 135]. For an influence of temperature on the coupled substitution Th + Si for RE + P in monazite, see p. 295. For a substitution of RE by Th and Ta, see p. 293. Since in Th-containing monazite the contents of Th, Si, and Ca are not correlated, a substitution of 4Ce by 3Th (as is observed in the RE phosphate rhabdophane) is assumed [2, p. 117]. A substitution of RE^{3+} by Th^{4+} and U^{4+} may be compensated by a replacement of a part of Ce^{3+} by Na^+ (besides a replacement of PO_4^{3-} by SiO_4^{4-}; see also below) [164]. Valency compensation during substitution of RE^{3+} by Th^{4+} and/or Ca^{2+} may also be attained by the incorporation of OH and/or F instead of O. Thus, during the isomorphous replacement of Ce^{3+} by Ca^{2+}, valency compensation occurs by the substitution of O^{2-} by OH^- [15]. Small amounts of Be, Zr, and Sr probably are isomorphous with Si, Th, and Ca, respectively (as, e.g., in monazite from granite, Kirovograd-Zhitomir complex, southeastern Ukraine) [160].

The following paragraphs give individual examples of the above types of substitution (see below), which also may be expressed by a solid solution of several isostructural components (see pp. 293/4).

In monazite from granites of the Rila Mountains, Bulgaria, Th is incorporated mainly according to the scheme ThSi → CeP, subordinately is the scheme 3Th → 4Ce. The incorporation of Th at the expense of Ce is confirmed by a constant sum of Ce and Th contents (with a variation of only ±5 to 6%), whereas the individual contents of Ce and especially of Th vary at ±8 to 9 and ±40 to 50%, respectively [46, p. 19]. The increased cationic charge owing to the introduction of Th is balanced solely by the replacement of P by Si, whereas Ca remains negligible, in a monazite from beach sands of Ratnapura, Sri Lanka [165]. In addition to Th and Y, only traces of Ca replace RE in monazite from calcite veinlets of the Mecsek Mountains, Hungary [85].

In graphite-bearing metashales of the Anakeesta Formation, North Carolina, virtually all Th present in monazite is accomodated by a charge-coupled substitution $Th^{4+} + Ca^{2+} = 2 RE^{3+}$ [33]. This substitution is responsible for the incorporation of Th into monazite from biotite granites of the Carnmenellis pluton, Cornwall, Great Britain, as is inferred from the approximate 1:1 atomic ratio between Th and Ca, whereas the atomic proportion of Si does not increase with increasing atomic proportion of Th in the range 0.20 to 0.35 Th [34, pp. 496, 498]. The incorporation of Th^{4+} into the monazite lattice in exchange for Ce^{3+} with a charge balance by Ca^{2+} seems to be stable during cooling of Nb-Ta-Sn pegmatites from Thailand [66, p. 24].

To maintain the structure type during incorporation of Ca and compensation of an equivalent amount of Ce^{3+} by Si^{4+} according to the scheme $(Ca^{2+} + P^{5+})^{7+} \to (Ce^{3+} + Si^{4+})^{7+}$, an equivalent amount of OH, which is twice that of Ca^{2+}, is incorporated instead of O into the lattice of green monazite of southern India [30, p. 84]. In monazite of the Bogatynia area, Lower Silesia, Poland, valency differences resulting from the substitution of RE by Th are compensated by a concomitant entrance of OH, F, as well as of Ca. This is indicated by Th:Si and RE:P ratios both differing markedly from 1:1 and by $Th:[Si + F_4 + (OH)_4]$ and (RE + Ca):P ratios close to 1:1 [90].

In monazite from pegmatite, southern Alps, Italy, the excess charge on the RE position owing to the presence of high contents of Th^{4+} and U^{4+} is essentially balanced by the presence of ions of lower charge distributed on the same lattice positions. An additional charge of −0.942 electrons per formula unit caused by U and Th (−0.541 and −0.401 electrons per formula unit, respectively) is balanced by an opposite charge of +1.022 electrons per formula unit caused by Ca (+0.742) and K (+0.140 × 2). The substitution of Si for P seems to be of minor importance, because the mineral contains only little SiO_2 (see analysis No. 16 in Table 14, p. 306) [69]. But note that substitution of RE by Th and U is accompanied by the presence of enough Si or Ca to maintain charge balance with no need to call in other possible causes (e.g., lattice defects), which are very improbable in large amounts, at least in fresh crystals, in monazites from Alpine fissures and pegmatites of the Alps [166]. These monazites reveal maximum substitution in specimens from Piona and Arvogno (see analyses No. 21 and 3, respectively, in Table 14, pp. 307 and 305), the latter being actually higher in SiO_2 than type cheralite. In all samples studied, the stoichiometric ratio between the lanthanides, the actinides, and Ca on the one side, and P plus Si on the other side is substantially respected, implying a correlation between the contents of Si and Ca and the actinides, and excluding any mechanism involving a considerable number of vacancies or other lattice defects [91].

Uranium in monazite isomorphously replaces Th and RE (mainly of the yttrian subgroup) [10], as is indicated, for example, by disseminations of U with a low tenor in the whole monazite crystals without concentration at certain points in two-mica granites of the Massif Saint-Sylvestre, Limousin region, France [67]. In monazite from granitoid rocks of the Kuramin zone, Central Asia, the entry of U into the lattice occurs by heterovalent substitution of RE^{3+} by U^{4+} with simultaneous substitution of a part of the PO_4^{3-} by SiO_4^{4-} [167]. In monazite from Alpine fissures and pegmatites of the Alps, the presence of U in the tetravalent state is confirmed from the stoichiometric balance, which is respected even in the highly uranoan samples. Also from crystal-chemical considerations it is difficult to imagine how any U^{6+} could enter the monazite lattice (see p. 344) [91].

The substitution of RE by Ta^{5+} and Th^{4+} in monazite from a stream placer of Ngahere, South Island, New Zealand, involves upset in electrostatic balance by substitution of Si for P, and electrical neutrality may be restored by substitution of Th and Ta for Ce, Y, and La (see analysis No. 140 in Table 14, p. 320). Possibly, Ta replaces ions in the 6-fold coordinated groups (since the Ta-O radius ratio lies within the limits for 6-fold coordination and no structures with a 4-fold coordination of Ta^{5+} are known) [5, pp. 668/9].

The **chemical composition** of, and hence the diadochic substitution in, monazite may also be **expressed by a solid solution of several isostructural components**, as is exemplified in the following paragraphs.

Besides a monazite component ($CePO_4$), mainly the huttonite ($ThSiO_4$) and the cheralite component ($Th_{0.5}Ca_{0.5}PO_4$) isomorphously form the monazite lattice [2, p. 115]. Note also that there is a major solid solution in the system $CePO_4-ThSiO_4-ThCa(PO_4)_2$ [38], and that monazite invariably contains the isomorphous molecules of huttonite and cheralite [7, p. 306]. An ideal-solution behavior of monazite is supported by unit-cell volumes of natural monazite, which are very similar to those of cheralite, huttonite, and synthetic end members (comprising $LaPO_4$, $CePO_4$, $PrPO_4$, $NdPO_4$, and $CaTh(PO_4)_2$), even for an extensive substitution of non-essential elements (such as Th and U) for RE, as is described for monazites from fissures and pegmatites of the Alps; for details, see original paper [91]. Eight monazite specimens from various sources of the Rila Mountains, Bulgaria, are composed of the three isomorphous components monazite, huttonite, and cheralite [46, pp. 13/4, 18/9]. Huttonite and cheralite are important components isomorphously replacing the $CePO_4$ component in monazite from granites of the Azov region and (based on data of [104]) from pegmatites of the Chokrak and Kal'chik rivers, Chudnov-Berdichev complex, all southeastern Ukraine, USSR [81]. Monazite of the Ilmen Mountains, Southern Urals, is composed of 79% $Ce_2P_2O_8$, 16% $Th_2Si_2O_8$, and 5% $ThCaP_2O_8$ in a sample from amazonite pegmatite (see analysis No. 5 in Table 14, pp. 305/30); and of 92 and 95% $Ce_2P_2O_8$, 3 and 0% $Th_2Si_2O_8$, and 5% $ThCaP_2O_8$ each in two samples from syenite pegmatite (see analyses No. 163 and 171). Monazite from albitite of the Vishnevye Gory, Middle Urals, contains 95% $Ce_2P_2O_8$ and 5% $ThCaP_2O_8$ (see analysis No. 203) [14]. Monazite contains little cheralite and somewhat more huttonite in granitic pegmatite from Siberia (see analysis No. 7) [94].

A solid-solution series between monazite and huttonite may result from the coupled replacement of RE and P by Th, U, and Si [93, pp. 229/30]. Huttonite is the end member of the coupled replacement of RE and P solely by Th and Si [165]. Probably, monazite forms a complete isomorphous series with $ThSiO_4$ but no such naturally occurring mineral series is observed [163]. Judging from the content of cations, the solid-solution series $CePO_4-ThSiO_4$ is almost continuous from the monazite end member up to 27% Th, and from the huttonite end member up to at least 20% RE. But it is continuous when judging from the content of anions [90]. However, the continuity of the solid-solution series between monazite and huttonite is questioned by the fact that monazite mostly contains 4 to 12% ThO_2, and that data on huttonite containing amounts of RE(Ce) are lacking [82]. The $REPO_4 - ThSiO_4$ series is only continuous if the contribution of divalent cations as well as of F and OH is taken into account in order to balance charge [168]. Since accounting for ThO_2 as cheralite and huttonite molecules yields an overestimation of ThO_2 in monazite from stream sediments of the southeastern United States, it is assumed that most solid solution is only between monazite and huttonite (as is also corroborated by factor analyses and correlation coefficients, see original paper) [70]. As is indicated by a direct relation between the contents of Si and Th, monazite from granites of the Novo-Ukrainkii massif, Ukraine, contains an isomorphous admixture of huttonite (see analyses No. 124, 150, and 174 in Table 14, pp. 305/30) [87], as does monazite from pegmatites of Eastern Siberia, whereas the cheralite component is absent [83], [134, p. 278]. In a quartz-Mo vein of the rare-metal deposit Kounrad, Kazakhstan, USSR, monazite mainly consists of $(La,Ce,Y)PO_4$ with an isomorphous admixture of huttonite (see analysis No. 115) [129, pp. 47, 136/7]. Some Th silicate isomorphously enters monazite from a terrigenous sediment of the interfluvial region of Dnestr and southern Bug rivers, Ukraine, USSR (see analysis No. 98) [126]. Silicomonazite from an alluvial placer of the Kyugyulyur Creek, Kular Ridge, Yakutia, is intermediate in composition between monazite and huttonite (see analysis No. 209) and indicates a continuous isomorphous series between the two end members caused by the replacement of Ce and PO_4 by Th and SiO_4 [32].

An absence of Si from monazite in hydrothermal veins of Alinci, southeastern Yugoslavia, indicates that Th is not bound as the isomorphous huttonite component, but rather is incorporated by atomic substitution [21, pp. 142/3, 148].

Probably, a component $(Ca,Mg)_2(Si,Zr,Al,P)O_4$ is present besides the huttonite component in monazite of Eastern Siberia [83].

In monazite containing an isomorphous admixture of huttonite, a camouflaged admixture of monoclinic $USiO_4$ apparently is dissolved in $ThSiO_4$ [79]. And the fact that U often is enriched in monazite containing a high amount of Th indicates that the admixture of $USiO_4$ is not dissolved in $CePO_4$ but mainly in $ThSiO_4$ [167].

For changes of refractive indices of monazite caused by the isomorphous entrance of the huttonite component, see p. 350.

Relationship Between Chemical Composition of Monazite and Its Geological Occurrence

The chemical composition of monazite may be influenced by the geological environment in which monazite crystallized, as is exemplified in the following paragraphs, which deal preferentially with the Th content; other elements are considered subordinately. The relationship between the chemical composition of monazite and type of igneous host rock is given for granites on pp. 295/7; for pegmatites on pp. 297/9; and for pneumatolytic and hydrothermal veins and for carbonatites on p. 299. Thereafter is described the relationship between the composition of monazite and the grade of metamorphic rocks (see pp. 299/301) and the type of metasomatism (see pp. 301/2). For sedimentary rocks, relationships between the composition of monazite and their source rocks are discussed on pp. 302/4. Finally, examples for the dependence of the amount of the cheralite or huttonite component in monazite from conditions of formation are cited (see p. 304).

General Trends. The ThO_2 content of monazite from igneous and metamorphic rocks is controlled by the p-T conditions prevailing during crystallization of monazite. Based on a concept of three p-T zones of rock formation it is shown that the ThO_2 content tends to be low in monazite from rocks of the epizone (formed at low p and T), increases in monazite from rocks of the mesozone (formed at intermediate p and T), and is highest in monazite from rocks of the katazone (formed at highest p and T) [6, pp. 12/6]. Note also that low contents of isomorphous admixtures (as, e.g., Th and Y) are typical for low-temperature monazite [151]. For a comprehensive compilation of literature data on the ThO_2 content of monazite in various geological environments, see [6, pp. 13/6].

In monazite samples derived from different stages of magmatic evolution, the content of ThO_2 decreases with decreasing temperature of formation, because Th^{4+} ions that enter the crystal lattice by a coupled substitution (as, e.g., $Th^{4+} + Si^{4+} = RE^{3+} + P^{5+}$) are not stable in the lattice at lower temperature [28, p. 59]. Lowest contents of Th as well as of Si characterize monazite from quartz veins, whereas the contents are somewhat higher in samples from granites and highest in samples from pegmatitic veins [15].

Besides p-T conditions during crystallization, also the chemical composition of the igneous host rock influences the composition of monazite [169]. In the southeastern United States, the Th content of monazite decreases from 6.2% ThO_2 in samples from synkinematic granites (containing 1.97% CaO) through 6.0% ThO_2 in samples from postkinematic granites (containing 2.09% CaO) to 5.2% ThO_2 in samples from synkinematic granites (containing 2.27% CaO). In contrast, in metamorphic rocks of the same region, no effect of the whole-rock CaO content on the contents of ThO_2 and ΣRE_2O_3 of monazite is evident [77, pp. 109/11, 115/6].

The composition of monazite in **granites** may be influenced by the type of granite (see below), its relative age (see p. 296), and the metamorphic grade of its wallrock (see pp. 296/7). There are also regional differences in composition for monazite, as are described, for example, for samples from Ukrainian granites (see pp. 287/8).

As is shown in the following table, the contents of ThO_2 and U_3O_8 and their average ratio (as well as the contents and composition of RE, see original paper) differ in monazite from three types of granitic rocks of the Dnepr area, Ukraine, USSR (for complete analyses of one monazite sample of each type, see analyses No. 36, 103, and 108 in Table 14, pp. 308 and 316) [109]:

source rock and location; n = number of analyses	range of content in % in monazite		average ratio in monazite
	ThO_2	U_3O_8	$ThO_2:U_3O_8$
rose-colored aplite-pegmatoid granites, Sukhen'ka; n = 3	8.25 to 9.30	0.09 to 0.095	102.3
granites and pegmatites, Mokromoskovsky massif; n = 3	6.61 to 7.19	0.248 to 0.40	21.3
plagiogranites, Zaporozh'e region; n = 4	5.36 to 6.88	0.076 to 0.134	74.3

Monazite differs slightly in chemical composition in various granitoid rocks of the Kola Peninsula, USSR. In alkali granites, it is most depleted in Th and most enriched in RE (see analyses No. 121 and 154 in Table 14, pp. 305/30), whereas in anatectic granites and alaskite it is enriched in Th (see analyses No. 22, 94 and 27, respectively) [29, pp. 131, 135/6]. Monazite

296

contains ~4.63% Th in a sample from alaskite and 4422 ppm Th on average in 35 samples from biotite granite, all of the USSR [118, p. 163]; see also [170]. Tin granites of southern Thailand (including as varieties muscovite, tourmaline, and medium-grained biotite granites) contain Th-poor monazite (Th < 5%), whereas Sn-free granites from southern and northern Thailand (including coarse-grained biotite and hornblende-biotite granites) contain Th-rich monazite (Th > 5%). The contents of Th and Ca in monazite from these various granite types are as follows [66, pp. 31/2]:

granite type; n = number of samples	content (range and mean) in % in monazite	
	Th	Ca
muscovite granite; n = 6	3.3 to 3.9; 3.6	0.0 to 0.5; 0.2
tourmaline granite; n = 7	3.4 to 6.0; 4.8	0.0 to 0.8; 0.4
medium-grained biotite granite; n = 11	3.0 to 7.2; 4.9	0.0 to 0.1; 0.1
coarse-grained biotite granite; n = 4	5.3 to 7.0; 6.2	1.0 to 2.6; 1.7
hornblende-biotite granite; n = 5	6.8 to 9.1; 8.3	1.0 to 4.5; 2.0

However, a real dependence between the content of ThO_2 in monazite and the amount of cassiterite in granites is questioned by [6, p. 22]. Monazite is poorer in Th in granites that crystallized at shallow depths than it is in plutonic granites [169].

The composition of monazite may depend on the relative age of its host granite. Thus, in the Kuu massif, central Kazakhstan, USSR, the ThO_2:CeO_2 ratio in magmatic monazite decreases from early to late generations of granite, whereas the contents of ThO_2 and ΣRE_2O_3 vary with decreasing age of the host rock as follows [171]:

host rock	content in % in monazite	
	ThO_2	ΣRE_2O_3
early coarse-grained porphyritic granites	5.55	53.45
	1.47	60.53
granite porphyry	5.20	55.30
unzoned veiny fine-grained granite	2.78	56.22
rhythmically zoned granite	5.07	63.93
late medium-grained granite	7.95	60.05

In the Kirovograd-Zhitomir complex, southeastern Ukraine, the content of Th is higher in monazite from granites of the 2nd intrusive phase as compared with monazite from those of the 1st intrusive phase [160]. And compared with monazite from early formed granites, the content of radioelements increases and the Th:U ratio decreases in monazite from later formed granites of the Megri pluton, Little Caucasus, southern Armenian SSR [172].

With increasing metamorphic grade of the wall rock of granites, the ThO_2 content of monazite from these granites increases, as is shown in the following table [6, p. 22]:

metamorphic grade of wall rock of granite; n = number of samples	% ThO$_2$ (range and mean) in monazite
greenschist facies	0.47
epidote-amphibolite facies	6.8
amphibolite facies	
lower and middle subfacies; n = 40	2.2 to 6.9; 4.2
middle and upper subfacies; n = 43	4.1 to 9.4; 6.0
upper subfacies; n = 92	2.48 to 13.66; 6.0
upper subfacies of amphibolite facies and granulite facies	7.8

Depending on regional p-T differences during the consolidation of granitic rocks (which are reflected by the metamorphic facies of the wall rocks), monazite is rich in Th in the Toluca quartz monzonite, North Carolina, and in granitic rocks of Georgia, all emplaced and consolidated under conditions of the sillimanite-almandine subfacies of amphibolite facies. On the other hand, monazite is low in Th in the Whiteside granite, North Carolina, and in granites of Virginia and Georgia, all emplaced and consolidated under conditions of the staurolite-kyanite subfacies and lower facies of amphibolite facies [77, pp. 106/9]. With increasing metamorphic grade of the wall rock of cassiterite-bearing granites, the average ThO$_2$ content of monazite increases from 1.8% in 30 samples from granites with a greenschist-facies wall rock through 4.2% in 17 samples from granites with an albite-epidote amphibolite-facies wall rock to 6.9% in 15 samples from granites with an amphibolite-facies wall rock [6, p. 22]. Monazite is devoid of Th in samples from cassiterite granites intruded into sandstone, shale, phyllite, or limestone but contains as much as 9% ThO$_2$ in samples from gneissic cassiterite granites intruded into amphibolite-facies rocks [169].

Regional differences in the composition of monazite from granites are described, for example, for Ukrainian granites. Thus, monazite differs in composition in Kirovograd and Zhitomir granites, as is evident from total analysis (see analyses No. 106 and 128, respectively, in Table 14, pp. 316 and 319) and from the ratios Ce$_2$O$_3$:RE(Ce), Ce$_2$O$_3$:RE(Y), and Ce$_2$O$_3$:ThO$_2$, which are for the two granite types 34.82 and 1.57, 100.50 and 37.95, and 7.95 and 5.29, respectively [86, pp. 141/3]. Or, monazite from Kirovograd-type granite has a higher content of Ce and considerably lower contents of cerian RE than that from Zhitomir-type granite [160]. Monazite from granitic rocks of the Bug complex is characterized by a high ThO$_2$ content, reaching 11.28 to 12.90% in aplite-pegmatitic granites and charnockites [89, p. 126].

As is exemplified in the following paragraphs, the composition of monazite in **pegmatites** may depend on the type of pegmatite which, partly, is related to its conditions of formation (see next paragraph), on the mineral paragenesis (see p. 298), and on the kind of surrounding rock (see p. 299). There are also differences in composition of monazites from pegmatite and from the corresponding source granite (see p. 299). The abundance of Th in monazite from pegmatites is related to the degree of fractionation of the fluid that formed the pegmatite [6, p. 22], and the Th:U ratio of monazite (which, e.g., averages 16.0 for three monazites of Alakurtti and 8.5 for two monazites of Tedino, both Karelia, USSR) reflects this ratio in the melt at the time of crystallization of pegmatites [173]. In pegmatites of the southern Alps, Italy-Switzerland-Austria, the composition of the solution or melt from which monazite grew (and not p and T) controls the composition of monazite, as is inferred, in spite of different physical conditions of formation, from nearly identical or similar compositions of various monazite specimens (see analyses No. 2, 3, 21, 81, 110, 131, 132, and 159 in Table 14, pp. 305/30). The peculiar composition of some monazite crystals derives from a series of previous enrichment processes, as, for example, fractional crystallization of other (possibly rock-forming) minerals [91]; see also [166].

In high-temperature granitic RE-niobate pegmatites from different locations (Norway; Karelia, USSR; and the USA), monazite is enriched in Th, U, Sc, Y, Nd, Sm, Gd, and other heavy RE [2, pp. 126, 128]. Monazite contains more Th, Ca, U, and Si in high-temperature granitic pegmatites (see analysis No. 8 in Table 14, pp. 305/30) than in syenitic pegmatites (no analyses are given in the original paper) at Kurumbapatti, southern India [95, pp. 94/5]. And in the Ilmen Mountains, Southern Urals, USSR, syenite pegmatites are characterized by Th-poor monazite (see analyses No. 163 and 171), whereas an amazonite pegmatite contains Th-rich monazite (see analysis No. 5) [14]. In low-temperature muscovite-plagioclase pegmatites formed during a final stage of pegmatite formation in the Chupa region, northern Karelia, USSR, metasomatically formed monazite (see analyses No. 130 and 157) is impoverished in isomorphous admixtures including Th, Ca, Si, Y, and RE(Y) (and is very close to the theoretical formula $(Ce,La)PO_4$) as compared to monazite that crystallized directly from a pegmatitic solution in earlier formed plagioclase-microcline and microcline-plagioclase pegmatites (see analyses No. 64 and 89, respectively) [116]. Note also that in the Chupa region, monazite from essentially muscovite-bearing pegmatite (see analysis No. 122) differs from monazite from plagioclase-microcline pegmatites (see analyses No. 52, 58, and 84) and from microcline-plagioclase pegmatite (see analysis No. 153). From the former it differs by an increase of the contents of ΣRE and RE(Ce) and a decrease of the contents of Y, RE(Y), Th, and Ca and from the latter by higher contents of Th, Y, and RE(Y) and lower contents of ΣRE and RE(Ce) [113, pp. 244, 247]. A low content of ThO_2 is typical for monazite from pegmatite of the Transangara alkaline massif, Enisei Ridge, Central Siberia (see analysis No. 192) [148].

The composition of monazite may be influenced by the mineral paragenesis of the host pegmatite. Thus, a higher average ThO_2 content of monazite correlates with a higher xenotime:monazite ratio in pegmatites from three districts of eastern Minas Gerais, Brazil. For a ratio of 0.30, 0.55, and 3.3 in the pegmatites, the corresponding content of ThO_2 in monazite is 9.4, 7.9 (average of 10 samples), and 11.6% (average of 6 samples) [174]. For paragenetically identical monazites from the same pegmatite district of New Mexico and Colorado, the contents of RE and, to a somewhat lesser degree, Th are remarkably uniform, as is shown in the following table. The contents of CaO, FeO, and PbO are nearly identical for all samples (0.5, 0.5 to 3.0, and 0.1 to 0.2%, respectively) [175]:

location of the pegmatite district	content in % in monazite	
	ThO_2	ΣRE_2O_3
New Mexico		
North Star, Petaca district	11.8; 11.2; 10.7	60.2; 57.5; 59.8
Coats, Petaca district	10.4; 9.5	61.5; 59.8
Fridlund, Petaca district	7.7	64.3
Pecos, Elk Mountain district	9.9	56.4
Colorado		
Brown Derby	4.0; 4.1; 3.8	70.2; 71.6; 63.5
Pass area	10.1; 9.2	57.3; 60.6
Baumer, Guffy area	7.1; 7.3	60.1; 66.0

The content of Th in monazite from pegmatites varies with the kind of surrounding rock. Thus, monazite from the eastern part of the Baltic Shield, USSR, contains only 0.22 to 1.48% ThO_2 in pegmatites within ultrametamorphic plagioclase or plagioclase-microcline granite gneisses, but contains 5.47% ThO_2 in pegmatite within anatectic plagioclase-microcline granites. For differences in content and composition of RE in these two monazite types, see original paper [176].

Monazite of pegmatites may differ in composition from monazite of granites. Thus, monazite from a variety of locations contains on average 8% ThO_2 for 152 specimens from pegmatites, whereas 254 specimens from granites only contain 5.1% ThO_2 [6, pp. 15, 22]. But note that in 9 pegmatites the average ThO_2 content of monazite is $\sim 4.58\%$ which practically does not deviate from the average of $\sim 4.43\%$ for monazite from 36 granitoid rocks (including granites, diorites, and granodiorites; mainly of the USSR?) [118, pp. 163/4]. In a pegmatite of the 1st intrusive phase of the Chudnov-Berdichev complex, southeastern Ukraine, USSR, monazite shows higher contents of ThO_2 and Ce_2O_3 (and CaO [160]) and lower contents of RE(Ce) and RE(Y) (and P_2O_5 [160]) as compared with monazite from granites of the same intrusive phase (see analyses No. 99 and 107, 109, 120, and 141 in Table 14, pp. 305/30). Monazite from pegmatites of the 2nd intrusive phase of the same complex is enriched in Ce and Th and depleted in RE(Ce) relative to monazite from rose-colored granites of the same intrusive phase [86, pp. 135/6, 141/3]. Note also that the average content of ThO_2 is lower (5.8%) in monazite from granites than in monazite from syngenetic pegmatites and pegmatitic granites (6.6%) of the Chudnov-Berdichev complex [89, pp. 126/7].

In **pneumatolytic and hydrothermal veins**, the Th content of monazite is low. Thus, in a pneumatolytic quartz vein (containing wolframite and Bi minerals) of the Ebisu Mine, Gifu prefecture, Japan, the ThO_2 content of monazite (see analysis No. 158 in Table 14, p. 322) is lower than in the usual monazite from pegmatites [93, pp. 225/6, 234]. In Alpine fissures (Austria), monazite is poorer in nonessential elements (comprising Si, Ca, Y, Th, and U) than monazite from pegmatites of the Alps (Italy-Switzerland) [91]. In hydrothermal formations, monazite has low contents of Th (mostly $\leqq 1\%$ ThO_2) and U [2, p. 127]; in low-temperature hydrothermal veins, monazite contains little Th [169]. The content of Th is unusually low ($\sim 1\%$ ThO_2) in monazite from Th veins of the Lemhi Pass area, Idaho—Montana [177]. Monazite of hydrothermal origin of the southeastern Ukraine is markedly depleted in Th, Sm, and Nd and is enriched in Ce as compared with monazite from gneiss and granite (and its pegmatite) of the same area [112]. Monazite is almost a pure phosphate of RE(Ce) with insignificant admixtures of Th, Y, and other elements (see analysis No. 199 in Table 14, p. 327) in a hydrothermal muscovite-quartz vein of the Kola Peninsula [151]. In three monazite specimens from hydrothermal quartz-carbonate veins of the Pamirs, the content of ThO_2 is lower (2.1, 2.6, and 3.0) than in a monazite from the enclosing intensely albitized quartzite (3.6% ThO_2) [178]. In monazite from Fe oxide-thorite-monazite veins of Kazakhstan, the low ThO_2 and U_3O_8 contents (1.03 and 0.23%, respectively) are explained by a simultaneous crystallization of Th minerals or of other Th-containing minerals [179].

The composition of monazite in **carbonatites** seems to be controlled by p and T. The content of Th in monazite is higher in deeper parts of carbonatite bodies [6, p. 23]. Generally, monazite has low Th contents in carbonatites [6, p. 23], [10, 169], as, for example, 1 to 3% ThO_2 in the Mountain Pass area, California, at Magnet Cove, Arkansas, and in the Bancroft area, Ontario, Canada [180]. In monazite from the Wigu Hill carbonatite, Tanzania, the distinctly low content of ThO_2 may be caused by a release of Th through incipient dissolution by ground water and subsequent adsorption of Th on the surface of associated zircon [181].

In **metamorphic rocks**, the type and grade of metamorphism may determine the Th content of monazite, which increases with increasing metamorphic grade. It is not known if this increase only affects monazite crystals forming at a given stage in the metamorphic evolution of the rock, or if there is a continuous reaction between early formed monazite and other Th-containing minerals in the host rock [6, pp. 17, 20]. As follows from a compilation of the ThO_2 content of 124 monazite samples from metamorphosed pelitic sediments of Africa, Asia (exclusive of the USSR), Australia, and South America, the proportion of Th in monazite, despite a wide range and certain inconsistencies, increases on average from 0.5% ThO_2 in samples from phyllites of the greenschist facies to as much as 10.72% ThO_2 in samples from rocks of the granulite facies (the maximum average content of 31.50% ThO_2 given in this compilation refers to cheralite) [169]. A decrease in the ThO_2 content of monazite appears to accompany the decrease in metamorphic grade in regional-metamorphic rocks from the northeastern (Oconee county) to the southwestern Piedmont of Georgia (Harris and Troup counties) (or of the Inner Piedmont belt, southeastern United States [182]), as is inferred from analyses of detrital monazite from fluviatile placers originating from amphibolite-facies metamorphic rocks (see pp. 302/3). However, the source rocks of these detrital monazites are rarely known with certainty, and thus they are far from ideal for the study of changes in the abundance of Th related to genetic aspects [75, pp. 64/5, 69/71]. The contents of Th, U, and RE in monazite from gneisses can indicate their source rocks. Thus, low contents of Th, U, and RE in monazite suggest a primarily sedimentary nature for amphibolite-facies gneisses of Tuva, the Urals, and the Kola Peninsula, all USSR [183]. The dependence of the Th content of monazite from metamorphic grade is used to identify the metamorphic source rocks of sedimentary rocks, see pp. 303/4.

As compared with monazite from granites, metamorphic monazite in paragneisses of the USSR is characterized by a lower content of Th, which possibly may be caused by a lower Th content in the sedimentary source rock [184]. Accessory monazite from schists and gneisses surrounding granites of the Kochkar region, Southern Urals, shows a more variable composition (higher contents of U, Y, and RE(Y) and lower contents of Th, RE(Ce), Mn, and Be) as compared with monazite from the enclosed granites. The content of Th (range and mean) in monazite from the above metamorphic rocks of the Southern Urals is as follows (for contents of U, Y, and RE and composition of RE, see original paper): 1.28 and 2.08 in two samples from kyanite schist; 2.31 to 6.05, 4.14 in four samples from gneiss and biotite schist; and 1.67 to 4.76, 3.31 in five samples from granite gneiss [185].

Changes of the composition of monazite caused by **metasomatic alteration** are very diverse and without regular trends, as can be seen from the following paragraphs.

Owing to postmagmatic metasomatic processes, Th may be partly removed from monazite. Thus, in monazite from metasomatic leucocratic granite of the Aktau massif, western Uzbekistan, the content of Th is lower by nearly half that in monazite from earlier formed biotite granite [186]. However, in the Kuu massif, central Kazakhstan, monazite is enriched in Th and Y, relative to monazite from granites (see p. 301), in samples from various postmagmatic metasomatites and quartz veins. The contents of ThO_2 and ΣRE_2O_3 in monazite from these postmagmatic formations (possibly reflecting a genetic sequence) are given in the table on top of p. 301 [171]. The content of Th (range and mean) is 3.3 to 3.7, 3.6% and 3.4 to 3.9, 3.7% in three samples each of magmatic and postmagmatic monazite, respectively, from metasomatically formed muscovite granite; and is 3.7 and 4.5% in two samples of magmatic and 3.4 to 6.0, 5.0% in five samples of postmagmatic monazite from metasomatically formed tourmaline granite, all Thailand [66, pp. 31/2].

host rock	content in monazite	
	% ThO$_2$	% ΣRE$_2$O$_3$
metasomatic albite-quartz rock in coarse-grained granite	5.35	63.65
quartz vein with wolframite (accompanied by biotite-quartz greisen) in coarse-grained granite		
vein-near biotite-quartz greisen	7.23	60.97
biotite segregation replacing greisen and vein quartz	4.15	60.85
ore-free quartz vein (accompanied by biotite-quartz greisen) in granite	5.60	64.60
ore-containing quartz vein (accompanied by phlogopite greisen) in serpentinite	4.00	63.75
feldspar metasomatite in coarse-grained granite	5.46	63.04
quartz vein with wolframite and cassiterite (accompanied by muscovite-quartz greisen) in granite		
greisen	8.50	59.8
vein filling	8.0	55.56

Owing to albitization, the content of ThO$_2$ in monazite from a pegmatite vein of Siberia is lower (5.16%) than in a sample from the adjacent unaltered vein (8.48%) [187]. The content of Th is higher and that of U is lower in monazite from epigenetic intensely albitized and biotitized pegmatites than in samples from syngenetic pegmatites of the southeastern Ukraine, as is shown in the following table [188]:

host rock	content in monazite	
	% Th	% U
Chudnov-Berdichev granite	4.6; 5.3	0.47; 0.55
syngenetic body of porphyritic and pegmatitic granite, Chudnov-Berdichev complex	6.11; 5.11; 6.02	0.54; 0.40; 0.34
epigenetic pegmatitic vein at the contact of charnockite and migmatite	8.63; 8.7	0.12; 0.18
epigenetic lens of biotite-containing rock	10.4; 9.78; 11.5	0.2; 0.195; 0.24
epigenetic pegmatite in charnockite	8.87; 8.17; 9.21; 10.3	0.11; 0.11; 0.1; 0.07

Generally, low contents of ThO$_2$ (0.00n to 5%) and heavy RE (comprising Nd, Sm, and Gd) are typical for monazite from greisens. Whereas in high-temperature greisens of South Africa, monazite contains 1 to 5% ThO$_2$, low-temperature greisens of Karelia, USSR, and of Bolivia have between 0.00n and 0.n% ThO$_2$ [2, pp. 126/7]. In pneumatolytic Sn greisens and Sn ores of the greisen impregnation type of Thailand, postmagmatic monazite is richer in Th and Ca than coexisting magmatic monazite, as is shown for samples of four mines in the following table [66, pp. 36/7]:

mine and monazite generation; n = number of samples	content (range and mean) in % in monazite	
	Th	Ca
Hoey Sinang; n = 22		
magmatic monazite	3.5 to 9.1; 6.2	0.0 to 4.5; 1.2
postmagmatic monazite	6.1 to 12.1; 9.0	1.7 to 6.3; 3.1
Chalerm; n = 14		
magmatic monazite	3.1 to 4.7; 3.7	0.0 to 0.4; 0.1
postmagmatic monazite	2.2 to 6.9; 4.9	0.0 to 0.7; 0.3
Tawat; n = 5		
magmatic monazite	2.8 to 5.7; 4.7	0.1 to 2.2; 1.2
postmagmatic monazite	8.2 to 8.8; 8.5	3.4 to 3.6; 3.5
Haad Som Pan; n = 7		
magmatic monazite	3.7 to 4.5; 4.1	0.0 to 0.1; 0.1
postmagmatic monazite	3.4 to 6.0; 5.0	0.1 to 0.8; 0.5

The composition of monazite in **sedimentary rocks,** mainly placers, may vary and depends on the kind of their source rock. The more plutonic the source rock, the more Th is contained in monazite. Mixing during transport leads to uniform mechanical blends of detrital monazite from diverse sources. Thus, the content of Th in monazite varies less in placers than in crystalline rocks [6, p. 25]. Depending on the source rock of fluviatile and beach placers, monazite contains more Th (9 to 11% ThO_2) in placers of India and Sri Lanka (see analyses No. 55, 70 and 23, 29, 32 to 35, 42, and 49, respectively, in Table 14, pp. 305/30), which formed by erosion of rocks of the charnockitic series, than in placers of the USA (see analyses No. 172, 177, 181, and 182), Brazil, Nigeria, and Malaya (3 to 6% ThO_2), which formed by erosion of granites [2, p. 126].

For monazite from fluviatile placers of the Piedmont of Georgia, the $U_3O_8:ThO_2$ ratio is thought to be indicative of the source rock: Low values may indicate granitic source rocks, whereas high values point to paraschists or -gneisses (see table below). Thus, monazite with an average ratio of 0.5 (and an average ThO_2 content of $\sim 6\%$) is derived, possibly, from granite with a wall rock of the sillimanite-almandine subfacies. Monazite with a ratio of ~ 0.09 (containing on average 4.8% ThO_2) may be derived mainly from metasedimentary rocks of the sillimanite-almandine subfacies, those with a ratio around 0.12 to 0.15 (containing around 3.6 to 4.2% ThO_2) from metasedimentary rocks of the staurolite-kyanite or lower subfacies, and those with a ratio around 0.08 (containing $\sim 4\%$ ThO_2) from granite intrusive into metamorphic rocks of dominantly lower grade than the staurolite-kyanite subfacies. Examples of the contents of ThO_2 and U_3O_8 and of the $U_3O_8:ThO_2$ ratio for monazite (partly investigated for different size fractions; for details, see original paper) from fluviatile placers with different underlying rocks of the Piedmont of Georgia are as follows [75, pp. 65, 69, 72/4]:

county, underlying rock, and metamorphic subfacies	monazite		
	content in % ThO$_2$	U$_3$O$_8$	ratio U$_3$O$_8$:ThO$_2$

Oconee
  biotite schist and biotite gneiss of staurolite-kyanite
  subfacies grading locally into sillimanite-almandine

	ThO$_2$	U$_3$O$_8$	U$_3$O$_8$:ThO$_2$
subfacies[a]	3.7	0.49	0.13
	4.8	0.62	0.13
granite intrusive into schist of sillimanite-almandine subfacies[a]	6.1	0.44	0.07

Harris
  biotite schist and biotite gneiss of staurolite-kyanite

subfacies and lower metamorphic grade[a]	3.7	0.61	0.17

Troup
  biotite schist and biotite gneiss isofacial with

staurolite-kyanite subfacies (but lacking kyanite)	3.6	0.48	0.13
granite (dominating) and biotite schist of staurolite-kyanite subfacies	4.1	0.33	0.08

Spalding
  biotite granite intrusive into biotite schist and

biotite gneiss of probable staurolite-kyanite subfacies	5.4	0.24	0.04

Pike
  biotite granite intrusive into biotite schist and

biotite gneiss of probable staurolite-kyanite subfacies[b]	5.6	0.16	0.03

[a] Average of two different size fractions. — [b] Average of four different size fractions.

According to plots of Th vs. the La or Ce content, the sources of monazite in conglomerates are determined as follows: Monazites with high Σ(La + Ce) and low Th contents in the Mozaan conglomerate of Denny Dalton and of Amsterdam, both R.S.A., possibly are derived from relatively low-grade metamorphic rocks (albite-epidote amphibolite facies and greenschist to albit-epidote amphibolite facies, respectively). Monazites with lower Σ(La + Ce) and higher Th contents of the same locations may be derived from high-grade metamorphic rocks, granites, and/or pegmatites. In conglomerates of the Dominion Reef, R.S.A., and of the Elliot Lake area, Ontario, Canada, monazites with a high ThO$_2$ content of >5% indicate pegmatites as their source rocks, whereas monazites with intermediate to low ThO$_2$ contents (<3%) indicate granites or metamorphic rocks of the albite-epidote amphibolite and lower facies as their source rocks [22, pp. 134/6].

As is shown in the following table, in saprolites derived from crystalline rocks of the Shelby Quadrangle and its vicinity, North Carolina, the amount of ThO$_2$ in monazite varies with the kind of source rock as follows [189]:

source rock of saprolite; n = number of samples	content of ThO$_2$ (range and mean) in % in monazite
sillimanite schist; n = 16	3.4 to 9.0; 4.7[*]
biotite schist; n = 31	2.1 to 6.9; 4.8
biotite gneiss; n = 9	3.7 to 8.8; 5.5[*]
Toluca quartz monzonite; n = 23	4.3 to 8.8; 6.2[*]
microcline-oligoclase-quartz pegmatite; n = 43	3.8 to 11.2; 6.1
quartz vein	6.1
Cherryville quartz monzonite; n = 3	5.6 to 6.9; 6.4

[*] Average value given incorrectly in the original paper.

The Th and U contents of monazite are higher in a sample from sandy letten (8.3% Th and 1.2% U) than in a sample from red sandstone (2.8% Th and 0.7% U) of the Fölschnitz and Ködnitz profiles, Frankenwald, F.R.G. [190].

The amount of the cheralite and/or huttonite component in monazite is used as an indicator for the environment of mineralization in rocks of the southeastern Ukraine, USSR. Thus, cheralite comprises 19% in monazite from pegmatite cutting pyroxene gneisses, but is only 4.8% in monazite from pegmatite cutting migmatites and granites. It reaches 18.1% in monazite from Zhitomir granite intruded into gneisses but is only 2% in monazite from veins of Zhitomir granite intersecting Berdichev granite. A higher content of cheralite in monazite from Berdichev granite, as compared with monazite from Zhitomir granite, is caused by a higher basicity of the geological environment. The amount of huttonite (calculated from the ThO$_2$ content) in monazite from granites is low in the Berdichev and Kirovograd-Zhitomir complexes (4.5 to 9%) and reaches a maximum in the Bug complex (10 to 20%) [89, p. 127].

Chemical Alteration

Thorium and apparently also Ce were leached selectively from yellowish green monazite of Taiwan, and were redeposited into veinlets of reddish brown, probably goethitic material, during alteration at temperatures <295°C (corresponding to the upper limit of the stability range of greigite, which occurs as minute spherical occlusions on the monazite grains) [17]. And hydrothermal alteration removed Th from monazite of biotite granite, Rila Mountains, Bulgaria [46, p. 20]. But, possibly as a result of hydrothermal alteration, two monazites (see analyses No. 9 and 18 in Table 14, pp. 305/6) out of 21 samples (containing mostly between ~6.4 and ~9.0% ThO$_2$) are enriched in Th, Ca, and Si in biotite granites of the Carnmenellis pluton, Cornwall, Great Britain [34, pp. 497/8, 503/4].

With increasing degree of supergene (hypergene) weathering of monazite in granites of the southeastern Ukraine, USSR, the amount of components captured from the enclosing rocks into the alteration products of the monazite increases. Based on data given in [131], it is shown that in a strongly kaolinitized granite of the Berdichev complex the increased content of Al$_2$O$_3$ in the alteration crust of monazite (1.76% against 0.20%) is caused by the decay of feldspar and garnet in the enclosing rock. The corresponding contents of Th and U are 5.11 and 0.91% in the Berdichev monazite and 3.78 and 0.75% in its alteration crust; a sample from the Zhitomir granite has 8.22 and 0.16% and 3.57 and 0.145%, respectively [89, pp. 124/6].

Text continued on p. 331.

Table 14
Chemical Composition and Density of Individual Monazite Samples (with ThO_2 contents $\geqq$ 1%) from Various Types of Occurrence. For the footnotes, see pp. 329/30. ΣRE_2O_3 includes oxides of various individual RE (for details, see original papers) and, if not listed separately, occasionally Y_2O_3. Besides Fe_2O_3, ΣFe_2O_3 may comprise only FeO (as in analyses No. 1, 16, 147, 178, 197, and 207) or FeO + Fe_2O_3 (as in analyses No. 17, 50, 57, 62, 88, 92, 164, 191, 198, and 203). tr. = traces.

	content in % in sample number								
	1	2	3	4	5	6	7	8	9
ThO_2	25.4	19.5	18.6	18.10	15.18	14.52	13.00	12.67	12.61
ΣRE_2O_3	47.2	43.6	50.4	48.05	43.46	53.51	50.70	54.53	47.49
Y_2O_3	–	2.7	3.1	1.70	–	–	2.5	–	1.08
CaO	2.2	3.3	0.6	2.00	0.93	0.82	1.33	2.21	3.50
ΣFe_2O_3	$\leqq$0.09	–	–	0.73	2.71	0.93	0.16	0.24	–
Al_2O_3	–	–	–	0.42	1.78	0.43	–	0.52	1.20
U_3O_8	–	–	–	–	–	0.32	–	0.26	–
UO_2	$\leqq$0.45	1.6	0.6	tr.	–	–	–	–	–
PbO	–	–	–	–	–	0.20	–	–	0.26
P_2O_5	17.9	30.0	25.4	25.64	19.09	26.84	27.34	28.24	25.64
SiO_2	2.4	0.8	2.8	2.67	6.65	2.46	3.50	1.36	2.97
H_2O	–[a]	–	–	–	tr.	–	0.63[f]	–	–
l.o.i.	–	–	–	0.20	2.10	0.35	–	–	–
others	6.1[b]	–	–	0.28[d]	8.56[e]	–	–	0.21[g]	–
sum	101.74[c]	101.5	101.5	99.79	100.46	100.38[c]	99.16	100.24	94.75
D in g/cm³	–	–	–	–	–	–	5.17	5.20	–

No. 1 = Monazite from Bogatynia area, Lower Silesia, Poland [90]. – No. 2 = Monazite (greenish) from pegmatite, Rio Graia, Val Vigezzo, Italy; average of 7 analyses of one sample (for the ranges, see original paper) [91]. – No. 3 = Monazite (yellow) from pegmatite, Arvogno, Val Vigezzo, Italy; average of 8 analyses of one sample (for the ranges, see original paper) [91]. – No. 4 = Monazite from China [92]. – No. 5 = Monazite from amazonite pegmatite, Ilmen Mountains, Southern Urals, USSR [14]. – No. 6 = Monazite from Ratnapura, Sri Lanka [93, p. 226]. – No. 7 = Monazite from granitic pegmatite, Siberia [94]. – No. 8 = Monazite from high-temperature granitic pegmatite, Kurumbapatti, India [95, p. 94]. – No. 9 = Monazite (outer zone showing marginal dissolution) within partially chloritized biotite from porphyritic biotite granite, Carnmenellis pluton, Cornwall, Great Britain [34, pp. 497/8].

Table 14 (continued)

	content in % in sample number								
	10	11	12	13	14	15	16	17	18
ThO_2	12.40	12.16	11.96	11.73	11.70	11.6	11.34	11.16	10.89
ΣRE_2O_3	53.07	63.24	52.00	55.41	52.99	59.24	34.23	52.26	53.72
Y_2O_3	–	0.07	–	–	–	–	1.01	5.19	1.15
CaO	0.73	0.29	0.96	1.11	0.52	1.0	4.45	1.12	2.61
ΣFe_2O_3	1.12	–	1.98	1.49	1.12	–	tr.	1.35	–
Al_2O_3	0.42	–	0.71	0.09	0.40	0.02	–	0.53	0.32
U_3O_8	0.20	–	0.21	–	0.20	0.4	–	–	–
UO_2	–	0.32	–	–	–	–	15.64	–	–
PbO	1.02[h]	–	0.92	–	1.02[h]	0.1[n]	–	–	0.21
P_2O_5	25.08	26.30	25.38	26.69	25.38	27.4	31.02	25.39	27.55
SiO_2	2.04	2.41	3.92	2.73	2.32	0.4	0.16	2.81	1.08
H_2O	–	–	–	0.20[l]	–	–	–	0.10[l]	–
l.o.i.	0.80	–	0.64	–	0.80	–	–	–	–
others	3.65[i]	0.09[j]	1.07[k]	–	3.84[m]	–	0.71[o]	0.07[p]	–
sum	100.53[c]	104.88	99.75	99.45	100.29[c]	100.16	98.56	99.98	97.53
D in g/cm³	–	–	–	–	–	5.247	5.15	–	–

No. 10 = Monazite (light) from granite, Rusava suite, Ukrainian Crystalline Shield, USSR [96]. – No. 11 = Monazite from biotite granite, Ririwai ring complex, Nigeria [97]. – No. 12 = Monazite from Anatoliiski granite, Azov region, Ukraine, USSR [81]. – No. 13 = Monazite from Ishikawa, Fukushima prefecture, Japan [98]. – No. 14 = Monazite (dark) from granite, Rusava suite, Ukrainian Crystalline Shield, USSR [96]. – No. 15 = Monazite from gem gravel, Sri Lanka [99]. – No. 16 = Monazite (rim) from pegmatite, Piona, southern Alps, Italy; average of 4 analyses of one sample [69]. – No. 17 = Monazite from mixed gneiss, Poiana Marului, southern Carpathian Mountains, Roumania [100]. – No. 18 = Monazite (outer zone showing marginal dissolution) within partially chloritized biotite from porphyritic biotite granite, Carnmenellis pluton, Cornwall, Great Britain [34, pp. 497/8].

307

Table 14 (continued)

	19	20	21	22	23	24	25	26	27
	content in % in sample number								
ThO_2	10.84	10.6	10.5	10.36	10.29	10.22	10.05	10.0	9.96
ΣRE_2O_3	49.76	55.5	39.6	53.29	57.50	60.36	54.16	56.2	51.72
Y_2O_3	–	1.7	1.5	–	2.14	–	–	0.9	–
CaO	1.40	–	4.7	2.79	0.41	–	2.87	0.2	3.18
ΣFe_2O_3	5.10	2.1	–	1.33	0.81	1.50	1.83	–	1.57
Al_2O_3	0.81	–	–	0.86	0.17	–	0.26	–	1.84
U_3O_8	–	–	–	0.06	–	–	–	0.5[r]	0.12
UO_2	–	–	12.0	–	–	–	–	–	–
PbO	–	–	–	0.27	–	–	–	0.7	–
P_2O_5	23.57	22.7	31.9	25.92	27.67	26.82	25.53	25.8	24.88
SiO_2	6.84	7.6	0.1	3.26	1.03	0.90	4.28	2.7	3.21
H_2O	0.24	–	–	1.29[l]	–	0.46	0.25	–	3.04[l]
l.o.i.	1.64	–	–	0.69	–	0.37	–	–	–
others	0.19[p]	–	–	0.31[p]	–	–	0.38[q]	0.3[s]	0.26[p]
sum	100.39	100.2	100.3	100.43	100.02	100.63	99.61[c]	97.3[c]	99.78
D in g/cm^3	4.687	5.7	–	4.921	–	5.1	–	–	4.790

No. 19 = Monazite from pegmatite of the 2nd intrusive phase, Chudnov-Berdichev complex, southeastern Ukraine, USSR [86, p. 140]. – No. 20 = Monazite from metasomatic rock, central Timan Ridge, European USSR [101]. – No. 21 = Monazite (dark green) from pegmatite, Piona, southern Alps, Italy; average of 9 analyses of one sample (for the ranges, see original paper) [91]. – No. 22 = Monazite from anatectic biotite granite, Kola Peninsula, USSR [29, pp. 136/7]. – No. 23 = Monazite from placer, Ratnapura, Sri Lanka [102]. – No. 24 = Monazite from Travancore, India [103]. – No. 25 = Monazite from pegmatite of the 2nd intrusive phase, Chudnov-Berdichev complex, southeastern Ukraine, USSR [104]. – No. 26 = Monazite from Dells granite, Arizona [105, p. 316]. – No. 27 = Monazite from alaskite, Kola Peninsula, USSR [29, pp. 136/7].

20 *

Table 14 (continued)

	content in % in sample number								
	28	29	30	31	32	33	34	35	36
ThO₂	9.80	9.77	9.74	9.64	9.64	9.51	9.49	9.47	9.30
ΣRE₂O₃	57.19	58.09	60.67	54.71	57.82	57.26	62.75[y]	59.00	55.13
Y₂O₃	–	0.74	–	–	0.74	1.05	–	0.87	–
CaO	0.41	0.44	tr.	0.57	0.76	0.89	–	0.25	1.13
ΣFe₂O₃	1.12	0.67	0.14	3.32	0.51	0.10	–	0.59	1.21
Al₂O₃	0.95	1.00	0.06	0.20	1.44	1.31	–	0.50	0.74
U₃O₈	–	–	–	0.29	–	–	0.43[h]	0.08	0.09
UO₂	–	–	–	–	–	–	–	–	–
PbO	–	0.11	–	0.89	0.05	–	–	0.22	1.26
P₂O₅	26.51	27.09	25.84	25.57	26.74	28.91	24.50	27.25	25.53
SiO₂	3.76	1.57	3.19	3.14	1.48	–	–	1.21	3.16
H₂O	0.46	–	–	–	–	–	–	–	1.40
l.o.i.	–	–	–	0.81	–	–	–	–	–
others	0.30[t]	0.57[u]	–	0.51[v]	0.91[w]	0.05[x]	2.56[z]	0.56[aa]	0.81[ab]
sum	100.50	100.05	99.64	99.65	100.09	99.08	99.73	100.00	99.76
D in g/cm³	–	–	5.286	–	–	–	–	–	5.16

No. 28 = Monazite from pegmatite of the 2nd intrusive phase, Chudnov-Berdichev complex, southeastern Ukraine, USSR [104]. – No. 29 = Monazite from beach sand, Beruwala, Sri Lanka [106]. – No. 30 = Monazite from Lake Baikal area, USSR [2, p. 116]. – No. 31 = Monazite from Anatoliiski granite, Azov region, Ukraine, USSR [81]. – No. 32 = Monazite from beach sand, Kaikawala, Sri Lanka [106]. – No. 33 = Monazite from beach placer, Dondra, Sri Lanka [102]. – No. 34 = Monazite from stream placer, Bentota River, Sri Lanka [107]; see also [108]. – No. 35 = Monazite from beach sand, Beruwala, Sri Lanka [106]. – No. 36 = Monazite from rose-colored pegmatitic granite, Sukhen'ka, Ukraine, USSR [109].

Table 14 (continued)

	37	38[de]	39	40	41	42	43	44	45
					content in % in sample number				
ThO_2	9.13	9.07	9.05	9.04	9.0	8.98	8.93	8.91	8.84
ΣRE_2O_3	55.45	47.71	58.82	55.21	57.5	55.13	59.74	58.61	56.91
Y_2O_3	4.16	0.53	0.89	–	1.2	1.37	0.94	1.11	–
CaO	1.34	1.73	1.71	0.20	0.9	0.72	1.65	1.73	–
ΣFe_2O_3	1.28	–	–	2.88	–	0.89	–	–	–
Al_2O_3	–	–	0.03	1.24	0.4[r]	0.96	0.03	0.18	0.44
U_3O_8	–	–	–	–	–	–	–	–	–
UO_2	–	–	0.55	–	0.7	–	2.28	–	–
PbO	–	0.40	0.16	–	–	0.09	0.20	0.21	0.04[ag]
P_2O_5	26.70	30.18	29.11	26.50	26.1	28.06	28.31	28.84	31.10
SiO_2	1.69	0.65	0.49	3.82	2.0	1.02	0.49	0.44	4.83
H_2O	0.25[l]	–	–	0.40	–	–	–	–	–
l.o.i.	–	–	–	–	–	–	–	–	–
others	–	–	–	0.29[ac]	0.4[ae]	2.22[af]	–	–	–
sum	100.00	90.27[c]	100.81	99.58[c]	98.2[c]	99.44	102.57	100.03	102.16
D in g/cm^3	–	–	–	–	–	–	–	–	–

No. 37 = Monazite from mixed gneiss, Poiana Marului, southern Carpathian Mountains, Romania [100]. – No. 38 = Monazite from thoriferous conglomerate (taken at 350.00 m depth), southeastern Gabon [110]. – No. 39 = Monazite from porphyritic biotite granite, Carnmenellis pluton, Cornwall, Great Britain [34, p. 497]. – No. 40 = Monazite from pegmatite of the 2nd intrusive phase, Chudnov-Berdichev complex, southeastern Ukraine, USSR [104]. – No. 41 = Monazite (clear) from two-mica granite, Lawler Peak pluton, Yavapai county, west-central Arizona [105, p. 112]. – No. 42 = Monazite from beach sand, Induruwa, Sri Lanka [106]. – Nos. 43 and 44 = Monazite from porphyritic biotite granite, Carnmenellis pluton, Cornwall, Great Britain [34, p. 497]. – No. 45 = Monazite from pegmatite, Valle Fértil region, San Juan, Argentina [111].

Table 14 (continued)

	content in % in sample number								
	46	47	48	49	50	51	52	53	54
ThO$_2$	8.8	8.75	8.71	8.65	8.63	8.6	8.59	8.57	8.53
ΣRE$_2$O$_3$	39.8	62.37	59.44	58.43	56.14	55.0	56.27	56.58	57.21
Y$_2$O$_3$	2.3	–	1.02	0.95	5.17	2.3	2.94	1.07	–
CaO	7.2	2.37	1.70	0.20	0.46	–	1.80	1.73	1.12
ΣFe$_2$O$_3$	0.1$^{r)}$	1.85	–	0.15	0.27	–	–	–	1.62
Al$_2$O$_3$	0.5	–	0.03	0.78	0.08	0.2$^{r)}$	0.31	0.03	1.42
U$_3$O$_8$	0.6$^{r)}$	–	–	–	–	–	–	–	0.16
UO$_2$	–	–	0.48	–	–	0.2	–	–	–
PbO	1.2	–	0.16	–	–	–	–	0.21	0.92
P$_2$O$_5$	27.1	22.83	28.25	27.50	27.36	28.0	29.85	28.48	27.34
SiO$_2$	0.3	1.65	0.44	1.60	1.39	1.3	0.24	0.46	1.48
H$_2$O	–	–	–	–	–	–	–	–	0.20
l.o.i.	–	–	–	–	–	–	–	–	1.02
others	1.7$^{ah)}$	0.18$^{ai)}$	–	0.15$^{x)}$	0.18$^{p)}$	1.7$^{aj)}$	–	–	0.37$^{ak)}$
sum	89.6$^{c)}$	100.00	100.23	98.41	99.98	97.3$^{c)}$	100.00	97.13	101.39
D in g/cm^3	–	5.18	–	–	–	–	5.16	–	5.15

No. 46 = Monazite (pale) from muscovite granite, Lawler Peak pluton, Yavapai county, west-central Arizona [105, p. 173]. – No. 47 = Monazite from aplitic biotite granite, southeastern Ukraine, USSR [112]. – No. 48 = Monazite from porphyritic biotite granite, Carnmenellis pluton, Cornwall, Great Britain [34, p. 497]. – No. 49 = Monazite from beach placer, Beruwala, Sri Lanka [102]. – No. 50 = Monazite (alluvial) from Pianul de Sus, southern Carpathian Mountains, Romania [100]. – No. 51 = Monazite from muscovite granite, Lawler Peak pluton, Yavapai county, west-central Arizona [105, p. 177]. – No. 52 = Monazite from plagioclase-microcline pegmatite Chernaya Salma No. 1, Chupa region, northern Karelia, USSR; analysis is recalculated to 100% by omitting Fe$_2$O$_3$ and H$_2$O [113, p. 247]. – No. 53 = Monazite from porphyritic biotite granite, Carnmenellis pluton, Cornwall, Great Britain [34, p. 497]. – No. 54 = Monazite from migmatite, Bug complex, Ukraine, USSR [114].

Table 14 (continued)

	content in % in sample number								
	55	56	57	58	59	60	61	62	63
ThO_2	8.51	8.48	8.48	8.46	8.45	8.45	8.44	8.35	8.33
ΣRE_2O_3	57.44	55.62	61.32	56.50	58.38	59.51	59.99	52.85	59.98
Y_2O_3	0.22	–	–	4.04 [ao]	1.17	0.91	0.81	3.44	0.85
CaO	2.49	0.24	–	1.37	1.66	1.65	1.55	1.43	1.60
ΣFe_2O_3	–	2.42	0.09	–	–	–	–	1.61	–
Al_2O_3	–	1.11	1.30	–	0.18	0.06	0.03	0.33	0.03
U_3O_8	0.35 [al]	–	0.04	–	–	–	–	0.56	–
UO_2		–	–	–	0.24	0.50	–	–	0.58
PbO	0.33	–	–	–	0.20	0.16	0.15	–	0.18
P_2O_5	30.67 [am]	26.09	26.35	28.06	28.44	28.91	28.55	26.87	28.70
SiO_2	— [am]	4.06	2.09	1.57	0.57	0.46	0.49	1.74	0.46
H_2O	–	0.44	0.40 [l]	–	–	–	–	0.86 [l]	–
l.o.i.	–	–	–	–	–	–	–	–	–
others	–	0.80 [an]	0.10 [p]	–	–	–	–	0.41 [ap]	–
sum	100.01	99.26 [c]	100.17	100.00	99.29	100.61	100.01	98.45	100.71
D in g/cm³	–	–	–	5.05	–	–	–	–	–

No. 55 = Monazite from heavy-mineral placer, southern India; contents are corrected for impurities (including zircon and garnet) [30, p. 81]. – No. 56 = Monazite from pegmatite of the 2nd intrusive phase, Chudnov-Berdichev complex, southeastern Ukraine, USSR [104]. – No. 57 = Monazite from pegmatite of field B, Siberia [88, p. 102]. – No. 58 = Monazite from plagioclase-microcline pegmatite Chernaya Salma No. 1, Chupa region, northern Karelia, USSR; analysis is recalculated to 100% by omitting Fe_2O_3 and H_2O [113, p. 247]. – Nos. 59 to 61 = Monazite from porphyritic biotite granite, Carnmenellis pluton, Cornwall, Great Britain [34, p. 497]. – No. 62 = Monazite from simple pegmatite, Krom Puts, Kenhardt district, Cape Province, R.S.A. [115]. – No. 63 = Monazite from porphyritic biotite granite, Carnmenellis pluton, Cornwall, Great Britain [34, p. 497].

Table 14 (continued)

content in % in sample number

	64	65	66	67	68	69	70	71	72
ThO_2	8.32	8.32	8.27	8.19	8.12	8.1	8.10	8.07	8.04
ΣRE_2O_3	57.38	60.67	59.30	57.36	45.07	54.0	60.00	58.00	59.14
Y_2O_3	–	–	0.78	3.96	1.71	2.1	0.40	1.06	–
CaO	1.75	0.82	1.59	0.20	1.66	1.9	0.10[ad]	1.57	0.34
ΣFe_2O_3	1.52	0.20	–	0.95	–	0.2[r]	–	–	0.35
Al_2O_3	–	–	0.06	0.50	–	–	0.30	0.06	0.16
U_3O_8	0.30	–	–	–	–	0.7[r]	–	–	0.06
UO_2	–	–	0.82	–	0.38	–	–	0.91	–
PbO	–	0.87	0.15	–	–	0.3	–	0.18	–
P_2O_5	28.93	28.04	28.87	26.26	31.78	29.7	26.20	28.63	26.51
SiO_2	1.32	1.32	0.49	2.06	0.30	0.62	2.40	0.49	3.53
H_2O	0.10	–	–	0.48[f]	–	–	–	–	0.53[l]
l.o.i.	–	–	–	–	–	–	–	–	–
others	0.90[aq]	–	–	0.32[ar]	–	0.4[s]	–	–	–
sum	100.52	100.24	100.33[c]	100.28	89.02	98.9[c]	97.50	98.97	98.66[c]
D in g/cm³	5.163	–	–	–	–	–	5.1	–	5.12

No. 64 = Monazite from plagioclase-microcline pegmatite Chkalov, Chupa region, northern Karelia, USSR [116]. – No. 65 = Monazite from Karelia, USSR [117]. – No. 66 = Monazite from porphyritic biotite granite, Carnmenellis pluton, Cornwall, Great Britain [34, p. 497]. – No. 67 = Monazite from alaskite, Uzbekistan [118, p. 106]. – No. 68 = Monazite from monzogranite, Carnmenellis pluton, Cornwall, Great Britain; average of 4 analyses of one sample (for standard deviations, see original paper) [119]. – No. 69 = Monazite (clear) from two-mica granite, Lawler Peak pluton, Yavapai county, west-central Arizona [105, p. 112]. – No. 70 = Monazite from beach sand, Quiton, India [120]. – No. 71 = Monazite from porphyritic biotite granite, Carnmenellis pluton, Cornwall, Great Britain [34, p. 497]. – No. 72 = Monazite from the Ushiroda pegmatite, Ishikawa, Fukushima prefecture, Japan [121].

Table 14 (continued)

					content in % in sample number				
	73[de]	74	75	76	77[de]	78	79	80	81
ThO_2	8.04	8.01	8.00	8.00	7.89	7.85	7.80	7.80	7.7
ΣRE_2O_3	60.64	56.40	56.52	59.30	51.42	59.60	52.40	59.18	59.2
Y_2O_3	0.76	4.77[at]	–	1.43	0.03	0.78	–	0.67	0.3
CaO	1.47	2.40	0.84	–	1.27	1.50	1.59	1.43	1.4
ΣFe_2O_3	–	0.33	3.28	0.81	–	–	1.40	–	–
Al_2O_3	–	1.14	0.82	0.20	0.08	0.06	0.14	0.06	–
U_3O_8	–	0.08	–	–	–	–	–	–	–
UO_2	0.64	–	–	–	1.02	–	–	0.30	0.4
PbO	0.16	0.05	–	–	–	0.20	–	0.15	–
P_2O_5	28.58	26.10	26.33	28.16	28.00	28.95	25.13	28.96	31.0
SiO_2	0.52	1.00	2.86	1.79	1.50	0.38	4.48	0.38	0.2
H_2O	–	0.20[f]	0.10	–	–	–	1.55[f]	–	–
l.o.i.	–	–	0.68	–	–	–	–	–	–
others	–	0.215[au]	0.29[p]	–	–	–	0.78[av]	–	–
sum	100.81	100.70	99.72	99.69	91.21	99.32[c]	95.27	98.93	100.2
D in g/cm³	–	–	4.945	–	–	–	5.30	–	–

No. 73 = Monazite from porphyritic biotite granite, Carnmenellis pluton, Cornwall, Great Britain [34, p. 497]. – No. 74 = Monazite from vein, Steenkampskraal, Vanrhynsdorp district, R.S.A. [122]; see also [50]. – No. 75 = Monazite from pegmatite of the 2nd intrusive phase, Chudnov-Berdichev complex, southeastern Ukraine, USSR [86, pp. 138, 140]. – No. 76 = Monazite from biotite granite, Jarawa ring complex, Nigeria [123]. – No. 77 = Monazite from thoriferous conglomerate (taken at 593.20 m depth), southeastern Gabon [110]. – No. 78 = Monazite from porphyritic biotite granite, Carnmenellis pluton, Cornwall, Great Britain [34, p. 497]. – No. 79 = Monazite from biotite gneiss, southeastern Ukraine, USSR [112]. – No. 80 = Monazite from porphyritic biotite granite, Carnmenellis pluton, Cornwall, Great Britain [34, p. 497]. – No. 81 = Monazite (yellow) from pegmatite, Nibbio, Val d'Ossola, Italy; average of 7 analyses of one sample (for ranges, see original paper) [91].

Table 14 (continued)

	content in % in sample number								
	82	83	84	85	86	87	88	89	90
ThO_2	7.63	7.59	7.53	7.5	7.47	7.4	7.37	7.35	7.33
ΣRE_2O_3	46.15	57.42	60.58	56.5	58.63	58.9	56.15[aw]	59.89	56.90
Y_2O_3	1.79	1.11	0.80	1.7	0.92	2.1	–	–	0.77
CaO	1.26	1.60	1.01	0.9	1.44	1.4	1.27	0.98	1.52
ΣFe_2O_3	–	–	–	–	–	–	1.77	1.48	–
Al_2O_3	–	0.03	0.24	0.3[r]	0.06	–	0.83	–	0.06
U_3O_8	–	–	–	–	–	–	0.08	0.24	–
UO_2	0.78	–	–	–	–	0.5	–	–	–
PbO	–	0.20	–	0.6	0.18	–	–	–	0.18
P_2O_5	29.57	28.93	29.36	29.5	29.21	29.2	24.87	28.64	28.55
SiO_2	–	0.41	0.48	1.2	0.44	–	4.22	1.26	1.48
H_2O	–	–	–	–	–	–	1.80[l]	0.13	–
l.o.i.	–	–	–	–	–	–	–	–	–
others	–	–	–	0.6[s]	–	0.5[s]	3.32[ax]	0.40[ay]	–
sum	87.18	97.29	100.00	98.8[c]	98.35[c]	100.00	101.68	100.37	96.79[c]
D in g/cm³	–	–	5.29	–	–	–	–	5.290	–

No. 82 = Monazite from monzogranite, Lamorna Cove, Cornwall, Great Britain; average of one sample (for standard deviations, see original paper) [119]. – No. 83 = Monazite from porphyritic biotite granite, Carnmenellis pluton, Cornwall, Great Britain [34, p. 497]. – No. 84 = Monazite from plagioclase-microcline pegmatite Kheto-Lambina No. 1, Chupa region, northern Karelia, USSR; analysis is recalculated to 100% by omitting Fe_2O_3 and H_2O [113, p. 247]. – No. 85 = Monazite from biotite granite, Lawler Peak pluton, Yavapai county, west-central Arizona [105, p. 163]. – No. 86 = Monazite from porphyritic biotite granite, Carnmenellis pluton, Cornwall, Great Britain [34, p. 497]. – No. 87 = Monazite from stream sediment, southeastern Piedmont, USA [18, pp. 7, 28, 35]; average of 3 points of one sample. – No. 88 = Monazite from pegmatite of field A, Siberia [88, p. 102]. – No. 89 = Monazite from microcline-plagioclase pegmatite Kheto-Lambina No. 1, Chupa region, northern Karelia, USSR [116]. – No. 90 = Monazite from porphyritic biotite granite, Carnmenellis pluton, Cornwall, Great Britain [34, p. 497].

Table 14 (continued)

					content in % in sample number				
	91	92	93	94	95	96	97	98	99
ThO_2	7.26	7.17	7.16	7.07	7.04	7.01	7.0	7.00	6.93
ΣRE_2O_3	64.80	56.86	58.10	58.82	55.82	58.11	59.6	60.47	56.29
Y_2O_3	0.07	–	0.75	–	–	–	0.8	–	–
CaO	0.43	1.65	1.45	2.03	1.80	2.52	0.5	0.88	0.84
ΣFe_2O_3	–	0.75	–	0.63	3.80	0.26	0.2	0.45	2.39
Al_2O_3	–	0.05	0.06	0.88	0.89	0.40	–	0.68	0.55
U_3O_8	–	0.32	–	0.01	–	0.19	1.3	–	–
UO_2	0.04	–	–	–	–	0.34	–	–	–
PbO	–	–	0.16	0.15	–	–	–	–	0.67
P_2O_5	28.24	27.70	28.68	26.91	26.53	26.64	26.6	26.85	25.94
SiO_2	1.45	3.96	0.49	2.73	3.36	1.83	–	3.04	3.40
H_2O	–	0.46[az]	–	0.87[f]	0.18	0.30[f]	–	–	0.07[f]
l.o.i.	–	–	–	–	0.74	–	–	0.48	0.90
others	–	0.32[ba]	–	0.07[p]	0.13[p]	0.454[bb]	–	–	2.14[bc]
sum	102.29	99.24[c]	96.85	100.17	100.29	98.054	96.00	99.85	100.12
D in g/cm^3	–	5.15 to 5.20	5.20	4.937	4.820	–	–	–	5.50

No. 91 = Monazite from biotite granite, Ririwai ring complex, Nigeria [97]. – No. 92 = Monazite from unaltered medium-grained granite, Zolotonosha massif, Orlinogor complex, northern Kazakhstan, USSR [124]. – No. 93 = Monazite from porphyritic biotite granite, Carnmenellis pluton, Cornwall, Great Britain [34, p. 497]. – No. 94 = Monazite from anatectic plagiomicrocline granite, Kola Peninsula, USSR [29, pp. 136/7]. – No. 95 = Monazite from pegmatite of the 2nd intrusive phase, Chudnov-Berdichev complex, southeastern Ukraine, USSR [86, p. 140]. – No. 96 = Monazite from simple pegmatite, Daberas, Kenhardt district, Cape Province, R.S.A. [115]. – No. 97 = Monazite from a W-Bi-Mo deposit in greisenized acid volcanic rock, Mount Pleasant, New Brunswick, Canada [125]. – No. 98 = Monazite from terrigenous sediment, interfluvial region of Dnestr and southern Bug rivers, Ukraine, USSR [126]. – No. 99 = Monazite (yellow) from pegmatite (of the 1st intrusive phase [86, pp. 133/4]), Chudnov-Berdichev complex, southeastern Ukraine, USSR [127].

Table 14 (continued)

	content in % in sample number								
	100	101	102	103	104	105	106	107	108
ThO_2	6.83	6.8	6.79	6.76	6.74	6.72	6.70	6.64	6.61
ΣRE_2O_3	49.16	64.2	56.42[aw]	56.30	59.29	55.56	55.33	57.89	56.76
Y_2O_3	–	–	2.95	–	–	–	–	–	–
CaO	3.22	1.1	2.48	0.86	0.67	2.32	0.62	0.30	0.67
ΣFe_2O_3	4.00	–	1.39	1.88	1.92	3.80	2.03	2.55	2.73
Al_2O_3	0.74	–	0.33	1.44	0.41	0.58	0.42	0.83	0.47
U_3O_8	0.59	1.7	–	0.28	0.51	–	–	–	0.088
UO_2	–	–	0.05	–	–	–	–	–	–
PbO	0.67	–	0.06	0.90	0.65[bg]	–	0.56	–	1.02
P_2O_5	29.62	29.9	27.82	25.66	26.23	25.79	25.62	28.86	28.24
SiO_2	4.12	–	1.52	2.66	1.13	4.44	5.60	2.64	1.51
H_2O	–	–	–	1.82	0.27	0.27	–	0.06[az]	0.90
l.o.i.	0.39	–	–	–	1.47	0.88	0.87	0.34	–
others	1.09[bd]	–	0.56[be]	1.01[bf]	1.38[bh]	0.17[p]	1.92[bi]	0.28[bj]	0.84[bk]
sum	100.43	103.7	100.37	99.57[c]	100.67	100.53[c]	99.67[c]	100.39	99.838[c]
D in g/cm³	–	–	–	5.06	5.06	4.800	5.29	5.259	–

No. 100 = Monazite (black-brown to black) from rose-colored aplite-pegmatitic granite, Zhitomir complex, southeastern Ukraine, USSR [127]. – No. 101 = Monazite from a W-Bi-Mo deposit in greisenized acid volcanic rock, Mount Pleasant, New Brunswick, Canada [125]. – No. 102 = Monazite from stream bed of Keelungho River, northern Taiwan [128]. – No. 103 = Monazite from gray granite, Mokromoskovsky massif, Dnepropetrovsk-Zaporozh'e region, Ukraine, USSR [109]. – No. 104 = Monazite (dark gray) from trachytic granite, Zhitomir complex, southeastern Ukraine, USSR [127]; see also [86, pp. 141/2]. – No. 105 = Monazite from pegmatite of the 2nd intrusive phase, Chudnov-Berdichev complex, southeastern Ukraine, USSR [86, pp. 138, 140]. – No. 106 = Monazite from Kirovograd granite, southeastern Ukraine, USSR [86, pp. 141/2]. – No. 107 = Monazite from granite of the 1st intrusive phase Chudnov-Berdichev complex, southeastern Ukraine, USSR [86, pp. 133/4]. – No. 108 = Monazite from plagiogranite, Zaporozh'e region, Ukraine, USSR [109].

Table 14 (continued)

	content in % in sample number								
	109	110	111	112	113	114	115	116	117
ThO_2	6.60	6.5	6.44	6.42	6.24	6.19	6.06	6.00	6.0
ΣRE_2O_3	58.70	59.0	59.68	62.27	59.06	63.03	61.44	56.60	68.18
Y_2O_3	–	2.3	–	0.34	–	–	2.00	–	1.0
CaO	0.28	0.8	0.24	1.11	0.64	0.52	–	0.85	1.5
ΣFe_2O_3	2.20	–	1.47	–	3.50	–	–	0.75	<0.35
Al_2O_3	0.67	–	0.57	0.06	0.51	0.47	0.40	0.46	<0.50
U_3O_8	–	–	–	0.77	–	–	–	–	–
UO_2	–	0.2	–		–	–	–	–	–
PbO	–	–	–	0.15	–	–	–	–	–
P_2O_5	28.57	30.3	27.56	27.87	26.89	28.53	27.80	26.75	26
SiO_2	2.14	0.9	3.44	0.58	2.40	1.39	2.20	6.34	1.6
H_2O	0.10[f]	–	0.13	–	0.20	0.29[f]	–	0.91[f]	–
l.o.i.	0.48	–	0.55	–	0.72	–	–	–	–
others	0.34[bl]	–	0.25[bm]	–	0.11[p]	–	–	1.78[bn]	<0.50[bo]
sum	100.08	100.0	100.33	99.57	100.27[c]	100.42	99.90	100.44	~105.63
D in g/cm³	5.123	–	5.171	5.084	5.084	–	–	5.43	–

No. 109 = Monazite from granite of the 1st intrusive phase, Chudnov-Berdichev complex, southeastern Ukraine, USSR [86, pp. 133/4]. – No. 110 = Monazite (yellow) from pegmatite, Val Codera, Italy; average of 7 analyses of one sample (for ranges, see original paper) [91]. – No. 111 = Monazite from rose-colored pegmatitic granite of the 2nd intrusive phase, Chudnov-Berdichev complex, southeastern Ukraine, USSR [86, pp. 138, 140]. – No. 112 = Monazite from porphyritic biotite granite, Carnmenellis pluton, Cornwall, Great Britain [34, p. 497]. – No. 113 = Monazite from pegmatite of the 2nd intrusive phase, Chudnov-Berdichev complex, southeastern Ukraine, USSR [86, pp. 139/40]. – No. 114 = Monazite from fractured metasomatite, Kola Peninsula, USSR [29, pp. 136/7]. – No. 115 = Monazite from quartz-Mo vein, eastern Kounrad deposit, Kazakhstan, USSR [129, pp. 47, 136/7]. – No. 116 = Monazite (yellow) from orthotectic pegmatite with biotite and garnet, southeastern Ukraine, USSR [112]. – No. 117 = Monazite (yellow) from alluvial beach placer, southwestern Taiwan [130, pp. 38, 42].

Table 14 (continued)

	content in % in sample number								
	118	119	120	121	122	123	124	125	126
ThO_2	5.95	5.92	5.88	5.86	5.82	5.80	5.78	5.77	5.73
ΣRE_2O_3	53.97	60.38	58.22	61.21	62.82	59.63	61.55	50.56	45.6
Y_2O_3	–	2.26	–	–	1.55	–	–	0.05	–
CaO	1.79	0.47	0.56	0.44	0.46	1.04	0.10	4.60	1.32
ΣFe_2O_3	2.42	0.44	2.40	0.62	–	0.64	0.07	6.10	0.8
Al_2O_3	1.24	–	1.23	0.44	–	0.20	–	0.06	1.8
U_3O_8	0.53	0.45	–	0.13	–	1.08	–	–	1.28
UO_2	–	–	–	–	–	–	–	–	–
PbO	0.58	0.10	–	0.57	–	0.96	–	–	0.4
P_2O_5	29.50	27.27	28.58	25.00	28.88	27.34	26.77	25.84	17.1
SiO_2	1.80	2.62	2.18	3.09	0.47	1.90	3.0	0.70	13.9
H_2O	–	–	–	0.87[l]	–	0.14[l]	–	3.50[l]	–
l.o.i.	1.02	–	0.44	0.70	–	0.20	–	–	3.4
others	0.82[bp]	0.22[bq]	0.66[br]	0.71[bs]	–	0.58[bt]	0.99[bu]	3.29[bv]	6.57[bw]
sum	99.62	100.13	100.15	99.64	100.00	99.51	98.26	100.47	97.9
D in g/cm³	4.9	–	–	4.946	5.03	–	–	4.54	3.95

No. 118 = Monazite (black-brown to black) from rose-colored aplite-pegmatitic granite, Zhitomir complex, southeastern Ukraine, USSR [127]. – No. 119 = Monazite from black sand, Rashîd (Rosetta), Nile delta, Egypt [82]. – No. 120 = Monazite from granite of the 1st intrusive phase, Chudnov-Berdichev complex, southeastern Ukraine, USSR [86, pp. 133/4]. – No. 121 = Monazite from alkali granite, Kola Peninsula, USSR [29, pp. 136/7]. – No. 122 = Monazite from muscovite-microcline-plagioclase pegmatite, Lake Loukhskoe area, Chupa region, northern Karelia, USSR; analysis is recalculated to 100% by omitting Fe_2O_3 and H_2O [113, p. 247]. – No. 123 = Monazite from kaolinitized Chudnov-Berdichev granite, southeastern Ukraine, USSR [89, pp. 124/5], [131]. – No. 124 = Monazite from pegmatitic granite, Novo-Ukrainkii massif, Ukraine, USSR [87]. – No. 125 = Sulfate monazite from gravel of carbonatite, Namo-Vara region, Kola Peninsula, USSR; the sample contains ~8 to 10% Fe hydroxide [84]; see also [132]. – No. 126 = Monazite from lujavrite, Kvanefjeld plateau, southern West Greenland [72].

Table 14 (continued)

	content in % in sample number								
	127	128	129	130	131	132	133	134	135
ThO_2	5.7	5.68	5.66	5.63	5.6	5.6	5.60	5.50	5.50
ΣRE_2O_3	61.2	59.01	57.60	62.23	57.3	59.0	64.00	47.70	58.00
Y_2O_3	0.4	–	–	–	3.0	0.7	–	4.26	–
CaO	1.25	0.16	–	0.44	1.0	0.3	0.22	0.55	1.30
ΣFe_2O_3	0.40	3.03	–	2.06	–	–	0.03	6.32	1.03
Al_2O_3	0.15	0.66	–	–	–	–	0.13	1.72	0.11
U_3O_8	0.1	–	–	–	–	–	0.05	–	–
UO_2	–	–	–	–	0.4	0.2	–	–	–
PbO	–	–	–	–	–	–	–	0.30	–
P_2O_5	27.83	28.45	27.95	27.92	30.6	30.0	27.68	23.50	28.58
SiO_2	1.37	2.80	1.33	0.45	0.5	0.8	1.86	6.26	2.68
H_2O	0.64[l)]	–	–	–	–	–	0.005	–	0.71[f)]
l.o.i.	–	0.6	–	–	–	–	–	0.90	–
others	0.194[bx)]	0.37[by)]	3.10[bz)]	–	–	–	0.539[ca)]	2.80[cb)]	1.58[cc)]
sum	99.234	100.76	95.64	98.73	98.4[c)]	96.6	100.114[c)]	99.81	99.49
D in g/cm³	5.15	4.995	–	5.027	–	–	–	–	5.02

No. 127 = Monazite from pegmatitic biotite-rich vein, Puumula, eastern Finland [133]. – No. 128 = Monazite (black-brown to black) from rose-colored aplite-pegmatitic granite, Zhitomir complex, southeastern Ukraine, USSR [127]. – No. 129 = Monazite from offshore bars, western Taiwan [128]. – No. 130 = Monazite from muscovite-plagioclase pegmatite, Lake Loukhskoe area, Chupa region, northern Karelia, USSR [116]. – No. 131 = Monazite (yellow) from pegmatite, Val Masino, Italy; average of 10 analyses of one sample (for ranges, see original paper) [91]. – No. 132 = Monazite (yellowish pink) from pegmatite, Mondei, Valle Antrona, Italy; average of 7 analyses of one sample (for ranges, see original paper) [91]. – No. 133 = Monazite from muscovite pegmatite, Mama region, Eastern Siberia [83], [134, p. 275]. – No. 134 = Monazite from Alakurtti, northern Karelia, USSR [92]. – No. 135 = Monazite (brown) from orthotectic pegmatite with biotite and garnet, southeastern Ukraine, USSR [112].

Table 14 (continued)

				content in % in sample number					
	136	137	138	139	140	141	142	143	144
ThO$_2$	5.47	5.47	5.45	5.37	5.32	5.28	5.26[cg)]	5.25	5.20
ΣRE$_2$O$_3$	61.92	65.23	57.44	60.92	55.51	58.96	62.3	60.38	64.80
Y$_2$O$_3$	–	–	1.45	–	–	–	2.5	–	–
CaO	0.82	0.16	–	0.02	–	0.20	–	1.97	0.50
ΣFe$_2$O$_3$	0.93	0.46	–	2.88	2.88	2.48	–	1.18	1.26
Al$_2$O$_3$	1.03	0.20	0.32	1.08	1.23	0.21	–	0.29	0.58
U$_3$O$_8$	–	–	–	tr.	–	–	0.62	–	–
UO$_2$	–	–	–	–	–	–	–	–	0.10
PbO	–	–	–	0.09	–	–	–	–	–
P$_2$O$_5$	28.57	25.90	26.70	28.57	27.12	28.27	29.1	27.31	25.66
SiO$_2$	1.18	1.94	–	1.18	2.46	3.44	–	2.69	1.50
H$_2$O	–	–	–	–	–	0.19	–	0.04	–
l.o.i.	–	0.50	4.8	0.46	–	0.49[f)]	–	–	0.38
others	0.08[p)]	–[cd)]	–	–	5.48[ce)]	0.32[cf)]	–[ch)]	0.46[ci)]	–[cj)]
sum	100.00	99.86	96.16	100.57	100.00	99.84	99.78	99.57	99.98
D in g/cm^3	5.23	5.184	–	–	5.26	5.056	–	–	5.1

No. 136 = Monazite from placer, Mudtown, Pegasus district, Stewart Island, New Zealand; contents are corrected for impurities (comprising ~6% gahnite, zircon, and ilmenite-titanhematite) [5, pp. 637, 668]. – No. 137 = Monazite from pegmatite Alakurtti No. 1, northern Karelia, USSR [31]. – No. 138 = Monazite from granite, Harwell ring complex, Nigeria [123]. – No. 139 = Monazite from Minas Gerais, Brazil [93, p. 226]. – No. 140 = Monazite from stream placer, Ngahere, South Island, New Zealand; contents are corrected for impurities (comprising ~15% zircon, tantalian cassiterite, and ilmenite) [5, pp. 637, 668]. – No. 141 = Monazite from granite of the 1st intrusive phase, Chudnov-Berdichev complex, southeastern Ukraine, USSR [86, pp. 133/4]. – No. 142 = Monazite from syenite, Megri pluton, Little Caucasus, southern Armenian SSR [135]. – No. 143 = Monazite from pegmatite of the 2nd intrusive phase, Chudnov-Berdichev complex, southeastern Ukraine, USSR [104]. – No. 144 = Monazite from granosyenite, Siberia [136].

Table 14 (continued)

				content in % in sample number					
	145	146	147	148	149	150	151	152	153
ThO_2	5.1	5.10	5.1	4.95	4.80[cg]	4.78	4.78	4.76	4.65
ΣRE_2O_3	61.9	62.82	66.9	60.22	65.9[cn]	58.80	61.33	61.41	63.93
Y_2O_3	1.8	–	–	–	0.05	–	–	–	0.52
CaO	1.0	0.14	0.7	0.26	–	0.70	1.68	0.24	0.71
ΣFe_2O_3	–	0.06	0.3	3.10	–	0.41	0.67	2.82	–
Al_2O_3	–	0.15	–	0.65	–	–	0.53	0.55	tr.
U_3O_8	0.7	0.03	–[cl]	–	–	–	0.88	–	–
UO_2	–	–	–[cl]	–	–	–	–	–	–
PbO	–	–	–	–	–	–	0.68	–	–
P_2O_5	29.0	28.32	27.3	27.15	28.82	29.41	26.64	26.77	29.97
SiO_2	–	2.24	–	3.00	–	1.57	1.41	2.80	0.22
H_2O	–	0.007	–	0.11	–	–	–	0.16	–
L.o.i.	–	–	–	0.55	–	–	–	0.44	–
others	0.4[s]	0.787[ck]	–	0.48[cm]	–	1.85[co]	1.05[cp]	0.42[cq]	–
sum	99.9	99.654	100.3	100.47	99.57	97.52	99.65	100.37	100.00
D in g/cm³	–	–	–	5.135	–	–	4.30	5.165	5.25

No. 145 = Monazite from stream sediment, southeastern Piedmont, USA [18, pp. 7, 28, 35]; average of 3 points of one sample. – No. 146 = Monazite from muscovite pegmatite, Gutar-Biryusa region, Eastern Siberia [83], [134, p. 275]. – No. 147 = Monazite from monazite-Fe-Ti oxide-apatite deposit in aplitic, graphic, and pegmatitic granite, Kulyk Lake area, northern Saskatchewan, Canada [20]. – No. 148 = Monazite from rose-colored pegmatitic granite of the 2nd intrusive phase, Chudnov-Berdichev complex, southeastern Ukraine, USSR [86, pp. 138, 140]. – No. 149 = Monazite from alaskite, northern Nuratau, western Uzbekistan, USSR [137]. – No. 150 = Monazite from a pegmatitic part in granite, Novo-Ukrainkii massif, Ukraine, USSR [87]. – No. 151 = Monazite (dark gray) from plagioclase granite, Chudnov-Berdichev complex, southeastern Ukraine, USSR [127]. – No. 152 = Monazite from rose-colored pegmatitic granite of the 2nd intrusive phase, Chudnov-Berdichev complex, southeastern Ukraine, USSR [86, pp. 138, 140]. – No. 153 = Monazite from microcline-plagioclase pegmatite Eki-Varaka No. 1, Chupa region, northern Karelia, USSR; analysis is recalculated to 100% by omitting Fe_2O_3 and H_2O [113, p. 247].

Table 14 (continued)

content in % in sample number

	154	155	156	157	158	159	160	161	162
ThO_2	4.62	4.61	4.60	4.56	4.51	4.5	4.5	4.50	4.49
ΣRE_2O_3	61.69	64.69	62.73	63.17	63.09	57.5	62.6	64.50	63.07
Y_2O_3	–	–	–	–	–	2.7	0.4	–	–
CaO	2.43	–	1.00	0.70	0.86	1.4	0.4	1.07	0.51
ΣFe_2O_3	0.81	–	0.06	0.60	0.48	–	–	0.04	0.37
Al_2O_3	0.53	–	0.15	–	0.36	–	0.2	0.14	3.16
U_3O_8	0.25	–	0.04	tr.	0.21	–	–	0.05	–
UO_2	–	–	–	–	–	2.5	–	–	–
PbO	0.22	–	–	–	0.026	–	0.4	–	–
P_2O_5	26.60	29.20	28.20	29.38	26.81	31.8	28.0	27.17	26.61
SiO_2	2.93	–	2.23	0.70	2.24	0.2	1.2	2.04	1.05
H_2O	0.78[l]	–	0.002	0.12	–	–	–	0.002	0.31
l.o.i.	0.22	[cs]	–	–	0.91	–	–	–	–
others	0.24[cr]	–	0.597[ct]	0.20[cu]	–	–	0.5[cv]	0.762[cw]	0.08[p]
sum	101.32[c]	98.50	99.609	99.43	99.496	100.6	98.2[c]	100.274	99.65
D in g/cm³	4.820	–	–	5.25	–	–	–	–	–

No. 154 = Monazite from alkali-granite vein, Kola Peninsula, USSR [29, pp. 136/7]. – No. 155 = Monazite (pale green) from porous sandstone-clay deposit at the border of the Western Siberian Lowlands [138]. – No. 156 = Monazite from muscovite pegmatite, Gutar-Biryusa region, Eastern Siberia [83], [134, p. 275]. – No. 157 = Monazite from muscovite-plagioclase pegmatite Eki-Varaka No. 1, Chupa region, northern Karelia, USSR [116]. – No. 158 = Monazite from wolframite- and Bi minerals-containing pneumatolytic quartz vein, Ebisu Mine, Gifu prefecture, Japan [93, pp. 225/6]. – No. 159 = Monazite (greenish) from pegmatite, Val Dombastone, Sondalo, Italy; average of 7 analyses of one sample (for ranges, see original paper) [91]. – No. 160 = Monazite (dark) from muscovite granite, Lawler Peak pluton, Yavapai county, west-central Arizona [105, p. 170]. – No. 161 = Monazite from muscovite pegmatite, Gutar-Biryusa region, Eastern Siberia [83], [134, p. 275]. – No. 162 = Monazite from quartz-fluorite pegmatite, Tarbagatai, Kazakhstan, USSR [139, pp. 139/40].

Table 14 (continued)

	content in % in sample number								
	163	164	165	166	167	168	169	170	171
ThO_2	4.39	4.36	4.30	4.3	4.3	4.24	4.1	4.1	3.94
ΣRE_2O_3	65.40	61.98	50.06	62.1	67.7	59.68	62.8	64.1	66.62
Y_2O_3	–	2.52	–	0.6	–	–	0.6	1.5	–
CaO	–	0.13	2.32	0.5	–	0.20	0.4	0.9	1.27
ΣFe_2O_3	0.62	0.38	1.14	0.2[r]	–	2.62	–	–	tr.
Al_2O_3	–	0.30	1.76	–	–	0.53	–	–	tr.
U_3O_8	–	0.34	0.88	–	–	–	–	–	–
UO_2	–	–	–	–	–	–	0.1	0.5	–
PbO	–	–	–	0.4	–	–	–	–	–
P_2O_5	28.31	27.60	25.67	29.1	27.4	28.17	31.0	28.9	27.89
SiO_2	0.39	1.44	6.64	0.6	–	3.84	0.4	–	0.95
H_2O	–	0.32[f]	0.25[f]	–	–	0.17	–	–	–
l.o.i.	–	–	2.35	–	–	0.51	–	–	–
others	0.62[p]	0.431[cx]	4.10[cy]	0.6[s]	–	0.32[cz]	–	0.5[s]	–
sum	99.73[c]	99.801[c]	100.45	98.4	99.4	100.28	99.4	100.4	100.67
D in g/cm^3	–	–	–	–	–	4.985	–	–	–

No. 163 = Monazite from syenite pegmatite, Ilmen Mountains, Southern Urals, USSR [14]. – No. 164 = Monazite from complex pegmatite, Styr-Kraal, Kenhardt district, Cape Province, R.S.A. [115]. – No. 165 = Monazite (altered rim) from kaolinitized Chudnov-Berdichev granite, southeastern Ukraine, USSR [89, pp. 124/5], [131]. – No. 166 = Monazite (salmon) from muscovite granite, Lawler Peak pluton, Yavapai county, west-central Arizona [105, p. 173]. – No. 167 = Monazite from the marginal zone of a pegmatite, Kulyk Lake area, northern Saskatchewan, Canada [140]. – No. 168 = Monazite from granite (termed rose-colored pegmatitic granite of the 2nd intrusive phase by [86, pp. 138, 140]), Chudnov-Berdichev complex, southeastern Ukraine, USSR [127]. – No. 169 = Monazite from Alpine fissure, Kollergraben, Binn, Switzerland; average of 7 analyses of one sample (for ranges, see original paper) [91]. – No. 170 = Monazite from stream sediment, southeastern Piedmont, USA [18, pp. 7, 28, 35]; average of 2 points of one sample. – No. 171 = Monazite from syenite pegmatite, Ilmen Mountains, Southern Urals, USSR [14].

Table 14 (continued)

	content in % in sample number								
	172	173	174	175	176	177	178	179	180
ThO$_2$	3.9	3.9	3.78	3.7	3.64	3.49	3.47	3.4	3.32
ΣRE$_2$O$_3$	63.9	65.6	61.00	65.5	60.62	64.8[dc]	55.72	63.5	68.87
Y$_2$O$_3$	1.1	–	–	0.6	–	–	0.09	0.4	–
CaO	0.7	0.4	0.80	0.5	1.50	0.47	0.96	0.3	–
ΣFe$_2$O$_3$	0.2	–	0.05	–	1.66	0.11	0.65	–	–
Al$_2$O$_3$	0.15	–[c]	–	–[c]	1.23	–	0.06	0.6	–
U$_3$O$_8$	0.2	–[c]	–	–[c]	–	0.25	–	–	–
UO$_2$	–	–	–	–	–	–	0.15	–	–
PbO	<0.01[da]	–	–	–	–	0.019[h]	–	0.3	–
P$_2$O$_5$	28.5	27.6	28.95	28.4	26.94	28.37	30.11	29.0	27.30
SiO$_2$	0.9	–	1.16	–	2.61	0.81	0.85	0.6	–
H$_2$O	0.15[az]	–	–	–	–	0.35[l]	–	–	–
l.o.i.	–	–	–	–	1.15	–	–	–	–
others	0.23[db]	0.9[s]	1.50[bn]	–	0.18[p]	<0.057[dd]	–	0.5[df]	–[dg]
sum	~99.94[c]	98.4	97.24	98.7	99.53	~98.726[c]	92.06[c]	98.6[c]	99.49
D in g/cm^3	5.21±0.02	–	–	–	–	5.21±0.02	–	–	–

No. 172 = Monazite from black-sand pay streak situated at storm level on beach south of the mouth of Año Nuevo Creek, California; additionally contains an appreciable amount of Cu [141]. – No. 173 = Monazite from monazite-Fe-Ti oxide-apatite deposit in aplitic, graphic, and pegmatitic granite, Kulyk Lake area, northern Saskatchewan, Canada [20]. – No. 174 = Monazite from granite, Novo-Ukrainkii massif, Ukraine, USSR [87]. – No. 175 = Monazite from monazite-Fe-Ti oxide-apatite deposit in aplitic, graphic, and pegmatic granite, Kulyk Lake area, northern Saskatchewan, Canada [20]. – No. 176 = Monazite from an undisclosed occurrence [142]. – No. 177 = Monazite from beach sand, Del Monte Properties, Pacific Grove, Monterey county, California [141]. – No. 178 = Monazite from thoriferous conglomerate (taken at 498.08 m depth), southeastern Gabon [110]. – No. 179 = Monazite (light) from muscovite granite, Lawler Peak pluton, Yavapai county, west-central Arizona [105, p. 170]. – No. 180 = Monazite (cream-colored) from porous sandstone-clay deposit at the border of the Western Siberian Lowlands [138].

Table 14 (continued)

					content in % in sample number				
	181	182	183	184	185	186	187	188	189
ThO_2	3.2	3.2	3.2	3.1	3.07	3.04	3.02	3.01	2.85
ΣRE_2O_3	60.99	64.90	65.4	66.9	62.53	63.74	64.86	67.83	63.35
Y_2O_3	0.65	1.6	1.4	0.7	0.70	–	–	–	–
CaO	0.49	0.51	0.6	0.5	0.10	0.15	1.31	–	0.72
ΣFe_2O_3	1.6	0.52	0.2	0.2	0.36	1.02	0.07	–	0.50
Al_2O_3	0.72	–	–	–	4.90	0.69	0.18	–	0.15
U_3O_8	–	–	0.5	0.4	–	0.11	0.15	–	–
UO_2	–	–	–	–	–	–	–	–	–
PbO	–	–	–	–	tr.	0.41	–	0.12	0.18
P_2O_5	24	22	30.4	30.4	25.32	28.45	28.32	29.23	28.11
SiO_2	2.1	2.3	–	–	1.99	0.99	1.84	0.70	2.14
H_2O	1.46	–	–	–	0.28[l]	0.10	0.004	–	–
l.o.i.	–	–	–	–	–	0.50	–	–	–
others	0.71[x]	0.11[p]	–	–	0.48[dh]	0.24[di]	0.348[dj]	–	2.12[dk]
sum	95.92	95.14	101.7	102.2	99.73	99.44	100.102	100.89	100.12
D in g/cm³	–	–	–	–	5.01	–	–	4.98	–

No. 181 = Monazite (dark, Eu-rich) from alluvial placer, Ruby, Alaska [130, p. 38]. – No. 182 = Monazite (yellow) from alluvial placer, Talkeetna, Alaska [130, p. 42]. – Nos. 183 and 184 = Monazite from a W-Bi-Mo deposit in greisenized acid volcanic rock, Mount Pleasant, New Brunswick, Canada [125]. – No. 185 = Monazite from hydrothermally altered gneiss xenolith, southeastern Ukraine, USSR [143]. – No. 186 = Monazite from pegmatite, Chokrak River, Ukraine, USSR [81]. – No. 187 = Monazite from muscovite pegmatite, Mama region, Eastern Siberia [83], [134, p. 275]. – No. 188 = Monazite from barite-carbonate (mainly dolomite) rock, Mountain Pass district, San Bernardino county, California [144]. – No. 189 = Monazite from placer sand, Baltic Sea, Poland [117].

Table 14 (continued)

	content in % in sample number								
	190	191	192	193	194	195	196	197	198
ThO_2	2.7	2.39	2.36	2.2	2.15	2.11	2.1	1.79	1.79
ΣRE_2O_3	51.4	53.98	65.88	62.2	59.98	63.88	68.6	62.33	63.46
Y_2O_3	0.2	–	–	3.1	4.69	–	–	–	5.13
CaO	0.9	2.86	–	0.6	2.89	1.72	0.1	–	0.83
ΣFe_2O_3	–	2.05	0.80	–	–	1.22	0.1	2.12	1.39
Al_2O_3	0.4	2.70	0.17	0.4	0.29	0.80	–	–	0.04
U_3O_8	tr.[r]	0.03	–	–	–	–	0.6	–	–
UO_2	–	–	–	–	–	–	–	–	–
PbO	–	–	–	–	–	–	–	–	–
P_2O_5	27.7	26.65	28.89	31.2	27.09	25.57	29.3	25.48	26.83
SiO_2	0.7	3.90	–	–	–	1.92	–	–	0.23
H_2O	–	1.37	0.66	–	–	0.44	–	–	0.21[f]
l.o.i.	–	–	–	–	–	2.18	–	–	–
others	0.3[s]	2.90[dl]	1.33[dm]	–	2.93[dn]	2.60[do]	–	–	0.07[p]
sum	84.3[c]	98.83	100.09[c]	99.7	100.02	102.44	100.8	91.72	99.98
D in g/cm³	–	4.683	5.06	–	5.195	–	–	–	–

No. 190 = Monazite (chalky) from muscovite granite, Lawler Peak pluton, Yavapai county, west-central Arizona [105, p. 175]. – No. 191 = Monazite from talc-actinolite-carbonate rock in crystalline rocks of the Urushten complex, left bank of Little Laba River, northern Caucasus, USSR [145, 146]; D is cited from [147]. – No. 192 = Monazite from pegmatite, Transangara alkaline massif, Enisei Ridge, Central Siberia [148]. – No. 193 = Monazite from a W-Bi-Mo deposit in greisenized acid volcanic rock, Mount Pleasant, New Brunswick, Canada [125]. – No. 194 = Monazite from pegmatite, Sierra de la Huerta, San Juan, Argentina [149]. – No. 195 = Monazite from hydrothermally altered granite, Central Asia [2, p. 116]. – No. 196 = Monazite from a W-Bi-Mo deposit in greisenized acid volcanic rock, Mount Pleasant, New Brunswick, Canada [125]. – No. 197 = Monazite from ferrocarbonatite, Fen complex, Telemark, southeastern Norway [150]. – No. 198 = Monazite from mixed gneiss, Sebes Mountains, southern Carpathian Mountains, Roumania [100].

Table 14 (continued)

	content in % in sample number								
	199	200	201	202	203	204	205	206	207
ThO_2	1.68	1.5	1.5	1.40	1.31	1.3	1.3	1.28[dv]	1.26
ΣRE_2O_3	67.19	56.4	66.0	62.46	65.78	45.3[dt]	64.7	68.90	68.07
Y_2O_3	—	—	0.3	—	—	—	0.7	—	—
CaO	0.20	—	0.2	0.29	—	0.64	0.8	—	0.74
ΣFe_2O_3	0.29	1.23	0.1[r]	—	0.87	1.0	—	—	1.41
Al_2O_3	0.25	—	—	—	—	1.3	—	—	—
U_3O_8	0.033	—	0.1[r]	—	—	—	—	—	—
UO_2	—	—	—	—	—	—	—	—	—
PbO	0.14	—	0.2	—	—	0.04	—	—	—
P_2O_5	28.94	28	28.3	30.00	29.36	22.6	31.6	28.87	26.92
SiO_2	0.38	4.42	1.1	0.15	0.20	25.0	0.2	—	—
H_2O	1.04[dp]	2.9[f]	—	—	—	—	—	—	—
l.o.i.	0.08	—	—	—	—	—	—	—	—
others	0.20[dq]	0.89[dr]	0.6[ds]	—	—	2.855[du]	—	—	—
sum	100.423[c]	95.34	98.4	94.30	97.52	100.035	99.3	99.05	98.40
D in g/cm³	5.142	—	—	—	5.20	—	—	—	—

No. 199 = Monazite from greisen (termed hydrothermal muscovite-quartz vein by [151]), Kola Peninsula, USSR [2, p. 116]; D is cited from [151]. – No. 200 = Monazite from Devonian Ti placer, Pizhma-Umba, central Timan Ridge, European USSR [101]. – No. 201 = Monazite from two-mica granite, Lawler Peak pluton, Yavapai county, west-central Arizona [105, p. 112]. – No. 202 = Monazite (kularite) from alluvial placer, Dzhatka River, Yakutia, USSR [36]. – No. 203 = Monazite from albitite in miaskite, Vishnevye Gory, Southern Urals, USSR [152]. – No. 204 = Monazite (black, Eu-rich) from beach placer, southwestern Taiwan [28, pp. 43, 53]. – No. 205 = Monazite from Alpine fissure, Rauris Plattenbrüche, Austria; average of 7 analyses of one sample (for ranges, see original paper) [91]. – No. 206 = Monazite from ferrocarbonatite, Fen complex, Telemark, southeastern Norway [150]. – No. 207 = Monazite from an eluvial pebble, Afu Hills, Nigeria [153].

Table 14 (continued)

	208	209	210	211	212	213	214	215
				content in % in sample number				
ThO_2	1.21	1.14	1.09	1.09	1.06	1.05	1.0	1.0
ΣRE_2O_3	69.59	54.01	54.11	69.08	57.58	69.13	62.5	68.55
Y_2O_3	–	–	–	–	–	0.30	3.4	0.57
CaO	–	0.80	1.28	0.78	0.82	0.08	0.2	0.28
ΣFe_2O_3	0.34	2.76	2.61	0.20	3.08	0.43	0.1	0.69
Al_2O_3	–	0.50	1.49	0.37	0.18	–	–	1.3
U_3O_8	–	0.41	–	0.35[dy]	–	–	–	–
UO_2	–	–	–	–	–	–	–	–
PbO	–	1.03	–	–	–	–	–	–
P_2O_5	29.56	24.08	21.39	28.60	25.11	27.79	30.1	23
SiO_2	0.50	12.04	16.78	0.40	12.48	0.66	–	7.1
H_2O	–	2.03[f]	0.82	–	–	0.34	–	0.70
l.o.i.	–	–	–	–	–	–	–	–
others	–	0.75[dw]	0.41[dx]	0.71[dz]	0.14[x]	–	–	0.73[as]
sum	101.20[c]	99.55	99.98[c]	101.58[c]	100.45	99.78	97.3	103.92
D in g/cm^3	–	4.3	–	5.06	4.30	5.2	–	–

No. 208 = Monazite from fenitized nepheline-syenite xenolith, Khibiny massif, Kola Peninsula, USSR [154]. – No. 209 = Silicomonazite (untreated) from alluvial placer, Kyugyulyur Creek, Kular Ridge, Yakutia, USSR; for the analysis of HCl-treated silicomonazite, see original paper [32]. – No. 210 = Monazite (Eu-rich) from placer, Verkhoyansk-Chukotka Fold Region, Eastern Siberia [155]. – No. 211 = Monazite from carbonatite, Samalpatti, Tamil Nadu, India [27, pp. 550, 553]; D is cited from [95, p. 94]. – No. 212 = Monazite from alluvial placer, basin of Yana River, northeastern Yakutia, USSR [36]. – No. 213 = Monazite from pegmatite, Tommot complex, Eastern Siberia [156]. – No. 214 = Monazite from a W-Bi-Mo deposit in greisenized acid volcanic rock, Mount Pleasant, New Brunswick, Canada [125]. – No. 215 = Monazite (dark, Eu-rich) from alluvial placer, Kivu, Zaire [130, p. 38].

Footnotes for Table 14:

a) Water is determined as OH (see footnote b)). — b) Comprises 5.1% OH (calculated by difference and by balancing anions with cations) and 1.0% F. — c) Summary content does not correspond to that given in the original paper. — d) Comprises 0.28% MgO and traces of MnO. — e) Comprises 0.51% MgO, 0.75% MnO, 6.10% ZrO_2, 0.12% TiO_2, and 1.08% (Ta_2O_5 + Nb_2O_5).

f) Only H_2O^+. — g) Comprises 0.09% MgO and 0.12% MnO. — h) Recalculated from atomic percentage. — i) Comprises 0.26% MgO, 3.34% ZrO_2, and 0.05% TiO_2. — j) Only ZrO_2.

k) Comprises 0.79% MgO and 0.28% TiO_2. — l) Comprises H_2O^+ and H_2O^-. — m) Comprises 0.30% MgO, 3.42% ZrO_2, and 0.12% TiO_2. — n) Given as Pb_3O_4. — o) Only K_2O.

p) Only MgO. — q) Comprises 0.30% MgO and 0.08% TiO_2. — r) Cation oxide state not determined. — s) Only F. — t) Comprises 0.14% MgO and 0.16% TiO_2.

u) Comprises 0.10% MgO, 0.14% ZrO_2, 0.32% TiO_2, and 0.01% SnO_2. — v) Comprises 0.39% MgO and 0.12% TiO_2. — w) Comprises 0.07% MgO, 0.29% ZrO_2, 0.48% TiO_2, and 0.07% SnO_2. — x) Only TiO_2. — y) Additionally are given 104 ppm Eu, 7500 ppm Gd, 437 ppm Tb, 1032 ppm Dy, 230 ppm Er, 123 ppm Yb, and 35 ppm Lu.

z) Comprises 1.35% ZrO_2 and 1.21% K_2O; additionally are given 5000 ppm Ba, 2000 ppm Cr, 2000 ppm Cl, 538 ppm Na, 100 ppm Co, and 85 ppm Mn. — aa) Comprises 0.09% MgO, 0.13% ZrO_2, 0.32% TiO_2, and 0.02% SnO_2. — ab) Comprises 0.30% MgO, 0.05% TiO_2, and 0.46% WO_3. — ac) Comprises 0.12% MgO and 0.17% TiO_2. — ad) Comprises $\Sigma(CaO + Al_2O_3 + Fe_2O_3)$.

ae) Comprises traces of ZrO_2 and 0.4% F. — af) Comprises 0.05% MgO, 0.01% MnO, 0.82% ZrO_2, and 1.34% TiO_2. — ag) Given as PbO_2. — ah) Comprises 0.3% ZrO_2 and 1.4% F. — ai) Only SrO.

aj) Comprises 1.2% MgO, 0.1% ZrO_2, and 0.4% F. — ak) Comprises 0.27% MgO and 0.10% TiO_2. — al) Given as UO_3. — am) Calculated by difference from 100%; includes SiO_2. — an) Comprises 0.10% MgO and 0.70% TiO_2.

ao) Comprises Y and RE(Y). — ap) Comprises 0.07% MgO, 0.08% TiO_2, 0.18% Na_2O, 0.07% K_2O, and 0.010% SnO_2. — aq) Comprises traces of MnO and 0.90% insoluble residue. — ar) Comprises 0.20% MgO, traces of MnO, 0.12% ZrO_2, and traces of TiO_2. — as) Comprises 0.12% MgO and 0.61% TiO_2.

at) Includes Er. — au) Comprises 0.19% MgO, 0.005% MnO, 0.01% BeO, and 0.01% ZrO_2. — av) Comprises 0.09% SrO and 0.69% SO_3. — aw) Ce given as CeO_2. — ax) Comprises 0.36% MgO, 0.03% SO_3, 1.13% CO_2, and 1.8% F.

ay) Comprises traces of MnO and 0.40% insoluble residue. — az) Only H_2O^-. — ba) Comprises 0.13% MgO, 0.12% Na_2O, and 0.07% K_2O. — bb) Comprises 0.10% MgO, 0.11% TiO_2, 0.16% Na_2O, 0.07% K_2O, and 0.014% SnO_2. — bc) Comprises 0.39% MgO, 1.53% ZrO_2, and 0.22% TiO_2.

bd) Comprises 0.62% MgO, 0.24% ZrO_2, and 0.23% TiO_2. — be) Comprises 0.01% MnO, traces of ZrO_2, and 0.55% TiO_2. — bf) Comprises 0.22% MgO, 0.06% TiO_2, and 0.73% WO_3. — bg) Given as 0.50% PbO by [86, pp. 141/2]. — bh) Comprises 0.11% MgO, 1.00% ZrO_2, and 0.27% TiO_2 [127].

bi) Comprises 0.39% MgO and 1.53% ZrO_2. — bj) Comprises 0.05% MgO and 0.23% TiO_2. — bk) Comprises 0.42% MgO, 0.03% TiO_2, and 0.39% WO_3. — bl) Comprises 0.17% MgO and 0.17% TiO_2. — bm) Comprises 0.12% MgO and 0.13% TiO_2.

bn) Only SO_3. — bo) Comprises <0.10% MgO and <0.40% TiO_2. — bp) Comprises 0.42% MgO, 0.40% ZrO_2, and traces of TiO_2. — bq) Comprises TiO_2; additionally contains 500 g/t Al, 50 g/t Be, 700 g/t Mg, 40 g/t Mn, 30 g/t Nb, 10 g/t Ni, 50 g/t Sn, and 80 g/t Zr. — br) Comprises 0.03% MgO and 0.63% TiO_2.

bs) Comprises 0.40% ZrO_2, 0.23% Na_2O, and 0.08% F. — bt) Comprises 0.40% MgO and 0.18% TiO_2. — bu) Comprises 0.03% TiO_2 and 0.96% SO_3. — bv) Comprises 0.08% MgO, 0.09% TiO_2, and 3.12% SO_3. — bw) Comprises 0.6% MnO, 0.03% TiO_2, 0.24% Nb_2O_5, 1.55% Na_2O, 0.48% K_2O, and 3.67% ZnO.

bx) Comprises 0.05% MgO, 0.004% MnO, and 0.14% TiO_2. — by) Comprises 0.07% MgO and 0.30% TiO_2. — bz) Comprises 2.95% ZrO_2 and 0.15% TiO_2. — ca) Comprises 0.15% MgO, 0.37% ZrO_2, and 0.019% Nb_2O_5. — cb) Comprises 0.10% MgO and 2.70% MnO.

cc) Comprises 0.16% SrO, 0.22% TiO_2, and 1.20% SO_3. — cd) Additionally contains small amounts of Tl, Bi, and Pb. — ce) Only $\Sigma(Nb_2O_5+Ta_2O_5)$. — cf) Comprises 0.17% MgO and 0.15% TiO_2. — cg) Erroneously given as Th in the original paper.

ch) Additionally contains 0.001 to 0.003% each of Sc and Mo, 0.001% Be, 0.003% Pb, and 0.01% Sr [135] as well as 0.3 to 1.0% Ti [157]. — ci) Comprises 0.17% MgO, 0.20% ZrO_2, and 0.09% TiO_2. — cj) Additionally contains traces of Mg and Sr. — ck) Comprises 0.13% MgO, 0.60% ZrO_2, 0.017% Nb_2O_5, and 0.04% SO_3. — cl) Given as $\leq$0.05% U.

cm) Comprises 0.20% MgO and 0.28% TiO_2. — cn) Erroneously given as ΣRE in the original paper. — co) Comprises 0.42% TiO_2 and 1.43% SO_3. — cp) Comprises 0.35% MgO, 0.70% ZrO_2, and traces of TiO_2. — cq) Comprises 0.18% MgO and 0.24% TiO_2.

cr) Comprises 0.04% MgO and 0.20% ZrO_2. — cs) In addition, the following ranges, determined spectral analytically and not included in the sum of chemically determined elements, are given: 0.1 to 0.3% each of Mg, Mn, Al, Fe, Pb, and Cr, 0.01 to 0.03% Si, and traces of Be, Zn, Cu, Sn, and U. — ct) Comprises 0.23% MgO, 0.31% ZrO_2, 0.027% Nb_2O_5, and 0.03% SO_3. — cu) Comprises traces of MnO and 0.20% insoluble residue. — cv) Comprises 0.1% ZrO_2 and 0.4% F.

cw) Comprises 0.19% MgO, 0.50% ZrO_2, 0.032% Nb_2O_5, and 0.04% SO_3. — cx) Comprises 0.08% MgO, 0.05% TiO_2, 0.21% Na_2O, 0.08% K_2O, and 0.011% SnO_2. — cy) Comprises 0.66% MgO and 3.44% TiO_2. — cz) Comprises 0.14% MgO and 0.18% TiO_2. — da) Given in atomic percentage in the original paper.

db) Comprises 0.1% MgO, traces of MnO, 0.04% TiO_2, and 0.09% Cu. — dc) Additionally are given 0.5 to 1.0% RE(Y). — dd) Comprises <0.01% ZrO_2, 0.04% TiO_2, and 0.007% Cu. — de) Total analysis erroneously given in atomic percentage. — df) Comprises 0.1% ZrO_2 and 0.4% F.

dg) In addition, the following ranges, determined spectral analytically and not included in the sum of chemically determined elements, are given: 0.1 to 0.3% each of Mn and Pb, 0.01 to 0.03% each of Si, Al, Fe, and Cr, and traces of Be, Cu, and U. — dh) Comprises 0.26% MgO, 0.02% Na_2O, and 0.20% F. — di) Comprises 0.15% MgO and 0.09% TiO_2. — dj) Comprises 0.13% MgO, 0.21% ZrO_2, and 0.008% Nb_2O_5. — dk) Comprises 0.12% MgO, 0.20% ZrO_2, and 1.80% TiO_2.

dl) Comprises 1.69% MgO, 0.07% TiO_2, 0.35% Na_2O, 0.24% K_2O, and 0.55% CO_2. — dm) Comprises 0.15% MgO, traces of UO_3, 0.98% TiO_2, and 0.20% CO_2. — dn) Comprises 1.51% ZrO_2 and 1.42% Nb_2O_5. — do) Comprises 0.06% MgO, 0.01% MnO, 1.65% CO_2, 0.86% F, and 0.02% S. — dp) Given as 0.38% H_2O^+ and 0.33% H_2O^- by [151].

dq) Comprises 0.12% MgO and 0.08% TiO_2. — dr) Comprises TiO_2 and traces of MgO. — ds) Comprises 0.1% TiO_2 and 0.5% F. — dt) Given as 44.24% ΣRE_2O_3 and 0.50% Y_2O_3 by [158]. — du) Comprises 0.66% MgO, 0.035% MnO (erroneously given as MnO_2 in the original paper), 0.66% ZrO_2, and 1.5% TiO_2.

dv) In the text of the original paper, the ThO_2 content is given as 1.13%. — dw) Comprises 0.42% MgO and 0.33% TiO_2. — dx) Comprises 0.18% MgO, 0.02% MnO, and 0.21% TiO_2. — dy) Given as 0.05% U_3O_8 by [95, p. 94]. — dz) Comprises 0.05% TiO_2, 0.02% SrO, and 0.64% unspecified components.

Partial oxidation of U^{4+} at the surface of monazite grains, with a consequent greater tendency for some components to be leached out or adsorbed, may enhance differences in the composition of rim and core of this mineral in pegmatites, northern Italy; the core is richer in Th and U and poorer in RE [69].

For H_2O contents in monazite caused by hydration and partial conversion of the mineral, see p. 288.

References for 2.4.1.1.2:

[1] Palache, C.; Berman, H.; Frondel, C. (Dana's System of Mineralogy, 7th Ed., Vol. 2, Wiley, New York 1963, pp. 1/1124).

[2] Semenov, E. I. (Mineralogiya Redkikh Zemel', Akad. Nauk SSSR, Moscow 1963, pp. 1/412).

[3] Perez Mateos, J.; Galvan Garcia, J. (Anales Edafol. Agrobiol. **26** [1967] 1227/44).

[4] Frondel, C. (Geol. Surv. Bull. [U.S.] No. 1064 [1958] 1/400, 151/3).

[5] Hutton, C. O. (Bull. Geol. Soc. Am. **61** [1950] 635/716).

[6] Overstreet, W. C. (Geol. Surv. Profess. Paper [U.S.] No. 530 [1967] 1/327).

[7] Rapp, R. P.; Watson, E. B. (Contrib. Mineral. Petrol. [Berlin] **94** [1986] 304/16).

[8] Kauffman, A. J., Jr.; Baber, K. D. (Inform. Circ. U.S. Bur. Mines No. 7767 [1956] 1/36, 5).

[9] Heinrich, E.W. (Mineralogy and Geology of Radioactive Raw Materials, McGraw-Hill, New York 1958, pp. 1/654, 75).

[10] Kaplan, G. E.; Uspenskaya, T. A.; Zarembo, Yu. I.; Chirkov, I. V. (Torii, Ego Syr'evye Resursy, Khimiya i Tekhnologiya, Atomizdat, Moscow 1960, pp. 1/224, 15/6).

[11] de Oliveira, A. I. (Engenharia Miner. Metalurgia **24** [1956] 163/4; Ref. Zh. Geol. **1959** 3328).

[12] Betechtin, A. G. (Lehrbuch der Mineralogie, Verlag Tech., Berlin 1953, pp. 1/681, 409 [Russian original, Moscow 1951]).

[13] Li, A. F.; Grebennikova, O. T. (Zap. Vost. Sibirsk. Otd. Vses. Mineral. Obshch. No. 4 [1962] 155/61).

[14] Makarochkin, B. A. (Geol. Geofiz. **1975** No. 2, pp. 155/9).

[15] Sokova, K. P. (Tr. Inst. Geol. Rudn. Mestorozhd. Petrogr. Mineral. Geokhim. No. 81 [1962] 3/22, 19/22).

[16] Platt, R. G.; Williams, C. T.; Woolley, A. R. (Mineral. Mag. **51** [1987] 253/63, 261).

[17] Matzko, J. J.; Overstreet, W. C. (Chung-Kuo Ti Chih Hsueh Hui Hui K'an No. 20 [1977] 16/35, 25, 31; C.A. **88** [1978] No. 194615).

[18] Karfunkel, B. S.; Fay, W. M.; Price, V., Jr. (DPST-81-141-8 [1981] 1/57; C.A. **98** [1983] No. 6530).

[19] Le, V. T.; Stussi, J.-M. (Sci. Terre **18** [1973] 353/79, 375).

[20] Watkinson, D. H.; Mainwaring, P. R. (Can. J. Earth Sci. **13** [1976] 470/5).

[21] Bermanec, V.; Tibljas, D.; Gessner, M.; Kniewald, G. (Mineral. Petrol. **38** [1988] 139/50).

[22] Stupp, H. D. (Diss. Univ. Köln 1984, pp. 1/280).

[23] Reimer, T. O. (Neues Jahrb. Mineral. Abhandl. **158** [1987] 13/46, 22).

[24] Glatthaar, C. W.; Feather, C. E. (Appl. Mineral. Proc. 2nd Intern. Congr. Appl. Mineral. Minerals Ind., Los Angeles 1984 [1985], pp. 617/28, 622/3).

[25] Ramdohr, P.; Strunz, H. (Klockmanns Lehrbuch der Mineralogie, 16th Ed., Enke, Stuttgart 1978, pp. 1/876, 625).

[26] Staatz, M. H. (Geol. Surv. Profess. Paper [U.S.] No. 1049-D [1983] D1/D52, D41).

[27] Semenov, E. I.; Upendran, R.; Subramanian, V. (J. Geol. Soc. India **19** [1978] 550/7).

[28] Soong, K.-L. (Acta Oceanogr. Taiwan. No. 8 [1978] 43/62).

332

[29] Bel'kov, I. V. (Aktsessornye Mineraly Granitoidov Kol'skogo Poluostrova, Nauka, Leningrad 1979, pp. 1/184).

[30] Chaudhuri, J. N. B. (Berlin. Geowiss. Abhandl. A **52** [1984] 1/123).

[31] Kalita, A. P. (Redkozemel'nye Pegmatity Alakurtti i Priladozh'ya, Akad. Nauk SSSR, Moscow 1961, pp. 1/119, 56/7).

[32] Nekrasov, I. Ya. (Dokl. Akad. Nauk SSSR **204** [1972] 941/3; Dokl. Acad. Sci. USSR Earth Sci. Sect. **204** [1972] 134/6).

[33] Mohr, D. W. (Am. Mineralogist **69** [1984] 98/103).

[34] Jefferies, N. L. (Mineral. Mag. **49** [1985] 495/504).

[35] Flinter, B. H.; Butler, J. R.; Harral, G. M. (Am. Mineralogist **48** [1963] 1210/26, 1220).

[36] Nekrasova, R. A.; Nekrasov, I. Ya. (Dokl. Akad. Nauk SSSR **268** [1983] 688/93; Dokl. Acad. Sci. USSR Earth Sci. Sect. **268** [1983] 151/5).

[37] Deer, W. A.; Howie, R. A.; Zussman, J. (Rock-Forming Minerals, Vol. 5, Non-Silicates, Longmans, London 1962, pp. 1/371, 339/40).

[38] Haaker, R. F.; Ewing, R. C. (Sci. Basis Nucl. Waste Management **2** [1980] 281/8, 283).

[39] Frondel, J. W.; Fleischer, M. (Geol. Surv. Bull. [U.S.] No. 1009-F [1955] 169/209, 185).

[40] Twenhofel, W. S.; Buck, K. L. (Proc. Intern. Conf. Peaceful Uses At. Energy, Geneva 1955 [1956], Vol. 6, pp. 562/7).

[41] Frondel, C. (Geol. Surv. Profess. Paper [U.S.] No. 300 [1956] 567/79, 570).

[42] Roubault, M. (Geologie de l'Uranium, Masson, Paris 1958, pp. 1/462, 411).

[43] Davidson, C. F. (Mining Mag. [London] **94** [1956] 197/208).

[44] Brown, J. C.; Dey, A. K. (India's Mineral Wealth, Oxford Univ. Press, London 1955, pp. 1/761, 277).

[45] Murata, K. J.; Rose, J. J., Jr.; Carron, M. K.; Glass, J. J. (Geochim. Cosmochim. Acta **11** [1957] 141/61, 144/5, 148).

[46] Aleksiev, E.; Tsvetkova, V. (Tr. Vurkhu Geol. Bulg. Ser. Geokhim. Polezni Izkop. **3** [1962] 5/24).

[47] Dar, K. K. (J. Geol. Soc. India **5** [1964] 112/20, 113).

[48] Wadia, D. N. (Sci. Cult. [Calcutta] **21** [1956] 561/5).

[49] Kelly, F. J. (Inform. Circ. U.S. Bur. Mines No. 8124 [1962] 1/38, 19).

[50] Kremers, H. E. (Soc. Mining Engineers Am. Inst. Mining Eng. Preprint No. 5819A18 [1958] 1/14 from [6, p. 43]).

[51] Rosenblum, S. (J. Res. U.S. Geol. Surv. **2** [1974] 689/92).

[52] Bonatti, S. (Atti 1st Conv. Geol. Nucl., Rome 1955, pp. 17/21).

[53] Bhola, K. L.; Chatterji, B. D.; Dar, K. K.; Mahadevan, C.; Mahadevan V.; Mehta, N. R.; Nagarajarao; Nandi, H.; Narayandas, G. R.; Sahasrabudhe, G. H.; Shirke, V. G.; Udas, G. R. (Proc. 2nd U.N. Intern. Conf. Peaceful Uses At. Energy, Geneva 1958, Vol. 2, pp. 100/2).

[54] Rayner, E. O. (Australian At. Energy Symp. Proc., Sydney 1958, pp. 27/34, 29).

[55] Whitworth, H. F. (New South Wales Dept. Mines Tech. Rept. **4** 1956 [1959] 7/60, 25).

[56] Bowie, S. H. U. (J. Roy. Soc. Arts **107** [1959] 704/18, 711/2).

[57] Cooper, P. F.; MacNevin, A. A.; Winward, K. (Mineral. Ind. New South Wales Nos. 38, 40, 44 [1979] 1/80, 20).

[58] Callaghan, C. C. (Ann. Geol. Surv. [South Africa] **17** 1983 [1984] 35/41).

[59] Adams, J. W.; Fish, G. E., Jr. (Geol. Surv. Profess. Paper [U.S.] No. 580 [1968] 430/5).

[60] Troxel, B. W. (Bull. Calif. Div. Mines Geol. No. 176 [1957] 635/40).

[61] de la Roche, H. (Serv. Geol. Madagascar **1958** 1/112 from C.A. **1958** 16136).

[62] de la Roche, H. (Document. Bur. Geol. [Malagasy] No. 140 [1958] 1/9 from C.A. **1959** 5989).

[63] Borrowman, S. R.; Rosenbaum, J. B. (Rept. Invest. U.S. Bur. Mines No. 5917 [1962] 1/8, 2).

[64] Pike, D. R. (Proc. 2nd U.N. Intern. Conf. Peaceful Uses At. Energy, Geneva 1958, Vol. 2, pp. 91/6).

[65] Oberli, F.; Sommerauer, J.; Steiger, R. H. (Schweiz. Mineral. Petrogr. Mitt. **61** [1981] 323/48, 331).

[66] Pluhar, E. (Berlin. Geowiss. Abhandl. A **12** [1979] 1/53).

[67] Ranchin, G. (Sci. Terre **13** [1968] 159/205, 163, 184).

[68] Kantor, J. (Geol. Prace Zpravy **11** [1957] 5/28, 13, 23/4).

[69] Gramaccioli, C. M.; Segalstad, T. V. (Am. Mineralogist **63** [1978] 757/61).

[70] Price, V.; Cook, J. R.; Fay, W. M.; Karfunkel, B. S. (Uranium Explor. Methods Rev. NEA/IAEA R&D Programme Proc. Symp., Paris 1982, pp. 431/44, 437).

[71] Silver, L. T.; Williams, I. S.; Woodhead, J. A. (GJBX 45-81 [1980] 1/380, 219; C.A. **95** [1981] No. 65599).

[72] Asmund, G. (RISO-253 [1971] 1/125, 62/4; N.S.A. **26** [1972] No. 47767).

[73] Köppel, V.; Grünenfelder, M. (Schweiz. Mineral. Petrogr. Mitt. **51** [1971] 385/409, 390, 394, 408).

[74] Overstreet, W. C.; Warr, J. J., Jr.; White, A. M. (Geol. Surv. Profess. Paper [U.S.] No. 700-D [1970] D207/D216, D207/D214).

[75] Overstreet, W. C.; Warr, J. J., Jr.; White, A. M. (Southeast. Geol. **10** [1969] 63/76).

[76] Lee, D. E.; Bastron, H. (Geochim. Cosmochim. Acta **31** [1967] 339/56).

[77] Overstreet, W. C.; Warr, J. J., Jr.; White, A. M. (Southeast. Geol. **13** [1971] 99/125).

[78] Murata, K. J.; Rose, H. J., Jr.; Carron, M. K. (Geochim. Cosmochim. Acta **4** [1953] 292/300, 293/4).

[79] Semenov, E. I. (Geokhimiya **1957** 626/37, 633/4).

[80] Shcherbak, N. P.; Alekseeva, K. N.; Gol'denfel'd, I. V.; Eliseeva, G. D. (Tr. Sess. Kom. Opred. Absolyut. Vozrasta Geol. Formatsii Akad. Nauk SSSR **10** 1961 [1962] 85/93, 88/90).

[81] Kuts, V. P. (Mineral. Sb. [Lvov] **20** [1966] 279/84).

[82] El-Hinnawi, E. E. (Beitr. Mineral. Petrogr. **9** [1964] 519/32, 528/9).

[83] Shmakin, B. M.; Shiryaeva, V. A. (Geokhimiya **1970** 1263/8; Geochem. Intern. **7** [1970] 908).

[84] Kukharenko, A. A.; Bulakh, A. G.; Baklanova, K. A. (Zap. Vses. Mineral. Obshch. **90** [1961] 371/81, 374, 376/7).

[85] Pantó, G. (Acta Geol. [Budapest] **19** [1975] 59/93, 80/2).

[86] Ivantishin, M. M. (Aktsesorni Ridkisni Minerali ta Rozsiyani Elementi v Granitakh i Pegmatitakh Ukrains'kogo Kristalichnogo Shchita, Akad. Nauk Ukr. SSR, Kiev 1960, pp. 1/243).

[87] Yalovenko, I. P.; Yur'eva, A. L. (Mineral. Sb. [Lvov] **21** [1967] 299/304).

[88] Kornetova, V. A. (Tr. Mineral. Muzeya Akad. Nauk SSSR No. 14 [1963] 96/107).

[89] Shcherbak, N. P. (Aktsessornye Mineraly Ukr. Shchita **1976** 118/32, 244/58).

[90] Kucha, H. (Mineral. Mag. **43** [1979/80] 1031/4).

[91] Mannucci, G.; Diella, V.; Gramaccioli, C. M. (Can. Mineralogist **24** [1986] 469/74).

[92] Saltykova, V. S. (Tr. Inst. Mineral. Geokhim. Kristallokhim. Redk. Elem. No. 2 [1959] 189/208, 204).

[93] Kato, T. (Mineral. J. [Tokyo] **4** [1958] 224/35).

[94] Kornetova, V. A.; Kazakova, M. E. (Nov. Dannye Mineral. No. 30 [1982] 191/4).

[95] Semenov, E. I. (Typochimizm Mineralov Shchelochnykh Massivov, Nedra, Moscow 1977, pp. 1/119).

[96] Slenzak, O. I. (Charnockity Pridnestrov'ya i Nekotorye Obshchie Voprosy Petrologii, Akad. Nauk Ukr. SSR, Kiev 1960, pp. 1/212, 71).

[97] Ixer, R. A.; Ashworth, J. R.; Pointer, C. M. (Geol. J. **22** (Winter Thematic Issue) [1987] 403/27, 407).

[98] Ueda, T. (Mem. Coll. Sci. Univ. Kyoto B **20** [1953] 227/46, 228).

[99] Jobbins, E. A.; Tresham, A. E.; Young, B. R. (J. Gemmol. **15** [1977] 295/9).

[100] Pavelescu, L.; Pavelescu, M. (TMPM Tschermaks Mineral. Petrogr. Mitt. [3] **17** [1972] 208/14).

[101] Serdyuchenko, D. P.; Kochetkov, O.S. (Dokl. Akad. Nauk SSSR **218** [1974] 1175/7; Dokl. Acad. Sci. USSR Earth Sci. Sect. **218** [1974] 124/5).

[102] Herath, J. W. (Econ. Bull. Sri Lanka Geol. Surv. Dept. No. 2 [1975] 1/72, 25).

[103] Kosterin, A. V.; Alekhina, K. N.; Kizyura, V. E. (Soobshch. Dal'nevost. Filiala Sibirsk. Otd. Akad. Nauk SSSR **1962** No. 15, pp. 23/6).

[104] Yurk, Yu. Yu. (Redkie Mineraly Pegmatitov Priazov'ya, Inst. Geol. Nauk Akad. Nauk USSR, Kiev 1956 from [81], [86, p. 139]).

[105] Silver, L. T.; Woodhead, J. A.; Williams, I. S.; Chappell, B. W. (GJBX-7-84 [1984] 1/412; C.A. **102** [1985] No. 152440).

[106] Cooray, P. G. (Geol. Surv. Ceylon Mem. No. 3 [1965] 1/110 from [107, p. 107]).

[107] Rupasinghe, M. S.; Gocht, W.; Dissanayake, C. B. (J. Natl. Sci. Counc. Sri Lanka **11** [1983] 99/110, 99, 102, 107).

[108] Rupasinghe, M. S.; Senaratne, A.; Dissanayake, C. B. (J. Gemmol. **20** [1986] 177/84, 177, 183/4).

[109] Orsa, V. I. (Dopov. Akad. Nauk Ukr. SSR B **31** [1969] 694/6).

[110] Gauthier-Lafaye, F. (Sci. Geol. Mem. No. 78 [1986] 1/206, 179).

[111] Linares, E.; Toubes, R. O. (Actas Jorn. Geol. Argent. **1** [1960/62] 191/205, 195).

[112] Marchenko, E. Ya. (Dokl. Akad. Nauk SSSR **176** [1967] 1153/5; Dokl. Acad. Sci. USSR Earth Sci. Sect. **176** [1967] 142/5).

[113] Nikitin, Yu. V.; Sal'e, M. E.; Leonova, V. A.; Safronova, G. P.; Shurkin, K. A. (in: Shurkin, K. A.; Gorlov, N. V.; Sal'e, M. E.; Duk, V. L.; Nikitin, Yu. V., Belomorskii Kompleks Severnoi Karelii i Yugo-Zapada Kol'skogo Poluostrova, Akad. Nauk SSSR, Moscow—Leningrad 1962, pp. 223/55).

[114] Kononov, Yu. V.; Nechaev, S. V. (Pitannya Geokhim. Mineral. Petrogr. **1963** 289/301, 291).

[115] Hugo, P. J. (Mem. South Africa Geol. Surv. No. 58 [1970] 1/94, 57).

[116] Leonova, V. A.; Nikitin, Yu. V. (Zap. Vses. Mineral. Obshch. **91** [1962] 136/45, 140/1).

[117] Lozinski, J. (Pr. Mineral. Pol. Akad. Nauk Oddzial Krakowie Kom. Nauk Mineral. No. 16 [1969] 43/9).

[118] Lyakhovich, V. V. (Redkie Elementy v Aktsessornykh Mineralakh Granitoidov, Nedra, Moscow 1973, pp. 1/309).

[119] Charoy, B. (J. Petrol. **27** [1986] 571/604, 576, 587).

[120] Titus, P. O.; Nair, N. R. (Plant Maint. Import Substitution **11** [1978] 86/102, 98).

[121] Hasegawa, S. (Ganseki Kobutsu Kosho Gakkaishi **46** [1961] 57/61).

[122] Coetzee, C. B. (Handb. South Africa Geol. Surv. No. 7 [1976] 199/201).

[123] Kinnaird, J. A. (J. African Earth Sci. **3** [1985] 229/51, 246/7).

[124] Krasil'nikova, A. V. (Geologiya [Alma-Ata] **11** [1976] 104/8).

[125] Petruk, W.; Owens, D. (Can. Mineralogist **13** [1975] 298/9).

[126] Dishchuk, V. L.; Druzhinin, L. N.; Smirnov, G. I.; Stasiv, T. V.; Yur'eva, A. L. (Mineral. Sb. [Lvov] **24** [1970] 431/4).

[127] Shcherbak, M. P. (Dopov. Akad. Nauk Ukr. SSR **1959** No. 2, pp. 188/91).

[128] Shen, J. T. (Proc. Intern. Conf. Peaceful Uses At. Energy, Geneva 1955 [1956], Vol. 6, pp. 147/51).

[129] Chukhrov, F. V. (Tr. Inst. Geol. Rudn. Mestorozhd. Petrogr. Mineral. Geokhim. No. 50 [1960] 3/237).

[130] Rosenblum, S.; Mosier, E. L. (Geol. Surv. Profess. Paper [U.S.] No. 1181 [1983] 1/67).

[131] Burkser, E. S.; Eliseeva, G. D.; Lechekhleb, V. R.; Shcherbak, N. P. (Byull. Kom. Opred. Absolyut. Vozrasta Geol. Formatsii Akad. Nauk SSSR No. 5 [1962] 48/52).

[132] Bulakh, A. G.; Kukharenko, A. A.; Knipovich, Yu. N.; Kondrash'eva, V. V.; Baklanova, K. A.; Baranova, E. N. (Mater. Sess. Uchen. Soveta VSEGEI Rezul'tatam Rabot **1959** [1961] 114/6).

[133] Haapala, I.; Ervamaa, P.; Löfgren, A.; Ojanperä, P. (Bull. Geol. Soc. Finland No. 41 [1969] 117/24, 117, 122).

[134] Shmakin, B. M.; Shiryaeva, V. A. (Zap. Vses. Mineral. Obshch. **100** [1971] 274/81).

[135] Meliksetyan, B. M. (in: Smorchkov, I.E., Aktsessornye Mineraly Izverzhennykh Porod, Nauka, Moscow 1968, pp. 95/108, 95, 103).

[136] Khomyakov, A. P. (Mineralogiya i Geneticheskie Osobennosti Shchelochnykh Massivov **1964** 56/82, 72).

[137] Khamrabaev, I. K.; Urunbaev, K. (in: Smorchkov, I. E., Aktsessornye Mineraly Izverzhennykh Porod, Nauka, Moscow 1968, pp. 141/9, 147).

[138] Yankovskii, V. A. (Sb. Statei Geol. Sibiri **1975** 30/4).

[139] Semenov, E. I.; Kostyunina, L. P.; Kulakov, M. P. (Mineral. Pegmatitov Gidroterm. Shchelochnykh Massivov Akad, Nauk SSSR Inst. Mineral. Geokhim. Kristallokhim. Redk. Elem. **1967** 137/49).

[140] Watkinson, D. H.; Mainwaring, P. R. (Can. Mineralogist **12** [1973] 148).

[141] Hutton, C. O. (Spec. Rept. Calif. Div. Mines Geol. No. 59 [1959] 1/32, 19/20).

[142] Wang, H.-C. (Kexue Tongbao **23** [1978] 743/5).

[143] Marchenko, E. Ya. (Dokl. Akad. Nauk SSSR **155** [1964] 349/52; Dokl. Acad. Sci. USSR Earth Sci. Sect. **155** [1964] 118/22).

[144] Jaffe, H. W. (Bull. Geol. Soc. Am. **66** [1955] 1247/56, 1249/50, 1253).

[145] Ploshko, V. V. (Izv. Akad. Nauk SSSR Ser. Geol. **1961** No. 1, pp. 86/93, 87, 89/90).

[146] Ploshko, V. V. (Urushtenskii Kompleks Severnogo Kavkaza (Geologiya, Petrografiya i Aktsessornaya Mineralizatsiya), Nauka, Moscow 1965, pp. 1/182, 118, 120).

[147] Ploshko, V. V. (Izv. Akad. Nauk SSSR Ser. Geol. **1958** No. 11, pp. 113/7).

[148] Sveshnikova, E. V.; Semenov, E. I.; Khomyakov, A. P. (Zaangarskii Shchelochnoi Massiv, Ego Porody i Mineraly, Nauka, Moscow 1976, pp. 1/80, 48).

[149] Kittl, E.; Tosi, A. (Rev. Fac. Ing. Quim. Univ. Nacl. Litoral Santa Fe Arg. **23** No. 36 [1955] 167/72).

[150] Andersen, T. (Mineral. Mag. **50** [1986] 503/9).

[151] Kupriyanova, I. I.; Volkova, M. I.; Goroshchenko, Z. I. (Tr. Mineral. Muzeya Akad. Nauk SSSR No. 15 [1964] 123/33, 123, 125).

[152] Es'kova, E. M.; Zhabin, A. G.; Mukhitdinov, G. N. (Mineralogiya i Geokhimiya Redkikh Elementov Vishnevykh Gor, Nauka, Moscow 1964, pp. 1/318, 165).

[153] Styles, M. T.; Young, B. R. (Mineral. Mag. **47** [1983] 41/6).

[154] Tikhonenkova, R. P. (Dokl. Akad. Nauk SSSR **266** [1982] 1236/9).

[155] Semenov, E. I.; Khomyakov, A. P. (Redk. Elem. Syr'e Ekon. No. 6 1971 [1972] 77/81).

[156] Nekrasov, I. Ya. (Magmatizm i Rudonosnost' Severo-Zapadnoi Chasti Verkhoyano-Chukotskoi Skladchatoi Oblasti, Akad. Nauk SSSR, Moscow 1962, pp. 1/412, 3, 101/2).

336

[157] Meliksetyan, B. M. (Zap. Arm. Otd. Vses. Mineral. Obshch. No. 2 [1963] 57/80, 69).

[158] Soong, K. L. (written communication 1978 from [130, p. 39]).

[159] Strunz (Mineralogische Tabellen, 5th Ed., Geest & Portig, Leipzig 1970, pp. 1/621, 313).

[160] Ivantishin, M. N. (Metallog. Dokembr. Shchitov Drevnikh Podvizhnykh Zon Dokl. 2nd Vses. Ob'edin. Sess., Kiev 1960, Pt. 1, pp. 116/31, 119/21).

[161] Ghouse, K. M. (Indian J. Pure Appl. Phys. **6** [1968] 265/8).

[162] Skrzat, Z. (Kwart. Geol. **4** [1960] 874/88, 881).

[163] Rowley, E. B. (Rocks Minerals **35** [1960] 328/30).

[164] Betekhtin, A. G. (Geol. Rudn. Mestorozhd. No. 1 [1959] 5/24, 9).

[165] Bowie, S. H. U.; Horne, J. E. T. (Mineral. Mag. **30** [1953/55] 93/9).

[166] Mannucci, G.; Diella, V.; Gramaccioli, C. M. (Papers Proc. General Meeting Intern. Mineral. Assoc. **14** [1986] 166).

[167] Smorchkov, I. E.; Tseitlin, S. G.; Batyreva, N. N. (Tr. Inst. Geol. Rudn. Mestorozhd. Petrogr. Mineral. Geokhim. No. 99 [1963] 60/7).

[168] Banas, M.; Kucha, H. (Proc. 6th Quadrenn. IAGOD Symp., Tbilisi 1982 [1984], Vol. 1, pp. 241/6).

[169] Overstreet, W. C. (Geol. Surv. Profess. Paper [U.S.] No. 400-B [1960] B55/B57).

[170] Lyakhovich, T. T.; Lyakhovich, V. V. (Geokhimiya **1983** 1616/33; Geochem. Intern. **20** No. 6 [1983] 91/108, 92, 98).

[171] Povilaitis, M. M.; Varshal, G. M. (Tipomorfizm Miner. **1969** 196/209, 202/3, 206/8).

[172] Meliksetyan, B. M. (Izv. Akad. Nauk Arm. SSR Geol. Geogr. Nauki **14** No. 2 [1961] 21/41, 31).

[173] Mineev, D. A. (Geokhimiya **1960** 131/8; Geochemistry [USSR] **1960** 156/66, 159/60).

[174] Murata, K. J.; Dutra, C. V.; Teixeira Da Costa, M.; Branco, J. J. R. (Geochim. Cosmochim. Acta **16** [1959] 1/14, 5, 7, 10/1).

[175] Heinrich, E. W.; Borup, R. A.; Levinson, A. A. (Geochim. Cosmochim. Acta **19** [1960] 222/31, 224/8).

[176] Kalita, A. P. (in: Kuz'menko, M. V., Osobennosti Raspredeleniya Redkikh Elementov Pegmatitakh, Nauka, Moscow 1969, pp. 79/100, 82/5, 91/2, 95).

[177] Austin, S. R.; Hetland, D. L.; Sharp, B. J. (Miner. Resour. Rept. Idaho Bur. Mines Geol. No. 11 [1970] 1/10, 4).

[178] Komov, I. L.; Mel'nikova, E. M.; Kokarev, G. N. (Tr. Mineral. Muzeya Akad. Nauk SSSR No. 23 [1974] 87/93, 91/2).

[179] Krol, O. F. (Geol. Geokhim. Mineral. Mestorozhd. Redk. Elem. Kaz. **1966** 60/5).

[180] Lombard, J. (Chron. Mines Coloniales **23** [1955] 310/6).

[181] McKie, D. (Records Geol. Surv. Tanganyika **6** 1956 [1958] 87/93).

[182] Overstreet, W. C.; White, A. M.; Whitlow, J. W.; Theobald, P. K., Jr.; Caldwell, D. W.; Cuppels, N. P. (Geol. Surv. Profess. Paper [U.S.] No. 568 [1968] 1/85, 25/6).

[183] Lyakhovich, V. V. (Mineral. Sb. [Lvov] **20** [1966] 199/208, 199, 205).

[184] Lyakhovich, V. V. (Litol. Polez. Iskop. **1984** No. 4, pp. 107/13; Lithol. Mineral. Resour. [USSR] **1984** No. 4, pp. 373/9).

[185] L'vov, B. K.; Zhangurov, A. A. (in: Smorchkov, I. E., Aktsessornye Mineraly Izverzhennykh Porod, Nauka, Moscow 1968, pp. 196/204, 199/201).

[186] Azimov, P. T. (Dokl. Akad. Nauk Uzbek. SSR **27** No. 2 [1970] 45/7).

[187] Kornetova, V. A.; Varshal, G. M. (Khim. Anal. Mineral. Ikh Khim. Sostav **1964** 83/9).

[188] Shcherbak, M. P.; Lavrov, D. A. (Pitannya Geokhim. Mineral. Petrogr. **1963** 223/35, 226/8).

[189] Overstreet, W. C.; Yates, R. G.; Griffitts, W. R. (Geol. Surv. Bull. [U.S.] No. 1162-F [1963] F1/F31, F13/F16).

[190] Berger, K.; Salger, M. (Geol. Bavarica No. 53 [1964] 174/93, 192).

2.4.1.1.3 Crystal Form and Crystal Structure

Crystal Form

Monazite is monoclinic prismatic, 2/m [1]; see also [2 to 5].

A compilation of observed **crystal forms** of monazite and their symbols is given in [1]; see also [6, 7]. Major crystal forms are: a{100}, v{$\bar{1}$11}, m{110}, x{$\bar{1}$01}, e{011}, w{101}, b{010}, u{021}, z{311}, and r{111} [1]; or a{100}, b{010}, m{110}, w{101}, x{$\bar{1}$01}, and v{$\bar{1}$11} [6]. Simple crystal faces predominate (as, e.g., pinacoids of the 1st and 2nd order and rhombohedral prisms); see, for example, [8 to 15], [16, pp. 141/2].

The crystal forms (and thus the habit, see below) of monazite are highly variable in different parageneses [5]. Simple crystal forms are common for monazite samples from vugs and low-temperature veins, whereas complex forms tend to be common for monazite samples from pegmatites. Monazites from plutonic gneisses and schists generally lack crystal forms and are globular [17]. Mostly simple forms predominate, as, for example, the pinacoids a, b, and w and the prisms m and e in monazite from syenite and amazonite pegmatites of the Ilmen Mountains, Southern Urals [18]. Or in monazite from the southeastern Ukraine, the pinacoids a and w and the prism e in samples from Chudnov-Berdichev granite; the prism and pinacoid of the 1st order and, less abundantly, the pinacoid of the 2nd order in samples from rose-colored pegmatitic granite and pegmatite of the 2nd intrusive phase of the Bug-Podolian group; and the pinacoids a and w and, less abundantly, the prism m in samples from Kirovograd and Zhitomir granites [19].

Habit. Most abundantly monazite occurs as subhedral rounded grains [17], sometimes as anhedral pods. Detrital grains of monazite are {100} tablets with rounded corners, variously rounded (001)-cleavage pieces, strongly rounded ovoid grains, or very slightly abraded euhedra of varying complexity. Monazite may also occur as radial fan-shaped masses [20, p. 75]. Idiomorphic crystals of monazite [6], [20, p. 75] or grains are mostly small-sized [1, 2, 4, 6, 7], rarely exceeding 0.02 in across [17]. Exceptions are large coarse crystals in pegmatites [6].

Common habits of idiomorphic monazite (see also **Fig. 12**a to c, p. 338) are listed in the standard reference literature as follows: tabular on {100} [1, 2, 6], [20, p. 75]; tabular elongated parallel to the b axis [1, 21], [20, p. 75]; flattened on {100} and somewhat elongated parallel to the c or a axis [6]; wedge-shaped by large development of {100} and {$\bar{1}$11} [1, 6], [20, p. 75]; equant [1, 6]; and prismatic [2] resulting from elongation of {111} [20, p. 75] or {$\bar{1}$11} [1]. More rarely, monazite is isometric or pyramidal [2] or platy on {100} [21]. An unusual plate-like development parallel to {$\bar{1}$01} is observed for Th-containing monazite from hydrothermal veins of Alinci, Yugoslavia [16, p. 141]. The habit of monazite may depend on the kind of host rock; see, for example, [17, 22 to 25]. For the habits of synthetic Th-containing $CePO_4$, see pp. 338/9.

Depending on the grade of development of the individual crystal forms, the following percentages of the morphological type of idiomorphic monazite crystals are distinguished for samples from alluvial sediments of the interfluvial region of Dnepr and southern Bug rivers, Ukraine: (1) 95% of the investigated crystals are tabular, strongly flattened on a and slightly elongated along the b axis with a predominance of the pinacoids a and w and the prism m (whereas x, r, and sometimes {252} are developed much less perfectly); (2) 4% are isometric with a predominance of a, m, and w, good development of v, and worse development of x; and (3) 1% are prismatic, elongated along the c axis with forms corresponding to those of the

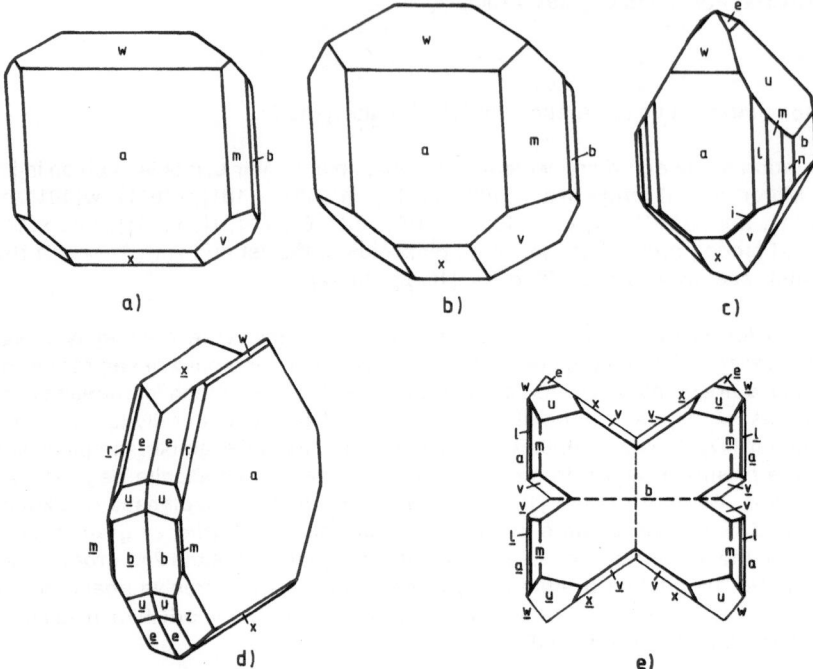

Fig. 12. Habits of monazite crystals from [1].

a and b = typical habits composed of the forms a{100}, b{010}, m{110}, w{101}, x{$\overline{1}$01}, v{$\overline{1}$11}; habit a is from the Ilmen Mountains, Southern Urals.

c = wedge-shaped crystal from New York composed of the forms a{100}, b{010}, l{210}, m{110}, n{120}, w{101}, e{011}, u{021}, i{$\overline{2}$11}, x{$\overline{1}$01}, v{$\overline{1}$11}.

d = twin on (100) from Tavetsch, Switzerland, composed of the forms a{100}, b{010}, e{011}, u{021}, r{111}, w{101}, m{110}, z{311}, x{$\overline{1}$01}.

e = cruciform twin on (100) from Binnental, Switzerland, composed of the forms a{100}, b{010}, e{011}, u{021}, w{101}, x{$\overline{1}$01}, v{$\overline{1}$11}, l{210}, m{110}.

type (2) and sometimes an additional development of b [14]. Note that ∼40% of monazite occurs as well developed, mostly tabular crystals flattened on a in the heavy-mineral fraction of clastic sediments of the above interfluvial region; the rest forms rounded slightly flattened grains without clear crystal forms [26].

The introduction of Th and Ca into the structure of synthetic $CePO_4$ (with a dominant habit controlled by a large development of x; for the symbols, see p. 337) probably induces a crystalhabit change by the development of the formerly unreported crystal forms j{313} (as important terminal form) and y{$\overline{4}$12}*), of the rare forms {212} and {041}, and of twins (see p. 339). In the resulting synthetic Th-containing monazite crystals (that are similar with natural monazites elongated by the development of v), three habits predominate: (1) crystals twinned on c and elongated on e having a square-appearing cross section; (2) thin rhombic-shaped

*) Note that [1] gives {310} for the symbol y.

crystals, tabular on x, with small bounding faces of e or m, or both; and (3) a thick but somewhat tabular habit controlled by extension of x with moderate development of e and m and minor development of b, r, {130}, {120}, {210}, and {041}. The portions of these three types vary with the amount of ThO_2 present (2.40 to 13.80%). In ThO_2-poor crystals with an initial ratio of ~2:1 for the habits (1) and (2), the portion of habit (1) increases as the ThO_2 content increases. In crystals containing slightly more than 5% ThO_2 (or 8%?, see table 2 in the original paper), habit (3) appears and is more abundant than (1) and (2) together; habits (1) and (2) abruptly attain about equal portions. In samples with the highest ThO_2 content, twinned crystals of habit (1) are rare. However, in natural monazite, probably such morphological changes would be masked by changes due to variations in the relative contents of RE [27].

The **crystal faces** of monazite often are rough, uneven, or striated [1, 6]; dull because of superficial alteration, massive granular [6]; wrinkled, fluted, or irregular [21]; and often show hatching [2]. Faces on larger crystals may be rough to irregular [20, p. 75].

Twinning of monazite is common on (100) [1 to 3, 5, 6, 21] (see Fig. 12d), and sometimes cruciform [1] (Fig. 12e). Lamellar twinning on (001) rarely occurs [1, 3]. Polysynthetic twinning is observed in monazites from granites of the southeastern Ukraine [28] (with a band width of 0.1 to 0.2 mm [29]), and from muscovite-quartz and albite-quartz veins of the European USSR [30]. Twins on (001) are caused by the introduction of Th and Ca into synthetic $CePO_4$ [27].

Cleavage of monazite is perfect parallel to (001), distinct parallel to (100) [5], [20, p. 79], and poor parallel to (010) [20, p. 79]; or is moderate parallel to (100) and variable parallel to (001) [3]. It also may be distinct parallel to (100), less distinct parallel to (010), and indistinct parallel to (110), (101), and (011) [1, 17]. The cleavage parallel to (100) is explained by the arrangement of the PO_4 groups in the monazite structure (see pp. 342/4) [31, p. 245].

Degree and kind of cleavage of monazite vary, apparently owing to alteration [1]. Thus, unaltered specimens of monazite of Chester, Morris county, north-central New Jersey, have no cleavage, whereas with increasing alteration cleavage appears at first in one, then in two, and finally in three directions [32, p. 511]. Note also that a good cleavage is observed in strongly altered monazite from granites of the Rusava suite, Ukraine, USSR [33].

Parting of monazite is marked parallel to (001) [1, 17] (which may be due to lamellar twinning [6], see above) and is rare parallel to ($\bar{1}$11) [17].

Crystal Structure

The **space group** of monazite is $P2_1/n$ [1, 3, 34] or C^5_{2h} (No. 14) [31, p. 229], [4, 5, 7, 35, 36]. $Z = 3.96$ [31, p. 229] or, rounded off, $Z = 4$ [3 to 5, 34].

The **lattice constants** of monazite are given in standard textbooks as a = 6.79, b = 7.04, c = 6.47 Å, β = 104°24′ [4, 5] or as a $\cong$ 6.78, b $\cong$ 7.00, c $\cong$ 6.45 Å, β $\cong$ 104° [3] with an axial ratio a:b:c of 0.970:1:0.923 [1] or 0.965:1:0.919 [4]. In basic studies of the crystal structure of Th-bearing monazite are given: a = 6.77 ± 0.01, b = 6.99 ± 0.01, c = 6.45 ± 0.01 Å, β = 103°38′ ± 12′, V = 297 Å³ [31, pp. 228/9]; a = 6.66, b = 6.98, c = 6.32 Å, β = 103°22′ [35]; or a = 6.77, b = 7.04, c = 6.46 Å [34] with β = 104°0′ [37]. The axial ratio is given as 0.969:1:0.923 [31, p. 229]. For a compilation of examples of lattice constants that can be assigned to a definite Th-bearing monazite sample, see Table 15, p. 340. For lattice constants that cannot be assigned to an individual Th-containing monazite sample, refer, for example, to [16, p. 144]. For the lattice constants of pure monoclinic La, Ce, Pr, or Nd phosphate, see, for example, [36].

Table 15

Lattice Constants of Monazite from Various Localities.

Samples are arranged by decreasing ThO$_2$ content; analysis numbers refer to those in Table 14, pp. 305/28, where a complete sample description is given.

monazite from	analysis No.	% ThO$_2$	a in Å	b in Å	c in Å	β	V in Å^3	Ref.
pegmatite, Italy	2	19.5	6.771(1)	6.982(1)	6.471(1)	103.81(1)°	297.07	[41]
pegmatite, Italy	3	18.6	6.782(1)	6.997(1)	6.485(1)	103.85(1)°	298.79	[41]
gem gravel, Sri Lanka	15	11.6	6.761	6.966	6.478	103°35'	–	[42]
pegmatite, Italy	16	11.34	6.78	6.96	6.48	103°54'	296.83	[43]
Thailand	–	8.50	6.813	7.030	6.503	103°47'	302.6	[38, p. 233]
Liberia	–	8.4	6.781	7.012	6.517	–	301.4	[44]
pegmatite, Italy	81	7.7	6.788(3)	7.009(2)	6.464(1)	103.57(1)°	298.95	[41]
India	–	6.54	6.820	7.039	6.485	104°00'	302.5	[38, p. 233]
carbonatite, USSR	125	5.77	6.73	6.93	6.45	103°35'	292	[45]
pegmatitic vein, Finland	127	5.7	6.778	7.003	6.479	103°34'	298.95	[46, pp. 117, 122]
pegmatite, Italy	131	5.6	6.776(1)	6.992(2)	6.475(1)	103.79(1)°	297.73	[41]
quartz vein, Japan	158	4.51	6.820	7.003	6.480	103°39'	299.6	[38, pp. 225/6, 233]
beach placer, Canada	–	1.5	6.76	6.96	6.46	103°50'10"	–	[47]
beach placer, Taiwan	204	1.3	6.80	7.02	6.46	–	305.3	[48, p. 53]
alluvial placer, USSR	209	1.14	6.87	6.91	6.51	104°36'	–	[49]

Table 16

Interplanar Spacings (d values) with Corresponding Relative Intensities in Parentheses (vs = very strong, s = strong) for Only the Strongest Observed X-Ray Powder Diffraction Reflections of Monazite from Various Localities.

Analysis number refers to that in Table 14, pp. 305/28, where a complete sample description is given. Note that no indices are given in the original paper for the d values of analyses number 24, 45, 57, 88, 155, 180, 185, and 191; n = total number of observed d values given in the original paper.

analysis No.	% ThO$_2$	d value (in Å) for plane indexed as			n	Ref.
		(200)	(120)	(012), (112)		
24 [a]	10.22	3.311(5)	3.115(10)	2.885(9)	25	[56]
45	8.84	3.27(s)	3.12(vs)	2.86(s)	21	[57]
57 [b]	8.48	3.27(7)	3.08(10)	2.85(9)	41	[11]
88 [c]	7.37	3.31(6)	3.11(10)	2.876(9)	30	[11]
125 [d]	5.77	3.260(70)	3.070(90)	2.860(100) [l]	24	[45]
127	5.7	3.296(80)	3.096(100)	2.876(50) [m]	25	[46, pp. 121/2]
155 [e]	4.61	3.300(7)	3.100(10)	2.860(4)	33	[58]
158	4.51	3.30(60)	3.09(100)	2.877(45) [l]	21	[38, pp. 226, 231]
180 [f]	3.32	3.290(8)	3.080(10)	2.860(3)	36	[58]
185	3.07	— [i]	3.10(10) [k]	2.86(8) [k]	19	[59]
191 [g]	2.39	3.27(7)	3.08(10)	2.87(7)	28	[60]
199 [h]	1.68	3.305(6)	3.109(10)	2.877(10)	23	[30]
204	1.3	3.30(140) [j]	3.10(100)	2.87(29) [l]	12	[48, pp. 51, 53]
209	1.14	3.31(9)	3.09(10)	2.86(10)	34	[49]

[a] Additional strong d values are 2.197(7), 2.152(8), 2.130(8), 1.968(6), 1.866(7), 1.739(6), and 1.694(7). — [b] Additional strong d values are 1.872(8), 1.236(8), and 1.049(8). — [c] Additional strong d values are 2.14(8), 1.976(7), 1.877(8), 1.283(7), 1.239(8), and 1.104(7). — [d] Additional strong d values are 4.13(80) for ($\bar{1}$11) and 2.128(80) for ($\bar{3}$11). — [e] Additional strong d values are 2.190(6), 2.140(6), 1.868(6), and 1.689(6).

[f] Additional strong d values are 2.180(7), 2.130(6), 1.964(4), 1.864(6), 1.686(5), 1.603(4), 1.339(4), 1.328(4), 1.232(4), and 1.097(6). — [g] Additional strong d values are 1.867(8) and 1.743(8). —

[h] Additional strong d values are 1.880(9) and 1.743(8). — [i] Only a weak reflection with a d value of 3.29 kX(1) is observed. — [j] The enhanced reflection is caused by a zircon impurity. —

[k] d values are given in kX. — [l] Indexed as (012). — [m] Indexed as (012), ($\bar{1}$12).

No systematic relationship is found between lattice constants and content of ThO$_2$ in monazite of the Gifu prefecture, Japan [38, p. 230].

Heat treatment does not change the space group and unit-cell dimensions of monazite from India [37].

The complex structure (see pp. 342/4) and the variety of substitutions possible among the RE and Th components of monazite (see pp. 291/3) create an intricate **X-ray diffraction pattern** with a few intense and many small and overlapping reflections [32, p. 517]. Alteration (not metamictization) may be responsible for diffuse diffraction patterns observed in some monazite samples [39]. The interplanar spacings (d values) of the three strongest X-ray powder diffraction reflections (with relative intensities of 100, 70, and 70) vary from 3.28 to 3.30, 3.07 to 3.09, and 2.87 to 2.88 Å, respectively, in ten Th-containing monazites from various localities

[32, pp. 511, 514/5, 519/20, 524]. Refer also to Table 16, p. 341, where d values are listed for monazites with assignable ThO_2 content. For d values of Th-containing monazite that cannot be assigned to individual samples, see, for example, [16, p. 144], [40].

Compared with the quite diffuse X-ray powder diffraction pattern obtained for unheated material, the pattern of Th-containing monazite is sharp for samples from Western Australia and Zimbabwe heated at 1100°C [50] and for a sample from black sand, Egypt, heated at ~1000°C for ~8 h (with d values closely agreeing with those given by [32, pp. 514/5, 519/20]; see above) [51]. The indexed d values (with corresponding relative intensities in parentheses) of the strongest reflections of monazite (from India?) heated at 1130°C for 24 h are 3.28 Å (strong) for (200); 3.09 Å (very strong) for (120); 2.86 Å (very strong) for (012), (112); 2.14 Å (strong) for (103), (311); 1.97 Å (strong) for ($\bar{2}$12); 1.87 Å (strong) for (032), ($\bar{1}$03); and 1.75 Å (strong) for (040) [52]. Guinier and Weissenberg diffraction patterns are sharper for Indian monazite heated at 1130°C for 48 h as compared with unheated material from the same source [37]. After heating at 850°C for 4 h, reflections of churchite or rhabdophane are visible in the X-ray powder diffraction pattern of silicomonazite from an alluvial placer, Yakutia, USSR [49].

Variations in content of RE, and particularly of Th, give no recognizable features on X-ray powder diffraction patterns of monazite, as is shown by a comparison of chemical and X-ray data of samples described in the literature [16, pp. 143/5]. A decrease in Th content from about 7% in 11 yellow monazites (from the USSR, USA, R.S.A., and Liberia) to <1% in 18 dark monazites (from Alaska, Montana, Taiwan, Eastern Siberia, and Zaire), which contain ~0.2 to 0.7% Eu, does not discernibly change the d values (for the mean d values of the above samples, see original paper) [44]. No difference in d values is found between Th-bearing yellow-green and Th-poor black monazites from Thailand and Th-free monazite from Arkansas [53]. And only slight differences in the d values are observed between Th-rich yellow monazites from Thailand and Australia (no ThO_2 contents given) and Th-poor black monazite (containing 1.3% ThO_2) from Thailand [48, p. 51]. A variable ThO_2 content does not affect the interplanar spacings of monazites from Malaya (whereas they increase with increasing La_2O_3 content and decrease with increasing contents of U_3O_8, Y_2O_3, Gd_2O_3, and, to a lesser extent, of Sm_2O_3; for details, see original paper) [54]. Shifts in the position of various X-ray diffraction peaks and differences in their relative height indicate differences in chemical composition of monazite from Th-bearing quartz veins, Lemhi Pass quadrangle, Idaho-Montana [55].

Crystal Structure. Monazite is isostructural with huttonite (see "Thorium" Suppl. Vol. A1b) [6, 61], cheralite (see pp. 364/5), synthetic $CaTh(PO_4)_2$ [6], and brabantite (see p. 369) [61]; see also p. 243. Silicomonazite of Yakutia, USSR, is isostructural with cheralite of Travancore, India [49]. Monazite has a dense and tightly packed structure without large vacant interstitial voids [62].

Descriptions of the monazite structure in the literature are rather inconsistent, especially as far as the coordination number of Ce is concerned as is shown in the following. Based on a two-dimensional Fourier synthesis on the planes (010) and (001), it is shown that the P atom is situated at the center of a more or less deformed tetrahedron of four oxygens with O-P distances of 1.66 and 1.62 Å for two edges each (mean 1.64 Å) and O-O distances of 2.96, 2.77, 2.68, 2.62, 2.57, and 2.43, mean 2.67 Å, respectively. The distance between neighboring oxygens belonging to different tetrahedra is $\leq$ 2.47 Å. Each Ce atom is surrounded by nine oxygens that form an irregular polyhedron. The Ce-O distances vary between 2.43 and 2.63, mean 2.53 Å. One oxygen is shared by one P atom and two Ce atoms. The strength of electrostatic bond is 5/4 between O and P atoms and 3/9 between O and Ce atoms. The sum

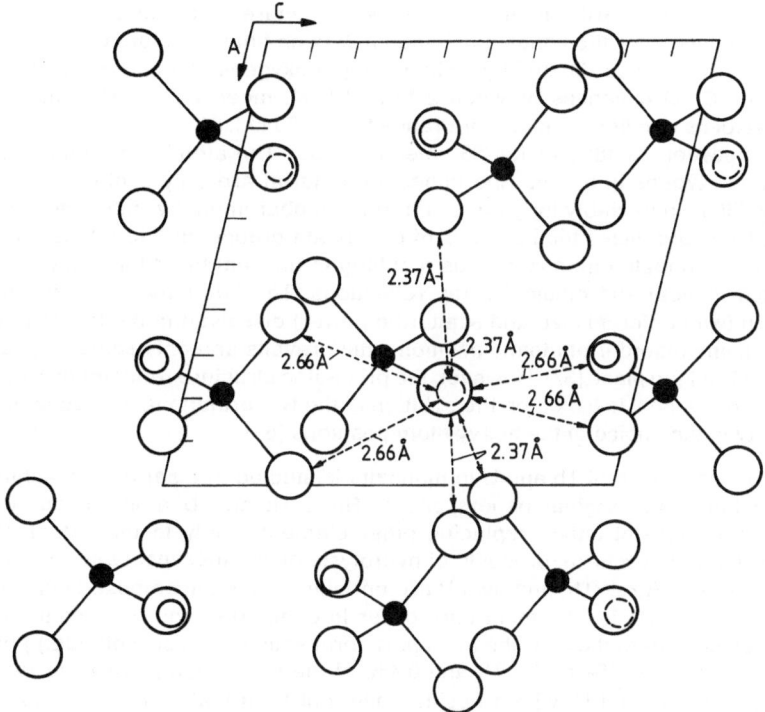

Fig. 13. Crystal structure of synthetic $CePO_4$ projected along the b axis from [36]. Included are only interatomic distances; for atomic parameters, see original paper. Black circles are P, small and big open circles are Ce and O, respectively.

of bond strengths is 23/12, which is nearly equal to 2 (i.e., the charge of an O atom) and thus satisfies the Pauling rule [31, p. 245]. The structure of synthetic $CePO_4$ (considered as an analog of naturally occurring monazite) is determined by difference Fourier synthesis and final least-squares refinement (with final R and R_w values of 0.0276 and 0.0280, respectively) as follows: Nine-coordinated Ce atoms are linked together by distorted phosphate tetrahedra with P-O distances varying between 1.518(7) and 1.534(6), mean 1.527(6) Å and O-P-O angles varying between 104.2(4)° and 113.5(4)°, mean 109.5°. The distortion of the phosphate tetrahedron may be due largely to the bonding of one oxygen (O_2) to three Ce atoms (with Ce-O distances of 2.640(6), 2.560(6), and 2.779(6) Å), whereas each of the other three oxygens (O_1, O_3, and O_4) is bonded to only two Ce atoms (with Ce-O distances of 2.517(6) and 2.445(6) Å for O_1, 2.468(6) and 2.573(6) Å for O_3, and 2.526(6) and 2.448(6) Å for O_4). The distances between neighboring oxygens of different tetrahedra vary between 2.403(9) and 2.556(9), mean 2.493(9) Å [63]. As a three-dimensional space-filling network, in which the cation has an irregular coordination of eight oxygens, the monazite structure is described for both synthetic $CePO_4$ (see **Fig. 13**) and for the natural mineral $(Ce, La, Dy)PO_4$. From each cation, four short bonds (with lengths of 2.37 Å in $CePO_4$ and of 2.46 Å in $(Ce, La, Dy)PO_4$) extend to the corners of four different phosphate groups. Four longer bonds of 2.66 Å connect the edges of two additional phosphate tetrahedra [36]. The structure of monazite heat-treated at 1130°C for 48 h, just like that of unheated monazite, consists of Ce^{3+} ions and discrete distorted PO_4^{3-} tetrahedra with P-O distances between 1.50 and 1.69, mean 1.60 Å, as is corroborated for monazite of India (preliminarily studied by [52]) by a structure refinement achieved by difference Fourier syntheses for the zones [001] and [100] (with final R factors of 10.0 and 10.2%, respectively). Cerium is coordinated by a shell of eight oxygens at the corners of a distorted polyhedron with Ce-O distances between 2.22 and 2.94, mean 2.59 Å. Distances between neighboring oxygens of different tetrahedra range from 2.35 to 2.81, mean 2.59 Å, with valency angles O-P-O between 93°18′ and 134°36′, mean 108°56′. The atomic coordinates (see also [34]) of the four oxygens in heated and unheated material are only slightly different for x, whereas they differ appreciably for y and z. Internal bombardment by radioactive α particles emitted from Th disorganizes the positions of phosphate groups and, to a lesser extent, the vibration of Ce in unheated monazite, thus resulting in unusual bond lengths (as compared with calculated values) and diffused X-ray reflections. The latter may also be due to the presence of Th (with different size and scattering power) dissolved in the lattice of Ce. Even after heating, some disorder of atomic positions still prevails and only some rearrangement of the structure is taking place [37]. However, the preceding structure refinement is considered to be incorrect because it is based on the solution of the two orthogonal projections resulting in an ambiguity in the choice of the heavy-atom positions [64].

The form of occurrence of Th and U in monazite is studied in a series of leaching experiments (for details, see original papers cited). Thus, Th and U atoms occur mainly as components of the crystal lattice replacing other elements; only to the extent of several percent both elements occur as products of hydrolysis or in other insoluble forms in lattice defects [65]. Whereas Ra, ^{228}Th, and also U are enriched in the decomposed superficial part of the lattice, RE and Th, for the most part, occur in capillaries and on vacant sites of the undestroyed lattice, where they, for the most part, form solid (probably colloidal) phases and are only subordinately (1.8% of the Th and 0.4% of the RE content) available for leaching solutions [66]. But note that U, with a position different from that of Th, does not enter the crystal lattice but is located on interstitial sites [67]. Uranium is present in the tetravalent state in monazite from a pegmatite of Piona, Italy (as is evident from the absorption spectrum) [43], as well as in various monazites from fissures and pegmatites of the Alps, Italy-Switzerland-Austria. From stoichiometric balance and from crystal-chemical considerations it is difficult to imagine how any U^{6+} could enter the monazite structure (owing to the strong difference in ionic radii and to the tendency to form UO_2^{2+} ions, which would result in a considerable deformation of the oxygen framework) [14].

As is shown by transmission and high-resolution electron microscopy of crushed monazite fragments and by transmission electron microscopy of ion-thinned foils of monazite, the microstructure of yellow, gray, and brown specimens from a paragneiss, Antarctica, is formed by tiny crystalline domains (with an approximate diameter of 100 Å) that show a uniform periodic lattice image and are ubiquitously misorientated from each other by no more than 2×10^{-3} radians. These domains are separated from each other by narrow confused boundary regions, where misorientation is probably accomodated by imperfect atomic arrangements. These narrow regions of misorientation (with a thickness of 3 Å and thus occupying 9% of the total volume of the mineral) form a continuous three-dimensional network throughout each monazite grain and represent potential zones of high permeability or diffusivity which are believed to be fundamental to the isotopic and color differences between the monazite grains (see pp. 357 and 348/9, respectively). Regions of aperiodic structure image (which would correspond to disordered, damaged or amorphous material) and regular defect structures (e.g., dislocations) are absent. It is supposed that this domain structure, which is equally developed and of indistinguishable size in all studied samples, is controlled by self-annealing of monazite, or that it is the final result of a slight but progressive lattice damage caused by radioactive decay that was activated by an input of thermal energy associated with the polyphase-metamorphic evolution of the hosting gneiss [68].

References for 2.4.1.1.3:

[1] Palache, C.; Berman, H.; Frondel, C. (Dana's System of Mineralogy, 7th Ed., Vol. 2, Wiley, New York 1963, pp. 1/1124, 691/6).

[2] Betechtin, A. G. (Lehrbuch der Mineralogie, Verlag Technik, Berlin 1953, pp. 1/681, 409/10 [Russian original, Moscow 1951]).

[3] Deer, W. A.; Howie, R. A.; Zussman, J. (Rock-Forming Minerals, Vol. 5, Non-Silicates, Longmans, London 1962, pp. 1/371, 339).

[4] Semenov, E. I. (Mineralogiya Redkikh Zemel', Akad. Nauk SSSR, Moscow 1963, pp. 1/412, 115).

[5] Ramdohr, P.; Strunz, H. (Klockmanns Lehrbuch der Mineralogie, 16th Ed., Enke, Stuttgart 1978, pp. 1/876, 625).

[6] Frondel, C. (Geol. Surv. Bull. [U.S.] No. 1064 [1958] 1/400, 154/6).

[7] Perez Mateos, J.; Galvan Garcia, J. (Anales Edafol. Agrobiol. **26** [1967] 1227/44, 1229/30).

[8] Kittl, E.; Tosi, A. (Rev. Fac. Ing. Quim. Univ. Nacl. Litoral Santa Fe Arg. **23** No. 36 [1955] 167/72).

[9] Leonova, V. A.; Nikitin, Yu. V. (Zap. Vses. Mineral. Obshch. **91** [1962] 136/45, 137/8).

[10] Nikitin, Yu. V.; Sal'e, M. E.; Leonova, V. A.; Safronova, G. P.; Shurkin, K. A. (in: Shurkin, K. A.; Gorlov, N. V.; Sal'e, M. E.; Duk, V. L.; Nikitin, Yu. V., Belomorskii Kompleks Severnoi Karelii i Yugo-Zapada Kol'skogo Poluostrova, Akad. Nauk SSSR, Moscow−Leningrad 1962, pp. 223/55, 243/4).

[11] Kornetova, V. A. (Tr. Mineral. Muzeya Akad. Nauk SSSR No. 14 [1963] 96/107, 99/100).

[12] Kuts, V. P. (Mineral. Sb. [Lvov] **20** [1966] 279/84).

[13] Yalovenko, I. P.; Yur'eva, A. L. (Mineral. Sb. [Lvov] **21** [1967] 299/304).

[14] Dishchuk, V. L.; Druzhinin, L. N.; Smirnov, G. I.; Stasiv, T. V.; Yur'eva, A. L. (Mineral. Sb. [Lvov] **24** [1970] 431/4).

[15] Krasil'nikova, A. V. (Geologiya [Alma-Ata] **11** [1976] 104/8).

[16] Bermanec, V.; Tibljas, D.; Gessner, M.; Kniewald, G. (Mineral. Petrol. **38** [1988] 139/50).

[17] Overstreet, W. C. (Geol. Surv. Profess. Paper [U.S.] No. 530 [1967] 1/327, 4/5).

[18] Makarochkin, B. A. (Geol. Geofiz. **1975** No. 2, pp. 155/9).

[19] Ivantishin, M. M. (Aktsesorni Ridkisni Minerali ta Rozsiyani Elementi v Granitakh i Pegmatitakh Ukrains'kogo Kristalichnogo Shchita, Akad. Nauk Ukr. SSR, Kiev 1960, pp. 1/243, 134, 136/7, 141).

[20] Heinrich, E. W. (Mineralogy and Geology of Radioactive Raw Materials, McGraw-Hill, New York 1958, pp. 1/654).

[21] Roubault. M. (Geologie de l'Uranium, Masson, Paris 1958, pp. 1/462, 411).

[22] Marchenko, E. Ya. (Dokl. Akad. Nauk SSSR **176** [1967] 1153/5; Dokl. Acad. Sci. USSR Earth Sci. Sect. **176** [1967] 142/5).

[23] L'vov, B. K.; Zhangurov, A. A. (in: Smorchkov, I. E., Aktsessornye Mineraly Izverzhennykh Porod, Nauka, Moscow 1968, pp. 196/204, 196/7, 199).

[24] Pavelescu, L.; Pavelescu, M. (TMPM Tschermaks Mineral. Petrogr. Mitt. [3] **17** [1972] 208/14).

[25] Bel'kov, I. V. (Aktsessornye Mineraly Granitoidov Kol'skogo Poluostrova, Nauka, Leningrad 1979, pp. 1/184, 127, 131).

[26] Bobrievich, A. P.; Druzhinin, L. N.; Smirnov, G. I.; Yur'eva, A. L. (Mater. Mineral. Petrogr. Geokhim. Osad. Porod Rud No. 2 [1974] 88/99, 95).

[27] Anthony, J. W. (Am. Mineralogist **50** [1965] 1421/31, 1423/30).

[28] Shcherbak, N. P. (Aktsessornye Mineraly Ukr. Shchita **1976** 118/32, 244/58, 128).

[29] Shcherbak, N. P.; Alekseeva, K. N.; Gol'denfel'd, I. V.; Eliseeva, G. D. (Tr. Sess. Kom. Opred. Absolyut. Vozrasta Geol. Formatsii Akad. Nauk SSSR **10** 1961 [1962] 85/93, 87).

[30] Kupriyanova, I. I.; Volkova, M. I.; Goroshchenko, Z. I. (Tr. Mineral. Muzeya Akad. Nauk SSSR No. 15 [1964] 123/33, 123/5).

[31] Ueda, T. (Mem. Coll. Sci. Univ. Kyoto B **20** [1953] 227/46).

[32] Molloy, M. W. (Am. Mineralogist **44** [1959] 510/32).

[33] Slenzak, O. I. (Charnockity Pridnestrov'ya i Nekotorye Obshchie Voprosy Petrologii, Akad. Nauk Ukr. SSR, Kiev 1960, pp. 1/212, 67).

[34] Ghouse, K. M. (Naturwissenschaften **52** [1965] 32/3).

[35] Shankar, J.; Khubchandani, P. G. (Proc. Indian Acad. Sci. A **44** [1956] 130/3).

[36] Mooney-Slater, R. C. L. (Z. Krist. **117** [1962] 371/85, 377/9).

[37] Ghouse, K. M. (Indian J. Pure Appl. Phys. **6** [1968] 265/8).

[38] Kato, T. (Mineral. J. [Tokyo] **4** [1958] 224/35).

[39] Floran, R. J.; Abraham, M. M.; Boatner, L. A.; Rappaz, M. (Sci. Basis Nucl. Waste Management **3** [1981] 507/14, 510).

[40] Aleksiev, W.; Tsvetkova, V. (Tr. Vurkhu Geol. Bulg. Ser. Geokhim. Polezni Izkop. **3** [1962] 5/24, 17).

[41] Mannucci, G.; Diella, V.; Gramaccioli, C. M. (Can. Mineralogist **24** [1986] 469/74).

[42] Jobbins, E. A.; Tresham, A. E.; Young, B. R. (J. Gemmol. **15** [1977] 295/9).

[43] Gramaccioli, C. M.; Segalstad, T. V. (Am. Mineralogist **63** [1978] 757/61).

[44] Rosenblum, S.; Mosier, E. L. (Geol. Surv. Profess. Paper [U.S.] No. 1 181 [1983] 1/67, 34/5).

[45] Kukharenko, A. A.; Bulakh, A. G.; Baklanova, K. A. (Zap. Vses. Mineral. Obshch. **90** [1961] 371/81, 374/9).

[46] Haapala, I.; Ervamaa, P.; Löfgren, A.; Ojanperä, P. (Bull. Geol. Soc. Finland No. 41 [1969] 117/24).

[47] Folinsbee, R. E. (Trans. Roy. Soc. Can. Sect. 4 [3] **49** [1955] 7/24, 14/5).

[48] Soong, K.-L. (Acta Oceanogr. Taiwan. No. 8 [1978] 43/62).

[49] Nekrasov, I. Ya. (Dokl. Akad. Nauk SSSR **204** [1972] 941/3; Dokl. Acad. Sci. USSR Earth Sci. Sect. **204** [1972] 134/6).

[50] Vance, E. R.; Pillay, K. K. S. (DOE/ET-41900-5 [1981] 1/39, 14; C.A. **96** [1982] No. 112068).

[51] El-Hinnawi, E. E. (Beitr. Mineral. Petrol. **9** [1964] 519/32, 528).

[52] Karkhanavala, M. D.; Shankar, J. (Proc. Indian Acad. Sci. A **40** [1954] 67/71).

[53] Matzko, J. J.; Overstreet, W. C. (Chung-Kuo Ti Chih Hsueh Hui Hui K'an No. 20 [1977] 16/35, 23/4).

[54] Flinter, B. H.; Butler, J. R.; Harral, G. M. (Am. Mineralogist **48** [1963] 1210/26, 1223/4).

[55] Staatz, M. H. (Geol. Surv. Bull. [U.S.] No. 1351 [1972] 1/94, 71).

[56] Kosterin, A. V.; Alekhina, K. N.; Kizyura, V. E. (Soobshch. Dal'nevost. Filiala Sibirsk. Otd. Akad. Nauk SSSR **1962** No. 15, pp. 23/6).

[57] Linares, E.; Toubes, R. O. (Actas Jorn. Geol. Argent. **3** [1960/62] 191/205, 195, 203).

[58] Yankovskii, V. A. (Sb. Statei Geol. Sib. **1975** 30/4).

[59] Marchenko, E. Ya. (Dokl. Akad. Nauk SSSR **155** [1964] 349/52; Dokl. Acad. Sci. USSR Earth Sci. Sect. **155** [1964] 118/22).

[60] Ploshko, V. V. (Izv. Akad. Nauk SSSR Ser. Geol. **1961** No. 1, pp. 86/93, 89).

[61] Speer, J. A. (Rev. Mineral. **5** [1980] 113/35, 123).

[62] Haaker, R. F.; Ewing, R. C. (Sci. Basis Nucl. Waste Management **2** [1980] 281/8, 283).

[63] Beall, G. W.; Boatner, L. A.; Mullica, D. F.; Milligan, W. O. (J. Inorg. Nucl. Chem. **43** [1981] 101/5).

[64] Finney, J. J.; Rao, N. N. (Am. Mineralogist **52** [1967] 13/9, 17).

[65] Starik, I. E.; Lazarev, K. F. (Radiokhimiya 1 [1959] 60/5; Radiochem. [USSR] 1 [1959] 174).
[66] Starik, I. E.; Lazarev, K. F.; Petryaev, E. P. (Radiokhimiya 3 [1961] 207/14, 209, 213; Radiochem. [USSR] 3 [1961] 129).
[67] Starik, I. E.; Starik, F. E.; Elizarova, A. N.; Petryaev, E. P.; Lazarev, K. F. (Tr. Sess. Kom. Opred. Absolyut. Vozrasta Geol. Formatsii Akad. Nauk SSSR 5 1956 [1958] 221/32, 232).
[68] Black, L. P.; Fitzgerald, J. D.; Harley, S. L. (Contrib. Mineral. Petrol. [Berlin] 85 [1984] 141/8).

2.4.1.1.4 Optical Properties

Diaphaneity. Usually, monazite is subtransparent to subtranslucent [1] or translucent [2, 3]. It is transparent to subtransparent in small grains. Superficially altered monazite may be opaque [4]. The degree of opacity (determined by photo voltmeter readings and expressed as the observed alteration in %) may be used as a measure of the degree of alteration of monazite [5]. Dark-colored monazite generally is opaque, as, for example, brown monazite from terrigenous sediments, Ukraine [6], and dark gray (rarely light gray and light brown) monazite from Central Asia, Siberia, and the Far East [7]. Intensely colored monazite from Chudnov-Berdichev granite, southeastern Ukraine, is semitransparent to opaque [8]. Caused by a fine dust of inclusions, monazite is opaque, for example, in pegmatites of the Transangara alkaline massif, Enisei Ridge, Central Siberia [9], and in gneisses and migmatites of the southeastern Ukraine [10].

The **color** of monazite is variable (see below) and may be influenced by chemical composition (see pp. 347/8), host minerals (see pp. 348/9), or mineral inclusions, degree of alteration, and source rock (see p. 349).

According to the standard reference literature, the color of monazite varies from light yellow to dark brown [3], from yellow to brown-red [11], or from light yellow to red-brown [12]. The most frequently described color is yellow [1, 2, 4, 13 to 15] and commonly includes shades of honey yellow to golden yellow and transparent pale yellow [4]; often given is brown [1, 4, 13, 15, 16], cinnamon brown [12], yellowish brown [4, 14, 16], reddish brown [1, 2, 4, 13 to 15], red [4, 16], green [1, 2, 4, 12, 14, 16], yellowish green, or greenish brown [4]. Sometimes monazite is white [1, 2, 4, 12], colorless [4, 12], gray, pale orange-yellow to yellowish gray [4], or black [4, 12]. Colors or color shadings of Th-containing monazite, differing from those cited in the standard reference literature, are given as follows: apple green [17], cream-colored, tan [18, 19], rose-brown [20, 21] or brownish rose [22], buff brownish orange [23], light to dark orange [18, 19, 24, 25], intensely orange-yellow [26], straw yellow through orange to turbid orange-brown [27], dark red [28], and pale rose-colored [29].

In transmitted light, the color of monazite is yellowish brown or yellow to colorless [1]. It also may be brown [40, p. 375], cinnamon brown [47], or pale greenish yellow [48].

Internal reflections are orange-yellow and cloudy diffuse in weakly metamict and gray-white in stronger metamict monazite from conglomerates, Denny Dalton, northern Zululand, R.S.A. [52].

The color of monazite may depend on chemical composition. But note that no chemical basis is known for the color of natural monazite, except the possible relation of a black color to the presence of carbon [4]. Differences in chemical composition of monazite (besides admixtures of foreign minerals; see p. 349) may cause some of the differences in color (cream, tan, brown, orange, and yellow) between samples from Th-bearing quartz veins of the Lemhi Pass quadrangle, Idaho-Montana [18, 19]. The content of ThO_2 and the

$ThO_2:U_3O_8$ ratio of monazite from conglomerates of the Blind River ore region, Canada, vary with color as follows [30]; see also [31]:

	color of monazite		
	orange	yellow	gray
%ThO_2	5.9	3.9	6.7 and 5.9
$ThO_2:U_3O_8$	19	4.1	4.1 and 2.0

The content of ThO_2 is 5.6% in green and 9.6% in yellow-brown monazite from quartz syenite of the western Garo Hills district, Meghalaya, India [33]. Brown and green monazite from beach placers of southern India differ in relative chemical composition as follows (for details, see original paper): As compared with the brown monazite, in the green variety the relative contents of Th, La, Nd, Yb, Y, and Pb are about half, the content of Ce is somewhat lower, Dy, Ho, Tm, and Lu are missing, and the content of Ca is about six times higer. The relative contents of Pr, Sm, Eu, Gd, and U are nearly the same in both varieties [32, pp. 81, 83]. Compared with a cream-colored monazite from a sandstone-clay deposit, Western Siberian Lowlands, the contents of ThO_2, Cr, and Fe in a pale green variety are somewhat higher (the presence of Fe, especially of Fe^{2+}, explaining the pale green color) and the content of ΣRE_2O_3 is somewhat lower [34]; for the contents, see analyses No. 180 and 155 and corresponding footnotes, respectively, in Table 14, pp. 324 and 322.

In contrast to yellow monazite, dark monazite is characterized by lower contents of ThO_2, Y_2O_3, P_2O_5, and CaO and higher contents of Eu_2O_3 (0.36% as compared to 0.05%), SiO_2, TiO_2, Al_2O_3, Fe_2O_3, MgO, and H_2O, as is shown by a comparison of the average composition (partly based on literature data) of 31 dark and 64 yellow monazites from various localities; deviating numbers of samples are given in parentheses [35, pp. 39/41]:

	mean content in % in			mean content in % in	
	31 dark monazites	64 yellow monazites		31 dark monazites	64 yellow monazites
ThO_2	0.81 (25)	7.15	Al_2O_3	2.27	0.17 (40)
ΣRE_2O_3	60.23	59.01	TiO_2	0.69 (25)	0.005 (40)
Y_2O_3	0.61	1.58	P_2O_5	22.07	26.00 (11)
CaO	0.33 (24)	1.98 (40)	SiO_2	9.90	1.70 (26)
MgO	0.23 (24)	0.01 (40)	H_2O^+	1.16 (12)	0.11 (40)
Fe_2O_3	1.89	0.57 (40)	rest	—	0.42 (37)

The composition of colorless to yellowish green monazite from offshore bars of Taiwan, containing 1.9 to 12.0, mean 5.1% ThO_2 and almost no Eu, differs markedly from that of associated black pellet-like monazite that lacks Th and is enriched in Eu [36, pp. 23, 25]. Note also that in monazites from Taiwan the respective contents of ThO_2 and ΣRE_2O_3 are 0.6 and 38.1% in black and 5.7 and 57.9% in yellow varieties [37]. Black monazite from Taiwan is lower in ThO_2 but higher in $(Ta,Nb)_2O_5$ and SiO_2 contents than associated yellow monazite [38].

A distinct correlation between color and host minerals is observed for clear (comprising yellow and cloudy gray) and dark brown monazite inclusions from a felsic paragneiss, northeastern Antarctica. Probably, these different colors are related to the accessibility of the host minerals to fluids. That means, the different colors depend on the capacity of these minerals to shield included monazite grains from fluids circulating in the rock system, and arise from either minor elements introduced with the fluids and/or from the presence of the fluids itself. Thus, clear samples are most common within recrystallized seams, brown samples predominate within garnet and quartz grains, and clear and brown samples are about equal within plagioclase as well as, in minor amounts, within biotite and sillimanite grains. The marked preference of brown monazite for quartz and garnet is explained by the relative impermeability of these minerals to fluids; the occasional presence of clear monazite in these minerals probably relates to a local decrease in the shielding properties of these host minerals, caused by the presence of cracks [39].

Sometimes inclusions of other minerals cause dark colors of monazite, as, for example, inclusions of goethite and hydrogoethite in brown sulfate-monazite from carbonatites and quartz-carbonate rocks of the Kola Peninsula [40, p. 375]; dusty inclusions of an ore mineral (as, e.g., dispersed inclusions of pyrite and other sulfides [42]) in samples from granitoid rocks, Ukraine [41, p. 121]; and inclusions of Fe oxide in samples from Th-bearing quartz veins of the Lemhi Pass quadrangle, Idaho-Montana [18, 19]. A reddish brown color in transmitted light of monazite from a talc-actinolite-carbonate rock, Urushten complex, northern Caucasus, USSR, is caused by Fe hydroxide [49]. As compared with light colors (e.g., yellow, light brown, greenish yellow) of monazite samples from the central part of the Chudnov-Berdichev granite, southeastern Ukraine, dark colors (e.g., dirty greenish brown, cinnamon brown, brownish red, gray, and black) of samples from the peripheral part are caused by a higher content of foreign admixtures (as, e.g., zircon and biotite) [43]. Whereas among alluvial monazites from various localities a light gray color is ascribed to the scatter of light by a sagenitic mesh of rutile, gray to black colors are caused by numerous mainly submicrometer-sized inclusions of amorphous carbon (which is proved by the replacement of the dark color by white to light gray or buff to pink tints in the outer parts of the crystals during heating at 1000°C for 3 h) [35, p. 25]. For a color zonation caused by a zonal arrangement of inclusions in monazite, see p. 272.

The color of monazite may depend on the degree of alteration, as is shown in the following examples. Thus, the color of monazite from granitoid rocks of the Kola Peninsula is transparent yellowish cinnamon brown with different intensity for fresh, unaltered samples and cloudy to opaque brick red for samples altered incipiently to rhabdophane and, rarely, bastnaesite [44, p. 131]. Alteration of monazite causes brownish, reddish, and grayish colors of samples from granitoid rocks of the Azov region, Ukraine [45]. As alteration to limonite and minute opaque particles proceeds, the color changes from greenish yellow to opaque brown in monazite samples from various localities (mostly of the USA) [5]. With increasing oxidation of RE, the color of alluvial monazite from Malaya varies from a clear deep canary yellow through cloudy white to cream-colored, and only rarely to dark gray or black [17].

The color of monazite may vary with the kind of host rock. Thus, commonly detrital monazite from stream and beach placers and accessory monazite from granulites and gneisses are yellow. Locally, various shades of brown, red, and green are observed in detrital monazite from stream and lake placers and in accessory monazite from pegmatites as well as from schists, gneisses, and granites. Pale orange-yellow to yellowish gray, white, gray, or nearly colorless monazite is rare and occurs in vugs and veins. Black monazite is found only locally in beach and stream placers and pegmatites [4]. Note also that a yellowish color is typical for monazite from beach sands, whereas darker colors characterize monazite from pegmatites or from alluvial inland deposits [46].

The **luster** of monazite is variable, usually resinous or waxy, but may incline towards vitreous or adamantine [1]; it also can be greasy (fatty) [12, 16]. Besides the kinds of luster cited above from the standard reference literature, Th-bearing monazite can show a dull luster [7, 20, 28], [35, p. 25], which is caused by weathering [53]; that of micropitted surfaces tends towards silky and pearly [35, p. 25]. The luster of monazite is more brilliant in transparent grains than in subtransparent grains [4]. It is vitreous in clear and transparent varieties, resinous in colored and translucent varieties, and weakly resinous in opaque varieties [54].

The **streak** of monazite is white or faintly colored [1], colorless, yellow, or pale brown [14]. Black monazite from Taiwan has a gray streak [36, p. 19].

Anisotropy. Usually, monazite is anisotropic; see, for example, [29, 56, 57]. Owing to metamictization, some monazite crystals from the South Carpathians, Roumania, are more or less isotropic and opaque [58].

Interference colors of monazite are moderately high [50] to high [51]. A varying intensity of interference colors is explained by the presence of a high amount of irregularly distributed minute inclusions [7].

The **optical sign, refractive indices, birefringence**, and/or **optical axial angle** of monazite are given in the standard reference literature as follows:
biaxial positive [1], see also [2, 11];
$n_\alpha = 1.774$ to 1.800, $n_\beta = 1.777$ to 1.801, $n_\gamma = 1.828$ to 1.851,
$n_\gamma - n_\alpha = 0.045$ to 0.075, $2 V_\gamma = 6°$ to $19°$ [15];
$n_\alpha = 1.785$, $n_\beta = 1.787$, $n_\gamma = 1.837$, $2 V = 1°$ to $14°$ [2]; or
$n_\alpha = 1.796$, $n_\beta = 1.797$, $n_\gamma = 1.841$ [3].

For a compilation of examples of refractive indices, birefringence, and optical axial angles, which can be assigned to a definite ThO_2-containing monazite sample, see Table 17, p. 351. For refractive indices and/or optical axial angle which cannot be assigned to an individual Th-containing monazite sample, refer, for example, to [5, 26, 34], [35, p. 27], [41, pp. 124/5], [57, 59, 60], [61, p. 199], [62, 63].

For monazite with a composition of 2.98% ThO_2, 67.23% $(Ce,La)_2O_3$, 0.12% PbO, 28.97% P_2O_5, and 0.70% SiO_2 and a density of 4.98 g/cm³, the measured mean index of refraction, $n_m = (n_\alpha + n_\beta + n_\gamma)/3 = 1.798$, deviates by -0.017 units from the calculated value obtained by the rule of Gladstone and Dale [79].

The refractive indices of monazite are increased by an incorporation of Th or of the Th-silicate (huttonite) component into the mineral [15]; low values of n correspond to monazites poor in ThO_2 and SiO_2 [54]. Low-refractive indices of monazite from Finland (see Table 17, p. 351) probably are due to its relatively high content of CaO (1.25%) [69]. High contents of CaO and SO_3 (4.60 and 3.12%, respectively), besides partial hydration (see below), may cause the low refractive indices ($n_\alpha \approx n_\beta = 1.730$, $n_\gamma = 1.776$), birefringence ($n_\gamma - n_\alpha = 0.046$), and optical axial angle ($2 V \leq 5°$) of sulfate-monazite from carbonatites and quartz-carbonate rocks of the Kola Peninsula [40, pp. 376/7].

With increasing degree of alteration, a decrease is observed for the refractive indices (see, e.g., [5, 80]) as well as for birefringence and for optical axial angle (see, e.g., [44, p. 131]) of monazite. Both the low refractive indices and birefringence of sulfate-monazite, Kola Peninsula (see above), partly are caused by its hydration, which may be accompanied by formation of rhabdophane [40, p. 377].

Table 17

Refractive Indices, Birefringence, and Optical Axial Angle of Monazite from Various Localities. Samples are arranged by decreasing ThO_2 content; analysis numbers refer to those in Table 14, pp. 305/28, where a complete sample description is given.

monazite from	analysis No.	%ThO₂	n_α	n_β	n_γ	$n_\gamma - n_\alpha$	2V	Ref.
Urals	–	12.60	1.789	–	1.840	–	–	[61, pp. 199, 201]
Sri Lanka	15	11.6	1.795a)	–	1.845a)	–	–	[64]
Italy	16	11.34	1.777a)	1.778a)	1.823a)	0.045	14.5°	[65]
Kola Peninsula	22	10.36	1.785	1.788	1.833	0.048	12°	[44, pp. 135/7]
Kola Peninsula	27	9.96	1.781	1.787	1.840	0.059	6°	[10]
Kola Peninsula	–	9.41	1.778	1.779	1.824	0.046	18°	[66]
Ukraine	47	8.75	1.785	–	1.834	0.049	–	[10]
Ukraine	54	8.53	1.767	–	1.780	–	–	[51]
Ukraine	79	7.80	1.785	–	1.830	0.045	–	[44, pp. 135/6]
Kazakhstan	92	7.17	1.795a)	–	1.846a)	0.051	–	[6]
Kola Peninsula	94	7.07	1.762	1.764	1.816	0.052	5°	[67]
Ukraine	98	7.00	1.785	1.787	1.836	–	–	[68]
Wyoming	–	~7	1.788a)	–	1.838a)	0.050	–	[10]
Kazakhstan	115	6.06	1.783	–	1.833	–	very small	[44, pp. 135/7]
Ukraine	116	6.00	1.785	–	1.838	0.053	–	[69]
Kola Peninsula	121	5.86	1.778	1.780	1.831	0.053	5° to 12°	[10]
Kola Peninsula	–	5.77	1.794	1.796	1.840	0.048	8°	[70]
Finland	127	5.7	1.778a)	1.780a)	1.831a)	0.053	12° to 14°	[71]
Ukraine	135	5.50	1.787	–	1.834	0.047	–	[44, pp. 135/6]
New Zealand	140	5.32	1.791a)	1.792a)	1.844a)	0.053	12° to 14°	
Little Caucasus	142	5.26	1.775	–	–	–	12° to 16°	
Kola Peninsula	154	4.62	1.783	1.787	1.840	0.058	8°	
Japan	158	4.51	1.790	1.791	1.840	–	10°	[72]
California	172	3.9	1.788a)	–	1.842	0.054	13°	[73]
California	177	3.49	1.787a)	–	1.840	0.053	13°	[74]
California	188	3.01	1.779a)	1.781a)	1.833a)	0.054	–b)	
Central Siberia	192	2.36	1.792	–	1.847	–	–	[9]
Canada	–	1.7	1.776a)	–	1.826a)	–	–	[75]
Central Timan Ridge	200	1.5	1.767a)	–	–	–	–	[76]
Yakutia	209	1.14	1.773a)	–	1.812a)	–	16° to 23°	[77]
Eastern Siberia	213	1.05	1.78	–	1.85	0.07	–	[78]

a) For the standard deviation, see original paper. — b) Calculated 2V is 26°.

Metamictization apparently causes a low birefringence of monazite inclusions in biotite from gneisses, Lepontine Alps, Switzerland [81].

In heated, as compared to unheated, monazite samples from India and Brazil, both average refractive index (1.83 as against 1.79) and birefringence are increased [82].

The **optical orientation** of monazite is $n_\alpha = b$ with $n_\gamma \wedge c = 2°$ to $6°$ [1, 2], $2°$ to $7°$ [15], or $1°$ to $10°$ [14]; the angle depends on the contents of ThO_2 and SiO_2 [54]. The optical axial plane is almost parallel to (100) [3, 54].

The **extinction** of monazite is oblique [54] (see, e.g., [81, 83]) to nearly straight (see, e.g., [32, p. 80]).

Monazite shows a weak horizontal **dispersion** $r < v$, sometimes $r > v$ [1]; see also [2, 14, 15]. Dispersion $r < v$ is distinct, for instance, in monazites from beach sands of California [73].

Pleochroism of monazite is faint or not perceptible [1], or very weak and varies from light yellow to very light yellow [15]. Monazite is pleochroic with α = light yellow, β = dark yellow, γ = greenish yellow [14, 54]. Dark-colored grains of sulfate-monazite from the Kola Peninsula, containing minute inclusions of goethite, sometimes are faintly pleochroic with yellow-brown colors with an absorption $\alpha \geq \gamma$ [40, p. 375]. Silicomonazite is clearly pleochroic from pale yellow to green [84].

The **IR spectrum** (measured in the 4000 to 400 cm^{-1} range) of monazite from Yugoslavia shows a steady ramplike increase in transmittance without absorption peaks or bands between 4000 and 1400 cm^{-1}. The broad and intense but slightly diffuse band, comprising the absorptions at 1089, 1044, and 993 cm^{-1}, belongs to the triple degenerate v_3 vibration of the phosphate anion, which is in accordance with the expected vibration modes for the tetrahedral coordination of the central P atom in the PO_4^{3-} ionic structure. The nondegenerate stretching vibration v_1 accounts for the sharp absorption peak at 948 cm^{-1}. The absorptions at 615 through 538 cm^{-1} are the result of the split v_4 vibrations. The peaks in the range 3400 to 1600 cm^{-1} certainly cannot be ascribed to PO_4^{3-} vibrations and are very likely caused by impurities, such as mineral inclusions or other ions and radicals hosted in the monazite structure [85].

The **absorption spectrum in the visible region** of monazite is dominated by the absorption bands of Nd and, to a lesser degree, of Pr. In low concentrations, Nd produces only a darkening of the yellow region of the spectrum; in high concentrations, the yellow region will contain distinct bands, and other bands will appear in the green and blue regions [55]. The absorption spectrum of green monazite from Piona, Italy, which is markedly different from that of "usual" yellow monazite, shows strong bands in the red at 640 to 680 nm, in the green with a maximum at 555 nm, and in the blue with a maximum at 510 nm. Probably owing to vibration of the ions in the crystal, these bands are clearly defined at the higher wavelength side and fade towards lower wavelengths. In addition, weak bands are observed in the yellow at about 570 to 590 nm and in the green at 522 nm. Except these two bands, all stronger bands observed in the green monazite are absent in the spectrum of yellow monazite, indicating an unusually low Nd content in the green sample. The other bands in the spectrum of the green monazite virtually coincide with the absorption spectrum of U^{4+} and are considered as a direct evidence of tetravalent U in the studied monazite (see p. 344) [65]. As against the spectra of parisite and didymium glass, the absorption spectrum of orange-reddish monazite of gem quality from Sri Lanka shows additional lines in the orange and red that are caused by U [64].

Luminescence. Under UV radiation, monazite neither fluoresces nor phosphoresces [4]. Monazite from black sands, Egypt, fluoresces in deep vermilion colors [23].

353

References for 2.4.1.1.4:

[1] Palache, C.; Berman, H.; Frondel, C. (Dana's System of Mineralogy, 7th Ed., Vol. 2, Wiley, New York 1963, pp. 1/1124, 693).

[2] Semenov, E. I. (Mineralogiya Redkikh Zemel', Akad. Nauk SSSR, Moscow 1963, pp. 1/412, 115).

[3] Ramdohr, P.; Strunz, H. (Klockmanns Lehrbuch der Mineralogie, 16th Ed., Enke, Stuttgart 1978, pp. 1/876, 625).

[4] Overstreet, W. C. (Geol. Surv. Profess. Paper [U.S.] No. 530 [1967] 1/327, 4/5).

[5] Molloy, M. W. (Am. Mineralogist **44** [1959] 510/32, 511/2).

[6] Dishchuk, V. L.; Druzhinin, L. N.; Smirnov, G. I.; Stasiv, T. V.; Yur'eva, A. L. (Mineral. Sb. [Lvov] **24** [1970] 431/4).

[7] Li, A. F.; Grebennikova, O. T. (Zap. Vost. Sibirsk. Otd. Vses. Mineral. Obshch. No. 4 [1962] 155/61, 155).

[8] Ivantishin, M. M. (Aktsesorni Ridkisni Minerali ta Rozsiyani Elementi v Granitakh i Pegmatitakh Ukrains'kogo Kristalichnogo Shchita, Akad. Nauk Ukr. SSR, Kiev 1960, pp. 1/243, 131).

[9] Sveshnikova, E. V.; Semenov, E. I.; Khomyakov, A. P. (Zaangarskii Shchelochnoi Massiv, Ego Porody i Mineraly, Nauka, Moscow 1976, pp. 1/80, 48/9).

[10] Marchenko, E. Ya. (Dokl. Akad. Nauk SSSR **176** [1967] 1153/5; Dokl. Acad. Sci. USSR Earth Sci. Sect. **176** [1967] 142/5).

[11] Roubault, M. (Geologie de l'Uranium, Masson, Paris 1958, pp. 1/462, 411).

[12] Kaplan, G. E.; Uspenskaya, T. A.; Zarembo, Yu. I.; Chirkov, I. V. (Torii, Ego Syr'evye Resursy, Khimiya i Tekhnologiya, Atomizdat, Moscow 1960, pp. 1/224, 16).

[13] Frondel, J. W.; Fleischer, M. (Geol. Surv. Bull. [U.S.] No. 1009-F [1955] 169/209, 185).

[14] Heinrich, E. W. (Mineralogy and Geology of Radioactive Raw Materials, McGraw-Hill, New York 1958, pp. 1/654, 79).

[15] Deer, W. A.; Howie, R. A.; Zussman, J. (Rock-Forming Minerals, Vol. 5, Non-Silicates, Longmans, London 1962, pp. 1/371, 339/46).

[16] Betechtin, A. G. (Lehrbuch der Mineralogie, Verlag Technik, Berlin 1953, pp. 1/681, 410 [Russian original, Moscow 1951]).

[17] Flinter, B. H.; Butler, J. R.; Harral, G. M. (Am. Mineralogist **48** [1963] 1210/26, 1211).

[18] Staatz, M.H. (Geol. Surv. Bull. [U.S.] No. 1351 [1972] 1/94, 71).

[19] Staatz, M. H.; Sharp, B. J.; Hetland, D. L. (Geol. Surv. Profess. Paper [U.S.] No. 1049-A [1979] A1/A90, A45).

[20] Kornetova, V. A. (Tr. Mineral. Muzeya Akad. Nauk SSSR No. 14 [1963] 96/107, 100).

[21] Arkhangel'skaya, V. V. (Redkometal'nye Shchelochnye Kompleksy Yuzhnogo Kraya Sibirskoi Platformy, Nedra, Moscow 1974, pp. 1/127, 42).

[22] Zhirova, V. V.; Zykov, S. I.; Tugarinov, A. I. (Geokhimiya **1957** 592/9, 593).

[23] Gindy, A. R. (Econ. Geol. **56** [1961] 436/41).

[24] Chistyakova, M. B.; Kazakova, M. E. (Tr. Mineral. Muzeya Akad. Nauk SSSR No. 18 [1968] 245/9).

[25] Chistyakova, M. B. (Tr. Mineral. Muzeya Akad. Nauk SSSR No. 23 [1974] 113/76, 137).

[26] Orsa, V. I. (Dopov. Akad. Nauk Ukr. SSR B **31** [1969] 694/6).

[27] Silver, L.T.; Williams, I. S.; Woodhead, J. A. (GJBX-45-81 [1980] 1/380, 193; C.A. **95** [1981] No. 65599).

[28] Kalita, A. P. (Redkozemel'nye Pegmatity Alakurtti i Priladozh'ya, Akad. Nauk SSSR, Moscow 1961, pp. 1/119, 56).

354

[29] Khomyakov, A. P. (Mineralogiya i Geneticheskie Osobennosti Shchelochnykh Massivov 1964 56/82, 72).

[30] Roscoe, S. M. (Can. Mining J. 80 No. 7 [1959] 65/6).

[31] Rudnitskaya, L. S. (Uran Drevnikh Konglomeratakh 1963 129/40, 135).

[32] Chaudhuri, J. N. B. (Berlin. Geowiss. Abhandl. A 52 [1984] 1/123).

[33] Basu, A. N.; Sharma, G. S.; Varma, H. M.; Dhana Raju, R.; Dougall, N. K.; Raju, B. N. V. (J. Geol. Soc. India 30 [1987] 267/72).

[34] Yankovskii, V. A. (Sb. Statei Geol. Sib. 1975 30/4).

[35] Rosenblum, S.; Mosier, E. L. (Geol. Surv. Profess. Paper [U.S.] No. 1 181 [1983] 1/67).

[36] Matzko, J. J.; Overstreet, W. C. (Chung-Kuo Ti Chih Hsueh Hui Hui K'an No. 20 [1977] 16/35).

[37] Chang, F.-C.; Tsai, H.-T.; Wu, S.-C. (J. Chinese Chem. Soc. [Taipei] [2] 22 [1975] 309/16, 315).

[38] Chen, M. C.; Li, K. T.; Wu, K. C. (K'uang Yeh 17 No. 3 [1973] 61/72 from C.A. 80 [1974] No. 39282).

[39] Black, L. P.; Fitzgerald, J. D.; Harley, S. L. (Contrib. Mineral. Petrol. [Berlin] 85 [1984] 141/8, 141/2, 146/7).

[40] Kukharenko, A. A.; Bulakh, A.G.; Baklanova, K. A. (Zap. Vses. Mineral. Obshch. 90 [1961] 371/81, 375/8).

[41] Shcherbak, N. P. (Aktsessornye Mineraly Ukr. Shchita 1976 118/32, 244/58).

[42] Komlev, L. V.; Danilevich, S. I.; Ivanova, K.S.; Kuchina, G. N.; Savonenkov, V. G.; Filippov, M. S.; Chupakhin, M. S. (Tr. Sess. Kom. Opred. Absolyut. Vozrasta Geol. Formatsii Akad. Nauk SSSR 5 1956 [1958] 159/75, 164).

[43] Ivantishin, M. N.; Kul'skaya, O. A.; Gornyi, G. Ya.; Eliseeva, G. D. (Tr. Inst. Geol. Nauk Akad. Nauk Ukr. SSR Ser. Petrogr. Mineralog. Geokhim. No. 21 [1964] 1/168, 19).

[44] Bel'kov, I. V. (Aktsessornye Mineraly Granitoidov Kol'skogo Poluostrova, Nauka, Leningrad 1979, pp. 1/184).

[45] Kuts, V. P. (Mineral. Sb. [Lvov] 20 [1966] 279/84).

[46] Davidson, C. F. (Mining Mag. [London] 94 [1956] 197/208, 198).

[47] Bulakh, A. G.; Kukharenko, A. A.; Knipovich, Yu. N.; Kondrash'eva, V. V.; Baklanova, K. A.; Baranova, E. N. (Mater. Sess. Uchen. Soveta VSEGEI Rezul'tatam Rabot 1959 [1961] 114/6).

[48] Mohr, D. W. (Am. Mineralogist 64 [1984] 98/103).

[49] Ploshko, V. V. (Izv. Akad. Nauk SSSR Ser. Geol. 1961 No. 1, pp. 86/93, 89).

[50] Andersen, T. (Mineral. Mag. 50 [1986] 503/9).

[51] Krasil'nikova, A. V. (Geologiya [Alma-Ata] 11 [1976] 104/8).

[52] Stupp, H. D. (Diss. Univ. Köln 1984, pp. 1/280, 85).

[53] Kupriyanova, I. I.; Volkova, M. I.; Goroshchenko, Z. I. (Tr. Mineral. Muzeya Akad. Nauk SSSR No. 15 [1964] 123/33, 124).

[54] Perez Mateos, J.; Galvan Garcia, J. (Anales Edafol. Agrobiol. 26 [1967] 1227/44, 1237/8).

[55] Aleksiev, E.; Tsvetkova, V. (Tr. Vurkhu Geol. Bulg. Ser. Geokhim. Polezni Izkop. 3 [1962] 5/24, 16/7).

[56] Makarochkin, B. A. (Geol. Geofiz. 1975 No. 2, pp. 155/9).

[57] Es'kova, E. M.; Zhabin, A. G.; Mukhitdinov, G. N. (Mineralogiya i Geokhimiya Redkikh Elementov Vishnevykh Gor, Nauka, Moscow 1964, pp. 1/318, 164).

[58] Pavelescu, L.; Pavelescu, M. (TMPM Tschermaks Mineral. Petrogr. Mitt. [3] 17 [1972] 208/14).

[59] Omel'yanenko, B. I.; Sirotinina, N. A. (Mater. Geol. Rudn. Mestorozhd. Petrogr. Mineral. **1959** 414/22, 422).

[60] Leonova, V. A.; Nikitin, Yu. V. (Zap. Vses. Mineral. Obshch. **91** [1962] 136/45, 137).

[61] L'vov, B. K.; Zhangurov, A. A. (in: Smorchkov, I. E., Aktsessornye Mineraly Izverzhennykh Porod, Nauka, Moscow 1968, pp. 196/204).

[62] Komov, I. L.; Mel'nikova, E. M.; Kokarev, G. N. (Tr. Mineral. Muzeya Akad. Nauk SSSR No. 23 [1974] 87/93, 90).

[63] Korneva, N. G.; Gavrusevich, I. B. (Geol. Zh. **47** No. 4 [1987] 57/63).

[64] Jobbins, E. A.; Tresham, A. E.; Young, B. R. (J. Gemmol. **15** [1977] 295/9).

[65] Gramaccioli, C. M.; Segalstad, T. V. (Am. Mineralogist **63** [1978] 757/61).

[66] Kononov, Yu. V.; Nechaev, S. V. (Pitannya Geokhim. Mineral. Petrogr. **1963** 289/301, 291, 293).

[67] Houston, R. S.; Murphy, J. R. (Bull. Geol. Surv. Wyoming No. 49 [1962] 1/120, 39/40).

[68] Chukhrov, F. V. (Tr. Inst. Geol. Rudn. Mestorozhd. Petrogr. Mineral. Geokhim. No. 50 [1960] 3/237, 136/7).

[69] Haapala, I.; Ervamaa, P.; Löfgren, A.; Ojanperä, P. (Bull. Geol. Soc. Finland No. 41 [1969] 117/24, 122).

[70] Hutton, C. O. (Bull. Geol. Soc. Am. **61** [1950] 635/716, 667/8).

[71] Meliksetyan, B. M. (in: Smorchkov, I. E., Aktsessornye Mineraly Izverzhennykh Porod, Nauka, Moscow 1968, pp. 95/108, 103).

[72] Kato, T. (Mineral. J. [Tokyo] **4** [1958] 224/35, 225/6).

[73] Hutton, C. O. (Spec. Rept. Calif. Div. Mines Geol. No. 59 [1959] 1/32, 20).

[74] Jaffe, H. W. (Bull. Geol. Soc. Am. **66** [1955] 1247/56, 1249/50).

[75] Folinsbee, R. E. (Trans. Roy. Soc. Can. IV [3] **49** [1955] 7/24, 15).

[76] Serdyuchenko, D. P.; Kochetkov, O. S. (Dokl. Akad. Nauk SSSR **218** [1974] 1175/7; Dokl. Acad. Sci. USSR Earth Sci. Sect. **218** [1974] 124/5).

[77] Nekrasova, R. A.; Nekrasov, I. Ya. (Dokl. Akad. Nauk SSSR **268** [1983] 688/93; Dokl. Acad. Sci. USSR Earth Sci. Sect. **268** [1983] 151/5).

[78] Nekrasov, I. Ya. (Magmatizm i Rudonosnost' Severo-Zapadnoi Chasti Verkhoyano-Chukotskoi Skladchatoi Oblasti, Akad. Nauk SSSR, Moscow 1962, pp. 1/412, 101/2).

[79] Jaffe, H. W. (Am. Mineralogist **41** [1956] 757/77, 767).

[80] El-Hinnawi, E. E. (Beitr. Mineral. Petrol. **9** [1964] 519/32, 528).

[81] Schwander, H.; Wenk, E. (Schweiz. Mineral. Petrogr. Mitt. **45** [1956] 797/806, 803).

[82] Karkhanavala, M. D.; Shankar, J. (Proc. Indian Acad. Sci. A **40** [1954] 67/71).

[83] Schermerhorn, L. J. G. (Comun. Serv. Geol. Port. No. 37 [1956] 1/617, 282).

[84] Nekrasov, I. Ya. (Dokl. Akad. Nauk SSSR **204** [1972] 941/3; Dokl. Acad. Sci. USSR Earth Sci. Sect. **204** [1972] 134/6).

[85] Bermanec, V.; Tibljas, D.; Gessner, M.; Kniewald, G. (Mineral. Petrol. **38** [1988] 139/50, 141, 145/6).

[86] Adams, J. W. (Am. Mineralogist **50** [1965] 356/66, 360).

2.4.1.1.5 Other Physical Properties

The Mohs **hardness** H_M of monazite is 5 to 5½ [1]; see also [2 to 4], [5, p. 79]. It is 3.5 in silicomonazite from Yakutia [6] and varies between 3.5 and 5.5 in dark Th-poor monazites from various localities [7, p. 27]. The microhardness (VHN) of monazite is 400 to 570 kg/mm^2 (corresponding to $H_M = 4.7$ to 5.1) [8]. It is 319 ± 8 kg/mm^2 at 150 g load in a sample from the Bogatynia area, Poland [9].

The **fracture** of monazite is conchoidal [4, 8] to uneven [1], [5, p. 79]. It is developed parallel to the c axis in monazite from rose-colored pegmatitic granites and pegmatites of the Chudnov-Berdichev complex, southeastern Ukraine [10]. Monazite is brittle [1, 4, 8].

The range of **density** of monazite is given in the standard reference literature as 4.6 to 5.4 (mostly 5 to 5.2) g/cm^3 [1]; 4.8 to 5.5 g/cm^3 [4]; 4.9 to 5.5 g/cm^3 [2, 8, 11]; or 5.0 to 5.3 g/cm^3 [3]. For densities of monazite, which can be assigned to a definite ThO_2-containing sample and, in a few cases, lie outside the ranges given above, see Table 14, pp. 305/28.

The density of monazite increases with increasing ThO_2 content [1], [5, p. 79] from 5.15 ± 0.05 g/cm^3 for pure $(Ce,La)PO_4$ up to at least 5.4 g/cm^3 for Th-containing monazite [12]. Monazite containing high contents of heavy radioactive elements (such as Th and U) possesses a somewhat higher density than monazite poor in these elements [11]. As compared with yellow monazite with a higher ThO_2 content and a density of 5.15 g/cm^3 (described in standard reference literature), dark monazites containing 0 to 3.2, mean 0.78% ThO_2 in 25 samples from various localities have a low density ranging from 4.13 to 4.79 g/cm^3 in 27 samples [7, pp. 27, 38/9]. The density is 4.18 g/cm^3 for Th-poor black and 5.03 g/cm^3 for yellow monazite from Taiwan [13]. However, the increase in density with increasing ThO_2 content is only a poorly developed trend and is difficult to demonstrate by actual analysis [14] (see also Table 14, pp. 305/28). Thus, for instance, no interrelation between D (ranging from 5.06 to 5.23 g/cm^3) and $\Sigma(ThO_2 + U_3O_8)$ (ranging from 2.19 to 11.52%) is found for 16 monazites from Malaya (for individual values, see original paper) [15, pp. 1215, 1221].

With increasing degree of alteration, the density of monazite decreases in samples from granitoid rocks of the Kola Peninsula [16].

On heat treatment at 1130°C for 24 h, the average density measured on different grains of monazite from India increases from 5.33 to 5.62 g/cm^3 [17]. During heating of Th-containing monazite from various localities (e.g., from Western Australia and southern Zimbabwe), the following, only slightly systematic, changes of density are observed [18]:

density in g/cm^3 of monazite				
unheated	3.66	3.97	5.13	5.24	5.28
heated at 950°C for 20 h	3.69	3.94	5.20	5.29	5.43
heated at 1100°C for 2 h	3.52	3.91	5.20	5.29	5.55

Magnetic Properties. Monazite is moderately paramagnetic [1]. It is less magnetic than xenotime [5, p. 75] and ilmenite [19]. Detrital monazite from New Zealand is decidedly paramagnetic [20]. The magnetic susceptibility χ of monazite is high (e.g., 38×10^{-6} cm^3/g for the mean mass magnetic susceptibility of monazite from a pegmatitic vein, Puumala, eastern Finland [21]) and derives from the paramagnetism of the RE [14] (see, e.g., [15, pp. 1222, 1224], [22]) or, partly, from extremely fine-grained unidentifiable inclusions, as is described for green monazite with a higher magnetic susceptibility than associated yellow-brown monazite without inclusions from quartz syenite, western Garo Hills district, Meghalaya, India [23]. Thorium does not affect χ in alluvial monazite from Malaya [15, p. 1224]. Dark Th-poor and yellow Th-rich monazite from various localities have about the same magnetic susceptibility [7, pp. 29, 33].

Radioactivity. Owing to the presence of ThO_2, monazite often is radioactive [2, 24]. The surface activity values of monazite from black sands of Egypt range from 1.447 to 7.500 $\alpha/cm^2/s$ (corresponding to an equivalent ThO_2 content of 2.720 to 14.087%) and are not influenced by shape, size, roundness, or degree of alteration of the monazite grains [25]. In dark Th-poor and yellow Th-rich monazites from various localities, the net radioactivity varies as follows [7, p. 34]:

monazite type; locality	content of ThO_2 in %	radioactivity in net counts/10 min
dark monazite		
Alaska	3.2; 0.79; 0.82	87; 25; 36
Taiwan	0.58; 0.59; 0.58	13; 6; 10
Zaire	1.0; 0.85; 0.75; 0.68; 0.65	9; 19; 17; 10; 21
yellow monazite		
Liberia	6.4; 8.4	183; 248

Inclusions of radioactive monazite in biotite from granites of northern Portugal are surrounded by uniformly black opaque ("overexposed") pleochroic haloes with relatively diffuse edges and with radii corresponding to the maximum distances (0.042 to 0.043 mm) travelled by the α particles emitted by ^{212}Po deriving from the radioactive decay of Th in monazite [26]. Note also that albite and albitized microcline in Siberian pegmatites have pink or brown colors in a zone of 5 to 6 cm width around inclusions of monazite [27]. For the occurrence of pleochroic haloes around inclusions of monazite in other minerals, see also p. 271.

Metamictization. In spite of a relatively high ThO_2 content, appreciable amounts of natural monazite have not been found in a metamict state. This may be due to the melting temperature of monazite (see p. 360), which is higher than that of other radioactive minerals [28]. Note also that in view of a ThO_2 content of 20%, natural monazite has an unusual stability to radiation damage [29], which is demonstrated by the extreme rarity of reported partially metamict occurrences of monazite. Totally metamict or fully amorphous examples have never been reported [30]. Thus, monazite is considered as a non-metamict [31] or almost non-metamict mineral [32]; see also [33 to 35]. Only a few occurrences of slightly metamict Th-bearing monazite are described, see, for example, [36, 37], [38, pp. 143/4, 147]. About 20 to 30% of silicomonazite from Yakutia is in a metamict state, which disappears after heating at 850 °C for 4 h [6]. As an exception, monazites from India and Brazil show a fairly high degree of metamictization, which is supported by a low birefringence of unheated material that increases after heating (see p. 352) [17]. At 450 °C, monazite from Madagascar (containing 11.2% ThO_2) is partly transformed from the amorphous isotropic into the crystalline state [39]. For an isotropization of monazite owing to metamictization, see also p. 350. Monazite from the Dominion Reef conglomerate, western Transvaal, R.S.A., sometimes is metamictized by radioactive inclusions [40]. Possibly, minor differences in radiation damage of the crystal lattice cause small-scale differences in isotopic composition of different size fractions of brown monazite from northeastern Antarctica (the generally observed complete inverse correlation of the $^{207}Pb:^{206}Pb$ ratio with the inferred average Th content is missing solely in the <215-μm fraction) [38, p. 142].

References for 2.4.1.1.5:

[1] Palache, C.; Berman, H.; Frondel, C. (Dana's System of Mineralogy, 7th Ed., Vol. 2, Wiley, New York 1963, pp. 1/1124, 693).

358

[2] Betechtin, A. G. (Lehrbuch der Mineralogie, Verlag Technik, Berlin 1953, pp. 1/681, 410 [Russian original, Moscow 1951]).

[3] Deer, W. A.; Howie, R. A.; Zussman, J. (Rock-Forming Minerals, Vol. 5, Non-Silicates, Longmans, London 1962, pp. 1/371, 339).

[4] Ramdohr, P.; Strunz, H. (Klockmanns Lehrbuch der Mineralogie, 16th Ed., Enke, Stuttgart 1978, pp. 1/876, 625).

[5] Heinrich, E. W. (Mineralogy and Geology of Radioactive Raw Materials, McGraw-Hill, New York 1958, pp. 1/654).

[6] Nekrasov, I. Ya. (Dokl. Akad. Nauk SSSR 204 [1972] 941/3; Dokl. Acad. Sci. USSR Earth Sci. Sect. 204 [1972] 134/6).

[7] Rosenblum, S.; Mosier, E. L. (Geol. Surv. Profess. Paper [U.S.] No. 1181 [1983] 1/67).

[8] Semenov, E. I. (Mineralogiya Redkikh Zemel', Akad. Nauk SSSR, Moscow 1963, pp. 1/412, 115).

[9] Kucha, H. (Mineral. Mag. 43 [1979/80] 1031/4).

[10] Ivantishin, M. M. (Aktsesorni Ridkisni Minerali ta Rozsiyani Elementi v Granitakh i Pegmatitakh Ukrains'kogo Kristalichnogo Shchita, Akad. Nauk Ukr. SSR, Kiev 1960, pp. 1/243, 136/7).

[11] Kaplan, G. E.; Uspenskaya, T. A.; Zarembo, Yu. I.; Chirkov, I. V. (Torii, Ego Syr'evye Resursy, Khimiya i Tekhnologiya, Atomizdat, Moscow 1960, pp. 1/224, 16/7).

[12] Frondel, C. (Geol. Surv. Bull. [U.S.] No. 1064 [1958] 1/400, 156/7).

[13] Chen, M. C.; Li, K. T.; Wu, K. C. (K'uang Yeh 17 No. 3 [1973] 61/72 from C.A. 80 [1974] No. 39282).

[14] Overstreet, W. C. (Geol. Surv. Profess. Paper [U.S.] No. 530 [1967] 1/327, 4/5).

[15] Flinter, B. H.; Butler, J. R.; Harral, G. M. (Am. Mineralogist 48 [1963] 1210/26).

[16] Bel'kov, I. V. (Aktsessornye Mineraly Granitoidov Kol'skogo Poluostrova, Nauka, Leningrad 1979, pp. 1/184, 131).

[17] Karkhanavala, M. D.; Shankar, J. (Proc. Indian Acad. Sci. A 40 [1954] 67/71).

[18] Vance, E. R.; Pillay, K. K. S. (DOE/ET-41900-5 [1981] 1/39, 15, 17; C.A. 96 [1982] No. 112068).

[19] Davidson, C. F. (Mining Mag. [London] 94 [1956] 197/208, 198).

[20] Hutton, C. O. (Bull. Geol. Soc. Am. 61 [1950] 635/716, 667).

[21] Haapala, I.; Ervamaa, P.; Löfgren, A.; Ojanperä, P. (Bull. Geol. Soc. Finland No. 41 [1969] 117/24, 122).

[22] Richartz, W. (Fortschr. Mineral. 39 [1961] 53/9).

[23] Basu, A. N.; Sharma, G. S.; Varma, H. M.; Dhana Raju, R.; Dougall, N. K.; Raju, B. N. V. (J. Geol. Soc. India 30 [1987] 267/72).

[24] Perez Mateos, J.; Galvan Garcia, J. (An. Edafol. Agrobiol. 26 [1967] 1227/44, 1237).

[25] Gindy, A. R. (Econ. Geol. 56 [1961] 436/41).

[26] Schermerhorn, L. J. G. (Comun. Serv. Geol. Port. No. 37 [1956] 1/617, 286).

[27] Kornetova, V. A. (Tr. Mineral. Muzeya Akad. Nauk SSSR No. 14 [1963] 96/107, 99).

[28] Hikichi, Y.; Nomura, T. (J. Am. Ceram. Soc. 70 [1987] C-252/C-253).

[29] Haaker, R. F.; Ewing, R. C. (Sci. Basis Nucl. Waste Management 2 [1980] 281/8, 283).

[30] Floran, R. J.; Abraham, M. M.; Boatner, L. A.; Rappaz, M. (Sci. Basis Nucl. Waste Management 3 [1981] 507/14, 509/10, 512).

[31] Povarennykh, A. S. (Zap. Vses. Mineral. Obshch. 85 [1956] 593/7).

[32] Bouška, V. (Acta Univ. Carolinae Geol. 1970 143/69, 160).

[33] Kato, T. (Mineral. J. [Tokyo] 4 [1958] 224/35, 231, 233).

[34] Es'kova, E. M.; Zhabin, A. G.; Mukhitdinov, G. N. (Mineralogiya i Geokhimiya Redkikh Elementov Vishnevykh Gor, Nauka, Moscow 1964, pp. 1/318, 165).

[35] Gramaccioli, C. M.; Segalstad, T. V. (Am. Mineralogist 63 [1978] 757/61).
[36] McKie, D. (Records Geol. Surv. Tanganyika 6 [1956/58] 87/93).
[37] El-Hinnawi, E. E. (Beitr. Mineral. Petrol. 9 [1964] 519/32, 528).
[38] Black, L. P.; Fitzgerald, J. D.; Harley, S. L. (Contrib. Mineral. Petrol. [Berlin] 85 [1984] 141/8).
[39] Noddack, W.; Jakobi, R. (Z. Anorg. Allgem. Chem. 284 [1956] 208/33, 222/5).
[40] Reimer, T. O. (Neues Jahrb. Mineral. Abhandl. 158 [1987] 13/46, 22).

2.4.1.1.6 Chemical and Thermal Behavior

Chemical Behavior. In the following paragraphs are compiled general data on the chemical behavior of monazite that are derived mainly from standard reference books and supplemented by a few examples of the chemical behavior of Th-containing monazite or of the behavior of Th during chemical treatment. For the chemical behavior of monazite, refer also to "Phosphor" A, 1965, p. 274. Chemical decomposition of monazite and its industrial utilization are described in "Thorium" 1955, pp. 37/40 and 66/72, respectively.

Monazite is slowly decomposed by acids [1 to 3]. It is difficultly soluble in HCl [4, 5]; if the solution is diluted, a phosphate of low solubility is again precipitated [6, p. 21]. Monazite is easily dissolved by H_2SO_4 [5], [6, p. 63]. Concentrated $HClO_4$ decomposes monazite more rapidly than concentrated H_2SO_4. Monazite may be decomposed by H_3PO_4 at 350°C [6, pp. 70, 75]. It is weakly soluble in HNO_3 without preliminary fusion with Na_2CO_3 [7].

A part of the ThO_2 content in monazite is leached by acids. Thus, for instance, after treatment with HCl, the content of ThO_2 of granular monazite and of a radial monazite aggregate from the Timan Ridge, USSR, decreases from 10.2 to 8.5% and from 8.7 to 8.15%, respectively; the corresponding content of Y_2O_3 changes from 1.29 to 2.5% and from 0.5 to 0.81% [8]. Leaching with HNO_3 of monazite from Quebec decreases the content of Th from 3.8 to 0.427% in the residue [9].

Monazite is slowly decomposed by fusion with $KHSO_4$ or Na_2CO_3 [1]; it decomposes easily when fused with alkali carbonate, and alkali phosphates pass into the aqueous extract. Monazite decomposes when fused with NaOH [6, pp. 102, 107].

In the borax pearl, monazite changes to a yellow or yellowish red glass which becomes colorless during cooling [4]. Before the blowpipe, monazite turns gray but does not fuse [1] or is almost infusible [4, 5]. Monazite powder moistened with H_2SO_4 colors the flame green during annealing [4]. Monazite is unchanged in the closed tube [1].

Thermal Behavior. The DTA curve of unaltered monazite from eastern Finland [10] and of pure monazite from Madison county, North Carolina, is nearly featureless. Other samples with high Fe content and partly extensive alteration from the USA, Brazil, Norway, and Russia display exothermic peaks caused by Fe-bearing alteration products (for details, see original paper). The DTA curve of one sample from New Jersey, altered superficially to goethite, shows endothermic reactions at 250 to 300°C (distorted by the effects of the superimposed exothermic reactions at 250 and 330°C) and at 560°C. This pattern is interpreted as a composite of the exothermic reactions at 250, 330, and 430°C (belonging to the monazite alteration) superimposed on the endothermic reactions of goethite and water at 200 to 400°C [11].

A TGA curve of monazite from eastern Finland shows a nearly linear weight loss up to a total of 0.425% between 200 and 1000°C [10].

The melting temperature is $2057 \pm 40\,°C$ for a natural monazite containing 11.73% ThO_2. It varies in almost pure synthetic $REPO_4$ for the individual rare earth elements as follows: $2072 \pm 20\,°C$ (La), $2045 \pm 20\,°C$ (Ce), $1938 \pm 20\,°C$ (Pr), $1975 \pm 20\,°C$ (Nd), and $1916 \pm 20\,°C$ (Sm) [12].

For an influence of heating on lattice constants, d values, crystal structure, average refractive index and birefringence, and density, see pp. 341, 342, 343/4, 352, and 356, respectively.

References for 2.4.1.1.6:

[1] Palache, C.; Berman, H.; Frondel, C. (Dana's System of Mineralogy, 7th Ed., Vol. 2, Wiley, New York 1963, pp. 1/1124, 694).

[2] Deer, W. A.; Howie, R. A.; Zussman, J. (Rock-Forming Minerals, Vol. 5, Non-Silicates, Longmans, London 1962, pp. 1/371, 339).

[3] Semenov, E. I. (Mineralogiya Redkikh Zemel', Akad. Nauk SSSR, Moscow 1963, pp. 1/412, 115).

[4] Betechtin, A. G. (Lehrbuch der Mineralogie, Verlag Technik, Berlin 1953, pp. 1/681, 410 [Russian original, Moscow 1951]).

[5] Ramdohr, P.; Strunz, H. (Klockmanns Lehrbuch der Mineralogie, 16th Ed., Enke, Stuttgart 1978, pp. 1/876, 625).

[6] Doležal, J.; Povondra, P.; Šulcek, Z. (Decomposition Techniques in Inorganic Analysis, Iliffe Books, London 1968, pp. 1/224 [Czechish original, Prague 1966]).

[7] Ploshko, V. V. (Izv. Akad. Nauk SSSR Ser. Geol. **1961** No. 1, pp. 86/93, 89).

[8] Kochetkov, O. S. (Redk. Elem. Porodakh Razlichnykh Metamorf. Fatsii **1967** 105/25, 110).

[9] Rimsaite, J. (Paper Geol. Surv. Can. No. 81-23 [1982] 19/30, 24).

[10] Haapala, I.; Ervamaa, P.; Löfgren, A.; Ojanperä, P. (Bull. Geol. Soc. Finland No. 41 [1969] 117/24, 122).

[11] Molloy, M. W. (Am. Mineralogist **44** [1959] 510/32, 512/3, 516/7, 531).

[12] Hikichi, Y.; Nomura, T. (J. Am. Ceram. Soc. **70** [1987] C-252/C-253).

2.4.1.2 Cheralite $(Ca, Ce, Th)[P, Si]O_4$

Cheralite from a kaolinized pegmatite dike near Trivandrum, Travancore, southern India, was described in 1953. It is named after the ancient Dravidian kingdom Chera which corresponded roughly to the territory of Travancore (now in Kerala) [1].

The name cheralite is applied to minerals of the monazite group with a composition near to $(RE, Th, Ca, U)[P, Si]O_4$ and intermediate in composition between monazite and $CaTh[PO_4]_2$ [1], or to members of a solid-solution series apparently extending between monazite $(CePO_4)$ and artificial $CaTh[PO_4]_2$ [2]; it represents a solid solution of the form $(Ce_{1-x}Ca_{0.5x}Th_{0.5x})PO_4$ [3], containing between 25 and 75 mol% $CaTh[PO_4]_2$ [1] (see also [36] and Fig. 11, p. 243), or is a solid solution toward $CaTh[PO_4]_2$ [4]. More recently it was proposed that type cheralite contains >50 mol% $CaTh[PO_4]_2$, and cheralite was redefined to include also the pure Ca-Th end member [5]. Note further that cheralite is considered as a phosphosilicate with composition $(Ca, Th)[(SiO_4), (PO_4)]$, forming an intermediate member of the series monazite− huttonite [6], or as a phosphate-silicate of Th, Ca, and RE [2]. Cheralite is also considered as an intermediate member of the system $ThSiO_4 - CePO_4 - CaTh[PO_4]_2$ [7], or as a member of an isostructural series of monoclinic minerals with the general formula ABO_4 (where A = RE, Th, U, Ca and B = P, Si) and with monazite and huttonite as end members [8].

ThO_2- and U_3O_8-rich monazite is used synonymously with cheralite in the older literature, see, for example, [9, 10], or it is viewed as a Ca-containing monazite without Y [11].

Whereas cheralite closely resembles monazite in all of its morphological, physical, and structural properties, it differs significantly from normal monazite ($CePO_4$) in composition, since it contains Ca, Th, and U besides lanthanides as major cationic constituents [12], see also pp. 362/4.

Occurrence and Paragenesis. As is shown in the following, cheralite occurs mainly in pegmatites (see below), rarely in alkaline or granitic igneous rocks (see below), and only subordinately in sedimentary or metamorphic rocks (see p. 362).

Grains of cheralite (1 to 5 µm in size and forming aggregates of 10 to 50 µm in diameter or veinlets) are present in a matrix of highly weathered alkaline igneous rocks, interlaced with a stockwork of magnetite dikes, from Morro do Ferro, Minas Gerais, southeastern Brazil. Cheralite, as the major host of Th, is associated with monazite and zircon. Owing to its habit, it is considered as a secondary mineral, although quite possibly it may be a late-stage phase in the intrusive cycle [13]. Cheralite occurs in less altered ("fresh") brecciated dark-colored tinguaite from Morro do Ferro, and is formed as a primary mineral during emplacement of the tinguaitic rocks and accumulated by subsequent hydrothermal activity (causing sericitization and kaolinization) [14, pp. 301, 308, 310/1]. Cheralite occurs in quartz-containing medium-grained syenites of the Arga-Yuryakhskii intrusion, northeastern Yakutia [15, pp. 80, 305]. It occurs as a relatively frequent accessory mineral in the grain-size fraction 5 to 60 µm of postkinematic real igneous red feldspar-bearing biotite granite of the Velence Mountains and as a main accessory mineral, coated by monazite or apatite, in the grain-size fraction 5 to 30 µm of synkinematic anatectic fine-grained veined granite aplite from Mórágy, Mecsek Mountains, both Hungary [16, pp. 59, 61/2, 80/1, 83]. Cheralite appears to be much more abundant in muscovite granite than in two-mica or biotite granites of the Lawler Peak pluton, Yavapai county, west-central Arizona. It occurs as separate euhedral grains or as crystal masses associated with xenotime and zircon, and also as euhedral prismatic crystals inter-grown with apatite and monazite (see pp. 270 and 362) [17, pp. 179, 191/7, 199/200, 202/6].

The type cheralite is pale to dark green, brittle, and occurs sparsely distributed throughout a kaolinized pegmatite dike outcropping at Kuttakuzhi, Kalkulam-taluk, east-southeast of Trivandrum, Travancore, southern India, where it is associated with black tourmaline, small greenish yellow crystals of chrysoberyl, partly metamict zircon, considerable smoky quartz, and a few crystals of brown monazite [1]; see also [18]. Rounded grains of grayish brown cheralite, at most 0.5 cm in diameter, occur as a great rarity in Li-stage replacement bodies and fracture fillings generated by fluids that later penetrated the Haapaluoma granitic pegmatite, Peräseinäjoki-Alavus region, southern Pohjanmaa, Finland. The cheralite is associated with albite (cleavelandite), lepidolite, red tourmaline, and columbite [19, pp. 30, 53]. Black cheralite seems to be of postmagmatic (hydrothermal-metasomatic) origin in a pegmatite vein of the Osnitsk complex, Volyn region, northwestern Ukraine. It occurs as segregations and nests and also, together with smirnovskite and xenotime, as aggregates [20]. Cheralite belongs to an Fe-Nb-Th-RE-PO_4 mineralization in pegmatites which recrystallized or originated metasomatically along supposed tectonic zones in the Rumburk granite near Bogatynia, southwestern Poland. Rejuvenated faults and surrounding fissures could facilitate the penetration of the later hydrothermal solutions causing crystallization of the Fe-Nb-Th-RE-PO_4 minerals [21]. Some of the minerals identified as monazite (see pp. 257/8) in the Th veins of the Lemhi Pass area, Idaho-Montana, might be cheralite [22]. Black, sometimes weakly greenish or dark brown cheralite ("cherolite" in the original paper) occurs in pegmatites of the Arga-Yuryakhskii intrusion, northeastern Yakutia. The average content of cheralite in three pegmatite samples is $2 \times 10^{-3}\%$ [15, pp. 301, 305].

Cheralite is associated with uraniferous minerals (as, e.g., uranothorite, uranothorianite, cyrtolite, monazite, euxenite, and aeschynite) in uraniferous placer deposits [23]. Pale yellowish brown rounded grains of cheralite with an average grain size of 0.18 mm occur in black-sand concentrate of beach placers between Visakhapatnam and Bhimunipatnam, eastern India [24]. Green cheralite occurs in surface wash of the Kuttakuzhi pegmatite, east-southeast of Trivandrum, Travancore, southern India, and in the gravel of adjacent streams [1].

Small-sized cheralite is disseminated in cavities of lateritic soil particles (e.g., of sericitized orthoclase) of the Morro do Ferro deposit, Minas Gerais, southeastern Brazil [14, pp. 298, 308]. Grains of cheralite, < 10 μm in size, occur in bauxite from the San Giovanni Rotondo deposit, central Italy, and are associated with monazite, zircon, and xenotime, suggesting, at least in part, an igneous source rock [25].

A radioactive U-RE phosphosilicate similar to cheralite is associated with augite, quartz, allanite, apatite, sulfides, monazite-bastnaesite intergrowths, and traces of uraninite in a discontinuous RE- and sulfide-rich contact zone between a magnetite ore body and amphibolite near Warwick, New York [26]. Cheralite occurs sparsely in the wall rock (comprising kaolinized granitic gneiss) adjacent to the Kuttakuzhi pegmatite, east-southeast of Trivandrum, Travancore, southern India [1].

Intergrowths of cheralite with apatite and monazite occur in the two-mica and muscovite granites of the Lawler Peak pluton, Yavapai county, west-central Arizona. In the muscovite granite, cheralite is much more common than thorite, which also is intergrown with apatite. Cheralite and thorite never form intergrowths with the same monazite species [17, pp. 100/2, 105/6, 109/11, 179, 183/90, 197].

Cheralite forms **inclusions** in feldspar and/or biotite from red feldspar-bearing biotite granite of the Velence Mountains and from granite aplite of the Mecsek Mountains, both Hungary [16, pp. 61/2]. In the Lawler Peak pluton, Yavapai county, west-central Arizona, cheralite occurs as inclusions in biotite and chlorite from two-mica granite, and possibly in monazite from two-mica and biotite granites [17, pp. 100/2, 114/6, 159]. Black medium-grained aggregates of cheralite with smirnovskite and xenotime occur as veinlets in quartz and feldspar from a pegmatite vein of the Osnitsk complex, Volyn region, northwestern Ukraine [20].

Huttonite and galena occur as inclusions in cheralite from biotite granite of the Velence Mountains, Hungary [16, pp. 65, 80]. Rarely, well-rounded pale olive detrital grains of cheralite enclose remnants of thorite and zircon (cyrtolite) in a matrix of quartz, sericite, and pyrite in the Dominion Reef, western Transvaal, R.S.A. [37].

Cheralite is refractory and therefore not as susceptible to **alteration** and leaching as are Th-rich RE silicates. Thus it should not contribute much to the formation of secondary minerals within the ore body of Morro do Ferro, Minas Gerais, southeastern Brazil [14, p. 309].

Cheralite occurs as a secondary alteration product of complex U minerals (e.g., of samarskite, fergusonite, annerodite, betafite (?), davidite, uraninite, or pitchblende [10]) in Indian pegmatites [27].

Chemistry. The composition of cheralite is extremely variable, even in samples of the same locality [28]; refer also to the Table 18. A partial analysis of cheralite from Indian pegmatite shows 31.4% ThO_2, 4.48% U_3O_8, 24.55% P_2O_5, and 3.12% SiO_2 [10, 27]. Cheralite in monazite sands of Travancore and Madras, India, contains 19 to 33% ThO_2 and 4 to 6% U_3O_8 [27]. Cheralite contains 1.05% Pb in pegmatite from Kerala, India [30].

The formula of cheralite is given as $Ca_{1.08}Th_{1.15}RE_{1.62}U_{0.14}[Si_{0.34}P_{3.64}]O_4$ by [4], and as $(Ca_{1.084}La_{0.780}Ce_{0.836}Th_{1.15}U_{0.139}Pb_{0.040})_{\Sigma 4.030}(Si_{0.337}P_{3.642})_{\Sigma 3.979}O_4$ for a sample from the Kuttakuzhi

Table 18

Chemical Composition and Density of Cheralite from Various Types of Occurrence and Localities of India.

	content in % in sample number					
	1	2	3	4	5	6
ThO_2	36.25	31.64	31.50	31.00	29.45	23.99
CaO	7.45	5.99	6.30	6.00	5.73	4.01
ΣRE_2O_3	19.45[a]	27.28[b]	27.56[c]	28.29	27.43[c]	38.48[a]
U_3O_8	5.82	4.33	4.05	3.99	6.56	3.08
PbO	0.95	1.15	0.92	—	0.07	0.51
P_2O_5	28.67	27.10	26.80	27.00	25.90	26.75
SiO_2	0.88	2.08	2.10	2.82	1.15	2.65
others	0.10[d]	—	0.06[e]	0.93[f]	2.80[g]	0.49[h]
sum	99.57	99.57	99.29	100.03	99.59	99.96[i]
D in g/cm^3	5.20	5.39[j]	5.3 ± 0.1	—	—	5.28

Nos. 1 to 3 = Cheralites from pegmatite, Kuttakuzhi, Travancore: No. 1 is a chemical analysis of material from the type locality [12]; No. 2 is a microprobe analysis [8] and No. 3 is the original chemical analysis of the type material [1, 18]. — No. 4 = Cheralite from placer, Kerala [29]. — No. 5 = Cheralite from pegmatite, Cootykad Pothay, Travancore [1] (later shown to be one and the same locality as Kuttakuzhi [28]). — No. 6 = Cheralite from placer, Visakhapatnam [24].

[a] Comprises Y, La, and Ce. — [b] Comprises Y and La to Tm. — [c] Comprises Y, La to Sm, and Gd. — [d] Comprises 0.05% Fe_2O_3 and 0.05% H_2O. — [e] Comprises traces of Al_2O_3 and Fe_2O_3 and 0.06% H_2O^+.
[f] Comprises 0.50% H_2O and 0.43% unspecified rest. — [g] Comprises 2.50% Fe_2O_3 2.50, 0.18% MnO, 0.32% TiO_2, and 0.30% H_2O^+. — [h] Only H_2O^+. — [i] Erroneously given as 100.06% in the original paper. — [j] Calculated from unit-cell contents and unit-cell dimensions.

pegmatite, southern India (based on the analysis given in [1] and neglecting U) [28]. Based on more recent analyses, the formula for the latter cheralite is calculated as $(Ce_{0.580}La_{0.532}Y_{0.012}Th_{1.304}Ca_{1.260}Pb_{0.040}U_{0.196})_{\Sigma3.924}(P_{3.836}Si_{0.140})_{\Sigma3.976}O_{16}$ [12], or $(RE_{1.58}Th_{1.15}Ca_{1.03}Pb_{0.05}U_{0.15})_{\Sigma3.96}(P_{3.67}Si_{0.33})_{\Sigma4.01}O_{16}$ [8].

In cheralite, the substitution of the RE(Ce) by Th and U is advanced to a much higher degree than in monazite. The increase in cationic charge is compensated for by a corresponding substitution of Ca for the RE(Ce) and of some Si for P. Thus, in cheralite the sum of Th, U, and Ca atoms exceeds that of RE atoms with a ratio $(Th + U + Ca):\Sigma RE$ of 3:2. A complete replacement of RE by Ca and Th would result in the (at the time of the original description) hypothetical end member $CaTh[PO_4]_2$ [1], that recently was found as brabantite (see pp. 368/70). Note also that in the type cheralite the presence of Th^{4+} apparently is accompanied largely by the concomitant substitution of Ca^{2+} for RE with a relatively small substitution of Si for P [31].

In cheralite (with the general formula ABO_4, refer to p. 360), the electronegativity is 137 and 165 kcal/g-atom for group-A and 320 kcal/g-atom for group-B cations. The difference in electronegativity between cations of group A and B is 183 to 155 kcal/g-atom [34].

The chondrite-normalized RE distribution in cheralite from the Kuttakuzhi pegmatite, Travancore, India, is similar to that in monazite, especially monazite from granitic pegmatites (see original paper) [8]. Cheralite from beach placers of Kerala, India, is enriched in intermediate and heavy RE (Gd, Sm, Dy) [29].

Crystal Form and Structure. Cheralite is monoclinic [12]. It rarely shows crystal faces and possesses cleavages distinct parallel to (010) and weak parallel to (100) [1].

The space group of cheralite is $P2_1/n$ [32] or C_{2h}^5 (No. 14) [28]; Z = 4 [12]. Variable unit-cell parameters of cheralite indicate its extremely variable chemical composition [28]. The lattice constants of cheralite from various localities are given as follows:

| cheralite from | lattice constants in Å and angle β | | | | Ref. |
	a	b	c	β	
Travancore	6.74	7.00	6.43	104.6°	[1]
	6.70	6.87	6.39	103°24′	[32]
	6.717 ±0.005	6.920 ±0.005	6.434 ±0.005	103°50′±5′	[28]
	6.751$_5$±0.0005	6.962$_5$±0.0005	6.468$_5$±0.0005	103°53′	[8]
eastern India	6.76	6.99	6.46	104°22′	[24]
Finland	6.72 ±0.02	6.95 ±0.02	6.44 ±0.02	103°55′±5′	[19, pp. 53/4]

The X-ray diffraction pattern of cheralite from Kerala is definitely poor in the region of small interplanar spacings (d values) with considerable line broadening, indicating at least partial disorder [30]. In the following table are given the interplanar spacings with corresponding relative intensities in parentheses (vs = very strong, s = strong, ms = medium strong) for only the strongest X-ray powder diffraction reflections of cheralite from various localities; n = total number of observed d values given in the original paper:

| cheralite from | d value (in Å) for plane indexed as | | | | | n | Ref. |
	($\bar{1}$11)	(200)	(120)	($\bar{1}$12), (012)	($\bar{3}$11)		
Travancore	4.14 (ms)	3.26 (s)	3.07 (vs)	2.86 (s)	—	19	[1]
	4.167(25)	3.277(58)	3.074(100)	2.862$_5$(65)	2.130(14)	45	[8]
eastern India[a]	4.199(50)	3.27(70)	3.1(100)	2.867(70)[b]	2.135(50)	30	[24]
Ukraine[c]	4.17(3)	—	3.05(10)	2.85(10)	2.14(4)	6	[20]

[a] Additional strong d value for (230) is 1.908 Å(70). — [b] Indexed as (012). — [c] No plane indices given in the original paper; additional strong d values are 3.74(4) and 1.872 Å(4).

As compared to unheated cheralite from the Osnitsk complex, Volyn region, northwestern Ukraine, a sample heated to 800°C shows a much greater number of d values (29 and 6, respectively). The strongest d values of the heated sample are 3.40(10), 3.20(10), 3.00(10), 2.80(10), and 2.49 Å (10) [20]. Also note that cheralite from beach placers between Visakhapatnam and Bhimunipatnam, eastern India, shows sharp X-ray peaks only after heating [24].

Cheralite is isotypic [4] or isostructural with monazite [2]; note also that it is isostructural with both the silicate huttonite and the phosphate monazite [7]; see also [11]. A refinement

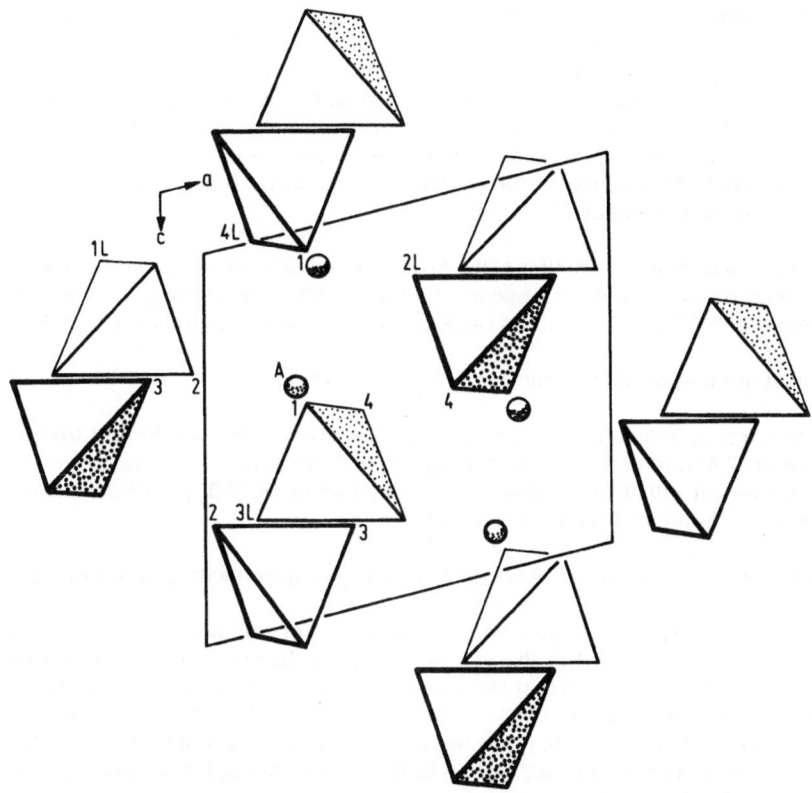

Fig. 14. Crystal structure of cheralite viewed along the b axis from [12]. Numbered oxygen atoms located at corners of phosphate tetrahedra belong to Ce atom A.

of the cheralite structure (with a final R factor of 0.084) confirms the model given for monazite (see p. 343). Each P atom is surrounded by four oxygens in tetrahedral coordination at an average distance of 1.541 Å from the P atom. The tetrahedron is nearly regular with respect to bond lengths but the bond angles O-P-O vary from 104.8° to 114.2°. Cerium atoms (represented by stippled circles in **Fig. 14**) are bonded to two each of the four oxygens at distances ranging from 2.403 to 2.564 Å. Longer Ce-O bonds (marked with an L in the figure) occur with distances ranging from 2.778 to 3.945 Å (see original paper for detailed bond lengths and angles). Each O atom is bonded to three Ce atoms by two short and one long bond. In addition, each O atom is bonded to one P atom. The coordination around each O atom is thus fourfold. However, the ionic charge of each O atom is satisfied by two contributions from Ce, assuming each at $+3/8$ position, and one from P, assuming it at $+5/4$ position. The increase of the long Ce-O bond with decreasing average short Ce-O bonds and P-O bond cannot be readily assessed with a purely ionic model [12].

Optical Properties. Cheralite is pale to dark green [1], pale yellowish brown [24], grayish brown [19, p. 53], dark brown [15, p. 305], or black [15, p. 305], [20]; under the microscope it is pale green. Cheralite has a resinous to vitreous luster and a white streak [1].

Cheralite is biaxial positive. The refractive indices are $n_\alpha = 1.779$, $n_\beta = 1.780$, $n_\gamma = 1.816$ for the type cheralite from Travancore, India [1], and $n_\alpha = 1.776 \pm 0.002$, $n_\beta = 1.778 \pm 0.002$, $n_\gamma = 1.831 \pm 0.003$ for cheralite in an Li-stage replacement body from the Haapaluoma pegmatite, Peräseinäjoki-Alavus region, southern Pohjanmaa, Finland [19, pp. 53/4]. The refractive indices range from 1.720 to 1.730 for cheralite from the Osnitsk complex, Volyn region, northwestern Ukraine [20]. An increase in refractive indices due to Th replacing Ce and La (as is also described for monazite, see p. 350) appears to be balanced in cheralite by the effect of Ca replacing Ce and La [33].

Note that cheralite from pegmatites of the Arga-Yuryakhskii intrusion, northeastern Yakutia, is isotropic with refractive indices between 1.99 and 2.00. The isotropy does not disappear after annealing at 800°C for 3 h, but the refractive indices increase to 2.07 [15, p. 305].

The optical orientation of cheralite is $\alpha = b$, $\gamma \wedge c = 7°$ [1].

The dispersion of optical axes is not perceptible. Measured on different fragments, 2 V varies between 17.4° and 19.0° (with an average of 18.1° for some twenty determinations) and is in good agreement with the calculated optical axial angle of 17.9° [1]. Cheralite from beach placers of eastern India shows a 2 Vγ of 16° [24].

Cheralite is weakly pleochroic: α and β = green, γ = green with yellow tinge [1].

The IR spectrum of type cheralite is very similar to that of monazite from Skardalen quarry, Evje, Norway, lacking the band in the 950 cm^{-1} region. In an absorption spectrum in the visible range, the bands in the 570 to 580 nm and 525 nm regions for the type cheralite are similar to those for gem monazite (see p. 352) and are ascribed to RE, notably to Nd. A broad band in the 640 to 660 nm region in type cheralite, which is not present in monazite, may be due to the significant content of 4.33% U_3O_8 in the former (as opposed to 0.4% U_3O_8 in the monazite) and to Tm (and Ho?) [8].

Other Physical Properties. The Mohs hardness of cheralite is 5. Fracture is uneven. The density D is 5.3 ± 0.1 g/cm^3 [1] or 5.20 ± 0.05 g/cm^3 for the type cheralite. The density calculated from the analysis given in [1] and from measured unit-cell dimensions is 5.47 g/cm^3 [28]. For other densities, see p. 363. Density decreases from 4.32 g/cm^3 before annealing to 4.27 g/cm^3 after annealing at 800°C for 3 h in cheralite from pegmatites of the Arga-Yuryakhskii intrusion, northeastern Yakutia [15, p. 305].

Metamictization is not observed in cheralite [34]; it is not totally, but possibly partially, metamict [35]. For isotropic cheralite, see above. Metamictization is not prominent in cheralite from beach placers between Visakhapatnam and Bhimunipatnam, eastern India. Depending on the variable U and Th contents, this cheralite is less radioactive (α = 3360 c/h/mg, β = 2100 c/g/min) than cheralite from Kerala (α = 4398 c/h/mg, β = 2725 c/g/min) [24].

Chemical and Thermal Behavior. The chemical behavior of cheralite probably is similar to that of monazite, but no data are given in the literature.

For changes of d values, refraction, and density of cheralite during annealing, see the foregoing paragraphs.

References for 2.4.1.2:

[1] Bowie, S. H. U.; Horne, J. E. T. (Mineral. Mag. **30** [1953/55] 93/9).
[2] Frondel, J. W.; Fleischer, M.; Jones, R. S. (Geol. Surv. Bull. [U.S.] No. 1250 [1967] 1/69, 15).
[3] Hikichi, Y.; Hukuo, K.; Shiokawa, J. (Nippon Kagaku Kaishi **1978** 1635/40; C.A. **90** [1979] No. 61971).
[4] Speer, J. A. (Rev. Mineral. **5** [1980] 113/35, 123).
[5] Fleischer, M.; Chao, G. Y.; Francis, C. A. (Am. Mineralogist **66** [1981] 878/9).
[6] Sidorenko, G. A. (Geol. Mestorozhd. Redk. Elem. No. 26 [1966] 5/23, 12/3).
[7] Frondel, C. (Proc. Intern. Conf. Peaceful Uses At. Energy, Geneva 1955 [1956], Vol. 6, pp. 568/77, 570).
[8] Bowles, J. F. W.; Jobbins, E. A.; Young, B. R. (Mineral. Mag. **43** [1980] 885/8).
[9] Davidson, C. F. (Mining Mag. [London] **94** [1956] 197/208, 198).
[10] Wadia, D. N. (Proc. Intern. Conf. Peaceful Uses At. Energy, Geneva 1955 [1956], Vol. 6, pp. 163/6).

[11] Roubault, M. (Geologie de l'Uranium, Masson, Paris 1958, pp. 1/462, 411).
[12] Finney, J. J.; Rao, N. N. (Am. Mineralogist **52** [1967] 13/9).
[13] Eisenbud, M.; Lei, W.; Ballad, R.; Krauskopf, K.; Franca, E. P.; Cullen, T. L. (PUC-TN-09-81 [1981] 1/34, 7).
[14] Barretto, P. M. C.; Fujimori, K. (Chem. Geol. **55** [1986] 297/312).
[15] Nekrasov, I. Ya. (Magmatizm i Rudonosnost' Severo-Zapadnoi Chasti Verkhoyano-Chukotskoi Skladchatoi Oblasti, Akad. Nauk SSSR, Moscow 1962, pp. 1/412).
[16] Pantó, G. (Acta Geol. Acad. Sci. Hung. **19** [1975] 59/93).
[17] Silver, L. T.; Woodhead, J. A.; Williams, I. S.; Chappell, B. W. (GJBX-7-84 [1984] 1/412).
[18] Bowie, S. H. U.; Horne, J. E. T. (Great Britain Geol. Surv. At. Energy Div. Rept. No. 134 [1952] 1/5) from Overstreet, W. C. (Geol. Surv. Profess. Paper [U.S.] No. 530 [1967] 1/327, 68).
[19] Haapala, I. (Bull. Comm. Geol. Finlande No. 224 [1966] 1/98).
[20] Vynar, O. N.; Remeshilo, B. G. (Mineral. Sb. [Lvov] **30** No. 2 [1976] 91/5).

[21] Banas, M.; Kucha, H. (Proc. 6th Quadrenn. IAGOD Symp., Tiflis 1982 [1984], Vol. 1, pp. 241/6).
[22] Staatz, M. H. (Geol. Surv. Bull. [U.S.] No. 1351 [1972] 1/94, 71).
[23] Page, L. R. (Intern. Geol. Congr. Rept. 21st Session Norden, Copenhagen 1960, Pt. 15, pp. 149/64, 155).
[24] Rao, A. T.; Rao, C. S. P. (Indian J. Mar. Sci. **9** [1980] 214/6).
[25] Bárdossy, G.; Pantó, G. (Congr. ICSOBA, Nice 1973, Vol. 3, pp. 47/53).
[26] Baillieul, T.; Indelicato, G. (Econ. Geol. **76** [1981] 167/71).
[27] Wadia, D. N. (Sci. Cult. [Calcutta] **21** [1956] 561/5).
[28] Rao, N. N.; Finney, J. J. (Am. Mineralogist **50** [1965] 507/8).
[29] Semenov, E. I.; Upendran, R.; Subramanian, V. (J. Geol. Soc. India **19** [1978] 550/7, 552/4, 556).
[30] Sivaramakrishnan, V.; Venkatasubramaniam, V. S. (Proc. Natl. Inst. Sci. India A **25** [1959] 278/80).

[31] Gramaccioli, C. M.; Segalstad, T. V. (Am. Mineralogist **63** [1978] 757/61).
[32] Khubchandani, P. G. (Current Sci. [India] **25** [1956] 323).
[33] Deer, W. A.; Howie, R. A.; Zussman, J. (Rock-Forming Minerals, Vol. 5, Non-Silicates, Longmans, London 1962, pp. 1/371, 341).

[34] Povarennykh, A. S. (Zap. Vses. Mineral. Obshch. **85** [1956] 593/7).

[35] Floran, R. J.; Abraham, M. M.; Boatner, L. A.; Rappaz, M. (Sci. Basis Nucl. Waste Management **3** [1981] 507/14, 512).

[36] Rose, D. (Neues Jahrb. Mineral. Monatsh. **1980** 247/57, 254/5).

[37] Smits, G. (MINTEK – M278 [1987] 1/33, 13, 20/1).

2.4.1.3 Brabantite $CaTh[PO_4]_2$

Brabantite was described in 1980 and is named for its type locality, the Brabant pegmatite in the Karibib district, South-West Africa (Namibia) [1, pp. 247/8]. The mineral "lingaitukuang" (or "cathophorite", as named in [2]) is synonymous with brabantite. The name brabantite has priority over "lingaitukuang" as the latter was published without approval by the International Mineralogical Association (I.M.A.), Commission on New Minerals and Mineral Names. However, both brabantite and "lingaitukuang" seem to be unnecessary, as cheralite, the intermediate member of the monazite-$CaTh[PO_4]_2$ series, could be redefined to include the pure Ca-Th end member (see p. 360) [3].

Occurrence and Paragenesis. Brabantite occurs as a primary mineral in the younger shell of a zoned pegmatite vein located on the farm Brabant, Karibib district, South-West Africa (Namibia). It occurs in a zone of microcrystalline muscovite and forms aggregates of elongated crystals reaching up to 1.5 cm in diameter. Other primary minerals are thorite and uraninite [1, pp. 248/9]. Granular aggregates and grains, 0.2 to 0.3 mm in size, of "lingaitukuang" occur in a rare-metal pegmatite in Xingjiang, northern P.R.C., and are associated with hafnian zircon, spodumene, mangano-columbite, and albite [4].

Alteration. Brabantite and other primary U and Th minerals are altered partially by later processes to brockite, thorogummite, uranophane, β-uranophane, phosphuranylite, and meta-autunite in the South-West African pegmatite vein [1, p. 249].

Synthesis. The compound $CaTh[PO_4]_2$ can be prepared from a homogenized mixture of stoichiometric amounts of $Th(NO_3)_4 \cdot 5 H_2O$, $(NH_4)_2HPO_4$, and $CaCO_3$ after heating at 1200°C for 13 h [1, p. 252]; see also [5]. Up to a temperature of 1600°C, the synthetic compound $Ca_{0.5}Th_{0.5}PO_4$ shows huttonite structure [5]. The compound $CaTh[PO_4]_2$ is also synthesized using Th orthophosphate $Th_3(PO_4)_4 \cdot 4 H_2O$, mixed in equimolecular proportions with tricalcium orthophosphate $Ca_3(PO_4)_2$. Heating of these reagents in an electrical furnace at 1400°C leads to a single phase isostructural with monazite. Fusion of the above reagents in an oxygen and coal-gas flame produces $CaTh[PO_4]_2$ and a ThO_2 phase, suggesting that dissociation occurs at higher temperature [6].

Chemistry. The chemical composition and density of brabantite (analysis No. 1) [1, p. 251] and "lingaitukuang" (No. 2) [4] is given in the table on p. 369.

The theoretical formula $CaTh[PO_4]_2$ of brabantite requires 12.14% CaO, 57.14% ThO_2, and 30.72% P_2O_5. The formula of the type brabantite (after deduction of 3% nonessential H_2O and recalculation to 100%) is given as $(Ca_{1.000}Mg_{0.065}Mn^{2+}_{0.021}Al_{0.068}Fe^{3+}_{0.003})_{\Sigma 1.157}Th_{0.936}$ $[P_{1.832}Si_{0.177}]_{\Sigma 2.009}O_8$ [1, p. 251]. The "lingaitukuang" formula is $[Ca_{0.52}Th_{0.43}RE_{0.04}U_{0.01}]_{\Sigma 1.00}$ $[(PO_4)_{0.89}(SiO_4)_{0.06}]_{\Sigma 0.95}$ [4].

Brabantite contains $\sim$10% huttonite in solid solution, which is comparable to cheralite. "Lingaitukuang" has a similar chemical composition but contains a small amount of $CaU(PO_4)_2$ [7, p. 132].

	content in % in sample number	
	1	2
ThO_2	52.65	51.12
CaO	11.94	13.33
MgO	0.56	tr.
Al_2O_3	0.74	—
Fe_2O_3	0.05	0.66
ΣRE_2O_3	—	3.05
P_2O_5	27.68	28.78
SiO_2	2.27	1.60
H_2O	3.07	—
others	0.32[a]	1.64[b]
sum	99.28	100.18
D in g/cm^3	4.72	5.20

[a] Only MnO. — [b] Comprises 0.12% TiO_2, 1.23% UO_2, and 0.29% UO_3.

Crystal Form and Structure. Brabantite is monoclinic. Two cleavages parallel to (100) and (001), in analogy to monazite, are observed [1, pp. 249/50]. "Lingaitukuang" shows no cleavage [4].

Brabantite is isostructural with monazite [7, p. 132]. The space group is $P2_1/n$ (No. 14) and Z = 2. The lattice constants are a = 6.726 ± 0.006, b = 6.933 ± 0.005, c = 6.447 ± 0.012 Å, β = 103°53′ ± 16′, V = 291.84 Å^3 for unheated brabantite and a = 6.718 ± 0.004, b = 6.916 ± 0.003, c = 6.442 ± 0.009 Å, β = 103°46′ ± 11′, V = 290.70 Å^3 for brabantite heated to 900°C. The X-ray powder diffraction patterns of unheated and heated material are nearly identical with those of monazite, cheralite, and synthetic $CaTh[PO_4]_2$ (see below and original paper). On heating at 900°C for 10 h the reflections become sharper and two reflections (at 6.02 and 4.40 Å) caused by admixed brockite disappear [1, pp. 249/50, 252, 254]. The X-ray powder diffraction pattern of "lingaitukuang" is clear and sharp and is essentially identical to that of synthetic $CaTh[PO_4]_2$ [4]. For the strongest X-ray powder diffraction reflections the following indexed interplanar spacings (d values), with their relative intensities in parentheses, are given for unheated and heated brabantite (from a total of 25 d values given) [1, p. 250] and for "lingaitukuang" (from a total of 18 d values given) [4]:

mineral	d value (in Å) for plane indexed as				
	($\bar{1}$11)	(200)	(120)	($\bar{1}$12), (012)	(312)
brabantite, unheated	4.15(30)	3.26(70)	3.06(100)	2.85(75)	1.947(30)
brabantite, heated	4.14(30)	3.26(70)	3.06(100)	2.84(75)	1.945(30)
"lingaitukuang"	4.13(5)[a]	3.25(6)	3.05(9)	2.84(10)[b]	1.94(4)[c]

[a] Indexed as (111). — [b] Indexed as (112), (012). — [c] Indexed as ($\bar{2}$12).

Optical Properties. Brabantite aggregates are grayish brown, altering to reddish brown at the rim of larger masses because of minute hematite platelets, and have a dull luster. In thin section, the color is brown-gray [1, pp. 249/50]. "Lingaitukuang" is pale yellow to brownish green with a greasy luster [4].

Brabantite is optically biaxial. Only the refractive index $n_\beta = 1.73$ and birefringence $n_\gamma - n_\alpha = 0.05$ could be measured owing to an extreme optical inhomogeneity, which is considered as an effect of different degrees of metamictization [1, p. 251]. "Lingaitukuang" is optically biaxial positive with $n_\alpha = 1.691$, $n_\beta = 1.696$, $n_\gamma = 1.725$, and $2V = 44°$ [4].

Pleochroism is not visible in brabantite [1, p. 250] and weak in "lingaitukuang" [4]. In polished section, the reflection of brabantite is higher ($\sim 7\%$) than that of surrounding mica (~ 4 to 5%), and the mineral shows gray color. Brabantite is nonfluorescent under short- and longwave UV radiation. IR absorption spectra of unheated and heated brabantite with a stretching-vibration peak at 1060 cm^{-1}, corresponding to the PO_4 group, are practically identical with that of monazite (see p. 352). The unique difference lies in some weak absorption peaks in the region from 400 to 460 cm^{-1}. Absorption peaks caused by OH^- or CO_3^{2-} groups are not observed [1, pp. 250/1, 255].

Other Physical Properties. The Mohs hardness is 5½ for brabantite [1, p. 250] and small for "lingaitukuang" [4]. For the densities of brabantite and "lingaitukuang", see p. 369. The density of brabantite varies and is 4.72 g/cm^3 for unheated and 5.02 g/cm^3 for heated material. Densities calculated from X-ray data are 5.26 and 5.28 g/cm^3, respectively [1, p. 250].

Brabantite is partially metamict [1, p. 249].

Chemical and Thermal Behavior. "Lingaitukuang" dissolves in HCl [4].

The DTA curve of brabantite shows a very broad exothermic reaction between 100 and 450°C which probably is due to recrystallization of the partially metamict mineral combined with dehydration of the nonessential H_2O content. The TGA curve shows a weight loss of 3% in the region between 80 and 450°C with a maximum between 100 and 150°C, which obviously is caused by admixed brockite, and there are no further significant changes up to 1100°C [1, p. 252].

References for 2.4.1.3:

[1] Rose, D. (Neues Jahrb. Mineral. Monatsh. **1980** 247/57).
[2] Wang, H.-C. (Kexue Tongbao **23** [1978] 743/5 from C.A. **91** [1979] No. 24051).
[3] Fleischer, M.; Chao, G. Y.; Francis, C. A. (Am. Mineralogist **66** [1981] 878/9).
[4] Wang, H.-C. (Kexue Tongbao **23** [1978] 743/5 from [3]).
[5] Schwarz, H. (Z. Anorg. Allgem. Chem. **334** [1964] 175/85, 177/9).
[6] Bowie, S. H. U.; Horne, J. E. T. (Mineral. Mag. **30** [1953/55] 93/9).
[7] Speer, J. A. (Rev. Mineral. **5** [1980] 113/35, 132).

2.4.1.4 Unnamed Fe-Th Phosphate $Fe^{2+}Th[PO_4]_2$

The unnamed Fe^{2+}-Th phosphate, described in 1980, belongs to the monazite group and seems to be an intermediate member of the solid-solution series extending between monazite, huttonite, cheralite, and $Fe^{2+}Th[PO_4]_2$.

Occurrence and Paragenesis. Small crystals of $Fe^{2+}Th[PO_4]_2$ are found in pegmatites and hydrothermal veins which occur in the Lużyce (Lusatia) granitoid rocks near Bogatynia, Lower Silesia, Poland, and which contain several Th minerals related to the monazite, rhabdophane, Th-ningyoite, and thorogummite groups as well as hydrothermal alteration and weathering products of these Th minerals. The amount of $Fe^{2+}Th[PO_4]_2$ varies in three samples from 18 to 30, mean 22.9 vol%. The mineral is intimately associated with hydrated Th phosphates (comprising the unnamed minerals $Fe^{2+}Th[PO_4]_2 \cdot H_2O$, see pp. 378/80, and $(Fe^{2+},Ca)_{1-x(?)}$ $Th_{1-x(?)}$ $(Fe^{3+},RE)_{2x(?)}$ $[PO_4]_2 \cdot 1-3\ H_2O$, see pp. 386/8) and other contaminants.

Intergrowths and Alteration. Unnamed $Fe^{2+}Th[PO_4]_2$ usually forms fine intergrowths with both the above given unnamed phosphates, and probably also with $Fe_3^{2+}(H_2O)[PO_4]_2$, which could not all be separated.

The alteration of unnamed $Fe^{2+}Th[PO_4]_2$ to submicrocrystalline $Fe^{2+}Th[PO_4]_2 \cdot H_2O$ by attaching water is often accompanied by the formation of fine-grained amorphous Fe hydroxides disseminated throughout the Fe-Th minerals. Probably $Th_3[PO_4]_4$ is formed during hydrothermal decay of $Fe^{2+}Th[PO_4]_2$ into goethite, thorogummite, and $Th(OH)_4$, which is observed at all stages of evolution for rocks penetrated by ground waters. Some PO_4 and SiO_4 are leached from Th veins and are encountered in secondary limonite cores and veinlets leading to the formation of $Fe_3^{2+}(H_2O)[PO_4]_2$. For crusts of $(Fe^{2+},Ca)_{1-x(?)}$ $Th_{1-x(?)}$ $(Fe^{3+},RE)_{2x(?)}$ $\cdot 1-3\ H_2O$ on altered $Fe^{2+}Th[PO_4]_2$, see p. 386.

Chemistry. The theoretical composition of unnamed $Fe^{2+}Th[PO_4]_2$ is calculated as 55.31% ThO_2, 15.05% FeO, and 29.79% P_2O_5. Contents of the main elements, recalculated from atomic percentages, of $Fe^{2+}Th[PO_4]_2$ from three rock samples of Bogatynia are given in the following table (for atomic proportions of OH and H_2O, calculated to complete analyses to 100%, see original paper); Nos. 1 to 5 are from rock samples S-9 and S-9B, No. 6 from rock sample G-1a:

	content in % in sample number					
	1	2	3	4	5	6
ThO_2	72.95	72.49	65.66	65.32	62.70	46.54
FeO	3.86	7.41	9.91	8.11	13.25	26.63
CaO	0.70	0.11	1.12	1.54	0.84	0.70
Ce_2O_3	$\leqq 0.23$	$\leqq 0.23$	$\leqq 0.23$	$\leqq 0.23$	$\leqq 0.23$	$\leqq 0.23$
P_2O_5	15.58	12.14	7.79	9.39	11.69	9.85
SiO_2	2.35	2.57	8.34	6.20	2.35	7.91
sum	~95.67	~94.95	~93.05	~90.79*)	~91.06	~91.86

*) Additionally contains 0.15% Cu.

Unnamed $Fe^{2+}Th[PO_4]_2$ contains traces of C and F.

The formula derived from the above analyses is $(Th, Fe^{2+}, Ca, Fe^{3+}, RE ...)[PO_4, SiO_4, OH] \cdot$ $x\ H_2O$ with $x \ll 0.5$. The possible oxidation of Fe^{2+} to Fe^{3+} requires additional valence compensation by a more extensive substitution of P by Si than is observed in cheralite. In addition, the incorporation of some OH may balance the anion-cation system. Both substitutions cause the variable $Th:M^{2+}$ ratio (with M^{2+} = bivalent metals) in the unnamed Fe^{2+}-Th phosphate, which decreases with decreasing $P:Si$ ratio.

Crystal Form and Structure. Unnamed $Fe^{2+}Th[PO_4]_2$ is assumed to be monoclinic as follows from its isostructural relation to monazite and cheralite.

Space group presumably is $P2_1/n$; Z = 4. The lattice constants are a = 6.68 ± 0.03, b = 6.97 ± 0.05, c = 6.41 ± 0.03 Å, $\beta = 103.88° \pm 0.18°$, and a:b:c = 0.957:1:0.919. For the strongest out of about 40 X-ray powder diffraction reflections, the interplanar spacings (d values) and corresponding relative intensities (I) of unnamed $Fe^{2+}Th[PO_4]_2$ are as follows:

sample from rock S-9		sample from rock G-1a	
d in Å	I	d in Å	I
4.69	6.5	4.73[*]	7.0
4.46	8.0	4.20[*]	10.0
4.36	9.0	3.56	7.0
3.08	10.0	3.24	8.0
2.86	10.0	3.24	7.0
2.16	6.5	2.68[*]	9.0

[*] Double line.

Optical Properties. Unnamed $Fe^{2+}Th[PO_4]_2$ is subtranslucent to opaque. Its color is reddish brown or brown; in transmitted light, it is red-brown to red-orange. The luster is variable, usually resinous or waxy.

Due to radioactive decay, $Fe^{2+}Th[PO_4]_2$ often is isotropic in transmitted light. Sometimes it shows radial-spherulitic anisotropy. Often it is birefringent. In reflected light, it shows a strong radial-spherulitic anisotropy and a distinct bireflectance. Internal reflections are brown, orange-red, sometimes red-pinkish. The reflectance R is ~14% in air and changes when $Fe^{2+}Th[PO_4]_2$ attaches H_2O.

$Fe^{2+}Th[PO_4]_2$ is nonfluorescent in UV light. An electron beam causes emission of a soft orange light.

Other Physical Properties. The Mohs hardness of $Fe^{2+}Th[PO_4]_2$ is 5, and the microhardness (VHN) varies between 270 and 520 kg/mm^2 at 150 g load, depending on the degree of alteration. Fracture is conchoidal to uneven. $Fe^{2+}Th[PO_4]_2$ is brittle.

Thermal Behavior. Data on the thermal behavior of pure $Fe^{2+}Th[PO_4]_2$ are not available since it forms fine intergrowths with the other two Fe-Th phosphates, and one mineral could not be separated from the other. The DTA, DTG, and TGA studies on mixtures of the Fe-Th phosphates (partly with $Fe_3^{3+}(H_2O)[PO_4]_2$ and thorogummite) undoubtedly show the presence of both OH and Fe^{2+} in the structure of the phosphates, but the presence of OH in the structure of $Fe^{2+}Th[PO_4]_2$ is not quite certain.

Reference for 2.4.1.4:

Kucha, H. (Mineral. Polonica **10** No. 1 [1979] 3/28, 4/7, 9/14, 16/7).

2.4.1.5 Unnamed Th-Y Phosphate and Th-Y-Ca-P Phase

A red-brown metamict and variably weakly birefringent unnamed Th-Y phosphate from two depositions of Th-U ore related paragenetically to an alkali (potassium) syenite (of the USSR?) was described in 1968. It is associated with thorite, huttonite, hypogene U minerals (including coffinite, uraninite and its variety nasturan), hydromica, apatite (dahllite), Fe^{3+}

phosphate, and quartz. The unnamed Th-Y phosphate is isostructural with monazite and contains 12.8% ThO_2, 13.8% RE_2O_3, 10.37% FeO, 1 to 6% CaO and MgO, 17.80% P_2O_5, 9.83% Al_2O_3, and 9.83% H_2O [1]; refer also to [2].

As a late-stage accessory primary mineral, or as an alteration product of other radioactive minerals, an unnamed Th-Y-Ca-P phase was mentioned in 1984 from muscovite granite of the Lawler Peak pluton, Yavapai county, west-central Arizona. Its formula is given as $(Y_{0.36}Th_{0.22}Ca_{0.21}U_{0.048}Yb_{0.04}Ce_{0.03}Zr_{0.03}Dy_{0.03}Fe_{0.03}Er_{0.03}Nb_{0.02}Gd_{0.01}Ti_{0.01}Sm_{0.01})$ $(P_{0.59}Si_{0.25}Al_{0.05})$-$O_4F_{0.40}$; the analysis gives 24.2% ThO_2, 17.2% Y_2O_3, 12.8% Σ (Ce, Nd, Sm, Gd, Tb, Dy, Er, Yb, Lu)$_2O_3$, 4.9% CaO, 5.4% UO_2 (oxidation state not determined), traces of PbO, 1.5% ZrO_2, 0.8% FeO (oxidation state not determined), 0.9% Nb_2O_5, 0.5% Ta_2O_5, 0.5% TiO_2, 0.1% MgO, 17.4% P_2O_5, 6.2% SiO_2, 1.0% Al_2O_3, 3.2% F, traces of H_2O, sum 96.6% [3].

References for 2.4.1.5:

[1] Rogova, V. P. (Geologiya i Voprosy Genezisa Endogennykh Uranovykh Mestorozhdenii, Nauka, Moscow 1968, pp. 319/33, 319, 321).
[2] Bonshtedt-Kupletskaya, E. M. (Zap. Vses. Mineral. Obshch. **100** [1971] 77/92, 88).
[3] Silver, L. T.; Woodhead, J. A.; Williams, I. S.; Chappell, B. W. (GJBX-7-84 [1984] 1/412, 175/6, 198, 209).

2.4.2 Brockite $(Ca, Th, Ce)[PO_4] \cdot H_2O$

Brockite from a vein in altered granitic rock of the Wet Mountains, Colorado, was described in 1962. It is a hydrous Th phosphate, related to rhabdophane [1, p. 1346], and belongs to the hexagonal rhabdophane group [2]. It is named for Maurice Brock of the U.S. Geological Survey [1, p. 1347].

Occurrence and Paragenesis. Aside from an occurrence as primary mineral in granite, brockite occurs mainly in veins, where it is often an alteration product of primary Th minerals, and is subordinate in disseminated deposits and sedimentary rocks, as is shown in the following paragraphs.

A mineral considered as brockite crystallized during the magmatic stage in the interspaces between quartz and lepidolite from peraluminous albitic topaz-lepidolite leucogranite of Echassières, Beauvoir, Brittany, France. It is associated with a poorly crystalline RE-rich calcic phosphate (possibly representing a more or less altered monazite) [3].

Brockite occurs in quartz-feldspar-Fe oxide-bearing Th veins occupying fractures, breccia zones, or small faults from widely scattered districts in the United States [4, pp. 495/6]. Type brockite is found as massive aggregates and as earthy coatings in Th veins and altered granitic rocks of the Wet Mountains, Custer county, Colorado. Massive aggregates of very fine-grained (on average 20 µm in length) reddish brown brockite form nodules up to 35 mm in diameter in deeply weathered granitic rocks near the Bassick mine. These nodules, along with blebs of brecciated quartz and albite and vein quartz, are enclosed in pale yellow cryptocrystalline earthy material which is largely brockite. Some cracks in apatite and/or quartz, filling the cores of irregular cavities in the ore, contain brockite and hematite. In addition, brockite occurs as yellow earthy coatings in aggregates of red kalifeldspar, typical of the potassic alteration zones cutting granitic rocks, from the Hardwick mine and from the Nightingale shaft. Thorite, barite, and hematite fill the spongy matrix. The finely crystalline brockite formed at the expense of thorite adjacent to small cavities [1, pp. 1347/8]. In other

Th veins, for example, at Monroe Canyon, Utah, and in two veins from the Laughlin Peak area, New Mexico, brockite is the only Th mineral identified, and it is the most abundant Th mineral in several veins of the Lemhi Pass district, Idaho-Montana [4, pp. 496, 498, 500]. Thus, in irregularly banded Black Rock Th veins, Montana, with coarsely crystalline quartz, limonite, and Mn oxides, fine-grained tan, yellow, or orange anhedral brockite occurs together with microcline, barite, apatite, thorite, chrysocolla, magnetite, hematite (specularite), and monazite [5, pp. A45, A68]. Fine-grained yellowish orange to reddish orange or tan brockite is moderately abundant in the Th-bearing quartz or quartz-microcline veins occupying small faults, fractures, and shears of the Lemhi Pass quadrangle. Brockite is associated with Fe-oxide minerals, thorite, and/or monazite [6, pp. 55, 57, 60/4, 71]; note also that brockite occurs in almost one fifth of the veins in the Lemhi Pass quadrangle [7] or is fairly common in nine veins in association with Fe oxides, thorite, monazite, allanite, and bastnaesite [8]. Variously colored brockite occurs especially in hematite-rich samples from five Th- and RE-containing veins near the middle of an oval-shaped intrusive rock body in the southern Bear Lodge Mountains, Crook county, Wyoming [9, pp. D2, D34/D36]. Dark reddish brown cryptocrystalline aggregates of a Th-bearing phase, similar to brockite (see [1, p. 1346]), are associated with Fe-containing minerals and minor amphibole and rutile in the heavy-mineral fraction of a small vein in Precambrian rocks southwest of Gunnison, Colorado [10]. As relics on the faces of columbite, dark or light red-brown quite fine-grained brockite usually forms equidimensional aggregates with indistinct crystal faces in a replacement body caused by Na-rich fluids penetrating the Haapaluoma granitic pegmatite of the Peräseinäjoki-Alavus region, southern Pohjanmaa, Finland [11, pp. 30, 36, 54]. Brockite occurs in Th-bearing feldspar-pyrochlore-Fe oxide pegmatites of the Rumburk granite near Bogatynia, southwestern Poland, which recrystallized or originated metasomatically along supposed tectonic zones of the granite. The rejuvenated faults and surrounding fissures could facilitate the penetration of later hydrothermal solutions causing the crystallization of the Fe-Nb-Th-RE-PO$_4$ minerals [12]. Admixed with brabantite (see pp. 369/70), brockite occurs as a later alteration product of primary Th minerals in the younger shell of a zoned pegmatite vein located on the farm Brabant, Karibib district, South-West Africa (Namibia) [13].

Brockite occurs in five disseminated Th-RE deposits within fractured Tertiary magmatic rocks (comprising mainly trachyte or phonolite) that form the core of the Bear Lodge dome, Crook county, Wyoming, and are cut by numerous thin, crisscrossing veinlets containing the Th-RE minerals [9, pp. D16, D32, D43, D46/D48].

Accessory brockite occurs in F-Ba-Th-RE deposits (related to alkaline volcanism) in Tertiary graywackes and pyroclastites from Kizilcaören, western Anatolia, Turkey [19].

In the cement of fresh-water and marine sedimentary U deposits of Stráž, northern Bohemia, Czechoslovakia, minute grains of authigenic brockite are closely associated with hydrozircon and pyrite as well as with marcasite, kaolinite, dolomite, a mineral structurally close to crandallite, and Ti oxides. These deposits also contain phosphates with high contents of Ca, U, and Th (with an atomic ratio U:Th amounting up to 1:1) which may be a very fine-grained mixture of ningyoite and brockite, a mixed gel (not yet entirely differentiated), or a new, so far unknown U-Th-Ca phosphate [14, pp. 113, 117]. Probably brockite, although not specified (see [1, p. 1346]), occurs together with disordered cristobalite, quartz, and montmorillonite in the fine-sized material (~1 μm) of the Limestone Member of the Pliocene Moonstone Formation, Fremont county, Wyoming [10].

Inclusions. Anhedral grains of thornasite are imbedded in patches (4 to 8 mm in diameter) of white powdery brockite from the floor of the De-Mix quarry, Mont St. Hilaire, Quebec, Canada [18].

Alteration. Brockite alters to a brown earthy material, containing less Ca and P and more Fe and Si than the fresh mineral, in an Na-stage replacement body of the Haapaluoma pegmatite, Peräseinäjoki-Alavus region, southern Pohjanmaa, Finland. The excess H$_2$O in

Table 19
Chemical Composition and Density of Brockite from Various Localities.

| | content in % in sample number | | | | |
	1	2	3	4	5
ThO_2	54.56	53.79	46.0	45.52	42.7
CaO	12.46	12.88	11.0	9.65	9.7
ΣRE_2O_3	1.75[a]	0.37[a]	5.6[h]	tr.[k]	6.6
UO_2	0.73	1.58	—	0.85	0.068
Fe_2O_3	0.12[b]	—	3.4	2.00	4.6
P_2O_5	26.10[c]	26.17[c]	29.0	28.64	23.6
H_2O	—	—	—[i]	—	7.5
others	0.92[d]	1.6[f]	5.0[j]	0.59[l]	7.5[m]
sum	96.64[e]	96.39[g]	100.00	87.25	102.2
D in g/cm^3	—	—	4.05 ± 0.05	—	3.9 ± 0.2

Nos. 1 and 2 = Brockite from albitic topaze-lepidolite granite, Beauvoir, Brittany, France from [3]. — No. 3 = Brockite from pegmatite, Haapaluoma, Finland [11, pp. 54, 56]. — No. 4 = Brockite from sediments (fresh-water and marine), Stráž, Czechoslovakia (contents are calculated from atomic percentages) [14, pp. 117/8]. — No. 5 = Brockite from vein, Wet Mountains, Colorado [1, pp. 1347, 1350/1].

[a] Comprises only Ce_2O_3. — [b] Calculated from FeO content. — [c] Given as P content in the original paper, but obviously intended as P_2O_5 content. — [d] Comprises 0.52% PbO, 0.15% MnO, and 0.25% TiO_2. — [e] Given as 98.06% in the original paper.
[f] Comprises 0.95% PbO and 0.65% MnO. — [g] Given as 96.45% in the original paper. —
[h] Comprises oxides of Y, La, Ce, Nd, Sm, and Gd. — [i] The l.o.i. at 900°C is 10.64%. Heated at 110°C, the mineral lost 3.67% H_2O^- which is not included in the l.o.i. — [j] Comprises 4.1% Bi_2O_3 and 0.9% SiO_2.
[k] Comprises oxides of Y, La, Ce, and Nd. — [l] Comprises 0.59% SrO and traces of Ba. —
[m] Comprises 2% insolubles (which could not be recovered for identification), 1.1% BaO, 1.3% SrO, and 3.1% CO_2.

this brockite (see footnote for analysis No. 3 in Table 19) evidently is due to this earthy alteration product [11, pp. 54, 56].

Brockite forms as an alteration product of monazite [15] or of primary Th minerals, as, for example, in the younger shell of a zoned pegmatite vein near the farm Brabant, South-West Africa, see p. 374, or in the potassic alteration zone of granites from the Wet Mountains, Custer county, Colorado, see p. 373. Brockite could be an alteration product of thornasite, or both minerals may be the alteration product of another mineral, in the De-Mix quarry, Mont St. Hilaire, Quebec, Canada [18].

Chemistry. The chemical analyses of brockite from different localities are given in Table 19. For semiquantitative analyses of a hydrous Th phosphate resembling brockite (see [1, p. 1346]), refer to [10].

The general formula of brockite is $(Ca, Th)PO_4 \cdot H_2O$ with a Ca:Th ratio of 1:1 [1, p. 1350]. For individual brockites, the following formulas have been calculated: $(Ca, Fe)Th[PO_4]_2 \cdot H_2O$ for a sample from Bogatynia, Poland, with a small amount of Fe (2%) [16]; $Ca_{0.43^-}$

$Sr_{0.03}Ba_{0.02}Th_{0.41}RE_{0.11}$ $[(PO_4)_{0.83}(CO_3)_{0.17}] \cdot 0.9\ H_2O$ for a purified brockite from a vein east of the Bassick mine, Custer county, Colorado (ferric oxide being regarded as impurity and omitted from calculation) [1, pp. 1350/1]; and $(Ca, Th, RE, Bi)_{1.0}[P, Si]_{1.0}O_4 \cdot 1.4\ H_2O$ from a semiquantitative analysis for brockite from an Na-stage replacement body of the Haapaluoma granitic pegmatite, southern Pohjanmaa, Finland (omitting Fe_2O_3 as a nonessential component and assuming that loss of ignition is only H_2O) [11, pp. 54/6].

Probably a solid solution exists between brockite and rhabdophane [6, p. 71]. Note also that brockite, saryarkite, and rhabdophane represent members of a series characterized by regular isomorphous substitutions [17]. In a purified brockite from the Wet Mountains, Colorado, CaO, ThO_2, and P_2O_5 are present in a molecular ratio of approximately 1:1:1. Thus, Ca and Th must substitute in nearly equal proportions for the RE in the rhabdophane structure. Minor constituents are held in the brockite structure, with CO_3^{2-} substituting for PO_4^{3-} to balance the valence difference between Th^{4+} and RE^{3+} [1, pp. 1350/1]. Some Si may substitute for P in a hydrous Th phosphate, similar to brockite (see [1, p. 1346]), from the Limestone Member of the Moonstone Formation, Fremont county, Wyoming, and from a vein southwest of Gunnison, Colorado [10]. Brockite with 2% Fe in Th-bearing pegmatites near Bogatynia, southwestern Poland, belongs to the brockite $(CaTh[PO_4]_2 \cdot H_2O)$ — $FeTh[PO_4]_2 \cdot H_2O$ series [12].

The contents of individual RE in brockite from the Wet Mountains, Colorado, show an unusual excess of Nd_2O_3 over Ce_2O_3, La_2O_3, and Y_2O_3 (for the data, see original paper) with an atomic ratio Ce:(La + Nd) of 0.58. The deficiency in Ce relative to Nd may reflect the prevalence of oxidizing conditions at the time of brockite formation [1, p. 1352]. Brockite contains very high amounts of Gd in nine veins in the Lemhi Pass district in Montana [5, p. A45]; note also that three samples of Th-bearing veins with high Gd content also contain moderate amounts of brockite, and these peaks may reflect the RE distribution in the mineral [6, p. 77]. Brockite is also a marker of high contents of Gd and other heavier RE in Th-RE veins of the Bear Lodge Mountains, where samples containing brockite have a relatively high content of heavy RE [9, p. D41].

Crystal Form and Structure. Brockite is hexagonal [1, p. 1346]. Elongated brockite crystals with well developed faces occur in albitic topaz-lepidolite granite from Beauvoir, Brittany, France [3]. Two morphologically different forms of brockite occur together in the cement of fresh-water and marine sediments of Stráž, northern Bohemia, Czechoslovakia: (1) Idiomorphic columnar crystals with hexagonal cross section and a maximum size of $1.3 \times 10\ \mu m$ (but often up to one order of magnitude smaller), usually forming sheaves of parallely intergrown crystals; and (2) colloform (papillary) aggregates. The ratio of these two morphological forms varies and probably is connected with the varying degrees of recrystallization of primary gels [14, pp. 117/8]. Brockite occurs predominantly as radial aggregates of fibrous grains, and also as stubby, imperfect hexagonal prisms (usually with granulated margins) in the Wet Mountains, Colorado [1, p. 1348].

Space group is P_622 or D_6^4 (No. 180); Z approximates 3 (as is required as a minimum by the space group) on a basis of a cell volume of 270 $Å^3$ and a cell weight of 651. The low value of Z may reflect structural vacancies resulting from substitution of CO_3 for PO_4 and of Ca and Th for RE. The lattice constants of type brockite are $a = 6.98 \pm 0.03$, $c = 6.40 \pm 0.03$ Å with $V = 270\ Å^3$, and are very close to those of rhabdophane [1, pp. 1353/4]. Comparable lattice constants are: $a = 6.982 \pm 0.003$, $c = 6.425 \pm 0.005$ Å with $V = 271.3\ Å^3$ for brockite from sediments of Stráž, Czechoslovakia [14, p. 118]; $a = 6.96_5 \pm 0.02$, $c = 6.41_5 \pm 0.02$ Å with $V = 269.5\ Å^3$ for brockite from the Haapaluoma pegmatite, Finland [11, pp. 54/5]; and $a = 7.08$, $c = 6.24$ Å for an Fe-containing brockite from Bogatynia, Poland [16].

The X-ray powder diffraction pattern of brockite is indistinguishable from that of rhabdophane [1, p. 1346], [6, p. 71]. The lines are rather broad, which may indicate poor crystallinity [1, p. 1352]. For the strongest reflections, the following interplanar spacings (d values), with relative intensities in parentheses, for brockite from various localities are given; n = total number of d values given in the original paper:

brockite from	d value (in Å) for plane indexed as						n	Ref.
	(101)	(110)	(200)	(102)	(211)	(212)		
Finland	4.39(80)	3.49(60)	3.02(100)	2.830(80)	2.147(80)	1.858(60)	20	[11, pp. 54/5]
Czecho-slovakia	4.44(80)*)	3.49(50)	3.03(100)	2.838(95)	2.152(50)	1.862(25)	9	[14, pp. 117/8]
Colorado	4.37	3.47	3.03	2.83	2.15	1.86	16	[1, pp. 1347,
	4.39	3.47	3.02	2.83	2.15	1.86		1353]

*) Line coincides with that of hydrozircon.

After heating at temperatures between 800 and 900°C, brockite exhibits an X-ray diffraction pattern of the monazite type [1, p. 1353], [11, pp. 55/6]; see also [10]. The interplanar spacings of heated brockite are close to those of synthetic cheralite [1, p. 1353], [11, pp. 55/6].

Brockite is isostructural with rhabdophane [1, p. 1346], [6, p. 71], [14, p. 117]. Note also that a brockite-like mineral (see [1, p. 1346]) has a structure closely resembling that of rhabdophane [10].

Optical Properties. Brockite is translucent with a greasy to vitreous luster [1, pp. 1347/8]. The color of brockite is variable and ranges from white [4, p. 500], [9, p. D34], [18], over pale yellow [1, p. 1347], [9, p. D34], yellow [1, p. 1347], [5, p. A45], yellowish orange [4, p. 500], [6, p. 71], orange [5, p. A45], orange-red [4, p. 500], reddish orange [6, p. 71], dark or light reddish brown [1, p. 1347], [9, p. D34], [10], [11, p. 54], yellowish brown [9, p. D34], tan [4, p. 500], [5, p. A45], [6, p. 71], [9, p. D34] to brown [9, p. D34]. This color variation apparently is due to a variation in grain size and in the amount of hematite stain (partly leading to opacity) [1, pp. 1346/8] or limonite stain [9, p. D34].

Type brockite is uniaxial positive, moderately birefringent, and shows a positive elongation and parallel undulatory extinction. Its refractive indices are $n_\omega = 1.680 \pm 0.002$ and $n_\epsilon = 1.695 \pm 0.002$ [1, pp. 1346, 1348]. A range of 1.67 to 1.69 is given for the refractive indices of brockite from the Haapaluoma pegmatite, Finland [11, p. 54], and of 1.66 to 1.69 for a brockite-like mineral (see [1, p. 1346]) from a vein southwest of Gunnison, Colorado [10].

Other Physical Properties. The Mohs hardness of brockite is 3 to 4 [11, p. 54]; see also [10]. Brockite has a conchoidal fracture. According to the Gladstone-Dale rule, type brockite has a calculated density of 4.0 g/cm^3 [1, pp. 1348, 1351], corresponding well to pycnometric measurements, see p. 375. A brockite-like mineral (see [1, p. 1346]) from a vein southwest of Gunnison, Colorado, has a range in density from 3.7 to 4.3 g/cm^3 [10].

Thermal Behavior. For the X-ray diffraction pattern and interplanar spacings of heated brockite, see above.

References for 2.4.2:

[1] Fisher, F. G.; Meyrowitz, R. (Am. Mineralogist **47** [1962] 1346/55).
[2] Kucha, H.; Wieczorek, A. (Mater. Konf. 6th Konf. Mikrosk. Elektron. Ciala Stalego, Krynica, Pol., 1981, pp. 449/52).
[3] Cuney, M.; Brouand, M. (Geol. France **1987** No. 2/3, pp. 247/57, 248/51).
[4] Staatz, M. H. (Econ. Geol. **69** [1974] 494/507).
[5] Staatz, M. H.; Sharp, B. J.; Hetland, D. L. (Geol. Surv. Profess. Paper [U.S.] No. 1049-A [1979] A1/A90).
[6] Staatz, M. H. (Geol. Surv. Bull. [U.S.] No. 1351 [1972] 1/94).
[7] Staatz, M. H. (Econ. Geol. **67** [1972] 240/8, 247).
[8] Staatz, M. H.; Shaw, V. E.; Wahlberg, J. S. (Econ. Geol. **67** [1972] 72/82, 72).
[9] Staatz, M. H. (Geol. Surv. Profess. Paper [U.S.] No. 1049-D [1983] D1/D52).
[10] Dooley, J. R., Jr.; Hathaway, J. C. (Geol. Surv. Profess. Paper [U.S.] No. 424-C [1961] C339/C341).

[11] Haapala, J. (Bull. Comm. Geol. Finlande No. 224 [1966] 1/98).
[12] Banas, M.; Kucha, H. (Proc. 6th Quadrenn. IAGOD Symp., Tiflis 1982 [1984], Vol. 1, pp. 241/6).
[13] Rose, D. (Neues Jahrb. Mineral. Monatsh. **1980** 247/57, 248/9).
[14] Scharm, B.; Burda, J.; Hofreitr, V.; Sulovský, P.; Scharmová, M. (Cas. Mineral. Geol. **25** [1980] 113/24).
[15] Floran, R. J.; Abraham, M. M.; Boatner, L. A.; Rappaz, M. (Sci. Basis Nucl. Waste Management **3** [1981] 507/14, 508).
[16] Kucha, H.; Wieczorek, A. (Mineral. Polonica **11** No. 1 [1980] 123/36 from C.A. **96** [1982] No. 146295).
[17] Shipovalov, Yu. V.; Krol, V. F. (Rentgenogr. Miner. Syr'ya No. 8 [1971] 41/4; C.A. **79** [1973] No. 33506).
[18] Ansell, V. E.; Chao, G. Y. (Can. Mineralogist **25** [1987] 181/3).
[19] Stumpfl, E. F.; Kirikoglu, M. S. (Mitt. Österr. Geol. Ges. **78** 1985 [1986] 193/200).

2.4.3 Unnamed Hydrous Fe-Th Phosphate $Fe^{2+}Th[PO_4]_2 \cdot H_2O$

Occurrence and Paragenesis. Small crystals of unnamed $Fe^{2+}Th[PO_4]_2 \cdot H_2O$ belonging to the rhabdophane group were described in 1980 from pegmatites and hydrothermal veins in the Lużyce (Lusatia) granitoid rocks near Bogatynia, Lower Silesia, Poland. The amount of the unnamed mineral varies in three samples from 3 to 30, mean 16 vol%; it occurs together with monazite-, rhabdophane-, Th ningyoite-, and thorogummite-group minerals as well as together with hydrothermal alteration and weathering products of these minerals.

Intergrowths and Alteration. Very small ($<1 \, \mu m$) crystals of $Fe^{2+}Th[PO_4]_2 \cdot H_2O$ often are intergrown with $(Fe^{2+},Ca)_{1-x(?)}Th_{1-x(?)}(Fe^{3+},RE)_{2x(?)}[PO_4]_2 \cdot 1-3 \, H_2O$ or $Fe^{2+}Th[PO_4]_2 \cdot$ Unnamed $Fe^{2+}Th[PO_4]_2 \cdot H_2O$ can be transformed into $(Fe^{2+},Ca)_{1-x(?)}Th_{1-x(?)}(Fe^{3+},RE_{2x(?)}$ $[PO_4]_2 \cdot 1-3 \, H_2O$ by addition of H_2O and oxidation of some Fe^{2+} to Fe^{3+} with formation of some $Fe_3^{2+}(H_2O)[PO_4]_2$. Unnamed $Fe^{2+}Th[PO_4]_2 \cdot 2 \, H_2O$ most likely is a secondary mineral and usually forms after $Fe^{2+}Th[PO_4]_2$ (see p. 371). For crusts of $(Fe^{2+},Ca)_{1-x(?)}Th_{1-x(?)}$ $(Fe^{3+},RE)_{2x(?)}[PO_4]_2 \cdot 1-3 \, H_2O$ on altered $Fe^{2+}Th[PO_4]_2 \cdot H_2O$, see p. 386.

Chemistry. The theoretical composition of $Fe^{2+}Th[PO_4]_2 \cdot H_2O$ is calculated as 53.26% ThO_2, 14.54% FeO, and 28.64% P_2O_5. The contents of the main elements, recalculated from atomic percentages, of $Fe^{2+}Th[PO_4]_2 \cdot H_2O$ from Bogatynia are as follows (for atomic proportions of OH and H_2O, calculated to complete the analyses to 100%, see original paper); Nos. 1, 2, and 3 are from rock samples S-9, S-9B, and G-1a, respectively:

	content in % in sample number		
	1	2	3
ThO$_2$	70.44	64.98	49.50
FeO	2.32	6.56	18.27
CaO	1.68	1.54	0.84
Ce$_2$O$_3$	$\leqq$0.23	0.59	0.82
P$_2$O$_5$	8.48	7.33	11.46
SiO$_2$	8.98	8.13	7.49
C	1.0	1.0	0.8
sum	$\sim$93.13	90.13	89.18

The unnamed $Fe^{2+}Th[PO_4]_2 \cdot H_2O$ seems to be an intermediate member of the solid-solution series apparently extending between rhabdophane, brockite, and hypothetical $FeTh[PO_4]_2 \cdot H_2O$. The presence of part of Fe in the trivalent state may cause a strong substitution of P by Si or OH. The presence of hydroxylic water in the structure of $Fe^{2+}Th[PO_4]_2 \cdot H_2O$ has not been proved. The real chemical formula is given as $(Th, Fe, Ca, RE\ldots)[PO_4, SiO_4, CO_3, (OH)?] \cdot 0.5 H_2O$. After ignition, part of Fe probably remains in the structure, but excess Fe unbalanced by the sum of PO$_4$ and SiO$_4$ is oxidized and yields Fe$_2$O$_3$.

Crystal Form and Structure. Unnamed $Fe^{2+}Th[PO_4]_2 \cdot H_2O$ is assumed to be hexagonal, as follows from an isostructural relationship to rhabdophane and brockite; after ignition, it has a monazite-like structure.

The lattice constants are a = 7.04 $\pm$ 0.06 and b = 6.36 $\pm$ 0.03 Å and are similar to those of rhabdophane and brockite. The interplanar spacings (d values) for the strongest out of a total of 20 X-ray powder diffraction reflections (with corresponding relative intensities (I)) for $Fe^{2+}Th[PO_4]_2 \cdot H_2O$ from rock sample G-1a are:

(hkl)	d in Å	I	(hkl)	d in Å	I
(101)	4.36	8	(211)	2.15	10
(200)	3.05	10	(212)	1.861	8
(102)	2.82	10			

For the material ignited in air, lacking interplanar spacings for ThO$_2$, the strongest from a total of 15 reflections with relative intensities (I) are:

d in Å	I	d in Å	I
3.06	10	1.85	10
2.82	10	1.693	9
2.14	10	1.447	9

Physical Properties. Unnamed $Fe^{2+}Th[PO_4]_2 \cdot H_2O$ is red- to green-brown. In transmitted light, it is subtranslucent white-yellow with a green hue or red-orange and shows a variably distinct to weak birefringence yellow-white to green-yellow, or orange to red-brown. In reflected light, bireflectance is weak or absent. Radial-anisotropy is usually distinct in transmitted and weak to distinct in reflected light. The reflectance in air is $\sim 12\%$. Nonfluorescent in UV light.

The Mohs hardness is ~ 4 and the microhardness (VHN) at 150 g load is 192 to 280 kg/mm². The mineral is brittle.

Thermal Behavior. For the behavior of Fe in the structure of $Fe^{2+}Th[PO_4]_2 \cdot H_2O$ and changes in structure and interplanar spacings of the mineral after ignition, see p. 379.

Reference for 2.4.3:

Kucha, H. (Mineral. Polonica **10** No. 1 [1979] 3/28, 4/5, 7, 17/20).

2.4.4 Eylettersite $(Th, Pb)_{1-x}Al_3[PO_4, SiO_4]_2(OH)_6$

The hydrated Th phosphate eylettersite, a member of the crandallite series [1, pp. 1375, 1380], or constituting the Th analog of crandallite [2, p. 104], [3, p. 146], was described in 1971 from the Kobokobo beryl pegmatite, Kivu province, Zaire. It is named in honor of the wife of L. van Wambeke [2, p. 104]. The mineral belongs to a group of phosphates with the general formula $A_xB_y[PO_4]_z(OH)_{n \geq 0}$ (with A = leachable cation or group of leachable and stable cations, B = Fe^{3+} and/or Al). To characterize the cation-deficient nature of eylettersite, the term "keno-eylettersite" is proposed [1, pp. 1366/7, 1381/2].

Occurrence and Paragenesis. Eylettersite occurs as pulverulent nodules [1, p. 1375] in feldspar, zircon (cyrtolite), columbite, and apatite, or is intimately associated with phosphuranylite, crandallite, rare plumbogummite, and very rarely with clay minerals in the Kobokobo beryl pegmatite, Kivu province, Zaire. It is formed during partial or total hydrothermal and supergene alteration of primary Th- and Pb-containing uraninite by phosphate solutions, leading to the formation of oxides and of phosphates of the phosphuranylite type. During the first stage of oxidation, Th occurs as a complex amorphous hydrate associated with U oxides (e. g., with schoepite). The subsequent supply of Al-rich phosphate solutions again causes partial leaching of U and its concentration in Al-rich phosphates and radioactive crandallite [2, pp. 99, 104].

Inclusions and Alteration. Mineral impurities in eylettersite are fluorapatite, some autunite, limonite, and highly altered zircon (cyrtolite) [1, pp. 1376, 1379], [2, pp. 101/2, 104]. Hydrothermal and supergene alteration causes considerable changes in the original chemical composition of eylettersite (see pp. 381/2) [1, p. 1382], [2, p. 99], whereas changes in structure are minor [1, p. 1382], and leads to the formation of autunite inclusions inside the eylettersite since U is leached out easily [2, p. 103].

Chemistry. The (average?) chemical analyses (columns a) and corresponding contents corrected for mineral impurities (columns b) of two eylettersite samples from the Kobokobo pegmatite are shown in the following [1, p. 1376], [2, p. 102]; for the densities, see p. 383:

| | content in % in | | | |
| | sample 1 | | sample 2 | |
	a [a]	b	a [b]	b
ThO_2	20.10	20.72	18.43	19.18
PbO	3.11	3.21	4.62	4.80
BaO	1.90	1.96	1.45	1.51
UO_3	3.10	2.42	3.05	2.09
CaO	0.55	0.26	0.72	0.11
Al_2O_3	34.9	35.98	35.7	37.13
P_2O_5	19.72	19.98	18.3	18.38
SiO_2	1.20	1.03	1.65	1.56
H_2O	14.4	14.51	15.0	15.18
others	1.4 [c]	0.31 [d]	1.34 [e]	0.32 [f]
sum	100.38	100.38	100.26	100.26

[a] Impurities comprise 1.8 to 1.9% autunite, 0.9% fluorapatite, 0.1% limonite, and highly altered zircon (cyrtolite). — [b] Impurities comprise 1.3% autunite, 0.4% fluorapatite, limonite, and highly altered zircon (cyrtolite). — [c] Comprises 0.2% SrO, 0.1% Fe_2O_3, 1.0% ZrO_2, and 0.1% CO_2. — [d] Comprises 0.21% SrO and 0.10% CO_2. — [e] Comprises 0.95% ZrO_2, 0.1% SrO, 0.09% Fe_2O_3, and 0.2% CO_2. — [f] Comprises 0.11% SrO and 0.21% CO_2.

The structural formulas derived from analyses in column a are $[Th_{0.355}Pb_{0.106}Ba_{0.049}U_{0.036}Ca_{0.010}Sr_{0.007}H_{0.172}]Al_{3.58}[(PO_4)_{1.270}(SiO_4)_{0.130}(CO_3)_{0.025}(H_4O_4)_{0.575}](OH)_{6.10}$ for sample No. 1 and $[Th_{0.389}Pb_{0.071}Ba_{0.063}U_{0.042}Ca_{0.023}Sr_{0.011}H_{0.168}]Al_{3.50}[(PO_4)_{1.395}(SiO_4)_{0.087}(CO_3)_{0.007}(H_4O_4)_{0.512}](OH)_{6.09}$ for sample No. 2. Hypothetical formulas for the unaltered equivalents are $Th_{0.30}A_{0.70}Al_3(PO_4)_{1.70}(SiO_4, H_4O_4)_{0.30}(OH)_6$ for sample No. 1 and $Th_{0.335}A_{0.665}HAl_3(PO_4)_{1.665}(SiO_4, H_4O_4)_{0.335}(OH)_6$ for sample No. 2; A = bivalent cations, H, and U [1, pp. 1377/8], [2, pp. 101, 103].

The contents of Th and Al show only slight variations which do not exceed a few percent (e.g., from 17.5 to 22.6% ThO_2 in sample No. 1). The distribution of P is more irregular, but anomalously high contents are due to the presence of apatite or U phosphate inclusions. The other variations in P content are small and probably are related to the degree of hydration. Generally, a slight increase in P content is accompanied by a relative decrease in the Al content. The CaO content is always low ($<1\%$ CaO); SiO_2, supposed to be distributed uniformly between Al and Th phosphate, may participate in the structure of the mineral. According to its SiO_2 content, eylettersite is close to a phosphosilicate [2, pp. 100/1, 104]. Note also that some of Si belongs to the structure of eylettersite, whereas some belongs to altered zircon [1, p. 1375].

As is indicated by the presence of SiO_2 and the relatively low content of P_2O_5 in eylettersite, the entry of tetravalent Th in the crandallite series is compensated by a partial substitution of PO_4 by SiO_4 groups and probably also by some tetrahedral hydroxyl groups [1, p. 1378].

In phosphates of the crandallite series, alteration often leads to deficiencies in cations and anionic groups, but with preservation of their structures with some changes in physical properties (see p. 383) [1, pp. 1366/7]; see also [3, p. 146]. Note also that in phosphates containing the leachable cations Na, RE, Ca, and U, late hydrothermal alteration may give rise not only to pronounced deficiencies (which range between 30 and 35%) in the A positions but also to partial replacement of PO_4 and SiO_4 by OH groups. Thus, the cations most resistant to leaching may become prevalent in the formula, as is observed in the Al phosphate eylettersite with a predominance of Th [3, pp. 125, 146]. As a result of hydrothermal and supergene alteration, eylettersite shows an unusual chemical composition compared to the unaltered equivalent, namely a sharp decrease in the contents of leachable A ions and of PO_4 and SiO_4 groups accompanied by a higher content of stable elements, in particular of Al [1, pp. 1376, 1379]; refer also to [2, p. 104]. Relatively stable A ions include Th, Ba, and Pb, whereas Ca, U, and, to a lesser extent, Sr are leached selectively by hydrothermal and supergene alteration. With increasing degree of alteration the atomic percentages of the stable elements (i.e., Al in B, and Th, Ba, and Pb in A positions) increase and those of the other A cations (namely Ca and U) and of PO_4 and SiO_4 radicals decrease, which is explained by partial substitution of PO_4 and SiO_4 by hydroxyl groups (H_4O_4). The approximate loss of PO_4 plus SiO_4 groups by alteration is 37 and 32% for eylettersite samples No. 1 and 2, respectively. About 53 and 48%, respectively, of the A ions (especially Ca and U) are leached by alteration [1, pp. 1377/9], [2, pp. 101, 103/4]. Assuming an approximately identical leaching rate for Ca and U, the unaltered equivalent of sample No. 1 has about the same atomic percentages for Th, U, and Ca, whereas in the unaltered equivalent of sample No. 2, the Th content remains prevalent, probably with a U content slightly higher than the Ca content [1, p. 1379]; or Th $\geq$ U $\cong$ Ca in the unaltered equivalent of eylettersite No. 1 and Th $>$ U $\geq$ Ca in the unaltered equivalent of eylettersite No. 2 [2, p. 103].

Crystal Form and Structure. Eylettersite is hexagonal (rhombohedral), space group R$\bar{3}$m (No. 166) [1, p. 1375]. The unit-cell parameters of the two analyzed specimens are: a = 6.99 $\pm$ 0.04 and c = 16.70 $\pm$ 0.10 Å with V = 706.6 Å^3 for sample No. 1; a = 6.98 $\pm$ 0.03 and c = 16.66 $\pm$ 0.08 Å with V = 702.9 Å^3 for sample No. 2 [1, p. 1375], [2, p. 100]. The entry of Th, U, and other A ions, substituting for Ca, increases the value of c [1, p. 1375]; note also that the presence of U and Th, besides an elevated degree of hydration, causes an increase of the parameter c of altered eylettersite [2, p. 104]. The X-ray powder diffraction pattern of eylettersite sample No. 1 is analogous to that of crandallite and shows the following strongest (from a total of 10) reflections for interplanar spacings (d values) with corresponding relative intensities (I) [2, pp. 99/100]:

(hkl)	d in Å	I	(hkl)	d in Å	I
(101)	5.70	55	(116), (107)	2.187	40
(110)	3.51	60	(033), (303)	1.899	30
(015), (113), (021)	2.95	100			

Material heated at 1000 °C shows a complex X-ray diffraction pattern with lines corresponding to ThO_2, U_3O_8, Al_2O_3 (corundum), and probably to an Al pyrophosphate. The d values for the three predominant reflections are 4.15, 4.36, and 2.53 Å [2, p. 104].

Optical Properties. Eylettersite is creamy white [1, p. 1375], [2, p. 99]. Microscopically, it is colorless or slightly brownish (owing to traces of limonite) [2, p. 99].

Eylettersite is nearly isotropic to uniaxial (probably negative) [2, p. 99]. The refractive indices are: n = 1.615 to 1.64, mean 1.635, for eylettersite sample No. 1; and n = 1.625 to 1.66, mean 1.645, for eylettersite sample No. 2 [1, p. 1379], [2, p. 99]. Generally, the outer, more altered parts have lower indices than the inner parts of the mineral. The indices vary as a function of chemical composition and degree of alteration; rather low values of n are caused by an elevated degree of hydration, the existence of vacancies in position A, and deficiencies in P. Eylettersite is weakly fluorescent in UV light (brown-creamy in the long-wave and yellow greenish in the short-wave region) [2, pp. 99, 104].

Other Physical Properties. Measured densities of eylettersite samples No. 1 and 2 are 3.38 ± 0.10 and 3.44 ± 0.10 g/cm^3, and calculated densities are 3.44 ± 0.05 and 3.50 ± 0.06 g/cm^3, respectively [2, p. 100]; see also [1, p. 1375]. The density varies as a function of chemical composition and degree of alteration; rather low values of D are caused by the same reasons as for refractive indices (see above). The presence of heavy elements such as Th, Pb, Ba, or U does not influence the density [2, pp. 99, 104].

Chemical and Thermal Behavior. Eylettersite is difficultly soluble in sulfuric acid [2, p. 99].

The DTA curve of eylettersite has an endothermic peak with a maximum at 165 to 175°C and two exothermic peaks at 730 to 740 and at 830°C. The second, rather large exothermic peak is bisected in some samples [2, pp. 103/4].

For the X-ray diffraction pattern of heated eylettersite, see p. 382.

References for 2.4.4:

[1] Van Wambeke, L. (Am. Mineralogist **56** [1971] 1366/84).
[2] Van Wambeke, L. (Bull. Soc. Franc. Mineral. Crist. **95** [1972] 98/105).
[3] Van Wambeke, L. (Neues Jahrb. Mineral. Abhandl. **112** [1970] 117/49).

2.4.5 Grayite (Th, Pb, Ca)[PO$_4$] · H$_2$O

Pale yellow powdery grayite with a chemical composition approximating to Th phosphate, associated with thorite in an Li-bearing pegmatite of the Mtoko district, Zimbabwe, was described in 1957 [1]; see also [2]. Description of grayite is incomplete, its etymology is not stated [6]; it occurs as an alteration product of monazite [3].

Chemical analyses of grayite are lacking, its formula is given as (Th, Pb, Ca)[PO$_4$] · H$_2$O [4].

Structurally, grayite is related to rhabdophane and other hexagonal RE phosphates [1, 2]. However, X-ray diffraction lines of grayite at 4.54 and 1.81 Å cannot be indexed on a hexagonal lattice, and the angle between axes a and b deviates from 60°. Grayite is thus considered as pseudohexagonal orthorhombic and belongs to the ningyoite group [5].

After heating to 850°C, grayite shows X-ray diffraction reflections of the monazite type [6].

References for 2.4.5:

[1] Bowie, S. H. U.; Horne, J. E. T.; Beer, K. E.; Branscombe, K. C.; Harrison, R. K.; Miller, J. M.; Ostle, D.; Ackermann, K. J.; Burke, K. C.; Cosgrove, M. E.; Peacock, J. D.; Rutland, R. W. R.; Taylor, J.; Taylor, K.; Tidy, A. J. L. (Summary Progr. Geol. Surv. Great Britain Museum Practical Geol. **1956/57** 66/7).

[2] Bowie, S. H. U. (Chron. Mines Outre-Mer Rech. Min. No. 279 [1959] 304/8).

[3] Floran, R. J.; Abraham, M. M.; Boatner, L. A.; Rappaz, M. (Sci. Basis Nucl. Waste Management 3 [1981] 507/14, 508).

[4] Fleischer, M. (Glossary of Mineral Species, 5th Ed., Mineral. Record Inc., Tucson, Arizona, 1987, pp. 1/227, 69).

[5] Fisher, F. G.; Meyrowitz, R. (Am. Mineralogist 47 [1962] 1346/55, 1346, 1353/4).

[6] Chudoba, K. F. (Hintze's Handbuch der Mineralogie, Erg.-Bd. 3, Neue Mineralien und Neue Mineralnamen, de Gruyter, Berlin 1968, pp. 1/684, 125/6).

2.4.6 Kivuïte $(Th, Ca, Pb)H_2(UO_2)_4[PO_4]_2(OH)_8 \cdot 7 H_2O$ (?)

The Th phosphate kivuïte from the Kobokobo pegmatite, Kivu province, Zaire, was described in 1958 and is named after the type locality [1].

The discovery of the Th phosphate eylettersite (see p. 380) questions the existence of the mineral kivuïte [2], which is considered to be a mixture of thorian crandallite or eylettersite and phosphuranylite [3].

Occurrence and Paragenesis. As a member of the phosphuranylite — renardite — kivuïte series, constituting the main part of the radioactive mineralization, earthy kivuïte occurs in pockets or as coatings as a secondary mineral in the Kobokobo pegmatite, Kivu province, Zaire. The minerals of the above series (comprising all intermediate stages between the three end members) are arranged by decreasing frequency as follows: thorophosphuranylite, kivuïte, phosphuranylite, thororenardite, and renardite. Kivuïte is associated with uraninite, uraniferous zircon (cyrtolite), columbotantalite, and, very rarely, with apatite. Kivuïte formed at the expense of thorian curite and uraninite which supplied Th and Pb. More or less altered apatite certainly supplied both P and Ca [4, pp. 384/5, 388, 396]. Note also that phosphates of the phosphuranylite — renardite — kivuïte series generally occur as nodules, and very rarely as veinlets, in the beryl pegmatite of Kobokobo. The interaction of late phosphate-rich phases with U, Th, and Pb contained in uraninite led to the formation of mineral associations where Th is present as secondary kivuïte [5].

Inclusions. Inclusions of phosphates of the phosphuranylite — renardite — kivuïte series are observed in frondelite and rockbridgeite [5]. Inclusions in kivuïte are cyrtolite and columbotantalite; and in a Pb-rich kivuïte quartz and phosphuranylite [4, p. 392].

Chemistry. Kivuïte is the thorian member of the phosphuranylite — renardite series. Its chemical composition is close to the general formula of this series, except for the presence of Th and an elevated H_2O content. The theoretical composition of kivuïte, calculated for a theoretical formula $ThH_2(UO_2)_4(PO_4)_2(OH)_8 \cdot 7 H_2O$, is 14.50% ThO_2, 62.87% UO_3, 7.80% P_2O_5, 7.91% H_2O^+, and 6.92% H_2O^-. The chemical analyses of original kivuïte containing 4% cyrtolite besides columbotantalite as impurities (analysis No. 1), and the analysis corrected for impurities (analysis No. 2) are given in the table on top of p. 385 [4, pp. 394/5].

The formula of kivuïte derived from analysis No. 2 (regarded as unsatisfactory, because it is not known how the impurities were deducted [1]) is $(Th, Ca, Pb)_{5.4}(UO_2)_{23.4}(PO_4)_{9.1}$-$(OH)_{47.8}H_2O_{36.9}$ [4, pp. 394/5].

As against the theoretical composition, the observed deficiency in P_2O_5 in kivuïte is not compensated by the presence of 0.1 to 0.2% As_2O_5 in the analyzed material and may be due to a partial substitution of PO_4 by SiO_4 groups; however, this substitution is not confirmed by the given analysis. The excess H_2O may be explained, to a certain degree, by dehydration

	content in % in analysis number	
	1	2
ThO_2	8.32	8.85
CaO	0.60	0.64
PbO	1.84	1.96
UO_3	62.90	66.89
P_2O_5	6.04	6.42
H_2O^+	8.09	8.60
H_2O^-	6.24	6.64
sum	94.03	100.00

of uraniferous cyrtolite at 450 °C. The major part of Th, which substitutes for Ca, is bound to PO_4 groups [4, p. 395].

Crystal Form and Structure. Kivuïte is orthorhombic and isostructural with phosphuranylite and renardite. It occurs as plates or tablets with cleavage parallel c (001). Space group is Bmbm (No. 66) or Cmcm (No. 63). The lattice constants for chemical analyzed kivuïte and for a second, Pb-rich kivuïte sample are identical with a = 15.88 ± 0.06 and c = 13.76 ± 0.06 Å, b is 17.24 ± 0.1 and 17.23 ± 0.1 Å, respectively. The X-ray diffraction pattern of kivuïte is identical to that of phosphuranylite. The strongest reflections of indexed interplanar spacings (d values), with corresponding relative intensities in parentheses (vs = very strong, s = strong, ms = medium strong), for kivuïte (from a total of 13 reflections) and for Pb-rich kivuïte (from a total of 16 reflections) are as follows [4, pp. 388, 390/1, 393]:

(hkl)	d in Å for	
	kivuïte	Pb-rich kivuïte
(101)	10.27 (vs)	10.28 (s)
(200)	7.96 (s to vs)	7.90 (vs)
(143), (501)	3.08 (s)	3.09 (ms to s)
(060)	2.87 (s)	2.864 (vs)

Optical Properties. Kivuïte is yellow; uniaxial or biaxial (2 V = 0° to 5°). The refractive indices are n_α = 1.618 ± 0.002, n_β = 1.654 to 1.655, and n_γ = 1.655 ± 0.003 for kivuïte, and n_α = 1.675 ± 0.003, n_β = 1.705 ± 0.005, and n_γ = 1.705 ± 0.005 for Pb-rich kivuïte; these values are higher than those for phosphuranylite and renardite. Possibly, a relation exists between the values of the refractive indices and the content of H_2O. The elongation of kivuïte is positive. Dispersion r > v [4, pp. 388, 392/4].

Kivuïte is pleochroic (β and γ = dark yellowish greenish, α = white) and nonfluorescent [4, p. 388].

Chemical Behavior. Kivuïte dissolves partly in nitric acid; the solution contains all of the U and only a small portion of Th and Pb. The insoluble residue does not dissolve in hydrofluoric and perchloric acid, indicating that Th is not present as silicate but as phosphate [4, pp. 388/9]. However, the behavior of kivuïte against nitric acid needs to be tested [1].

References for 2.4.6:

[1] Fleischer, M. (Am. Mineralogist **44** [1959] 1326/7).
[2] Van Wambeke, L. (Bull. Soc. Franc. Mineral. Crist. **95** [1972] 98/105, 99).
[3] Van Wambeke, L. (Am. Mineralogist **56** [1971] 1366/84, 1380).
[4] Van Wambeke, L. (Bull. Soc. Belge Geol. Paleontol. Hydrol. **67** [1958] 383/403).
[5] Safiannikoff, A.; van Wambeke, L. (Mineralium Deposita **2** [1967] 119/30, 124, 126).

2.4.7 Unnamed Hydrous Ca-Th-RE Phosphate $Ca_{1-x}Th_{1-x}RE_{2x}[PO_4]_2 \cdot 2 H_2O$

An unnamed mineral $Ca_{1-x}Th_{1-x}RE_{2x}[PO_4]_2 \cdot 2 H_2O$, structurally related to ningyoite, was described in 1980 from Bogatynia, Poland, where it is associated with minerals of the monazite, rhabdophane (e.g., brockite), and ningyoite groups as well as with $Fe_3^{2+}(H_2O)[PO_4]_2$. It formed as a member of the transformation series of monazite-group through rhabdophane-group to ningyoite-group minerals (see p. 242) and is orthorhombic with lattice constants a = 6.67, b = 12.06, and c = 6.40 Å.

Reference for 2.4.7:

Kucha, H.; Wieczorek, A. (Mineral. Polonica **11** No. 1 [1980] 123/36 from C.A. **96** [1982] No. 146295).

2.4.8 Unnamed Hydrous Fe-Ca-Th-RE Phosphate $(Fe^{2+}, Ca)_{1-x(?)}Th_{1-x(?)}(Fe^{3+}, RE)_{2x(?)}[PO_4]_2 \cdot 1-3 H_2O$

Occurrence, Paragenesis, Intergrowths, and Alteration. Fine aggregates of unnamed $(Fe^{2+}, Ca)_{1-x(?)}Th_{1-x(?)}(Fe^{3+}, RE)_{2x(?)}[PO_4]_2 \cdot 1-3 H_2O$ belonging to the ningyoite group were described in 1980 from pegmatites and hydrothermal veins of the Lużyce (Lusatia) granitoid rocks near Bogatynia, Lower Silesia, Poland. The amount of the mineral varies in three samples from 14 to 30, mean 23 vol%; it occurs together with several Th minerals related to the monazite, rhabdophane, Th-ningyoite, or thorogummite group as well as together with hydrothermal alteration and weathering products of these Th minerals. The unnamed mineral usually is intergrown with the unnamed phosphates $Fe^{2+}Th[PO_4]_2$ and $Fe^{2+}Th[PO_4]_2 \cdot H_2O$ and forms opaline to microcrystalline crusts on altered $Fe^{2+}Th[PO_4]_2$ and $Fe^{2+}Th[PO_4]_2 \cdot H_2O$. For its occurrence as an alteration product of $Fe^{2+}Th[PO_4]_2 \cdot H_2O$, see also p. 378.

Chemistry. The contents of the main elements, recalculated from atomic percentages, of the unnamed $(Fe^{2+}, Ca)_{1-x(?)}Th_{1-x(?)}(Fe^{3+}, RE)_{2x(?)}[PO_4]_2 \cdot 1-3 H_2O$ are as follows (for atomic proportions of OH and H_2O, calculated to complete analyses to 100%, see original paper); Nos. 1 to 3, 5, and 6 are from rock samples S-9 and S-9 B, Nos. 4 and 7 from rock sample G-1a:

	content in % in sample number						
	1	2	3	4	5	6	7
ThO_2	61.22	58.72	58.38	58.15	56.22	52.12	50.53
FeO	8.36	9.91	2.44	1.42	10.68	6.82	15.18
CaO	1.96	1.82	0.42	1.68	1.26	0.98	0.70
Ce_2O_3	$\leqq 0.23$	$\leqq 0.23$	$\leqq 0.23$	1.17	$\leqq 0.23$	$\leqq 0.23$	0.23
P_2O_5	8.02	6.65	9.39	11.69	8.25	12.83	11.46
SiO_2	8.34	7.91	3.42	7.27	5.99	1.28	10.27
sum	~ 88.13	~ 85.24	~ 74.28	81.38	~ 82.63	~ 74.26	88.37

Primary variations in the chemical composition of unnamed $(Fe^{2+}, Ca)_{1-x(?)}Th_{1-x(?)}$ $(Fe^{3+}, RE)_{2x(?)}[PO_4]_2 \cdot 1-3\ H_2O$ involve the substitution of Fe^{2+} chiefly by Ca and RE as well as, to a minor extent, by Fe^{3+}. Usually small amounts of SiO_4 substitute for PO_4 but some highly silicious varieties exist with an Si:P ratio of 1.2:1, which may be considered as P-rich varieties of thorogummite. Sometimes a small amount of As substitutes for P (e.g., 0.5% As in sample No. 3). A ratio (P + Si): total sum of cations <1:1, and the presence of essential H_2O in the studied samples (see below), suggest a continuous substitution of OH for PO_4 and SiO_4.

The idealized formula of the unnamed mineral is $Fe^{2+}_{1-x}Th_{1-x}Fe^{3+}_{2x}[PO_4]_2 \cdot 2\ H_2O$. Owing to solid solution and secondary alteration, all analyzed samples deviate from this formula, and the general formula should be $(Fe^{2+}, Ca)_{1-x}Th_{1-x}(Fe^{3+}, RE)_{2x}[PO_4, SiO_4, (OH)]_2 \cdot 1-2\ H_2O$.

Crystal Form and Structure. Unnamed $(Fe^{2+}, Ca)_{1-x(?)}Th_{1-x(?)}(Fe^{3+}, RE)_{2x(?)}[PO_4]_2 \cdot 1-3\ H_2O$ is orthorhombic. Space group presumably P222 or D^1_2 (No. 16); Z = 3. Lattice constants are a = 6.77 ± 0.03, b = 12.06 ± 0.07, and c = 6.41 ± 0.04 Å. For the calculated d values and corresponding relative intensities, see original paper. The mineral is isostructural with ningyoite.

Physical Properties. Unnamed $(Fe^{2+}, Ca)_{1-x(?)}Th_{1-x(?)}(Fe^{3+}, RE)_{2x(?)}[PO_4]_2 \cdot 1-3\ H_2O$ is subtranslucent reddish brown. In transmitted light, it is brown-red to orange-yellow.

In both transmitted and reflected light, the mineral often is isotropic due to radioactive decay and sometimes is weakly anisotropic. In transmitted light at high magnifications, the mineral is resolved into formless birefringent specks. In reflected light, bireflectance is absent. The reflectance in air is $\sim 10\%$. Internal reflections are brown-red, red-yellow, or red-orange and brown in isotropic components.

The mineral is nonfluorescent in UV light. Electron beam causes emission of soft orange light.

The Mohs hardness is $\sim 3½$ to 4. The mineral is brittle. The density equals 4.11 ± 0.03 g/cm^3 for a 1:1 mixture of unnamed $Fe^{2+}Th[PO_4]_2$ and $(Fe^{2+}, Ca)_{1-x(?)}Th_{1-x(?)}(Fe^{3+}, RE)_{2x(?)}[PO_4]_2 \cdot 1-3\ H_2O$.

Thermal Behavior. Data on the thermal behavior of pure $(Fe^{2+}, Ca)_{1-x(?)}Th_{1-x(?)}(Fe^{3+}, RE)_{2x(?)}[PO_4]_2 \cdot 1-3\ H_2O$ are not available. DTG, DTA, and TGA studies show the presence of essential water in mixtures of the described mineral with other Fe^{2+}-Th phosphates, $Fe_3^{2+}(H_2O)[PO_4]_2$, or thorogummite, indicating that water is associated chiefly with the described mineral.

Reference for 2.4.8:

Kucha, H. (Mineral. Polonica **10** No. 1 [1979] 3/28, 4, 9/14, 20/2).

2.4.9   Althupite $AlTh(UO_2)[(UO_2)_3O(OH)(PO_4)_2]_2(OH)_3 \cdot 15 H_2O$

The **aluminium-thorium-uranyl** phosphate althupite from the uraniferous zone of the beryl-columbite pegmatite of Kobokobo, Kivu province, Zaire, was described in 1987. Its name is given in allusion to the chemical composition.

Occurrence and Paragenesis. Althupite occurs as microcrystalline crusts and as very small (maximum length 0.1 mm) yellow-amber crystals, which often encrust crystals of kaolinized feldspar, in the pegmatite of Kobokobo. It is associated with numerous other uranyl phosphates (including ranunculite, upalite, phuralumite, threadgoldite, mundite, triangulite, and meta-autunite).

Chemistry. The theoretical composition of althupite is 9.05% ThO_2, 1.75% Al_2O_3, 68.65% UO_3, 9.73% P_2O_5, and 10.81% H_2O. Range and mean for ten analyses of althupite are as follows: 9.61 to 11.43, 10.25% ThO_2; 1.62 to 2.21, 1.75% Al_2O_3; 66.80 to 68.93, 68.22% UO_3; 9.57 to 10.19, 9.86% P_2O_5; and 9.92% H_2O (calculated by difference). The somewhat higher ThO_2 content against the theoretical value is due to the difficulty to find a suitable analytical standard for this element.

Crystal Form and Structure. Althupite is triclinic. It occurs as thin pentagonal tablets flattened on (100), elongated along [001], and terminated by (011) and (021); the crystal top is sometimes truncated by (001). Cleavage after (100) is rather distinct.

The space group is $P\bar{1}$ (No. 2); Z = 2. Unit-cell parameters are a = 10.953(3), b = 18.567(4), c = 13.504(3) Å, $\alpha = 72.64(2)°$, $\beta = 68.20(2)°$, and $\gamma = 84.21(2)°$, with V = 2434(1) $Å^3$. The X-ray powder diffraction pattern gives the following interplanar spacings (d values) and corresponding relative intensities (I) for the strongest (from a total of 39) reflections:

(hkl)	d in Å	I
(100)	10.2	100
($\bar{1}$21)	5.80	50
(200)	5.09	70
(2$\bar{2}$1), (123)	4.41	50
(044), (324)	2.896	50

The crystal structure (for figures, see original paper) of althupite consists of $[(UO_2)_3O(OH)(PO_4)_2]_n^{3n-}$ layers and intercalated additional UO_2 groups and Th and Al atoms, which are surrounded by O atoms, OH groups, and H_2O molecules. The layers consist of pentagonal bipyramids around 7-fold coordinated U^{6+}, of hexagonal bipyramids around 8-fold coordinated U^{6+}, and of PO_4 tetrahedra; the U^{6+} cations belong to uranyl groups UO_2^{2+}. The intercalated additional UO_2 groups are surrounded almost linearly by five oxygens which are located approximately in one plane. Two of these oxygens belong to the PO_4 groups of the layers, the other three are part of the H_2O molecules. Each Al atom is surrounded by six oxygens (belonging to four OH^- and two H_2O) forming a relatively regular octahedron $Al_2(OH)_6(H_2O)_4$. These octahedra share in one edge ($2\,OH^-$) around a symmetry center. Thorium atoms are surrounded by nine oxygens (each 3 O, OH, and H_2O) and are connected with the $[(UO_2)_3O(OH)(PO_4)_2]_n^{3n-}$ layers by the three oxygens, two of which are from PO_4 groups and one from the UO_2 group. The $[(UO_2)_3O(OH)(PO_4)_2]_n^{3n-}$ layers are interlocked on both sides of the symmetry center by two $Al_2(OH)_6(H_2O)_4$ octahedra through three OH^- and thus form a complex centrosymmetric structure. This structure is the main link between the layers with three O^{2-} belonging to a layer and the other three to the next layer. For interatomic distances and their dependence on the charge of oxygen and for atomic positions, see original paper.

Optical Properties. Althupite is yellow. It is transparent to translucent with a vitreous luster.

Althupite is biaxial negative with 2 V = 34° (calculated) and 31(3)° (measured). The refractive indices are n_α = 1.620(3), n_β = 1.661(2), and n_γ = 1.665(2). Optical orientation is X ~ a, Z ∧ c ~ 15°. Dispersion is strong, r < v.

Althupite is pleochroic with α = very pale yellow, β and γ = yellow, and it is nonfluorescent in UV light.

Other Physical Properties. The Mohs hardness of althupite is close to 4 and cannot be measured exactly owing to the minute size of the crystals. The measured density is 3.9 ± 0.1 g/cm^3.

Reference for 2.4.9:

Piret, P.; Deliens, M. (Bull. Mineral. **110** [1987] 65/72).

Physical Constants and Conversion Factors

Avogadro constant N_A (or L) = 6.02214×10^{23} mol^{-1}

Faraday constant F = 9.64853×10^4 C/mol

molar gas constant R = 8.31451 J·mol^{-1}·K^{-1}

molar volume (ideal gas) V_m = 2.24141×10^1 L/mol
(273.15 K, 101325 Pa)

Planck constant h = 6.62608×10^{-34} J·s

elementary charge e = 1.60218×10^{-19} C

electron mass m_e = 9.10939×10^{-31} kg

proton mass m_p = 1.67262×10^{-27} kg

1 kg = 2.205 pounds

1 m = 3.937×10^1 inches = 3.281 feet

1 m^3 = 2.642×10^2 gallons (U.S.)

1 m^3 = 2.200×10^2 gallons (Imperial)

Force	N	dyn	kp
1 N	1	10^5	1.019716×10^{-1}
1 dyn	10^{-5}	1	1.019716×10^{-6}
1 kp	9.80665	9.80665×10^5	1

Pressure	Pa	bar	kp/m²	at	atm	Torr	lb/in²
1 Pa =1N/m²	1	10^{-5}	1.019716×10^{-1}	1.019716×10^{-5}	9.86923×10^{-6}	7.50062×10^{-3}	1.450378×10^{-4}
1 bar =10^6 dyn/cm²	10^5	1	1.019716×10^4	1.019716	9.86923×10^{-1}	7.50062×10^2	1.450378×10^1
1 kp/m²=1mm H$_2$O	9.80665	9.80665×10^{-5}	1	10^{-4}	9.67841×10^{-5}	7.35559×10^{-2}	1.422335×10^{-3}
1 at (technical)	9.80665×10^4	9.80665×10^{-1}	10^4	1	9.67841×10^{-1}	7.35559×10^2	1.422335×10^1
1 atm = 760 Torr	1.01325×10^5	1.01325	1.033227×10^4	1.033227	1	7.60×10^2	1.469595×10^1
1 Torr=1mmHg	1.333224×10^2	1.333224×10^{-3}	1.359510×10^1	1.359510×10^{-3}	1.315789×10^{-3}	1	1.933678×10^{-2}
1 lb/in²=1psi	6.89476×10^3	6.89476×10^{-2}	7.03069×10^2	7.03069×10^{-2}	6.80460×10^{-2}	5.17149×10^1	1

Work, Energy, Heat	J	kW·h	kcal	Btu	eV
1 J = 1 W·s = 1 N·m = 10⁷ erg	1	2.778×10^{-7}	2.39006×10^{-4}	9.4781×10^{-4}	6.242×10^{18}
1 kW·h	3.6×10^{6}	1	8.604×10^{2}	3.41214×10^{3}	2.247×10^{25}
1 kcal	4.1840×10^{3}	1.1622×10^{-3}	1	3.96566	2.6117×10^{22}
1 Btu (British thermal unit)	1.05506×10^{3}	2.93071×10^{-4}	2.5164×10^{-1}	1	6.5858×10^{21}
1 eV	1.602×10^{-7}	4.450×10^{-14}	3.8289×10^{-11}	1.51840×10^{-10}	1

$1\,\text{cm}^{-1} = 1.239842 \times 10^{-4}\,\text{eV}$
$1\,\text{hartree} = 27.2114\,\text{eV}$

$1\,\text{Hz} = 4.135669 \times 10^{-15}\,\text{eV}$
$1\,\text{eV} \triangleq 23.0578\,\text{kcal/mol}$

Power	kW	hp	kp·m·s⁻¹	kcal/s
1 kW = 10³ J/s	1	1.35962	1.01972×10^{2}	2.39006×10^{-1}
1 hp (horsepower, metric)	7.3550×10^{-1}	1	7.5×10^{1}	1.7579×10^{-1}
1 kp·m·s⁻¹	9.80665×10^{-3}	1.333×10^{-2}	1	2.34384×10^{-3}
1 kcal/s	4.1840	5.6886	4.26650×10^{2}	1

References:

International Union of Pure and Applied Chemistry, Manual of Symbols and Terminology for Physicochemical Quantities and Units, Pergamon, London 1979; Pure Appl. Chem. **51** [1979] 1/41.

The International System of Units (SI), National Bureau of Standards Spec. Publ. 330 [1972].

Landolt-Börnstein, 6th Ed., Vol. II, Pt. 1, 1971, pp. 1/14.

ISO Standards Handbook 2, Units of Measurement, 2nd Ed., Geneva 1982.

Cohen, E. R., Taylor, B. N., Codata Bulletin No. 63, Pergamon, Oxford 1986.

Key to the Gmelin System
of Elements and Compounds

System Number	Symbol	Element
1		Noble Gases
2	H	Hydrogen
3	O	Oxygen
4	N	Nitrogen
5	F	Fluorine
6	**Cl**	**Chlorine**
7	Br	Bromine
8	I	Iodine
8a	At	Astatine
9	S	Sulfur
10	Se	Selenium
11	Te	Tellurium
12	Po	Polonium
13	B	Boron
14	C	Carbon
15	Si	Silicon
16	P	Phosphorus
17	As	Arsenic
18	Sb	Antimony
19	Bi	Bismuth
20	Li	Lithium
21	Na	Sodium
22	K	Potassium
23	NH_4	Ammonium
24	Rb	Rubidium
25	Cs	Caesium
25a	Fr	Francium
26	Be	Beryllium
27	Mg	Magnesium
28	Ca	Calcium
29	Sr	Strontium
30	Ba	Barium
31	Ra	Radium
32	**Zn**	**Zinc**
33	Cd	Cadmium
34	Hg	Mercury
35	Al	Aluminium
36	Ga	Gallium

System Number	Symbol	Element
37	In	Indium
38	Tl	Thallium
39	Sc, Y La—Lu	Rare Earth Elements
40	Ac	Actinium
41	Ti	Titanium
42	Zr	Zirconium
43	Hf	Hafnium
44	Th	Thorium
45	Ge	Germanium
46	Sn	Tin
47	Pb	Lead
48	V	Vanadium
49	Nb	Niobium
50	Ta	Tantalum
51	Pa	Protactinium
52	**Cr**	**Chromium**
53	Mo	Molybdenum
54	W	Tungsten
55	U	Uranium
56	Mn	Manganese
57	Ni	Nickel
58	Co	Cobalt
59	Fe	Iron
60	Cu	Copper
61	Ag	Silver
62	Au	Gold
63	Ru	Ruthenium
64	Rh	Rhodium
65	Pd	Palladium
66	Os	Osmium
67	Ir	Iridium
68	Pt	Platinum
69	Tc	Technetium[1]
70	Re	Rhenium
71	Np,Pu...	Transuranium Elements

HCl

$CrCl_2$

$ZnCrO_4$

$ZnCl_2$

Material presented under each Gmelin System Number includes all information concerning the element(s) listed for that number plus the compounds with elements of lower System Number.

For example, zinc (System Number 32) as well as all zinc compounds with elements numbered from 1 to 31 are classified under number 32.

---

[1] A Gmelin volume titled "Masurium" was published with this System Number in 1941.

A Periodic Table of the Elements with the Gmelin System Numbers is given on the Inside Front Cover